Approaches in Enhancing Antioxidant Defense in Plants

Approaches in Enhancing Antioxidant Defense in Plants

Editors

Mirza Hasanuzzaman
Masayuki Fujita

MDPI • Basel • Beijing • Wuhan • Barcelona • Belgrade • Manchester • Tokyo • Cluj • Tianjin

Editors

Mirza Hasanuzzaman
Department of Agronomy,
Faculty of Agriculture
Sher-e-Bangla Agricultural
University
Dhaka
Bangladesh

Masayuki Fujita
Laboratory of Plant Stress
Responses, Faculty of
Agriculture
Kagawa University
Kagawa
Japan

Editorial Office
MDPI
St. Alban-Anlage 66
4052 Basel, Switzerland

This is a reprint of articles from the Special Issue published online in the open access journal *Antioxidants* (ISSN 2076-3921) (available at: www.mdpi.com/journal/antioxidants/special_issues/antioxidant_defense_plants).

For citation purposes, cite each article independently as indicated on the article page online and as indicated below:

LastName, A.A.; LastName, B.B.; LastName, C.C. Article Title. *Journal Name* **Year**, *Volume Number*, Page Range.

ISBN 978-3-0365-4186-0 (Hbk)
ISBN 978-3-0365-4185-3 (PDF)

Contents

About the Editors

Mirza Hasanuzzaman

Dr. Mirza Hasanuzzaman is a Professor of Agronomy at Sher-e-Bangla Agricultural University, Bangladesh. He received his Ph.D. with a dissertation on 'Plant Stress Physiology and Antioxidant Metabolism' from the United Graduate School of Agricultural Sciences, Ehime University, Japan. Later, he completed his postdoctoral research at the University of the Ryukyus, Okinawa, Japan. Subsequently, he became an Adjunct Senior Researcher at the University of Tasmania with an Australian Government's Endeavour Research Fellowship. Prof. Hasanuzzaman has over 200 publications in the Web of Science. He has edited over 20 books and written over 30 book chapters. His publications are cited over 11,000 times as per Scopus with an h-index of 54 (as of April 2022). He is an Editor and a reviewer for more than 50 peer-reviewed international journals and the recipient of 'Publons Peer Review Award 2017, 2018, and 2019'. He is acting Associate Editor of *Plant Signaling & Behavior* (Taylor and Francis), *Frontiers in Plant Science* (Switzerland), *Biocell* (Tech Science Press, USA), *Phyton* (Tech Science Press, USA); *Academic Editor of PeerJ* (USA), *Plants* (MDPI), *Vegetos* (Springer); and Member of the Editorial Board of *Plant Physiology and Biochemistry* (Elsevier), *Annals of Agricultural Sciences* (Elsevier), etc. He is an active member of 40 professional societies and is the acting Research and Publication Secretary of the Bangladesh JSPS Alumni Association. He received theWorld Academy of Science (TWAS) Young Scientist Award 2014, University Grants Commission (UGC) Gold Medal 2018, Global Network of Bangladeshi Biotechnologists (GNOBB) Award 2021, and Distinguished Scientists Award (2020) from AETDS, India. He is a fellow of the Bangladesh Academy of Sciences (BAS), Royal Society of Biology, and a Foreign fellow of The Society for Science of Climate Change and Sustainable Environment. He is named a Highly Cited Researcher by Clarivate AG.

Masayuki Fujita

Dr. Masayuki Fujita is Professor in the Laboratory of Plant Stress Responses, Faculty of Agriculture, Kagawa University, Kagawa, Japan. He received his B.Sc. in Chemistry from Shizuoka University, Shizuoka, and M.Agr. and Ph.D. in plant biochemistry from Nagoya University, Nagoya, Japan. His research interests include physiological, biochemical, and molecular biological responses based on secondary metabolism in plants under various abiotic and biotic stresses; phytoalexin, cytochrome P450, glutathione S-transferase, and phytochelatin; and redox reaction and antioxidants. In the last decade, his works were focused on oxidative stress and antioxidant defense in plants under environmental stress. His group investigates the role of different exogenous protectants in enhancing antioxidant defense and methylglyoxal detoxification systems in plants. He has supervised 4 M.S. students and 13 Ph.D. students as the main supervisor. He has about 150 publications in journals and books, and has edited 10 books.

Editorial

Approaches to Enhancing Antioxidant Defense in Plants

Masayuki Fujita [1] and Mirza Hasanuzzaman [2,*]

[1]   Laboratory of Plant Stress Responses, Faculty of Agriculture, Kagawa University, Miki-cho, Kita-gun, Takamatsu 761-0795, Japan; fujita@ag.kagawa-u.ac.jp

[2]   Department of Agronomy, Faculty of Agriculture, Sher-e-Bangla Agricultural University, Dhaka 1207, Bangladesh

*   Correspondence: mhzsauag@yahoo.com; Tel.: +880-17165-87711

Citation: Fujita, M.; Hasanuzzaman, M. Approaches to Enhancing Antioxidant Defense in Plants. *Antioxidants* **2022**, *11*, 925. https://doi.org/10.3390/antiox11050925

Received: 28 April 2022

Accepted: 5 May 2022

Published: 8 May 2022

Publisher's Note: MDPI stays neutral with regard to jurisdictional claims in published maps and institutional affiliations.

In the era of global climate change, plants are exposed to various adversities in field conditions [1]. Water stress (drought and waterlogging), salinity, metal/metalloid toxicity, extreme temperatures, xenobiotics, and other abiotic stressors have a significant impact on plant growth, development, and sustainable crop production [2]. Different reactive oxygen species (ROS) including free radicals, (superoxide anion, $O_2^{\bullet-}$; hydroperoxyl radical, $HO_2^{\bullet}$; alkoxy radical, $RO^{\bullet}$; and hydroxyl radical, $^{\bullet}OH$) and non-radical molecules (hydrogen peroxide, H_2O_2, and singlet oxygen, $^1O_2$) are naturally produced in plants as a result of the cellular metabolism [2–4]. However, the overproduction of ROS occurs in plant cells under stress. ROS are highly reactive, interfering with plant metabolism and causing significant damage to essential cellular components such as carbohydrates, lipids, proteins, DNA, and others [5]. Therefore, this disrupts the balance between normal ROS generation and antioxidant activity, leading to oxidative stress in plants [2].

Enhancing the capacity of the antioxidant defense system in plants is the main adaptive response to oxidative stress [6]. The system is mainly maintained by some low-molecular-weight nonenzymatic antioxidants, such as ascorbic acid (AsA), glutathione (GSH), α-tocopherol, phenolic compounds, flavonoids, alkaloids, and nonprotein amino acids along with other antioxidant enzymes [2]. Likewise, a bunch of antioxidant enzymes are associated with this defense system, including peroxidase (POD), superoxide dismutase (SOD), catalase (CAT), ascorbate peroxidase (APX), glutathione reductase (GR), monodehydroascorbate reductase (MDHAR), dehydroascorbate reductase (DHAR), glutathione peroxidase (GPX) and glutathione *S*-transferase (GST) [7]. For example, SOD removes $O_2^{\bullet-}$, CAT converts H_2O_2 into H_2O and O_2, POD scavenges H_2O_2 in the vacuoles, GST combines GSH with the electrophilic or hydrophobic compounds, and MDHAR and DHAR control the ascorbate pool [8]. Other xenobiotics and electrophilic compounds are converted into less hazardous molecules by the GST and GPX enzymes and eventually sequestered in extracellular spaces [9]. One of the vital defense systems is the AsA-GSH cycle, which regulates the levels of H_2O_2 [4]. Żur et al. [10] reported that microspore embryogenesis in triticale was induced through various stress treatments of tillers, and its effectiveness was analyzed in terms of AsA and GSH contents, the total activity of low-molecular-weight antioxidants, and the activities of AsA-GSH cycle enzymes. The exogenous application of melatonin (MEL) was found to be effective in enhancing plant defense under a low temperature [10].

The importance of an antioxidant defense system in plants is crucial for the plants in stressful environments as it delays programmed cell death. In the absence of sufficient antioxidant enzymes in plants to scavenge excessive ROS, cellular organelles cannot properly continue their activities, resulting in lipid peroxidation, protein damage caused by oxidation, breakdown of DNA molecules and nucleic acids, and several enzyme inhibitions [11]. The antioxidant defense system ensures efficient ROS detoxification, reduced lipid peroxidation in membranes, and the prevention of protein damage by delaying oxidation and controlling DNA and nucleic acid damages under stress. Thus, by providing overall cellular protection, this enables the plants' stress tolerance abilities.

However, antioxidant defense efficiency differs between plant species and genotypes, as well as between various stress conditions and their frequencies [2]. El-Badri et al. [12] found that the ROS level and malondialdehyde (MDA) content were minimized in tolerant genotypes due to the activation of antioxidant enzymes, such as SOD, POD, CAT, and APX to scavenge over-accumulated ROS under salinity stress. Most of the contributors to stress tolerance are relevant to amino acids, sucrose, flavonoid metabolism, and tricarboxylic acid cycle, which accumulated as a response to salinity stress. The tolerant cultivar showed improved antioxidant enzyme activity and higher metabolite accumulation, which enhances its tolerance against salinity. Their study is of great reference value for plant breeders to develop salt-tolerant rapeseed cultivars [12]. Begum et al. [13] investigated the antioxidant metabolism in four soybean cultivars, viz., PI408105A, PI567731, PI567690, and PI416937 exposed to drought (5, 10, and 15% polyethylene glycol, PEG-6000), salinity (50, 100, and 150 mM NaCl), and their combination, particularly at the seed germination stage. They found that the ROS accumulation was accompanied by improved enzymatic antioxidant activity, such as SOD, POD, CAT, and APX. However, the enhancement was most noticeable in PI31 and PI90 under both salinity and drought conditions. They concluded that the stress tolerance and improved seed germination of soybean are mainly regulated by its superior antioxidative enzyme activity and secondary metabolites [13].

The production of ROS in plant cells displays both detrimental and beneficial effects. However, the exact pathways of ROS-mediated stress alleviation have yet to be fully elucidated. Sachdev et al. [14] summarized the status of known production sites, signaling mechanisms/pathways, effects, and the management of ROS within plant cells under stress. Moreover, they discussed the role played by advancements in modern techniques such as molecular priming, systems biology, phenomics, and crop modeling in preventing oxidative stress, as well as diverting ROS into signaling pathways [14]. Cross-tolerance is one of the plant adaptive responses to abiotic stress associated with ROS signaling. Rani et al. [15] reported that cold acclimation remarkably enhanced enzymatic (SOD, CAT, APX, GR) and non-enzymatic (AsA and GSH) activity, which resulted in reduced low-temperature-induced leaf damage under cold stress in tolerant chickpea genotypes. This information will be useful in directing efforts to increase cold tolerance [15].

Phytohormones have a significant function in stress signaling and oxidative stress mitigation, along with their direct involvement in growth control. The use of probiotic bacteria from the genera *Bacillus*, *Pseudomonas*, *Enterobacter*, *Micrococcus*, *Lysobacter*, and others improves plant tolerance to a variety of stressors [6]. Qadir et al. [16] revealed that phytohormone-producing, plant-growth-promoting rhizobacteria *Acinetobacter bouvetii* P1 restored the sunflower growth under Cr^{+6} by strengthening the host antioxidant system and triggering the higher production of enzymatic antioxidants, including CAT, APX, SOD, and POD. Moreover, P1 also promoted a higher production of nonenzymatic antioxidants, such as flavonoids, phenolics, proline (Pro), and GSH [16]. Fatma et al. [6] found that exogenous application of methyl jasmonate (MeJA) resulted in an increased enzymatic antioxidant activity that reduced the H_2O_2 content and thiobarbituric acid reactive substances and enhanced the photosynthetic efficiency of wheat under heat stress (42 °C).

The addition of signaling molecules such as nitric oxide (NO), hydrogen sulfide (H_2S), and H_2O_2 stimulates the antioxidant system. Some stress-tolerant plant species and halophytes showed a better ability to synthesize signaling molecules such as NO, which is correlated with stress tolerance. Hasanuzzaman et al. [17] observed that the enhanced salt tolerance in halophyte *Kandelia obovata* was strongly associated with the antioxidant defense, which was governed by NO. The treatment of salt-stressed plants with sodium nitroprusside (SNP) increased endogenous NO levels, reduced ion toxicity, and improved nutrient homeostasis while further increasing Pro levels. SNP treatment also improved the activities of antioxidant enzymes (CAT, APX, MDHAR, DHAR). However, treatment with NO scavengers or inhibitors reversed these beneficial SNP effects and exacerbated salt damage, confirming that SNP promoted stress recovery and improved plant growth under salt stress [17]. Rahman et al. [18] observed that the NO donor could sustain alfalfa plants from iron (Fe) deficiencies. Exogenous NO donor restored Fe-homeostasis and oxidative status in Fe-deficient alfalfa. Specifically, the increase in antioxidant genes and their related enzymes (Fe-SOD, APX) in response to SNP treatment suggests that Fe-SOD and APX are key contributors to reductions in H_2O_2 accumulation and oxidative stress in alfalfa. Furthermore, the elevation of AsA-GSH pathway-related genes (GR and MDAR) in Fe-deficiency with SNP implies that the presence of NO relates to an enhanced antioxidant defense against Fe-deficiency stress [18].

Recently, MEL has been widely tested in abiotic stress situations including drought, waterlogging, and salinity [19]. In their review article, Moustafa-Farag et al. [19] summarized that MEL controls the ROS levels and reactive nitrogen species, and positively changes the molecular defense to improve plant tolerance against water stress. Moreover, the crosstalk between MEL and other phytohormones is a key element of plant survival under drought stress, while this relationship needs further investigation under waterlogging stress [19].

Qi et al. [20] revealed that improvements in endogenous GABA levels in leaf and root by GABA pretreatment could significantly alleviate the damage to white clover during high-temperature stress. The GABA significantly enhanced the gene expression and enzyme activities involved in antioxidant defense, including SOD, CAT, POD, and key enzymes of the AsA–GSH cycle, thus reducing the accumulation of ROS and the oxidative injury to membrane lipids and proteins. In addition, the expression and decline in the GABA-induced aquaporin in endogenous abscisic acid levels could improve the heat dissipation capacity by maintaining a higher stomatal opening and transpiration in white clovers under high-temperature stress [20].

Phytohormone-based seed priming also proved to be effective against abiotic stress. Basit et al. [21] concluded that seed priming with brassinosteroids (EBL) could be adopted as a promising strategy to enhance rice growth by copping the toxic effect of chromium (Cr). They [21] revealed that seed priming with a low dose (0.01 μM) of EBL could alleviate the adverse effects of Cr in two different rice cultivars. Seed priming with EBL stimulatingly increased antioxidant enzyme activities to scavenge ROS production under Cr stress. The gene expression of SOD and POD in EBL-primed rice plants followed a similar increasing trend, as observed in the case of enzymatic activities of SOD and POD compared to water-primed rice plants. Simultaneously, Cr uptake was observed to be higher in the water-primed control compared to plants primed with EBL.

Antioxidant metabolism has been improved by a variety of plant nutrients [8]. According to Imran et al. [22], molybdenum (Mo) supply was found to strengthen plant metabolism at prominent growth stages through an improved enzymatic and non-enzymatic antioxidant defense system, thereby increasing the grain yield and quality characteristics of aromatic rice under cadmium (Cd) toxicity. Importantly, Mo supply enhanced photosynthesis, Pro, and soluble protein content, and also strengthened plant metabolism and antioxidant defense by maintaining higher activities and transcript abundance of ROS-detoxifying enzymes at the vegetative, reproductive, and maturity stages of aromatic rice plants under Cd toxicity [22].

Researchers have used a variety of techniques to reduce the negative consequences of oxidative damage by increasing antioxidant defenses [6]. Approaches ranging from genetic manipulation to the introduction of exogenous protectants in plants. Characterization and profiling antioxidant enzymes are among the targeted approaches. In their study, Su et al. [23] performed a genome-wide investigation to identify the rapeseed *SOD* genes. They recognized 31 *BnSOD* genes in the rapeseed genome, including 14 *BnCSDs*, 11 *BnFSDs*, and six *BnMSDs*. In brief, gene ontology annotation outcomes confirm the *BnSODs* role under different stress stimuli, cellular oxidant detoxification processes, metal ion binding activities, SOD activity, and different cellular components. Moreover, the expression profiling showed that eight genes (*BnCSD1, BnCSD3, BnCSD14, BnFSD4, BnFSD5, Bn*-

FSD6, *BnMSD2*, and *BnMSD10*) were significantly up-regulated under different hormones (abscisic acid, gibberellic acid, indole acetic acid, and kinetin) and abiotic stress (salinity, cold, waterlogging, and drought) treatments. Their findings deliver the foundation for additional functional investigations into the *BnSOD* genes in rapeseed breeding programs. Rudíc et al. [24] have analyzed eight functional *SOD* genes from potato, three *StCuZnSODs*, one *StMnSOD*, and four *StFeSODs*. The quantitative analysis revealed a higher induction of *StCuZnSODs* (the major potato *SODs*) and *StFeSOD3* in thermotolerant cultivars than in thermosensitive cultivars during long-term exposure to elevated temperature. *StMnSOD* was constitutively expressed, while the expression of *StFeSODs* was cultivar-dependent. Their results provide the basis for further research on *StSODs* and their regulation in potato, particularly in response to elevated temperatures [24].

This Special Issue, "Approaches in Enhancing Antioxidant Defense in Plants" published 13 original research works and a couple of review articles that discuss the various aspects of plant oxidative stress biology and ROS metabolism, as well as the physiological mechanisms and approaches to enhancing antioxidant defense and mitigating oxidative stress. These papers will serve as a foundation for plant oxidative stress tolerance and, in the long term, provide further research directions in the development of crop plants' tolerance to abiotic stress in the era of climate change.

Author Contributions: Conceptualization, M.F. and M.H.; writing—original draft preparation, M.H. and M.F.; writing—review and editing, M.F. and M.H. All authors have read and agreed to the published version of the manuscript.

Funding: This work received no external funding.

Acknowledgments: We would like to thank all contributors for their submission to this Special Issue and the reviewers for their valuable input in improving the articles. We also thank the editorial office for their helpful support during the compilation of this Special Issue. The authors acknowledge Farzana Nowroz, Ayesha Siddika, Shamima Sultana, and Md. Rakib Hossain Raihan for their help during the organization of the literature and necessary formatting.

Conflicts of Interest: The authors declare no conflict of interest.

References

1. Chaki, M.; Begara-Morales, J.C.; Barroso, J.B. Oxidative stress in plants. *Antioxidants* **2020**, *9*, 481. [CrossRef] [PubMed]
2. Hasanuzzaman, M.; Bhuyan, M.H.M.B.; Zulfiqar, F.; Raza, A.; Mohsin, S.M.; Mahmud, J.A.; Fujita, M.; Fotopoulos, V. Reactive oxygen species and antioxidant defense in plants under abiotic stress: Revisiting the crucial role of a universal defense regulator. *Antioxidants* **2020**, *9*, 681. [CrossRef] [PubMed]
3. Mehla, N.; Sindhi, V.; Josula, D.; Bisht, P.; Wani, S.H. An introduction to antioxidants and their roles in plant stress tolerance. In *Reactive Oxygen Species and Antioxidant Systems in Plants: Role and Regulation under Abiotic Stress*; Khan, M.I.R., Khan, N.A., Eds.; Springer: Singapore, 2017; pp. 1–23.
4. Hasanuzzaman, M.; Bhuyan, M.; Anee, T.I.; Parvin, K.; Nahar, K.; Mahmud, J.A.; Fujita, M. Regulation of ascorbate-glutathione pathway in mitigating oxidative damage in plants under abiotic stress. *Antioxidants* **2019**, *8*, 384. [CrossRef] [PubMed]
5. Raja, V.; Majeed, U.; Kang, H.; Andrabi, K.I.; John, R. Abiotic stress: Interplay between ROS, hormones and MAPKs. *Environ. Exp. Bot.* **2017**, *137*, 142–157. [CrossRef]
6. Fatma, M.; Iqbal, N.; Sehar, Z.; Alyemeni, M.N.; Kaushik, P.; Khan, N.A.; Ahmad, P. Methyl jasmonate protects the PS II system by maintaining the stability of chloroplast D1 protein and accelerating enzymatic antioxidants in heat-stressed wheat plants. *Antioxidants* **2021**, *10*, 1216. [CrossRef]
7. Hasanuzzaman, M.; Mahmud, J.A.; Anee, T.I.; Nahar, K.; Islam, M.T. Drought stress tolerance in wheat: Omics approaches in understanding and enhancing antioxidant defense. In *Abiotic Stress-Mediated Sensing and Signaling in Plants: An Omics Perspective*; Zargar, S.M., Zargar, M.Y., Eds.; Springer: Singapore, 2018; pp. 267–307. [CrossRef]
8. Nahar, K.; Rhaman, M.S.; Parvin, K.; Bardhan, K.; Marques, D.N.; García-Caparrós, P.; Hasanuzzaman, M. Arsenic-induced oxidative stress and antioxidant defense in plants. *Stresses* **2022**, *2*, 179–209. [CrossRef]
9. Hasanuzzaman, M.; Raihan, M.R.H.; Khojah, E.; Samra, B.N.; Fujita, M.; Nahar, K. Biochar and chitosan regulate antioxidant defense and methylglyoxal detoxification systems and enhance salt tolerance in jute (*Corchorus olitorius* L.). *Antioxidants* **2021**, *10*, 2017. [CrossRef]
10. Żur, I.; Kopeć, P.; Surówka, E.; Dubas, E.; Krzewska, M.; Nowicka, A.; Janowiak, F.; Juzoń, K.; Janas, A.; Barna, B.; et al. Impact of ascorbate—Glutathione cycle components on the effectiveness of embryogenesis induction in isolated microspore cultures of barley and triticale. *Antioxidants* **2021**, *10*, 1254. [CrossRef]

11. Dumanović, J.; Nepovimova, E.; Natić, M.; Kuča, K.; Jaćević, V. The significance of reactive oxygen species and antioxidant defense system in plants: A concise overview. *Front. Plant Sci.* **2021**, *11*, 552969. [CrossRef]
12. El-Badri, A.M.; Batool, M.; Mohamed, I.A.A.; Wang, Z.; Khatab, A.; Sherif, A.; Ahmad, H.; Khan, M.N.; Hassan, H.M.; Elrewainy, I.M.; et al. Antioxidative and metabolic contribution to salinity stress responses in two rapeseed cultivars during the early seedling stage. *Antioxidants* **2021**, *10*, 1227. [CrossRef]
13. Begum, N.; Hasanuzzaman, M.; Li, Y.; Akhtar, K.; Zhang, C.; Zhao, T. Seed germination behavior, growth, physiology and antioxidant metabolism of four contrasting cultivars under combined drought and salinity in soybean. *Antioxidants* **2022**, *11*, 498. [CrossRef] [PubMed]
14. Sachdev, S.; Ansari, S.A.; Ansari, M.I.; Fujita, M.; Hasanuzzaman, M. Abiotic stress and reactive oxygen species: Generation, signaling, and defense mechanisms. *Antioxidants* **2021**, *10*, 277. [CrossRef]
15. Rani, A.; Kiran, A.; Sharma, K.D.; Prasad, P.V.V.; Jha, U.C.; Siddique, K.H.M.; Nayyar, H. Cold tolerance during the reproductive phase in chickpea (*Cicer arietinum* L.) is associated with superior cold acclimation ability involving antioxidants and cryoprotective solutes in anthers and ovules. *Antioxidants* **2021**, *10*, 1693. [CrossRef] [PubMed]
16. Qadir, M.; Hussain, A.; Hamayun, M.; Shah, M.; Iqbal, A.; Irshad, M.; Ahmad, A.; Lodhi, M.A.; Lee, I.-J. Phytohormones producing *Acinetobacter bouvetii* P1 mitigates chromate stress in sunflower by provoking host antioxidant response. *Antioxidants* **2021**, *10*, 1868. [CrossRef] [PubMed]
17. Hasanuzzaman, M.; Inafuku, M.; Nahar, K.; Fujita, M.; Oku, H. Nitric oxide regulates plant growth, physiology, antioxidant defense, and ion homeostasis to confer salt tolerance in the mangrove species, *Kandelia obovata*. *Antioxidants* **2021**, *10*, 611. [CrossRef] [PubMed]
18. Rahman, M.A.; Kabir, A.H.; Song, Y.; Lee, S.-H.; Hasanuzzaman, M.; Lee, K.-W. Nitric oxide prevents Fe deficiency-induced photosynthetic disturbance, and oxidative stress in alfalfa by regulating Fe acquisition and antioxidant defense. *Antioxidants* **2021**, *10*, 1556. [CrossRef] [PubMed]
19. Moustafa-Farag, M.; Mahmoud, A.; Arnao, M.B.; Sheteiwy, M.S.; Dafea, M.; Soltan, M.; Elkelish, A.; Hasanuzzaman, M.; Ai, S. Melatonin-induced water stress tolerance in plants: Recent advances. *Antioxidants* **2020**, *9*, 809. [CrossRef]
20. Qi, H.; Kang, D.; Zeng, W.; Jawad Hassan, M.; Peng, Y.; Zhang, X.; Zhang, Y.; Feng, G.; Li, Z. Alterations of endogenous hormones, antioxidant metabolism, and aquaporin gene expression in relation to γ-aminobutyric acid-regulated thermotolerance in white clover. *Antioxidants* **2021**, *10*, 1099. [CrossRef]
21. Basit, F.; Chen, M.; Ahmed, T.; Shahid, M.; Noman, M.; Liu, J.; An, J.; Hashem, A.; Fahad Al-Arjani, A.-B.; Alqarawi, A.A.; et al. Seed priming with brassinosteroids alleviates chromium stress in rice cultivars via improving ROS metabolism and antioxidant defense response at biochemical and molecular levels. *Antioxidants* **2021**, *10*, 1089. [CrossRef]
22. Imran, M.; Hussain, S.; He, L.; Ashraf, M.F.; Ihtisham, M.; Warraich, E.A.; Tang, X. Molybdenum-induced regulation of antioxidant defense-mitigated cadmium stress in aromatic rice and improved crop growth, yield, and quality traits. *Antioxidants* **2021**, *10*, 838. [CrossRef]
23. Su, W.; Raza, A.; Gao, A.; Jia, Z.; Zhang, Y.; Hussain, M.A.; Mehmood, S.S.; Cheng, Y.; Lv, Y.; Zou, X. Genome-wide analysis and expression profile of superoxide dismutase (SOD) gene family in rapeseed (*Brassica napus* L.) under different hormones and abiotic stress conditions. *Antioxidants* **2021**, *10*, 1182. [CrossRef] [PubMed]
24. Rudić, J.; Dragićević, M.B.; Momčilović, I.; Simonović, A.D.; Pantelić, D. In Silico study of superoxide dismutase gene family in potato and effects of elevated temperature and salicylic acid on gene expression. *Antioxidants* **2022**, *11*, 488. [CrossRef] [PubMed]

Review

Abiotic Stress and Reactive Oxygen Species: Generation, Signaling, and Defense Mechanisms

Swati Sachdev [1], Shamim Akhtar Ansari [2], Mohammad Israil Ansari [3,*], Masayuki Fujita [4,*] and Mirza Hasanuzzaman [5,*]

[1] Department of Environmental Science, School for Environmental Sciences, Babasaheb Bhimrao Ambedkar University, Vidya Vihar, Rae Bareli Road, Lucknow 226 025, India; swati_sachdev2003@yahoo.com
[2] Institute of Forest Research and Productivity, Ranchi 835 303, India; shamimansari_1@yahoo.co.uk
[3] Department of Botany, University of Lucknow, Lucknow 226 007, India
[4] Laboratory of Plant Stress Responses, Department of Applied Biological Science, Faculty of Agriculture, Kagawa University, 2393 Ikenobe, Miki-cho, Kita-gun, Kagawa 761-0795, Japan
[5] Department of Agronomy, Faculty of Agriculture, Sher-e-Bangla Agricultural University, Dhaka 1207, Bangladesh
* Correspondence: ansari_mi@lkouniv.ac.in (M.I.A.), fujita@ag.kagawa-u.ac.jp (M.F.); mhzsauag@yahoo.com (M.H.)

Citation: Sachdev, S.; Ansari, S.A.; Ansari, M.I.; Fujita, M.; Hasanuzzaman, M. Abiotic Stress and Reactive Oxygen Species: Generation, Signaling, and Defense Mechanisms. *Antioxidants* **2021**, *10*, 277. https://doi.org/10.3390/antiox10020277

Academic Editor: Francisco Corpas
Received: 3 December 2020
Accepted: 1 February 2021
Published: 11 February 2021

Abstract: Climate change is an invisible, silent killer with calamitous effects on living organisms. As the sessile organism, plants experience a diverse array of abiotic stresses during ontogenesis. The relentless climatic changes amplify the intensity and duration of stresses, making plants dwindle to survive. Plants convert 1–2% of consumed oxygen into reactive oxygen species (ROS), in particular, singlet oxygen ($^1O_2$), superoxide radical ($O_2{}^{\bullet-}$), hydrogen peroxide (H_2O_2), hydroxyl radical ($^{\bullet}OH$), etc. as a byproduct of aerobic metabolism in different cell organelles such as chloroplast, mitochondria, etc. The regulatory network comprising enzymatic and non-enzymatic antioxidant systems tends to keep the magnitude of ROS within plant cells to a non-damaging level. However, under stress conditions, the production rate of ROS increases exponentially, exceeding the potential of antioxidant scavengers instigating oxidative burst, which affects biomolecules and disturbs cellular redox homeostasis. ROS are similar to a double-edged sword; and, when present below the threshold level, mediate redox signaling pathways that actuate plant growth, development, and acclimatization against stresses. The production of ROS in plant cells displays both detrimental and beneficial effects. However, exact pathways of ROS mediated stress alleviation are yet to be fully elucidated. Therefore, the review deposits information about the status of known sites of production, signaling mechanisms/pathways, effects, and management of ROS within plant cells under stress. In addition, the role played by advancement in modern techniques such as molecular priming, systems biology, phenomics, and crop modeling in preventing oxidative stress, as well as diverting ROS into signaling pathways has been canvassed.

Keywords: abiotic stress; antioxidant; biomolecules; climate change; reactive oxygen species

1. Introduction

Climate change has drastically reduced the environmental services, enhancing plants' vulnerability to various abiotic stresses during ontogenesis [1] that disparages their struggle for survival, growth, and economic output [2]. Abiotic stresses encompassing heat shock, chilling/freezing, water-deficit, waterlogging, salinity, nutrient imbalance, heavy metals, and xenobiotic stress account for 50% productivity loss [3]. The contributory environmental factors are extreme temperature events (low or high), excess irradiation (UV-A and UV-B), fluctuation in light intensities (low or high), strong storm events, non-uniformity in the rainfall pattern (deficit or excess), discharge and accumulation of heavy metals, and other xenobiotic compounds (pesticides, fertilizers, hydrocarbons) [4–6]. In a natural environment, the abiotic stresses often occur in combination [7] due to their interrelated pathways and show unparalleled and compounded effects on plants, impinging their cellular, metabolic, and physiological activities [1,5].

Reactive oxygen species (ROS) such as superoxide radical ($O_2{}^{\bullet-}$), hydrogen peroxide (H_2O_2), hydroxyl radical ($^\bullet OH$), singlet oxygen ($^1O_2$), peroxy radical ($ROO^\bullet$), and alkoxyl radicals ($RO^\bullet$) are produced at low temperature within a threshold concentration in the plant cell under ambient environmental conditions. However, the extreme environmental conditions trigger excessive production of ROS [8]. ROS damage molecular and cellular components due to the oxidation of biomolecules (lipid, carbohydrates, proteins, enzymes, DNA) and cause plant death [6,9]. To avert the damages, plants tightly regulate ROS production via the recruitment of enzymatic and non-enzymatic antioxidants. The enzymatic antioxidant system comprising superoxide dismutase (SOD), catalase (CAT), ascorbate peroxidase (APX), glutathione reductase (GR), peroxidase (POX), etc. and non-enzymatic antioxidants such as vitamins, flavonoids, stilbenes, and carotenoids quench the excess ROS, thereby providing a shield against oxidative stress [7,10,11]. Unfettered propagation of oxygen (O_2) derived reactive species is detrimental to the plant health. However, a controlled ROS production participates in redox signaling, plant growth, and development during stress [12]. Fine-tuned ROS production mediates cell to cell communication by magnifying signals via the nicotinamide adenine dinucleotide phosphate (NADPH) oxidase, also called respiratory burst oxidase homolog (RBOH) and retaliating stress by modulating the protein structure and activating defense responsive genes [9].

The occurrence of abiotic stresses either individually or simultaneously triggers the overproduction of ROS in plant cells that becomes a major challenge for optimal plant growth and productivity. Exploring the underlying molecular mechanisms of ROS signaling pathways assumes a great significance to mitigate stress, or promote signaling under current and future climatic scenarios, as well as retain tolerance and economic productivity in plants of economic importance. The present review provides a critical analysis of the accumulated knowledge on the impact of plant fitness under abiotic stresses as well as explores antioxidant-based defense mechanisms regulating ROS accumulation and dissipating oxidative stress. The review unravels the dual role of ROS as a signaling molecule triggering plant acclimatization and development under stress(es). Moreover, the implication of scientific and technological applications such as molecular priming, systems biology, phenomics, and crop modeling to fortify plants' tolerance against oxidative stress has also been discussed.

2. Climate Change Triggers Abiotic Stress and ROS Generation

Climate change has escalated the prevalence of abiotic stress and their debt on plants (Figure 1), which is witnessed on a broad geographical scale. FAO (2019) has reported that 96.5% of the global cultivation area experiences one or the other kind of stress [13]. The atmospheric enrichment of greenhouse gases has raised the mean global surface temperature (0.85 °C from 1880 to 2012) and changed rainfall patterns [14]. It has been anticipated that a 1 °C hike in temperature enhances 4–4.5% more water requirement in plants [9,15] making plant cultivation a more denting task in drought-affected areas. An increase in temperature is suspected to bio-transform chemical pollutants into more toxic or bioactive forms that will aggravate environmental nuisance and perniciously affect plant homeostasis [16]. A rise in temperature coupled with precipitation reduction promotes the volatilization of xenobiotic compounds as persistent organic pollutants, exacerbating air pollution. On the other hand, the excess precipitation enhances deposits of air pollutants on land and reinforces the leaching of soil nutrients and pollutants to groundwater causing soil pollution, aquifer contamination, nutrient imbalance, and salinity [16–18]. In normal circumstances, fluctuation in temperature and rainfall, nutrient imbalance, waterlogging, etc. temporarily and competitively restrict plant growth. However, due to extreme climatic events and fluctuations in routine weather conditions, both the severity and duration of stresses prolong and get amplified, drastically eclipsing the plant performance beyond recovery [13,19]. Climate change has been geared up to create adverse conditions that plants cannot escape and face several vandalizing impacts of abiotic stresses (Table 1). Improving plant growth and productivity to feed the existing global population is not the only challenge, but to fulfill the nutritional needs of the future generation is equally important. Therefore, it is crucial to review the extent of impinging effects of various persisting abiotic stresses on plants. Coupling these data with simulation models could help chalk out sustainable strategies for crop protection in accordance with the projected change in environmental conditions.

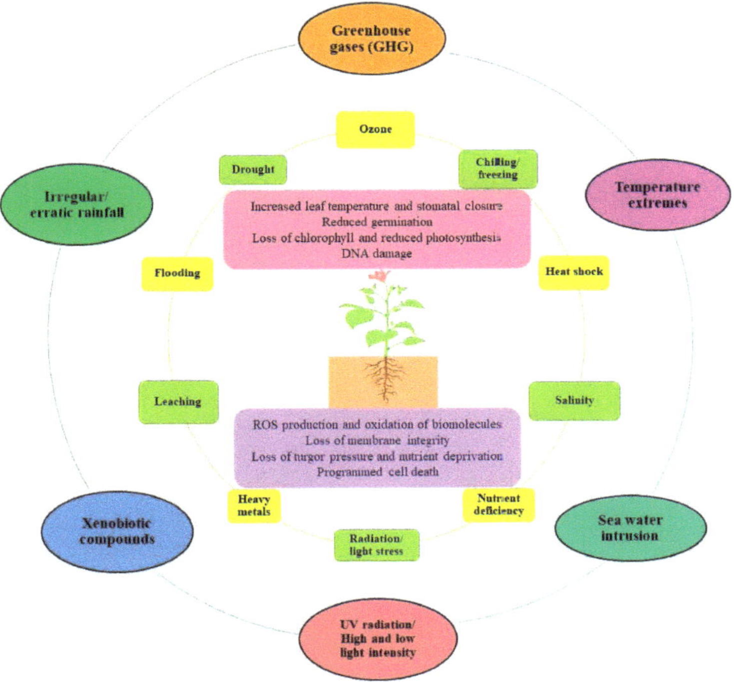

Figure 1. Climate change intensifies the magnitude of abiotic stresses that severely affect plant growth and physiological activities. Various abiotic factors modified by environmental conditions (outer circle) leads to abiotic stresses (the inner circle) that hamper plant physiological and metabolic activities, biomolecules, cellular structure, growth, and productivity (rectangular innermost blocks).

Table 1. Abiotic stresses triggered secondary stresses and their damaging effects on plant growth and activity.

Abiotic Stress	Induced Secondary Stresses	Effects in Plant	References
Chilling/freezing stress	Nutritional imbalance, osmotic and oxidative stress	• Accumulation of ROS and oxidative damage; inhibition of enzymes' activities and metabolic imbalance. • Increased cell dehydration and starvation, senescence, delayed maturation, damage of PS II, and reduced photosynthetic activity. • Decreased growth and productivity.	[20,21]

Table 1. *Cont.*

Abiotic Stress	Induced Secondary Stresses	Effects in Plant	References
Drought	Osmotic, heavy metal, and oxidative stress	• Increased ROS production and ion leakage; induced dehydration and turgor loss. • Decrease in absorption and translocation of mineral nutrients. • Protein denaturation, loss of enzyme activities, reduced photosynthetic activity due to abridged chlorophyll content and CO_2 assimilation. • Increase in leaf temperature, premature abscission, necrosis, and stunted plant growth.	[22,23]
Flooding/waterlogging	Water and nutrient deficiency stress, oxidative stress	• Increased ROS and ethylene production and decreased antioxidants level. • Reduced stomatal conductance; abridged water and nutrient uptake. • Reduced gaseous exchange, anoxia/hypoxia, increased anaerobic metabolism and inhibited root respiration; reduced photosynthetic activity due to the decreased chlorophyll content and damage of PS II. • Stunted growth and senescence of leaf and inflorescence.	[24,25]
Heat stress	Water scarcity, osmotic and oxidative stress	• Enhanced ROS production and oxidative damage, protein misfolding, and denaturation. • Growth inhibition, foliar senescence, and abscission, leaf and fruit discoloration, reduced CO_2 fixation, PS I and PS II disruption, disturbed ion transport.	[26–28]
Heavy metals/xenobiotic compounds	Nutrient and oxidative stress	• Increased ROS production and oxidative damage. • Disruption of function and structure of enzymes; reduced stomatal conductance, CO_2 assimilation, and net photosynthesis rate. • Reduced biomass accumulation, inhibition of seed germination, and impaired nutrient uptake.	[29]

Table 1. *Cont.*

Abiotic Stress	Induced Secondary Stresses	Effects in Plant	References
Light/radiation stress	Oxidative stress		[30]
		• Increased ROS production and oxidative damage, disrupted photosynthesis ETC, and/or increased activity of membrane-bounded NADPH oxidase, chlorophyll degradation, reduced photosynthetic activity, and epidermal cell expansion inhibition. • Leaf senescence, reduced rosette diameter, condensed inflorescence stem with a boosted number of flowering stems.	
Nutrient imbalance	Oxidative stress		[31,32]
		• ROS accumulation with reduced antioxidants; increased leakage of ion and solutes, reduced activities of metalloenzymes, declined photosynthesis. • Susceptibility to other biotic and abiotic stresses. • Stunted growth, chlorosis, necrosis, poor flowering and fruiting, declined productivity.	
Ozone (O_3) stress	Oxidative stress		[33,34]
		• ROS production inducing oxidative damage, inhibited enzyme activities, chlorophyll and xanthophyll degradation, diminished stomatal conductance, and decreased photosynthesis. • Leaf chlorosis and necrosis, early senescence, and reduced plant biomass and productivity.	
Salinity	Water scarcity, ionic imbalance, nutrient, osmotic and oxidative stress	• ROS production causing oxidative damage, restricted uptake and translocation of water and mineral nutrients causing Na^+ toxicity and decreased K^+, Ca^{2+}, and Mg^{2+} content, reduced soil water potential. • Decreased stomatal opening, disorganized thylakoid ultrastructure, and reduced photosynthesis. • Reduced seed germination, immature leaf senescence, and abridged growth and productivity.	[7,35,36]

2.1. Temperature Stress

The extreme variation in temperature (10–15 °C deviation) above or below an optimum condition induces heat or chilling/freezing stress that impairs photosynthesis, plant architecture, reproduction, and productivity [37]. A plant encountered with heat stress undergoes morphological, cellular, and metabolic changes that decrease the function of photosynthetic and respiratory apparatus, reduce enzymatic activity, upregulate transcription, and translation of heat shock proteins (HSP), increase calcium (Ca^{2+}) influx, and intensify ROS production [37]. Heat stress inhibits the cell differentiation process, therefore, affecting the leaf area [38]. Exposure of hyacinth bean (*Lablab purpureus* L.) to a high temperature significantly affects membrane permeability, increases ROS production, and lipid peroxidation; abridges plant growth, productivity, and leaf area; reduces leaf chlorophyll and carotenoid content; and causes an imbalance between the generation and scavenging of H_2O_2 and $O_2^{\bullet-}$ [39]. The imposition of cucumber plant to heat stress reduces growth, yield, chlorophyll content, photosynthesis, stomatal conductance, transpiration rate, antioxidants, and membrane stability index, while increasing ROS production, lipid peroxidation, intercellular carbon dioxide (CO_2) concentration, and non-photochemical quenching (NPQ) [38]. A high temperature elevates the production of ozone (O_3) in the troposphere which imposes oxidative stress on plants [40]. Chilling stress characterized by low-temperature events facilitates solubility and the accumulation of O_2 and electron leakage from the photosynthetic electron transport chain (ETC)/reduction of respiratory ETC that together enhances ROS production in plant cells [41], affecting membrane fluidity and enzymes activities [42]. Under chilling stress, an enhanced electrolyte leakage with reduced chlorophyll and tissue water content has been reported in cucumber seedlings [43]. Increased malondialdehyde (MDA) content, *RBOH1* expression, and accumulation of H_2O_2 and $O_2^{\bullet-}$ in leaves, and reduced net photosynthesis rate, as well as chlorophyll fluorescence, has been observed in tomato under low-temperature stress [44].

2.2. Water Stress

During the last decades, change in climatic scenarios has tremendously affected the rainfall patterns causing erratic precipitation with an altered magnitude and seasonal variations [45]. The situation fosters extremes of drought and flooding in different parts of the globe.

2.2.1. Water Deficit (Drought)

Drought imposing water deficit stress leads to water scarcity, restricted growth, and yield in plants [23,46,47]. Water deficit stress sets a reduction in the plant water potential and turgor to the level that impairs the normal functioning of cells [45]. The physiological impact of water deficit conditions varies with the severity and duration of stress. Water deficit stress reduces stomatal opening, abridges CO_2 fixation, accelerates photoreduction of O_2 in the chloroplast, and increases photorespiration, eventually leading to ROS accumulation and oxidative damage in plants [42]. The reduced number of spells coupled with a high temperature has aggravated drought conditions in many parts of the world. According to a World Bank report (2006), India ranks second among the most severely drought-affected Asian countries [48]. Due to drought, worldwide productivity has reduced by 21% in wheat and 40% in maize during the past few years [23]. Lee et al. [49] have reported a decrease in dry mass, enhanced accumulation of ROS, and increased MDA content in white clover leaves under water deficit conditions.

2.2.2. Waterlogging and Flooding

The excessive accumulation of water in soil due to heavy precipitation over a period of time, poor drainage, etc. causes soil flooding or waterlogging [50]. Nearly 10% of the world's total land has been detrimentally affected by waterlogging [51]. During 2006–2016, two-thirds of the total global crop loss and damage has been attributed to floods [50]. Waterlogging covers plant roots and is characterized by low light, impaired gaseous exchange, hypoxia, and anoxia [50]. It reduces O_2 diffusion by 10,000 times compared to air, thereby suppressing aerobic activity, including root respiration in soil [52]. The anoxic condition inhibits ETC of chloroplast and mitochondria that consequently results in the production of ROS [53,54]. Sesame plants subjected to waterlogging conditions show increased lipid peroxidation, ROS accumulation, and methylglyoxal content that induce oxidative stress [55]. In the case of clear flooded water, light easily reaches the submerged plant parts and induces photorespiration, and produces peroxisomal H_2O_2 [54]. Flooding also leaches out essential nutrients from the soil, accumulates salts, and increases the availability of heavy metals owing to the change in soil pH. These adverse changes ultimately induce nutrient deficiency and other stresses (salinity, heavy metal) in plants [56].

2.3. Salt Stress

Soil salinity has globally degraded nearly 20% of total arable and 33% of the irrigated land [57]. Excess sodium (Na^+) and chloride (Cl^-) ions present in the saline soil are transported and accumulated to the toxic level at the expense of other essential ions in plant cells [36]. Salt stress plants desisting water absorption experience drought-like conditions [36]. Therefore, salinity reduces stomatal conductance and disrupts photosystem (PS) and photosynthetic enzymes that lead to ROS production in plants [57]. The accumulation of ROS in plant cells under salinity is also mediated through the plasma membrane NADPH oxidase and apoplast (all parts beyond the plasma membrane including the cell wall) diamine oxidases (DAOs) [58]. The exposure of wheat cultivars to salinity stress increases ROS accumulation that induces lipid peroxidation and electrolyte leakage thereby reducing membrane stability [59]. The effects of salinity have been more pronounced on sensitive wheat cultivar HD2329. Similarly, the higher H_2O_2 accumulation and MDA content under salinity stress have been reported in a salt-sensitive cultivar of *Brassica juncea* as compared to its tolerant cultivar [60].

2.4. Nutrient Deficiency

Accessibility to essential nutrients, ensuring proper plant growth and development, has become a major challenge owing to the persistently changing attributes of global climate. The scarcity of essential plant nutrients in soil adversely affects their physiological activities particularly ETC, water relation, and gaseous exchange that contribute to ROS production and trigger oxidative stress in plants [61]. Plasma membrane-bounded NADPH oxidase is one of the major sources of ROS generation in plant cells [62]. Plant nutrients such as zinc (Zn^{2+}) and potassium (K^+) regulate the activity of NADPH oxidase and therefore, their scarcity elevates the enzyme activity which catalyzes the production of $O_2^{\bullet-}$ [61] or H_2O_2 [63]. Nutrient starved plants elicit ROS production via the ethylene signaling cascade. The low availability of K^+ prompts ethylene biosynthesis that, in turn, up-streams ROS production [64]. Mineral nutrients such as nitrogen (N), magnesium (Mg), copper (Cu), manganese (Mn), Zn, etc. are an integral part of various enzymes (Cu/Zn-SOD, Mn-SOD, etc.) and antioxidants that participate in energy metabolism or scavenge ROS [61,63,65]. The deficiency of nutrients impairs the ROS scavenging capacity of plants and indirectly results in ROS production [61,65]. For instance, the diminished potential of enzymes to scavenge H_2O_2 and $O_2^{\bullet-}$ within plant cells increases the level of $^\bullet OH$ via the Heber-Weiss reaction [61]. Further deprivation of elements such as Mg which is a major constituent of chlorophyll impairs the photosynthetic activity resulting in ROS generation [66].

2.5. Heavy Metal and Xenobiotics Stress

The accumulation of non-essential metals shows toxicity in plants via ROS generation. However, the unrestricted uptake of essential nutrients also induces ROS production [32]. Heavy metals such as iron (Fe), chromium (Cr), and Cu are major redox-active metals that impose oxidative stress in plants owing to their high concentrations in soil [67]. Heavy metal stress triggers ROS production mediated through ETC of chloroplast, mitochondria, apoplast, and peroxisome [68,69]. Cadmium (Cd) is a non-essential metal that causes toxicity in plants. Cd supersedes Cu or Fe ions in antioxidant metalloenzymes with their impeded activities, indirectly inducing ROS production, impairing respiratory ETC, and interfering with the redox status in cells [63]. Despite being an essential micronutrient, the excess accumulation of Fe also initiates the production of ROS in plants through a series of reactions [70] and causes damage to the lipid membrane and chlorophyll [61]. The reduced form of Fe oxidizes to produce H_2O_2 and $O_2^{\bullet-}$. In turn, H_2O_2 oxidizes the reduced Fe compounds to generate highly toxic $^{\bullet}OH$ [61]. This auto-oxidation of redox-active metals such as Fe and Cu consequently results in ROS formation, mediated by the Fenton-type reaction [67]. Homologous to heavy metals, xenobiotic compounds such as pesticides also trigger ROS production leading to oxidative stress [71]. Out of the total pesticides applied, only 1% reaches the target, the remaining very large proportion accumulates in soil and non-target living organisms [72]. Pesticides retard plant growth, abridge photosynthetic efficiency, induce molecular alterations, increase ROS production, and modify the antioxidant status [71,73]. The degradation of chlorophyll with an increase in H_2O_2 and MDA level has been reported in tomato leaves treated with thiram [73]. In another study, imidacloprid declines the chlorophyll content in *B. juncea* seedlings. The reduction in chlorophyll is attributed to an enhanced expression of gene *CHLASE* encoding chlorophyllase enzyme that catalyzes chlorophyll degradation [71]. Moreover, insecticides enhance the RBOH transcript level and ROS accumulation.

2.6. Co-Occurrence of Multiple Abiotic Stresses

Plants growing in natural conditions are exposed to multiple stresses at the same time. For example, an increase in temperature enhances evapotranspiration that induces stresses of water-deficit and soil salinization simultaneously and has a dramatic impact on growth and productivity. A combination of abiotic stress induces a unique and complex set of responses at the physiological, metabolic, and molecular levels, which are different than what is being observed under individual stress scenarios [28]. The confluence of heat and drought stress induces the closure of stomata, whereas the individual heat stress effect prompts the opening of stomata for transpiration and assists cooling in *Arabidopsis* [74]. Rizhsky et al. [75] have demonstrated differential physiological responses during heat shock, drought, and combined stress (heat+drought) in the tobacco plant. Drought reduces the respiration rate and photosynthesis, whereas heat shock increases the respiration rate without a significant change in the photosynthesis as compared to the control. The combined stress treatment reduces the process of photosynthesis compared to the individual drought stress but significantly enhances respiration compared to the heat shock stress. The stomatal conductance and leaf temperature significantly alter during the combined stress conditions. Stomatal conductance gets reduced and the leaf temperature, increased by 2–3 °C in plants, is exposed to stress combination due to the closed stomata and negligible transpiration. Analogously, Semwal and Khanna-Chopra [76] have reported that jointly operating heat and water deficit stress leads to ROS production, oxidative damage, and attenuates the antioxidant defense capacity (CAT activity and higher redox pool) in *Chenopodium album*.

Correspondingly, the combined stress conditions also provoke a dissimilar alteration at the molecular level in many cases. For instance, the individual gene in the *Arabidopsis* ROS gene network follows differential expressions under dissimilar stresses [75] due to different sets of responses being required under various stress conditions. As a result, the combination of stresses shows an independent and unique set of responses [77]. On exposure to the combined stresses of heat and drought, 770 specific transcripts have been recorded compared to the individual stress of either heat or drought, indicating elicitation of a unique acclimation response under stress combination [74]. Similarly, the combined effect of heat, drought, and biotic (viruses) stress induce molecular reprogramming leading to a significant reorganization of defense response [78]. Plants, to survive under the persisting combination of environmental cues, tailor their defense responses resulting in a cross-talk between various mechanisms. Several studies highlight that the cross-talks of regulatory molecules with signaling pathways trigger tolerance to multiple stresses [79]. The concurrent occurrence of stresses may have complementary or detrimental consequences on plants [76]. For example, in comparison to individual stress, the combined episode of heat and drought stress induces detrimental effects on physiological activities, growth, and productivity of several crops (maize, barley, sorghum) and grasses such as bluegrass [77]. It is difficult to predict the strategies adopted by plants to cope with the concert of diverse environmental stresses due to their tailored responses. However, the elucidation of cross-talk mechanisms among cellular pathways responsible for differential responses of various plant species under the concert of stresses can augment crop breeding programs to develop tolerant varieties.

3. Abiotic Stress-Induced Oxidative Stress in Cellular Compartments

Oxidative stress is an unparalleled and intricate phenomenon of imbalance in cellular redox homeostasis that arises due to an exponential increase in ROS [80]. Under stress conditions, the activity of antioxidants declines to aid in ROS accumulation at an uncompensated level, leading to oxidative burst and oxidative damage [81]. The generation of a particular ROS in a cell is highly localized and regulated by a particular compartment depending upon the operating enzymatic and non-enzymatic pathways [82,83]. The photosynthetic and respiratory ETC, plasma membrane-localized NADPH oxidases, and apoplast POXs are major pathways, which are mainly involved in ROS production in the plant cell [82]. The major events leading to ROS production in a plant cell under the influence of unfavorable abiotic conditions trigger either retrograde signaling or oxidative burst (Figure 2). ROS generated in different organelles affect ETC, chlorophyll, proteins, and enzymes. However, inducing a mechanism that curbs ROS at the initial point of generation in cell organelles can prevent further damage. Additionally, channelizing ROS into signaling pathways averts the oxidative damage and induces tolerance to an individual or, may be, to a set of stresses.

3.1. Photosynthetic Apparatus (Chloroplast)

The photosynthetic apparatus (chloroplast) is an extremely important plant cell organelle that generates energy to drive life on earth. The chloroplast is susceptible to hostile conditions and a prime site for ROS generation (Figure 3). ROS produced within the chloroplast reduces the photosynthetic efficiency leading to dwindling growth and productivity. Exploring molecular processes affecting the photosynthetic activity and excess ROS generation may prevent deleterious effects. Adverse environmental conditions reduce stomatal conductance, decrease CO_2 assimilation, and/or result in the formation of excited triplet chlorophyll ($^3Chl^*$) that disturbs photosynthetic ETC, induces overproduction of ROS, and prompts photo-oxidation [84]. ROS are generated at the reaction center of PS I and II mainly due to the presence of excess high energy-intermediates, reductants, and O_2 [85,86]. Upon illumination, light-harvesting complexes (LHC) absorb energy (photon) and produce an excited singlet chlorophyll ($^1Chl^*$), which is a long-lived molecule and participates in the conversion of excitation energy into electrochemical energy via charge separation. In the presence of excess light, energy absorbed by LHC at the acceptor side of PS II exceeds over its utilization threshold limit and results in the formation of $^3Chl^*$ [87]. $^3Chl^*$ reacts with O_2 leading to the generation of highly oxidizing $^1O_2$. Apart from the excess light, other stresses such as drought induce disequilibrium between the light capture and its utilization, resulting in the production of $^1O_2$ [88]. Abiotic stresses limit the availability of CO_2 to Calvin's cycle due to the reduced stomatal conductance, causing an over-reduction of plastoquinone Q_A and Q_B (photosynthetic ETC component of PS II) that hinders the charge separation between P680 (chlorophyll molecules present at PS II) and pheophytin. The phenomenon triggers the formation of triplet chlorophyll (^{3}P680) at the PS II reaction center, which ultimately leads to the production of $^1O_2$ [89]. Due to the low concentration of CO_2 (final electron acceptor) under abiotic stress conditions, the decreased availability of $NADP^+$ prompts excessive electron leakage from the photosynthetic electron transport and reduces O_2 at the acceptor side of PS I via ferredoxin into $O_2^{\bullet-}$ known as

Mehler's reaction [89,90]. The over-reduction (overloading) of photosynthetic ETC causes electron leakage from plastoquinone Q_A and Q_B to O_2 resulting in the generation of $O_2^{\bullet-}$ at the reaction center of PS II [91]. An increased thylakoid membrane electron leakage to O_2 under drought has been reported in sunflower by Sgherri et al. [92]. Excitation of O_2 by highly energized chlorophyll pigments also results in the formation of $O_2^{\bullet-}$ [8]. The $O_2^{\bullet-}$ is then converted into a stable H_2O_2 either spontaneously or by dismutation via the action of thylakoid membrane-bounded/stromal membrane Cu/Zn-SOD [86,93]. The H_2O_2 generated is a potential photo-inhibitor that causes oxidation of cysteine (Cys) or methionine (Met) residues [89] and thiol modulated enzymes of Calvin's cycle inhibiting CO_2 fixation by 50% even at a concentration of 10 µM [93]. H_2O_2 has also been reported to mediate the signaling pathway, hence modulation of H_2O_2 into the signaling can avert oxidative damages. H_2O_2 undergoes further transformation leading to the formation of highly reactive and most toxic $^{\bullet}OH$ through the Fenton reaction mediated by redox metals (Fe^{2+} or Cu^+) [87]. However, quenching excess redox metals from chloroplast can prevent the Fenton reaction and production of $^{\bullet}OH$ that can bridge associated damages. Therefore, studies need to be carried out to elucidate mechanisms for intrinsically sequestering excess redox metals.

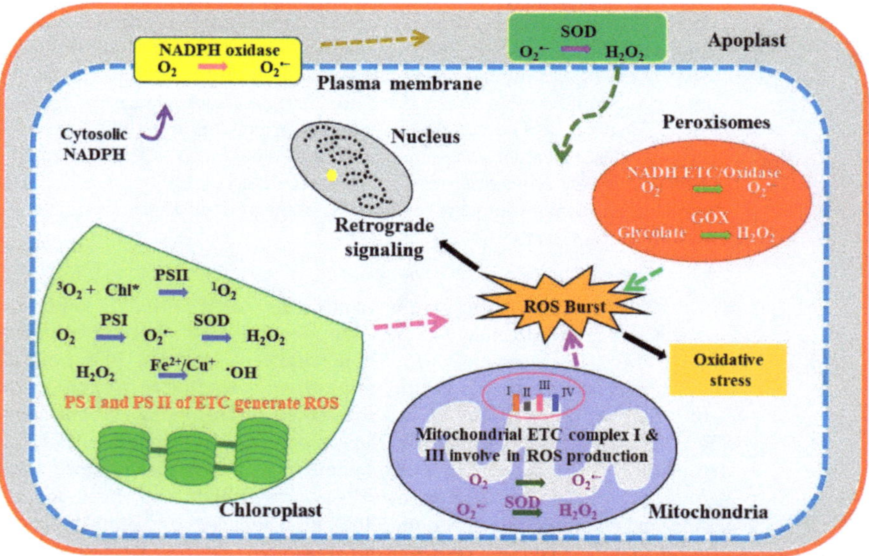

Figure 2. Abiotic stresses induced the production of ROS in different plant cell organelles which either initiate signaling (retrograde) or cause oxidative stress. In chloroplast singlet oxygen ($^1O_2$), superoxide radical ($O_2^{\bullet-}$), hydrogen peroxide (H_2O_2), and hydroxyl radical ($^{\bullet}OH$) are produced by an excited chlorophyll (Chl*), via the electron transport chain (ETC) at PS I and II (Mehler's reaction), dismutation of $O_2^{\bullet-}$ by superoxide dismutase (SOD) and via the Fenton reaction catalyzed by reduced iron (Fe^{2+}) and copper (Cu^+), respectively. At peroxisomes, photorespiration (glycolate), enzymes, and NADH (nicotinamide adenine dinucleotide) dependent small ETC induce the production of $O_2^{\bullet-}$ and H_2O_2. Mitochondrial ETC participates in the generation of $O_2^{\bullet-}$ which on dismutation by SOD produces H_2O_2. Cytosolic NADPH induces conversion of O_2 into $O_2^{\bullet-}$ by the action of NADPH oxidase of the plasma membrane which further dis-mutates into H_2O_2 in the apoplast by SOD. ROS produced in different cell organelles under the duress of abiotic stresses mediate signaling pathways at a low/moderate concentration or induce oxidative stress at a high concentration.

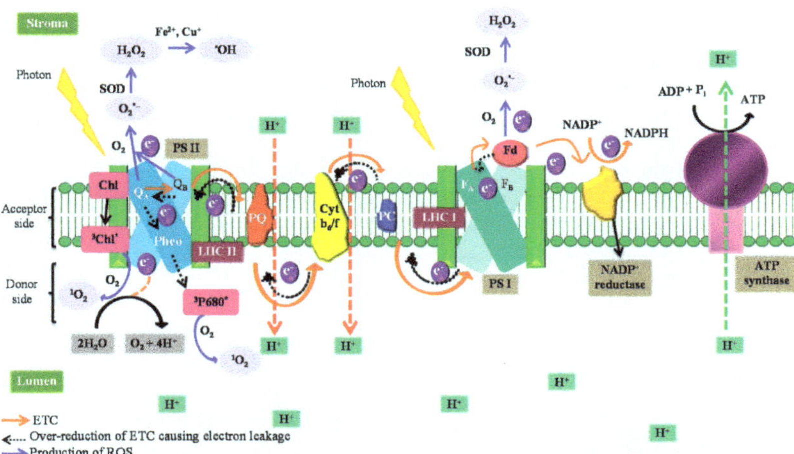

Figure 3. Photosynthetic electron transport chain under abiotic stress gets over-reduced and triggers the production of ROS. Photons striking at light-harvesting complex I and II (LHC I and II) result in electron (e^-) generation and hydrogen or proton (H^+) gradient, which initiates the electron transport chain (ETC) at photosystem (PS) I and II (through photolysis of H_2O) and production of NADPH and ATP by NADPH reductase and ATP synthase, respectively. However, the excess illumination of photons at LHC II converts the chlorophyll (Chl) molecule into an excited triplet form ($^3Chl^*$), which reduces O_2 to $^1O_2$. The reduced activity of Calvin's cycle due to low CO_2, leads to the over-reduction of ETC causing electron leakage. The electron moves in reverse from PS I to II and at PS II from Q_B to Q_A and then to the pheophytin forming an excited triplet chlorophyll ($^3P680$), which reduces O_2 to $^1O_2$. Over-reduction of Q_B and Q_A also directly reduces O_2 to $O_2^{\bullet-}$. At PS I, the over-reduction of ETC prompts electron leakage from ferredoxin (Fd) to O_2 forming $O_2^{\bullet-}$ via Mehler's reaction. The $O_2^{\bullet-}$ generated is dis-mutated either spontaneously or by the action of superoxide dismutase (SOD) to H_2O_2, which in the presence of reduced redox metals (Fe^{2+}, Cu^+) changed to a highly toxic $^\bullet OH$.

ROS produced in the chloroplast results in photo-oxidative stress leading to lipid peroxidation, damage to the membrane protein that affects the PS II reaction center, and ultimately cell death [94,95]. For instance, herbicides such as bentazon, paraquat, and 3-acetyl-5-isopropyltetramic acid inhibit photosynthesis and trigger ROS generation by competing with the D1 binding site of plastoquinone and blocking photosynthetic ETC from PS II [96] and/or by inhibiting the ultimate electron acceptor of PS I, i.e., $NADP^+$ and accepting an electron from PS I, which finally actuates the production of $O_2^{\bullet-}$, H_2O_2, and $^\bullet OH$ [85]. Analogously, the availability of $NADP^+$ to electrons reduces the under chilling stress that disrupts ETC and elicits ROS generation [42]. The chilling stress also induces overexcitation of the thylakoid membrane, which causes photo-inhibition and impairs the functioning of the photosynthetic machinery [42,97]. Yamane et al. [98] and Shu et al. [99] have reported damage to the chloroplast ultrastructure, i.e., destruction of chloroplast membrane, swelling of thylakoid, and aberrations in the thylakoid membrane, which is attributed to the production of ROS such as H_2O_2 and $O_2^{\bullet-}$ under salinity stress. Pandey et al. [68] have reported increased production of $O_2^{\bullet-}$, H_2O_2, and $^\bullet OH$ in the pea plant chloroplast exposed to Cr (VI). Similarly, the inhibition of PS II, ATP synthetase, enzymes of Calvin's cycle, disruption of photosynthetic ETC, and ROS production in the presence of metals such as nickel (Ni), Cd, Cu, Zn, and Cr has been reported by Dietz et al. [100] and Shahzad et al. [29]. Shakirova et al. [84] have observed the oxidative stress in wheat exposed to Cd resulting in the production of MDA and increased electrolyte leakage.

3.2. Peroxisomes

Peroxisomes are another major site for intracellular H_2O_2 production [101]. They also operate several important cellular functions, including high oxidative metabolic pathways in most of the eukaryotic cells [95,102] (Figure 4). Apart from H_2O_2, $O_2^{\bullet-}$ are also produced in the matrix and/or at the membrane of peroxisomes and are released into the cytosol [103]. The processes such as photorespiration, fatty acid β-oxidation mediated by acyl CoA oxidase (ACX), and the activity of enzymes such as flavin oxidase, urate oxidase (UO), xanthine oxidase (XOD), etc. in peroxisomes partake in ROS generation [104,105]. Under abiotic stress such as flooding, drought, salinity, high irradiance, heavy metals, xenobiotic compounds, high temperature, or chilling, the process of photorespiration initiates in the chloroplast due to the limited availability of CO_2 and increased solubility of O_2 that competitively accelerate the oxygenation of ribulose-1,5-biphosphate [106,107] to produce glycolate, which then gets exported to peroxisomes where glycolate oxidase (GOX) oxidizes it, generating H_2O_2 [90,95].

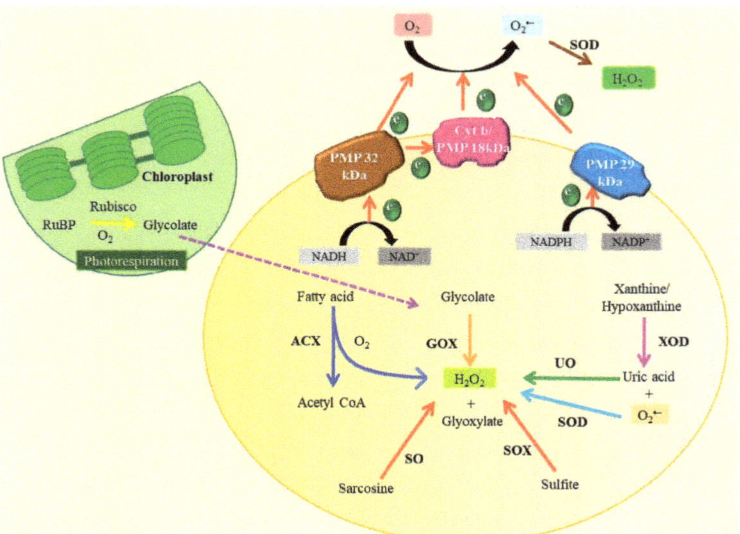

Figure 4. Different pathways for ROS production in peroxisomes under abiotic stress. ROS in the peroxisomes matrix is generated via the action of different enzymes. Glycolate produced in the chloroplast during photorespiration moves to peroxisomes where the action of glycolate peroxidase (GOX) generates glyoxylate and H_2O_2. The fatty acid undergoes β-oxidation in the presence of enzyme acyl-CoA oxidase (ACX) leading to the production of acetyl CoA and H_2O_2 [104]. Xanthine oxidase (XOD) catalyzes xanthine and/or hypoxanthine into the uric acid and $O_2^{\bullet-}$. The uric acid gets catalyzed by urate oxidase (UO) resulting in the production of H_2O_2 [105]. Other compounds such as sarcosine and sulfite undergo oxidation in the presence of enzymes sarcosine oxidase (SOX) and sulfite oxidase (SO) in peroxisomes and generate H_2O_2 [101]. NAD(P)H dependent small ETC consisting of three peroxisome membrane polypeptides (PMPs)—32, 18, and 29kDa generate ROS through electron leakage. NADH releases an electron to PMP 32kDa (NADH ferricyanide reductase) and forms NAD^+ (oxidized nicotinamide adenine dinucleotide), the electron either reduces O_2 to $O_2^{\bullet-}$ in cytosol or moves to cytochrome b (Cyt b/PMP 18kDa) where it reduces O_2 to $O_2^{\bullet-}$ in the cytosol. At PMP 29kDa, NADPH regenerates $NADP^+$ releasing electron which reduces O_2 to $O_2^{\bullet-}$ in the cytosol. $O_2^{\bullet-}$ forms dis-mutate either spontaneously or in the presence of superoxide dismutase (SOD) into H_2O_2.

Yamane et al. [98] have reported that salinity stress enhances the photorespiration and H_2O_2 level in peroxisomes. The increased lipid peroxidation and reduced activity of the ascorbic acid (AsA) and glutathione (GSH) in tomato plants subjected to salt stress have been reported by Mittova et al. [108]. The salt stress-induced oxidative damage probably arises from the production of ROS by the activity of peroxisomal GOX [106]. During drought conditions, photorespiration is estimated to contribute to >70% of H_2O_2 generation [106]. Further, β-oxidation of fatty acids, activities of enzymes such as flavin oxidases, XOD, UO, and disproportionation of $O_2^{\bullet-}$ trigger the production of H_2O_2 in peroxisomes [102,103,105]. Under abiotic stress characterized by prolonged darkness, chloroplasts release fatty acids which subsequently get metabolized by the peroxisomal β-oxidation [109]. Ortega-Galisteo et al. [103] have reported that the Cd and 2,4-dichlorophenoxyacetic acid (2,4-D) induced the production of H_2O_2 in pea leaves. Cd increases the H_2O_2 level due to the increased activity of GOX and reduces the CAT activity, whereas 2,4-D elevates ACX (β-oxidation of fatty acids) and XOD activities. The number of peroxisomes in plant cells also proliferate in the presence of abiotic stress including xenobiotic compounds, salinity, O_3, heavy metals, salinity, and high light [102]. Another important ROS, $O_2^{•-}$ is produced in peroxisomes on the action of salinity, Cd, herbicides, and other xenobiotics [110]. Peroxisomal $O_2^{•-}$ is generated via two different mechanisms. The first mechanism involves peroxisome membrane-localized NADH dependent small ETC comprising peroxisomes membrane polypeptide (PMP)-NADH: Ferricyanide reductase and cytochrome (Cyt) b of molecular masses 32 and 18kDa, respectively. NADH dependent ETC oxidizes NADH and Cyt b as well as reduces O_2 to $O_2^{•-}$ which is released into the cytosol [8,95]. In addition to PMP 32kDa and PMP 18kDa, another PMP of about 29kDa molecular mass generates $O_2^{•-}$ using NADPH as an electron donor and reduces Cyt c [95,102]. The second mechanism includes the oxidation of xanthine and hypoxanthine to uric acid with a simultaneous production of $O_2^{•-}$ mediated by XOD present in the peroxisomal matrix [8,95]. *A. thaliana* seedlings exposed to Cd stress overproduce $O_2^{•-}$ in peroxisomes [111]. Similarly, pea plants exposed to Cd stress exhibit an increased number of peroxisomes, $O_2^{•-}$ and H_2O_2 overproduction, and alteration in some endogenous proteins [112,113].

3.3. Mitochondria

Mitochondria are the other potential site for the production of $O_2^{•-}$, H_2O_2, and $^•OH$ in plants (Figure 5). Mitochondrial ETC (mtETC) and photorespiration favor ROS formation under abiotic stress. The mtETC or respiratory ETC operates in the inner membrane of mitochondria through two pathways, i.e., cytochrome oxidase (COX) with the ATP synthesis and alternative oxidase (AOX)-cyanide insensitive pathway without the ATP synthesis [114]. The mtETC comprises four oxido-reductase complexes I-IV (complex I-NADH dehydrogenase; complex II-succinate dehydrogenase; complex III-Cyt c reductase; complex IV-COX), two interior alternatives (NDin), and two exterior alternatives (NDex) NAD(P)H dehydrogenases (rotenone), one ATP synthase (complex V), mobile ubiquinone (UQ), mobile Cyt c, AOX, and uncoupling proteins (UCPs) [115]. A constraint on respiration during stress causes an over-reduction of mtETC that stimulates electron leakage to O_2 and ROS production [94,116]. The input of electron to mtETC when it exceeds more than its ability to utilize, over-reduces the UQ pool accelerating ROS generation [117]. Complex I and III of mtETC partake in ROS generation [95], whereas alternative NDs, AOX, and UCP are known to reduce the ROS production under stress [114,118]. $O_2^{•-}$ gets produced through the reduction of O_2 at the flavoprotein region and iron-sulfur (Fe-S) center of NADH dehydrogenase and/or by Cyt c reductase due to the reduction of UQ, which favors leakage of an electron to O_2 by generating highly reducing ubisemiquinone radicals [95]. Under drought and/or salinity stress, the over-reduction of the UQ pool in mitochondria due to the perturbation of ETC favors the production of ROS [94,95,98,119]. Hu et al. [120] have obtained similar results under chilling stress. Exposure to stress results in the over-reduction of mtETC and electron leakage to O_2 forming $O_2^{•-}$. Concomitantly, heat stress-induced hyperpolarization of the mitochondrial inner membrane of winter wheat cells due to the high potential gradient accelerates the over-reduction of the respiratory electron chain and actuates the production of ROS [121]. Complex II (succinate dehydrogenase) indirectly contributes to the ROS load in mitochondria by reversing the electron flow towards complex I due to the dearth of NAD^+ (oxidized nicotinamide diamine dinucleotide)–linked substrate [122]. This reverse electron flow from complex II to I is regulated by ATP hydrolysis [109]. $O_2^{•-}$ is the major ROS produced in the mitochondria, which disproportionates into H_2O_2 by the activity of Mn-SOD and APX [8,89]. H_2O_2 formed in the presence of reduced Fe^{2+} or Cu^+ yields a highly toxic ROS radical $^\bullet$OH via the Fenton reaction [116]. Photorespiration occurring in peroxisomes under stress conditions produces glycine which enters the mitochondria where it gets converted to serine and reduces NAD^+ to NADP by the action of glycine dehydrogenase complex (GDC) in the mitochondrial matrix [115]. In an excess light condition, GDC is probably the main substrate that produces NADP and donates an electron to complex I which initiates ETC [115] and may induce $O_2^{\bullet-}$ formation. The ROS generated in the mitochondria under stress affect its structure [123] and function, sometimes even leading to programmed cell death (PCD). Yamane et al. [98] have suggested that H_2O_2 generated under salinity stress is probably responsible for the degradation of mitochondrial cristae. Overproduction of ROS in the mitochondria leads to lipid peroxidation and PCD. This results in an alteration in the membrane potential, prompting the release of intermembrane space localized Cyt c to the cytosol [107,124,125]. The translocation of Cyt c from the mitochondria to cytosol has been observed in cucumber under heat stress [126]. Gao et al. [125] have reported the activation of caspase-like protease, DNA laddering, nucleus fragmentation, and PCD in *A. thaliana* due to the mitochondrial transmembrane potential loss and ROS formation after exposure to excess UV radiation.

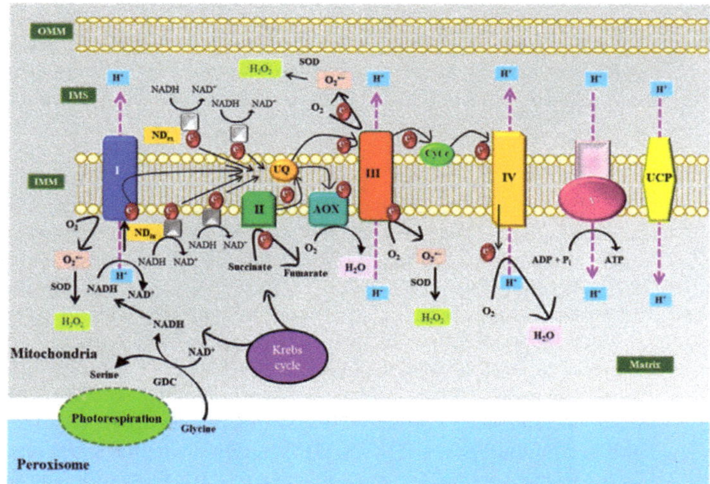

Figure 5. Mitochondrial electron transport chain (mtETC) mediated ROS production and alternative pathway under abiotic stress. The process, such as photorespiration and Krebs cycle, results in the generation of NADH and/or succinate which enters mtETC at complex I or complex II, respectively. At complex I, NADH converts into NAD^+ and H^+ with the generation of an electron. At complex II, succinate is changed to fumarate with the electron generation. An electron from both complex I and II get transferred to UQ from where they move to complex III and then to complex IV via Cyt c. The electron at complex I and III reduces O_2 to generate ROS ($O_2^{\bullet-}$ and H_2O_2), whereas, at complex IV, O_2 oxidized to H_2O. H^+ generates at complex I, III, and IV pumped to IMS and then moves to complex V or ATP synthase to form ATP from ADP. The mtETC also comprises an alternative pathway consisting of two each NDex and NDin with AOX and UCP which limit ROS generation. NDex and NDin function in stress conditions and transfer electrons to UQ. The AOX present between UQ and complex III accepts an electron from UQ and reduces O_2 to H_2O, thus terminating the electron transport to complex III. OMM: Outer mitochondrial membrane; IMS: Inter-mitochondrial space; IMM: Inner mitochondrial membrane; e^-: electron; UQ: Ubiquinone; I–V: Complex I–V; Cyt c: Cytochrome c; NDex and NDin: NAD(P)H dehydrogenase on the exterior and interior side of IMM, respectively; AOX: Alternative oxidase; UCP: Uncoupling protein.

3.4. Plasma Membrane, Cell Wall, and Apoplast

The plasma membrane and apoplast envelope the cell organelles and maintain cell activity, fluidity, rigidity, ion transport, as well as secure its integrity [83,127]. Plasma membrane-localized NADPH oxidases are major ubiquitous enzymes that catalyze reactions generating

ROS [82,83]. NADPH oxidases mediate the transfer of an electron from cytosolic NADPH to O_2 which results in the production of $O_2^{\bullet-}$ in the apoplast [128] that undergoes dismutation either spontaneously or by the action of antioxidant enzyme SOD, yielding H_2O_2 [95,129] (Figure 6). During oxygen depriving stress conditions (hypoxia), the plasma membrane located NADPH oxidase partakes in the production of H_2O_2 in the apoplastic space [130]. The apoplast produces extracellular ROS such as H_2O_2 under sub-optimal conditions. The pathway for ROS production operates under cell wall-associated enzymes including pH dependent extracellular POXs, quinine reductase, lipoxygenases, amine oxidases (AO), polyamine oxidases (PAO), and germin-like oxalate oxidases (OXOs) [83,129,131] (Figure 6). H_2O_2 is constantly generated in the apoplast on the combined action of abscisic acid and stress signals [132]. Voothuluru and Sharp [133] have reported an increase in apoplastic H_2O_2 content in the primary root of maize, experiencing a water-deficient condition which is mediated by the activity of the OXO enzyme. Lin and Kao [134] have also recorded a reduced root growth of rice seedlings grown under salinity stress impacted by increased activity of cell-wall POX, NADH peroxidase, and DAO which promote the accumulation of H_2O_2 in the cell wall. Other abiotic stress such as the presence of ground-level O_3 induces oxidative burst in a plant cell by actuating the production and accumulation of H_2O_2 and $O_2^{\bullet-}$ in the apoplast which inflicts necrosis and cell death [135]. The production of H_2O_2 by the cell wall-associated POX in *Arabidopsis* under the K^+ deficient condition has been reported [136]. Stress conditions such as salinity/osmotic stress activate NADPH oxidases, apoplastic DAO, and PAO enzymes which promote the production of ROS [137]. PAO catabolizes polyamines such as spermidine and produces/releases H_2O_2 as a byproduct in the apoplast under high salinity stress [138].

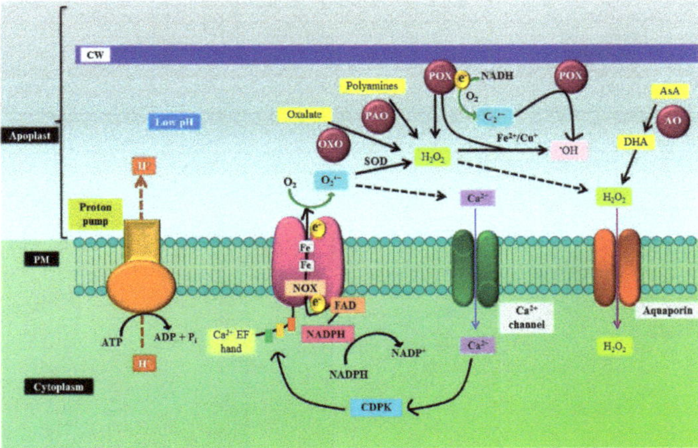

Figure 6. ROS production by the plasma membrane, apoplast, and cell wall under abiotic stress. The plasma membrane (PM) localized NADPH oxidase consists of two cytoplasmic binding sites: 1) Flavin adenine dinucleotide (FAD) and nicotinamide adenine dinucleotide phosphate (NADPH) and 2) Ca^{2+} binding EF-hand motifs. The NADPH oxidase transfer electron from the cytosolic NADPH to the apoplast via cytochrome (Fe) present in the channel is formed by NADPH oxidase transmembrane domains and reduces O_2 to $O_2^{\bullet-}$. In the apoplast, $O_2^{\bullet-}$ either spontaneously (due to low pH maintained through the proton pump) or by the action of SOD disproportionates into H_2O_2. $O_2^{\bullet-}$ induces the Ca^{2+} influx through the Ca^{2+} channel which moves to the Ca^{2+} binding EF-hand motif of NADPH oxidase via the calcium-dependent protein kinase (CDPK) and activates the NADPH oxidase leading to ROS production. Other enzymes such as cell wall (CW) bound peroxidases (POX) and apoplast localized amine oxidases (AO), polyamine oxidases (PAO), and oxalate oxidases (OXO) in the presence of specific substrates result in ROS generation. POX in the presence of NADP reduces O_2 to $O_2^{\bullet-}$ which dis-mutates to H_2O_2. Similarly, AO breaks down AsA to dehydroascorbate (DHA), which in turn generates H_2O_2. OXO and PAO partake in H_2O_2 formation in the presence of oxalate and polyamine, respectively. H_2O_2 is converted into $^{\bullet}OH$ either through the Fenton reaction in the presence of redox metals or by the action of POX. H_2O_2 produced in the apoplast also moves to the cytoplasm through aquaporins and participates in signaling

4. Biomolecules Targeted by ROS and Oxidative Damage

ROS overproduction leads to oxidative burst and damage to biomolecules under adverse environmental conditions (Figure 7). The damaged biomolecules comprise the product of protein oxidation, inactivation of enzymes, lipid peroxidation, increase membrane fluidity, chlorophyll degradation, nucleic acid damage, and commencement of the apoptosis pathway and PCD in severe conditions [9,80]. These damages affect the growth, development, and ultimately plant survival. The extent of damage to biomolecules depends on various factors including the concentration of particular biomolecule(s), location of the target biomolecule(s) in relation to the site of ROS generation, the rate constant for the reaction between target biomolecule(s) and ROS, the occurrence of secondary damaging incidents and ROS scavenging or detoxifying repair system [139].

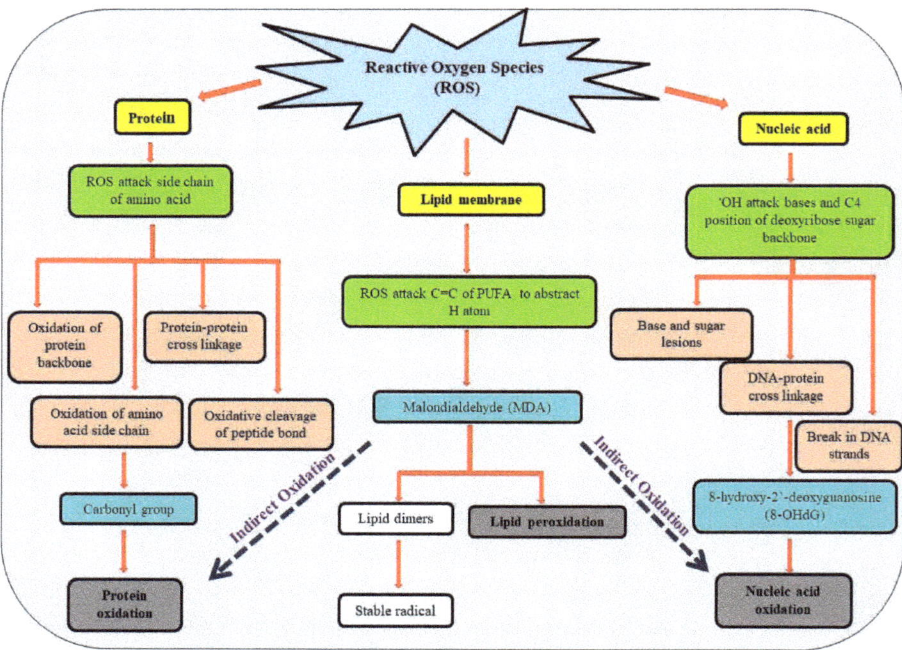

Figure 7. Reactive oxygen species attack biomolecules (proteins, membrane lipids, and nucleic acids) at different sites leading to oxidation that alters their structural and functional activities. Oxidation of biomolecules results in the formation of carbonyl group, malondialdehyde, and 8-hydroxy-2′-deoxyguanosine, which are considered as a best marker of protein, lipid, and nucleic acid oxidation, respectively.

4.1. Lipid Membrane

The oxidative burst in a cell under stress conditions damages the lipid membrane. Lipid peroxidation reactions involve lipoxygenase activity, $^1O_2$ generation, and radical catalyzed mechanism, which differ quantitatively between underground and aboveground tissues depending on the type of ROS [140]. ROS targets unsaturated C-C double bond polyunsaturated fatty acids (PUFA), e.g., arachidonic acid, linolenic acid, and linoleic acid. The •OH radical attacks on the methylene group of fatty acid and abstracts the hydrogen (H) atom forming carbon-center lipid radical [141]. ROS also breaks the ester linkage between glycerol and fatty acids, disintegrating membrane phospholipids [89,95]. The ROS radical, •OH initiates the cyclic reaction resulting in peroxidation of PUFA [95]. The process of lipid peroxidation involves three stages: Initiation, propagation, and termination (cleavage) [141]. Initiation involves the production of ROS by the reduction of O_2. The ROS generated trigger a cascade of reactions leading to the formation of lipid radicals (lipid peroxyl radicals, hydroperoxides, etc.) and MDA conforming the second stage, i.e., propagation. Finally, lipid radicals end up as the formation of lipid dimmers [95]. Lipid peroxidation perpetrates membrane destabilization with regards to permeability, electrolyte leakage, deactivation of enzymes and receptors as well as enhances the oxidation of nucleic acids and proteins [89,95,141]. Lipid radicals generated during lipid oxidation undergo enzymatic or non-enzymatic degradation and yield compounds such as reactive carbonyl species (RCS) [142]. These RCS selectively react with proteins via the lipoxidation reaction and result in the loss of functional activities of proteins. The alleviating degradation of lipid radicals by their direct elimination from a cell can prevent the lipoxidation reaction and further oxidative damage to the cell. For example, Gram-positive bacterium *Deinococcus radiodurans* possesses the ability to reduce ROS induced injury to Fe-S proteins by eliminating the cellular iron outside the cytosol [140]. Studying the underlying mechanisms for sequestration of susceptible or damaging molecules can provide avenues to enhance tolerance in plant cells against oxidative damage.

Abiotic stresses such as salinity [143], temperature [144], metals and metalloids [145,146], drought [147,148], xenobiotic compounds including pesticides [149], ground-level O_3 [150], and UV radiation [151] converge oxidative stress to accelerate cellular and organelle lipid peroxidation. ROS in roots of rice seedlings exposed to excess aluminum exhibit lipid peroxidation, as well as DNA damage [152]. Arsenic (As) stress also prompts H_2O_2 accumulation, lipid peroxidation, and electrolyte leakage in common bean seedlings [153]. The lipid peroxidation product, MDA, indicates the degree of oxidative damage in the cell and hence a marker for the degree of the damage [89,95]. Martinez et al. [154] have reported an over-accumulation of H_2O_2 followed by a high MDA content and lipid peroxidation in tomato plants exposed to salinity, heat, and combined stresses. Kumari et al. [150] have also demonstrated a significant increase in lipid peroxidation/electrolyte leakage and reduction in the chlorophyll content and photosynthetic rate in *Solanum tuberosum* L. cv. Kufri chandramukhi grown under ambient CO_2 and elevated O_3. Membrane lipid peroxidation in *Phalaenopsis* due to the exposure to an elevated temperature induces the loss of membrane integrity and K^+ leakage [144].

4.2. Proteins

Proteins play a crucial role in mediating tolerance to abiotic stress by adjusting the physiological characters of plants [155]. Protein aggregation or change in conformation affects their enzymatic, binding, and other functional activities [141,156]. Proteins are more susceptible to oxidation than other biological molecules due to their abundance in the living system and high-rate constants for the reaction [140]. Both radical and non-radical oxidants cause protein oxidation either directly or indirectly. Some ROS cause little and selective damage to certain residues, while others such as $^\bullet OH$ induce widespread and non-selective (non-specific) damages [140]. The protein backbone attacked by non-radical oxidants results in limited damage due to the slow oxidation rate. However, the extensive or widespread damage to the protein backbone is induced by radicals that react rapidly primarily through the abstraction of the H atom at the α-carbon site resulting in the formation of stabilized carbon-center radicals [140]. The direct oxidation by ROS involves both, oxidation of side chains of amino acid specifically those containing sulfur (S) and thiol groups (e.g., oxidation of Cys and Met residue by $^1O_2$ and $^\bullet OH$) and degradation of peptide backbone resulting in carbonylation, nitrosylation, disulfide bond formation, and glutathionylation, which alters the protein activity [95]. On the contrary, indirect oxidation is mediated via products formed during lipid peroxidation [95,157]. The oxidation of protein enhances their susceptibility towards proteolytic digestion [158] by getting prepared for ubiquitination-mediated degradation by the proteasome [95]. Protein oxidation by ROS is either irreversible or reversible. ROS such as $O_2^{\bullet-}$ can irreversibly damage enzymes that contain the Fe-S center [95]. The irreversible damage to protein such as carbonylation, protein-protein cross-linking, etc. causes functional loss. On the other hand, reversible changes such as glutathionylation and *S*-nitrosylation can mediate the redox regulation [141]. Carbonylation of the protein is irreversible and an unrepairable damage mediated by the oxidative cleavage of proteins and is considered as the best marker for estimation of oxidative damage under stress [158]. Oxidation of heat shock proteins and late embryogenesis abundant (LEA) proteins reduces the capacity to maintain protein functioning in dehydrated seeds [159]. Karuppanapandian and Kim [160] has noted a significant increase in the carbonylated protein in cobalt-stressed *B. juncea* leaves. Carbonylated proteins occur in plant cell organelles including chloroplast, mitochondria, nucleus, cytosol, and peroxisomes [141]. Under drought conditions, the protein carbonyl level elevated by several folds has been detected in the mitochondria as compared to the chloroplast and peroxisomes in leaves of the wheat plant [161]. An increase in protein oxidation has been demonstrated in cashew plants subjected to salinity stress [162]. Exposure of *A. thaliana* and *Glycine max* to excessive CO_2 also induces protein carbonylation [163]. Oxidative damage to proteins under chilling, paraquat, and O_3 stress leads to functional loss, which has been demonstrated in several studies [164–166].

4.3. Nucleic Acid

Nucleic acids undergo oxidation on the ROS attack that affect protein synthesis and may lead to mutation [89,167]. The DNA present in the chloroplast and mitochondria are more susceptible to oxidation than the DNA present in the nucleus due to their proximity to the ROS production site and lack of protective histones and associative proteins [95]. ROS imperil oxidation of nucleic acid that includes the oxidation of sugar residue, alteration of nucleotide bases (insertion or deletion), and the abstraction of nucleotide break in the DNA strand, cross-linking the DNA and protein [95]. Intersomal nDNA fragmentation has also been reported in the sensitive genotype of wheat with PCD in leaves under drought [148]. The ROS subtract H-atom from the C4 position of deoxyribose sugar backbone forming the deoxyribose radical, further causes a break in DNA strand [168]. Among all ROS, the $^{\bullet}OH$ radical has been reported to cause maximum damage to DNA due to its ability to react with purines/pyrimidine bases, and even deoxyribose sugar [169]. Apart from $^{\bullet}OH$, $^1O_2$ reacts only with guanine, while $O_2^{\bullet-}$ and H_2O_2 do not react with any purine or pyrimidine bases [169]. The $^{\bullet}OH$ radical attacks the double bond of purines and pyrimidine bases [170] developing DNA lesions and forming 8-hydroquinine and some other less common products such as hydroxyl methyl urea, thymine glycol, etc. [95]. The cross-linking between DNA and protein is also facilitated by the $^{\bullet}OH$ radical by reacting either with the DNA or associated proteins. The repairing of this cross-linkage is a difficult task, and if not repaired before replication or transcription, can cause a lethal effect on the plant cell [95]. In addition to the direct oxidation of DNA, the lipid radicals obtained from lipid peroxidation prompt an indirect DNA oxidation [171]. MDA, a major product of lipid peroxidation reacts with guanine (G) residues in the DNA to form M_1G, i.e., pyrimidopurinone adduct [172]. RNAs are also susceptible to the ROS attack [173]. Oxidation of RNA results in the formation of 8-oxo-7,8-dihydroguanosine (8-OHG), which is used as a marker for the determination of the intensity of RNA oxidation. The Cd-induced oxidation of RNA in soybean seedling [174] and degradation of mRNA during water deficit stress (desiccation) in *Lindernia subracemosa* [175] have been documented. ROS affect the DNA replication and transcription that may abnormally affect the protein synthesis, membrane stability, as well as signal transduction pathways in a cell, reducing metabolic efficiency, genetic instability, and compromising cell homeostasis [169]. The accumulation of radicals formed due to the oxidization of the biomolecule shows the potential to oxidize other biomolecules, which may elevate oxidative damage in plant cells and result in PCD under a severe condition. To avoid extensive damage, the continuous elimination or repairing of damaged biomolecules is necessary. Fortifying plants' intrinsic mechanisms to remove or repair damaged biomolecules may induce the resistance towards stress and prevent productivity losses.

5. Antioxidants: Oxidative Stress Defense Mechanism

ROS at a low or moderate concentration act as a secondary messenger and participate in a signaling cascade within the cell that elicit a response to tide over stress situations [8,89]. Ironically in stress conditions, ROS are generated in high concentrations that become toxic and are responsible for PCD [116]. The activity of ROS in the plant cell (regulative, damaging, or signaling) depends on the equilibrium between their production and detoxification system [176]. Enzymatic and non-enzymatic antioxidants of plants act as ROS detoxifying machinery, which limit their concentration and maintain their steady-state level inside cellular compartments [154,177]. Enhancing the antioxidant level of plant cells either endogenously through genetic engineering or by an exogenous application can strengthen the defense system of the plant and rescue them from the debt of environmental stress.

5.1. Enzymatic Antioxidants

An enzymatic antioxidant such as SOD, CAT, APX, POX, monodehydroascorbate reductase (MDHAR), dehydroascorbate reductase (DHAR), glutathione *S*-transferase (GST), glutathione peroxidase (GPX), AOXs, peroxiredoxin (Prx), and thioredoxin (Trx) alleviates the ROS level by breaking them down and removing them from the system through various steps including conversion of ROS to H_2O_2 and then to the H_2O molecule in the presence of metallic co-factors [178]. SOD (EC: 1.15.1.1) are metalloenzymes that are found in three isoforms viz. Cu, Zn-SOD localized in the cytosol, chloroplast, peroxisomes, nuclei, mitochondria, and apoplast; Fe-SOD in chloroplast, peroxisomes, and mitochondria; and Mn-SOD in peroxisomes, mitochondria, and vascular tissues [116,179,180]. SODs provide an initial or first line of defense against toxic ROS [116]. They catalyze the disproportionation of $O_2^{\bullet-}$ free radicals by reducing one radical into H_2O_2 and oxidizing another into O_2 thereby eliminating the risk of production of more toxic free radical $^\bullet OH$ [116]. CAT (EC: 1.11.1.6), APX (EC: 1.11.1.11), and GR (EC: 1.6.4.2) catalyze the decomposition of H_2O_2 antioxidant into H_2O and O_2 [181,182]. CAT and APX are metalloenzymes localized in peroxisomes and mitochondria [182]. Apart from these, APX is also found in the cytosol, chloroplast, microbodies, and peroxisomes/glyoxysomes [86,183] and participates in the AsA-GSH (ascorbate-glutathione) pathway as a key enzyme [184,185]. The enzyme requires AsA as a reducing substrate for its stability and proper functioning [186]. AsA-GSH or Foyer-Halliwell-Asada pathway [184], comprising enzymatic (APX, MDHAR, DHAR, GPX) and non-enzymatic (AsA, GSH) antioxidant components, operates in chloroplast, plastids, mitochondria, and peroxisomes [185] to combat the overproduction of H_2O_2 [187].GR is a flavoprotein oxido-reductase that is localized in the chloroplast (where it displays 70–80% of the activity) [116,188], mitochondria, cytosol, peroxisomes, and in non-photosynthetic tissues and organelles [189]. Its activity is dependent on the combined action of pH and concentration of NADPH and glutathione disulfide (GSSG) at the site of action [188]. GR catalyzes the conversion of oxidized GSH (GSSG) into the reduced form-GSH using NADPH as an electron donor [116] and maintains a balance between the GSH/GSSG ratio necessary for the detoxification of H_2O_2 [190]. Prxs are thiol peroxide enzymes that detoxify peroxidase substrates such as H_2O_2 and alkyl hydroperoxide and reduce oxidative damage [191,192]. Prx participates in ROS dependent signaling by modulating the concentration of H_2O_2, processing alkyl hydroperoxide, switching to chaperone function, etc. [192]. Prxs contain one or two catalytic Cys in a conserved sequence and are classified into four groups, (1) 1-Cys Prx, (2) 2-Cys Prx, (3) YLR109-related Prx, or type II Prx, and (4) bacterioferritin-comigratory protein or Prx Q [191,193]. Trxs are small thiol-disulfide regulatory proteins (around 14kDa) that reduce the disulfide bond and participate in ROS regulation [192]. Trxs contain a pair of cysteinyl residues in a highly conserved amino acid motif WC[G/P]PC, which are involved in the catalytic activity of the enzyme [194].

5.2. Non-Enzymatic Antioxidants

Non-enzymatic antioxidants detoxify ROS by interrupting a free-radical chain reaction [179]. The non-enzymatic compounds such as AsA, GSH, compatible solutes, phenolics, α-tocopherol, carotenoids, flavonoids, and even proline counteract the uncontrolled cascade of ROS produced during stress [123,137,195]. GSH is a ubiquitous thiol tripeptide that participates in the degradation of H_2O_2 in a reaction catalyzed by GPX [196]. It takes part in the AsA-GSH pathway as a reductant for DHAR and aids in the scavenging of H_2O_2 [187,196] and/or degradation of H_2O_2 and lipid peroxides by forming a conjugate through a reaction catalyzed by the GPX and GST, respectively [196]. AsA or vitamin C participates in the AsA-GSH pathway as an electron donor for APX [168] and is a co-factor of POXs [197]. AsA helps in the regeneration of tocopherol and xanthophyll production that partakes in quenching of the excitation energy [173]. Carotenoids are a light-harvesting pigment [198,199] that alleviates high light illumination induced oxidative stress by quenching excessive energy as heat dissipation [198,199]. Carotenoids also avert the over-excitation of PS II in the thylakoid membrane by efficiently scavenging $^1Chl^*$, $^3Chl^*$, and $^1O_2$ [199].

The gamma-aminobutyric acid (GABA) is a ubiquitous non-protein amino acid that accumulates in plant cells under stress conditions and provides tolerance by scavenging free radicals and regulating the enzyme activity [36,200]. GABA acts as an osmolyte or encourages the production of other osmolytes such as proline under conditions such as the drought that aids in osmotic adjustment for acclimatization during stress [36]. GABA is metabolized by a GABA shunt pathway that comprises GABA transaminase (GABA-T), glutamate dehydrogenase (GDH), and succinic semialdehyde dehydrogenase [200]. Jalil et al. [41] have reported that the mutant of *A. thaliana* lacking the GABA-T gene reduces GABA and chlorophyll content, lowers photosynthesis, and GDH activity but increases membrane ion leakage, MDA content, and early leaf senescence under various abiotic stresses. GABA also participates in the signal transduction pathway under stress via the increased cytosolic calmodulin-dependent activity of the enzyme glutamate decarboxylase [36].

It is apparent that the enzymatic and non-enzymatic antioxidants acquire crucial pathways for tight regulation of ROS within plant cells and are responsible for efficient amelioration of abiotic stress-induced oxidative stress. Many researchers have documented the potential role of antioxidants in the alleviation of oxidative stress (Table 2). The tolerant genotypes of *B. juncea* alleviate the heat stress via the increasing activity of enzymatic POX and non-enzymatic GSH antioxidant [201]. Subjection to temperature stress, the tolerant wheat genotypes (HD 2815 and HDR 77) maintain a high activity of antioxidant enzymes SOD, CAT, and APX with the least reduction in the chlorophyll content and lower membrane damage in comparison to that of its susceptible genotypes. The investigation comprehensively establishes alleviating role of antioxidants for the maintenance of structural and functional characteristics in plants [202]. Further, the production of enzymatic antioxidants (SOD, GPX, APX, and GR) with non-enzymatic antioxidants (AsA, GSH) and proline confer tolerance to rice plants against excessive Cu induced oxidative stress [203]. The tolerant lentils to heat stress exhibit elevated SOD and other antioxidants and a negative correlation between MDA and H_2O_2, confirming their role in the alleviation of oxidative stress [204]. These studies show that the efficient and coordinated working of antioxidants confer a protective effect on plants under harsh environmental cues.

Plants often encounter multiple stresses simultaneously under field conditions that show a discrete antioxidant activity. For example, *Portulaca oleracea* subjected to combined heat and drought stress exhibits a higher activity of SOD and POX [205]. The cytosolic enzyme APX1 decomposes H_2O_2 and plays a significant role in the acclimatization of plants exposed to drought and high-temperature stress concurrently. The *Apx1* deficient mutant of *Arabidopsis* sensitive towards the combined stresses corroborates the findings [206]. Similarly, Zandalinas et al. [190] have observed an enhanced ROS detoxification and resilience to combined heat and drought stress in citrus genotypes Carrizo citrange exhibiting the effective activation of antioxidant machinery. The tolerance ability of Carrizo has been prompting to efficiently coordinate activities of SOD, CAT, APX, and GR with a maintained favorable ratio of GSH/GSSG. While the Cleopatra mandarin subjected to similar stress conditions displays sensitivity due to the increased SOD activity with inefficient activation of GR, diminished CAT activity, and lack of APX activity enhancing oxidative stress.

Table 2. Antioxidant activity in plants in response to abiotic stress-induced oxidative stress.

Abiotic Stress(es)	Plant Exposed	Antioxidant(s) Activity	References
Chilling stress	*Cucumis sativus* (Cucumber)	• The activity of SOD, APX, GR, and GP increased and CAT activity decreased in leaves.	[207]
Chilling stress	*Zea mays* (Maize) seedling	• Exogenous application of nitric acid before the onset of stress increased the activity of SOD and POX. • ROS level and lipid peroxidation alleviated.	[20]
Drought	*Triticum aestivum* (Wheat)	• The upregulated APX and balanced redox pool of AsA and GSH fortified photosynthetic apparatus and mitochondria in acclimatized plants.	[208]
Heavy metal (Cu) stress	*Oryza sativa* (Rice)	• The activity of SOD, guaiacol peroxidase (GP), APX, GR, AsA, GSH with proline increased. • CAT activity remained unaltered. • H_2O_2 level and lipid peroxidation declined.	[203]
High-temperature stress	*Triticum aestivum* (Wheat)	• The activity of SOD, APX, CAT, POX, and GR increased in tolerant genotype C306.	[209]
High-temperature stress	*Spinacia oleracea* (Spinach) seedling	• Overexpression of the gene encoding cytosolic heat shock 70 protein (SoHSC70) increased the activity of SOD, POX, CAT, and APX enzymes. • Oxidative membrane damage and ROS accumulation reduced.	[210]
Metalloid (Boron) stress	*Artemisia annua*	• The activity of SOD, POX, and CAT increased.	[211]

Table 2. *Cont.*

Abiotic Stress(es)	Plant Exposed	Antioxidant(s) Activity	References
Salinity stress	*Oryza sativa* (Rice) seedling	• Exogenous application of manganese to seedlings exposed to stress increased non-enzymatic antioxidants (phenolic compounds, flavonoids, and AsA), and enzymatic antioxidants (MDHAR, DHAR, SOD, and CAT) content. • ROS level reduced.	[35]
UV-B radiation	*Helianthus annuus* (Sunflower) cotyledons	• The activity of CAT, glutathione dehydrogenase, GP, and the ratio of GSH/GSSG increased. • The AsA/DHA ratio, APX, and GR activity remained unaltered. • Lipid peroxidation and oxidative damage in cotyledons reduced.	[212]
Low temperature + herbicide (isoproturon) stress	*Triticum aestivum* (Wheat) seedling	• Foliar application of AsA increased activity of antioxidants SOD, CAT, and POX. • MDA content and ROS production rate declined.	[213]
Salinity + herbicide (2,4 dichlorophenoxyacetic acid) stress	*Oryza sativa* (Rice)	• Enzymatic (SOD, CAT, APX, and POX) and non-enzymatic (phenolic compounds, total soluble phenols, proline, and sugars) antioxidants level modulated. • H_2O_2 and $O_2^{\bullet-}$ content decreased; oxidative stress and lipid peroxidation alleviated.	[214]

6. ROS as Signaling Molecules

Plants are equipped with an arsenal of adaptive strategies to endure harsh conditions [1,5]. The chief strategy includes initiation of systemic signals from an area under stress to an unstressed region that consequently alerts and activates defense or increases resilience [7,206] arising from signal transducers, including ROS [215]. In addition to oxidative damage, the ROS role has been well recognized as a signaling molecule that prompts tolerance against unfavorable conditions [176]. Inefficient scavenging capacities of antioxidants result in an oxidative burst within plant cells. Under such conditions, the activation or modulation of ROS into signaling transducing pathways could avert damaging consequences of the stress. The imposition of biotic and abiotic stress conditions compels cell organelles to switch the transient ROS production [3,216,217] that offsets ROS homeostasis and initiates the signal transduction cascade [218], involving specific feedback and feed-forward responses facilitating stress tolerance [7]. The spatio-temporal production of ROS is a critical factor that determines the ROS mediated cellular and intracellular signaling [219]. The systemic signaling against ROS generation arises as an auto-propagating wave to an adjacent cell [220] that confers stress tolerance through spatio-temporal communication. For this purpose, plants engage phytohormones and/or amino acids as specific signals to indicate a stress situation [7]. For instance, ROS generated under stress initiate signaling by oxidizing proteins that result in the production of peptides which in turn maintain signaling as a secondary messenger [7]. Among various ROS, H_2O_2, a non-ionic, relatively stable yet reactive molecule, diffuses through membranes via aquaporins and initiates signaling. Therefore, H_2O_2 acts as a perfect candidate for the signal transduction pathway [7,221]. ROS operates signaling in a highly coordinated manner to regulate stress. It activates antioxidants, kinases, defense genes, the influx of Ca^{2+} ions, protein phosphorylation, increasing synthesis of plant hormones such as salicylic acid, jasmonic acid, ethylene, etc. In the case of biotic stress, it elicits early defense responses such as the synthesis of phytoalexins and pathogenesis-related proteins, as well as cell wall strengthening/PCD promotion, restricting invasion/multiplication/spread of pathogens in plant cells [7,9,176,222]. For instance, GDH that participates in ammonia production/accumulation in stressed cells instead starts synthesizing glutamate and se-quentially leading to the production of proline (well known to partake in stress tolerance) in tobacco [7,221]. ROS signaling arbitrates transcription of the gene encoding for GDH α-subunit [7]. ROS such as $O_2^{\bullet-}$ and H_2O_2 also reportedly participate in plant growth and development, as well as in plant protection against biotic and abiotic stress condi-tions [3,178]. Therefore, ROS production below the stress threshold induces developmental changes such as the formation of tracheary elements, lignification, and cross-linking in the cell wall leading to PCD and ameliorates abiotic stress [67,223].

The ROS production in cell organelles mediates retrograde signals to the nucleus. The signals move with a speed of 8.4 cm min^{-1} under stress conditions and play a pivotal role as a secondary messenger to alleviate abiotic stress in plants [224,225]. The retrograde signaling assists the nucleus to modulate the anterograde control for the acclimatization of plants exposed to abiotic perturbation [226]. During abiotic stress, the ROS burst elicits the upstream transcription of stress-responsive genes such as heat shock gene (HSG) [225]. HSPs, for example, act as molecular chaperones, partake in the prevention of protein aggre-gation, misfolding, denaturation, and degradation, as well as facilitate protein refolding particularly during heat stress [225,227]. Apart from heat stress, the role of HSPs in the reg-ulation of light, anoxia, cold, and other abiotic stress has also been documented [225,228]. H_2O_2 in the *Arabidopsis* cell culture under heat stress also modulates HSG expression, which induces the production of APX2, HSP17.6, and HSP18.2 [229]. Similarly, the H_2O_2 burst in *Arabidopsis* cells upregulates the production of HSPs, APX1 that scavenge H_2O_2, and provides tolerance to light stress [230], as well as acclimatizes the plant exposed to the combination of heat and drought stress [206]. The onset of low oxygen stress (hypoxia) consequently leads to the production of ROS in a regulated manner via RBOHs [54]. The regulated production and signaling of ROS are considered an important factor in the man-agement of hypoxic stress [54,225,231–233]. Under oxygen deprivation (anoxia/hypoxia stress), H_2O_2 upregulates the expression of genes encoding HSPs and genes responsible for fermentation such as *ALCOHOL DEHYDROGENASE*, as well as ROS regulated tran-scription factor including *ZAT10* and *ZAT12* and proteins, which subsequently facilitate acclimatization to stress [54,225,234,235]. ROS also display systemic signaling in plants via auto-propagation as a wave to adjacent cells [54]. The systemic signaling to neighboring cells by $O_2^{\bullet-}$ and H_2O_2 in stagnant rice roots and *Arabidopsis*, respectively have been reported [220,224,233]. NADPH oxidase genes viz. *AtrbohF* and *AtrbohD* also trigger the production of ROS during salinity stress that consequently initiates signaling and provide tolerance by regulating Na^+/K^+ homeostasis in cells [236]. Jiang et al. [237] have reported that the *soil-salinity sensitive 1-1* mutant of *Arabidopsis* lacking functioning of the NADPH oxidase gene *AtrbohF* does not accumulate ROS in root vasculature and displays hyper-sensitivity towards salinity stress. Moreover, under nutrient deprivation conditions, ROS induces signaling pathways. ROS in low K^+ availability upregulates calcium signaling in cells [64,238]. In response to K^+ deficiency, the H_2O_2 concentration increases in plant roots, which enhances the expression of *HAK5* genes [239]. An understanding and extensive investigation of molecular mechanisms of ROS mediated signaling and cross-talk with other pathways could help develop more tolerant plant varieties that could easily sustain under extremely adverse conditions.

7. Strategies and Accomplishments

Improving plants' ability to adapt and tolerate abiotic stresses under changing climate scenarios is a potential strategy to lessen the oxidative stress-induced damages. Inhibition of pathways that partake in the overproduction of ROS, fortifying the plants' defense system through recruitment of antioxidants and modulation of ROS into the signaling pathway can boost plant survival under the stressful scenario. Pre-conditioning of plants to non-lethal stress [76] and molecular priming using agents such as micronutrients (β-sitosterol), osmolytes (GABA), etc. can fortify the plant defense mechanism and reduces oxidative stress [36,240,241]. Semwal and Khanna-Chopra [76] have reported that water-deficit pre-conditioning induced the tolerance to subsequent heat stress due to the recovery that escalates activities of antioxidants (SOD, CAT, POX, GSH, DHAR, AsA/DHA ratio, and GSH/GSSH ratio). The exogenous pretreatment with trehalose prompts H_2O_2 and the nitric oxide level in tomato leaves under cold stress that mediates signaling, upregulated Cu/Zn SOD, and CAT1 transcripts thereby the enhancement of defense capacity induced tolerance to stress, improvement in growth, and prevention of lipid membrane peroxidation [242]. Analogously, the exogenous application of melatonin in tea and AsA, GSH, and proline in chickpea plants increases the activity of enzymatic antioxidants (SOD, POX, CAT, APX) with the amplified accumulation of GSH and AsA under cold, salt, and/or drought stress [243,244]. The over-reduction of ETC and activities of certain enzymes/redox-active metals are major culprits for ROS overproduction in cell organelles. Consequently, the prevention of ETC over-reduction, inhibiting enzymes, and redox metals can arrest excessive ROS formation. Proline can bind with the redox-active metal ions, thus preventing the production of $^\bullet OH$ via the Fenton reaction and safeguard plant cells from oxidative damages [245]. Proline also maintains cellular redox homeostasis by maintaining the $NADP^+/NADPH$ balance [246]. In the chloroplast, proline is synthesized when glutamate is reduced by NADPH. Consequently, $NADP^+$ having been produced prevents the over-reduction of PS I by accepting electrons during stress conditions [94]. In addition, certain compounds such as nitric oxide have been reported to reduce activities of enzymes mediating the ROS production. The activity of XOD in peroxisomes of the *Phalaenopsis* flower that participates in $O_2^{\bullet-}$ production is downregulated by nitric oxide, leading to alleviation in the ROS level and oxidative stress [247]. Therefore, the priming using potential agents could alleviate ROS overproduction under stress.

Molecular priming is an efficient tool for improving plant tolerance to abiotic stress and its linkage with systems biology can strengthen the potential by unraveling the plants' complex defense and tolerance mechanism at the molecular level [248]. Systems biology deals with the omics study, i.e., genomics, transcriptomics, proteomics, and metabolomics to understand the functionality of the biological system altogether and facilitates in finding new genes, RNAs, proteins, and metabolites, deciphering their regulatory functions and intracellular interactions [248,249]. Several studies at the molecular level have provided deep insight into the regulatory network controlling response to abiotic stress in plants [250]. Some genes encode for functional proteins or products that directly partake in the regulation of stresses, while some regulate the expression of other stress-responsive genes [251]. The genomic approach focusing on identifying genes encoding for enzymatic antioxidants (APX, GPX, SOD, and CAT) in four resurrection species reveals their major role in the regulation of ROS homeostasis under desiccation (extreme water deficit condition) stress [252]. The study also highlights the ROS detoxification mechanism to be species-specific as having been evidenced through dissimilar expression patterns of all the studied antioxidant gene families. Similar results have been documented by Dubouzet et al. [253] who have reported the expression of dehydration responsive element binding (DREB) transcription factor homolog *OsDREB1A* and *OsDREB1B* gene under low-temperature stress and expression of *OsDREB2A* gene under dehydration and high salinity in rice (*Oryza sativa* L.). The integration of different omics study data elucidates the function and shared pathways of key molecular processes related to the multitude of stress and crops [223] facilitating the development of synthetic biology, which in turn aids in genetic manipulation to develop long-lasting stress-tolerant species [249]. The expression of DREB transcription factor (TF) genes *OsDREB1A* and *OsDREB1B* in transgenic rice improved plant tolerance to drought, high salinity, and cold stresses [254]. These genes encode proteins that might partake in stress tolerance and are associated with an increase in osmoprotectants such as free proline and soluble sugars in transgenic rice plants. A concomitant increase in the amount of osmolytes (free proline and soluble sugars), elevated expression of defense-related genes *OsDREB1A* with enhanced tolerance to drought and salt stress, and alleviated electrolyte leakage have been reported in rice seedlings [251]. Transgenic *Arabidopsis* over-expressing genes encoding for Cu/Zn-SOD demonstrate enhanced resistance against oxidative stress due to the escalated activities of SOD and POD [255]. The over-expression of zinc finger protein gene *OsZFP252* in rice seedlings elevates the expression of defense-related genes *OsDREB1A*, enhances tolerance to drought and salt stress, increases the number of osmolytes (free proline and soluble sugars), and alleviates electrolyte leakage [251]. The indispensable role of *OsZFP252* in the stress-responsive signaling transduction pathway has also been reported. Other studies have also reported the activation of oxidative signaling pathway with an expression of small HSP by the *Nicotiana* protein kinase (NPK1), tobacco mitogen-activated protein kinase kinase kinase (MAPKKK) in transgenic maize that confers protection to photosynthetic machinery after exposure to drought stress [256], and increases tolerance to freezing stress [257].

Recent advances in systems biology have added new impetus to improve plant tolerance. However, the expression and function of RNAs, proteomes, and metabolites in several genotypes are dynamic and still largely unknown. Traditional biotechnological approaches are unable to establish their niche in the field of plant stress management due to hindrance in the translation of identified agronomic traits into phenotypes [250]. The quest to understand the function, interaction, and response of molecular components under an environmental perturbation can be solved by quantifying and characterizing genotype to phenotype relationships [248]. Phenotypic attributes represent a response to abiotic stress. For example, plants growing under excess light possess thick leaves owing to the expanded palisade tissues and high stomatal density, and a lower number of thylakoids per chloroplast compared to plants growing in low light [258]. Phenotypic characters can be studied through a phenomics approach that spans a detailed study of physiological parameters influenced by the plant genetic layout, the spatio-temporal impact of the environment, and agricultural management practices [259]. The non-invasive method using cameras and sensor-based imaging (fluorescence, visible light, and infrared imaging, X-ray computed tomography, etc.) and advanced instruments such as fluorometers together with robust software systems are emerging techniques for studying plant morphological and developmental responses under a prevalent environment [258–260]. Such techniques enable us to explore information regarding the chemical composition and function that can be accessed from the cell to plant canopy level [260]. Under stress, the measurement of green and yellow areas of the leaf facilitates the determination of leaf senescence and tissue tolerance, corresponding to salt accumulation [260]. Similarly, studying chlorophyll fluorescence to monitor the impact of abiotic stress on plant photosynthesis and overall performance is a potential phenotyping technique [258]. Furthermore, chlorophyll fluorescence may be used to detect acclimatization mechanisms among genotypes under a defined set of stresses [258]. In particular, the variation in NPQ and leaf development has been observed in different accessions of *Arabidopsis* under similar environmental conditions [258]. In addition to the amalgamation of systems biology with plant physiology, crop modeling approaches further outline the plant responses by designing multiple simulations for farming practices and predicted climate change [261]. Crop models are expected to assist extrapolation of the complexity of climate change [261]. The climate-resilient barley (*Hordeum vulgare* L.) ideotypes designed through an assemblage of eight barley simulation models for the boreal and Mediterranean climate have revealed that specific ideotypes with a particular set of traits such as longer reproductive growing phase, higher radiation use efficiency/maximum assimilation rate, lower leaf senescence rate, and drought tolerance, in addition to a long (for boreal climate)/short (Mediterranean climate) photoperiod and vernalization sensitivity makes them promising cultivars for a projected future climate compared to other genotypes and confers better yield with a lower inter-annual yield variation [262]. A well-spun amalgamation of new and advanced scientific technologies could generate vast information that can be exploited to improve the plant tolerance capacity and resilience against hostile situations.

8. Conclusions and Future Prospects

Intensified abiotic stresses have perturbed ecological fitness via the production of ROS in plant cells. In the coming decades, the crisis will aggravate, challenging the plants' survival. ROS are similar to a double-edged sword that induces oxidative stress in plants when their production exceeds threshold levels, but at low or moderate concentrations, mediates the signal transduction that assists in maintaining cellular homeostasis and facilitates plant acclimatization to stress(es). To maintain equilibrium between ROS generation and their quenching, plants recruit antioxidants. Nevertheless, their potential diminishes during stress. Devising techniques that could avert damaging aspects of ROS under stress conditions and improving plants' tolerance mechanisms can unlock avenues for designing new generation stress-resilient crops. Recently, the techniques such as molecular priming or pre-conditioning have demonstrated the immense potential to improve plant resistance against abiotic stresses, however, some gaps still exist. Molecular priming requires a particular timing for the application of priming agents, for instance, just before the onset of stress, thus it requires continuous monitoring of environmental conditions to fortify plants against stress at both local and global levels. Furthermore, molecular priming has shown promising results in hydroponic or controlled conditions but the estimation of their efficiency in a real field under a present and projected climatic scenario is a necessity. In addition, the elucidation of pathways and impact of cross-talk between different priming agents and cellular compounds such as signaling agents, phytohormones, etc., within plant cells is important to gain insight into the fate of priming agents under variable conditions.

To fill the gaps, the integration of various disciplines such as systems biology, phenomics, and crop modeling with molecular priming is required (Figure 8). The identification of key genes, transcripts, proteins, and metabolites governing multiple pathways (signaling as well as the oxidation of biomolecules) through the systems biology approach assist in the discovery of new avenues. For example, the identification of genes responsible for the synthesis of priming agents endogenously in accordance with changing environmental conditions could eliminate the problem of continuous environmental monitoring at a global scale. Therefore, there is a need for intensive and dedicated research for the development of resilient crops via the application of molecular tools such as QTL mapping for the identification of the genomic network of metabolite biosynthesis and genome editing tools such as the regularly clustered interspaced short palindromic repeats and CRISPR-associated proteins (CRISPR/Cas). In addition to deciphering and improving the plant genetic network and underlying mechanisms, their vigilant amalgamation with phenomics and crop modeling is also necessary to maintain the unabridged potential of crops, while maintaining the ecosystem sustainability under climate change scenarios. The molecular responses of plants are highly influenced by environmental dynamics and demonstrate unique characters under each set of conditions. Therefore, quantification of the relationship between the genotype and phenotype under changing environmental conditions is important. This could be achieved through an integration of phenomics and crop modeling studies that furnish data related to the behavior of a particular or set of genetic networks in consonance to the dynamic environment. This integrated research framework could be highly obliged for laying out new improved management techniques that sustainably lead to the development of climate-smart crop cultivation with long-lasting tolerance to oxidative stress and boosts economical productivity.

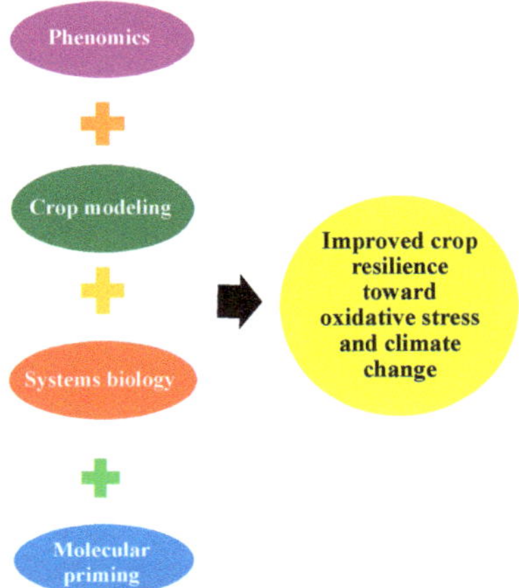

Figure 8. Integration of systems biology, phenomics, crop modeling, and molecular priming as a holistic approach for the development of climate-smart crops to improve growth and productivity.

Author Contributions: Conceptualization, M.I.A. and M.H.; writing—original draft preparation, M.I.A. and S.S.; writing—review and editing, S.S., M.H., M.I.A., M.F., and S.A.A.; supervision, M.I.A. and M.H. All authors have read and agreed to the published version of the manuscript.

Funding: This research received no external funding.

Acknowledgments: We acknowledge the past and present members of my laboratory as well as my scientific collaborators in the field of plant stress physiology.

Conflicts of Interest: The authors declare no conflict of interest.

References

1. Bulgari, R.; Franzoni, G.; Ferrante, A. Biostimulants application in horticultural crops under abiotic stress conditions. *Agronomy* **2019**, *9*, 306. [CrossRef]
2. Sachdev, S.; Singh, R.P. Root colonization: Imperative mechanism for efficient plant protection and growth. *MOJ Ecol. Environ. Sci.* **2018**, *3*, 240–242.
3. Saini, P.; Gani, M.; Kaur, J.J.; Godara, L.C.; Singh, C.; Chauhan, S.S.; Francies, R.M.; Bhardwaj, A.; Kumar, N.B.; Ghosh, M.K. Reactive oxygen species (ROS): A way to stress survival in plants. In *Abiotic Stress-Mediated Sensing and Signaling in Plants: An Omics Perspective*; Zargar, S.M., Zargar, M.Y., Eds.; Springer: Singapore, 2018; pp. 127–153.
4. Pereira, A. Plant abiotic stress challenges from the changing environment. *Front. Plant Sci.* **2016**, *7*, 1123. [CrossRef]
5. He, M.; He, C.Q.; Ding, N.Z. Abiotic stresses: General defenses of land plants and chances for engineering multi stress tolerance. *Front. Plant Sci.* **2018**, *9*, 1771. [CrossRef]
6. Bhuyan, M.B.; Hasanuzzaman, M.; Parvin, K.; Mohsin, S.M.; Al Mahmud, J.; Nahar, K.; Fujita, M. Nitric oxide and hydrogen sulfide: Two intimate collaborators regulating plant defense against abiotic stress. *Plant Growth Regul.* **2020**, *90*, 409–424. [CrossRef]
7. Kumar, V.; Khare, T.; Sharma, M.; Wani, S.H. ROS-induced signaling and gene expression in crops under salinity stress. In *Reactive Oxygen Species and Antioxidant Systems in Plants: Role and Regulation under Abiotic Stress*; Khan, M.I.R., Khan, N.A., Eds.; Springer: Singapore, 2017; pp. 159–184.
8. Maurya, A.K. Oxidative Stress in Crop Plants. In *Agronomic Crops: Stress Responses and Tolerance*; Hasanuzzaman, M., Ed.; Springer: Singapore, 2020; pp. 349–380.
9. Shah, K.; Chaturvedi, V.; Gupta, S. Climate Change and Abiotic Stress-Induced Oxidative Burst in Rice. In *Advances in Rice Research for Abiotic Stress Tolerance*; Hasanuzzaman, M., Fujita, M., Nahar, K., Biswas, J.K., Eds.; Woodhead Publishing: Cambridge, MA, USA, 2019; pp. 505–535.

10. Hasanuzzaman, M.; Bhuyan, M.H.M.; Zulfiqar, F.; Raza, A.; Mohsin, S.M.; Mahmud, J.A.; Fujita, M.; Fotopoulos, V. Reactive oxygen species and antioxidant defense in plants under abiotic stress: Revisiting the crucial role of a universal defense regulator. *Antioxidants* **2020**, *9*, 681. [CrossRef]

11. Wongshaya, P.; Chayjarung, P.; Tothong, C.; Pilaisangsuree, V.; Somboon, T.; Kongbangkerd, A.; Limmongkon, A. Effect of light and mechanical stress in combination with chemical elicitors on the production of stilbene compounds and defensive responses in peanut hairy root culture. *Plant Physiol. Biochem.* **2020**, *157*, 93–104. [CrossRef] [PubMed]

12. Florez-Sarasa, I.; Fernie, A.R.; Gupta, K.J. Does the alternative respiratory pathway offer protection against the adverse effects resulting from climate change? *J. Exp. Bot.* **2020**, *71*, 465–469. [CrossRef]

13. Raza, A.; Razzaq, A.; Mehmood, S.S.; Zou, X.; Zhang, X.; Lv, Y.; Xu, J. Impact of climate change on crops adaptation and strategies to tackle its outcome: A review. *Plants* **2019**, *8*, 34. [CrossRef] [PubMed]

14. Gray, S.B.; Brady, S.M. Plant developmental responses to climate change. *Dev. Biol.* **2016**, *419*, 64–77. [CrossRef] [PubMed]

15. Ye, Q.; Yang, X.; Dai, S.; Chen, G.; Li, Y.; Zhang, C. Effects of climate change on suitable rice cropping areas, cropping systems and crop water requirements in southern China. *Agric. Water Manag.* **2015**, *159*, 35–44. [CrossRef]

16. Noyes, P.D.; McElwee, M.K.; Miller, H.D.; Clark, B.W.; Van Tiem, L.A.; Walcott, K.C.; Ervin, K.N.; Levin, E.D. The toxicology of climate change: Environmental contaminants in a warming world. *Environ. Int.* **2009**, *35*, 971–986. [CrossRef] [PubMed]

17. Jarsjö, J.; Andersson-Sköld, Y.; Fröberg, M.; Pietroń, J.; Borgström, R.; Löv, Å.; Kleja, D.B. Projecting impacts of climate change on metal mobilization at contaminated sites: Controls by the groundwater level. *Sci. Total Environ.* **2020**, *712*, 135560. [CrossRef]

18. Zheng, W.; Wang, S.; Tan, K.; Lei, Y. Nitrate accumulation and leaching potential is controlled by land-use and extreme precipitation in a headwater catchment in the North China Plain. *Sci. Total Environ.* **2020**, *707*, 136168. [CrossRef]

19. Saijo, Y.; Loo, E.P.I. Plant immunity in signal integration between biotic and abiotic stress responses. *New Phytol.* **2020**, *225*, 87–104. [CrossRef]

20. Esim, N.; Atici, O. Nitric oxide improves chilling tolerance of maize by affecting apoplastic antioxidative enzymes in leaves. *Plant Growth Regul.* **2014**, *72*, 29–38. [CrossRef]

21. Megha, S.; Basu, U.; Kav, N.N. Regulation of low temperature stress in plants by microRNAs. *Plant Cell Environ.* **2018**, *41*, 1–15. [CrossRef] [PubMed]

22. Ansari, M.I.; Lin, T.P. Molecular analysis of dehydration in plants. *Int. Res. J. Plant Sci.* **2010**, *1*, 21–25.

23. Iqbal, M.S.; Singh, A.K.; Ansari, M.I. Effect of Drought Stress on Crop Production. In *New Frontiers in Stress Management for Durable Agriculture*; Rakshit, A., Singh, H.B., Singh, A.K., Singh, U.S., Fraceto, L., Eds.; Springer: Singapore, 2020; pp. 35–47.

24. Bailey-Serres, J.; Voesenek, L.A.C.J. Flooding stress: Acclimations and genetic diversity. *Annu. Rev. Plant Biol.* **2008**, *59*, 313–339. [CrossRef] [PubMed]

25. Ashraf, M.A. Waterlogging stress in plants: A review. *Afr. J. Agric. Res.* **2012**, *7*, 1976–1981.

26. Nievola, C.C.; Carvalho, C.P.; Carvalho, V.; Rodrigues, E. Rapid responses of plants to temperature changes. *Temperature* **2017**, *4*, 371–405. [CrossRef]

27. Sarkar, J.; Chakraborty, B.; Chakraborty, U. Plant growth promoting rhizobacteria protect wheat plants against temperature stress through antioxidant signalling and reducing chloroplast and membrane injury. *J. Plant Growth Regul.* **2018**, *37*, 1396–1412. [CrossRef]

28. Demirel, U.; Morris, W.L.; Ducreux, L.J.; Yavuz, C.; Asim, A.; Tindas, I.; Campbell, R.; Morris, J.A.; Verrall, S.R.; Hedley, P.E.; et al. Physiological, biochemical, and transcriptional responses to single and combined abiotic stress in stress-tolerant and stress-sensitive potato genotypes. *Front. Plant Sci.* **2020**, *11*, 169. [CrossRef]

29. Shahzad, B.; Tanveer, M.; Che, Z.; Rehman, A.; Cheema, S.A.; Sharma, A.; Song, H.; Rehman, S.; Zhaorong, D. Role of 24-epibrassinolide (EBL) in mediating heavy metal and pesticide induced oxidative stress in plants: A review. *Ecotoxicol. Environ. Saf.* **2018**, *147*, 935–944. [CrossRef]

30. Noshi, M.; Hatanaka, R.; Tanabe, N.; Terai, Y.; Maruta, T.; Shigeoka, S. Redox regulation of ascorbate and glutathione by a chloroplastic dehydroascorbate reductase is required for high-light stress tolerance in *Arabidopsis*. *Biosci. Biotechnol. Biochem.* **2016**, *80*, 870–877. [CrossRef] [PubMed]

31. Bisht, N.; Tiwari, S.; Singh, P.C.; Niranjan, A.; Chauhan, P.S. A multifaceted rhizobacterium *Paenibacillus lentimorbus* alleviates nutrient deficiency-induced stress in *Cicer arietinum* L. *Microbiol. Res.* **2019**, *223*, 110–119. [CrossRef]

32. Saleem, M.H.; Fahad, S.; Khan, S.U.; Din, M.; Ullah, A.; Sabagh, A.E.; Hossain, A.; Llanes, A.; Liu, L. Copper-induced oxidative stress, initiation of antioxidants and phytoremediation potential of flax (*Linum usitatissimum* L.) seedlings grown under the mixing of two different soils of China. *Environ. Sci. Pollut. Res.* **2020**, *27*, 5211–5221. [CrossRef] [PubMed]

33. Ainsworth, E.A.; Yendrek, C.R.; Sitch, S.; Collins, W.J.; Emberson, L.D. The effects of tropospheric ozone on net primary productivity and implications for climate change. *Annu. Rev. Plant Biol.* **2012**, *63*, 637–661. [CrossRef]

34. Ueda, Y.; Uehara, N.; Sasaki, H.; Kobayashi, K.; Yamakawa, T. Impacts of acute ozone stress on superoxide dismutase (SOD) expression and reactive oxygen species (ROS) formation in rice leaves. *Plant Physiol. Biochem.* **2013**, *70*, 396–402. [CrossRef]

35. Rahman, A.; Hossain, M.S.; Mahmud, J.A.; Nahar, K.; Hasanuzzaman, M.; Fujita, M. Manganese-induced salt stress tolerance in rice seedlings: Regulation of ion homeostasis, antioxidant defense and glyoxalase systems. *Physiol. Mol. Biol. Plants* **2016**, *22*, 291–306. [CrossRef] [PubMed]

36. Jalil, S.U.; Ansari, M.I. Physiological role of Gamma-aminobutyric acid in salt stress tolerance. In *Salt and Drought Stress Tolerance in Plants*; Hasanuzzaman, M., Tanveer, M., Eds.; Springer Nature: Cham, Switzerland, 2020; pp. 337–350.

37. Bita, C.; Gerats, T. Plant tolerance to high temperature in a changing environment: Scientific fundamentals and production of heat stress-tolerant crops. *Front. Plant Sci.* **2013**, *4*, 273. [CrossRef]
38. Hussain, H.A.; Hussain, S.; Khaliq, A.; Ashraf, U.; Anjum, S.A.; Men, S.; Wang, L. Chilling and drought stresses in crop plants: Implications, cross talk, and potential management opportunities. *Front. Plant Sci.* **2018**, *9*, 393. [CrossRef]
39. Balal, R.M.; Shahid, M.A.; Javaid, M.M.; Iqbal, Z.; Anjum, M.A.; Garcia-Sanchez, F.; Mattson, N.S. The role of selenium in amelioration of heat-induced oxidative damage in cucumber under high temperature stress. *Acta Physiol. Plant.* **2016**, *38*, 158. [CrossRef]
40. Rai, K.K.; Rai, N.; Rai, S.P. Salicylic acid and nitric oxide alleviate high temperature induced oxidative damage in Lablab purpureus L. plants by regulating bio-physical processes and DNA methylation. *Plant Physiol. Biochem.* **2018**, *128*, 72–88. [CrossRef]
41. Coates, J.; Mar, K.A.; Ojha, N.; Butler, T.M. The influence of temperature on ozone production under varying NOx conditions–a modelling study. *Atmos. Chem. Phys.* **2016**, *16*, 11601–11615. [CrossRef]
42. Jalil, S.U.; Ahmad, I.; Ansari, M.I. Functional loss of GABA transaminase (GABA-T) expressed early leaf senescence under various stress conditions in *Arabidopsis thaliana. Curr. Plant Biol.* **2017**, *9*, 11–22. [CrossRef]
43. Zhang, Y.; Jiang, W.; Yu, H.; Yang, X. Exogenous abscisic acid alleviates low temperature-induced oxidative damage in seedlings of *Cucumis sativus* L. *Trans. Chin. Soc. Agric. Eng.* **2012**, *28*, 221–228.
44. Liu, T.; Hu, X.; Zhang, J.; Zhang, J.; Du, Q.; Li, J. H_2O_2 mediates ALA-induced glutathione and ascorbate accumulation in the perception and resistance to oxidative stress in *Solanum lycopersicum* at low temperatures. *BMC Plant Biol.* **2018**, *18*, 1–10. [CrossRef]
45. Feng, X.; Porporato, A.; Rodriguez-Iturbe, I. Changes in rainfall seasonality in the tropics. *Nat. Clim. Chang.* **2013**, *3*, 811–815. [CrossRef]
46. Shao, H.B.; Chu, L.Y.; Jaleel, C.A.; Zhao, C.X. Water-deficit stress-induced anatomical changes in higher plants. *C. R. Biol.* **2008**, *331*, 215–225. [CrossRef]
47. Misra, V.; Solomon, S.; Ansari, M.I. Impact of drought on post-harvest quality of sugarcane crop. *Adv. Life Sci.* **2016**, *20*, 9496–9505.
48. Sam, A.S.; Padmaja, S.S.; Kachele, H.; Kumar, R.; Muller, K. Climate change, drought and rural communities: Understanding people's perceptions and adaptations in rural eastern India. *Int. J. Disaster Risk Reduct.* **2020**, *44*, 101436. [CrossRef]
49. Lee, B.R.; Li, L.S.; Jung, W.J.; Jin, Y.L.; Avice, J.C.; Ourry, A.; Kim, T.H. Water deficit-induced oxidative stress and the activation of antioxidant enzymes in white clover leaves. *Biol. Plant.* **2009**, *53*, 505–510. [CrossRef]
50. Fukao, T.; Barrera-Figueroa, B.E.; Juntawong, P.; Peña-Castro, J.M. Submergence and waterlogging stress in plants: A review highlighting research opportunities and understudied aspects. *Front. Plant Sci.* **2019**, *10*, 340. [CrossRef] [PubMed]
51. Hossain, M.A.; Uddin, S.N. Mechanisms of waterlogging tolerance in wheat: Morphological and metabolic adaptations under hypoxia or anoxia. *Aust. J. Crop Sci.* **2011**, *5*, 1094–1101.
52. Yamauchi, T.; Abe, F.; Tsutsumi, N.; Nakazono, M. Root cortex provides a venue for gas-space formation and is essential for plant adaptation to waterlogging. *Front. Plant Sci.* **2019**, *10*, 259. [CrossRef] [PubMed]
53. Chang, R.; Jang, C.J.; Branco-Price, C.; Nghiem, P.; Bailey-Serres, J. Transient MPK6 activation in response to oxygen deprivation and reoxygenation is mediated by mitochondria and aids seedling survival in *Arabidopsis. Plant Mol. Biol.* **2012**, *78*, 109–122. [CrossRef]
54. Sasidharan, R.; Hartman, S.; Liu, Z.; Martopawiro, S.; Sajeev, N.; van Veen, H.; Yeung, E.; Voesenek, L.A. Signal dynamics and interactions during flooding stress. *Plant Physiol.* **2018**, *176*, 1106–1117. [CrossRef] [PubMed]
55. Anee, T.I.; Nahar, K.; Rahman, A.; Mahmud, J.A.; Bhuiyan, T.F.; Alam, M.U.; Fujita, M.; Hasanuzzaman, M. Oxidative damage and antioxidant defense in Sesamum indicum after different waterlogging durations. *Plants* **2019**, *8*, 196. [CrossRef] [PubMed]
56. Steffens, B. The role of ethylene and ROS in salinity, heavy metal, and flooding responses in rice. *Front. Plant Sci.* **2014**, *5*, 685. [CrossRef]
57. Hasanuzzaman, M.; Oku, H.; Nahar, K.; Bhuyan, M.B.; Al Mahmud, J.; Baluska, F.; Fujita, M. Nitric oxide-induced salt stress tolerance in plants: ROS metabolism, signaling, and molecular interactions. *Plant Biotechnol. Rep.* **2018**, *12*, 77–92. [CrossRef]
58. Zhang, M.; Smith, J.A.C.; Harberd, N.P.; Jiang, C. The regulatory roles of ethylene and reactive oxygen species (ROS) in plant salt stress responses. *Plant Mol. Biol.* **2016**, *91*, 651–659. [CrossRef]
59. Kaur, H.; Bhardwaj, R.D.; Grewal, S.K. Mitigation of salinity-induced oxidative damage in wheat (*Triticum aestivum* L.) seedlings by exogenous application of phenolic acids. *Acta Physiol. Plant* **2017**, *39*, 221. [CrossRef]
60. Kumar, M.; Kumar, R.; Jain, V.; Jain, S. Differential behavior of the antioxidant system in response to salinity induced oxidative stress in salt-tolerant and salt-sensitive cultivars of *Brassica juncea* L. *Biocatal. Agric. Biotechnol.* **2018**, *13*, 12–19. [CrossRef]
61. Hajiboland, R. Effect of micronutrient deficiencies on plants stress responses. In *Abiotic Stress Responses in Plants: Metabolism, Productivity and Sustainability*; Ahmad, P., Prasad, M.N.V., Eds.; Springer: New York, NY, USA, 2012; pp. 283–329.
62. Gupta, D.K.; Pena, L.B.; Romero-Puertas, M.C.; Hernandez, A.; Inouhe, M.; Sandalio, L.M. NADPH oxidases differentially regulate ROS metabolism and nutrient uptake under cadmium toxicity. *Plant Cell Environ.* **2017**, *40*, 509–526. [CrossRef]
63. Shin, R.; Berg, R.H.; Schachtman, D.P. Reactive oxygen species and root hairs in Arabidopsis root response to nitrogen, phosphorus and potassium deficiency. *Plant Cell Physiol.* **2005**, *46*, 1350–1357. [CrossRef] [PubMed]
64. Jung, J.Y.; Shin, R.; Schachtman, D.P. Ethylene mediates response and tolerance to potassium deprivation in *Arabidopsis. Plant Cell* **2009**, *21*, 607–621. [CrossRef]

65. Tewari, R.K.; Kumar, P.; Sharma, P.N. Magnesium deficiency induced oxidative stress and antioxidant responses in mulberry plants. *Sci. Hortic.* **2006**, *108*, 7–14. [CrossRef]

66. Guo, W.; Nazim, H.; Liang, Z.; Yang, D. Magnesium deficiency in plants: An urgent problem. *Crop J.* **2016**, *4*, 83–91. [CrossRef]

67. Schutzendubel, A.; Polle, A. Plant responses to abiotic stresses: Heavy metal-induced oxidative stress and protection by mycorrhization. *J. Exp. Bot.* **2002**, *53*, 1351–1365. [CrossRef]

68. Pandey, V.; Dixit, V.; Shyam, R. Chromium effect on ROS generation and detoxification in pea (*Pisum sativum*) leaf chloroplasts. *Protoplasma* **2009**, *236*, 85–95. [CrossRef] [PubMed]

69. Trinh, N.N.; Huang, T.L.; Chi, W.C.; Fu, S.F.; Chen, C.C.; Huang, H.J. Chromium stress response effect on signal transduction and expression of signaling genes in rice. *Physiol. Plant.* **2014**, *150*, 205–224. [CrossRef] [PubMed]

70. Becanne, M.; Moran, J.F.; Iturbe-Ormaetxe, I. Iron dependent oxygen free radical generation in plants subjected to environmental stress: Toxicity and antioxidant protection. *Plant Soil* **1998**, *201*, 137–147. [CrossRef]

71. Sharma, A.; Yuan, H.; Kumar, V.; Ramakrishnan, M.; Kohli, S.K.; Kaur, R.; Thukral, A.K.; Bhardwaj, R.; Zheng, B. Castasterone attenuates insecticide induced phytotoxicity in mustard. *Ecotoxicol. Environ. Saf.* **2019**, *179*, 50–61. [CrossRef] [PubMed]

72. Sachdev, S.; Singh, R.P. Isolation, characterisation and screening of native microbial isolates for biocontrol of fungal pathogens of tomato. *Clim. Chang. Environ. Sustain.* **2018**, *6*, 46–58. [CrossRef]

73. Yuzbasıoglu, E.; Dalyan, E. Salicylic acid alleviates thiram toxicity by modulating antioxidant enzyme capacity and pesticide detoxification systems in the tomato (*Solanum lycopersicum* Mill.). *Plant Physiol. Biochem.* **2019**, *135*, 322–330. [CrossRef] [PubMed]

74. Rizhsky, L.; Liang, H.; Shuman, J.; Shulaev, V.; Davletova, S.; Mittler, R. When defense pathways collide: The response of *Arabidopsis* to a combination of drought and heat stress. *Plant Physiol.* **2004**, *134*, 1683–1696. [CrossRef]

75. Rizhsky, L.; Liang, H.; Mittler, R. The combined effect of drought stress and heat shock on gene expression in tobacco. *Plant Physiol.* **2002**, *130*, 1143–1151. [CrossRef]

76. Semwal, V.K.; Khanna-Chopra, R. Enhanced oxidative stress, damage and inadequate antioxidant defense contributes towards insufficient recovery in water deficit stress and heat stress combination compared to either stresses alone in *Chenopodium album* (Bathua). *Physiol. Mol. Biol. Plants* **2020**, *26*, 1331–1339. [CrossRef]

77. Mittler, R. Abiotic stress, the field environment and stress combination. *Trends Plant Sci.* **2006**, *11*, 15–19. [CrossRef]

78. Prasch, C.M.; Sonnewald, U. Simultaneous application of heat, drought, and virus to Arabidopsis plants reveals significant shifts in signaling networks. *Plant Physiol.* **2013**, *162*, 1849–1866. [CrossRef]

79. Woldesemayat, A.A.; Modise, D.M.; Gamieldien, J.; Ndimba, B.K.; Christoffels, A. Correction: Cross-species multiple environmental stress responses: An integrated approach to identify candidate genes for multiple stress tolerance in sorghum (*Sorghum bicolor* (L.) Moench) and related model species. *PLoS ONE* **2018**, *13*, e0197017. [CrossRef]

80. Roychowdhury, R.; Khan, M.H.; Choudhury, S. Physiological and molecular responses for metalloid stress in rice—A Comprehensive Overview. In *Advances in Rice Research for Abiotic Stress Tolerance*; Hasanuzzaman, M., Fujita, M., Nahar, K., Biswas, J.K., Eds.; Woodhead Publishing: Cambridge, MA, USA, 2019; pp. 341–369.

81. Apel, K.; Hirt, H. Reactive oxygen species: Metabolism, oxidative stress, and signal transduction. *Annu. Rev. Plant Biol.* **2004**, *55*, 373–399. [CrossRef]

82. Podgorska, A.; Burian, M.; Szal, B. Extra-cellular but extra-ordinarily important for cells: Apoplastic reactive oxygen species metabolism. *Front. Plant Sci.* **2017**, *8*, 1353. [CrossRef] [PubMed]

83. Janku, M.; Luhova, L.; Petrivalsky, M. On the origin and fate of reactive oxygen species in plant cell compartments. *Antioxidants* **2019**, *8*, 105. [CrossRef]

84. Shakirova, F.M.; Allagulova, C.R.; Maslennikova, D.R.; Klyuchnikova, E.O.; Avalbaev, A.M.; Bezrukova, M.V. Salicylic acid-induced protection against cadmium toxicity in wheat plants. *Environ. Exp. Bot.* **2016**, *122*, 19–28. [CrossRef]

85. Chen, S.; Yin, C.; Strasser, R.J.; Yang, C.; Qiang, S. Reactive oxygen species from chloroplasts contribute to 3-acetyl-5-isopropyltetramic acid-induced leaf necrosis of *Arabidopsis thaliana*. *Plant Physiol. Biochem.* **2012**, *52*, 38–51. [CrossRef]

86. Asada, K. Production and scavenging of reactive oxygen species in chloroplasts and their functions. *Plant Physiol.* **2006**, *141*, 391–396. [CrossRef] [PubMed]

87. Pospisil, P. Production of reactive oxygen species by photosystem II as a response to light and temperature stress. *Front. Plant Sci.* **2016**, *7*, 1950. [CrossRef] [PubMed]

88. Mattos, L.M.; Moretti, C.L. Oxidative stress in plants under drought conditions and the role of different enzymes. *Enzym. Eng.* **2015**, *5*, 1–6. [CrossRef]

89. Sharma, P.; Jha, A.B.; Dubey, R.S.; Pessarakli, M. Reactive oxygen species, oxidative damage, and antioxidative defense mechanism in plants under stressful conditions. *J. Bot.* **2012**. [CrossRef]

90. Cruz, D.C.M. Drought stress and reactive oxygen species: Production, scavenging and signaling. *Plant Signal. Behav.* **2008**, *3*, 156–165. [CrossRef] [PubMed]

91. Xie, X.; He, Z.; Chen, N.; Tang, Z.; Wang, Q.; Cai, Y. The roles of environmental factors in regulation of oxidative stress in plant. *BioMed Res. Int.* **2019**. [CrossRef]

92. Sgherri, C.L.; Pinzino, C.; Navari-Izzo, F. Sunflower seedlings subjected to increasing stress by water deficit: Changes in $O_2^{\bullet-}$ production related to the composition of thylakoid membranes. *Physiol. Plant.* **1996**, *96*, 446–452. [CrossRef]

93. Foyer, C.H.; Shigeoka, S. Understanding oxidative stress and antioxidant functions to enhance photosynthesis. *Plant Physiol.* **2011**, *155*, 93–100. [CrossRef] [PubMed]

94. Miller, G.A.D.; Suzuki, N.; Ciftci-Yilmaz, S.; Mittler, R.O.N. Reactive oxygen species homeostasis and signalling during drought and salinity stresses. *Plant Cell Environ.* **2010**, *33*, 453–467. [CrossRef]

95. Das, K.; Roychoudhury, A. Reactive oxygen species (ROS) and response of antioxidants as ROS-scavengers during environmental stress in plants. *Front. Environ. Sci.* **2014**, *2*, 53. [CrossRef]

96. Radwan, D.E.M.; Mohamed, A.K.; Fayez, K.A.; Abdelrahman, A.M. Oxidative stress caused by Basagran®herbicide is altered by salicylic acid treatments in peanut plants. *Heliyon* **2019**, *5*, e01791. [CrossRef] [PubMed]

97. Jung, S.; Steffen, K.L. Influence of photosynthetic photon flux densities before and during long-term chilling on xanthophyll cycle and chlorophyll fluorescence quenching in leaves of tomato (*Lycopersicon hirsutum*). *Physiol. Plant.* **1997**, *100*, 958–966. [CrossRef]

98. Yamane, K.; Taniguchi, M.; Miyake, H. Salinity-induced subcellular accumulation of H_2O_2 in leaves of rice. *Protoplasma* **2012**, *249*, 301–308. [CrossRef]

99. Shu, S.; Yuan, L.Y.; Guo, S.R.; Sun, J.; Yuan, Y.H. Effects of exogenous spermine on chlorophyll fluorescence, antioxidant system and ultrastructure of chloroplasts in *Cucumis sativus* L. under salt stress. *Plant Physiol. Biochem.* **2013**, *63*, 209–216. [CrossRef]

100. Dietz, K.J.; Baier, M.; Kramer, U. Free radicals and reactive oxygen species as mediators of heavy metal toxicity in plants. In *Heavy Metal Stress in Plants: From Molecules to Ecosystems*; Prasad, M.N.V., Hagemeyer, J., Eds.; Springer: Berlin/Heidelberg, Germany, 1999; pp. 73–97. [CrossRef]

101. Nyathi, Y.; Baker, A. Plant peroxisomes as a source of signalling molecules. *Biochim. Biophys. Acta (BBA)Mol. Cell Res.* **2006**, *1763*, 1478–1495. [CrossRef]

102. del Río, L.A.; Lopez-Huertas, E. ROS generation in peroxisomes and its role in cell signaling. *Plant Cell Physiol.* **2016**, *57*, 1364–1376. [CrossRef]

103. Ortega-Galisteo, A.P.; Rodriguez-Serrano, M.; Pazmino, D.M.; Gupta, D.K.; Sandalio, L.M.; Romero-Puertas, M.C. S-Nitrosylated proteins in pea (*Pisum sativum* L.) leaf peroxisomes: Changes under abiotic stress. *J. Exp. Bot.* **2012**, *63*, 2089–2103. [CrossRef]

104. del Río, L.A.; Pastori, G.M.; Palma, J.M.; Sandalio, L.M.; Sevilla, F.; Corpas, F.J.; Jimenez, A.; Lopez-Huertas, E.; Hernández, J.A. The activated oxygen role of peroxisomes in senescence. *Plant Physiol.* **1998**, *116*, 1195–1200. [CrossRef]

105. Corpas, F.J.; Del Río, L.A.; Palma, J.M. Plant peroxisomes at the crossroad of NO and H_2O_2 metabolism. *J. Integr. Plant Biol.* **2019**, *61*, 803–816. [PubMed]

106. Voss, I.; Sunil, B.; Scheibe, R.; Raghavendra, A.S. Emerging concept for the role of photorespiration as an important part of abiotic stress response. *Plant Biol.* **2013**, *15*, 713–722. [CrossRef]

107. Suo, J.; Zhao, Q.; David, L.; Chen, S.; Dai, S. Salinity Response in Chloroplasts: Insights from Gene Characterization. *Int. J. Mol. Sci.* **2017**, *18*, 1011. [CrossRef]

108. Mittova, V.; Tal, M.; Volokita, M.; Guy, M. Up-regulation of the leaf mitochondrial and peroxisomal antioxidative systems in response to salt-induced oxidative stress in the wild salt-tolerant tomato species *Lycopersicon pennellii*. *Plant Cell Environ.* **2003**, *26*, 845–856. [CrossRef] [PubMed]

109. Su, T.; Li, W.; Wang, P.; Ma, C. Dynamics of peroxisome homeostasis and its role in stress response and signaling in plants. *Front. Plant Sci.* **2019**, *10*, 705. [CrossRef] [PubMed]

110. Demidchik, V. Reactive oxygen species and oxidative stress in plants. In *Plant Stress Physiology*; Shabala, S., Ed.; CAB International: London, UK, 2012; pp. 24–58.

111. Corpas, F.J.; Barroso, J.B. Peroxynitrite (ONOO−) is endogenously produced in Arabidopsis peroxisomes and is overproduced under cadmium stress. *Ann. Bot.* **2014**, *113*, 87–96. [CrossRef]

112. Romero-Puertas, M.C.; McCarthy, I.; Sandalio, L.M.; Palma, J.M.; Corpas, F.J.; Gomez, M.; Del Rio, L.A. Cadmium toxicity and oxidative metabolism of pea leaf peroxisomes. *Free Radic. Res.* **1999**, *31*, 25–31. [CrossRef]

113. Romero-Puertas, M.C.; Palma, J.M.; Gómez, M.; Del Rio, L.A.; Sandalio, L.M. Cadmium causes the oxidative modification of proteins in pea plants. *Plant Cell Environ.* **2002**, *25*, 677–686. [CrossRef]

114. Keunen, E.; Remans, T.; Bohler, S.; Vangronsveld, J.; Cuypers, A. Metal-induced oxidative stress and plant mitochondria. *Int. J. Mol. Sci.* **2011**, *12*, 6894–6918. [CrossRef]

115. Schertl, P.; Braun, H.P. Respiratory electron transfer pathways in plant mitochondria. *Front. Plant Sci.* **2014**, *5*, 163. [CrossRef] [PubMed]

116. Gill, S.S.; Tuteja, N. Reactive oxygen species and antioxidant machinery in abiotic stress tolerance in crop plants. *Plant Physiol. Biochem.* **2010**, *48*, 909–930. [CrossRef]

117. Rhoads, D.M.; Umbach, A.L.; Subbaiah, C.C.; Siedow, J.N. Mitochondrial reactive oxygen species. Contribution to oxidative stress and interorganellar signaling. *Plant Physiol.* **2006**, *141*, 357–366. [CrossRef]

118. Cvetkovska, M.; Vanlerberghe, G.C. Alternative oxidase impacts the plant response to biotic stress by influencing the mitochondrial generation of reactive oxygen species. *Plant Cell Environ.* **2013**, *36*, 721–732. [CrossRef]

119. Bose, J.; Rodrigo-Moreno, A.; Shabala, S. ROS homeostasis in halophytes in the context of salinity stress tolerance. *J. Exp. Bot.* **2014**, *65*, 1241–1257. [CrossRef]

120. Hu, W.H.; Song, X.S.; Shi, K.; Xia, X.J.; Zhou, Y.H.; Yu, J.Q. Changes in electron transport, superoxide dismutase and ascorbate peroxidase isoenzymes in chloroplasts and mitochondria of cucumber leaves as influenced by chilling. *Photosynthetica* **2008**, *46*, 581–588. [CrossRef]

121. Fedyaeva, A.V.; Stepanov, A.V.; Lyubushkina, I.V.; Pobezhimova, T.P.; Rikhvanov, E.G. Heat shock induces production of reactive oxygen species and increases inner mitochondrial membrane potential in winter wheat cells. *Biochemistry* **2014**, *79*, 1202–1210. [CrossRef]

122. Turrens, J.F. Mitochondrial formation of reactive oxygen species. *J. Physiol.* **2003**, *552*, 335–344. [CrossRef]

123. Hernandez, J.A.; Corpas, F.J.; Gomez, M.; del Rio, L.A.; Sevilla, F. Salt-induced oxidative stress mediated by activated oxygen species in pea leaf mitochondria. *Physiol. Plant.* **1993**, *89*, 103–110. [CrossRef]

124. Vacca, R.A.; Valenti, D.; Bobba, A.; Merafina, R.S.; Passarella, S.; Marra, E. Cytochrome c is released in a reactive oxygen species-dependent manner and is degraded via caspase-like proteases in tobacco Bright-Yellow 2 cells en route to heat shock-induced cell death. *Plant Physiol.* **2006**, *141*, 208–219. [CrossRef]

125. Gao, C.; Xing, D.; Li, L.; Zhang, L. Implication of reactive oxygen species and mitochondrial dysfunction in the early stages of plant programmed cell death induced by ultraviolet-C overexposure. *Planta* **2008**, *227*, 755–767. [CrossRef]

126. Balk, J.; Leaver, C.J.; McCabe, P.F. Translocation of cytochrome c from the mitochondria to the cytosol occurs during heat-induced programmed cell death in cucumber plants. *FEBS Lett.* **1999**, *463*, 151–154. [CrossRef]

127. Schroeder, W.H.; Breuer, U.; Stelzer, R.; Gierth, M. The apoplast and its significance for plant mineral nutrition. *New Phytol.* **2009**, *182*, 284.

128. Choudhury, F.K.; Rivero, R.M.; Blumwald, E.; Mittler, R. Reactive oxygen species, abiotic stress and stress combination. *Plant J.* **2017**, *90*, 856–867. [CrossRef] [PubMed]

129. Choudhary, A.; Kumar, A.; Kaur, N. ROS and oxidative burst: Roots in plant development. *Plant Divers.* **2020**, *42*, 33–43. [CrossRef]

130. Wang, F.; Chen, Z.H.; Liu, X.; Colmer, T.D.; Shabala, L.; Salih, A.; Zhao, M.; Shabala, S. Revealing the roles of GORK channels and NADPH oxidase in acclimation to hypoxia in *Arabidopsis*. *J. Exp. Bot.* **2017**, *68*, 3191–3204. [CrossRef]

131. Mika, A.; Minibayeva, F.; Beckett, R.; Luthje, S. Possible functions of extracellular peroxidases in stress-induced generation and detoxification of active oxygen species. *Phytochem. Rev.* **2004**, *3*, 173–193. [CrossRef]

132. Hu, X.; Zhang, A.; Zhang, J.; Jiang, M. Abscisic acid is a key inducer of hydrogen peroxide production in leaves of maize plants exposed to water stress. *Plant Cell Physiol.* **2006**, *47*, 1484–1495. [CrossRef]

133. Voothuluru, P.; Sharp, R.E. Apoplastic hydrogen peroxide in the growth zone of the maize primary root under water stress. I. Increased levels are specific to the apical region of growth maintenance. *J. Exp. Bot.* **2013**, *64*, 1223–1233. [CrossRef] [PubMed]

134. Lin, C.C.; Kao, C.H. Cell wall peroxidase activity, hydrogen peroxide level and NaCl-inhibited root growth of rice seedlings. *Plant Soil* **2001**, *230*, 135–143. [CrossRef]

135. Langebartels, C.; Wohlgemuth, H.; Kschieschan, S.; Grün, S.; Sandermann, H. Oxidative burst and cell death in ozone-exposed plants. *Plant Physiol. Biochem.* **2002**, *40*, 567–575. [CrossRef]

136. Martinez, C.; Montillet, J.L.; Bresson, E.; Agnel, J.P.; Dai, G.H.; Daniel, J.F.; Geiger, J.P.; Nicole, M. Apoplastic peroxidase generates superoxide anions in cells of cotton cotyledons undergoing the hypersensitive reaction to *Xanthomonas campestris* pv. *malvacearum* race 18. *Mol. Plant-Microbe Interact.* **1998**, *11*, 1038–1047. [CrossRef]

137. Petrov, V.; Hille, J.; Mueller-Roeber, B.; Gechev, T.S. ROS-mediated abiotic stress-induced programmed cell death in plants. *Front. Plant Sci.* **2015**, *6*, 69. [CrossRef]

138. Moschou, P.N.; Paschalidis, K.A.; Delis, I.D.; Andriopoulou, A.H.; Lagiotis, G.D.; Yakoumakis, D.I.; Roubelakis-Angelakis, K.A. Spermidine exodus and oxidation in the apoplast induced by abiotic stress is responsible for H_2O_2 signatures that direct tolerance responses in tobacco. *Plant Cell* **2008**, *20*, 1708–1724. [CrossRef]

139. Davies, M.J. The oxidative environment and protein damage. *Biochim. Biophys. Acta Proteins Proteom.* **2005**, *1703*, 93–109. [CrossRef] [PubMed]

140. de Dios Alché, J. A concise appraisal of lipid oxidation and lipoxidation in higher plants. *Redox Biol.* **2019**, *23*, 101136. [CrossRef]

141. Anjum, N.A.; Sofo, A.; Scopa, A.; Roychoudhury, A.; Gill, S.S.; Iqbal, M.; Lukatkin, A.S.; Pereira, E.; Duarte, A.C.; Ahmad, I. Lipids and proteins—major targets of oxidative modifications in abiotic stressed plants. *Environ. Sci. Pollut. Res.* **2015**, *22*, 4099–4121. [CrossRef]

142. Slade, D.; Radman, M. Oxidative stress resistance in Deinococcus radiodurans. *Microbiol. Mol. Biol. Rev.* **2011**, *75*, 133–191. [CrossRef] [PubMed]

143. Katsuhara, M.; Otsuka, T.; Ezaki, B. Salt stress-induced lipid peroxidation is reduced by glutathione S-transferase, but this reduction of lipid peroxides is not enough for a recovery of root growth in Arabidopsis. *Plant Sci.* **2005**, *169*, 369–373. [CrossRef]

144. Ali, M.B.; Hahn, E.J.; Paek, K.Y. Effects of temperature on oxidative stress defense systems, lipid peroxidation and lipoxygenase activity in Phalaenopsis. *Plant Physiol. Biochem.* **2005**, *43*, 213–223. [CrossRef]

145. Singh, S.; Eapen, S.; D'souza, S.F. Cadmium accumulation and its influence on lipid peroxidation and antioxidative system in an aquatic plant, *Bacopa monnieri* L. *Chemosphere* **2006**, *62*, 233–246. [CrossRef]

146. Singh, H.P.; Batish, D.R.; Kohli, R.K.; Arora, K. Arsenic-induced root growth inhibition in mung bean (*Phaseolus aureus* Roxb.) is due to oxidative stress resulting from enhanced lipid peroxidation. *Plant Growth Regul.* **2007**, *53*, 65–73. [CrossRef]

147. Hameed, A.; Bibi, N.; Akhter, J.; Iqbal, N. Differential changes in antioxidants, proteases, and lipid peroxidation in flag leaves of wheat genotypes under different levels of water deficit conditions. *Plant Physiol. Biochem.* **2011**, *49*, 178–185. [CrossRef] [PubMed]

148. Hameed, A.; Goher, M.; Iqbal, N. Drought induced programmed cell death and associated changes in antioxidants, proteases, and lipid peroxidation in wheat leaves. *Biol. Plant.* **2013**, *57*, 370–374. [CrossRef]

149. Majid, U.; Siddiqi, T.O.; Iqbal, M. Antioxidant response of *Cassia angustifolia* Vahl. to oxidative stress caused by Mancozeb, a pyrethroid fungicide. *Acta Physiol. Plant.* **2014**, *36*, 307–314. [CrossRef]
150. Kumari, S.; Agrawal, M.; Singh, A. Effects of ambient and elevated CO_2 and ozone on physiological characteristics, antioxidative defense system and metabolites of potato in relation to ozone flux. *Environ. Exp. Bot.* **2015**, *109*, 276–287. [CrossRef]
151. Li, X.; Zhang, L.; Li, Y.; Ma, L.; Bu, N.; Ma, C. Changes in photosynthesis, antioxidant enzymes and lipid peroxidation in soybean seedlings exposed to UV-B radiation and/or Cd. *Plant Soil* **2012**, *352*, 377–387. [CrossRef]
152. Meriga, B.; Reddy, B.K.; Rao, K.R.; Reddy, L.A.; Kishor, P.K. Aluminium-induced production of oxygen radicals, lipid peroxidation and DNA damage in seedlings of rice (*Oryza sativa*). *J. Plant Physiol.* **2004**, *161*, 63–68. [CrossRef]
153. Talukdar, D. Arsenic-induced oxidative stress in the common bean legume, *Phaseolus vulgaris* L. seedlings and its amelioration by exogenous nitric oxide. *Physiol. Mol. Biol. Plants* **2013**, *19*, 69–79. [CrossRef] [PubMed]
154. Martinez, V.; Mestre, T.C.; Rubio, F.; Girones-Vilaplana, A.; Moreno, D.A.; Mittler, R.; Rivero, R.M. Accumulation of flavonols over hydroxycinnamic acids favors oxidative damage protection under abiotic stress. *Front. Plant Sci.* **2016**, *7*, 838. [CrossRef]
155. Kosova, K.; Vitamvas, P.; Urban, M.O.; Prasil, I.T.; Renaut, J. Plant abiotic stress proteomics: The major factors determining alterations in cellular proteome. *Front. Plant Sci.* **2018**, *9*, 122. [CrossRef] [PubMed]
156. Oracz, K.; Bouteau, H.E.M.; Farrant, J.M.; Cooper, K.; Belghazi, M.; Job, C.; Job, D.; Corbineau, F.; Bailly, C. ROS production and protein oxidation as a novel mechanism for seed dormancy alleviation. *Plant J.* **2007**, *50*, 452–465. [CrossRef]
157. Moller, I.M.; Rogowska-Wrzesinska, A.; Rao, R.S.P. Protein carbonylation and metal-catalyzed protein oxidation in a cellular perspective. *J. Proteom.* **2011**, *74*, 2228–2242. [CrossRef]
158. Moller, I.M.; Jensen, P.E.; Hansson, A. Oxidative modifications to cellular components in plants. *Annu. Rev. Plant Biol.* **2007**, *58*, 459–481. [CrossRef]
159. Sweetlove, L.J.; Møller, I.M. Oxidation of proteins in plants—mechanisms and consequences. *Adv. Bot. Res.* **2009**, *52*, 1–23.
160. Karuppanapandian, T.; Kim, W. Cobalt-induced oxidative stress causes growth inhibition associated with enhanced lipid peroxidation and activates antioxidant responses in Indian mustard (*Brassica juncea* L.) leaves. *Acta Physiol. Plant.* **2013**, *35*, 2429–2443. [CrossRef]
161. Bartoli, C.G.; Gómez, F.; Martínez, D.E.; Guiamet, J.J. Mitochondria are the main target for oxidative damage in leaves of wheat (*Triticum aestivum* L.). *J. Exp. Bot.* **2004**, *55*, 1663–1669. [CrossRef]
162. Ferreira-Silva, S.L.; Voigt, E.L.; Silva, E.N.; Maia, J.M.; Aragão, T.C.R.; Silveira, J.A.G. Partial oxidative protection by enzymatic and non-enzymatic components in cashew leaves under high salinity. *Biol. Plant.* **2012**, *56*, 172–176. [CrossRef]
163. Qiu, Q.S.; Huber, J.L.; Booker, F.L.; Jain, V.; Leakey, A.D.; Fiscus, E.L.; Yau, P.M.; Donald, R.O.; Huber, S.C. Increased protein carbonylation in leaves of *Arabidopsis* and soybean in response to elevated $[CO_2]$. *Photosynth. Res.* **2008**, *97*, 155. [CrossRef]
164. Prasad, T.K. Mechanisms of chilling-induced oxidative stress injury and tolerance in developing maize seedlings: Changes in antioxidant system, oxidation of proteins and lipids, and protease activities. *Plant J.* **1996**, *10*, 1017–1026. [CrossRef]
165. Kingston-Smith, A.H.; Foyer, C.H. Bundle sheath proteins are more sensitive to oxidative damage than those of the mesophyll in maize leaves exposed to paraquat or low temperatures. *J. Exp. Bot.* **2000**, *51*, 123–130. [CrossRef]
166. Feng, Y.; Komatsu, S.; Furukawa, T.; Koshiba, T.; Kohno, Y. Proteome analysis of proteins responsive to ambient and elevated ozone in rice seedlings. *Agric. Ecosyst. Environ.* **2008**, *125*, 255–265. [CrossRef]
167. Hollosy, F. Effects of ultraviolet radiation on plant cells. *Micron* **2002**, *33*, 179–197. [CrossRef]
168. Evans, M.D.; Dizdaroglu, M.; Cooke, M.S. Oxidative DNA damage and disease: Induction, repair and significance. *Mutat. Res. Rev. Mutat. Res.* **2004**, *567*, 1–61. [CrossRef] [PubMed]
169. Soares, C.; Carvalho, M.E.; Azevedo, R.A.; Fidalgo, F. Plants facing oxidative challenges—A little help from the antioxidant networks. *Environ. Exp. Bot.* **2019**, *161*, 4–25. [CrossRef]
170. Halliwell, B. Reactive species and antioxidants. Redox biology is a fundamental theme of aerobic life. *Plant Physiol.* **2006**, *141*, 312–322. [CrossRef] [PubMed]
171. Roldan-Arjona, T.; Ariza, R.R. Repair and tolerance of oxidative DNA damage in plants. *Mutat. Res. Rev. Mutat. Res.* **2009**, *681*, 169–179. [CrossRef]
172. Fink, S.P.; Reddy, G.R.; Marnett, L.J. Mutagenicity in Escherichia coli of the major DNA adduct derived from the endogenous mutagen malondialdehyde. *Proc. Natl. Acad. Sci. USA* **1997**, *94*, 8652–8657. [CrossRef] [PubMed]
173. Mittler, R. Oxidative stress, antioxidants and stress tolerance. *Trends Plant Sci.* **2002**, *7*, 405–410. [CrossRef]
174. Chmielowska-Bak, J.; Izbiańska, K.; Ekner-Grzyb, A.; Bayar, M.; Deckert, J. Cadmium stress leads to rapid increase in RNA oxidative modifications in soybean seedlings. *Front. Plant Sci.* **2018**, *8*, 2219. [CrossRef] [PubMed]
175. Dinakar, C.; Bartels, D. Light response, oxidative stress management and nucleic acid stability in closely related Linderniaceae species differing in desiccation tolerance. *Planta* **2012**, *236*, 541–555. [CrossRef]
176. Camejo, D.; Guzmán-Cedeno, Á.; Moreno, A. Reactive oxygen species, essential molecules, during plant–pathogen interactions. *Plant Physiol. Biochem.* **2016**, *103*, 10–23. [CrossRef]
177. Dumont, S.; Rivoal, J. Consequences of oxidative stress on plant glycolytic and respiratory metabolism. *Front. Plant Sci.* **2019**, *10*, 166. [CrossRef]
178. Nimse, S.B.; Pal, D. Free radicals, natural antioxidants, and their reaction mechanisms. *RSC Adv.* **2015**, *5*, 27986–28006. [CrossRef]
179. Ahmad, P.; Jaleel, C.A.; Salem, M.A.; Nabi, G.; Sharma, S. Roles of enzymatic and nonenzymatic antioxidants in plants during abiotic stress. *Crit. Rev. Biotechnol.* **2010**, *30*, 161–175. [CrossRef]

180. Zafra, A.; Castro, A.J.; de Dios Alche, J. Identification of novel superoxide dismutase isoenzymes in the olive (*Olea europaea* L.) pollen. *BMC Plant Biol.* **2018**, *18*, 114. [CrossRef] [PubMed]

181. Chen, N.; Teng, X.L.; Xiao, X.G. Subcellular localization of a plant catalase-phenol oxidase, AcCATPO, from Amaranthus and identification of a non-canonical peroxisome targeting signal. *Front. Plant Sci.* **2017**, *8*, 1345. [CrossRef]

182. Su, T.; Wang, P.; Li, H.; Zhao, Y.; Lu, Y.; Dai, P.; Ren, T.; Wang, X.; Li, X.; Shao, Q.; et al. The *Arabidopsis* catalase triple mutant reveals important roles of catalases and peroxisome-derived signaling in plant development. *J. Integr. Plant Biol.* **2018**, *60*, 591–607. [CrossRef]

183. Anjum, N.A.; Sharma, P.; Gill, S.S.; Hasanuzzaman, M.; Khan, E.A.; Kachhap, K.; Mohamed, A.A.; Thangavel, P.; Devi, G.D.; Vasudhevan, P.; et al. Catalase and ascorbate peroxidase—representative H_2O_2-detoxifying heme enzymes in plants. *Environ. Sci. Pollut. Res.* **2016**, *23*, 19002–19029. [CrossRef] [PubMed]

184. Ishikawa, T.; Shigeoka, S. Recent advances in ascorbate biosynthesis and the physiological significance of ascorbate peroxidase in photosynthesizing organisms. *Biosci. Biotechnol. Biochem.* **2008**, *72*, 1143–1154. [CrossRef] [PubMed]

185. Pandey, P.; Singh, J.; Achary, V.; Reddy, M.K. Redox homeostasis via gene families of ascorbate-glutathione pathway. *Front. Environ. Sci.* **2015**, *3*, 25. [CrossRef]

186. Pandey, S.; Fartyal, D.; Agarwal, A.; Shukla, T.; James, D.; Kaul, T.; Negi, Y.K.; Arora, S.; Reddy, M.K. Abiotic stress tolerance in plants: Myriad roles of ascorbate peroxidase. *Front. Plant Sci.* **2017**, *8*, 581. [CrossRef] [PubMed]

187. Bartoli, C.G.; Buet, A.; Grozeff, G.G.; Galatro, A.; Simontacchi, M. Ascorbate-glutathione cycle and abiotic stress tolerance in plants. In *Ascorbic Acid in Plant Growth, Development and Stress Tolerance*; Hossain, M.A., Munne-Bosch, S., Burritt, D.J., Diaz-Vivancos, P., Fujita, M., Lorence, A., Eds.; Springer: Cham, Switzerland, 2017; pp. 177–200.

188. Rao, A.C.; Reddy, A.R. Glutathione reductase: A putative redox regulatory system in plant cells. In *Sulfur Assimilation and Abiotic Stress in Plants*; Khan, N.A., Singh, S., Umar, S., Eds.; Springer: Berlin/Heidelberg, Germany, 2008; pp. 111–147.

189. Edwards, E.A.; Rawsthorne, S.; Mullineaux, P.M. Subcellular distribution of multiple forms of glutathione reductase in leaves of pea (*Pisum sativum* L.). *Planta* **1990**, *180*, 278–284. [CrossRef] [PubMed]

190. Zandalinas, S.I.; Balfagón, D.; Arbona, V.; Gomez-Cadenas, A. Modulation of antioxidant defense system is associated with combined drought and heat stress tolerance in citrus. *Front. Plant Sci.* **2017**, *8*, 953. [CrossRef]

191. Kohli, S.K.; Khanna, K.; Bhardwaj, R.; Abd_Allah, E.F.; Ahmad, P.; Corpas, F.J. Assessment of subcellular ROS and NO metabolism in higher plants: Multifunctional signaling molecules. *Antioxidants* **2019**, *8*, 641. [CrossRef] [PubMed]

192. Dietz, K.J. Peroxiredoxins in plants and cyanobacteria. *Antioxid. Redox Sign.* **2011**, *15*, 1129–1159. [CrossRef]

193. Ratajczak, E.; Dietz, K.J.; Kalemba, E.M. The occurrence of peroxiredoxins and changes in redox state in Acer platanoides and Acer pseudoplatanus during seed development. *J. Plant Growth Regul.* **2019**, *38*, 298–314. [CrossRef]

194. Gelhaye, E.; Rouhier, N.; Navrot, N.; Jacquot, J.P. The plant thioredoxin system. *Cell Mol. Life Sci.* **2005**, *62*, 24–35. [CrossRef]

195. Pandey, P.; Irulappan, V.; Bagavathiannan, M.V.; Senthil-Kumar, M. Impact of combined abiotic and biotic stresses on plant growth and avenues for crop improvement by exploiting physio-morphological traits. *Front. Plant Sci.* **2017**, *8*, 537. [CrossRef]

196. Szalai, G.; Kellős, T.; Galiba, G.; Kocsy, G. Glutathione as an antioxidant and regulatory molecule in plants under abiotic stress conditions. *J. Plant Growth Regul.* **2009**, *28*, 66–80. [CrossRef]

197. Noctor, G.; Reichheld, J.P.; Foyer, C.H. ROS-related redox regulation and signaling in plants. *Semin. Cell Dev. Biol.* **2018**, *80*, 3–12. [CrossRef]

198. Hix, L.M.; Lockwood, S.F.; Bertram, J.S. Bioactive carotenoids: Potent antioxidants and regulators of gene expression. *Redox Rep.* **2004**, *9*, 181–191. [CrossRef] [PubMed]

199. Uarrota, V.G.; Stefen, D.L.V.; Leolato, L.S.; Gindri, D.M.; Nerling, D. Revisiting carotenoids and their role in plant stress responses: From biosynthesis to plant signaling mechanisms during stress. In *Antioxidants and Antioxidant Enzymes in Higher Plants*; Gupta, D.K., Palma, J.M., Corpas, F.J., Eds.; Springer Nature: Cham, Switzerland, 2018; pp. 207–232.

200. Ansari, M.I.; Hasan, S.; Jalil, S.U. Leaf senescence and GABA shunt. *Bioinformation* **2014**, *10*, 734–736. [CrossRef] [PubMed]

201. Wilson, R.A.; Sangha, M.K.; Banga, S.S.; Atwal, A.K.; Gupta, S. Heat stress tolerance in relation to oxidative stress and antioxidants in *Brassica juncea*. *J. Environ. Biol.* **2014**, *35*, 383–387. [PubMed]

202. Almeselmani, M.; Deshmukh, P.S.; Sairam, R.K.; Kushwaha, S.R.; Singh, T.P. Protective role of antioxidant enzymes under high temperature stress. *Plant Sci.* **2006**, *171*, 382–388. [CrossRef]

203. Thounaojam, T.C.; Panda, P.; Mazumdar, P.; Kumar, D.; Sharma, G.D.; Sahoo, L.; Sanjib, P. Excess copper induced oxidative stress and response of antioxidants in rice. *Plant Physiol. Biochem.* **2012**, *53*, 33–39. [CrossRef]

204. Chugh, P.; Kaur, J.; Grewal, S.K.; Singh, S.; Agrawal, S.K. Upregulation of superoxide dismutase and catalase along with proline accumulation mediates heat tolerance in lentil (*Lens culinaris* Medik.) Genotypes during Reproductive Stage. *Ind. J. Agric. Biochem.* **2017**, *30*, 195–199. [CrossRef]

205. Jin, R.; Wang, Y.; Liu, R.; Gou, J.; Chan, Z. Physiological and metabolic changes of purslane (*Portulaca oleracea* L.) in response to drought, heat, and combined stresses. *Front. Plant Sci.* **2016**, *6*, 1123. [CrossRef]

206. Koussevitzky, S.; Suzuki, N.; Huntington, S.; Armijo, L.; Sha, W.; Cortes, D.; Shulaev, V.; Mittler, R. Ascorbate peroxidase 1 plays a key role in the response of *Arabidopsis thaliana* to stress combination. *J. Biol. Chem.* **2008**, *283*, 34197–34203. [CrossRef] [PubMed]

207. Lee, D.H.; Lee, C.B. Chilling stress-induced changes of antioxidant enzymes in the leaves of cucumber: In gel enzyme activity assays. *Plant Sci.* **2000**, *159*, 75–85. [CrossRef]

208. Selote, D.S.; Khanna-Chopra, R. Drought acclimation confers oxidative stress tolerance by inducing co-ordinated antioxidant defense at cellular and subcellular level in leaves of wheat seedlings. *Physiol. Plant.* **2006**, *127*, 494–506. [CrossRef]
209. Almeselmani, M.; Deshmukh, P.; Sairam, R. High temperature stress tolerance in wheat genotypes: Role of antioxidant defence enzymes. *Acta Agron. Hung.* **2009**, *57*, 1–14. [CrossRef]
210. Qi, C.; Lin, X.; Li, S.; Liu, L.; Wang, Z.; Li, Y.; Bai, R.; Xie, Q.; Zhang, N.; Ren, S.; et al. SoHSC70 positively regulates thermotolerance by alleviating cell membrane damage, reducing ROS accumulation, and improving activities of antioxidant enzymes. *Plant Sci.* **2019**, *283*, 385–395. [CrossRef] [PubMed]
211. Aftab, T.; Khan, M.M.A.; Idrees, M.; Naeem, M.; Ram, M. Boron induced oxidative stress, antioxidant defence response and changes in artemisinin content in *Artemisia annua* L. *J. Agron. Crop Sci.* **2010**, *196*, 423–430. [CrossRef]
212. Costa, H.; Gallego, S.M.; Tomaro, M.L. Effect of UV-B radiation on antioxidant defense system in sunflower cotyledons. *Plant Sci.* **2002**, *162*, 939–945. [CrossRef]
213. Wang, X.; Wu, L.; Xie, J.; Li, T.; Cai, J.; Zhou, Q.; Dai, T.; Jiang, D. Herbicide isoproturon aggravates the damage of low temperature stress and exogenous ascorbic acid alleviates the combined stress in wheat seedlings. *Plant Growth Regul.* **2018**, *84*, 293–301. [CrossRef]
214. Islam, F.; Ali, B.; Wang, J.; Farooq, M.A.; Gill, R.A.; Ali, S.; Wang, D.; Zhou, W. Combined herbicide and saline stress differentially modulates hormonal regulation and antioxidant defense system in *Oryza sativa* cultivars. *Plant Physiol. Biochem.* **2016**, *107*, 82–95. [CrossRef]
215. Tsaniklidis, G.; Pappi, P.; Tsafouros, A.; Charova, S.N.; Nikoloudakis, N.; Roussos, P.A.; Paschalidis, K.A.; Delis, C. Polyamine homeostasis in tomato biotic/abiotic stress cross-tolerance. *Gene* **2020**, *727*, 144230. [CrossRef] [PubMed]
216. Gilroy, S.; Suzuki, N.; Miller, G.; Choi, W.G.; Toyota, M.; Devireddy, A.R.; Mittler, R. A tidal wave of signals: Calcium and ROS at the forefront of rapid systemic signaling. *Trends Plant Sci.* **2014**, *19*, 623–630. [CrossRef] [PubMed]
217. Huang, S.; Van Aken, O.; Schwarzlander, M.; Belt, K.; Millar, A.H. The roles of mitochondrial reactive oxygen species in cellular signaling and stress response in plants. *Plant Physiol.* **2016**, *171*, 1551–1559. [CrossRef]
218. Foyer, C.H.; Noctor, G. Redox homeostasis and antioxidant signaling: A metabolic interface between stress perception and physiological responses. *Plant Cell* **2005**, *17*, 1866–1875. [CrossRef]
219. Bailey-Serres, J.; Mittler, R. The roles of reactive oxygen species in plant cells. *Plant Physiol.* **2006**, *141*, 311. [CrossRef] [PubMed]
220. Mittler, R.; Vanderauwera, S.; Suzuki, N.; Miller, G.A.D.; Tognetti, V.B.; Vandepoele, K.; Gollery, M.; Shulae, V.; Van Breusegem, F. ROS signaling: The new wave? *Trends Plant Sci.* **2011**, *16*, 300–309. [CrossRef]
221. Sewelam, N.; Kazan, K.; Schenk, P.M. Global plant stress signaling: Reactive oxygen species at the cross-road. *Front. Plant Sci.* **2016**, *7*, 187. [CrossRef] [PubMed]
222. Andersen, E.J.; Ali, S.; Byamukama, E.; Yen, Y.; Nepal, M.P. Disease resistance mechanisms in plants. *Genes* **2018**, *9*, 339. [CrossRef]
223. Jacobson, M.D. Reactive oxygen species and programmed cell death. *Trends Biochem. Sci.* **1996**, *21*, 83–86. [CrossRef]
224. Miller, G.; Schlauch, K.; Tam, R.; Cortes, D.; Torres, M.A.; Shulaev, V.; Dangl, J.L.; Mittler, R. The plant NADPH oxidase RBOHD mediates rapid systemic signaling in response to diverse stimuli. *Sci. Signal.* **2009**, *2*, ra45. [CrossRef] [PubMed]
225. Pucciariello, C.; Banti, V.; Perata, P. ROS signaling as common element in low oxygen and heat stresses. *Plant Physiol. Biochem.* **2012**, *59*, 3–10. [CrossRef]
226. Woodson, J.D.; Chory, J. Coordination of gene expression between organellar and nuclear genomes. *Nat. Rev. Genet.* **2008**, *9*, 383–395. [CrossRef]
227. Ahuja, I.; de Vos, R.C.; Bones, A.M.; Hall, R.D. Plant molecular stress responses face climate change. *Trends Plant Sci.* **2010**, *15*, 664–674. [CrossRef]
228. Heidarvand, L.; Amiri, R.M. What happens in plant molecular responses to cold stress? *Acta Physiol. Plant.* **2010**, *32*, 419–431. [CrossRef]
229. Volkov, R.A.; Panchuk, I.I.; Mullineaux, P.M.; Schoffl, F. Heat stress-induced H_2O_2 is required for effective expression of heat shock genes in Arabidopsis. *Plant Mol. Biol.* **2006**, *61*, 733–746. [CrossRef]
230. Pnueli, L.; Liang, H.; Rozenberg, M.; Mittler, R. Growth suppression, altered stomatal responses, and augmented induction of heat shock proteins in cytosolic ascorbate peroxidase (Apx1)-deficient *Arabidopsis* plants. *Plant J.* **2003**, *34*, 187–203. [CrossRef]
231. Yamauchi, T.; Watanabe, K.; Fukazawa, A.; Mori, H.; Abe, F.; Kawaguchi, K.; Oyanagi, A.; Nakazono, M. Ethylene and reactive oxygen species are involved in root aerenchyma formation and adaptation of wheat seedlings to oxygen-deficient conditions. *J. Exp. Bot.* **2014**, *65*, 261–273. [CrossRef]
232. Gonzali, S.; Loreti, E.; Cardarelli, F.; Novi, G.; Parlanti, S.; Pucciariello, C.; Bassolino, L.; Banti, V.; Licausi, F.; Perata, P. Universal stress protein HRU1 mediates ROS homeostasis under anoxia. *Nat. Plants* **2015**, *1*, 1–9. [CrossRef] [PubMed]
233. Yamauchi, T.; Yoshioka, M.; Fukazawa, A.; Mori, H.; Nishizawa, N.K.; Tsutsumi, N.; Yoshioka, H.; Nakazono, M. An NADPH oxidase RBOH functions in rice roots during lysigenous aerenchyma formation under oxygen-deficient conditions. *Plant Cell* **2017**, *29*, 775–790. [CrossRef] [PubMed]
234. Banti, V.; Mafessoni, F.; Loreti, E.; Alpi, A.; Perata, P. The heat-inducible transcription factor HsfA2 enhances anoxia tolerance in *Arabidopsis*. *Plant Physiol.* **2010**, *152*, 1471–1483. [CrossRef]
235. Yang, C.Y.; Hong, C.P. The NADPH oxidase Rboh D is involved in primary hypoxia signalling and modulates expression of hypoxia-inducible genes under hypoxic stress. *Environ. Exp. Bot.* **2015**, *115*, 63–72. [CrossRef]

236. Ma, L.; Zhang, H.; Sun, L.; Jiao, Y.; Zhang, G.; Miao, C.; Hao, F. NADPH oxidase AtrbohD and AtrbohF function in ROS-dependent regulation of Na$^+$/K$^+$ homeostasis in Arabidopsis under salt stress. *J. Exp. Bot.* **2012**, *63*, 305–317. [CrossRef]

237. Jiang, C.; Belfield, E.J.; Mithani, A.; Visscher, A.; Ragoussis, J.; Mott, R.; Smith, J.A.C.; Harberd, N.P. ROS-mediated vascular homeostatic control of root-to-shoot soil Na delivery in *Arabidopsis*. *EMBO J.* **2012**, *31*, 4359–4370. [CrossRef]

238. Li, L.; Kim, B.G.; Cheong, Y.H.; Pandey, G.K.; Luan, S. A Ca^{2+} signaling pathway regulates a K$^+$ channel for low-K response in *Arabidopsis*. *Proc. Natl. Acad. Sci. USA* **2006**, *103*, 12625–12630. [CrossRef]

239. Shin, R.; Schachtman, D.P. Hydrogen peroxide mediates plant root cell response to nutrient deprivation. *Proc. Natl. Acad. Sci. USA* **2004**, *101*, 8827–8832. [CrossRef] [PubMed]

240. Kerchev, P.; van der Meer, T.; Sujeeth, N.; Verlee, A.; Stevens, C.V.; Van Breusegem, F.; Gechev, T. Molecular priming as an approach to induce tolerance against abiotic and oxidative stresses in crop plants. *Biotechnol. Adv.* **2020**, *40*, 107503. [CrossRef] [PubMed]

241. Elkeilsh, A.; Awad, Y.M.; Soliman, M.H.; Abu-Elsaoud, A.; Abdelhamid, M.T.; El-Metwally, I.M. Exogenous application of β-sitosterol mediated growth and yield improvement in water-stressed wheat (*Triticum aestivum*) involves up-regulated antioxidant system. *J. Plant Res.* **2019**, *132*, 881–901. [CrossRef]

242. Liu, T.; Ye, X.; Li, M.; Li, J.; Qi, H.; Hu, X. H$_2$O$_2$ and NO are involved in trehalose-regulated oxidative stress tolerance in cold-stressed tomato plants. *Environ. Exp. Bot.* **2020**, *171*, 103961. [CrossRef]

243. Li, J.; Yang, Y.; Sun, K.; Chen, Y.; Chen, X.; Li, X. Exogenous melatonin enhances cold, salt and drought stress tolerance by improving antioxidant defense in tea plant (*Camellia sinensis* (L.) O. Kuntze). *Molecules* **2019**, *24*, 1826. [CrossRef]

244. El-Beltagi, H.S.; Mohamed, H.I.; Sofy, M.R. Role of ascorbic acid, glutathione and proline applied as singly or in sequence combination in improving chickpea plant through physiological change and antioxidant defense under different levels of irrigation intervals. *Molecules* **2020**, *25*, 1702. [CrossRef]

245. Matysik, J.; Alia Bhalu, B.; Mohanty, P. Molecular mechanisms of quenching of reactive oxygen species by proline under stress in plants. *Curr. Sci.* **2002**, *82*, 525–532.

246. Nadarajah, K.K. ROS Homeostasis in Abiotic Stress Tolerance in Plants. *Int. J. Mol. Sci.* **2020**, *21*, 5208. [CrossRef] [PubMed]

247. Tewari, R.K.; Kumar, P.; Kim, S.; Hahn, E.J.; Paek, K.Y. Nitric oxide retards xanthine oxidase-mediated superoxide anion generation in Phalaenopsis flower: An implication of NO in the senescence and oxidative stress regulation. *Plant Cell Rep.* **2009**, *28*, 267–279. [CrossRef] [PubMed]

248. Cramer, G.R.; Urano, K.; Delrot, S.; Pezzotti, M.; Shinozaki, K. Effects of abiotic stress on plants: A systems biology perspective. *BMC Plant Biol.* **2011**, *11*, 1–14. [CrossRef]

249. Liu, D.; Hoynes-O'Connor, A.; Zhang, F. Bridging the gap between systems biology and synthetic biology. *Front. Microbiol.* **2013**, *4*, 211. [CrossRef] [PubMed]

250. Jogaiah, S.; Govind, S.R.; Tran, L.S.P. Systems biology-based approaches toward understanding drought tolerance in food crops. *Crit. Rev. Biotechnol.* **2013**, *33*, 23–39. [CrossRef]

251. Xu, D.Q.; Huang, J.; Guo, S.Q.; Yang, X.; Bao, Y.M.; Tang, H.J.; Zhang, H.S. Overexpression of a TFIIIA-type zinc finger protein gene ZFP252 enhances drought and salt tolerance in rice (*Oryza sativa* L.). *FEBS Lett.* **2008**, *582*, 1037–1043. [CrossRef]

252. Gupta, S.; Dong, Y.; Dijkwel, P.P.; Mueller-Roeber, B.; Gechev, T.S. Genome-wide analysis of ROS antioxidant genes in resurrection species suggest an involvement of distinct ROS detoxification systems during desiccation. *Int. J. Mol. Sci.* **2019**, *20*, 3101. [CrossRef] [PubMed]

253. Dubouzet, J.G.; Sakuma, Y.; Ito, Y.; Kasuga, M.; Dubouzet, E.G.; Miura, S.; Seki, M.; Shinozaki, K.; Yamaguchi-Shinozaki, K. OsDREB genes in rice, Oryza sativa L., encode transcription activators that function in drought-, high-salt-and cold-responsive gene expression. *Plant J.* **2003**, *33*, 751–763. [CrossRef]

254. Ito, Y.; Katsura, K.; Maruyama, K.; Taji, T.; Kobayashi, M.; Seki, M.; Shinozaki, K.; Yamaguchi-Shinozaki, K. Functional analysis of rice DREB1/CBF-type transcription factors involved in cold-responsive gene expression in transgenic rice. *Plant Cell Physiol.* **2006**, *47*, 141–153. [CrossRef] [PubMed]

255. Li, Z.; Han, X.; Song, X.; Zhang, Y.; Jiang, J.; Han, Q.; Liu, M.; Qiao, G.; Zhuo, R. Overexpressing the *Sedum alfredii* Cu/Zn superoxide dismutase increased resistance to oxidative stress in transgenic *Arabidopsis*. *Front. Plant Sci.* **2017**, *8*, 1010. [CrossRef] [PubMed]

256. Shou, H.; Bordallo, P.; Wang, K. Expression of the Nicotiana protein kinase (NPK1) enhanced drought tolerance in transgenic maize. *J. Exp. Bot.* **2004**, *55*, 1013–1019. [CrossRef]

257. Shou, H.; Bordallo, P.; Fan, J.B.; Yeakley, J.M.; Bibikova, M.; Sheen, J.; Wang, K. Expression of an active tobacco mitogen-activated protein kinase kinase kinase enhances freezing tolerance in transgenic maize. *Proc. Natl. Acad. Sci. USA* **2004**, *101*, 3298–3303. [CrossRef] [PubMed]

258. Rungrat, T.; Awlia, M.; Brown, T.; Cheng, R.; Sirault, X.; Fajkus, J.; Trtilek, M.; Furbank, B.; Badger, M.; Tester, M.; et al. Using phenomic analysis of photosynthetic function for abiotic stress response gene discovery. *Arab. Book* **2016**, *14*, e0185. [CrossRef] [PubMed]

259. Großkinsky, D.K.; Syaifullah, S.J.; Roitsch, T. Integration of multi-omics techniques and physiological phenotyping within a holistic phenomics approach to study senescence in model and crop plants. *J. Exp. Bot.* **2018**, *69*, 825–844. [CrossRef]

260. Singh, B.; Mishra, S.; Bohra, A.; Joshi, R.; Siddique, K.H. Crop phenomics for abiotic stress tolerance in crop plants. In *Biochemical, Physiological and Molecular Avenues for Combating Abiotic Stress Tolerance in Plants*; Wani, S.H., Ed.; Elsevier, Academic Press: New York, NY, USA, 2018; pp. 277–296.

261. Palit, P.; Kudapa, H.; Zougmore, R.; Kholova, J.; Whitbread, A.; Sharma, M.; Varshney, R.K. An integrated research framework combining genomics, systems biology, physiology, modelling and breeding for legume improvement in response to elevated CO_2 under climate change scenario. *Curr. Plant Biol.* **2020**, *22*, 100149. [CrossRef] [PubMed]

262. Tao, F.; Rötter, R.P.; Palosuo, T.; Díaz-Ambrona, C.G.H.; Minguez, M.I.; Semenov, M.A.; Kersebaum, K.C.; Nendel, C.; Cammarano, D.; Hoffmann, H.; et al. Designing future barley ideotypes using a crop model ensemble. *Eur. J. Agron.* **2017**, *82*, 144–162. [CrossRef]

 *antioxidants*

Article

Impact of Ascorbate—Glutathione Cycle Components on the Effectiveness of Embryogenesis Induction in Isolated Microspore Cultures of Barley and Triticale

Iwona Żur [1,*], Przemysław Kopeć [1], Ewa Surówka [1], Ewa Dubas [1], Monika Krzewska [1], Anna Nowicka [1], Franciszek Janowiak [1], Katarzyna Juzoń [1], Agnieszka Janas [1], Balázs Barna [2] and József Fodor [2]

[1] The Franciszek Górski Institute of Plant Physiology Polish Academy of Sciences, Niezapominajek 21, 30-239 Kraków, Poland; p.kopec@ifr-pan.edu.pl (P.K.); e.surowka@ifr-pan.edu.pl (E.S.); e.dubas@ifr-pan.edu.pl (E.D.); m.krzewska@ifr-pan.edu.pl (M.K.); a.nowicka@ifr-pan.edu.pl (A.N.); f.janowiak@ifr-pan.edu.pl (F.J.); k.juzon@ifr-pan.edu.pl (K.J.); a.janas@ifr-pan.edu.pl (A.J.)
[2] Plant Protection Institute, Centre for Agricultural Research, Herman Ottó út 15, 1022 Budapest, Hungary; barna.balazs@agrar.mta.hu (B.B.); fodor.jozsef@atk.hu (J.F.)
* Correspondence: i.zur@ifr-pan.edu.pl; Tel.: +48-12-425-33-01; Fax: +48-12-425-18-44

Abstract: Enhanced antioxidant defence plays an essential role in plant survival under stress conditions. However, excessive antioxidant activity sometimes suppresses the signal necessary for the initiation of the desired biological reactions. One such example is microspore embryogenesis (ME)—a process of embryo-like structure formation triggered by stress in immature male gametophytes. The study focused on the role of reactive oxygen species and antioxidant defence in triticale ($\times$ *Triticosecale* Wittm.) and barley (*Hordeum vulgare* L.) microspore reprogramming. ME was induced through various stress treatments of tillers and its effectiveness was analysed in terms of ascorbate and glutathione contents, total activity of low molecular weight antioxidants and activities of glutathione–ascorbate cycle enzymes. The most effective treatment for both species was a combination of low temperature and exogenous application of 0.3 M mannitol, with or without 0.3 mM reduced glutathione. The applied treatments induced genotype-specific defence responses. In triticale, both ascorbate and glutathione were associated with ME induction, though the role of glutathione did not seem to be related to its function as a reducing agent. In barley, effective ME was accompanied by an accumulation of ascorbate and high activity of enzymes regulating its redox status, without direct relation to glutathione content.

Keywords: antioxidant defence; ascorbate; glutathione; microspore embryogenesis; reactive oxygen species; *Hordeum vulgare*; $\times$ *Triticosecale* Wittm.

Citation: Żur, I.; Kopeć, P.; Surówka, E.; Dubas, E.; Krzewska, M.; Nowicka, A.; Janowiak, F.; Juzoń, K.; Janas, A.; Barna, B.; et al. Impact of Ascorbate—Glutathione Cycle Components on the Effectiveness of Embryogenesis Induction in Isolated Microspore Cultures of Barley and Triticale. *Antioxidants* **2021**, *10*, 1254. https://doi.org/10.3390/antiox10081254

Academic Editors: Masayuki Fujita and Mirza Hasanuzzaman

Received: 9 June 2021
Accepted: 3 August 2021
Published: 5 August 2021

Publisher's Note: MDPI stays neutral with regard to jurisdictional claims in published maps and institutional affiliations.

1. Introduction

Each stress that disturbs cellular homeostasis and oxygen metabolism leads to the generation of reactive oxygen species (ROS) and induces the so-called 'oxidative stress' [1]. Effective antioxidant protection in cells is needed for survival under stress conditions, especially for plants that do not have the possibility of the 'fight-or-flight' response.

However, ROS are no longer considered only the inevitable, toxic by-products of aerobic metabolism—their role proved to be essential in many life processes [1–3]. Depending on the subtle and dynamic homeostasis between production and scavenging, ROS can be considered as life-threatening or life-saving molecules, and the latter possibility is based on their involvement in stress signalling and initiation of defence reactions.

Our earlier studies [4,5] clearly showed that the initiation of microspore embryogenesis (ME) was also associated with ROS generation. This is fully justified as the process—highly resembling zygotic embryo formation in planta, although induced in immature male gametophyte cells in response to stress—involves very stressful procedures of anther or microspore isolation and transfer to in vitro culture conditions. Our recent results have suggested that ROS generation is not merely a side effect of stress treatment and a drastic change of the cell environment, but the first requirement for microspore reprogramming towards sporophytic development. At the same time, high activity of the antioxidant system is necessary for microspore survival and successful induction of embryogenic development. Furthermore, it was presumed that the next stages of embryo-like structure (ELS) development were also regulated by cellular redox homeostasis. The observed variability between two cereal species (barley and triticale) suggests that in addition to basic antioxidant enzymes, such as superoxide dismutase (SOD) and catalase (CAT), other enzymatic and non-enzymatic components of antioxidative defence are of great importance, including those involved in the ascorbate–glutathione cycle [5].

Ascorbate and glutathione are among the most important low molecular weight (LMW) antioxidants, and they are present in the majority of plant cells at high concentrations. Both molecules in the cellular environment are mostly found in reduced form, which enables them to interact with many cellular components and play a multifunctional role in plant metabolism, growth and development [6,7]. Both molecules scavenge ROS and are elements of the very important antioxidant Foyer–Halliwell–Asada pathway also known as the ascorbate–glutathione cycle. Reduced ascorbate (ASC) acts as an electron donor for ROS decomposition by ascorbate peroxidase (APX) in this cycle, and it is simultaneously oxidized to mono- and dehydroascorbate (MDHA, DHA). They can be reduced back in reactions catalysed by specific enzymes: monodehydroascorbate reductase (MDHAR) and dehydroascorbate reductase (DHAR) using reduced glutathione (GSH) and reduced nicotinamide adenine dinucleotide (NADH) as the source of electrons. The produced oxidized glutathione (glutathione disulphide, GSSG) is in turn reduced back with electrons transferred from nicotinamide adenine dinucleotide phosphate (NADPH) by glutathione reductase (GR). However, both antioxidants can also act independently and their functions in plant cells are generally determined by their cellular compartmentalization and intracellular environment. The role of glutathione in plant development, with particular emphasis on its function in plant embryo formation both in in vivo and in vitro systems, was discussed in details in our earlier reports [4,5]. Moreover, it plays a role in assimilation, transport and storage of sulphur as a thiol compound. It is also used to detoxify certain toxic electrophilic compounds (e.g., heavy metals, oxo-aldehydes or formaldehyde) by forming a reversible covalent bond in reactions catalysed by glutathione transferases, glyoxalases and formaldehyde dehydrogenase, or as a precursor of phytochelatins. The ratio of GSH to GSSG determines cell redox potential, and thus can regulate gene expression patterns, signalling as well as the synthesis and activity of some proteins [6,8].

In contrast, some data indicate that enhanced ROS generation has little effect on the reduced/oxidized ascorbate ratio. This is probably connected with the apoplastic location of DHA under oxidative stress (reviewed in [6]). It is possible that the intracellular ascorbate pool is largely maintained in a reduced state with a simultaneous increased DHA accumulation in the apoplast. This apoplastic fraction of ascorbate is mainly responsible for stress perception, signalling and defence initiation, while cytoplasmic ascorbate acts as a cofactor of several enzymes catalysing biosynthesis of phytohormones (abscisic acid, gibberellins, ethylene), pigments (anthocyanins) and secondary metabolites (flavonoids, glucosinolate) (reviewed by [6,9]). Ascorbate also regulates plant growth by controlling the biosynthesis of hydroxyproline-rich proteins involved in cell cycle progression, and by modulating the energetic state of plasmalemma and cell wall structure [10]. It is also involved in apical meristem formation, development of root architecture, regulation of flowering time and leaf senescence. Recent reports also postulated its participation in epigenetic regulation (review in [11]).

Several studies have indicated the possibility that ascorbic acid or glutathione application increases ME effectiveness. In the majority of cases, both compounds were supplemented to the induction or regeneration media, which resulted in enhanced ELS formation and green plant regeneration [12–15].

As mentioned above, recent results of our group [4,5] have suggested the involvement of various enzymatic or non-enzymatc components of antioxidant defence in ME regulation. Therefore, in the present study, we examined the role of the Foyer–Halliwell–Asada cycle in ME induction. Using two DH lines of triticale and two barley cultivars that significantly differed in ME responsiveness, we analysed the effect of various tiller stress pre-treatments on the effectiveness of ME induction, ROS generation, reduced and oxidized ascorbate and glutathione levels, as well as the total activity of LMW antioxidants and the activity of ascorbate–glutathione cycle enzymes: APX, DHAR, MDHAR and GR. The results broaden our understanding of the mechanisms determining ME effectiveness and the role of antioxidant system in the initiation of microspore reprogramming towards sporophytic development.

2. Materials and Methods

2.1. Plant Material and Growth Conditions

Two DH lines of winter triticale ($\times$*Triticosecale* Wittm.): DH19 and DH28, derived from the F1 generation of a cross between the German inbred line 'Saka 3006' and Polish cv. 'Modus' and two cultivars of barley (*Hordeum vulgare* L.): winter cv. 'Igri' and spring cv. 'Golden Promise', highly differentiated in terms of ME responsiveness [16–20], were used in the study. The seeds of DH lines of triticale were obtained from the State Breeding Institute at the University of Hohenheim (Germany), whereas the seeds of barley cultivars were obtained from Leibniz Institute of Plant Genetics and Crop Plant Research (Gatersleben, Germany).

Winter forms of barley and triticale were vernalized for 7 weeks at 4 °C and 8/16 h day/night photoperiod in perlite moistened with Hoagland's nutrient solution (HS; according to Wędzony [21]). Spring barley, cv. Golden Promise, was grown in the same conditions for 2 weeks. Subsequently, vernalized plantlets were transferred to a mixture of soil and sand (3/1; *v/v*) and grown at 20 °C and 16 h/8 h photoperiod. Additional illumination with an irradiance of 400 μmol m^{-2} s^{-1} was provided by high pressure sodium (HPS) lamps SON-T + AGRO (Philips, Brussels, Belgium) during unfavourable weather conditions.

The experiment was repeated twice, and microspore isolation was started each time in January (barley) and March (triticale).

2.2. ME Inducing Treatment, Microspore Isolation and In Vitro Culture Conditions for Antioxidant System Analysis

The procedure of ME induction was developed at the Plant Cell Biology Dept. IPP PAS as a combination of the requirements for triticale and barley, and was described in details in Żur et al. [5]. Briefly, tillers were harvested at the mid- to late uninucleate stage of microspore development—optimal for ME induction, placed in HS solution and stored for 3 weeks at a low temperature (4 °C) in the dark.

Four days before microspore isolation, tillers were transferred to fresh HS solution (control, low temperature treatment, LT), HS supplemented with 0.3 mM GSH (LT+GSH), HS supplemented with 0.3 M mannitol (LT+MAN) or HS supplemented with both GSH and mannitol (LT+MAN+GSH).

Depending on the plant species, about 20 (triticale) and from 20 to 60 spikes (for cv. 'Igri' and cv. 'Golden Promise', respectively) were used for one isolation procedure.

The microspore isolation procedure was previously described in Żur et al. [4,5]. After isolation, microspores were collected and sampled for in vitro culture and biochemical analysis.

Microspore suspensions with a density of approximately 70,000 microspores per ml (triticale) and 100,000 microspores per ml (barley) were cultured in the induction medium 190-2 [22], modified according to Pauk et al. [23], and KBP [24] for triticale and barley, respectively. Triticale microspores were co-cultured with simultaneously dissected ovaries (10 ovaries per 1 mL of suspension). All cultures were incubated in the dark at 26 °C.

2.3. Parameters Describing Productivity, Viability and Embryogenic Potential of Microspores

Microspore suspensions isolated from pre-treated tillers were evaluated using the following parameters:

(i) Microspore yield—mean number of isolated microspores obtained per one spike of a donor plant calculated using a Neubauer chamber.

(ii) Microspore viability—percentage of fully viable microspores in the whole population of isolated cells, assessed based on its reaction with fluorescein diacetate (FDA; 0.01%; λ_{Ex} = 465 nm, λ_{Em} = 515 nm, green fluorescence; according to Heslop-Harrison and Heslop-Harrison [25].

Microspore samples were collected at the end of the isolation procedure and examined using a Nikon Eclipse E600 epifluorescence microscope equipped with a Zyla 4.2 (Andor) camera and NIS-Elements AR 4.00 software. At least 500 microspores from 10 fields of view (magnification ×100) for each individual sample were analysed.

(iii) Effectiveness of ME induction—the number of developed embryo-like structures (ELS) observed after 6 weeks of in vitro culture, calculated per one spike of a donor plant [ELS/spike].

2.4. In Situ Histochemical Detection of Superoxide Anions and Hydrogen Peroxide in Microspores

Samples of isolated microspores were used for the detection of superoxide anions and hydrogen peroxide according to the protocols of Żur et al. [26] and Wohlgemuth et al. [27] with modifications.

Samples were collected with special handmade sieves prepared from pipette tips (5 mm in diameter) wrapped with 30 μm Nylon mesh (CellTrics, Sysmex Partec, Goerlitz, Germany). Sieves were placed in Eppendorf tubes in freshly prepared staining solutions containing 0.1% (*w/v*) nitroblue tetrazolium (NBT; for superoxide anion analysis) or 0.1% (*w/v*) 3,3′-diaminobenzidine-4-HCl (DAB; for hydrogen peroxide analysis) in potassium phosphate buffer (PPB, pH 7.0) containing 0.005% (*w/v*) Triton X-100. Then, the microspores were vacuum-infiltrated at room temperature (RT) for 120 min in the dark (NBT) or in the light (DAB). Afterwards, the microspores were treated with 50% (*v/v*) ethanol. Hydrogen peroxide was visualized as a reddish-brown colour, whereas superoxide anion through the conversion of NBT to blue formazan. Negative controls were in order to exclude false positive results.

Microscopic observations were performed under a Nikon Eclipse E600. Images were recorded using a digital camera (Nikon DS-Ri1, Tokyo, Japan) and processed with NIS-Elements AR 2.10 (Tokyo, Japan).

2.5. Low Molecular Weight Antioxidant Activity by DPPH Assay

To determine LMW antioxidant activity, microspores were incubated for 1 h at 26 °C in the dark, then frozen in liquid N_2 and stored at −60 °C.

Plant material (microspores with a total FW of about 10 mg) was freeze-dried, ground with an MM400 ball mill (Retsch, Haan, Germany), and shaken for 2 h at room temperature (RT) after adding 1 mL of 50% ethanol. The extracts were then centrifuged for 20 min in a refrigerated centrifuge at 18,000× *g* (MPW-350R, Warszawa, Poland) and the supernatant was used for the measurements. Total antioxidant content (free radical scavenging activity) was measured using 0.5 mM solution of stable free radical 1,1-diphenyl-2-picrylhydrazyl (DPPH, SIGMA, Munich, Germany) in methanol according to the method by Brand-Williams et al. [28] with some modifications adapting the protocol to 96-well microtitre plates, and absorbance measurements in said microtitre plates [29]. Each sample was measured three times on separate plates by pipetting 50 μL of its supernatant with the addition of 250 μL of 0.5 mM DPPH solution into 3 wells for each measurement. After 30 min of reaction incubation at 37 °C, absorbance at 515 nm was determined using a Model 680 reader (Bio-Rad Laboratories, Hercules, CA, USA). Trolox was used as a standard in the following eight concentrations: 0.75, 1, 1.25, 1.5, 1.75, 2, 2.5, and 3 nM.

17.5 µL of each concentration was pipetted into three wells on each microtitre plate with the addition of 250 µL of 0.5 mM DPPH solution. Trolox equivalents for each measurement were calculated using the linear regression equation from a calibration curve representing linear relationships between absorbance and different Trolox concentrations. The results were expressed as µmoles of Trolox equivalents g^{-1} DW. For each DH line/cultivar and treatment, at least four measurements were performed on two independent samples collected from at least 6 different spikes.

2.6. Sample Preparation for Other Assays

The collected samples (1 mL of microspore suspension) were incubated for one hour at 26 °C in the dark, then centrifuged (2 min; 10,000× g; 4 °C) and washed with 50 mM Tris HCl buffer (pH 7.8) containing 1 mM EDTA-Na$_2$. After the second centrifugation, pelleted microspores were used for further analysis. The activity of all antioxidant enzymes and the levels of reduced and oxidized forms of ascorbate and glutathione were detected spectrophotometrically in isolated microspore extracts using an Ultrospec 2100 pro UV/visible spectrophotometer (Amersham, Umeå, Sweden). At least three independent measurements were carried out for all assays.

2.7. Sample Preparation for Ascorbate and Glutathione Assays

Microspores were suspended in a five-fold (*v/w*) volume of ice-cold 6% meta-phosphoric acid and ground in liquid N$_2$ in a mortar until the samples were thawed. After centrifugation (15 min, 12,000× g, 4 °C), the supernatant was used for the analyses.

2.8. Reduced and Oxidized Glutathione Measurements

Reduced (GSH) and oxidized (GSSG) glutathione were determined by an enzymatic recycling assay using GR according to Law et al. [30]. The procedure was described in detail in Żur et al. [4]. Briefly, the samples (50 µL of metaphosphoric acid extracts) were neutralised with 9 µL of 1.5 M triethanolamine. Total glutathione content was measured in an assay mix containing 59 µL of neutralized sample and 50 mM potassium phosphate buffer (pH 7.4), 2.5 mM EDTA-Na$_2$, 1 mM 5,5′-dithio-bis(2-nitrobenzoic acid) (DNTB), 0.6 U of GR from baker's yeast, 0.2 mM NADPH in a total volume of 600 µL. Total glutathione content was estimated by recording the absorbance at 412 nm. GSSG measurement was carried out as for total glutathione, except that GSH was derivatised by adding 2 µL of 2-vinylpyridine to the neutralized samples and incubated for 1h at RT. Total glutathione and GSSG contents were estimated from the standard curves based on a dilution series of GSH and GSSG in 6% metaphosphoric acid. GSH was calculated as the difference between total glutathione and GSSG concentrations.

2.9. Reduced and Oxidized Ascorbate Measurements

Reduced ascorbate (ASC) concentration was determined using the method of Foyer et al. [31]. The samples (62.5 µL metaphosphoric acid extracts) were neutralized with 7.5 µL of 1.5 M triethanolamine and 75 µL of 150 mM sodium phosphate buffer (pH 7.4) and subsequently mixed with 500 µL of 100 mM sodium phosphate buffer (pH 5.6). To oxidize ASC, 1 U ascorbate oxidase (AO) from *Cucurbita* sp. (Sigma-Aldrich, Saint Louis, MO, USA) was added to the assay mix. ASC content was determined based on the difference between the initial and final absorbance measured at 265 nm. Total ascorbate levels were determined after dehydroascorbate (DHA) reduction to ascorbate using dithiothreitol (DTT). A 10 mM DTT solution (37.5 µL) was added to the reaction mixture, prior to the addition of sodium phosphate buffer (pH 5.6) and AO, and incubated for 15 min at RT. Ascorbate concentration was calculated using the extinction coefficient of ascorbic acid (14.7 mM^{-1} cm^{-1}). DHA content was obtained from the difference between total and reduced ascorbate concentrations.

2.10. Sample Preparation and Enzyme Activity Assays

Microspores were suspended in ice-cold 50 mM Tris HCl buffer (pH 7.8) containing 1 mM EDTA-Na$_2$, 3% (*w/v*) soluble polyvinylpyrrolidone K25 and 0.5 mM Pefabloc SC protease inhibitor (Roche, 11429876001), and ground in liquid N$_2$ in a mortar until thawed. After centrifugation (15 min, 12,000× *g*, 4 °C), the supernatant was used for the assays.

APX activity was detected according to the method of Nakano and Asada [32]. The assay mixture (570 μL) contained: 50 mM Tris-HCl buffer (pH 7.8), 0.25 mM ascorbic acid, 0.5 mM H$_2$O$_2$ and 20 μL of enzyme extract. Ascorbic acid oxidation was followed by measuring the absorbance at 290 nm, and APX activity was calculated using the extinction coefficient of ascorbic acid (2.8 mM^{-1} cm^{-1}). A control reaction was carried out for each sample in the absence of H$_2$O$_2$. The enzyme activity was expressed in nmol ASC mg^{-1} protein min^{-1}.

DHAR activity determination was based on the method of Klapheck et al. [33] by measuring the absorbance at 265 nm following DHA reduction. The assay mixture consisted of 30 μL of enzyme extract, 550 μL of 50 mM sodium phosphate buffer (pH 6.5), 1 mM EDTA-Na$_2$, 1 mM GSH and 0.5 mM DHA. Reaction mixture without extract was used as a negative control. DHAR activity was calculated using the extinction coefficient of ascorbic acid (14.7 mM^{-1} cm^{-1}) and expressed in nmol ASC mg^{-1} protein min^{-1}.

MDHAR activity was estimated by measuring the decrease in absorbance at 340 nm due to NADH oxidation [34]. The reaction mixture consisted of 30 μL of enzyme extract, 575 μL of 50 mM Tris-HCl buffer (pH 7.8), 1 mM ASC, 0.1 mM NADH and 0.2 U AO. A control reaction was carried out for each sample in the absence of AO. MDHAR activity was calculated using the extinction coefficient of NADH (6.2 mM^{-1} cm^{-1}) and expressed in nmol NADH mg^{-1} protein min^{-1}.

GR activity was assayed by monitoring NADPH oxidation at 340 nm according to the method of Klapheck et al. [33]. The assay mixture and) consisted of 600 μL 50 mM Tris-HCl buffer (pH 7.8), 0.1 mM NADPH, 5.8 mM GSSG and 30 μL of enzyme extract. Control reactions were conducted in the absence of GSSG. GR activity was calculated using the extinction coefficient of NADPH (6.2 mM^{-1} cm^{-1}) and expressed in nmol NADPH mg^{-1} protein min^{-1}.

Enzyme activities were normalized to soluble protein content of microspore homogenates. Protein content in enzyme extracts was determined according to Bradford [35] using bovine serum albumin as a standard.

2.11. Statistical Analysis

All data after testing for normal distribution using the Shapiro–Wilk test were subject to two-way analysis of variance (ANOVA) followed by post hoc comparison using Duncan's multiple range test ($p \leq 0.05$). Variables with non-normal data distribution were analysed using non-parametric Kolmogorov–Smirnov tests ($p \leq 0.001$). All statistical analyses were performed using the STATISTICA package version 12 (Stat Soft Inc., Tulsa, OK USA).

3. Results

3.1. The Effect of Tiller Pre-Treatments on the Effectiveness of ME Induction

Our results confirmed a significant variation in ME effectiveness between the two studied cereal species, as well as the intraspecific differentiation between two DH triticale lines and two barley cultivars (Table 1).

The results shown in Table 1 clearly indicated that microspore yield obtained from a single spike after isolation was almost 4-fold higher in triticale compared to barley. In this respect, the number of microspores obtained from DH19 (about 105,000 per spike) was significantly higher than that of DH28 (about 89,000 per spike), whereas the process of microsporogenesis was similarly effective in the studied barley cultivars, producing an average of 26,500 microspores per spike. At the same time, variation in microspore yields after various tiller pre-treatments was insignificant.

In contrast, microspore viability was higher in microspore suspensions of both barley cultivars (51–65%) in comparison to triticale (6–30%; Table 1). Moreover, the number of fully viable cells was higher in both responsive genotypes (DH28 and cv. Igri) in comparison to recalcitrant ones (DH19 and cv. Golden Promise). Particularly low cell viability (approximately 12%) was found in isolated DH19 microspores. Tiller pre-treatment with MAN resulted in a lower number of viable cells in triticale, although a decrease in the number of fully viable cells from about 30% to less than 18% in DH28 was a statistically significant effect.

Unfortunately, the number of ELS produced in isolated microspore cultures of triticale was relatively low (Table 1), not only in the standard recalcitrant DH19, but also in the usually moderately responsive DH28. The best effect was obtained in the combination of the LT treatment with MAN and GSH application (LT+MAN+GSH). This was the only treatment that allowed any ELS formation in DH19, although the effectiveness was below 1 ELS per spike. The results obtained with highly recalcitrant spring barley (Golden Promise) were only slightly better after the LT treatment in combination with MAN or after MAN and GSH application. The positive effects of LT+MAN and LT+MANGSH treatments could also be observed in isolated microspore cultures of cv. Igri characterized by a high embryogenic potential (Table 1).

Table 1. The effect of tillers pre-treatment on microspore yield ($\times 10^3$), viability (%) and the effectiveness of microspore embryogenesis (ME) induction (ELS/spike) in isolated microspore cultures of two DH lines of triticale and two cultivars of barley. Data represent means from 3–5 biological replications (isolations) $\pm$ SE. Statistical analysis was performed separately for triticale and barley. Data marked with different letters differ significantly according to the Duncan test ($p \leq 0.05$); ns, not significant.

Plant Material	Treatment	Microspore Yield	Viability	ME Induction
DH19	LT (control)	111.0 ± 11 [ns]	14.3 ± 1 [cd]	0
	LT+MAN	103.5 ± 11	6.9 ± 1 [d]	0
	LT+GSH	88.5 ± 6	17.8 ± 2 [bc]	0
	LT+MAN+GSH	116.5 ± 6	5.9 ± 1 [d]	0.7 ± 0.7
	mean	104.8 ± 5	11.7 ± 1	0.2 ± 0.2
DH28	LT (control)	99.2 ± 6 [ns]	30.3 ± 3 [a]	17 ± 6 [ns]
	LT+MAN	87.1 ± 10	17.4 ± 5 [bc]	32 ± 11
	LT+GSH	86.3 ± 67	29.4 ± 2 [a]	12 ± 5
	LT+MAN+GSH	81.2 ± 5	24.0 ± 3 [ab]	31 ± 18
	mean	88.4 ± 3	26.1 ± 2	22 ± 5
cv. Golden Promise	LT (control)	29 ± 4 [ns]	51 ± 3 [b]	2 ± 0.9 [c]
	LT+MAN	27 ± 5	57 ± 3 [ab]	15 ± 7 [c]
	LT+GSH	26 ± 5	53 ± 3 [b]	2 ± 1 [c]
	LT+MAN+GSH	25 ± 4	58 ± 3 [ab]	12 ± 8 [c]
	mean	26.1 ± 2	54.7 ± 1	8 ± 2
cv. Igri	LT (control)	21 ± 6 [ns]	65 ± 2 [a]	414 ± 121 [bc]
	LT+MAN	25 ± 7	64 ± 2 [a]	1069 ± 407 [b]
	LT+GSH	28 ± 8	57 ± 3 [ab]	543 ± 296 [bc]
	LT+MAN+GSH	32 ± 9	59 ± 2 [ab]	2047 ± 622 [a]
	mean	30.4 ± 5	61.6 ± 1	1073 ± 277

ELS—embryo-like structures; LT—low-temperature tiller pre-treatment (21 days at 4 °C); LT+GSH—low-temperature tiller pre-treatment combined with the application of 0.3 mmol·dm^{-3} reduced glutathione during the last 4 days before microspore isolation; LT+MAN—low-temperature tiller pre-treatment combined with the application of 0.3 mol·dm^{-3} mannitol during the last 4 days before microspore isolation; LT+MAN+GSH—low-temperature tiller pre-treatment combined with the application of reduced 0.3 mmol·dm^{-3} glutathione and 0.3 mol·dm^{-3} mannitol during the last 4 days before microspore isolation.

3.2. In Situ Histochemical Detection of Superoxide Anions and Hydrogen Peroxide in Microspores

Superoxide anion and hydrogen peroxide generation was observed in isolated microspores of triticale and barley after all types of tiller pre-treatments. It can be seen that both superoxide anion (Figure S1) and hydrogen peroxide (Figure S1D–F) were accumulated mainly in the cell cytoplasm. In general, the signal was diffused but, in some areas, particularly near the nucleus enhanced ROS generation was detected. It was observed more often it the case of superoxide anion.

3.3. Total Activity of Low Molecular Weight Antioxidants in Microspores after Various ME-Inducing Pre-Treatments of Tillers

Total activity of LMW antioxidants was estimated using a method based on the scavenging capacity of a stable free radical (DPPH) compared to antioxidant capacity of Trolox, used as a reference. We found significant variation between various plant genotypes, tiller pre-treatments and their interactions (Table 2). Antioxidant activity also depended on tiller pre-treatments and was significantly higher in microspores isolated from tillers treated with LT+GSH (3.2 µM Trolox g^{-1} DW) and LT+MAN+GSH (2.9 µM Trolox g^{-1} DW) in comparison to other treatments (2.1−2.3 µM Trolox g^{-1} DW). The LT+GSH treatment was particularly effective for the responsive DH28 and cv. Igri.

Table 2. The sources of variance for content of reduced ascorbate (ASC), reduced glutathione (GSH), total activity of low molecular weight (LMW) antioxidants, activity od ascorbate peroxidase (APX), monodehydroascorbate reductase (MDHAR), dehydroascorbate reductase (DHAR) and glutathione reductase (GR) were as follows: four plant genotypes, four tillers pre-treatments, and interaction between plant genotype and the treatment.

Parameter	Variable	MS	F	p
ASC	(1) Plant genotype	118 ^{E4}	179.5	***
	(2) Tillers pre-treatment	147 ^{E3}	22.4	***
	(1) × (2)	164 ^{E3}	24.9	***
GSH	(1) Plant genotype	199 ^{E2}	172.0	***
	(2) Tillers pre-treatment	252	2.2	ns
	(1) × (2)	826	7.1	***
LMW antioxidants	(1) Plant genotype	3.564	22.13	***
	(2) Tillers pre-treatment	5.635	34.99	***
	(1) × (2)	2.002	12.43	***
APX	(1) Plant genotype	887 ^{E3}	216.6	***
	(2) Tillers pre-treatment	294 ^{E3}	70.2	***
	(1) × (2)	279 ^{E3}	66.4	***
MDHAR	(1) Plant genotype	107 ^{E2}	61.9	**
	(2) Tillers pre-treatment	216 ^{E2}	125.2	**
	(1) × (2)	9349	54.3	**
DHAR	(1) Plant genotype	468 ^{E2}	576.4	***
	(2) Tillers pre-treatment	2113	26.0	***
	(1) × (2)	2181	26.8	***
GR	(1) Plant genotype	7468	91.21	***
	(2) Tillers pre-treatment	1298	15.85	***
	(1) × (2)	398	4.86	***

MS—mean squares, an estimate of variance of a variable or interaction between variables; F—F calculated as variance of a variable or interaction between variables divided by error variance of the experiment; p—error probability for rejection of the null hypothesis about the significance of variability within a variable or interaction between variables: ***, **, ns—significant at $p \leq 0.001$, $p \leq 0.01$, not significant, respectively; E—scientific notation (times ten raised to the power of two, three, etc.)

It was observed that antioxidant activity was in most cases significantly higher in microspores of responsive genotypes (DH28, cv. Igri) than in the recalcitrant DH19 and cv. Golden Promise (Figure 1A,B). The only exception was the LT+MAN+GSH treatment associated with enhanced activity of LMW antioxidants in microspores of both DH19 and cv. Golden Promise. The same treatment had no effect or reduced total antioxidant activity in microspores of cv. Igri and DH28. Only responsive genotypes were affected by the LT+MAN treatment, but its effect was genotype-specific and both increased and decreased antioxidant activity was detected in microspores of cv. Igri and DH28, respectively.

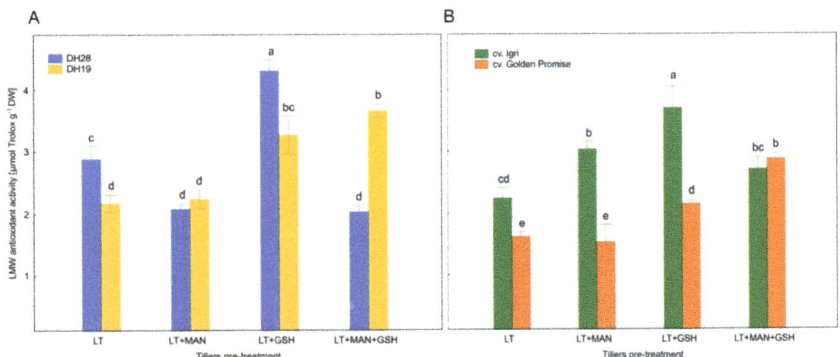

Figure 1. Total activity of low molecular weight (LMW) antioxidants in microspores of (**A**) two DH triticale lines (responsive DH28 and recalcitrant DH19) and (**B**) two barley cultivars of (responsive cv. Igri and recalcitrant cv. Golden Promise) isolated after various tiller pre-treatments. Data represent means of at least four measurements of two independent samples, each collected from at least 6 different spikes ± SE. Values marked with the same letter do not differ significantly according to Duncan's multiple range test ($p \leq 0.05$). Please see Figure 3, LT—low temperature tiller pre-treatment (21 days at 4 °C); LT+MAN—low temperature tiller pre-treatment combined with the application of 0.3 M mannitol; LT+GSH—low temperature tiller pre-treatment combined with the application of 0.3 mM reduced glutathione; LT+MAN+GSH—low temperature tillers pre-treatment combined with the application of 0.3 M mannitol and 0.3 mM reduced glutathione.

3.4. Glutathione and Ascorbate Levels and Their Reduction State in Triticale and Barley Microspores after Various ME-Inducing Pre-Treatments of Tillers

The level of ASC in microspores varied from 208 to 918 nmol g⁻¹ FW (Figure 2A,B) and was significantly affected by plant genotype, tiller pre-treatment and their interactions (Table 2). Compared to ASC, GSH was detected in a significantly lower amount, ranging from 39 to 128 nmol g⁻¹ FW (Figure 2C,D) and its content was influenced by plant genotype, but not tiller pre-treatment (Table 2). The effect of tiller pre-treatments also depended on the plant species and individual DH line/cultivar.

In general, variation in ASC and GSH levels, as well as the redox potential of ascorbate (ASC/ASC+DHA) and glutathione (GSH/GSH+GSSG) was low in triticale microspores (Figure 2A,C and Figure 3A,B). Only the LT+GSH treatment resulted in a significant decrease in ASC level in microspores of DH28 in comparison to control (LT, Figure 2A). However, this effect did not change the redox status of ascorbate (Figure 3A). On the contrary, the LT+MAN+GSH treatment did not affect ASC content, but was associated with a lower ascorbate redox potential (Figures 2A and 3A). This parameter was also relatively low in DH19 microspores after the LT+GSH treatment.

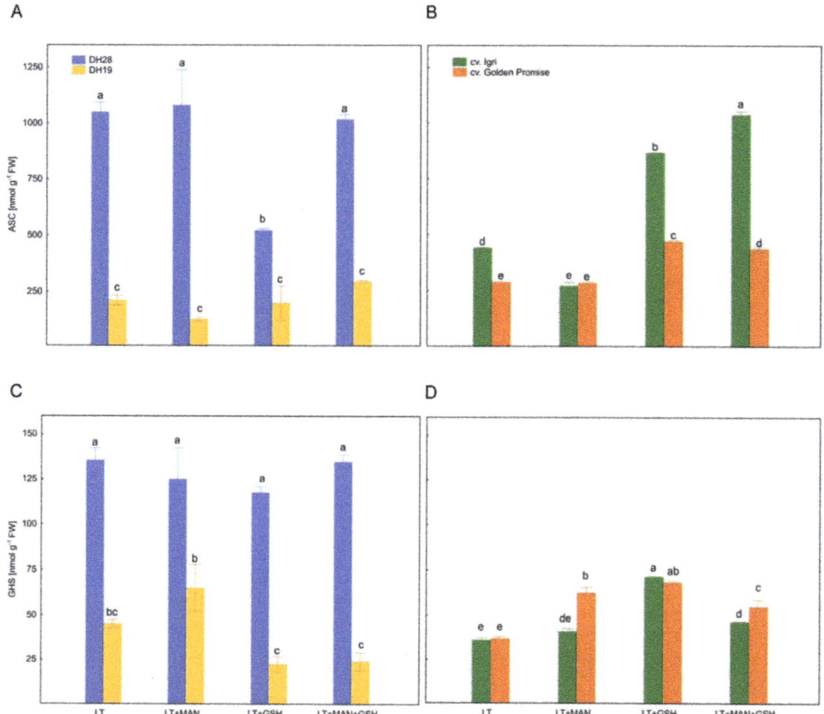

Figure 2. Reduced ascorbate (ASC) (**A,B**) and reduced glutathione (GSH) (**C,D**) contents in microspores of two DH triticale lines (responsive DH28 and recalcitrant DH19) and two barley cultivars (responsive cv. Igri and recalcitrant cv. Golden Promise) isolated after various tiller pre-treatments. Data represent means of three biological replications ± SE. Values marked with the same letter do not differ significantly according to Duncan's multiple range test ($p \leq 0.05$). Please see Figure 3, LT—low temperature tiller pre-treatment (21 days at 4 °C); LT+MAN—low temperature tiller pre-treatment combined with the application of 0.3 M mannitol; LT+GSH—low temperature tiller pre-treatment combined with the application of 0.3 mM reduced glutathione; LT+MAN+GSH—low temperature tillers pre-treatment combined with the application of 0.3 M mannitol and 0.3 mM reduced glutathione.

Two treatments (LT+GSH and LT+MAN+GSH) stimulated the accumulation of ASC in both barley cultivars and changed the ascorbate redox state (ASC/ASC+DHA) in microspores of one or both cultivars (Figure 2B; Figure 3C). The effect of the LT+MAN treatment was genotype-specific as it reduced ASC accumulation exclusively in microspores of cv. Igri. All treatments increased GSH content in microspores of cv. Golden Promise (Figure 2D), but only LT+MAN application also increased the glutathione redox potential (GSH/GSH+GSSG; Figure 3D). Similarly, an increased level of GSH was detected in microspores of cv. Igri after LT+GSH and LT+MAN+GSH treatments (Figure 2D), but it was not associated with a change in the GSH/GSH+GSSG ratio (Figure 3D).

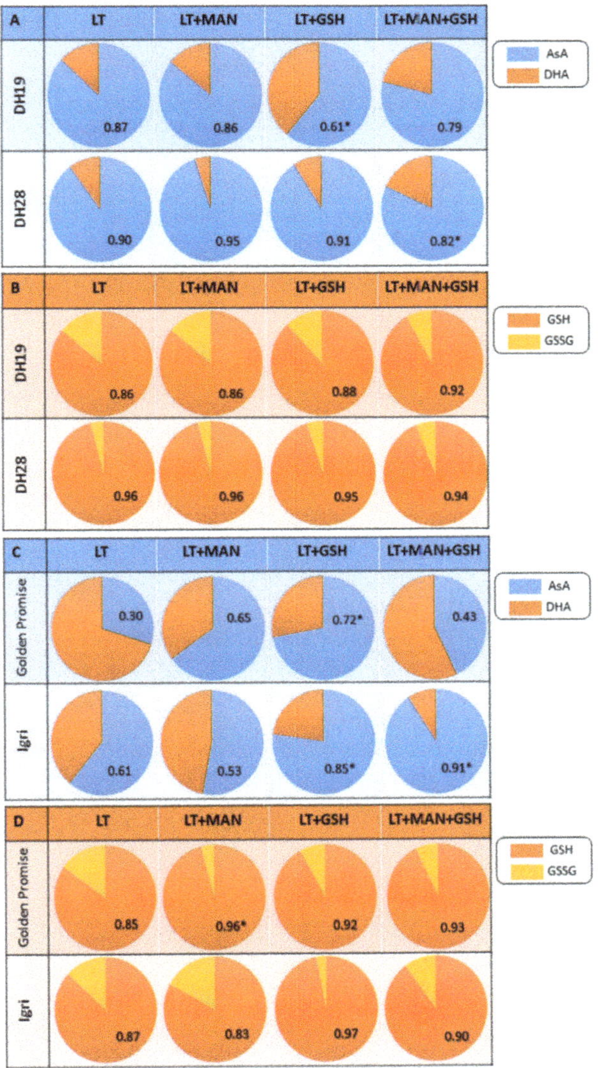

Figure 3. The proportion reduced (ASC, GSH) and oxidized (DHA, GSSG) forms of ascorbate and glutathione in their total pools in microspores of two DH triticale lines (responsive DH28 and recalcitrant DH19) (**A,B**) and two barley cultivars (responsive cv. Igri and recalcitrant cv. Golden Promise) (**C,D**) isolated after various tiller pre-treatments. The numbers represent the redox status of ascorbate (ASC/ASC+DHA) and glutathione (GSH/GSH+GSSG). Values significantly different from control (LT) according to the Kruskal–Wallis test (* $p \leq 0.001$) are marked with an asterisk. LT—low temperature tiller pre-treatment (21 days at 4 °C); LT+MAN—low temperature tiller pre-treatment combined with the application of 0.3 M mannitol; LT+GSH—low temperature tiller pre-treatment combined with the application of 0.3 mM reduced glutathione; LT+MAN+GSH—low temperature tillers pre-treatment combined with the application of 0.3 M mannitol and 0.3 mM reduced glutathione.

3.5. The Activity of Enzymes of the Glutathione–Ascorbate Cycle in Microspores after Various ME-Inducing Pre-Treatments of Tillers

The activity of APX, which catalyses H_2O_2 reduction via ASC oxidation to monodehydroascorbate (MDHA) significantly depended on plant genotype, tiller pre-treatment and their interactions (Table 2). The same sources of variation were found to be statistically significant for all other enzymes (MDHAR, DHAR, and GR).

It should be noted that the average APX activity was significantly higher in the ME-responsive DH28 and cv. Igri in comparison to the recalcitrant DH19 and cv. Golden Promise (Figure 4A,B).

All modified tiller treatments resulted in a lower APX activity in triticale microspores of both DH lines (Figure 4A) compared to the LT treatment, with the strongest effect of LT+GSH application. In contrast, all modified treatments increased APX activity in microspores of cv. Golden Promise and two of them (LT+MAN and LT+MAN+GSH) increased APX activity in microspores of cv. Igri (Figure 4B). Similar activities of this enzyme were detected in microspores of both barley cultivars after the LT+MAN treatment and in both DH triticale lines after LT+MAN and LT+MAN+GSH treatments.

The activity of MDHAR, which can reduce MDHA to ASC, was significantly higher after all modified tiller treatments in microspores of two responsive genotypes (DH28 and cv. Igri), but the effect of the LT+MAN+GSH treatment was especially prominent in comparison to control (LT; Figure 4C,D). A higher activity of this enzyme was detected in microspores of cv. Golden Promise only after the LT+MAN treatment, whereas the only change observed in DH19 microspores was a decrease in MDHAR activity after the LT+GSH treatment.

The produced MDHA is further oxidized to DHA, which can be converted back to ASC in the reaction catalysed by DHAR. In this reaction, GSH is used as an electron donor that is simultaneously oxidised to GSSG. Generally, modifications of tiller pre-treatment applied in this study resulted in a decrease of DHAR activity in triticale microspores, with the exception of DH28 tillers, which showed significantly increased DHAR activity after the LT+MAN+GSH pre-treatment (Figure 4E). On the other hand, all treatment modifications resulted in a slight but significant increase in DHAR activity in microspores of cv. Golden Promise, whereas a slight increase (after LT+MAN+GSH) as well as a slight decrease (after LT+GSH) could also be observed in DHAR activity in cv. Igri microspores. The activity of this enzyme was almost always higher in ME-responsive genotypes in comparison to recalcitrant ones (Figure 4E,F).

In comparison with other enzymes of the glutathione–ascorbate cycle, the average activity of GR which catalyses the conversion of GSSG to its reduced form, was low in triticale microspores, especially in the ME-recalcitrant line DH19 (8.7 nmol mg^{-1} protein min^{-1}). Microspores of DH28 showed more than twice higher average GR activity (21 nmol mg^{-1} protein min^{-1}) as compared to DH19. Similarly, the average GR activity in microspores of ME-recalcitrant cv. Golden Promise was about 1.5-fold lower in comparison to responsive cv. Igri (40.3 versus 62.9 nmol mg^{-1} protein min^{-1}). Among various tiller pre-treatments, LT+MAN+GSH was associated with a decreased GR activity in microspores of both triticale DH lines and barley cv. Igri (Figure 4G,H). The activity of GR in DH19 fell below the level of detection. Strikingly, the LT+GSH treatment resulted in a large decrease in GR activity in microspores of barley cv. Golden Promise and in both DH triticale lines. Variation in the reactions between responsive and recalcitrant genotypes was observed after the LT+MAN treatment in both species and after the LT+GSH treatment in barley (Figure 4G,H).

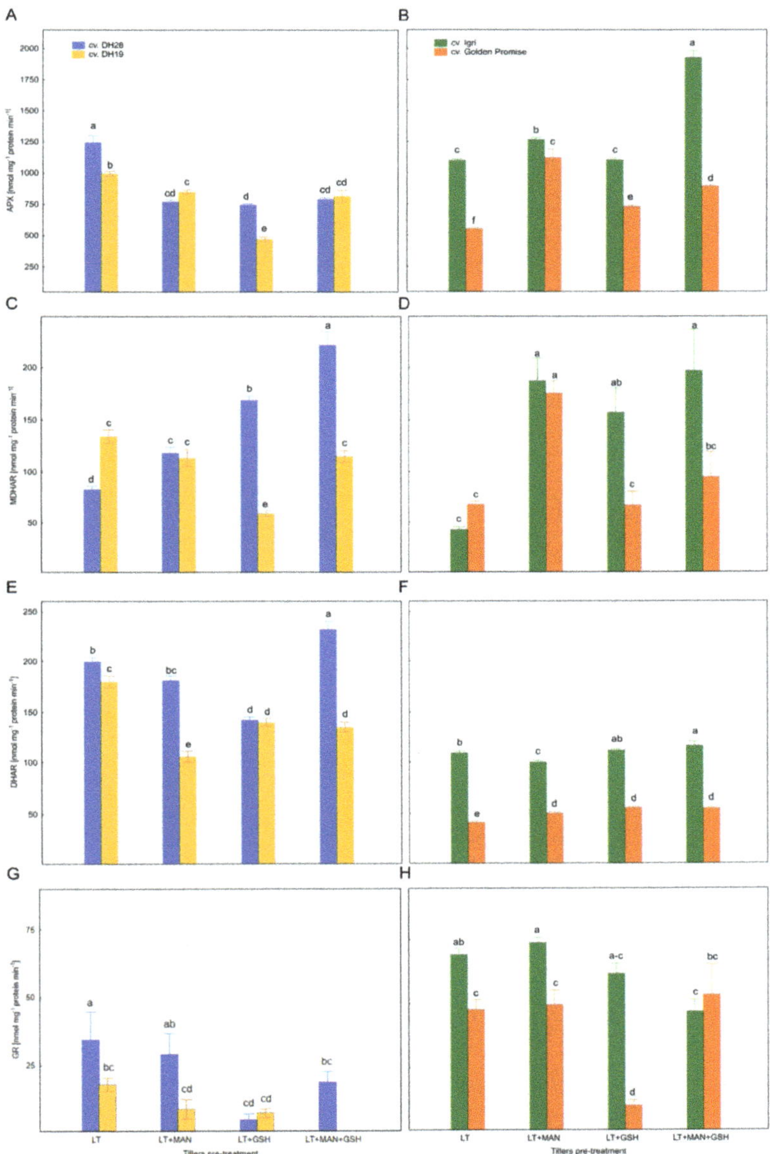

Figure 4. Activity of enzymes of the ascorbate–glutathione cycle: (**A,B**) ascorbate peroxidase (APX), (**C,D**) monodehydroascorbate reductase (MDHAR), (**E,F**) dehydroascorbate reductase (DHAR), (**G,H**) glutathione reductase (GR) in microspores of two DH triticale lines (responsive DH28 and recalcitrant DH19) and two barley cultivars (responsive cv. Igri and recalcitrant cv. Golden Promise) isolated after various tiller pre-treatments. Data represent means of three biological replications ± SE. Values marked with the same letter do not differ significantly according to Duncan's multiple range test ($p \leq 0.05$). Please see Figure 3, LT—low temperature tiller pre-treatment (21 days at 4 °C); LT + MAN—low temperature tiller pre-treatment combined with the application of 0.3 M mannitol; LT + GSH—low temperature tillers pre-treatment combined with the application of 0.3 mM reduced glutathione; LT + MAN + GSH—low temperature tillers pre-treatment combined with the application of 0.3 M mannitol and 0.3 mM reduced glutathione.

4. Discussion

The process of ME is both scientifically fascinating as a manifestation of cell totipotency and commercially important due to the possibility of instant production of completely homozygous doubled haploid (DH) plants. DH technology complements conventional plant breeding and can be used in areas such as genome engineering and gene mapping, often applied for substantial crop improvement [36–39]. However, the possibility of its application on a commercial scale is often limited by the low effectiveness of ME induction, which is strongly dependent on plant genotype, environmental conditions and complex system of interactions between internal and external factors.

Substantial progress in recent years has significantly increased our knowledge about ME mechanism and its regulation, but many issues remain unexplored or unclear. The use of stress treatments to induce efficient microspore reprogramming was a very important discovery [40]. In addition to being a trigger redirecting the pathway of development, stress also exerts a considerable adverse effect on cells, which can result in their death or genome rearrangements [41]. Thus, the type, intensity and duration of stress treatment must be carefully optimized to the requirements of each individual genotype and its stress tolerance. Prior experimental data suggested that exposure to certain mild primary stresses (e.g., LT, conventionally used as an ME-inducing factor in many plant species) can activate defence mechanisms and increase tolerance to more intensive stresses that occur during microspore isolation and transfer to in vitro culture conditions [42]. Such a cross-tolerance phenomenon is well known in plants, and in many cases, it is associated with increased ROS generation, which may also serve as signalling molecules activating cellular defence responses [43]. Several studies have confirmed that ME initiation is associated with the upregulation of genes encoding enzymes involved in antioxidative defence, as well as increased activity and/or accumulation of LMW antioxidants (reviewed in [20,44,45] and references therein). Therefore, different types of antioxidants were added to media used for in vitro cultures in order to achieve positive effects on ELS formation and plant regeneration ([5] and references therein).

The main aim of our research has been to determine the role of ROS and antioxidant defence in microspore reprogramming and ME induction. Our earlier investigations concerning several DH triticale lines significantly differed in ME effectiveness. Żur et al. [26] not only confirmed ROS generation in anthers after LT tiller pre-treatment, but also suggested that a certain threshold level of ROS is required for ME initiation; however, ROS generation had to go hand in hand with an efficient antioxidant defence for effective reprogramming of microspores. Furthermore, the effects of GSH application during the LT tiller pre-treatment [4] suggested its dual role: protection of microspores from oxidative stress and stimulation of ELS development, probably through modulation of cellular redox homeostasis. Our data indicated that intensive ROS elimination could suppress the signal necessary for microspore reprogramming [4]. To test this hypothesis in a broader range of taxa, the present study, in addition to two DH triticale lines (DH19 and DH28), also used two barley cultivars markedly heterogeneous in terms of ME capacity (Igri and Golden Promise) [16–20]. The experimental work was also extended to include tiller pre-treatments with MAN, a sugar alcohol that is often used to simulate osmotic stress and commonly applied for ME induction in barley. A strong positive correlation ($r = 0.85$) between the generation of hydrogen peroxide (H_2O_2) and ME effectiveness confirmed the important role of ROS in microspore reprogramming [5]. It was also revealed that the efficiency of ME induction in both triticale and barley was significantly modified by seasonal effects associated with fluctuations in the activity of the main antioxidant enzymes—SOD and CAT [5]. The observed genetic specificity of antioxidant defence triggered in response to various ME-inducing treatments suggested an important role of other enzymatic or non-enzymatic elements of antioxidative defence. Continuing our investigations, in the present study, we examined cellular ROS generation, the role of the main LMW antioxidants (ascorbate and glutathione) and enzymes involved in the ascorbate–glutathione cycle in determining the effectiveness of ME induction.

The effects of various ME-inducing stress pre-treatments of tillers were assessed by analysing several parameters related to isolated microspore condition, antioxidant activity and embryogenic potential. In line with previous results [5], the analysis of microspore yield confirmed that barley spikes produced a lower number of microspores in comparison to triticale and that the number of produced microspores was not significantly affected by stress pre-treatments. The low viability of microspores detected in the population of triticale microspores isolated in early spring (March) was also consistent with earlier observations [5]. It was probably the main cause of low effectiveness of ELS formation in isolated microspore cultures of DH28, which is usually significantly more productive [4,5,19,26]. Interestingly, the negative effect of MAN on triticale microspore viability was associated with a higher effectiveness of ELS production, which also confirmed previously published results [5].

Low effectiveness of ME induction in microspore suspension of cv. Golden Promise probably resulted from the specificity of tiller pre-treatments, not sufficiently meeting the requirements of this genotype. On the contrary, highly successful ME initiation in isolated microspore cultures of cv. Igri indicated a significant difference in response to various tiller pre-treatments (observed only as a tendency in other plant materials tested). The observed positive effect induced by the combined LT+MAN+GSH treatment on ELS formation was also in accordance with earlier results on triticale and rye [5,46].

In contrast to animal cells, in which ROS generation is associated mainly with mitochondrial activity, ROS biogenesis in plant cells can be induced by multiple metabolic processes in all cellular compartments [3]. Depending on the cell origin, phase of development and environmental conditions, ROS generation was detected in the plant cell wall, apoplast, plasma membrane, cytoplasm, and organelles involved in oxygen metabolism [47]. However, in light-deprived or non-green tissues (cells) such as microspores, the mitochondrion is postulated to be the major site of ROS production. Mitochondria-originating ROS affect many cell functions through various signalling cascades, retrograde signalling, and interaction with plant hormones [48]. However, the involvement of other ROS-generating pathways cannot be excluded. For example, ROS generation in the cell wall, apoplast, or plasma membrane plays an important role in the interactions between the plant and its environment as well as in plant development [3]. The diffuse patterns of superoxide anion and hydrogen peroxide distribution in the microspore's cytoplasm observed in our study suggest, that microspore isolation and transfer to in vitro culture conditions induced oxidative stress of high intensity in various cell compartments or that the induced signal was transported across the cell area. The more intense accumulation of superoxide anion observed near the nucleus seems to confirm its role in transmitting the signal to the nucleus, which probably leads to an activation of various redox signalling pathways [49]. It could be supposed that the strong NBT staining is probably related to extensive oxidative damage near the nucleus, and thus the modification of some nuclear factors followed by changes in the expression of genes linked to antioxidant responses, cellular transformation, and cell proliferation [3,49]. All these events ending finally in ME induction in the presented model.

The results of recalcitrant genotypes (DH19, cv. Golden Promise) suggested that the high total activity of LMW antioxidants might be one of the factors important for ME induction. It could be attributed to the compensation of inefficient enzymatic defence [5,26]. Such high activity of LMW antioxidants was not necessary for microspores of responsive genotypes equipped with more active antioxidant enzymes [5]. Furthermore, the relatively high total activity of LMW antioxidants induced by GSH pre-treatment of tillers was associated with reduced effectiveness of ELS formation in responsive DH lines/cultivars.

Based on the results, it can be concluded that the embryogenic potential of triticale seems to be associated with ascorbate and glutathione metabolism (Figure 5). Such a conclusion could be drawn by comparing the level of ASC and GSH accumulation in microspores of responsive and recalcitrant DH lines. The responsive DH28 line was also characterized by high activity of MDHAR and DHAR, enzymes that catalyse the conversion of oxidised ascorbate (MDHA and DHA) to a reduced form (ASC). However, the relatively low activity of APX in triticale microspores after the most effective ME-inducing treatments suggests that reduced ascorbate may play other roles than antioxidative in microspore reprogramming. The role of GR is questionable, because its activity, although relatively low in DH28 microspores, was significantly higher in comparison to recalcitrant DH19. Contrary to the previous report [4], exogenous GSH application during the LT pre-treatment of tillers did not improve ELS formation in isolated microspore cultures of the responsive DH28 line. Such a divergent effect of GSH was also observed earlier [5], and then it was explained as a result of excessive decomposition of H_2O_2, whose amount in the cells declined below the threshold necessary to initiate microspore reprogramming.

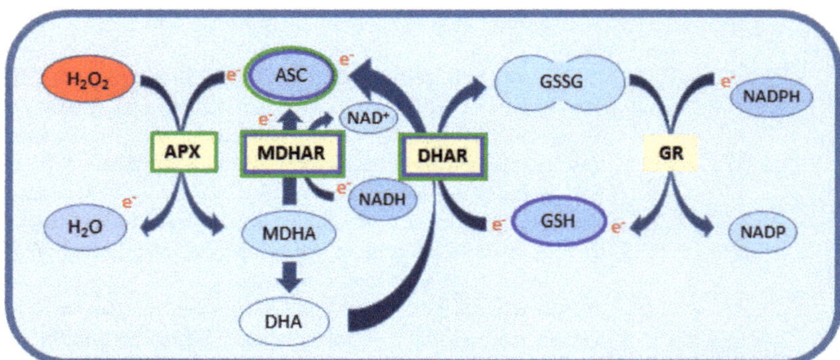

Figure 5. Summary of the work showing the role of the ascorbate–glutathione cycle in the scavenging of ROS (H_2O_2) generated during ME-induction in microspores of the responsive triticale line (DH28) and barley cultivar (cv. Igri). The generated H_2O_2 is initially reduced via the oxidation of reduced ascorbate (ASC) in a reaction catalysed by APX. Two MDHA molecules spontaneously disproportionate into ASC and dehydroascorbate (DHA). Both MDHA and DHA can be reduced back to ASC in reactions catalysed by MDHAR and DHAR, where GSH and NADH are used as electron donors. Oxidized glutathione (GSSG) is converted back to its reduced form by GR using NADPH as an electron donor. Ascorbate–glutathione cycle components, which are largely involved in the process of ME-induction in triticale and barley are marked with the following colour outlines: dark blue for triticale and green for barley. APX—ascorbate peroxidase; ASC—reduced ascorbate; DHA—dehydroascorbate; DHAR—dehydroascorbate reductase; GR—glutathione reductase; GSH—reduced glutathione; GSSG—oxidized glutathione, glutathione disulphide; MDHA—monodehydroascorbate; MDHAR—monodehydroascorbate reductase; NAD(P)H/NAD(P)+—nicotinamide adenine dinucleotide (phosphate) reduced/oxidized; e⁻—transferred electrons.

In microspores of highly responsive barley (cv. Igri), the most effective ME-induction treatment (LT+MAN+GSH) was associated with a very intensive accumulation of ascorbate and higher activity of enzymes involved in ascorbate regeneration (APX, MDHAR and DHAR) compared to the recalcitrant cultivar (cv. Golden Promise). Conversely to triticale, changes in GSH level and GR activity induced by various tiller pre-treatments in the studied barley cultivars suggested that glutathione did not appear to be directly involved in ME induction (Figure 5). It was confirmed by the fact that a positive effect of exogenously applied GSH on ME effectiveness has not been observed in barley, neither previously [5] nor in this study.

Ascorbate and glutathione have been localized in several cellular organelles (chloroplasts, mitochondria, peroxisomes, nuclei, cytosol and apoplast) acting independently or in concert in the Foyer–Halliwell–Asada pathway [9]. Although their essential roles are related to their antioxidant properties, and they usually respond in a compensatory manner, their functions in controlling cell division, plant growth and development seem to be interdependent and not interchangeable [6,50–52].

The strong reducing potential of ASC (physiologically active, anionic form of ascorbic acid) is based on the electron donating/accepting properties of the 2,3-enediol moiety [53].

It is located mainly in the cell cytoplasm, but usually about 5% of its pool is transported across the plasma membrane to the apoplast (reviewed in [9]). This fraction is believed to play a crucial role in oxidative stress signalling. The dominant role of ASC in antioxidant defence during ME induction could be explained by its substantially higher content detected in microspores of all studied plants in comparison to the level of GSH. Moreover, due to the higher affinity of APX for H_2O_2 in comparison to other peroxidases, the process of H_2O_2 removal by ASC may be more efficient than of GSH-utilising enzymes. APX activation was observed in a variety of plant species grown under stress conditions (reviewed in [9,54]). In addition, the previously observed strong seasonal effect on ME effectiveness [5] could be at least partially explained by the fact that ascorbate biosynthesis is highly sensitive to changing light conditions, especially red/far-red ratios [55,56].

The best recognized role of glutathione is also connected with its antioxidant properties determined by the cysteine sulfhydryl group (-SH) [7,57]. Acting as an electron donor/acceptor, it prevents uncontrolled, irreversible oxidation of other cellular components. However, it also plays a very important role in the protein protection through the formation of reversible, covalent disulphide bonds between the sulphur atom of GSH and protein cysteinyl residues [58]. This process of the so-called S-glutathionylation is promoted by oxidative stress, but it also occurs in cells under physiologically optimal conditions and causes specific changes in protein functions (activation or deactivation). The process is catalysed by a large family of glutathione S-transferases (GSTs), which are very diverse in structure and physiological functions. Some GSTs may also play a role as antioxidants, having peroxidase or DHAR activity ([7] and references therein). The importance of these enzymes may be underlined by the fact that certain GST transcripts have almost always been detected as upregulated in response to ME-inducing treatments [45,59–62]. Two of GST coding genes (*GSTF2, GSTA2*) were shown to be significantly upregulated in embryogenic microspores of triticale (including DH28) during the first 8 days of in vitro culture [63].

High cellular concentrations of glutathione and ascorbate have made them the major elements of redox buffering systems determining the cellular redox environment. Studies on several GSH-deficient Arabidopsis mutants [64] revealed GSH-responsive genes encoding transcription factors and other proteins involved in the regulation of cell division, redox potential and auxin signalling. Other functions of glutathione are related to redox signalling and regulation of the expression of genes determining plant growth and development. The role of glutathione in redox regulation has also been demonstrated in in vitro culture systems [65–68]. It was observed that the high redox potential (GSH/GSH+GSSG) promoted cell proliferation, whereas subsequent stages of embryo development required a more oxidized environment [65,69–71].

The role of the ASC/DHA ratio in cell redox state is still under discussion due to their spatial separation, since the DHA pool is mainly accumulated in the cell apoplast [6]. The results of this experiment confirmed that the level of ASC in terms of signalling and regulation seemed to be more important than the ASC/DHA ratio. This was probably due to other important roles of ASC, which is a cofactor of several enzymes involved in photoprotection, biosynthesis of plant growth regulators, pigments and secondary metabolites [8].

5. Conclusions

These observations support our hypothesis that ROS generation and antioxidant defence are critical for successful ME induction. The analysis of our data suggests that the very subtle, dynamic homeostasis between ROS production and scavenging determines the effectiveness of microspore reprogramming towards sporophytic development. Stress treatments effective for ME induction induce ROS generation and simultaneously enhance antioxidant defence in a balanced manner. Genetic and physiological determination of the antioxidant defence system could explain the observed variation in plant responsiveness to various stresses applied as triggers for ME induction. It appears to be based on stress tolerance mechanisms that have developed during the evolution of individual species in response to the most common environmental challenges. In addition to the function of basic antioxidant enzymes like SOD and CAT, enzymes involved in the Foyer–Halliwell–Asada cycle also seem to play an important role in the successful microspore reprogramming. The roles played by the main LMW antioxidants—ascorbate and glutathione—are much more complex and genotype-specific. They are based not only on their antioxidant capacity, but also on their potential key role in cell redox signalling and regulation of cell growth and differentiation. The highly complex multifactorial network of mutual interactions makes the development of effective ME induction procedures in certain economically important crop species still a great challenge for researchers and breeders. To overcome this challenge, more research is needed into the mechanism of microspore reprogramming. A very interesting and almost non-recognized problem is the source and subcellular localization of ROS generation and its interaction with other signalling molecules like Ca^{2+}, plant growth regulators or reactive nitrogen species [47]. Other various possibilities are the involvement of polyamine oxidases (PAOs) and/or NADPH-oxidase as enzymes contributing to H_2O_2 accumulation in response to abiotic stress [72] and generative development of microspores [73]. The direction of further research will be based however, on the results of RNAseq analysis which will show the most probable and promising objects of interest.

Supplementary Materials: The following are available online at https://www.mdpi.com/article/10.3390/antiox10081254/s1, Figure S1: In situ histochemical detection of superoxide anion (A–C) and hydrogen peroxide (D–F) in randomly selected samples of isolated microspores at the early stages of microspore embryogenesis.

Author Contributions: Conceived and managed the project, designed and performed the experiments and wrote the manuscript, I.Ż.; supervised the analysis of antioxidant enzymes, glutathione and ascorbate, B.B.; conducted the analysis of antioxidant enzymes, glutathione and ascorbate, J.F., P.K. and E.S.; measured the total activity of low molecular weight antioxidants, F.J.; performed the experiments, prepared isolated microspore cultures, conducted microscopic observations and analysed all cytological parameters, E.D., M.K., A.N., K.J. and A.J. All authors have read and agreed to the published version of the manuscript.

Funding: This research was funded by the NATIONAL SCIENCE CENTER, Poland, grant number 2015/18/M/NZ3/00348.

Institutional Review Board Statement: Not applicable.

Informed Consent Statement: Not applicable.

Data Availability Statement: Data is contained within the article and supplementary material.

Acknowledgments: The seeds of DH triticale lines were obtained from the State Breeding Institute at the University of Hohenheim (Germany), courtesy of H.P. Maurer. We would also like to thank Jochen Kumlehn of the Leibniz Institute of Plant Genetics and Crop Plant Research (Gatersleben, Germany) for the seeds of two barley cultivars, 'Igri' and 'Golden Promise'.

Conflicts of Interest: The authors declare that they have no conflict of interest.

Abbreviations

ASC—reduced ascorbate; APX—ascorbate peroxidase; CAT—catalase; DHA—dehydroascorbate; DHAR—dehydroascorbate reductase; ELS—embryo-like structure; GR—glutathione reductase; GSH—reduced glutathione; GSSG—glutathione disulphide; HS—Hoagland's nutrient solution; LMW—low molecular weight antioxidants; LT—low temperature tillers treatment; LT+GSH—tillers treatment with low temperature and 0.3 mM GSH; LT+MAN—tillers treatment with low temperature and 0.3 M mannitol; LT+MAN+GSH— tillers treatment with low temperature, 0.3 M mannitol and 0.3 mM GSH; MDHA—monodehydroascorbate; MDHAR—monodehydroascorbate reductase; ME—microspore embryogenesis; NADH—reduced nicotinamide adenine dinucleotide; NADPH—nicotinamide adenine dinucleotide phosphate; ROS—reactive oxygen species; SOD—superoxide dismutase.

References

1. Mittler, R. ROS Are Good. *Trends Plant Sci.* **2017**, *22*, 11–19. [CrossRef] [PubMed]
2. Dumanović, J.; Nepovimova, E.; Natić, M.; Kuča, K.; Jaćević, V. The Significance of Reactive Oxygen Species and Antioxidant Defense System in Plants: A Concise Overview. *Front. Plant Sci.* **2021**, *11*, 552969. [CrossRef] [PubMed]
3. Janků, M.; Luhová, L.; Petřivalský, M. On the Origin and Fate of Reactive Oxygen Species in Plant Cell Compartments. *Antioxidants* **2019**, *8*, 105. [CrossRef] [PubMed]
4. Żur, I.; Dubas, E.; Krzewska, M.; Zieliński, K.; Fodor, J.; Janowiak, F. Glutathione provides antioxidative defence and promotes microspore-derived embryo development in isolated microspore cultures of triticale (×*Triticosecale* Wittm.). *Plant Cell Rep.* **2019**, *38*, 195–209. [CrossRef]
5. Żur, I.; Dubas, E.; Krzewska, M.; Kopeć, P.; Nowicka, A.; Surówka, E.; Gawrońska, K.; Gołębiowska, G.; Juzoń, K.; Malaga, S. Triticale and barley microspore embryogenesis induction requires both reactive oxygen species generation and efficient system of antioxidative defence. *Plant Cell Tissue Organ Cult.* **2021**, *145*, 347–366. [CrossRef]
6. Foyer, C.H.; Noctor, G. Ascorbate and glutathione: The heart of the redox hub. *Plant Physiol.* **2011**, *155*, 2–18. [CrossRef] [PubMed]
7. Noctor, G.; Mhamdi, A.; Chaouch, S.; Han, Y.; Neukermans, J.; Marquez-Garcia, B.; Queval, G.; Foyer, C.H. Glutathione in plants: An integrated overview. *Plant Cell Environ.* **2012**, *35*, 454–484. [CrossRef]
8. Hasanuzzaman, M.; Bhuyan, M.H.M.B.; Anee, T.I.; Parvin, K.; Nahar, K.; Mahmud, J.A.; Fujita, M. Regulation of Ascorbate-Glutathione Pathway in Mitigating Oxidative Damage in Plants under Abiotic Stress. *Antioxidants* **2019**, *8*, 384. [CrossRef] [PubMed]
9. Akram, N.A.; Shafiq, F.; Ashraf, M. Ascorbic acid: A potential oxidant scavenger and its role in plant development and abiotic stress tolerance. *Front. Plant Sci.* **2017**, *8*, 613. [CrossRef]
10. Zhang, Y. Biological role of ascorbate in plants. In *Ascorbic Acid in Plants*; SpringerBriefs in Plant Science; Springer: New York, NY, USA, 2013; pp. 7–33. [CrossRef]
11. Xiao, M.; Li, Z.; Zhu, L.; Wang, J.; Zhang, B.; Zheng, F.; Zhao, B.; Zhang, H.; Wang, Y.; Zhang, Z. The multiple roles of ascorbate in the abiotic stress response of plants: Antioxidant, cofactor, and regulator. *Front. Plant Sci.* **2021**, *12*, 592. [CrossRef]
12. Elhiti, M.; Yang, C.; Belmonte, M.F.; Gulden, R.H.; Stasolla, C. Transcriptional changes of antioxidant responses, hormone signalling and developmental processes evoked by the *Brassica napus* SHOOTMERISTEMLESS during in vitro embryogenesis. *Plant Physiol. Biochem.* **2012**, *58*, 297–311. [CrossRef]
13. Asif, M.; Eudes, F.; Goyal, A.; Amundsen, E.; Randhawa, H.; Spaner, D. Organelle antioxidants improve microspore embryogenesis in wheat and triticale. *Vitro Cell. Dev. Biol.* **2013**, *49*, 489–497. [CrossRef]
14. Hoseini, M.; Ghadimzadeh, M.; Ahmadi, B.; da Silva, J.A.T. Effects of ascorbic acid, alpha-tocopherol, and glutathione on microspore embryogenesis in *Brassica napus* L. *Vitro Cell. Dev. Biol.* **2014**, *50*, 26–35. [CrossRef]
15. Yerzhebayeva, R.S.; Abekova, A.M.; Ainebekova, B.A.; Urazaliyev, K.R.; Bazylova, T.A.; Daniyarova, A.K.; Bersimbayeva, G.K. Influence of different concentrations of ascorbic and gibberellic acids and pH of medium on embryogenesis and regeneration in anther culture of spring triticale. *Cytol. Genet.* **2017**, *51*, 448–454. [CrossRef]
16. Cistue, L.; Ziauddin, A.; Simion, E.; Kasha, K.J. Effects of culture conditions on isolated microspore response of barley cultivar Igri. *Plant Cell Tissue Organ Cult.* **1995**, *42*, 163–169. [CrossRef]
17. Coronado, M.J.; Hensel, G.; Broeders, S.; Otto, I.; Kumlehn, J. Immature pollen-derived doubled haploid formation in barley cv. Golden Promise as a tool for transgene recombination. *Acta Physiol. Plant* **2005**, *27*, 591–599. [CrossRef]
18. Lippmann, R.; Friedel, S.; Mock, H.-P.; Kumlehn, J. The low molecular weight fraction of compounds released from immature embryogenesis. *Front. Plant Sci.* **2015**, *6*, 498. [CrossRef]
19. Żur, I.; Dubas, E.; Krzewska, M.; Waligórski, P.; Dziurka, M.; Janowiak, F. Hormonal requirements for effective induction of microspore embryogenesis in triticale (×*Triticosecale* Wittm.) anther cultures. *Plant Cell Rep.* **2015**, *34*, 47–62. [CrossRef]
20. Krzewska, M.; Gołebiowska-Pikania, G.; Dubas, E.; Gawin, M.; Żur, I. Identification of proteins related to microspore embryogenesis responsiveness in anther cultures of winter triticale (×*Triticosecale* Wittm.). *Euphytica* **2017**, *213*, 192. [CrossRef]
21. Wędzony, M. Protocol for doubled haploid production in hexaploid Triticale (×*Triticosecale* Wittm.) by crosses with maize. In *Doubled Haploid Production in Crop Plants. A Manual*; Maluszynski, M., Kasha, K.J., Forster, B.P., Szarejko, I., Eds.; Springer: Dordrecht, The Netherlands, 2003; pp. 135–140. [CrossRef]
22. Zhuang, J.; Xu, J. Increasing differentiation frequencies in wheat pollen callus. In *Cell and Tissue Culture Techniques for Cereal Crop Improvement*; Hu, H., Vega, M., Eds.; Science Press: Beijing, China, 1983; p. 431.
23. Pauk, J.; Mihály, R.; Monostori, T.; Puolimatka, M. Protocol of triticale (×*Triticosecale* Wittmack) microspore culture. In *Doubled Haploid Production in Crop Plants. A Manual*; Maluszynski, M., Kasha, K.J., Forster, B.P., Szarejko, I., Eds.; Springer: Dordrecht, The Netherlands, 2003; pp. 129–134. [CrossRef]
24. Kumlehn, J.; Serazetdinova, L.; Hensel, G.; Becker, D.; Loerz, H. Genetic transformation of barley (*Hordeum vulgare* L.) via infection of androgenetic pollen cultures with *Agrobacterium tumefaciens*. *Plant Biotechnol. J.* **2006**, *4*, 251–261. [CrossRef]
25. Heslopharrison, J.; Heslopha, Y. Evaluation of pollen viability by enzymatically induced fluorescence—Intracellular hydrolysis of fluorescein diacetate. *Stain Technol.* **1970**, *45*, 115–120. [CrossRef]
26. Żur, I.; Dubas, E.; Krzewska, M.; Janowiak, F.; Hura, K.; Pociecha, E.; Bączek-Kwinta, R.; Płażek, A. Antioxidant activity and ROS tolerance in triticale (×*Triticosecale* Wittm.) anthers affect the efficiency of microspore embryogenesis. *Plant Cell Tissue Organ Cult.* **2014**, *119*, 79–94. [CrossRef]

27. Wohlgemuth, H.; Mittelstrass, K.; Kschieschan, S.; Bender, J.; Weigel, H.-J.; Overmyer, K.; Kangasjärvi, J.; Sandermann, H.; Langebartels, C. Activation of an oxidative burst is a general feature of sensitive plants exposed to the air pollutant ozone. *Plant Cell Environ.* **2002**, *25*, 717–726. [CrossRef]

28. Brand-Williams, W.; Cuvelier, M.E.; Berset, C. Use of a free-radical method to evaluate antioxidant activity. *Food Sci. Technol.-Lebensm.-Wiss. Technol.* **1995**, *28*, 25–30. [CrossRef]

29. Laskoś, K.; Pisulewska, E.; Waligórski, P.; Janowiak, F.; Janeczko, A.; Sadura, I.; Polaszczyk, S.; Czyczyło-Mysza, I.M. Herbal additives substantially modify antioxidant properties and tocopherol content of cold-pressed oils. *Antioxidants* **2021**, *10*, 781. [CrossRef]

30. Law, M.Y.; Charles, S.A.; Halliwell, B. Glutathione and ascorbic acid in spinach (*Spinacia oleracea*) chloroplasts. The effect of hydrogen peroxide and of paraquat. *Biochem. J.* **1983**, *210*, 899–903. [CrossRef]

31. Foyer, C.; Rowell, J.; Walker, D. Measurements of the ascorbate content of spinach leaf protoplasts and chloroplasts during illumination. *Planta* **1983**, *157*, 239–244. [CrossRef]

32. Nakano, Y.; Asada, K. Hydrogen peroxide is scavenged by ascorbate-specific peroxidase in spinach chloroplasts. *Plant Cell Physiol.* **1981**, *22*, 867–880. [CrossRef]

33. Klapheck, S.; Zimmer, I.; Cosse, H. Scavenging of hydrogen peroxide in the endosperm of *Ricinus communis* by ascorbate peroxidase. *Plant Cell Physiol.* **1990**, *31*, 1005–1013. [CrossRef]

34. Hossain, M.A.; Nakano, Y.; Asada, K. Monodehydroascorbate reductase in spinach chloroplasts and its participation in regeneration of ascorbate for scavenging hydrogen peroxide. *Plant Cell Physiol.* **1984**, *25*, 385–395. [CrossRef]

35. Bradford, M.M. A rapid and sensitive method for the quantitation of microgram quantities of protein utilizing the principle of protein–dye binding. *Anal. Biochem.* **1976**, *72*, 248–254. [CrossRef]

36. Thomas, W.; Forster, B.; Gertsson, B. Doubled haploids in breeding. In *Doubled Haploid Production in Crop Plants. A Manual*; Maluszynski, M., Kasha, K.J., Forster, B.P., Szarejko, I., Eds.; Springer: Dordrecht, The Netherlands, 2003; pp. 337–349. [CrossRef]

37. Forster, B.; Thomas, W. Doubled haploids in genetic mapping and genomics. In *Doubled Haploid Production in Crop Plants. A Manual*; Maluszynski, M., Kasha, K.J., Forster, B.P., Szarejko, I., Eds.; Springer: Dordrecht, The Netherlands, 2003; pp. 367–390. [CrossRef]

38. Szarejko, I.; Forster, B.P. Doubled haploidy and induced mutation. *Euphytica* **2007**, *158*, 359–370. [CrossRef]

39. Humphreys, D.G.; Knox, R.E. Doubled haploid breeding in cereals. In *Advances in Plant Breeding Strategies: Breeding, Biotechnology and Molecular Tools*; Al-Khayri, J.M., Jain, S.M., Johnson, D.V., Eds.; Springer: Cham, Switzerland; Heidelberg, Germany; New York, NY, USA; Dordrecht, The Netherlands; London, UK, 2015; Volume 1, pp. 241–290.

40. Touraev, A.; Vicente, O.; HeberleBors, E. Initiation of microspore embryogenesis by stress. *Trends Plant Sci.* **1997**, *2*, 297–302. [CrossRef]

41. Shariatpanahi, M.E.; Belogradova, K.; Hessamvaziri, L.; Heberle-Bors, E.; Touraev, A. Efficient embryogenesis and regeneration in freshly isolated and cultured wheat (*Triticum aestivum* L.) microspores without stress pretreatment. *Plant Cell Rep.* **2006**, *25*, 1294–1299. [CrossRef]

42. Zoriniants, S.; Tashpulatov, A.S.; Heberle-Bors, E.; Touraev, A. The role of stress in the induction of haploid microspore embryogenesis. In *Haploids in Crop Improvement II*; Don Palmer, C., Keller, W.A., Kasha, K.J., Eds.; Springer: Berlin/Heidelberg, Germany, 2005; pp. 35–52. [CrossRef]

43. Bartoli, C.G.; Casalongue, C.A.; Simontacchi, M.; Marquez-Garcia, B.; Foyer, C.H. Interactions between hormone and redox signalling pathways in the control of growth and cross tolerance to stress. *Environ. Exp. Bot.* **2013**, *94*, 73–88. [CrossRef]

44. Hosp, J.; de Faria Maraschin, S.; Touraev, A.; Boutilier, K. Functional genomics of microspore embryogenesis. *Euphytica* **2007**, *158*, 275–285. [CrossRef]

45. Jacquard, C.; Mazeyrat-Gourbeyre, F.; Devaux, P.; Boutilier, K.; Baillieul, F.; Clement, C. Microspore embryogenesis in barley: Anther pre-treatment stimulates plant defence gene expression. *Planta* **2009**, *229*, 393–402. [CrossRef]

46. Zieliński, K.; Krzewska, M.; Żur, I.; Juzoń, K.; Kopeć, P.; Nowicka, A.; Moravcikova, J.; Skrzypek, E.; Dubas, E. The effect of glutathione and mannitol on androgenesis in anther and isolated microspore cultures of rye (*Secale cereale* L.). *Plant Cell Tissue Organ Cult.* **2020**, *140*, 577–592. [CrossRef]

47. Mignolet-Spruyt, L.; Xu, E.; Idänheimo, N.; Hoeberichts, F.A.; Mühlenbock, P.; Brosché, M.; Van Breusegem, F.; Kangasjärvi, J. Spreading the news: Subcellular and organellar reactive oxygen species production and signalling. *J. Exp. Bot.* **2016**, *67*, 3831–3844. [CrossRef]

48. Huang, S.; Van Aken, O.; Schwarzländer, M.; Belt, K.; Millar, A.H. The Roles of Mitochondrial Reactive Oxygen Species in Cellular Signaling and Stress Response in Plants. *Plant Physiol.* **2016**, *171*, 1551–1559. [CrossRef]

49. Lennicke, C.; Rahn, J.; Lichtenfels, R.; Wessjohann, L.A.; Seliger, B. Hydrogen peroxide—Production, fate and role in redox signaling of tumor cells. *Cell Commun. Signal.* **2015**, *13*, 39. [CrossRef] [PubMed]

50. Noctor, G.; Foyer, C.H. Ascorbate and glutathione: Keeping active oxygen under control. *Annu. Rev. Plant Physiol. Plant Mol. Biol.* **1998**, *49*, 249–279. [CrossRef]

51. Potters, G.; De Gara, L.; Asard, H.; Horemans, N. Ascorbate and glutathione: Guardians of the cell cycle, partners in crime? *Plant Physiol. Biochem.* **2002**, *40*, 537–548. [CrossRef]

52. Noctor, G. Metabolic signalling in defence and stress: The central roles of soluble redox couples. *Plant Cell Environ.* **2006**, *29*, 409–425. [CrossRef] [PubMed]

53. Bilska, K.; Wojciechowska, N.; Alipour, S.; Kalemba, E.M. Ascorbic acid: The little-known antioxidant in woody plants. *Antioxidants* **2019**, *8*, 645. [CrossRef] [PubMed]
54. Smirnoff, N. Ascorbic acid metabolism and functions: A comparison of plants and mammals. *Free Radic. Biol. Med.* **2018**, *122*, 116–129. [CrossRef]
55. Bartoli, C.G.; Yu, J.; Gomez, F.; Fernandez, L.; McIntosh, L.; Foyer, C.H. Inter-relationships between light and respiration in the control of ascorbic acid synthesis and accumulation in *Arabidopsis thaliana* leaves. *J. Exp. Bot.* **2006**, *57*, 1621–1631. [CrossRef]
56. Bartoli, C.G.; Tambussi, E.A.; Diego, F.; Foyer, C.H. Control of ascorbic acid synthesis and accumulation and glutathione by the incident light red/far red ratio in *Phaseolus vulgaris* leaves. *Febs Lett.* **2009**, *583*, 118–122. [CrossRef]
57. Ulrich, K.; Jakob, U. The role of thiols in antioxidant systems. *Free Radic. Biol. Med.* **2019**, *140*, 14–27. [CrossRef]
58. Dalle-Donne, I.; Rossi, R.; Colombo, G.; Giustarini, D.; Milzani, A. Protein S-glutathionylation: A regulatory device from bacteria to humans. *Trends Biochem. Sci.* **2009**, *34*, 85–96. [CrossRef]
59. Vrinten, P.L.; Nakamura, T.; Kasha, K.J. Characterization of cDNAs expressed in the early stages of microspore embryogenesis in barley (*Hordeum vulgare*) L. *Plant Mol. Biol.* **1999**, *41*, 455–463. [CrossRef] [PubMed]
60. Maraschin, S.d.F.; Caspers, M.; Potokina, E.; Wuelfert, F.; Graner, A.; Spaink, H.P.; Wang, M. CDNA array analysis of stress-induced gene expression in barley androgenesis. *Physiol. Plant.* **2006**, *127*, 535–550. [CrossRef]
61. Munoz-Amatriain, M.; Svensson, J.T.; Castillo, A.-M.; Cistue, L.; Close, T.J.; Valles, M.-P. Transcriptome analysis of barley anthers: Effect of mannitol treatment on microspore embryogenesis. *Physiol. Plant.* **2006**, *127*, 551–560. [CrossRef]
62. Belanger, S.; Marchand, S.; Jacques, P.-E.; Meyers, B.; Belzile, F. Differential expression profiling of microspores during the early stages of isolated microspore culture using the responsive barley cultivar Gobernadora. *G3-Genes Genom. Genet.* **2018**, *8*, 1603–1614. [CrossRef]
63. Żur, I.; Dubas, E.; Krzewska, M.; Sanchez-Diaz, R.A.; Castillo, A.M.; Valles, M.P. Changes in gene expression patterns associated with microspore embryogenesis in hexaploid triticale (×*Triticosecale* Wittm.). *Plant Cell Tissue Organ Cult.* **2014**, *116*, 261–267. [CrossRef]
64. Schnaubelt, D.; Queval, G.; Dong, Y.; Diaz-Vivancos, P.; Makgopa, M.E.; Howell, G.; De Simone, A.; Bai, J.; Hannah, M.A.; Foyer, C.H. Low glutathione regulates gene expression and the redox potentials of the nucleus and cytosol in *Arabidopsis thaliana*. *Plant Cell Environ.* **2015**, *38*, 266–279. [CrossRef]
65. Stasolla, C.; Yeung, E.C. Recent advances in conifer somatic embryogenesis: Improving somatic embryo quality. *Plant Cell Tissue Organ Cult.* **2003**, *74*, 15–35. [CrossRef]
66. Stasolla, C.; Belmonte, M.F.; van Zyl, L.; Craig, D.L.; Liu, W.B.; Yeung, E.C.; Sederoff, R.R. The effect of reduced glutathione on morphology and gene expression of white spruce (*Picea glauca*) somatic embryos. *J. Exp. Bot.* **2004**, *55*, 695–709. [CrossRef]
67. Belmonte, M.F.; Ambrose, S.J.; Ross, A.R.S.; Abrams, S.R.; Stasolla, C. Improved development of microspore-derived embryo cultures of *Brassica napus* cv Topaz following changes in glutathione metabolism. *Physiol. Plant.* **2006**, *127*, 690–700. [CrossRef]
68. Cistue, L.; Romagosa, I.; Batlle, F.; Echavarri, B. Improvements in the production of doubled haploids in durum wheat (*Triticum turgidum* L.) through isolated microspore culture. *Plant Cell Rep.* **2009**, *28*, 727–735. [CrossRef]
69. De Gara, L.; de Pinto, M.C.; Moliterni, V.M.C.; D'Egidio, M.G. Redox regulation and storage processes during maturation in kernels of *Triticum durum*. *J. Exp. Bot.* **2003**, *54*, 249–258. [CrossRef] [PubMed]
70. Belmonte, M.E.; Macey, J.; Yeung, E.C.; Stasolla, C. The effect of osmoticum on ascorbate and glutathione metabolism during white spruce (*Picea glauca*) somatic embryo development. *Plant Physiol. Biochem.* **2005**, *43*, 337–346. [CrossRef] [PubMed]
71. Stasolla, C. Glutathione redox regulation of in vitro embryogenesis. *Plant Physiol. Biochem.* **2010**, *48*, 319–327. [CrossRef]
72. Gémes, K.; Kim, Y.J.; Park, K.Y.; Moschou, P.N.; Andronis, E.; Valassaki, C.; Roussis, A.; Roubelakis-Angelakis, K.A. An NADPH-oxidase/polyamine oxidase feedback loop controls oxidative burst under salinity. *Plant Physiol.* **2016**, *172*, 1418–1431. [CrossRef]
73. Wu, J.; Shang, Z.; Wu, J.; Jiang, X.; Moschou, P.N.; Sun, W.; Roubelakis-Angelakis, K.A.; Zhang, S. Spermidine oxidase-derived H_2O_2 regulates pollen plasma membrane hyperpolarization-activated Ca^{2+}-permeable channels and pollen tube growth. *Plant J.* **2010**, *63*, 1042–1053. [CrossRef]

antioxidants

MDPI

Article

Seed Germination Behavior, Growth, Physiology and Antioxidant Metabolism of Four Contrasting Cultivars under Combined Drought and Salinity in Soybean

Naheeda Begum [1], Mirza Hasanuzzaman [2], Yawei Li [1], Kashif Akhtar [3], Chunting Zhang [1] and Tuanjie Zhao [1,*]

[1] National Center for Soybean Improvement, Key Laboratory of Biology and Genetics and Breeding for Soybean, Ministry of Agriculture, State Key Laboratory for Crop Genetics and Germplasm Enhancement, Nanjing Agricultural University, Nanjing 210095, China; t2020106@njau.edu.cn (N.B.); 2018201083@njau.edu.cn (Y.L.); 2018801229@njau.edu.cn (C.Z.)

[2] Department of Agronomy, Faculty of Agriculture, Sher-e-Bangla Agricultural University, Dhaka 1207, Bangladesh; mirzahasanuzzaman@sau.edu.bd

[3] State Key Laboratory for Conservation and Utilization of Subtropical Agro-Bio-Resources, College of Life Science and Technology, Guangxi University, Nanning 530004, China; kashif@zju.edu.cn

* Correspondence: tjzhao@njau.edu.cn

Abstract: Drought and salinity stresses are persistent threat to field crops and are frequently mentioned as major constraints on worldwide agricultural productivity. Moreover, their severity and frequency are predicted to rise in the near future. Therefore, in the present study we investigated the mechanisms underlying plant responses to drought (5, 10 and 15% polyethylene glycol, PEG-6000), salinity (50, 100, and 150 mM NaCl), and their combination, particularly at the seed germination stage, in terms of photosynthesis and antioxidant activity. in four soybean cultivars, viz., PI408105A (PI5A), PI567731 (PI31), PI567690 (PI90), and PI416937 (PI37). Results showed that seed germination was enhanced by 10% PEG and decreased by 15% PEG treatments compared to the control, while seed germination was drastically decreased under all levels of NaCl treatment. Furthermore, combined drought and salinity treatment reduced plant height and root length, shoot and root total weights, and relative water content compared with that of control. However, the reductions were not similar among the varieties, and definite growth retardations were observed in cultivar PI5A under drought and in PI37 under salinity. In addition, all treatments resulted in substantially reduced contents of chlorophyll pigment, anthocyanin, and chlorophyll fluorescence; and increased lipid peroxidation, electrolyte leakage, and non-photochemical quenching in all varieties of soybean as compared to the control plants. However, proline, amino acids, sugars, and secondary metabolites were increased with the drought and salinity stresses alone. Moreover, the reactive oxygen species accumulation was accompanied by improved enzymatic antioxidant activity, such as that of superoxide dismutase, peroxidase, catalase, and ascorbate peroxidase. However, the enhancement was most noticeable in PI31 and PI90 under both treatments. In conclusion, the cultivar PI31 has efficient drought and salinity stress tolerance mechanisms, as illustrated by its superior photosynthesis, osmolyte accumulation, antioxidative enzyme activity, and secondary metabolite regulation, compared to the other cultivars, when stressed.

Keywords: *Glycine max*; seed germination; photosynthesis; antioxidant enzymes; secondary metabolites; drought; salinity

Citation: Begum, N.; Hasanuzzaman, M.; Li, Y.; Akhtar, K.; Zhang, C.; Zhao, T. Seed Germination Behavior, Growth, Physiology and Antioxidant Metabolism of Four Contrasting Cultivars under Combined Drought and Salinity in Soybean. *Antioxidants* **2022**, *11*, 498. https://doi.org/10.3390/antiox11030498

Academic Editors: Stanley Omaye

Received: 24 December 2021
Accepted: 26 February 2022
Published: 3 March 2022

Publisher's Note: MDPI stays neutral with regard to jurisdictional claims in published maps and institutional affiliations.

1. Introduction

Abiotic stress is a serious environmental issue that restricts agricultural production, and therefore, global climate change is of great relevance [1]. The most common abiotic stressors that adversely affect plant development and growth are drought and soil salinity [2]. It has been noted that plants in around 10% and 40% of the world's arable land suffer from soil salinity and drought stress [3]. Drought stress inhibits the performance of essential biochemical mechanisms, such as the photosynthetic process [4], nutritional absorption, and osmotic adjustments by influencing the functions of essential enzymes [5]. Drought reduces the water potency in cells, resulting in plant water loss and wilting. Water deficiency for an extended period disrupts membrane integrity, removing the membrane's selective permeability and causing other severe damage. In plants, photosynthesis is one of the most critical metabolic processes that regulate photosynthetic efficiency, thereby regulating plant growth and development. Drought and salt stress have significant impacts on photosynthetic performance, causing several biochemical and physiological changes in plants [3,6,7]. Therefore, detecting the changes in photosynthesis to environmental factors, especially in drought and salinity conditions, is helpful for facilitating plants to adjust to environmental stresses, such as extreme salt and lack of water [3].

Similarly, salinity is another critical environmental stressor that due to toxicity has a negative impact on plants' productivity globally [8]. Accumulation of Na^+ in plant tissues, along with the proliferation of various other toxic ions, causes biological imbalances in plant cells, which in turn result in desperate destruction to the structural and functional permanence of plants' photosynthesis [5]. The processes through which salinity stress affects plant photosynthetic activity include: triggering the over-excitation of photosystem II, which tends to result in more non-radioactive energy loss [9]; lowering the apparent quantum efficiency of photosynthetic activity [9,10]; accumulating reactive oxygen species (ROS), resulting in oxidative damage, which disrupts cellular homeostasis and decreases photosynthetic efficiency and plant growth [11]. Plants, in general, have established natural morpho-physiological protective mechanisms to conquer stress conditions [12], the most significant of which is the antioxidant defense system, which plays an important role in scavenging overproduced ROS [13]. To counteract the oxidative damage induced by ROS, plants normally upregulate antioxidative defense and antioxidant enzymes [14]. Previous research revealed that enzymatic antioxidants play important roles in stress resistance and respond as a defense mechanism in plants [15]. Hence, plants with a stronger antioxidant defense system may be more stress resistant.

In addition, previously, research has focused on elucidating the mechanisms underlying drought and salt stress in plants, and screening and producing stress-resistant cultivars. The soybean (*Glycine max* (L.) Merr), as a key commercial crop, is an important legume widely adopted and cultivated worldwide for its protein content, oil (for consumption), and fatty acids. However, several abiotic variables, such as temperature, flooding, drought, salt, and acidity, significantly affect soybean crop's overall efficiency [16]. Consequently, crop yields must be protected from increasing and more frequent abiotic stresses, particularly drought and salinity, in current and future climates. For soybean crops, drought stress mainly occurs during the growing season, leading to considerable yield loss and quality deterioration, mainly in arid and semi-arid zones. Drought stress also disturbs carbon assimilation and plant phenology in the soybean [17]. Therefore, in order to fulfill future food demands, breeding efforts will need to improve not only current yields but also salinity and drought resistance.

To counteract the detrimental effects of stress on soybeans, a variety of approaches have been established and implemented, the most common of which include agricultural practices and genetic improvements to soybean cultivars [18]. Rainwater and irrigation water are also being used increasingly; although irrigation system adaption is region-specific and significantly increases soybean production costs [19]. Therefore, the most economic way of sustainable soybean cultivation is to develop soybean genotypes having the ability to germinate and survive under severe stress conditions [20,21]. Evidently, enhancing soybean cultivars' stress tolerance is critical for protecting yield gains [22]. Previously the PI 416937 soybean has been shown to have a substantially wider root system than other lines, which could be one component in improving drought resistance. Another investigation recently discovered that PI 416937 is implicated in soybean resistance to aluminum (Al) stress [23]. In addition, two soybean germplasms, PI 567690 and PI 567731, were recently found to consistently express drought tolerance in the field compared to two drought-sensitive genotypes [24]. Although these studies evaluated that these soybean genotypes show drought stress tolerance, the mechanisms underlying the seed germination traits and early seedling growth of these PIs under drought, salinity, and combined stresses require further investigation.

Drought and high salinity conditions are becoming more severe and frequent in agricultural areas, hampering plant growth and development and resulting in significant crop loss. Therefore, identifying the stress tolerant soybean cultivars and significant regulators involved in organized responses to concurrent stresses will be useful. The soybean is an important factor of global food security for both human beings and animals, and a significant means of supplying edible protein and cooking oil. The food demand has increased as the world's population has grown, but agricultural land has reduced. As a result, soybean cultivation will be critical in utilizing arid soil. Soybean performance under combined salinity and drought stress has received less attention to date, even though understanding the combined salinity and drought stress tolerance mechanisms and improving resistance are critical for soybean production.

Therefore, the main objective of the present study was to evaluate physiological and biochemical modulations in four contrasting soybean genotypes concerning germination and early seedling growth under drought, high salinity, and both. Physiological and biochemical analyses were used to investigate the mechanisms of drought and salinity stress tolerance in contrasting soybean genotypes. In particular, we showed that antioxidant enzymes' activities, osmolyte accumulation, and secondary metabolite regulation all have significant roles in salinity and drought stress responses. The present observations could serve as a conceptual framework for selecting and breeding drought and salinity-resistant soybean cultivars.

2. Materials and Methods

2.1. Experimental Plant Materials

This study used genetically diverse soybean genotypes, PI567731 (PI31), PI416937 (PI37) PI567690 (PI90), and PI408105A (PI5A). Accession numbers PI90, PI31, PI37, and PI5A were selected as contrasting genotypes with drought tolerance [25,26], aluminum resistance [23,27], and flooding tolerance. Other information about soybean cultivars that were used in the present study is provided in Table 1. The respective soybean cultivars seeds were allowed to dry naturally before being cleaned and placed in a tightly sealed glass container, which was then kept at 4 °C in the refrigerator until needed for experiments.

Table 1. Details of soybean cultivars used in the study.

Sl. No.	Accession Name of the Cultivar	Cultivar	Origin/Country	Stress Tolerance	References
1	PI5A	KAS 633-19	Korea, South	Flooding tolerance	[27]
2	PI31	Fu yang (56)	China	Drought tolerance	[26]
3	PI90	Fu yang (7)	China	Drought tolerance	[25]
4	PI37	Houjaku Kuwazu	Japan	Aluminum-resistant	[23]

2.2. Seeds' Germination Treatment

Mature seeds were sterilized for 5 min in a sodium hypochloride solution before being rinsed with distilled water. For each cultivar, 30 healthy seeds were placed in Petri dishes (9 cm length diameter) with two layers of Whatman no. 1 filter paper with six different treatments as follows: (i) distilled water (control, CK), (ii) 5% polyethylene glycol-6000, PEG, (iii) 10% PEG, (iv) 15% PEG, (v) 50 mM NaCl, and (vi) 100 mM NaCl and (vii) 150 mM NaCl respectively. Germination and growth analyses of four different available soybean cultivars (PI31, PI90, PI37 and PI5A) were conducted in a growth chamber, with a 16 h light/8 h dark day/night pattern at temperatures of 25 °C. There were three replications for each cultivar. Germinated seeds were removed after counting to avoid inaccuracies in germination records.

2.3. Seedlings' Treatments

Seeds of four soybean materials were sown in the growth chamber under the same conditions described in the previous paragraph. Seedlings were transplanted into pots filled with a Vermiculite, peat (2:1: w/w) mixture when they were 5 days old. The pots were irrigated with half-strength Hoagland's nutrient solution every other day. After 20 days of growth with regular management, seedlings were exposed to one of four treatments—viz., (1) control, (2) 15% PEG-6000, (3) 150 mM NaCl, or (4) 15% PEG-6000 + 150 mM NaCl (Table S1). The study used a completely randomized design with four factors (genotypes, drought, salinity, and combined stress), four genotypes (PI31, PI90, PI37, and PI5A), and four treatments (control; drought, 15% PEG-6000; salinity, 150 Mm NaCl; and combined stress, 15% PEG-6000 + 150 mM NaCl) (Table S1). Salt stress was applied by adding NaCl to the Hoagland solution to achieve a final concentration of 150 mM NaCl. Drought stress was addressed by dissolving 15% PEG in half-strength Hoagland's nutrient solution. For combined treatments, 15% PEG (150 mM combined) was added to the Hoagland solution, and the plants were irrigated with the Hoagland solution. The leaf samples were collected from each of the four treatments after the seedlings had been treated for seven days. After that, the samples were immediately frozen with liquid nitrogen and maintained in the fridge until measurements of chlorophyll (Chl) pigment, anthocyanin, Chl fluorescence, lipid peroxidation, electrolyte leakage, osmolytes, antioxidative enzymes' activity levels, and soluble protein content could be performed.

2.4. Seed Germination Calculation

Seed germination was tracked daily during the first five days of the experiment. When the radicle length was more than 2 mm, seeds were considered germinated [28]. The number of germinated seeds by the last day of the total number of seeds was used for the germination percentage (GP) [29]. Germination energy (GE) was calculated by using the formula below:

$$(GE, \%) = \frac{\text{No. of germinated seeds within 3 days}}{\text{total no. of seeds planted}} \times 100 \tag{1}$$

The Germination index (GI) was computed using Equation,

$$\text{Germination index (GI)} = \frac{\Sigma Gt}{Dt} \tag{2}$$

In the formula, number of seeds germinated on a given day is represented by Gt, and Dt represents the number of days from the start of the experiment.

2.5. Fresh and Dry Biomass, Plant Height, and Relative Water Content (RWC) Measurements

After harvesting, the plant height was measured with a measuring tape. After that, the seedlings were divided into two groups: shoots and roots. The fresh weight (FW) of the shoots and that of the roots were calculated using an electronic weight balance, and the dry masses of the shoot and roots were determined after oven drying at 60 °C for 72 h.

The RWC was determined using the WEATHERLEY [30] method. The 8 mm diameter leaf discs were weighed and immersed for 4 h in deionized water, following which the turgid weight was recorded (TW). After the discs had been dried in the oven, the dry biomass (DW) was estimated. For calculation, we utilized the formula below.

$$RWC(\%) = \frac{(FW - DW)}{(TW - DW)} \times 100 \tag{3}$$

2.6. Chlorophyll Fluorescence, Photosynthetic Pigments, and Anthocyanin Content

The plants' uppermost completely grown, fresh, and healthy leaves were selected after 7 days of stress treatments to assess Chl fluorescence. To ensure that all maximum yield of photosystem II (PSII, Fv/Fm) were dark-adapted, the leaves were dark-acclimated for at least 30 min prior to collecting these measurements. Then, a beam of saturating red light triggered the Chl fluorescence fluorometer transients (MINI-Imaging-PAM, (HeinZWalZ GmbH Fffeltrich, Germany). In addition, fresh leaves were homogenized in acetone at a concentration of 80%. A spectrophotometer was used to determine the contents of Chl a, Chl b, and carotenoids at 645, 663, and 480 nm, respectively, according to [31].

For determining anthocyanin content, approximately 500 mg of the leaf sample was taken in a 5 mL methanol solution, 6 M HCL, and drained at 4 °C for 24 h in an airtight container. The extract was then mixed with 1 mL distilled water and 2 mL chloroform in a 2 mL aliquot. A spectrophotometer was used to evaluate the absorbance of the upper chloroform layer containing extracted anthocyanins at 530 nm, by following [32]

2.7. Determination of Proline, Sugars, and Free Amino Acid and Soluble Protein

The proline activity was calculated according to the protocols established by Bates et al. [33]. The plant tissues were homogenized with 3% sulphosalicylic acid and centrifuged for 10 min at 3600 rpm before being tested. The 2 mL solution was mixed with ninhydrin and glacial acetic acid before being incubated at 100 °C for 1 h before being discarded, after which the optical density of the solution was measured at 520 nm. The sugar content in leaves was determined by mixing 0.1 g dried ground samples in 80 percent ethanol and measuring the extract absorbance at 620 nm, as described by Fong et al. [34]. Furthermore, the contents of amino acids were assayed by following the methods of Sadasivam and Manickam [35].

Fresh leaf tissue (0.5 g) was pulverized in a 1 mL extract solution that contained 1 mM ascorbic acid, 1 mM KCl, 0.5M KP buffer (pH 7.0), mercaptoethanol, and glycerol in an ice-cold mortar and pestle. The mixture was centrifuged for 15 min at $11,500\times g$. The supernatant was used to assess enzyme activity as a soluble protein solution. Each sample's protein concentration was measured using the [36] method.

2.8. Lipid Peroxidation and Electrolyte Leakage (EL)

According to the protocols developed by Heath and Packer [37], the lipid peroxidation content was determined. To obtain the final product, fresh tissues were extracted with 1 percent trichloroacetic acid and centrifuged at $10,000\times g$ for 5 min. The supernatant was then heated to 95 °C for 30 min before being diluted with 0.5% thiobarbituric acid. After cooling, the samples were centrifuged at $5000\times g$ for 15 min to measure absorbance at 532 and 600 nm.

The EL was estimated using the procedure provided by Sullivan & Ross [38], after immersing leaf disks in deionized water, electrical conductivity (EC1) was observed. The EC2 was determined by immersing the tubes in a water bath for 30 min at 45 °C and 55 °C, respectively. To determine the EC3, samples were further boiled at 100 °C for 10 min.

The EL values were determined using the following formulae:

$$EL(\%) = \frac{(EC2 - EC1)}{(EC3)} \times 100 \tag{4}$$

2.9. Determination of Phenol, Flavonoids

In order to assess the phenolic content, 500 mg of dry leaf was homogenized in 80% ethanol. When the supernatant was centrifuged at 10,000 rpm for 10 min, it was treated with the reagent 100 µL Folin Ciocalteu (0.1 mL) immediately afterward. The sample's absorbance was observed at 765 nm [39]. The method of Zhishen et al. [40] was used to assess the content of flavonoids. In this procedure, 500 µL of leaf extracts were combined with 300 µL of 5% $NaNO_2$ and 450 µL deionized water to make a solution. In a subsequent

6 min period, 100 µL of 10% $AlCl_3$ $6H_2O$ was added. After 5 min, deionized water and 200 µL of 1 M NaOH were added to get the correct 10 mL mixture. After 5 min of incubation, the absorbance at 500 nm was measured with a spectrophotometer.

2.10. Determination of the Activity Levels of Antioxidant Enzymes

The leaf samples were ground in a pestle and mortar that had been pre-chilled to soak in phosphate buffer (100 mM, pH 7.8) containing thylene diaminetetra acetic acid (EDTA) (1 mM), polyvinyl pyrrolidine (1%), and phenylmethylsulfonyl fluoride (0.1 m). The samples were centrifuged at 12,000 rpm for 10 min at 4 °C. This was done to collect the supernatant for use in an enzyme assay. Superoxide dismutase (SOD) activity was determined by the ability of an enzyme extract to prevent the photochemical reduction of nitroblue tetrazolium [41]. The optical density at 560 nm was determined after incubating the assay mixture for 15 min under fluorescent illumination. In addition, the catalase (CAT) activity was determined by monitoring a reduction in optical density at 240 nm for 2 min while using H_2O_2 as a substrate, and the extinction coefficient (ε) of $0.036 \, mM^{-1} \, cm^{-1}$ was utilized for computation [42]. Furthermore, the method published by Nakano and Asada [43] was utilized to evaluate ascorbate peroxidase (APX) activity, and H_2O_2-dependent oxidation of ascorbate was detected as a change in absorbance at 290 nm for 2 min. Using the method described by Hori et al. [44], peroxidase (POD) activity was evaluated in a reaction mixture including enzyme extract phosphate buffer (50 mM, pH 7.0), 1 mM H_2O_2, and 1 mM guaiacol. For 3 min, the absorbance of the sample was measured at 470 nm. The activity was measured in enzyme unit mg^{-1} protein.

2.11. Statistical Analysis

Using the Statistix 8.1 software, the data were subjected to two-factor analysis of variance (ANOVA). The least significant differences (LSD) were used to compare means at the 0.05 probability level. Sigma plot software was used for data processing and plotting figures. Canoco 5 and GraphPad Prism 7 software were used to test the principal component analysis (PCA) and heat map analysis to differentiate between the different cultivars with different treatments.

3. Results

3.1. Seed Germination Behavior under Different Drought and Salinity Conditions

Soybean seeds germination percentage (SGP) and germination rate index (GRI) were ominously impacted by drought and salinity stress in all cultivars (Figures1A,B and 2). A 10% PEG treatment resulted in a significant increase in the rate of seed germination compared to CK (Figures1A,B and 2). Soybean GP characteristics were modified by PEG content and cultivar; however, the interaction was not significant (Figure 1A,B). The GP was highest under the 10% drought treatment, followed by the 5% drought treatment, for all soybean cultivars compared to CK (Figures1A,B and 2, Table S2). Among the four genotypes, PI31, PI37, and PI90 resulted in better germination performances than the PI5A genotype. The PI31, PI37, and PI90 cultivars exhibited especially good performances. However, the GP decreased steadily with the increase in salt concentration except for PI31 (Figure 1A,B). In a control, the highest GP was recorded, but among the treatment groups, the highest GP was recorded with 100 mM of salt, followed by 50 mM and 150 mM (Figure 1A,B, Table S3).

Furthermore, the results revealed that the maximum GRI were at 5% and 10% of PEG (Figure 1A,B). However, at the 15% PEG level, the lowest GRI were found. The PI31 variety had the highest GRI among all varieties, followed by PI90 and PI37 varieties (PI5A having the lowest of all genotypes). As the PEG content increased, the GRI of all genotypes decreased (Figure 1A,B).

In addition, as the salt content increased, a gradual decrease in GRI was observed ($p < 0.05$). The GRI index was negatively affected by salinity levels ($p < 0.05$, Figure 1A,B); i.e., seed GRI steadily dropped in all varieties as salinity increased (Figure 1A,B, Table S3). The germination rate indexes of PI31 and PI90 were greater than those of PI5A and PI37 under normal conditions. In contrast, the GRI of PI31, PI90, and PI37 under salinity stress were considerably higher than that of FI5A (Figure 1A,B). The above results indicate substantial differences in the sensitivity of PI31, PI90, and PI5A to salinity, and that PI31 was more tolerant to salinity than PI37 (Figure 1A,B).

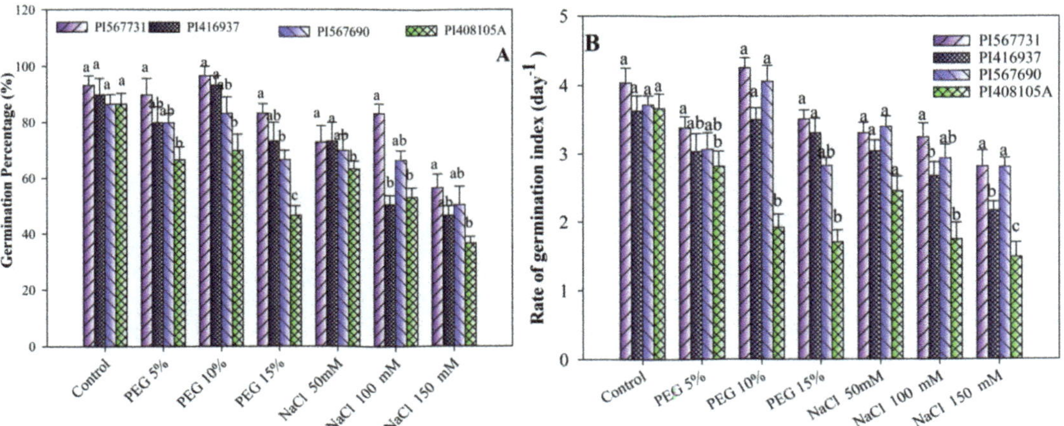

Figure 1. (**A**) Seed germination percentages (SGP) and (**B**) rate of germination indexes (GRI) of four soybean cultivars grown under drought and salinity stress conditions. Data presented are means $\pm$ SD. Different letters denote a significant difference at $p < 0.05$ based on the least significant difference (LSD) test. Letter a is highly significant than b, and ab means no significant differences between a with ab and b with ab.

3.2. Germination Energy

PEG and NaCl treatments significantly impacted all cultivars' GE, with PI31 having higher GE than the others in the PEG/NaCl treatment (Figure 3). The present results show that drought and salinity stress increments decreased GE in all cultivars. When all cultivars were exposed to PEG 15% and 150 mM NaCl, their GE was reduced dramatically, but there were no changes when exposed to PEG 10% and 50 mM NaCl. PI5A showed more germination energy retardation than PI31 under drought stress. However, under salinity stress, the lowest GE was observed in PI37. The results showed that the highest GE (Figure 3) was observed in PI31 at all PEG and NaCl levels. According to these findings, the degree of reduction was not the same for all soybean genotypes evaluated at the PEG and NaCl concentrations studied. A negative effect difference across all genotypes investigated for all germination traits, as shown in (Figure 3), indicates potential genetic variation in response to PEG-induced drought stress and salinity (Tables S2 and S3).

3.3. Plant Height and Biomass and RWC under Drought and Salinity Conditions

The results show that drought stress, salinity, and combined stress treatments reduced the plant height, biomass, and RWC (Tables 2 and S4–S7). The plant height of all cultivars was considerably reduced by drought stress and salinity. Under salinity treatment, the reduction was more in PI37 (64%) and less in PI31 (52%) than in control treatments. Under drought treatment, the reduction was greatest in PI5A (179.44%) and least in PI31 (49%), whereas it was intermediate in the other two varieties compared to the control treatments (Table 2).

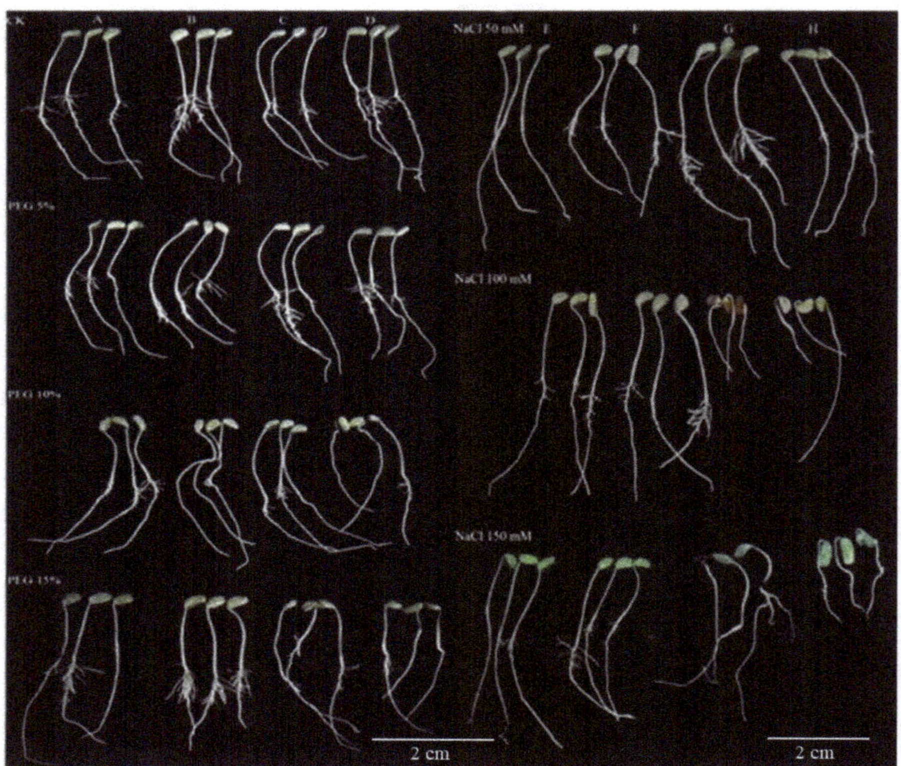

Figure 2. Seed germination of four soybean cultivars grown under different drought and salinity stress conditions. (**A**) PI31, (**B**) PI90, (**C**) PI37, (**D**) PI5A, (**E**) PI31, (**F**) PI90, (**G**) PI5A, (**H**) PI37.

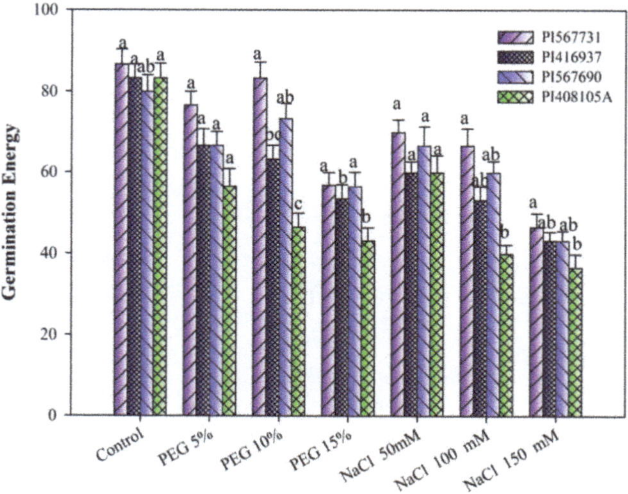

Figure 3. Germination energy of four soybean cultivars grown under drought and salinity stress conditions. Data presented are means ± SD. Different letters denote significant a difference at $p < 0.05$ based on the least significant difference (LSD) test. Letter a is highly significant than b, and ab means no significant differences between a with ab and b with ab.

Table 2. Plant height, shoot fresh weight, shoot dry weight, root dry weight, and leaf relative water content (LRWC) of four soybean cultivars grown under drought and salinity stress conditions. Data presented are means ± SD. Different letters denote a significant difference at $p < 0.05$ based on the least significant difference (LSD) test.

Cultivars	Treatments	Plant Height (cm)	Shoot Fresh Weight (mg g^{-1} FW)	Shoot Dry Weight (mg g^{-1} FW)	Root Dry Weight (mg g^{-1} FW)	LRWC (%)
PI31	Control	76.663 ± 0.95 a	19.603 ± 0.65 a	8.937 ± 0.30 a	3.297 ± 0.40 a	81.376 ± 2.37 a
	Drought	51.410 ± 3.20 a	12.173 ± 0.37 a	7.840 ± 0.32 a	2.217 ± 0.06 a	73.443 ± 2.55 a
	Salinity	45.840 ± 0.76 a	9.743 ± 0.36 a	7.037 ± 0.76 a	1.55 ± 0.28 a	63.09 ± 2.62 a
	drought + Salinity	33.93 ± 1.06 a	6.800 ± 0.46 a	3.270 ± 0.03 a	1.043 ± 0.07 a	58.45 ± 1.56 a
PI90	Control	66.567 ± 2.27 b	18.233 ± 0.32 a	9.433 ± 0.65 a	3.20 ± 0.07 a	83.119 ± 2.57 a
	Drought	43.91 ± 1.02 b	10.300 ± 0.34 a	7.633 ± 0.55 ab	1.280 ± 0.55 b	70.96 ± 1.89 a
	Salinity	40.85 ± 2.40 ab	8.933 ± 0.40 a	5.27 ± 0.17 b	0.920 ± 0.18 b	65.831 ± 1.57 a
	Drought + Salinity	27.68 ± 0.76 b	5.657 ± 0.18 ab	2.867 ± 0.18 ab	0.727 ± 0.18 ab	55.514 ± 2.81 ab
PI37	Control	69.233 ± 1.12 ab	18.100 ± 0.22 a	8.833 ± 0.22 a	3.100 ± 0.11 a	83.92 ± 3.21 a
	Drought	39.43 ± 1.06 b	7.927 ± 0.30 b	4.143 ± 0.64 b	1.190 ± 0.14 bc	50.51 ± 1.27 b
	Salinity	36.267 ± 1.20 b	6.7733 ± 0.38 b	4.410 ± 0.46 b	1.243 ± 0.12 ab	60.79 ± 1.35 a
	Drought + Salinity	20.367 ± 0.03 c	4.680 ± 0.20 b	2.263 ± 0.26 b	0.497 ± 0.26 c	48.47 ± 2.09 bc
PI5A	Control	62.93 ± 2.90 b	16.74 ± 0.90 a	8.267 ± 0.40 a	2.897 ± 0.08 a	80.39 ± 4.44 a
	Drought	22.03 ± 1.11 c	5.480 ± 0.30 c	2.573 ± 0.33 c	0.887 ± 0.03 c	61.78 ± 3.52 ab
	Salinity	25.43 ± 1.93 c	6.28 ± 0.23 b	4.023 ± 0.86 c	1.030 ± 0.08 ab	42.48 ± 2.45 b
	Drought + Salinity	17.33 ± 1.40 c	4.657 ± 0.23 b	1.633 ± 0.22 c	0.347 ± 0.22 c	41.39 ± 1.53 c

The results show that combined stress further decreased the plant height by 126%, 140, 240%, and 252%, in PI31, PI37, PI90, and PI5A, respectively. Moreover, under salinity treatment, the lowest shoot fresh weight (SFW), shoot dry weight (SDW), and root dry weight (RDW) were recorded in PI37 (103%, 79%, and 60% of the control) and the highest in PI31 (101%, 87%, and 79% of the control). However, under drought treatment, the lowest SFW, SDW, and RDW were recorded in PI5A (67%, 68% and 69% of the control) and the highest in PI31 (61%, 13.99%, and 30.68% of the control). On average, 76%, 24%, 158%; and 76%, 130%, 10% of SFW, SDW, and RDW were recorded in the PI37 and PI90 soybean cultivars (Table 2). Similarly, combined stress treatments decreased SFW, SDW, and RDW (Table 2). The lowest SFW, SDW, and RDW were found in PI5A and the highest in PI31 under the combined treatment (Tables 2 and S4–S7).

3.4. Chlorophyll, Carotenoid, and Anthocyanin Content, and Chlorophyll Fluorescence

The results show that the Chl, carotenoids, anthocyanin contents, and the Chl fluorescence were reduced in all varieties due to drought stress, salinity, and combined stress treatments compared with the controls (Tables 3 and S4–S7). The results revealed that Chl a, Chl b, Chl (a + b), and carotenoid contents were significantly affected by drought, salinity, and combined treatments. Soybean cultivar PI31 had higher Chl a, Chl b, Chl (a + b), and carotenoid contents under drought, salinity, and combined stress treatments. In the others, compared to the control, Chl a, Chl b, Chl (a + b), and carotenoids were decreased due to combined stress treatment. The reductions were lower in PI31 (68%, 66%, 69%, and 70%) and higher in PI5A (89%, 70%, 61%, and 80%) compared to the control treatments. Furthermore, drought, salinity, and combined stress significantly reduced the anthocyanin content in all cultivars (Table 3). Under combined stress treatment, the respective contents were reduced by 39%, 51%, 46%, and 61% (Tables 3 and S4–S7).

Moreover, the maximum yield of photosystem II (PSII, Fv/Fm) and photochemical quenching coefficient (qP) of soybean leaves reduced dramatically under drought, salinity, and combined stress, whereas the non-photochemical quenching coefficient (NPQ) increased (Figure 4A–C, Tables S4–S7). The Fv/Fm values evaluated under drought and salinity conditions were considerably reducedby 18% and 50%; 37% and 47%; 28% and 44%; and

62% and 57% in PI31, PI90, PI37, and PI5A, respectively, when compared with control treatment. The qP did not differ significantly in the CK and drought treatments, but the salinity and combined treatments reduced the qP. However, when compared to the control treatment, the salinity and drought treatments increased the qP in PI31 (Figure 4A–C). In all soybean cultivars, the NPQ in the drought and salinity treatment was considerably enhanced as compared to the control treatment. Compared to CK treatment, the NPQ in the combined treatment further increased in PI31, PI90, PI5A, and PI37 (Figure 4A–C, Tables S4–S7).

Table 3. Leaf chlorophyll pigments, carotenoid contents, and anthocyanin contents of four soybean cultivars grown under drought and salinity stress conditions. Data presented are means $\pm$ SD, and different letters denote a significant difference at $p < 0.05$ based on the least significant difference (LSD) test.

Cultivars	Treatments	Chl a (mg g^{-1} FW)	Chl b(mg g^{-1} FW)	Total Chl (mg g^{-1} FW)	Carotenoid (mg g^{-1} FW)	Anthocyanin (µg g^{-1} FW)
PI31	Control	1.414 ± 0.67 a	5.278 ± 0.16 a	6.692 ± 1.09 a	1.283 ± 0.16 a	71.490 ± 1.9 a
	Drought	0.816 ± 0.015 a	3.202 ± a 0.18 a	4.018 ± 1.21 a	0.476 ± 0.05 a	55.385 ± 4.46 a
	Salinity	0.774 ± 0.02 a	2.964 ± 0.2 a	3.738 ± 0.72 a	0.475 ± 0.22 a	56.956 ± 1.87 a
	Drought + Salinity	0.449 ± 0.04 a	1.982 ± 0.19 a	2.431 ± 0.12 a	0.451 ± 0.34 a	48.267 ± 3.24 a
PI37	Control	1.388 ± 0.88 a	5.149 ± 0. 10 a	6.537 ± 0.43 a	0.944 ± 0.27a	76.596 ± 1.12 a
	Drought	0.625± 0.03 a	2.380 ± 0.09 bc	3.005 ± 0.64 b	0.2887 ± 0.18 ab	57.676 ± 1.33 a
	Salinity	0.475 ± 0.01 b	2.231± 0.36 b	2.706± 0.57c	0.256 ± 0.45 b	42.815 ± 4.48 b
	Drought + Salinity	0.231± 0.07 b	1.728 ± 0.17 ab	1.959 ± 0.54c	0.174 ± 0.33 bc	32.144 ± 4.67 b
PI90	Control	1.437 ± 0.44 a	4.994 ± 0.14 a	6.431 ± 1.13 a	1.0172 ± 1.11a	78.691 ± 2.85 a
	Drought	0.316 ± 0.06 b	2.771 ± 0.28 ab	3.087 ± 0.74 a	0.337 ± 0.55 ab	52.544 ± 0.79 ab
	Salinity	0.251 ± 0.08 c	2.883 ± 0.19 a	3.134 ± 0.22 b	0.423 ± 0.13 ab	51.784 ± 2.74 ab
	Drought + Salinity	0.114 ± 0.03 b	1.922 ± 0.11 a	2.063 ± 0.29 b	0.272 ± 0.65 b	42.750 ± 3.53 ab
PI5A	Control	1.388 ± 0.055a	4.902 ± 0.29 a	6.29 ± 1.08 b	0.967 ± 0.88 a	80.262 ± 6.17 a
	Drought	0.215± 0.03b	2.225 ± 0.15 c	2.44 ± 0.98b	0.181 ± 0.57 b	46.94 ± 3.88 b
	Salinity	0.233 ± 0.21c	2.356 ± 0.17 b	2.589 ± 0.65c	0.254 ± 0.14 b	50.082 ± 5.57 ab
	Drought + Salinity	0.158 ±0.14 b	1.421 ± 0.12 b	1.579 ± 0.54ab	0.1284 ± 0.44 c	30.311 ± 0.74 b

3.5. Sugars, Proline, Free Amino Acids, and Protein

Under control conditions, all osmolytes (sugars, proline, and free amino acid) concentrations were statistically similar in all soybean varieties (Figure 5A–D, Tables S8–S11). However, drought, salinity, and combined stress significantly increased osmolytes' accumulation in the soybean seedlings of all varieties. Soybean cultivar PI31 had by far the highest soluble sugar, proline, and free amino acid concentrations compared to the other cultivars under drought and salinity stress treatments, respectively (Figure 5A–D). Upon exposure to combined stress treatment, proline, sugar, and free amino acid contents increased by 399%, 125%, and 174% in PI31; 255%, 101%, and 113% in PI90; 128%, 73%, and 11% in PI05A; and 195%, 111%, and 110% in PI37, compared with the control conditions (Figure 5A–D). In addition, drought, salinity, and combined stress resulted in a substantial decrease in protein content in all soybean varieties compared to the control (Figure 5A–D, Tables S8–S11). Protein content decreased by 32%, 70%, 79%, and 60% in PI31, PI567690, PI5A and PI37, respectively, under combined stress conditions compared to the controls.

3.6. Total Phenolic and Flavonoid Contents

In this study, the accumulation of phenol and flavonoids was significantly augmented with drought, salinity, and combined stress treatments (Figure 6A,B, Tables S8–S11). Compared to the control, drought treated plants showed substantial improvements in their contents of phenolic compounds and flavonoids: 111% and 231.40% in PI31; 145% and 147% in PI90; 85% and 86% in PI5A; and 135% and 215% in PI37, respectively. Similar to drought treatment, the contents of total phenolic compounds and flavonoids were also increased under high salinity by 134% and 241%; 117% and 119%; 133% and 135%; and 86%

and 152% in PI31, PI90, PI5A, and PI37 compared to the controls, respectively. Likewise, combined stress treatment further increased the phenol and flavonoid contents compared to the control (Figure 6A,B, Tables S8–S11).

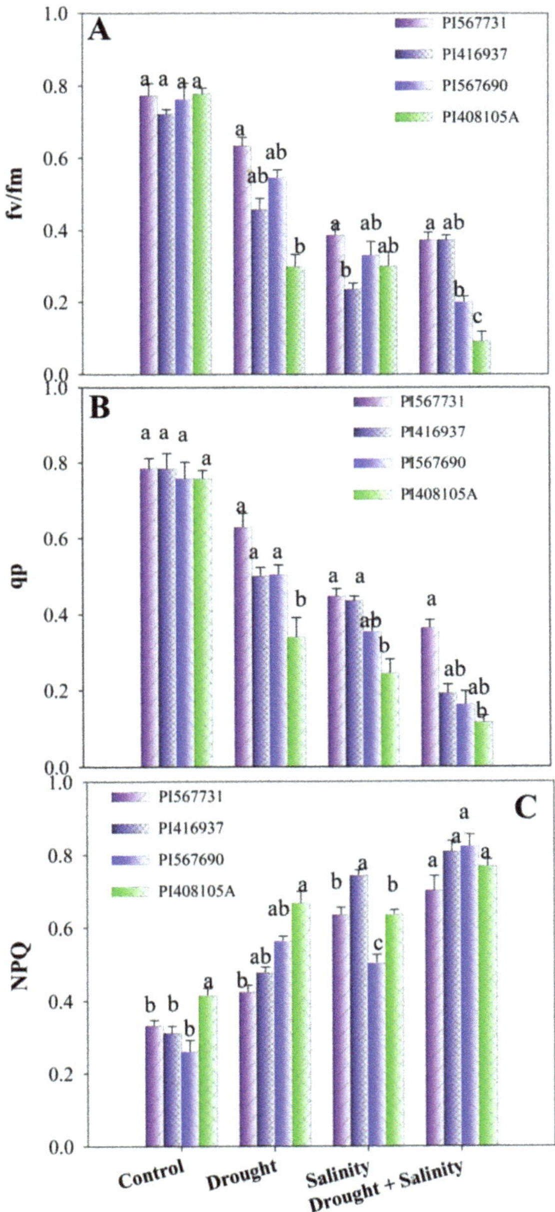

Figure 4. (**A**) Fv/Fm, (**B**) non-photochemical quenching (NPQ), and (**C**) photochemical quenching (qP) of four soybean cultivars grown under drought and salinity stress conditions. Data presented are means ± SD, and different letters denote a significant difference at $p < 0.05$ based on the least significant difference (LSD) test. Letter a is highly significant than b, and ab means no significant differences between a with ab and b with ab.

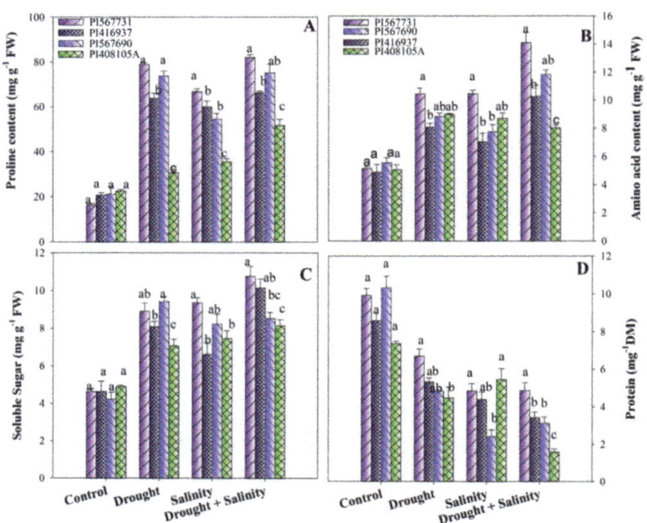

Figure 5. (**A**) Proline, (**B**) free amino acids, (**C**) soluble sugars, and (**D**) protein of four soybean cultivars grown under drought and salinity stress conditions. Data presented are means ± SD, and different letters denote a significant difference at $p < 0.05$ based on the least significant difference (LSD) test. Letter a is highly significant than b, and ab means no significant differences between a with ab and b with ab.

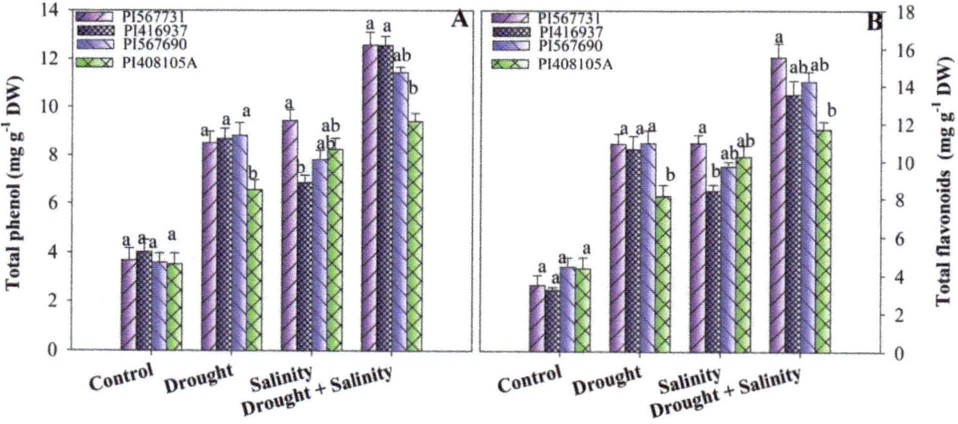

Figure 6. (**A**) Total phenols and (**B**) total flavonoids of four soybean cultivars grown under drought and salinity stress conditions. Data presented are means ± SD, and different letters denote a significant difference at $p < 0.05$ based on the least significant difference (LSD) test. Letter a is highly significant than b, and ab means no significant differences between a with ab and b with ab.

3.7. Oxidative Stress Indicators

Under drought, salinity, and combined stress treatment, lipid peroxidation (MDA) and electrolyte leakage (EL) significantly ($p \leq 0.05$) increased compared with the controls (Figure 7A,B, Tables S8–S11). Compared with its control, PI37 possessed higher MDA and EL contents under high salinity. Under drought stress treatment, PI5A has displayed the highest MDA and EL values. PI31 showed the least lipid peroxidation and electrolyte leakage under all stress conditions. Meanwhile, among the cultivars, PI37, PI90, and PI5A displayed maximum MDA and EL concentration reductions of 173%, 250%, and 174% and

281%, 686%, and 217%, respectively, compared to PI31 under combined stress conditions (Figure 7A,B).

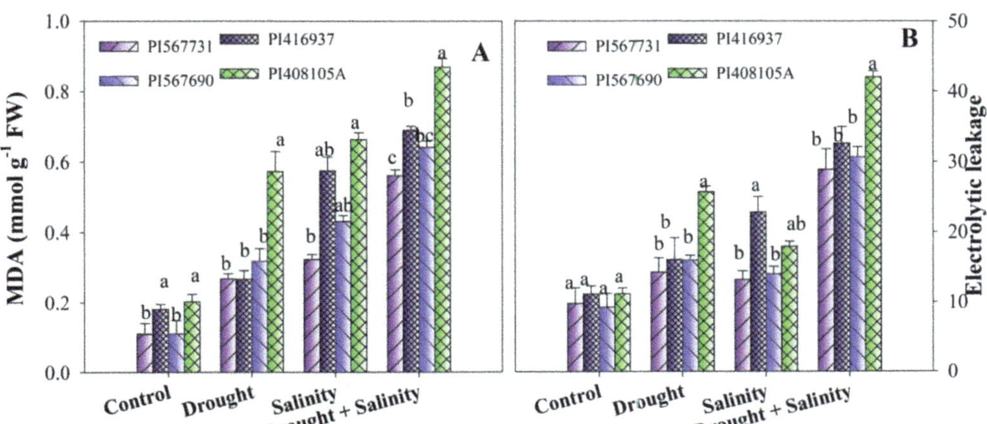

Figure 7. (**A**) Lipid peroxidation and (**B**) electrolyte leakage of four soybean cultivars grown under drought and salinity stress conditions. Data presented are means ± SD, and different letters denote a significant difference at *p* < 0.05 based on the least significant difference (LSD) test. Letter a is highly significant than b, and ab means no significant differences between a with ab and b with ab.

3.8. Activity Levels of Antioxidant Enzymes

Antioxidant enzymes' activity levels, such as those of SOD, POD, CAT, and APX, were examined in soybean cultivars and found to significantly differ (Figure 8A–D, Tables S8–S11). Under separate drought and salinity treatments, SOD and CAT activity levels were significantly increased in PI31, PI37, PI90, and PI5A (Figure 8A–D). Plants exposed to combined stress treatment further augmented their activity, having highest improvements of 153%, 179%, 190%, and 90% in SOD; and 161%, 108%, 162%, 121% in CAT in PI31, PI90, PI37, and PI5A, respectively, compared to the controls. Moreover, in drought and salinity treatments, POD and APX activity levels increased in PI31, PI90, PI5A, and PI37 compared to their control treatments (Figure 8A–D). Likewise, combined stress further significantly increased the APX and POD activity levels by 83% and 206% in PI567731; 71% and 198% in PI90; 57% and 97% in PI5A; and 38% and 121% in PI37, respectively, compared to their controls (Figure 8A–D, Tables S8–S11).

3.9. Relationships among Physiological Traits under Drought Stress, Salinity, and Combined Stress Conditions

Across drought, salinity, and combined stress conditions, the principal component analysis (PCA) with two major components described 44% of the total variation in four different soybean cultivars PI31, PI90, PI5A, and PI37 (Figure 9A). The first PCA axis was used to differentiate no salinity treatments (well-watered and drought-stressed), salinity treatments (salinity, well-watered), and combined stress treatments (salinity + drought) (Figure 9A). Each treatment was segregated well under stressful conditions, and the four soybean cultivars were separated along the second PCA axis. Interestingly, the PCA results revealed that SOD, POD, carotenoid, APX, CAT, phenol, flavonoids, and total Chl were the significant contributors in PC1 and were strongly related to "PI31." EL, MDA, and NPQ were substantially connected to "PI37" under salinity and "PI5A" under drought treatments (Figure 9A). Furthermore, "Chl a" was closely linked to "PI90." The contents of Chl a and Chl b were found to be the most closely linked to "PI31" (Figure 9A).

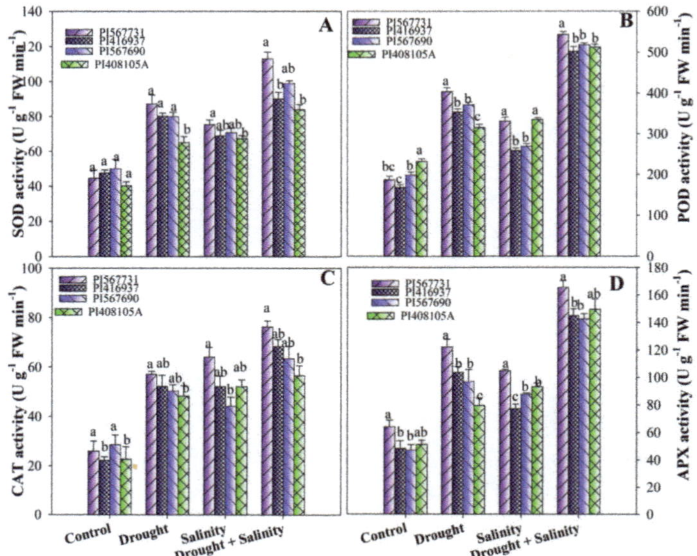

Figure 8. (**A**) Superoxide dismutase, (**B**) peroxidase, (**C**) catalase, and (**D**) ascorbate peroxidase activity levels of four soybean cultivars grown under drought and salinity stress conditions. Data presented are means ± SD, and different letters denote a significant difference at $p < 0.05$ based on the least significant difference (LSD) test. Letter a is highly significant than b, and ab means no significant differences between a with ab and b with ab.

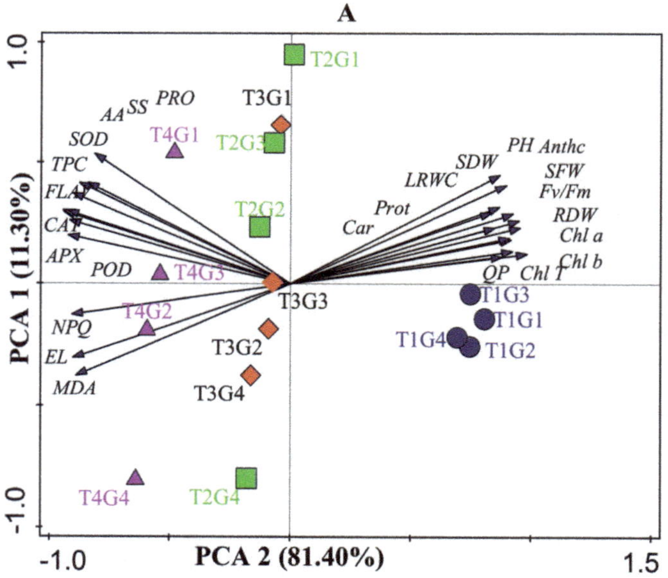

Figure 9. *Cont.*

B

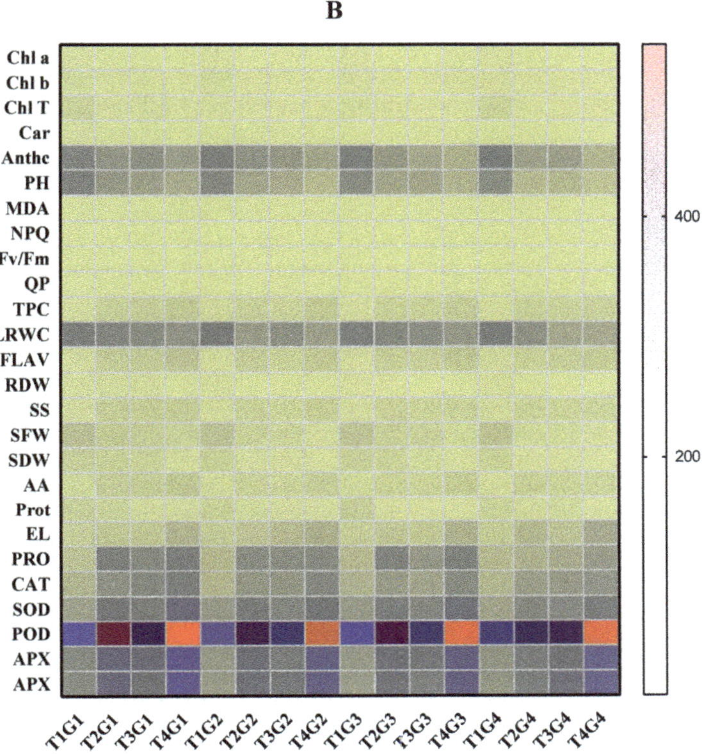

Figure 9. Multivariate statistical analysis indicates associations between treatments, variables, and cultivars. (**A**) Principal component analysis (PCA) based on eco-physiological traits in four soybean cultivars affected by drought, salinity, and combined stress. Blue circles, PI567731; green symbols, PI416937; red symbols, PI567690; and purple symbols, PI408105A in control, drought, salinity, and drought + salinity treatments; (**B**) Heatmap of correlation. T1, control: T2, drought: T3 salinity: T4 drought + salinity; G1, PI567731: G2, PI416937: G3, PI567690: G4, PI408105A. Chl: chlorophyll; Car: carotenoids Chl T: total chlorophyll; PH; plant height; SFW: shoot fresh weight: SDW: shoot dry weight: QP: photochemical quenching; NPQ: non-photochemical quenching; maximal photochemical efficiency (Fv/Fm), RLWC: relative leaf water content; MDA: malondialdehyde; EL: electrolyte leakage; PRO: proline; SS: soluble sugar; CAT: catalase; POD: peroxidase; SOD: superoxide dismutase; APX: ascorbate peroxidase; FLAV: flavonoid content; TPC: total phenolic content; AA: amino acid.

3.10. Correlation Analysis in Soybean Cultivars under Drought Stress, Salinity, and Combined Stress Conditions

Furthermore, the heat map elucidates the physiological responses of the four soybean cultivars under various stress conditions, including drought, salinity, and combined stress (Figure 9B). According to our data, under drought, salinity and combined treatments, plant weights (SFW and SDW), Chl content, qp, RWC, protein, Fv/Fm activity, and proline content were correlated positively with secondary metabolites (phenol and flavonoids) and antioxidative enzymes (CAT, POD, APX, and SOD). The stress treatments are grouped apart from the other soybean treatments because of their increased NPQ, MDA, and EL contents.

4. Discussion

Environmental stresses, particularly drought and salinity stress, are the two most important abiotic factors limiting plants' growth and productivity by restricting their physiological, biochemical, and molecular mechanisms, such as photosynthetic performance, metabolism of protein, and synthesis of lipids [7,45]. To date, most findings have focused on plant responses to each of these stressors separately [46], and these investigations have not provided predicted the strategies plants utilize to respond to combined stresses. Moreover, seedling growth may be a useful trait for the early detection of drought and salinity. Therefore, the present research explored the adverse effects of drought, salinity stress, and their combination on seed germination, growth, photosynthesis, and enzymatic antioxidation in soybeans.

Drought is well-known abiotic stress that causes major losses in plant growth, leading to considerable yield loss among many crops worldwide [6]. Many researchers have shown that simulating drought stress with PEG hypertonic solution is an effective tool during seed germination [47]. In this study, contrasting soybean cultivars were employed, including PI5A, PI31, PI90, and PI37 (Table 1), to evaluate the drought and salt resistance of the soybean germplasms and to comprehend the genetic influences on drought and salt stress tolerance. The present results showed that the seed germination attributes were enhanced by 10% PEG and decreased by 15% PEG treatments compared to the controls. A previous study reported that some plants could adapt to drought, and that low stress could enhance the seed germination, whereas high stress inhibited growth [48]. The effects of drought stress simulated by PEG on seed germination parameters, i.e., GP, GE, and GI, exhibited that seed germination attributes of soybean varied depending on the concentration of PEG600. The relative GP, GE, and GI were reduced as PEG concentrations increased [49]. We observed that 10% PEG could increase seed germination, but 15% PEG could greatly inhibit seed germination. Similar results to our findings were found in the study of Guo et al. [50]. Furthermore, salt concentration increases seed germination efficiency and decreases germination rate [51]. Salt stress can cause osmotic stress and ionic toxicity, which result in poor germination [52]. Our results showed that the seed germination performance was poor with a high salt concentration compared to the control conditions. Among the salinity treatments, he maximum germination performance was recorded at 100 mM NaCl, followed by 50 mM NaCl. The current findings are similar to those of Talei et al. [53].

Furthermore, a recent study stated that, during stressful situations, compared to seed germination, seedling growth is more affected [52]. The present findings demonstrate that salt and drought reduced plant fresh and dry weights in all soybean cultivars, but the responses differed. Due to drought stress, the decrease in plant biomass was greater in the sensitive cultivar (PI5A) than in the tolerant cultivar (PI31). In contrast, PI31 plants grown under combined stress exhibited produced more biomass than the others, indicating that PI567731 soybeans are the most drought and salinity tolerant of these cultivars. These findings demonstrate that the four genotypes have different growth strategies under stressful situations, which is consistent with recent findings [54,55]. In addition, according to previous research, plants can modify their growth distributions among roots and shoots to deal with stress conditions, such as water stress and salinity [56,57]. The present findings support previous research, as under drought, salinity, and combined stress, PI31 had more root biomass than the others (Table 2), which could be due to the roots' enhanced nutrition and water adsorption capability. On the other hand, drought, salinity, and combined stress treatment reduced root biomass in PI37, PI90, and PI5A. This complex mode of growth trait retardation in different soybean genotypes under combined stress could be owing to distinct, particularly opposed signaling pathways caused through combined stress. [58].

Alexieva et al. [59] observed that a reduction in RWC was the primary cause of osmotic stress, which can also be caused by salt and drought. The sensitive PI5A cultivar was more influenced by the reduction in RWC than the PI31 and PI37 cultivars in our study under combined stress (Table 2). We found that when those cultivars were subjected to combined stress, they displayed different levels of sensitivity. Additionally, to determine the performance of their photosynthetic systems under drought and salinity stress conditions, the photosynthetic pigments and Chl fluorescence were measured after drought, salinity, and combined stress treatments. Results showed that drought and salinity stress considerably affected the photosynthetic parameters in the soybean leaves, but the reductions were not similar in the cultivars (Table 3; Figure 4A–C). Generally, alterations in Fv/Fm are frequently corroborated by changes in the concentrations of particular photosynthetic pigments and cell structure modulations, which can be affected by a wide range of environmental factors, including water deficit, nutritional status, and temperature, that affect PSII activity [60]. In the present study, we found that drought and combined stress more adverse effects on the Fv/Fm of the drought-sensitive cultivar (PI5A) than the Fv/Fm of the drought-tolerant cultivars (PI31 and PI37) (Figure 4A–C), demonstrating that PI408105A is more sensitive to drought stress than the other cultivars. On the other hand, salinity stress has a more severe adverse effect on the Fv/Fm of the salt-sensitive cultivar (PI37) than the Fv/Fm of the other cultivars. A recent study has shown that drought can alter the reaction center and impede the electron transport system, decreasing the Fv/Fm [61]. Photosynthetic pigment concentrations, such as those of Chl a, Chl b, and Car, are widely employed to directly check the photosynthesis capabilities of plants in stressful situations [62]. Reductions in Chl and Car concentration have previously been documented to cause significant photosynthetic capacity impairments [63]. In this study, Chl content was observed to be reduced in all stress treatments, and the reductions in Chl in salt treatment groups were more severe than in drought treatment groups (Table 3). In addition, significant reductions in all Chl pigments occurred after seven days of salt stress treatment, demonstrating that salt stress disrupted the synthesis and expedited the breakdown of leaf pigments. Previous studies on hybrid *Pennisetum* [64], *Populus euphratica*, and *P. pruinosa* seedlings [57], supported this hypothesis when salt-treated plants were subjected to drought stress. Their Chl a, Chl b, total Chl, and carotenoid contents were reduced significantly compared to the control plants.

Plants also accumulate a wide range of organic solutes in response to external osmotic impending changes to deal with environmental factors. Proline, an organic solute, is well-known for its osmotic adaptation activity and role in stress responses enhancement by inhibiting cellular membranes and enzyme integrity [65]. According to our results, stress-treated plants had significantly higher proline content than non-treated plants. Furthermore, plants treated with 15% PEG and 200 mM stress exhibited similar adjustments to salt stress when proline levels were nearly equal. Previous findings demonstrated that plant survival, stress tolerance, and biochemical changes under stressed conditions, such as drought and salinity, are all dependent on non-structural carbohydrates [66,67]. Several studies have found that sugars play a vital role in plants' osmotic regulation, including cell turgor maintenance, and absorption and transportation of water under stress [67,68]. In the present study, under combined stress treatment, PI31, PI90, and PI37 soybean plants had higher leaf soluble sugar contents than their controls (Figure 5A–D), indicating that PI31, PI90, and PI37 soybeans may have superior osmotic regulation. Besides this, free amino acids (FAA) are involved in osmotic regulation, helping cells maintain their osmotic potential [69]. We observed that, under all stressful conditions, PI31 and PI90 accumulated more FAA than other soybeans used in this study. Therefore, PI31 and PI90 are more efficient in osmotic regulation, maintaining cell osmotic potential, and water absorption under stressful conditions than the other cultivars (Figure 5A–D).

Furthermore, phenols and flavonoids are crucial for improving cellular homeostasis during drought and salt stress. The amounts of phenolic compounds and flavonoids in our study increased as the concentrations of NaCl and PEG increased. Increased phenolic and flavonoid contents in drought and salt-treated plants could increase ROS production. Increased phenolic and flavonoid contents could also be helpful due to their scavenging activity, protecting plants from ROS by deactivating the free radicals, quenching the ROS, and the decomposing peroxides that are ultimately generated during stress. Some medicinal and aromatic plants, such as rosemary, basil, and pennyroyal [70], have shown that phenolic content increases as stress concentration increases. Surprisingly, in *Nigella sativa* [71], phenolic content decreases as the salt concentration rises.

Environmental conditions such as drought and high salinity lead to ROS production, leading to lipid peroxidation, antioxidant deprivation, and eventually, gene expression alterations [72]. The present investigation found that all soybean cultivars had greater ROS concentrations during drought and salinity stress and combined stress; however, the EL content was highest in the sensitive cultivars (PI5A) under drought and combined treatments and PI37 under salinity and combined treatments, implying higher oxidative stress under stressed conditions. The generation of ROS has been proposed as a major symptom of phytotoxicity, and this process has been extensively studied in plants in various environmental conditions [73,74]. Malondialdehyde (MDA), a key lipid peroxidation product, is linked to oxidative damage in severe stress situations [75]. Our data demonstrated that drought and salinity augmented lipid peroxidation levels in all soybean cultivars, indicating greater oxidative injuries [76], although PI37 and PI5A had more severe oxidative damage than other soybeans. These results revealed that PI31 had less damage to MDA than the other cultivars when exposed to drought, salinity, and combined stress treatments. Antioxidant enzymes are essential for scavenging superoxide ions and lipid peroxidation tolerance. According to previous research, stressful environments can cause substantial increases in the activity of enzymes such as POD, SOD, CAT, and APX, which are well associated with the scavenging ability of ROS and serve as essential protective measures to deal with stressful environmental factors [12,77]. SOD is the first enzyme in the antioxidant system, and it transforms the highly reactive $OH^{\bullet}$ radical and superoxide ($O_2^{\bullet-}$) to less harmful H_2O_2, reducing DNA, protein, and damage to the membrane [78]. The enzyme activity levels in all soybeans increased dramatically in this study. Under drought stress, the SOD, POD, CAT, and APX activity levels were higher in PI31, PI90, and PI37 than in PI5A. At the same time, POD contents were lower in PI37 and greater in PI5A in terms under high salinity. Under combined stress conditions, all four showed higher levels of SOD and POD APX, and CAT as compared to their controls (Figure 8A–D). In summary, PI31 and PI90 appear to have highly effective antioxidant defense mechanisms to cope with stress, explaining their higher stress tolerance. These results are in agreement with other findings: other authors have also found enhanced antioxidant activity in alfalfa [79], maize [80], sugarcane species [15], *Populus yunnanensis* [81], *T. aestivum* [82], *Populus euphratica*, and *P. pruinosa* [57] under stress. Overall, the antioxidant enzyme complex's improved efficacy found under stress conditions could be linked to tolerance mechanisms based on fine-tuned redox state management.

5. Conclusions

The responses of different soybean cultivars to drought and salinity stress were evaluated at the germination and seedling stages. Under drought treatment, compared to the control and 15% PEG treatment, seed germination was improved with the 10% PEG treatment. However, seed germination was reduced under salt treatment as the salt concentration increased. Furthermore, the present investigation revealed that under drought, salinity, and combined stress especially, PI37 and PI5A soybeans suffered more severe inhibitory effects and showed less tolerance than PI31 soybeans. Under stressful conditions, the antioxidant enzyme activity and osmoregulation of the PI31 soybean were higher, and the ROS accumulation was lower than in PI37 and PI5A. The present information collectively demonstrates that the PI31 soybean is more stress-tolerant whether exposed to drought stress, salt, or both. The results of this study are novel and advance our understanding of the plant responses to drought, salt, and combined stressors. These findings are especially valuable for cultivation management due to the frequency of droughts and worsened soil salinity in many areas. Having a better understanding of the specific responses and tolerances to drought and salinity of soybean cultivars is critical and will support environmental movements to re-green agricultural lands.

Supplementary Materials: The following supporting information can be downloaded at: https://www.mdpi.com/article/10.3390/antiox11030498/s1, Table S1: Different drought and salt treatments which used for imposing stress for 7days; Table S2. Seed germination percentage (SGP), rate of germination index (GRI) and Germination energy (GE) of four soybean cultivars grown under drought stress conditions. Data presented are means $\pm$ SD. Different letters denote significant difference at $p < 0.05$ based on the least significant difference (LSD) test. Abbreviation; CK = control, D1 = PEG6000 5% Drought; D2 = PEG6000 10% Drought; D3, PEG6000 15% Drought; Table S3. Seed germination percentage (SGP), rate of germination index (GRI) and Germination energy (GE) of four soybean cultivars grown under Salinity stress conditions. Data presented are means $\pm$ SD. Different letters denote significant difference at $p < 0.05$ based on the least significant difference (LSD) test. Abbreviation; CK = control, S1 = NaCL 50 mM; S2 = NaCL 100 mM; S3 = NaCL 150 Mm; Table S4. Plant height, Shoot fresh weight, Shoot dry weight, Root dry weight, and leaf relative water content (LRWC), Leaf chlorophyll pigments, carotenoid contents, and Anthocyanin contents, Fv/Fm, non-photochemical quenching (NPQ) and photochemical quenching (qP) of PI567731 soybean cultivars grown under drought and salinity stress conditions. Data presented are means $\pm$ SD. Different letters denote significant difference at $p < 0.05$ based on the least significant difference (LSD); Table S5. Plant height, Shoot fresh weight, Shoot dry weight, Root dry weight, and leaf relative water content (LRWC), Leaf chlorophyll pigments, carotenoid contents, and Anthocyanin contents, Fv/Fm, non-photochemical quenching (NPQ) and photochemical quenching (qP) of PI416937 soybean cultivars grown under drought and salinity stress conditions. Data presented are means $\pm$ SD. Different letters denote significant difference at $p < 0.05$ based on the least significant difference (LSD); Table S6. Plant height, Shoot fresh weight, Shoot dry weight, Root dry weight, and leaf relative water content (LRWC), Leaf chlorophyll pigments, carotenoid contents, and Anthocyanin contents, Fv/Fm, non-photochemical quenching (NPQ) and photochemical quenching (qP) of PI567690 soybean cultivars grown under drought and salinity stress conditions. Data presented are means $\pm$ SD. Different letters denote significant difference at $p < 0.05$ based on the least significant difference (LSD); Table S7. Plant height, Shoot fresh weight, Shoot dry weight, Root dry weight, and leaf relative water content (LRWC), Leaf chlorophyll pigments, carotenoid contents, and Anthocyanin contents, Fv/Fm, non-photochemical quenching (NPQ) and photochemical quenching (qP) of PI408105A soybean cultivars grown under drought and salinity stress conditions. Data presented are means $\pm$ SD. Different letters denote significant difference at $p < 0.05$ based on the least significant difference (LSD); Table S8. Proline, Free Amino acids, Soluble Sugars, Protein, Total phenols, Total flavonoids, lipid peroxidation and Electrolyte leakage, Superoxide dismutase, Peroxidase, Catalase, and Ascorbate peroxidase of PI567731 soybean cultivars grown under drought and salinity stress conditions. Data presented are means $\pm$ SD. Different letters denote significant difference at $p < 0.05$ based on the least significant difference (LSD); Table S9. Proline, Free Amino acids, Soluble Sugars, Protein, Total phenols, Total flavonoids, lipid peroxidation and Electrolyte leakage, Superoxide dismutase, Peroxidase, Catalase, and Ascorbate peroxidase of PI416937 soybean cultivars grown under drought and salinity stress conditions. Data presented are means $\pm$ SD. Different letters denote significant difference at $p < 0.05$ based on the least significant difference (LSD); Table S10. Proline, Free Amino acids, Soluble Sugars, Protein, Total phenols, Total flavonoids, lipid peroxidation and Electrolyte leakage, Superoxide dismutase, Peroxidase, Catalase, and Ascorbate peroxidase of PI567690 soybean cultivars grown under drought and salinity stress conditions. Data presented are means $\pm$ SD. Different letters denote significant difference at $p < 0.05$ based on the least significant difference (LSD); Table S11. Proline, Free Amino acids, Soluble Sugars, Protein, Total phenols, Total flavonoids, lipid peroxidation and Electrolyte leakage, Superoxide dismutase, Peroxidase, Catalase, and Ascorbate peroxidase of PI408105A soybean cultivars grown under drought and salinity stress conditions. Data presented are means $\pm$ SD. Different letters denote significant difference at $p < 0.05$ based on the least significant difference (LSD).

Author Contributions: Conceptualization, N.B. and T.Z.; methodology, N.B. and M.H.; formal analysis, N.B.; investigation, N.B.; writing—original draft preparation, N.B. and M.H.; writing—review and editing, N.B., M.H., Y.L., C.Z., K.A. and T.Z; visualization, N.B.; supervision, T.Z. All authors have read and agreed to the published version of the manuscript.

Funding: This work was supported by the National Key R & D Program of China (No. 2021YFD1201603), Science and Technology Innovation Team of Soybean Modern Seed Industry In Hebei (21326313D), the National Natural Science Foundation of China (grant numbers 31871646, 31571691), and the Jiangsu Collaborative Innovation Center for Modern Crop Production (JCIC-MCP) Program.

Institutional Review Board Statement: Not applicable.

Informed Consent Statement: Not applicable.

Data Availability Statement: Data is contained within the article.

Conflicts of Interest: Mirza Hasanuzzaman is a Guest Editor of this special issue. This fact did not affect the peer-review process or any decisions. All other authors declare no conflict of interest.

References

1. Hasanuzzaman, M.; Bhuyan, M.H.M.B.; Zulfiqar, F.; Raza, A.; Mohsin, S.M.; Al Mahmud, J.; Fujita, M.; Fotopoulos, V. Reactive oxygen species and antioxidant defense in plants under abiotic stress: Revisiting the crucial role of a universal defense regulator. *Antioxidants* **2020**, *9*, 681. [CrossRef]

2. Begum, N.; Ahanger, M.A.; Zhang, L. AMF inoculation and phosphorus supplementation alleviates drought induced growth and photosynthetic decline in *Nicotiana tabacum* by up-regulating antioxidant metabolism and osmolyte accumulation. *Environ. Exp. Bot.* **2020**, *176*, 104088. [CrossRef]

3. Sahin, U.; Ekinci, M.; Ors, S.; Turan, M.; Yildiz, S.; Yildirim, E. Effects of individual and combined effects of salinity and drought on physiological, nutritional and biochemical properties of cabbage (*Brassica oleracea* var. capitata). *Sci. Hortic.* **2018**, *240*, 196–204. [CrossRef]

4. Chastain, D.R.; Snider, J.L.; Collins, G.D.; Perry, C.D.; Whitaker, J.; Byrd, S.A. Water deficit in field-grown *Gossypium hirsutum* primarily limits net photosynthesis by decreasing stomatal conductance, increasing photorespiration, and increasing the ratio of dark respiration to gross photosynthesis. *J. Plant Physiol.* **2014**, *171*, 1576–1585. [CrossRef] [PubMed]

5. Ahanger, M.A.; Alyemeni, M.N.; Wijaya, L.; Alamri, S.A.; Alam, P.; Ashraf, M.; Ahmad, P. Potential of exogenously sourced kinetin in protecting *Solanum lycopersicum* from NaCl-induced oxidative stress through up-regulation of the antioxidant system, ascorbate-glutathione cycle and glyoxalase system. *PLoS ONE* **2018**, *13*, e0202175. [CrossRef]

6. Khan, A.; Anwar, Y.; Hasan, M.; Iqbal, A.; Ali, M.; Alharby, H.; Hakeem, K.; Hasanuzzaman, M. Attenuation of drought stress in *Brassica* seedlings with exogenous application of Ca^{2+} and H_2O_2. *Plants* **2017**, *6*, 20. [CrossRef]

7. Begum, N.; Qin, C.; Ahanger, M.A.; Raza, S.; Khan, M.I.; Ashraf, M.; Ahmed, N.; Zhang, L. Role of arbuscular mycorrhizal fungi in plant growth regulation: Implications in abiotic stress tolerance. *Front. Plant Sci.* **2019**, *10*, 1068. [CrossRef]

8. Pan, Y.J.; Souissi, A.; Souissi, S.; Hwang, J.S. Effects of salinity on the reproductive performance of *Apocyclops royi* (Copepoda, Cyclopoida). *J. Exp. Mar. Bio. Ecol.* **2016**, *475*, 108–113. [CrossRef]

9. Baker, N.R. Chlorophyll fluorescence: A probe of photosynthesis in vivo. *Annu. Rev. Plant Biol.* **2008**, *59*, 89–113. [CrossRef]

10. Oukarroum, A.; Bussotti, F.; Goltsev, V.; Kalaji, H.M. Correlation between reactive oxygen species production and photochemistry of photosystems I and II in *Lemna gibba* L. plants under salt stress. *Environ. Exp. Bot.* **2015**, *109*, 80–88. [CrossRef]

11. Khan, M.I.R.; Asgher, M.; Khan, N.A. Alleviation of salt-induced photosynthesis and growth inhibition by salicylic acid involves glycinebetaine and ethylene in mungbean (*Vigna radiata* L.). *Plant Physiol. Biochem.* **2014**, *80*, 67–74. [CrossRef]

12. Ahmad, P.; Latef, A.A.A.; Hashem, A.; Abd Allah, E.F.; Gucel, S.; Tran, L.S.P. Nitric oxide mitigates salt stress by regulating levels of osmolytes and antioxidant enzymes in chickpea. *Front. Plant Sci.* **2016**, *7*, 11. [CrossRef]

13. Hanin, M.; Ebel, C.; Ngom, M.; Laplaze, L.; Masmoudi, K. New insights on plant salt tolerance mechanisms and their potential use for breeding. *Front. Plant Sci.* **2016**, *7*, 1787. [CrossRef]

14. Wang, B.; Zhang, J.; Pei, D.; Yu, L. Combined effects of water stress and salinity on growth, physiological, and biochemical traits in two walnut genotypes. *Physiol. Plant.* **2021**, *172*, 176–187. [CrossRef]

15. Zhang, Y.B.; Yang, S.L.; Dao, J.M.; Deng, J.; Shahzad, A.N.; Fan, X.; Li, R.D.; Quan, Y.J.; Bukhari, S.A.H.; Zeng, Z.H. Drought-induced alterations in photosynthetic, ultrastructural and biochemical traits of contrasting sugarcane genotypes. *PLoS ONE* **2020**, *15*, e0235845. [CrossRef]

16. Le Gall, H.; Philippe, F.; Domon, J.-M.; Gillet, F.; Pelloux, J.; Rayon, C. Cell wall metabolism in response to abiotic stress. *Plants* **2015**, *4*, 112–166. [CrossRef]

17. Du, Y.; Zhao, Q.; Chen, L.; Yao, X.; Zhang, W.; Zhang, B.; Xie, F. Effect of drought stress on sugar metabolism in leaves and roots of soybean seedlings. *Plant Physiol. Biochem.* **2020**, *146*, 1–12. [CrossRef]

18. Turner, N.C.; Saxena, N.P.; Johansen, C.; Chauhan, Y.S.; Rao, R.C. Drought resistance: A comparison of two frameworks. In *Management of Agriculture and Drought: Agronomic and Genetic Options*; Saxena, N.P., Johansen, C., Chauhan, Y.S., Rao, R.C.N., Eds.; Science Publishers: Enfield, CT, USA, 2000; pp. 89–102.

19. Kebede, H.; Fisher, D.K.; Sui, R.; Reddy, K.N. Irrigation methods and scheduling in the Delta region of Mississippi: Current status and strategies to improve irrigation efficiency. *Am. J. Plant Sci.* **2014**, *5*, 2917–2928. [CrossRef]

20. Margenot, A. The State of Soybean in Africa: Soybean Diseases. *Farmdoc Dly.* **2019**, *9*, 1–4.

21. Lopez, M.A.; Xavier, A.; Rainey, K.M. Phenotypic variation and genetic architecture for photosynthesis and water use efficiency in soybean (*Glycine max* l. Merr). *Front. Plant Sci.* **2019**, *10*, 680. [CrossRef]

22. Hudak, C.M.; Patterson, R.P. Vegetative growth analysis of a drought-resistant soybean plant introduction. *Crop Sci.* **1995**, *35*, 464–471. [CrossRef]

23. Li, W.; Sun, Y.; Wang, B.; Xie, H.; Wang, J.; Nan, Z. Transcriptome analysis of two soybean cultivars identifies an aluminum respon-sive antioxidant enzyme GmCAT1. *Biosci. Biotechnol. Biochem.* **2020**, *84*, 1394–1400. [CrossRef]

24. Pathan, S.M.; Lee, J.D.; Sleper, D.A.; Fritschi, F.B.; Sharp, R.E.; Carter, T.E., Jr.; Nelson, R.L.; King, C.A.; Schapaugh, W.T.; Ellersieck, M.R.; et al. Two soybean plant introductions display slow leaf wilting and reduced yield loss under drought. *J. Agron. Crop Sci.* **2014**, *200*, 231–236. [CrossRef]

25. Mutava, R.N.; Prince, S.J.K.; Syed, N.H.; Song, L.; Valliyodan, B.; Chen, W.; Nguyen, H.T. Understanding abiotic stress tolerance mechanisms in soybean: A comparative evaluation of soybean response to drought and flooding stress. *Plant Physiol. Biochem.* **2015**, *86*, 109–120. [CrossRef]

26. Ye, H.; Song, L.; Schapaugh, W.T.; Ali, M.L.; Sinclair, T.R.; Riar, M.K.; Mutava, R.N.; Li, Y.; Vuong, T.; Valliyodan, B.; et al. The importance of slow canopy wilting in drought tolerance in soybean. *J. Exp. Bot.* **2020**, *71*, 642–652. [CrossRef]

27. Wu, C.; Chen, P.; Hummer, W.; Zeng, A.; Klepadlo, M. Effect of flood stress on soybean seed germination in the field. *Am. J. Plant Sci.* **2017**, *8*, 53–68. [CrossRef]

28. Saleem, M.H.; Ali, S.; Seleiman, M.F.; Rizwan, M.; Rehman, M.; Akram, N.A.; Liu, L.; Alotaibi, M.; Al-Ashkar, I.; Mubushar, M. Assessing the correlations between different traits in copper-sensitive and copper-resistant varieties of jute (*Corchorus capsularis* L.). *Plants* **2019**, *8*, 545. [CrossRef]

29. International Seed Testing Association. *The Germination Test in: International Rules for Seed Testing*; International Seed Testing Association: Bassersdorf, Switzland, 2015; pp. 5-1–5-6.

30. Weatherley, P.E. Studies in the water relations of the cotton plant: I. the field measurement of water deficits in leaves. *New Phytol.* **1950**, *49*, 81–97. [CrossRef]

31. Arnon, D.I. Copper enzymes in isolated chloroplasts. Polyphenoloxidase in *Beta vulgaris*. *Plant Physiol.* **1949**, *24*, 1–15. [CrossRef]

32. Hughes, N.M.; Smith, W.K. Attenuation of incident light in *Galax urceolata* (Diapensiaceae): Concerted influence of adaxial and abaxial anthocyanic layers on photoprotection. *Am. J. Bot.* **2007**, *94*, 784–790. [CrossRef]

33. Bates, L.S.; Waldren, R.P.; Teare, I.D. Rapid determination of free proline for water-stress studies. *Plant Soil* **1973**, *39*, 205–207. [CrossRef]

34. Fong, J.; Schaffer, F.L.; Kirk, P.L. The ultramicrodetermination of glycogen in liver. A comparison of the anthrone and reducing-sugar methods. *Arch. Biochem. Biophys.* **1953**, *45*, 319–326. [CrossRef]

35. Sadasivam, S.; Manickam, A. *Biochemical Methods*; New Age International (P) Limited Publishers: New Delhi, India, 2004.

36. Bradford, K.J. A water relations analysis of seed germination rates. *Plant Physiol.* **1990**, *94*, 840–849. [CrossRef] [PubMed]

37. Heath, R.L.; Packer, L. Photoperoxidation in isolated chloroplasts. I. Kinetics and stoichiometry of fatty acid peroxidation. *Arch. Biochem. Biophys.* **1968**, *125*, 189–198. [CrossRef]

38. Sullivan, C.Y.; Ross, W.M. Selecting for drought and heat resistance in grain sorghum. In *Stress Physiology in Crop Plants*; Mussell, H., Staples., R.C., Eds.; John Wiley and Sons: New York, NY, USA, 1979; pp. 263–281.

39. Singleton, V.L.; Rossi, J.A., Jr. Colorimetry of total phenolics with phosphomolybdic-phosphotungstic acid reagents. *Am. J. Enol. Vitic.* **1965**, *16*, 144–158.

40. Zhishen, J.; Mengcheng, T.; Jianming, W. The determination of flavonoid contents in mulberry and their scavenging effects on superoxide radicals. *Food Chem.* **1999**, *64*, 555–559. [CrossRef]

41. Dhindsa, R.S.; Plumb-dhindsa, P.; Thorpe, T.A. Leaf Senescence: Correlated with increased levels of membrane permeability and lipid peroxidation, and decreased levels of superoxide dismutase and catalase. *J. Exp. Bot.* **1981**, *32*, 93–101. [CrossRef]

42. Aebi, H. Catalase in vitro. *Methods Enzymol.* **1984**, *105*, 121–126.

43. Nakano, Y.; Asada, K. Hydrogen peroxide is scavenged by ascorbate-specific peroxidase in spinach chloroplasts. *Plant Cell Physiol.* **1981**, *22*, 867–880.

44. Hori, M.; Kondo, H.; Ariyoshi, N.; Yamada, H.; Hiratsuka, A.; Watabe, T.; Oguri, K. Changes in the hepatic glutathione peroxidase redox system produced by coplanar polychlorinated biphenyls in Ah-responsive and -less-responsive strains of mice: Mechanism and implications for toxicity. *Environ. Toxicol. Pharmacol.* **1997**, *3*, 267–275. [CrossRef]

45. Ilyas, M.; Nisar, M.; Khan, N.; Hazrat, A.; Khan, A.H.; Hayat, K.; Fahad, S.; Khan, A.; Ullah, A. Drought tolerance strategies in plants: A mechanistic approach. *J. Plant Growth Regul.* **2021**, *40*, 926–944. [CrossRef]

46. Kakar, N.; Jumaa, S.H.; Redoña, E.D.; Warburton, M.L.; Reddy, K.R. Evaluating rice for salinity using pot-culture provides a systematic tolerance assessment at the seedling stage. *Rice* **2019**, *12*, 57. [CrossRef]

47. Wang, Z.; Yuan, L.I.; Xin-Ming, W.U.; Gao, H.W.; Sun, G.Z. Study on germination characteristics and drought-resistance evaluation of *Dactylis glomerata* L. under osmotic stress. *Chin. J. Grassl.* **2008**, *30*, 50–54.

48. ZhenPeng, X.U.; Wan, T.; Cai, P.; Zhang, Y.R.; Jing, Y.U.; Meng, C. Effects of PEG simulated drought stress on germination and physiological properties of *Apocynum venetum* seeds. *Chin. J. Grassl.* **2015**, *37*, 75–80.

49. Yang, Z.Y.; Zhou, B.Z.; Zhou, Y.; Ge, X.G.; Wang, X.M.; Cao, Y.H. Effects of drought stress simulated by PEG on seed germination and growth physiological characteristics of *Phyllostachys edulis*. *For. Res.* **2018**, *31*, 47–54.

50. Guo, J.M.; Liu, J.; Dong, K.H. Effect of PEG-6000 stress on seed germination of *Bothriochica ischaemum*. *Chin. J. Grassl.* **2015**, *37*, 58–62.

51. Kumar, A.; Rodrigues, V.; Verma, S.; Singh, M.; Hiremath, C.; Shanker, K.; Shukla, A.K.; Sundaresan, V. Effect of salt stress on seed germination, morphology, biochemical parameters, genomic template stability, and bioactive constituents of *Andrographis paniculata* Nees. *Acta Physiol. Plant.* **2021**, *43*, 1–14. [CrossRef]

52. Bajji, M.; Kinet, J.M.; Lutts, S. Osmotic and ionic effects of NaCl on germination, early seedling growth, and ion content of *Atriplex halimus* (Chenopodiaceae). *Can. J. Bot.* **2002**, *80*, 297–304. [CrossRef]

53. Talei, D.; Yusop, M.K.; Kadir, M.A.; Valdiani, A.; Abdullah, M.P. Response of king of bitters (*Andrographis paniculata* Nees.) seedlings to salinity stress beyond the salt tolerance threshold. *Aust. J. Crop Sci.* **2012**, *6*, 1059–1067.

54. Grieco, M.; Roustan, V.; Dermendjiev, G.; Rantala, S.; Jain, A.; Leonardelli, M.; Neumann, K.; Berger, V.; Engelmeier, D.; Bachmann, G. Adjustment of photosynthetic activity to drought and fluctuating light in wheat. *Plant. Cell Environ.* **2020**, *43*, 1484–1500. [CrossRef]

55. Shawon, R.A.; Kang, B.S.; Sang, G.L.; Kim, S.K.; Yang, G.K. Influence of drought stress on bioactive compounds, antioxidant enzymes and glucosinolate contents of Chinese cabbage (*Brassica rapa*). *Food Chem.* **2019**, *308*, 125657. [CrossRef] [PubMed]

56. Dong, T.; Duan, B.; Zhang, S.; Korpelainen, H.; Niinemets, U.; Li, C. Growth, biomass allocation and photosynthetic responses are related to intensity of root severance and soil moisture conditions in the plantation tree *Cunninghamia lanceolata*. *Tree Physiol.* **2016**, *36*, 807–817. [CrossRef] [PubMed]

57. Yu, L.; Dong, H.; Li, Z.; Han, Z.; Korpelainen, H.; Li, C. Species-specific responses to drought, salinity and their interactions in *Populus euphratica* and *P. pruinosa* seedlings. *J. Plant Ecol.* **2020**, *13*, 563–573. [CrossRef]

58. Suzuki, N.; Rivero, R.M.; Shulaev, V.; Blumwald, E.; Mittler, R. Abiotic and biotic stress combinations. *New Phytol.* **2014**, *203*, 32–43. [CrossRef]

59. Alexieva, V.; Sergiev, I.; Mapelli, S.; Karanov, E. The effect of drought and ultraviolet radiation on growth and stress markers in pea and wheat. *Plant Cell Environ.* **2001**, *24*, 1337–1344. [CrossRef]

60. Miao, Y.; Bi, Q.; Qin, H.; Zhang, X.; Tan, N. Moderate drought followed by re-watering initiates beneficial changes in the photosynthesis, biomass production and Rubiaceae-type cyclopeptides (RAs) accumulation of *Rubia yunnanensis*. *Ind. Crops Prod.* **2020**, *148*, 112284. [CrossRef]

61. Hussain, T.; Koyro, H.W.; Zhang, W.; Liu, X.; Gul, B.; Liu, X. Low salinity improves photosynthetic performance in *Panicum antidotale* under drought stress. *Front. Plant Sci.* **2020**, *11*, 1–13. [CrossRef]

62. Gitelson, A.A.; Gritz, Y.; Merzlyak, M.N. Relationships between leaf chlorophyll content and spectral reflectance and algorithms for non-destructive chlorophyll assessment in higher plant leaves. *J. Plant Physiol.* **2003**, *160*, 271–282. [CrossRef]

63. Nezhadahmadi, A.; Prodhan, Z.H.; Faruq, G. Drought tolerance in wheat. *Sci. World J.* **2013**, *2013*, 610721. [CrossRef]

64. Li, P.; Zhu, Y.; Song, X.; Song, F. Negative effects of long-term moderate salinity and short-term drought stress on the photosynthetic performance of *Hybrid Pennisetum*. *Plant Physiol. Biochem.* **2020**, *155*, 93–104. [CrossRef]

65. Suresh, K.; Beena, A.S.; Monika, A.; Archana, S. Physiological, biochemical, epigenetic and molecular analyses of wheat (*Triticum aestivum*) genotypes with contrasting salt tolerance. *Front. Plant Sci.* **2017**, *8*, 1151.

66. Martínez-Vilalta, J.; Sala, A.; Asensio, D.; Galiano, L.; Hoch, G.; Palacio, S.; Piper, F.I.; Lloret, F. Dynamics of non-structural carbohydrates in terrestrial plants: A global synthesis. *Ecol. Monogr.* **2016**, *86*, 495–516. [CrossRef]

67. Tomasella, M.; Nardini, A.; Hesse, B.D.; MacHlet, A.; Matyssek, R.; Häberle, K.H. Close to the edge: Effects of repeated severe drought on stem hydraulics and non-structural carbohydrates in European beech saplings. *Tree Physiol.* **2019**, *39*, 717–728. [CrossRef]

68. De Baerdemaeker, N.J.F.; Salomón, R.L.; De Roo, L.; Steppe, K. Sugars from woody tissue photosynthesis reduce xylem vulnerability to cavitation. *New Phytol.* **2017**, *216*, 720–727. [CrossRef]

69. Boldizsár, Á.; Simon-Sarkadi, L.; Szirtes, K.; Soltész, A.; Szalai, G.; Keyster, M.; Ludidi, N.; Galiba, G.; Kocsy, G. Nitric oxide affects salt-induced changes in free amino acid levels in maize. *J. Plant Physiol.* **2013**, *170*, 1020–1027. [CrossRef]

70. Waskiewicz, A.; Muzolf-Panek, M.; Goliński, P. *Phenolic Content Changes in Plants under Salt Stress*; Ahmad, P., Azooz, M., Prasad, M.N.V., Eds.; Springer: New York, NY, USA, 2013; pp. 283–314.

71. Bourgou, S.; Bettaieb, I.; Tounsi, M.S.; Marzouk, B. Fatty acids, essential oil, and phenolics modifications of black cumin fruit under NaCl stress conditions. *J. Agric. Food Chem.* **2010**, *58*, 12399–12406. [CrossRef]

72. Choudhury, S.; Panda, P.; Sahoo, L.; Panda, S.K. Reactive oxygen species signaling in plants under abiotic stress. *Plant Signal. Behav.* **2013**, *8*, e23681. [CrossRef]

73. Ray, R.L.; Ampim, P.A.Y.; Gao, M. Crop protection under drought stress. In *Crop Protection under Changing Climate*; Jabran, K., Florentine, S., Chauhan, B., Eds.; Springer: Cham, Switzerland, 2020; pp. 145–170.

74. Farooq, M.A.; Ali, S.; Hameed, A.; Ishaque, W.; Mahmood, K.; Iqbal, Z. Alleviation of cadmium toxicity by silicon is related to elevated photosynthesis, antioxidant enzymes; suppressed cadmium uptake and oxidative stress in cotton. *Ecotoxicol. Environ. Saf.* **2013**, *96*, 242–249. [CrossRef]

75. Song, J.; Wang, Y.; Pan, Y.; Pang, J.; Zhang, X.; Fan, J.; Zhang, Y. The influence of nitrogen availability on anatomical and physiological responses of *Populus alba* × *P. glandulosa* to drought stress. *BMC Plant Biol.* **2019**, *19*, 1–2. [CrossRef]

76. Cao, X.; Jia, J.; Zhang, C.; Li, H.; Liu, T.; Jiang, X.; Polle, A.; Peng, C.; Luo, Z. Bin Anatomical, physiological and transcriptional responses of two contrasting poplar genotypes to drought and re-watering. *Physiol. Plant.* **2014**, *151*, 480. [CrossRef]

77. Petrov, V.; Hille, J.; Mueller-Roeber, B.; Gechev, T.S. ROS-mediated abiotic stress-induced programmed cell death in plants. *Front. Plant Sci.* **2015**, *6*, 69. [CrossRef]
78. Goswami, B.; Rankawat, R.; Gadi, B.R. Physiological and antioxidative responses associated with drought tolerance of *Lasiurus sindicus* Henr. endemic to Thar desert, India. *Rev. Bras. Bot.* **2020**, *43*, 761–773. [CrossRef]
79. Zhang, C.; Shi, S.; Wang, B.; Zhao, J. Physiological and biochemical changes in different drought-tolerant alfalfa (*Medicago sativa* L.) varieties under PEG-induced drought stress. *Acta Physiol. Plant.* **2018**, *40*, 25. [CrossRef]
80. Anjum, S.A.; Ashraf, U.; Tanveer, M.; Khan, I.; Hussain, S.; Shahzad, B.; Zohaib, A.; Abbas, F.; Saleem, M.F.; Ali, I.; et al. Drought induced changes in growth, osmolyte accumulation and antioxidant metabolism of three maize hybrids. *Front. Plant Sci.* **2017**, *8*, 69. [CrossRef] [PubMed]
81. Chen, L.; Zhang, S.; Zhao, H.; Korpelainen, H.; Li, C. Sex-related adaptive responses to interaction of drought and salinity in *Populus yunnanensis*. *Plant Cell Environ.* **2010**, *33*, 1767–1778. [CrossRef] [PubMed]
82. Kirova, E.; Pecheva, D.; Simova-Stoilova, L. Drought response in winter wheat: Protection from oxidative stress and mutagenesis effect. *Acta Physiol. Plant.* **2021**, *43*, 1–11. [CrossRef]

 *antioxidants*

Article

Antioxidative and Metabolic Contribution to Salinity Stress Responses in Two Rapeseed Cultivars during the Early Seedling Stage

Ali Mahmoud El-Badri [1,2], Maria Batool [1], Ibrahim A. A. Mohamed [1,3], Zongkai Wang [1], Ahmed Khatab [1,2], Ahmed Sherif [1,2], Hasan Ahmad [4], Mohammad Nauman Khan [1], Hamada Mohamed Hassan [2], Ibrahim M. Elrewainy [2], Jie Kuai [1], Guangsheng Zhou [1] and Bo Wang [1,*]

[1] MOA Key Laboratory of Crop Ecophysiology and Farming System in the Middle Reaches of the Yangtze River, College of Plant Science & Technology, Huazhong Agricultural University, Wuhan 430070, China; alyelbadry@webmail.hzau.edu.cn (A.M.E.-B.); maria.batool@webmail.hzau.edu.cn (M.B.); iaa04@fayoum.edu.eg (I.A.A.M.); wangzongkai@webmail.hzau.edu.cn (Z.W.); ahmedkhatab@webmail.hzau.edu.cn (A.K.); Sherif@webmail.hzau.edu (A.S.); nauman@webmail.hzau.edu.cn (M.N.K.); 105042014160@mail.hzau.edu.cn (J.K.); zhougs@mail.hzau.edu.cn (G.Z.)
[2] Field Crops Research Institute, Agricultural Research Center (ARC), Giza 12619, Egypt; hhassan997@yahoo.com (H.M.H.); imelrewainy@gmail.com (I.M.E.)
[3] Botany Department, Faculty of Agriculture, Fayoum University, Fayoum 63514, Egypt
[4] National Gene Bank, Agricultural Research Center (ARC), Giza 12619, Egypt; hasanngb@webmail.hzau.edu.cn
*   Correspondence: wangbo@mail.hzau.edu.cn; Tel.:+86-027-8728-2130 or +86-137-0719-2880

Citation: El-Badri, A.M.; Batool, M.; A. A. Mohamed, I.; Wang, Z.; Khatab, A.; Sherif, A.; Ahmad, H.; Khan, M.N.; Hassan, H.M.; Elrewainy, I.M.; et al. Antioxidative and Metabolic Contribution to Salinity Stress Responses in Two Rapeseed Cultivars during the Early Seedling Stage. *Antioxidants* **2021**, *10*, 1227. https://doi.org/10.3390/antiox10081227

Academic Editors: Masayuki Fujita and Mirza Hasanuzzaman

Received: 6 June 2021
Accepted: 16 July 2021
Published: 30 July 2021

Publisher's Note: MDPI stays neutral with regard to jurisdictional claims in published maps and institutional affiliations.

Abstract: Measuring metabolite patterns and antioxidant ability is vital to understanding the physiological and molecular responses of plants under salinity. A morphological analysis of five rapeseed cultivars showed that Yangyou 9 and Zhongshuang 11 were the most salt-tolerant and -sensitive, respectively. In Yangyou 9, the reactive oxygen species (ROS) level and malondialdehyde (MDA) content were minimized by the activation of antioxidant enzymes such as superoxide dismutase (SOD), peroxidase (POD), catalase (CAT), and ascorbate peroxidase (APX) for scavenging of over-accumulated ROS under salinity stress. Furthermore, Yangyou 9 showed a significantly higher positive correlation with photosynthetic pigments, osmolyte accumulation, and an adjusted Na^+/K^+ ratio to improve salt tolerance compared to Zhongshuang 11. Out of 332 compounds identified in the metabolic profile, 225 metabolites were filtrated according to $p < 0.05$, and 47 metabolites responded to salt stress within tolerant and sensitive cultivars during the studied time, whereas 16 and 9 metabolic compounds accumulated during 12 and 24 h, respectively, in Yangyou 9 after being sown in salt treatment, including fatty acids, amino acids, and flavonoids. These metabolites are relevant to metabolic pathways (amino acid, sucrose, flavonoid metabolism, and tricarboxylic acid cycle (TCA), which accumulated as a response to salinity stress. Thus, Yangyou 9, as a tolerant cultivar, showed improved antioxidant enzyme activity and higher metabolite accumulation, which enhances its tolerance against salinity. This work aids in elucidating the essential cellular metabolic changes in response to salt stress in rapeseed cultivars during seed germination. Meanwhile, the identified metabolites can act as biomarkers to characterize plant performance in breeding programs under salt stress. This comprehensive study of the metabolomics and antioxidant activities of *Brassica napus* L. during the early seedling stage is of great reference value for plant breeders to develop salt-tolerant rapeseed cultivars.

Keywords: *Brassica napus*; salinity stress; antioxidant enzymes; osmolytes; ROS; metabolites

1. Introduction

For thousands of years, rapeseed (*Brassica species*) has been planted for its high production of edible oils and its economic and significant nutritional value [1]. Canada is the largest rapeseed producer, followed by China and India [2]. Different *Brassica* species are grown or adapted to different climates; in particular, brassica crops are extensively cultivated in arid and semi-arid regions, where the accumulation of salts negatively affects germination, early seedling growth, and productivity [3].

Climate changes such as drought, salinity, and temperature are a threat to food production by limiting crop productivity [4]. Salinity is one of the main abiotic stresses that negatively affects agricultural crop productivity, by impairing germination, plant vigor, and crop yield [5]. All over the world, salinity causes damage to more than 20% of cultivated land in addition to 33% of irrigated agricultural land. Every year, about 1.5 million hectares are not cultivated due to high salinity levels in the soil. There are many reasons for increased salinity, including low precipitation, high surface evaporation, weathering of native rocks, and poor cultural practices [6]. It is expected that, by 2050, more than 50% of agricultural land could be damaged by salinity [7]. During the plant life cycle, seed germination and seedling vigor are complicated and critical phenomena that are highly affected by various environmental stresses, especially salinity [8]. Osmotic stress and ion toxicity are among the main reasons behind the restriction of plant growth in salinized soils, due to the higher levels of salt in the soil, which restricts plants from extracting water from the soil and inside the plants themselves, which then causes nutritional imbalance and oxidative stress [9,10]. Additionally, Na^+ can replace ions, particularly K^+, in key enzymatic reactions, which affects cytosol and organelle metabolism due to the Na^+/K^+ ratio, which is critical for cell performance under salinity [11,12]. Tolerant plants use ions as an alternative to organic compounds for osmotic modification, which requires the synthesis of more energy (ATP) [13].

Moreover, osmolytes also protect plant cells, as they act as antioxidants, buffer the cellular redox potential, stabilize membranes and macromolecules, and function as immediate sources of energy during recovery from stress [14], which maintains the functional balance of the cell [15]. Furthermore, defense through protective enzymes superoxide dismutase, peroxidase, and catalase against salt-induced ROS over-generation and membrane lipid peroxidation is attributed to the protection of cellular membranes, which leads to salt tolerance [16].

Under salinity stress, plants adapt by initiating multiple moleculeare and physiochemical changes, which results in modifications to metabolic pathways to reach a new homeostatic equilibrium [17]. In *Brassica napus* L., higher salt stress decreases the germination parameters and biomass during the early seedling stage [18]. Additionally, in *Brassica* spp., salinity decreases nutrient absorption [19], electrolyte leakage, biomass, RWC [18,20], root length, total chlorophyll content, hypocotyl, and leaf growth with increasing POD activity and IAA oxidase [21].

Furthermore, salinity is assumed to activate the alternative gene expression patterns, which may synthesize, degrade, or decorate the metabolites from related pathways. The process is attributed to retrotransposon mobilization over-inducing salt-induced transcription factors, binding the promoter of special retrotransposons [14]. Moreover, retrotransposition bursts were reported to be critical for the reformation of gene regularity networks and for creating new metabolite patterns to tolerate and adapt to salinity stress [12]. Measuring the metabolite patterns is very important to understand the physiological and molecular responses of plants under salinity in order to illustrate the functions of genes as vital tools in functional genomics and systems biology to develop new breeding and selection strategies to improve salt tolerance in crops [6].

In the long term, metabolic disturbances are beneficial to plants, as plants use them as adaptive mechanisms, but an imbalance in Na^+ and Cl^- levels in the metabolic compartments becomes toxic. The isolation of ions in the vacuoles expresses one mechanism for avoiding Na^+ and Cl^- toxicity, and the levels in the leaves increase over time [22]. Previous studies were conducted on the metabolic contribution of the stress responses in rice, maize, wheat, and barley [23]. During grain filling in wheat, heat stress was shown to increase sucrose and reduced sugar phosphates and starch [24], and reducing sugar and sucrose was shown to cause reduced starch and rice seed weight [25].

During plant growth, several changes have been noted, such as metabolite changes, which can be correlated with physiological and environmental responses and genetic perturbations [23]. The significance of metabolite accumulation, such as amino acids, indicates general cellular damage in salt-sensitive cultivars. In contrast, salt-tolerant barley cultivars were shown to have an accumulation of organic acids, polyols, hexose phosphates, sugars, and tricarboxylic acid cycle (TCA) intermediates under salt stress [26].

Identifying the key metabolites and gaining a comprehensive understanding of salt-related antioxidant responses can improve the selection of desirable phenotypes of salt tolerance. Therefore, we aimed to identify metabolites that could work as biomarkers for tolerant and sensitive rapeseed cultivars that are differently adapted to salinity stress by appllying metabolite profiling and examining the diverse salinity tolerance ability of five common rapeseed cultivars through changes in their morpho-physiological traits. Our investigation contributes to the understanding of various metabolic components and morpho-physiological attributes in salinity tolerance during the early seedling stage, which can be used for further analysis.

2. Materials and Methods

2.1. Determination of Optimum Salt Stress Concentration

Five cultivars with diverse genetic backgrounds developed at the Oil Crops Research Institute, Chinese Academy of Agricultural Science, Huazhong Agriculture University, Wuhan, China, with $\geq$90% seed viability (Yangza 11, Zhongshuang 11, Huayouza 62, Fengyou 520, and Yangyou 9), were used in this study to determine the optimum salt concentration among concentrations of 50, 100, 150, and 200 mM L^{-1} NaCl, along with a control group (CK). Seeds were sterilized with 70% ethanol for 5 min and washed with ddH$_2$O 3–5 times. The sterilized seeds were dried with blotted paper and kept at room temperature for complete redrying. Fifty uniform and healthy seeds (to decrease errors in seed germination and seedling vigor) were selected from each of the 5 cultivars and sown in germination boxes (3 technical replications with 4 biological replications) with a triple layer of germination paper containing 15 mL of NaCl solution (50, 100, 150, and 200 mM L^{-1} NaCl) or 15 mL ddH$_2$O (CK). The seeds were cultured under optimal conditions (day/night temperature at $25 \pm 1/20 \pm 1\,^\circ$C) with 12 h light (13,000 lx) and 12 h dark (HP250GS-C, Ruihua Instrument and Equipment Co., Ltd., Wuhan, China). Seed germination was recorded daily and seeds were considered to be germinated when the radical length was $\geq$ 2 mm. All samples were collected on the seventh day of sowing in 3 replicates, then kept at $-80\,^\circ$C for further analysis.

2.2. Phenotypic Trait Measurement

After the seventh day of germination, the final germination percentage (FG%), germination rate (GR), vigor index I (VI (I)), and vigor index II (VI (II)) were measured according to the equations reported in [27] as follows:

$$\text{FG\%} = (n/n_t) \times 100,$$

where n is the number of germinated seeds at the end of the experiment and n_t is the total number of seeds.

$$\text{GR} = (a/1) + (b - a/2) + (c - b/3) + \ldots + (n - n - 1/n)$$

where $a, b, c, \ldots, n$ are numbers of germinated seeds after 1, 2, 3, $\ldots$, N days from the start of imbibition.

$$\text{VI (I)} = \text{FG\%} \times \text{seedling length;}$$

$$\text{VI (II)} = \text{FG\%} \times \text{seedling fresh weight;}$$

$$\text{Seedling fresh weight stress index} = \frac{\text{Seedling fresh weight stressed}}{\text{Seedling fresh weight nonstressed}} \times 100;$$

$$\text{Seedling length stress index} = \frac{\text{Seedling length stressed}}{\text{Seedling length nonstressed}} \times 100. \tag{1}$$

After 7 days of treatment, seedlings of all cultivars were harvested, and 50 random seedlings were used to measure shoot and root length by ImageJ software. Then, the fresh and dry biomass were measured from the same seedlings after removing surface water by blotting, using 10 seedlings in each replicate according to [28].

2.3. Estimation of Photosynthetic Pigments, Total Soluble Sugar, and Protein Contents

Chlorophyll a (chl a), chlorophyll b (chl b), and carotenoids in fresh leaves were determined after 7 days of treatment. First, 0.5 g of fresh leaf tissue was ground and mixed with 10 mL of 80% acetone, then incubated in the dark at room temperature overnight. The absorption values of the extract at 665, 649, and 470 nm were measured using an ultraviolet spectrophotometer (UV-2100, UNIC, Shanghai, China). The contents of chl a, chl b, and carotenoids were measured according to the equations presented in [29]. Total soluble sugar content was analyzed using the anthrone sulfuric acid method. The absorbance of the samples and the standard solution was determined at 620 nm, while protein content was assessed by using bovine serum albumin as a standard, and the absorbance was estimated at 595 nm, accordingly [30].

2.4. Malondialdehyde Analysis and Proline Content

Malondialdehyde (MDA) content represents lipid peroxidation, which was measured according to [31]. In this investigation, 0.5 g of fresh shoots was homogenized in 5 mL of 10% TCA and 0.65% of 2-thiobarbituric acid (TBA). Afterward, the mixture was heated for 1 h at 100 °C and cooled at room temperature, followed by centrifugation at 10,000 rpm for 10 min. The absorbance was quantified at 450, 532, and 600 nm. Proline content was estimated by the indene triketone method [32]. Then, 0.5 g of fresh shoots was ground by a rapid automatic sample grinding instrument (JXFSTPRP24, Shanghai Jingxin Industrial Development, Shanghai, China), followed by digestion in 5 mL of 3% aqueous sulfosalicylic acid (3 g/100 mL ddH$_2$O). Afterward, 2 mL of extract solution was mixed with 2 mL ninhydrin reagent and 2 mL glacial acetic acid, then boiled at 100 °C for 60 min, and the reaction was stopped by placing in an ice bath for 5 min. The mixture was extracted with 4 mL toluene and mixed vigorously using a vortex for 15–20 s, then cooled at room temperature. Free toluene was measured at 520 nm with a spectrophotometer (Beckman Coulter Inc., Fullerton, CA, USA).

2.5. Histochemical Analysis of O_2^- and H_2O_2

The accumulated H_2O_2 and O_2^- were recognized by staining leaves in 3,3-diaminobenzidine (DAB) and nitro blue tetrazolium (NBT) solution, respectively. Both dyes, with a weight of 0.025 g, were dissolved in 50 mL phosphate-buffered saline (PBS) and allowed to incubate for 2 h with slow shaking. Using a high-powered microscope attached to a high-resolution digital camera (Leica DM—2500), DAB and NBT stained leaves were photographed according to [33], and images were quantified using ImageJ (http://www.imagej.nih.gov/ij/) according to [34].

2.6. Measurement of Antioxidant Enzyme Activity

To determine the enzyme activity, leaf samples were ground using a rapid automatic sample grinding instrument (JXFSTPRP24, Shanghai Jingxin Industrial Development, Shanghai, China), and 0.5 g FW was homogenized in 8 mL of 50 mM potassium phosphate buffer (PPB) (pH 7.8) under cooling conditions. This homogenized solution was centrifuged at 10,000 rpm for 20 min at 4 °C, and a crude enzyme extract was obtained to measure SOD, POD, APX, and CAT according to [35,36].

Superoxide dismutase (SOD; EC 1.15.1.1) activity was determined by inhibiting photochemical reduction by nitro blue tetrazolium (NBT). The reaction mixture comprised 50 mM PPB (pH 7.8), 13 mM methionine, 75 mM NBT, 2 mM riboflavin, 0.1 mM EDTA, and 0.1 mL of enzyme extract in a 3 mL volume. One unit of SOD activity was measured as the amount of enzyme required to cause 50% inhibition of NBT reduction measured at 560 nm.

To assay peroxidase (POD; EC 1.11.1.7) activity, 0.1 mL enzyme extract was mixed with 50 mM PPB (pH 7.0), 1% (m/v) guaiacol, and 0.4% (v/v) H_2O_2. The absorbance was quantified at a wavelength of 470 nm.

The assay for ascorbate peroxidase (APX, EC 1.11.1.11) was done by a reaction mixture (3 mL) containing 100 mM phosphate (pH 7), 0.1 mM EDTA-Na2, 0.3 mM ascorbic acid, 0.06 mM H_2O_2, and 0.1 mL enzyme extract. The change in absorption was read at 290 nm for 30 s after the addition of H_2O_2.

The method to measure catalase (CAT; EC 1.11.1.6) activity used H_2O_2 (extinction co-efficient 39.4 mM^{-1} cm^{-1}), 3 mL reaction mixture containing 50 mM PPB (pH 7.0), 2 mM EDTA-Na2, 10 mM H_2O_2, and 0.1 mL enzyme extract measured at 240 nm.

2.7. Determination of Na$^+$ and K$^+$ in Leaves

The determination of Na$^+$ and K$^+$ in rapeseed seedlings was carried out by following the method of [37] using a flame photometer (FP6431,Shanghai Yidian Analysis Instrument Co., Ltd. Shanghai, China). The dried powder of shoot samples (0.1 g) was digested with 2 mL of H_2SO_4-H_2O_2 mixture, filtered, and then diluted with ddH$_2$O. The acid mixture (2 mL) containing ddH$_2$O was considered blank. A standard curve of Na$^+$ and K$^+$ (10–100 µg mL^{-1}) was used as a reference.

2.8. Metabolite Extraction and Detection

Seeds were collected after 12 and 24 h after being sown under salt treatment. Freeze-dried seeds were crushed using a mixer mill (MM 400, Retsch GmbH, Haan, Germany), and the powder was weighed and extracted overnight at 4 °C with 1.0 mL of 70% aqueous methanol containing 0.1 mg L^{-1} lidocaine for water-soluble metabolites. Following centrifugation at 10,000 rcf for 10 min, the extracts were absorbed (CNWBOND Carbon-GCB SPE Cartridge, 250 mg, 3 mL; ANPEL, Shanghai, China, www.anpel.com.cn/cnw) and filtrated (SCAA-104, 0.22 mm pore size; ANPEL http://www.anpel.com.cn/) before LC-MS analysis. Samples were subjected to metabolite detection using high-performance liquid chromatography (HPLC) and high-performance liquid chromatography mass spectrometry (LC-MS/MS) [38]. The relative contents of each of these 332 identified metabolites were quantified using the scheduled multiple reaction monitoring (sMRM) method described previously [39]. The sMRM algorithm was used with an MRM detection window of 90 s and a target scan time of 1.0 s using Analyst 1.5 software. Given that biological variance is considerably higher than technical variance, we chose not to carry out technical replication. The analytical conditions were taken from a previous report [40] with minor modifications as follows: HPLC: column, shim-pack VP-ODS C18 (pore size 5.0 µm, length 2 × 150 mm); solvent system, water (0.04% acetic acid): acetonitrile (0.04% acetic acid); gradient program, 95:5 v/v at 0 min, ramping to 0:100 v/v at 15 min, 0:100 v/v at 15–17 min, 95:5 v/v at 17–17.1 min, 95:5 v/v at 17.1–22 min; flow rate, 0.35 mL min^{-1}; temperature, 40 °C; injection volume: 5 µL for one run.

2.9. Metabolite Analysis and Identification

The metabolite fold changes (FCs) were calculated and volcano plots were generated using MetaboAnalyst [23]. An increase in FC (ratio of metabolites in salt-stressed samples to control samples) was considered significant when FC ≥ 1 and the concentration of the metabolite significantly different ($p < 0.05$) between control and stressed plants. Metabolites were identified by comparing with the standards (wherein the "identifications" column was labeled as "standard") or taken from previous experiments (labeled as "putative") (supplementary data file 2) [41]. The metabolic pathway map was constructed based on the relevant literature and the KEGG database (https://www.kegg.jp/kegg-bin/show_pathway?161345586717248/vvi01100.args, accessed on 1 March 2021).

2.10. Statistical Analysis

The phenotypic data collected from the experiment were analyzed statistically. Two-way analysis of variance (ANOVA) of germination and seedling traits for all accessions was conducted using the Statistix 8.1 software package. The significance of differences between groups was further validated and determined by Duncan's multiple range test (DMRT) at a significance level of $p < 0.05$. The graphical representation was constructed using GraphPad prism (V: 8.0.1) (RStudio software, San Diego, CA, USA). The standard error is mentioned in the figures.

3. Results

3.1. Impact of NaCl Treatment on Germination Parameters, Phenotypic Appearance Traits, and Vegetative Biomass of Five Rapeseed Cultivars

To estimate the different responses of five common rapeseed cultivars to salinity stress during the seed germination and early seedling growth stage, we investigated the germination and growth parameters under various concentrations of NaCl (0, 50, 100, 150, and 200 mM L^{-1}) during 7-day treatments. Based on the morphological analysis (Table S1, Figures S1–S3), we used 150 mM NaCl to complete the study and evaluate the salt influence on salt-tolerant and -sensitive rapeseed cultivars using physiochemical parameters. Among the studied cultivars, Zhongshuang 11 displayed the lowest growth rate (Figure 1b), while Yangyou 9 showed higher growth (Figure 1e) compared to other cultivars with 150 and 200 mM of NaCl.

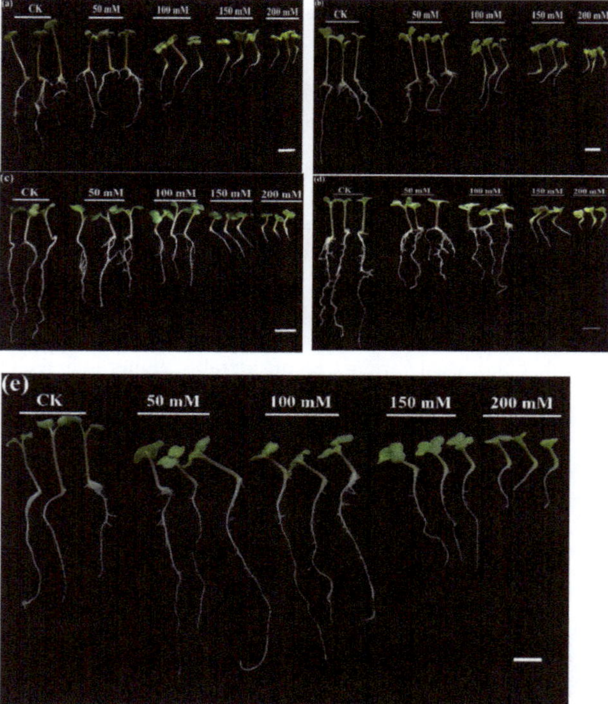

Figure 1. Impact of NaCl treatment on seedling growth of (**a**) Yangza 11, (**b**) Zhongshuang 11, (**c**) Huayouza 62, (**d**) Fengyou 520, and (**e**) Yangyou 9 rapeseed cultivars during the early seedling stage. Scale bar: 1 cm.

3.2. Alterations in Photosynthetic Pigments under Salt Stress

Salt stress caused a marked dose-dependent decline in photosynthetic pigments (chl a, chl b, and carotenoids) in both rapeseed cultivars. The response of the two cultivars with regard to photosynthetic pigments was not the same under 150 mM NaCl: the salt-induced decrease in photosynthetic pigments was 36.05 and 38.88% (chl a), and 39.26 and 39.77% (chl b) in Yangyou 9 and Zhongshuang 11, respectively (Figure 2a,b). A similar trend was observed for total chlorophyll in both cultivars; in Yangyou 9 and Zhongshuang 11, it decreased by 38.06 and 39.42%, respectively, under 150 mM NaCl versus normal conditions (CK) (Figure 2c). Additionally, salt caused significant perturbations in carotenoids in Yangyou 9 and Zhongshuang 11, which were reduced by 40.58 and 34.89%, respectively, under stress conditions versus CK (Figure 2d).

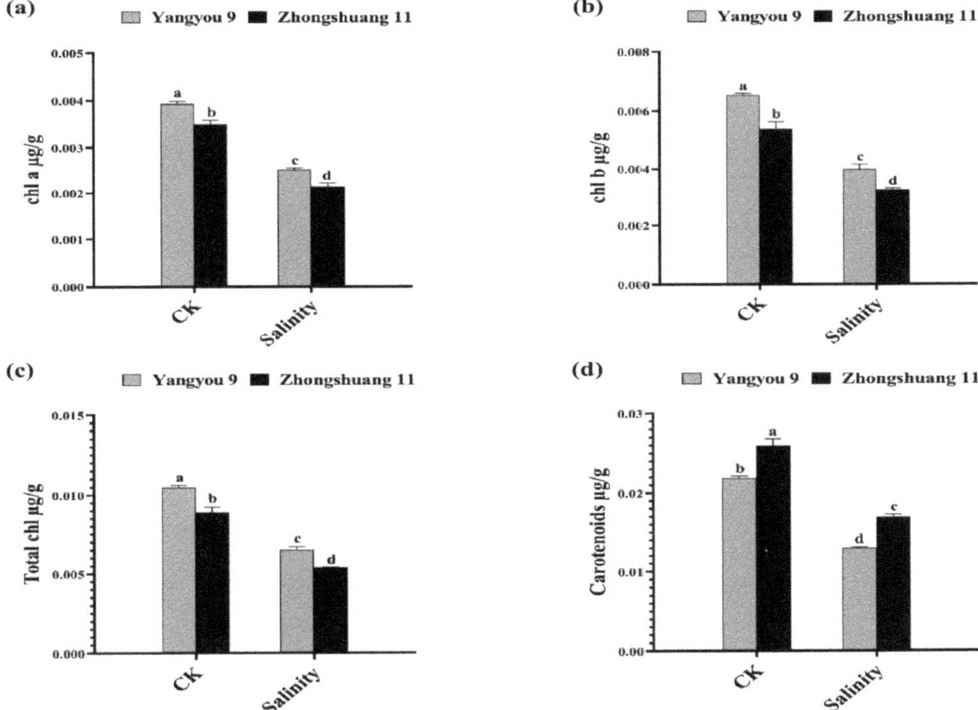

Figure 2. Modifications in (**a**) chlorophyll a, (**b**) chlorophyll b, (**c**) total chlorophyll, and (**d**) carotenoid content under normal and salt conditions induced by 150 mM NaCl. Bars represent ±SE of three replicates. Letters (a, b, c and d) on vertical bars represent significant differences between cultivars and treatments according to Duncan's multiple range test (DMRT) at $p < 0.05$.

3.3. Alterations in Total Soluble Sugar, Total Soluble Protein, MDA, and Proline Content under Salt Stress

Total soluble sugar (TSS) and protein (TSP) are the most significant osmolytes that actively participate in osmoregulation under stress conditions. Unstressed Zhongshuang 11 seedlings showed higher TSS and TSP content of 5.88 and 13.69 mg/g, while Yangyou 9 recorded values of 5.01 and 12.09 mg/g, respectively. However, salinity stress increased TSS and TSP content by 40.68 and 89.04% (Yangyou 9) and 43.96 and 88.89% (Zhongshuang 11), respectively, versus CK (Figure 3a,b).

On the other hand, under normal growth conditions (CK), Zhongshuang 11 showed slightly higher MDA and percent of proline content (0.0009 µmol g^{-1} FW and 0.219%) compared to Yangyou 9 (0.0004 µmol g^{-1} FW and 0.200%). Notably, salt stress increased MDA and proline content by 123.7 and 71.35% (Yangyou 9) and 201.6 and 108.4% (Zhongshuang 11), respectively (Figure 3c,d).

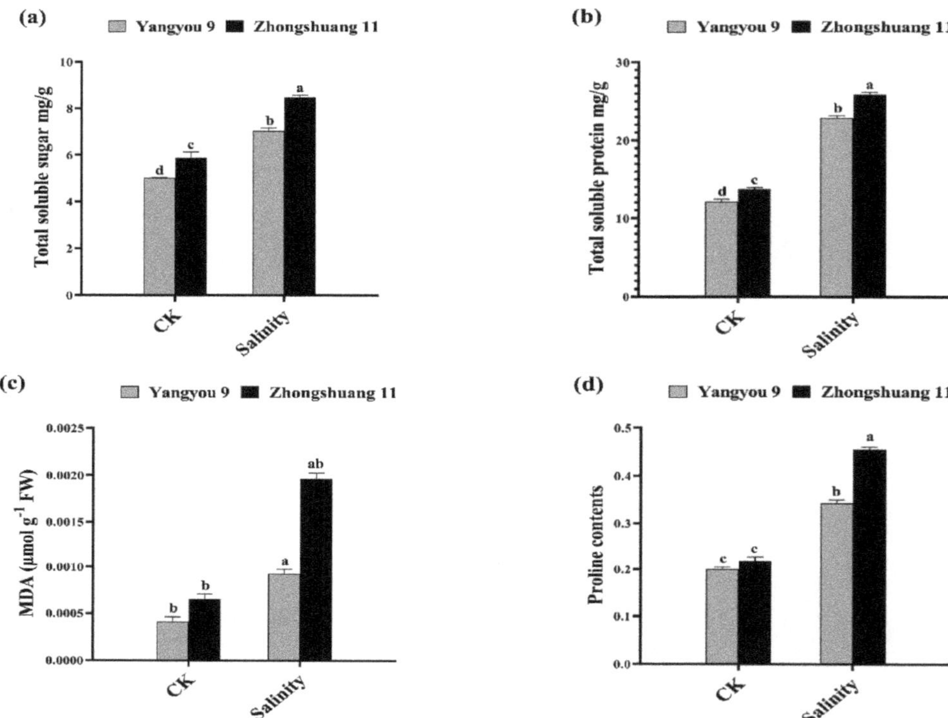

Figure 3. Modifications in (**a**) total soluble sugar, (**b**) total soluble protein, (**c**) MDA, and (**d**) proline content under normal and salt conditions induced by 150 mM NaCl. Bars represent ±SE of three replicates. Letters (a, b, c and d) on vertical bars represent significant differences between cultivars and treatments according to Duncan's multiple range test (DMRT) at $p < 0.05$.

3.4. Accumulation of $O_2{}^-$ and H_2O_2 under Salt Stress in Yangyou 9 and Zhongshuang 11

After 7 days of salt treatment, H_2O_2 and $O_2{}^-$ accumulation was examined in leaves of rapeseed seedlings. Histochemical detection of ROS ($O_2{}^-$ and H_2O_2) using NBT and DAB staining revealed that seedlings subjected to salt stress accumulated larger amounts of ROS. Our results show weaker staining in Yangyou 9 leaves (tolerant cultivar) for both $O_2{}^-$ and H_2O_2 under stress compared with Zhongshuang 11 (sensitive cultivar), with the salinized leaves of the latter being more damaged than leaves of Yangyou 9 seedlings (Figure 4a,b). Additionally, DAB and NBT intensity was higher under salt stress relative to control (%), which was higher in Zhongshuang 11 (190.4 and 147.8%, respectively) as compared to Yangyou 9 (161.1 and 134.2%) (Figure 4c,d).

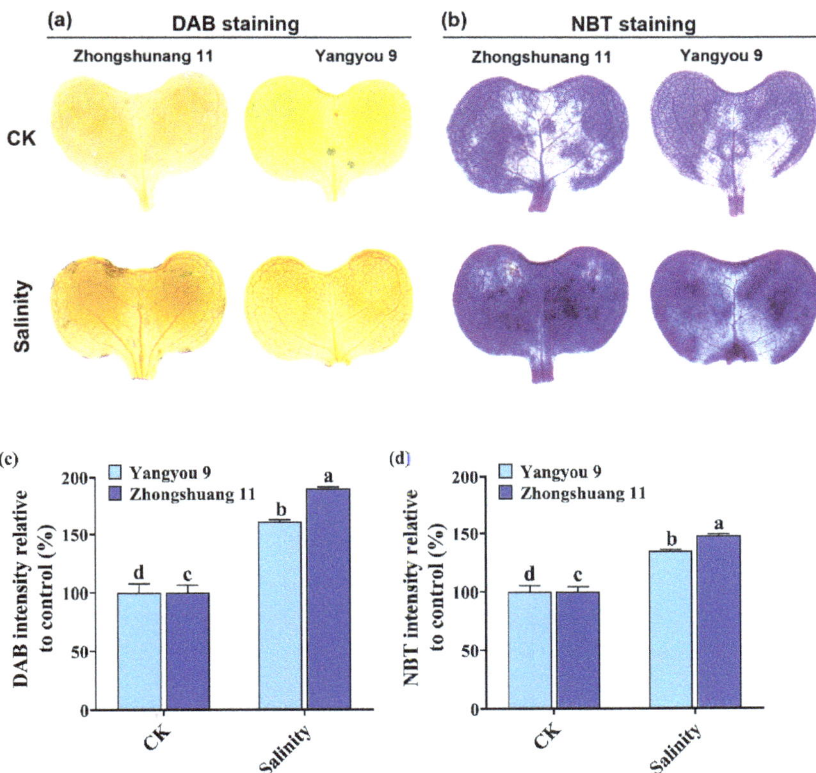

Figure 4. ROS accumulation in Yangyou 9 and Zhongshuang 11 leaves with (**a**) 3, 3-diaminobenzidine (DAB), (**b**) nitro blue tetrazolium (NBT), (**c**) DAB intensity relative to control (%), and (**d**) NBT intensity relative to control (%) under NaCl (150 mM). Bars represent ±SE of three replicates. Letters (a, b, c and d) on vertical bars represent significant differences between cultivars and treatments according to Duncan's multiple range test (DMRT) at $p < 0.05$.

3.5. Alterations in Antioxidant Enzyme Activity under Salt Stress

Adding NaCl to the growth medium resulted in a marked change in antioxidant enzyme activity in both rapeseed cultivars (Yangyou 9 and Zhongshuang 11). Salt stress caused marked dose-dependent changes in antioxidant enzyme activity (SOD, POD, APX, and CAT). Under normal conditions, Yangyou 9 recorded values of 1464 μg^{-1} FW, 601.2 $\mu min^{-1}\ g^{-1}$ FW, 0.058 $\mu min^{-1}\ g^{-1}$ FW, and 535.3 $\mu min^{-1}\ g^{-1}$ FW, while Zhongshuang 11 recorded values of 1400 μg^{-1} FW, 664.6 $\mu min^{-1}\ g^{-1}$ FW, 0.087 $\mu min^{-1}\ g^{-1}$ FW, and 552.6 $\mu min^{-1}\ g^{-1}$ FW, on SOD, POD, APX, and CAT, respectively (Figure 5a–d). The enzyme response of the two cultivars was not the same under NaCl stress: salt increased SOD, POD, and APX by 15.27, 45.46, and 214.8% in Yangyou 9 and by 13.96, 38.21, and 191.1% in Zhongshuang 11, respectively. Meanwhile, CAT activity decreased under salinity stress by 53.40 and 58.75% in Yangyou 9 and Zhongshuang 11, respectively, versus CK (Figure 5a–d).

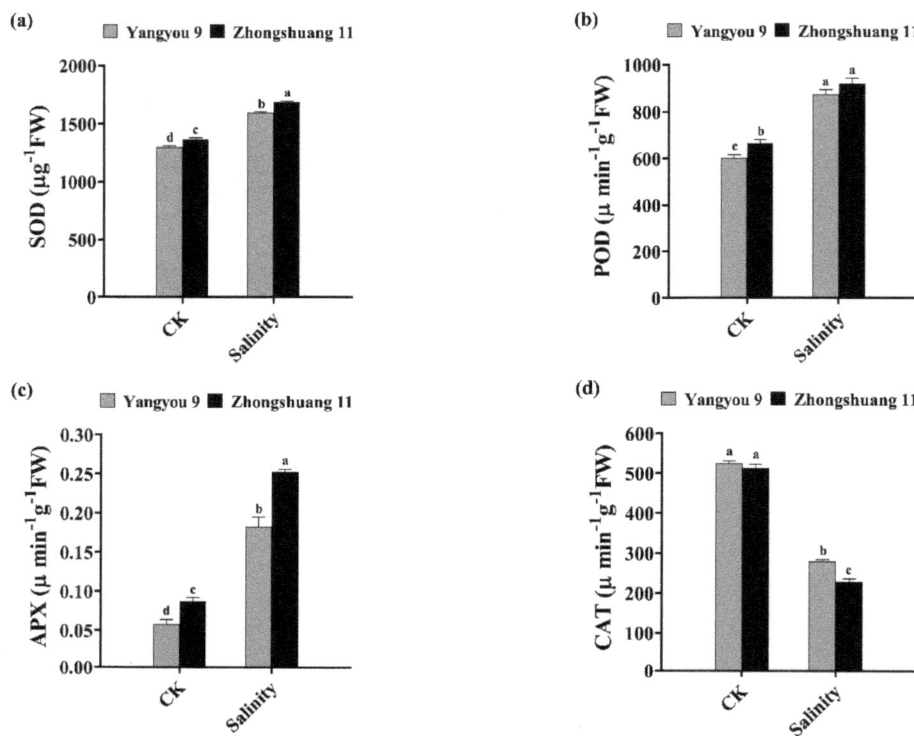

Figure 5. Modifications in (**a**) superoxidase (SOD), (**b**) peroxidase (POD), (**c**) ascorbate peroxidase (APX), and (**d**) catalase (CAT) activity under normal and salt conditions induced by 150 mM NaCl on fresh samples. Bars represent ±SE of three replicates. Letters (a, b, c and d) on vertical bars represent significant differences between cultivars and treatments according to Duncan's multiple range test (DMRT) at $p < 0.05$.

3.6. Impact of NaCl on Na⁺, K⁺, and Na⁺/K⁺ Ratio in Shoots

Under salinity stress, Na^+ uptake was decreased, and K^+ uptake was increased in the tolerant cultivar (Yangyou 9) compared to the sensitive cultivar (Zhongshuang 11). Under normal conditions, Na^+ content decreased by 24.95% and K^+ content increased by 16.39% in the Yangyou 9 versus Zhongshuang 11 shoots. Meanwhile, unstressed Zhongshuang 11 seedlings showed higher Na^+ content of 48.76 mg/g and lower K^+ content of 4.87 mg/g, while Yangyou 9 recorded values of 30.47 and 6.83 mg/g, respectively. On the other hand, the Na^+/K^+ ratio in Yangyou 9 shoots decreased by 35.52% (normal conditions) and 55.48% (stress conditions) compared to Zhongshuang 11 (Table 1).

Table 1. Na^+, K^+, and Na^+/K^+ ratio in shoots under salt stress during the early seedling stage.

Traits	Na⁺ (mg/g)		K⁺ (mg/g)		Na⁺/K⁺ (mg/g)	
Cultivars	Yangyou 9	Zhongshuang 11	Yangyou 9	Zhongshuang 11	Yangyou 9	Zhongshuang 11
CK	3.82 ± 0.21 [c]	5.09 ± 0.25 [c]	7.88 ± 0.30 [a]	6.77 ± 0.21 [ab]	0.48 ± 0.05 [c]	0.75 ± 0.06 [c]
NaCl	30.47 ± 0.55 [b]	48.76 ± 0.61 [a]	6.83 ± 0.41 [a]	4.87 ± 0.34 [b]	4.46 ± 0.02 [b]	10.01 ± 0.07 [a]

Data are the mean (± SE) of three replicates. Letters (a, b and c) on vertical bars represent significant differences between cultivars and treatments according to Duncan's multiple range test (DMRT) at $p < 0.05$.

3.7. Relationships and Variation among Growth and Biochemical Attributes of Two Rapeseed Cultivars

The score and loading plots of principal component analysis (PCA) were used to evaluate the performance of two *B. napus* cultivars under salt stress (150 mM NaCl). All 24 traits were loaded into two major principal components, PC1 (Dim1) and PC2 (Dim2), which showed a cumulative variance of about 96.9% in the dataset, where PC1 explained 76.6% of variation and PC2 revealed the difference of 20.3%, indicating variation of salt treatment applied on rapeseed. The distribution of the two cultivars displayed a clear signal of salinity stress, indicating significant effects on the studied characteristics of rapeseed. Specifically, Yangyou 9 was more displaced than Zhongshuang 11 under salt conditions, indicating that Yangyou 9 could alleviate the salt toxicity on the seeds and enhance their germination and early seedling growth (Figure 6a).

The PCA loading plot shows clear visualization and variation of the studied growth-related parameters. The members of the first group of variables with PC1 (ShFW, proline, SOD, TSS, TSP, Na$^+$, POD, APX, Na$^+$/K$^+$, and MDA) are positively correlated with each other but negatively correlated with chl, ShL, RL, RFW, GR, VI (I), VI (II), FG%, and GR. In contrast, a positive correlation is noticed in the remaining attributes, aligned with PC2: chl, GR, VI (I), RFW, ShL, and RL (Figure 6b).

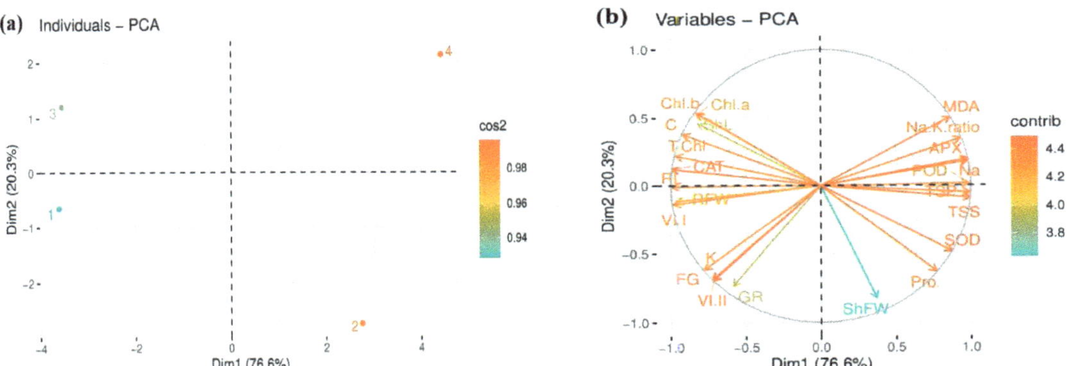

Figure 6. Principal component analysis (PCA) of salt treatment and relationships of variable traits in two rapeseed cultivars: (**a**) PCA score plot of salt treatment on rapeseed seedlings, and (**b**) PCA loading plot of PC1 and PC2 of examined variable traits; circles indicate most correlated variables. Score plot represents separation of treatments as (1) Ck and (2) salinity treatment for Yangyou 9, and (3) CK and (4) salinity treatment for Zhongshuang 11. Tested variables include final germination (FG %); germination rate (GR); vigor index I (VI (I)); vigor index II (VI (II)); shoot length (ShL); root length (RL); shoot fresh weight (ShFW); root fresh weight (RFW); chlorophyll a (Chl a); chlorophyll b (Chl b); total chlorophyll (T. Chl); carotenoid (C) content; total soluble sugar (TSS); total soluble protein (TSP); proline (pro. %); malondialdehyde (MDA) content; sodium ions (Na$^+$); potassium ions (K$^+$); superoxidase (SOD) activity; peroxidase (POD) activity; ascorbate peroxidase (APX) activity, and catalase (CAT) activity.

3.8. Metabolic Changes of Yangyou 9 (T) and Zhengsheng 11 (S) in Response to Salt Stress

For further clarification, the physiological mechanisms of salt tolerance underlying the salt-tolerant Yangyou 9 and Zhongshuang 11 metabolic changes during the early germination stage (12 and 24 h after sowing) were studied under salinity stress (150 mM NaCl) compared to control (no NaCl). Using a metabolomic approach based on HPLC-QQQ mass spectrometry, 332 compounds were detected. Statistical analysis was conducted to minimize the data complexity and get significant differences. The metabolites list came from [41]. A total of 225 metabolites were filtrated according to $p < 0.05$, as shown in Table S2. According to heatmap, principal component analysis (PCA), and cluster-based analysis, there were clear distinctions between samples within treatments and genotypes. All replications within each treatment clustered together, indicating that the changes induced by salinity were hierarchically greater than biological and technical variability (Figure S4a–d).

To further clarify the differential metabolites between Yangyou 9 (T) and Zhengsheng 11 (S) during 12 and 24 h of germination under salinity stress (150 mM NaCl), a Venn diagram was drawn to illustrate discriminating metabolites common to and distinct between the two cultivars (Figure 7a) by fold change (FC > 1) and Student's t-test ($p < 0.05$). Our results show that the accumulation of 47 metabolic compounds was involved in responding to salt stress in the two cultivars during the studied times (12 and 24 h) (Figure 7b). Interestingly, Yangyou 9, the tolerant cultivar, is characterized by accumulating some important metabolic compounds; 16 metabolic compounds were accumulated after 12 h: MAG (18:2), cholesterol, L-aspartic acid, L-asparagine, ornithine, beta-homothreonine, 5-hydroxytryptophan, N-p-coumaroylserotonin, N-feruloylserotonin, trans-zeatin N-glucoside, pyridoxine, delphinidin O-rutinoside, N-acetylneuraminic acid, isobornyl methacrylate, 2-aminoisobutyric acid, and diethylpyrocarbonate (Figure 7c). In comparison, nine metabolite compounds were accumulated after 24 h: LPE (18:2), linolenic acid, xanthosine, inosine 5′-monophosphate, adenosine 3′-monophosphate, niacinamide, oleamide, phosphoric acid, and etamiphylline (Figure 7d), indicating that they might be involved in enhancing salt tolerance. Heatmaps of the normalized intensity of these metabolites are presented in Figure 7b–d.

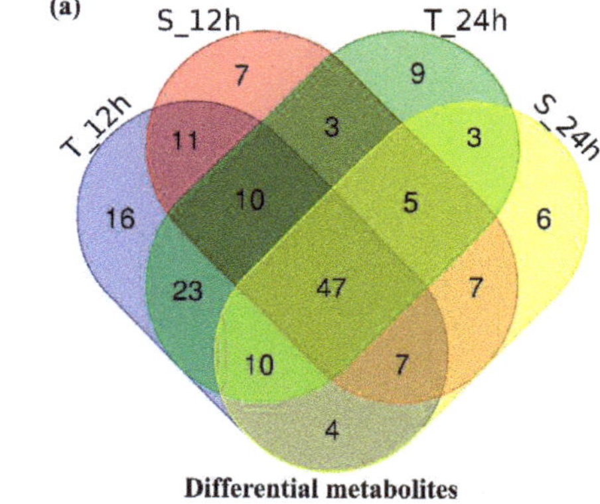

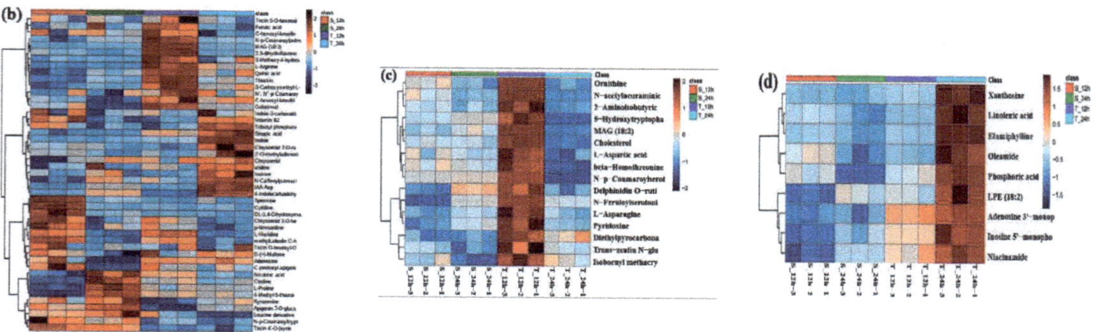

Figure 7. Differential metabolites of Yangyou 9 (T) and Zhengsheng 11 (S), fold change >1. (**a**) Venn diagram and (**b**) heatmaps of differential metabolites during 12 and 24 h of germination under salt treatment for tolerant (T) and sensitive (S) cultivar. Heatmaps of differential metabolites in tolerant cultivar after (**c**) 12 h and (**d**) 24 h of germination under salt treatment.

3.8.1. Amino and Polyamine-Related Metabolites

We found an increase in some amino acids and their derivatives, the most important of which were L-histidine, L-arginine, and L-proline, which increased by 1.43-, 1.74-, and 1.11-fold (12 h) and 1.08-, 1.38-, and 1.11-fold (24 h) in Yangyou 9, and by 1.37-, 1.13-, and 1.07-fold (12 h) and 1.06-, 1.15-, and 1.17-fold (24 h) in Zhongshuang 11. Moreover, a leucine derivative, kynurenine, and N-p-coumaroyltryptamine were increased in both cultivars under salt treatment during 12 and 24 h of seed germination; we noted an increase in the accumulation of serotonin and its derivatives. Of note, it increased the accumulation of trigonelline and betaine (alkaloid compounds) by 1.26- and 1.10-fold (12 h) and 1.15-, and 1.05-fold (24 h) in Yangyou 9, and decreased it in Zhongshuang 11. However, spermine (polyamine) increased by 1.20- and 1.55-fold (12 h) and 1.23-, and 1.25-fold (24 h) in Yangyou 9 and Zhongshuang 11, respectively (Table S2).

3.8.2. Polyphenolic-Related Metabolites

Phenolic compounds were altered in salt-treated Yangyou 9 and Zhongshuang 11 cultivars; most of them were highly accumulated. Alterations in polyphenol compounds were evident during the time of germination. Quinic acid, sinapic acid, and ferulic acid were accumulated in both Yangyou 9 and Zhongshuang 11 at 12 and 24 h of germination, while catechin and N-feruloylserotonin were increased only in a Yangyou 9 (tolerant cultivar) at both times. Moreover, coniferyl aldehyde increased during both times, while benzamidine accumulated only at 24 h of seed germination in Zhongshuang 11. In addition, phenol amine compounds such as N-caffeoylputrescine, N-p-coumaroylputrescine, and N′, N″-p-coumaroyl feruloylspermidine accumulated in the two cultivars, whereas p-coumaroy l-2-hydroxyputrescine accumulated at 12 h of germination, then decreased at 24 h. Additionally, N′, N″-di-p-coumaroylspermidine increased in Zhongshuang 11 only during 12 and 24 h of seed germination (Table S2).

3.8.3. Flavonoid-Related Metabolites

There was a significant change in the accumulation of flavonoid compounds based on the time of seed germination and the cultivar, and there was a clear difference between 12 and 24 h of germination and increased accumulation of flavonoid compounds in Yangyou 9 (tolerant cultivar) compared to Zhongshuang 11 (sensitive cultivar) during those times. The amounts of several flavonoids (apigenin 7-O-glucoside, C-pentosyl-apigenin O-hexoside, C-hexosyl-luteolin O-hexoside, C-hexosyl-luteolin O-p-coumaroylhexoside, chrysoeriol 7-O-hexoside, chrysoeriol 7-O-rutinoside, tricin O-hexosyl-O-hexoside, and tricin 4′-O-(syringyl alcohol) ether 5-O-hexoside, and methyl luteolin C-hexoside) were enhanced under salt stress in both cultivars at both time points. Meanwhile, some flavonoids (chrysoeriol C-hexoside, C-pentosyl-apeignin O-feruloylhexoside, luteolin 6-C-glucoside, tricin 7-O-hexoside, and chrysoeriol C-hexoside) showed an opposite trend. However, the level of chrysoeriol was accumulated in the tolerant and sensitive cultivar, but was higher in the former. Some flavonoids (chrysoeriol O-malonylhexoside, tricin, Selgin 5-O-hexoside, and cyanidin 3,5-di-O-hexoside) showed accumulation in Yangyou 9 and alleviation in Zhongshuang 11 under stress at both 12 and 24 h of seed germination under salt treatment (Table S2).

Interestingly, vitamin B2 was accumulated by 1.23-fold in Zhongshuang 11 after 12 h, whereas it increased by 1.20-fold in Yangyou 9 after 24 h. Moreover, pyridoxine O-glucoside, thiamin, 4-methyl-5-thiazoleethanol, and choline accumulated by 1.28-, 1.21-, 1.26-, and 1.23-fold at 12 h and 1.18-, 1.09-, 1.23-, and 1.17-fold at 24 h in Yangyou 9, whereas it decreased in Zhongshuang 11 by 1.08-, 1.04-, 1.02-, and 1.05-fold at 12 h and 0.80-, 1.07-, 1.43-, and 1.48-fold at 24 h. Additionally, the accumulation of carbachol and niacinamide was increased by 1.24- and 1.15-fold after 24 h in Yangyou 9, higher than in Zhongshuang 11 (Table S2).

3.8.4. Carbohydrate-Related Metabolites

In this investigation, there was increased sugar compound accumulation during seed germination in the studied cultivars. We found increases in fructose 1, 6-diphosphate, sucrose, D-(+)-maltose, and α-lactose by 1.20-, 1.05-, 2.56-, and 1.08-fold at 12 h and 1.16-, 1.06-, 2.59-, and 1.06-fold at 24 h in Yangyou 9. All compounds were decreased after 12 h in Zhongshuang 11 except for D-(+)-maltose, which increased by 2.58-fold at 12 h and 2.40-fold at 24 h and fructose 1, 6-diphosphate and sucrose by 1.13-, and 1.05-fold at 24 h. In addition, a-L-rhamnose was decreased in both cultivars at 12 and 24 h of seed germination under salt treatment. On the other hand, we noticed an increase in polygodial (terpene compounds) at 12 and 24 h by 1.33- and 1.10-fold, respectively, in Yangyou 9 and a decrease in Zhongshuang 11; additionally, diosgenin was increased at 12 and 24 h in Zhongshuang 11 and increased only at 24 h in Yangyou 9 (Table S2).

3.8.5. Fatty Acid-Related Metabolites in the Two Cultivars

Fatty acid content was decreased at 12 and 24 h of seed germination in response to salinity stress, and the decrease was more significant in the sensitive cultivar (Zhongshuang 11) than the tolerant cultivar (Yangyou 9). Among fatty acid compounds, MAG (18:3) was increased by 1.47- and 1.05-fold at 12 h and 1.08- and 1.07-fold at 24 h in Yangyou 9 and Zhongshuang 11, respectively. Furthermore, the accumulation of some fatty acids increased after 24 h of germination in the resistant cultivar, including LPE (18:2), linolenic acid, and 14,15-dehydrocrepenynic acid, while there was an increase in lysoPE 18:2 only at 24 h of germination in the sensitive cultivar (Table S2).

3.8.6. Nucleic Acid-Related Metabolites

The resistant cultivar showed a noticeable change in accumulation of nucleic acid and its derivatives at 12 and 24 h of seed germination under salt treatment, including crotonoside, guanosine, N2, N2-dimethylguanosine, and N-(9H-purin-6-ylcarbamoyl) threonine. In contrast, uridine, inosine, uracil, cytidine, and adenosine were accumulated in both cultivars (Table S2).

3.8.7. Other Metabolites

Several other metabolites that fall into a variety of biochemical classes were altered in the two cultivars: p-Nitroaniline, 2'-O-methyladenosine, golotimod, 1-[5-(2,3,4-trihydroxybutyl)-2-pyrazinyl]-1,2,3,4-butanetetrol, DL-3,4 dihydroxymandelic acid, S-carboxymethyl-L-cysteine, 2,3-dihydroflavone, 3-methoxy-4-hydroxybenzoic acid O-hexoside, 4-Indolecarbaldehyde, and tributyl phosphate. These were significantly enhanced under salinity stress at 12 and 24 h. In addition, hinokinin showed higher levels under salt stress at 24 h in both cultivars. Many other metabolites, including DIMBOA glucoside, pinoresinol 4-O-glucoside, N-acetyl-L-2-aminoadipic acid, and 4-hydroxybenzoic acid O-hexoside, were accumulated in the salt-tolerant cultivar (Yangyou 9), and decreased in salt-sensitive cultivar (Zhongshuang 11), at 12 and 24 h of seed germination under salinity stress (150 mM NaCl) (Table S2).

Furthermore, the resistant cultivar showed a noticeable change in hormone accumulation at 12 and 24 h of seed germination under salt treatment, indicating the critical role of hormones in plant tolerance to saline stress. For example, in Yangyou 9, the accumulation of gibberellin A14 was increased by 1.07- and 1.20-fold after 12 and 24 h of germination, and we found an increase in IAA-Asp, IAA-Glu, indole, and melatonin by 1.46-, 1.13-, 1.26-, and 1.19-fold at 12 h and 2.44-, 1.07-, 1.80-, and 1.23-fold at 24 h, while melatonin was decreased in Zhongshuang 11 at 12 and 24 h. The accumulation of methoxy indoleacetic acid increased in Zhongshuang 11 only at 12 and 24 h (Table S2).

4. Discussion

4.1. Differences in Morpho-Physiological Alterations in Response to Salinity between Yangyou 9 and Zhongshuang 11

Seed germination is a very sensitive process in the plant's life cycle, as it supports seedling development and survival, which is largely affected by genetic traits, moisture availability, and soil quality [42]. Our findings indicate that salt stress reduced the studied attributes of germination and seedling growth, especially in the Zhongshuang 11 cultivar, indicating that it was the less tolerant cultivar under high salt concentrations. Previous studies confirmed a negative relationship between salt concentration and germination parameters, ultimately leading to delayed germination in rapeseed [43], which is consistent with our study. Salinity affects the development process by causing an imbalance in the hormonal system and cellular functions of seeds, which alter enzymatic activity, changes metabolism, reduces the use of seed reserves, and reduces water uptake (osmotic effect) or ionic imbalance (ionic effect), ultimately slowing the growth rate [44,45].

It is always difficult to know whether a reduced photosynthesis rate is the cause of growth reduction or the result. Salinity causes an imbalance of chlorophyll and its intermediates (toxic photodynamic molecules that produce ROS, such as singlet oxygen), thus the chlorophyll metabolism must be strictly regulated under abiotic stresses [46]. Chloroplasts are the most important organelles responsible for photosynthesis, and an increased Na^+ or Cl^- concentration under salt stress leads to an imbalance in the photosynthesis process and chlorophyll production. Thus, chlorophyll is a substantial indicator of metabolite changes in salinized plants [47]. A study on *Brassica napus* [48] explained that the growth rate decreased with increased salt concentration through an imbalance in the photosynthesis and ET rates and a decrease in PS II efficiency.

Osmotic adjustment decreased the osmotic potential by increasing external osmolality, maintaining water absorption and various physiological processes in the plants [6,49], and including total soluble sugars and proteins, proline, amino acids, glycine betaine, etc., where they reduce osmotic stress. The increased proline and total soluble sugar content plays a vital role in protecting the cells under salt stress by maintaining the osmotic pressure and ionic balance in the cytosol and outside the cell, which increases the absorption of water and nutrients and balance of relative water content, which leads to increased protein function and cellular membrane stability [50,51]. Our investigation showed that the tolerant cultivar exhibited higher proline accumulation and elevated total soluble sugar and protein under salinity stress, which is in line with [30,38,52].

Under salinity stress, there is increased accumulation of toxic ions, especially sodium ions (Na^+), which leads to an imbalance in the ion and hyperosmosis levels in plants. As a result of this imbalance, there is weakness in the physiochemical processes of plant cells, which negatively affects development [1]. Moreover, in salinized plants with higher Na^+ levels, K^+ uptake is inhibited, which elevates the Na^+/K^+ ratio in the cellular tissues [53]. Our findings show that the tolerant cultivar worked to save and increase the K^+ in leaves and decrease Na^+ accumulation, which modulates Na^+ and K^+ uptake, to improve seedling growth, and showed an association with the restricted accumulation of toxic Na^+ ions; thus, it reduced the generation of increased oxidative stress due to salt stress (NaCl), which agrees with the results of [54,55].

Salinized seedlings displayed excessive accumulation of H_2O_2, hence oxidative stress, which also resulted in higher lipid peroxidation, thereby causing leakage of cellular components [56]. Abiotic stress, especially salinity, causes oxidative damage to plants due to excessive accumulation of ROS, which increases the damage to cellular components and injures cell structure and function, including membrane lipids, DNA damage, and enzyme inactivation [57]. Plants have comprehensive antioxidative systems, which play a critical role in scavenging ROS accumulation, whereas CAT and SOD mitigate the harmful effects of oxidative stress; likewise, SOD catalyzed $O_2^{\bullet-}$ is dismutated into $O_2^{\bullet-}$ and H_2O_2 [58,59].

Furthermore, the osmotic effects of salinity lead to metabolic changes, which disrupt the plant's ability to absorb and retain water, thereby increasing ROS levels [48]. Previous studies showed increased ROS accumulation in plants grown under salinized conditions and emphasized the role of ROS in damage to cell membranes and decreased crop yield, as it is one of the main causes of cytotoxicity under salt stress [6]. Conclusive results of a previous study in wheat and barley showed that antioxidant enzymes are involved in salt stress tolerance [60] in rapeseed [30]. Additionally, osmolytes reduced salt-induced oxidative stress by direct or indirect ROS degeneration and induced antioxidant enzyme activity [49], which agrees with our results showing that antioxidant enzymes and osmolyte accumulation were enhanced under salt stress.

4.2. Differences in Accumulation of Metabolites in Response to Salinity in Yangyou 9 and Zhongshuang 11

Metabolomics analysis is a helpful tool to understand the mechanisms underlying salt tolerance in plants. Recently, metabolic studies of the variations between salt-sensitive and salt-tolerant plants have revealed an interesting finding, that the levels of constitutive metabolites in tolerant varieties differ from those in sensitive varieties in the same species, and various species have shown both conserved and divergent metabolite responses. Moreover, specific metabolite synthesis is limited to a few plants. Thus, it would be helpful to perform further metabolic studies between salt-sensitive and salt-tolerant plants in the same species. In the current study, there was a clear distinction among the metabolites in response to salt stress between the two cultivars (Yangyou 9 and Zhongshuang 11) and treatments (12 and 24 h after sowing).

Under different salinity levels, there are changes in amino acid (Aa) content where the reorganization occurs, such as increased proline, amides, glutamate, and alanine content, while there are no changes in others [61]. The Aa content in this study decreased at 12 and 24 h of germination under salinity stress, and the decrease was more significant in the sensitive cultivar (Zhongshuang 11) than the tolerant cultivar (Yangyou 9); additionally, we found an accumulation of Aa derivatives that might act as compatible osmolytes under salt stress. In particular, Aas found at low concentrations in the cells increased under salinity pressure, especially branched-chain Aas, which can function as alternative electron givers in the electron transport chain and as compatible compounds in the mitochondria [62,63]. In *Arabidopsis*, after 12 h germination under salt treatment, there was an increase in the biosynthesis of aromatic Aas, which are primary factors stimulating the biosynthesis of lignin, which works to strengthen the cellular walls [64].

Osmolytes are one of the most important compatible solutes with low weight, which enables them to help remove excessive ROS and stabilize proteins by replacing water on the surface of the protein membrane [65]. The obtained results indicate an increased level of proline with increased exposure time during seed germination under salt stress, especially in Yangyou 9 (tolerant cultivar), and these results correspond to [65,66]; those studied suggested that the increased accumulation of osmolytes in the cells is associated with decreased accumulation of ROS, thus plants are better able to grow and develop under salinity stress. Fatty acid and lipid regulation are an essential means of protection in the plant under salt stress [67], as it works to maintain the safety of cellular membranes [68]. Our results show a change in the content of nitrogen-containing compounds, such as amines that relieve oxidative stress, the balance of enzymes, and the safety of cell membranes under salt stress due to increased Na^+ and Cl^- levels in the cells [69]. We also noted an increased accumulation of serotonin and its derivatives, which have a known role in plant resistance to various stressors [70].

Carbohydrates act as osmolytes, maintain cellular balance under salt stress, and have an important role in a host of biological functions [70]. An increase in sugar content was observed under the salt stress (100 mM) during 1–24 h in rice tissue cultures, including fructose, phosphorylated sugars, glucose, and galactose [71], whereas, under long-term NaCl exposure, co-induction of glycolytic metabolites and sucrose in *Arabidopsis* tissue cultures was observed [72]. The content of soluble sugars and other unknown putative complex glycans increased under high salinity levels in *Arabidopsis*, *Thellungiella salsuginea*, and durum wheat. Trehalose application decreased the adverse effect of salinity in rice [38].

There is a link between increased mannitol, sorbitol, cyclic polyols, and myo-inositol, and its methylated derivatives and salt tolerance; additionally, application of trehalose and polyols prevents oxidation of salt-bound lipids and protects the cells from damage by coating large particles in a protective shell, as well as reducing ROS accumulation [73].

The polyamines (Pas) have an essential role in modulating plant defense responses to various stresses, such as salinity, and include putrescine, spermidine, and spermine (free, conjugated, and bound polyamines) [74]. Due to their antioxidant properties, the Pas play a role in stabilizing cell walls and maintaining cell membrane function, in addition to having anti-aging and anti-stress effects. Moreover, Pas regulate the processes of gene expression and the formation of ionic bonds between anions i.e., DNA, RNA, and protein. Exogenous Pas (putrescine) have positive effects under salt stress conditions, by elevating the salinity tolerance [75]. In addition, Pa conjugates such as N-caffeoylputrescine, N-p-coumaroylputrescine, and N', N''-p-coumaroyl feruloylspermidine affect the response to salt treatment; they increased significantly in Yangyou 9 as compared to Zhongshuang 11. Previous studies confirmed the role of these compounds in maintaining the balance of secondary metabolic pathways. They also have an influential role in the life cycle and development of plants by participating in the processes of cell division and different physiological responses under stress conditions [74,76].

Under salt stress, flavonoids are considered to be one of the most crucial plant phenolics, playing an essential role in the plant's tolerance against salt stress, as these are potent antioxidants [66]. In this investigation, the obtained results show a difference in flavonoid content, which shows the effect of salt on compounds of flavonoids and other phenolics in rapeseed seeds. As the increase in Yangyou 9 was greater than Zhongshuang 11 at 12 and 24 h under treatment during seed germination, we agree that the increased accumulation of these compounds has a role in reducing the oxidative damage caused by increased ROS accumulation under salt stress [66,75].

Terpenes are natural plant products found in a simple, undefined form and often accumulate as accompaniments to carbohydrates. They participate in important functions such as pest resistance and contribute to crop quality and yield, and are accumulated in the epicuticular and intracuticular wax layers of stem and leaf surfaces, protecting the plant under abiotic stresses [77]. In this study, the content of terpenes in the resistant cultivar (Yangyou 9) was increased compared to the sensitive cultivar (Zhongshuang 11) during 12 and 24 h of seed germination under salt treatment. Additionally, terpenes participate in interactions with various transcription factors concerning apoptosis, cell cycle, and DNA reform due to oxidative imbalance under different abiotic stresses, especially salinity stress [70].

In this study, we found an increase in the level of plant hormones, including cytokinins (CK), gibberellins (GA), and indole acetic acid (IAA), in Yangyou 9 compared to Zhongshuang 11. The difference in GA and CK content confirms their role in different responses to oxidative stress caused by salinity stress [38]. Previous studies verified the role of GA in response to osmotic stress in *Arabidopsis* seedlings [78], confirming the role of plant hormones in different physiochemical responses during oxidative stress caused by abiotic stress [79]. Moreover, our results show accumulated IAA-Asp and IAA-Glu during seed germination under salt stress. In *Pisum sativum*, IAA-Asp has a protective role under salt and cadmium stress, demonstrating its relation with plant response to environmental stresses, which possibly indicates the signaling function, and IAA-Asp has a direct biological function in abiotic stresses [80].

4.3. Metabolic Pathway Analysis

Salt stress induces molecular reprogramming, which plays a significant role in rapeseed seeds under stress. Reduced water availability and/or ionic imbalance due to salinity stress might influence different pathways. A wide range of metabolite compounds are accumulated in response to salinity stress from a functional viewpoint; additionally, amino acids, sucrose, and flavonoids are the most common metabolites that accumulate in response to salinity stress in two cultivars (Yangyou 9 and Zhongshuang 11), as shown in Figure 8.

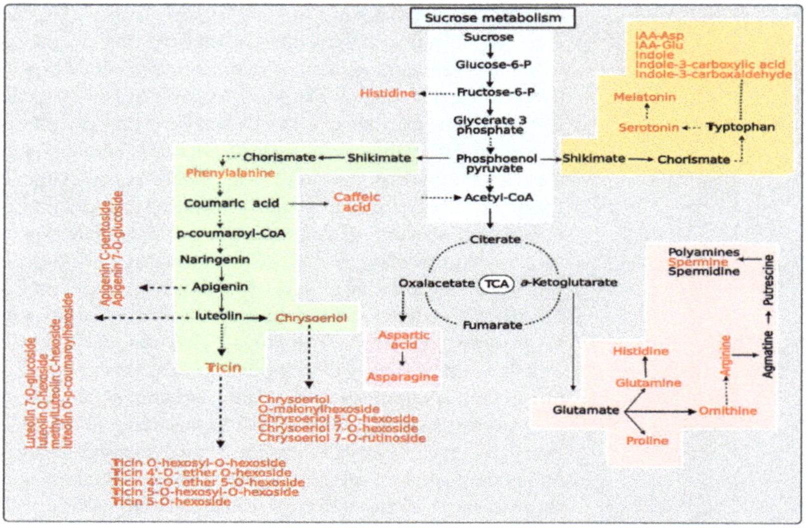

Figure 8. Proposed schematic of top differential metabolites of Yangyou 9 (T) and Zhengsheng 11 (S) exposed to salinity conditions.

5. Conclusions

Among the studied rapeseed cultivars, Yangyou 9 showed higher growth, while Zhongshuang 11 displayed a remarkably low growth rate under stress induced by 150 and 200 mM of NaCl. Higher osmolyte accumulation in Yangyou 9 reduced salt-induced oxidative stress by direct or indirect ROS degeneration and induced antioxidant enzyme activity. A comprehensive examination of metabolite changes under salt stress showed that, out of 332 compounds detected in the metabolic profile, 225 metabolites were filtrated. While the concentrations of inosine, spermine, vitamin B2, tricin O-hexosyl-O-hexoside, D-(+)-maltose, IAA-Asp, and indole were increased within both cultivars, several key metabolites—cholesterol, L-aspartic acid, L-asparagine, ornithine, N-feruloyl serotonin, pyridoxine, linolenic acid, inosine 5′-monophosphate and phosphoric acid were accumulated in Yangyou 9 only. Amino acid, sucrose, and flavonoids are the most common metabolites that can accumulate due to salt stress. Our study provides a sound basis for examining the salt tolerance of various rapeseed cultivars, and Yangyou 9 exhibited higher tolerance against salinity stress, which might be a significant germplasm resource for breeding programs that aim to develop salt-tolerant rapeseed.

Moreover, the identified metabolites can act as biomarkers to characterize plant performance under salt stress in breeding programs (as one promising strategy). Finally, the present study provides a framework for fundamental biochemical analyses and an informed database that can be used to recognize salt-tolerant and -sensitive characteristics before developing a salinity tolerance breeding program for rapeseed cultivars. Accordingly, our findings could be used for further phenotypic and genotypic association studies. This knowledge is of great value for plant breeders aiming at developing new rapeseed cultivars under stress conditions.

Supplementary Materials: The following are available online at https://www.mdpi.com/article/10.3390/antiox10081227/s1, Supplementary file (1), Figure S1: Impact of NaCl treatments on (a) shoot length (cm), (b) root length (cm), (c) shoot fresh weight (g), (d) root fresh weight (g), I shoot dry weight (g), and (f) root dry weight (g) of five rapeseed cultivars during early seedling stage, Figure

S2: Impact of NaCl treatments on (a) seedling fresh weight stress index, and (b) seedling length stress index of five rapeseed cultivars during the germination stag, Figure S3: Correlation among different growth and biochemical attributes of rapeseed cultivars, Figure S4: Up- and down-regulation of differential metabolites of Yangyou 9 (T) and Zhengsheng 11 (S). (a) Heat-map; (b) Correlation; (c) Scoring plot and (d) Loading plot of principal component analysis (PCA) of differential metabolites during 12, 24h of germination upon salt treatment for tolerant cultivar (T) and sensitive cultivar (S), Table S1: Means of FG%, GR, VI (I), and VI (II) of the studied cultivars in the germination stage under salinity stress during the seed germination stage, Table S2: Metabolites differentially accumulated in Yangyou 9 and Zhongshuang 11 seeds under salt stress (150 mM NaCl). Supplementary file (2), Identified metabolites list.

Author Contributions: Data curation, formal analysis, visualization, conceptualization, investigation, methodology, software, writing—review & editing, validation, writing–original draft, A.M.E.-B.; software, writing–original draft, M.B.; software, LA.A.M.; data curation, methodology, Z.W., A.K., and A.S.; formal analysis, H.A.; methodology, M.N.K., and H.M.H.; formal analysis, methodology, validation, I.M.E.; validation, J.K.; funding acquisition, conceptualization, project administration, supervision, writing—review & editing, B.W.; resources, conceptualization, project administration supervision, G.Z. All authors have read and agreed to the published version of the manuscript.

Funding: This research was funded by "National Key Research and Development Program of China, 2018YFD1000900" and "Technical Innovation Project in Hubei Province, 2020BBB061, and 2020BBB062)".

Institutional Review Board Statement: Not applicable.

Informed Consent Statement: Not applicable.

Data Availability Statement: Data is contained within the article.

Acknowledgments: We want to express our deep and sincere gratitude to Wei Chen and Jie Chen for helping us with data analysis and G. El-Rewainy for her continued support.

Conflicts of Interest: The authors declare no conflict of interest.

Abbreviations

AAs	Amino acid
APX	Ascorbate peroxidase
CAT	Catalase
Chl	Chlorophyll
DAB	3,3-diaminobenzidine
FC	Fold change
FW	Fresh weight
MDA	Malondialdehyde
NBT	Nitro blue tetrazolium
POD	Peroxidase
PA	Polyamine
SOD	Superoxide dismutase
TCA	Tricarboxylic acid
TSS	Total soluble sugar
TSP	Total soluble protein

References

1. Ashraf, M.; McNeilly, T. Salinity tolerance in brassica oilseeds. *Crit. Rev. Plant Sci.* **2004**, *23*, 157–174. [CrossRef]
2. Wu, W.; Ma, B.L.; Whalen, J.K. Chapter three-enhancing rapeseed tolerance to heat and drought stresses in a changing climate: Perspectives for stress adaptation from root system architecture. In *Advances in Agronomy*; Sparks, D.L., Ed.; Academic Press: Cambridge, MA, USA, 2018; Volume 151, pp. 87–157.
3. Shah, N.; Anwar, S.; Xu, J.; Hou, Z.; Salah, A.; Khan, S.; Gong, J.; Shang, Z.; Qian, L.; Zhang, C. The response of transgenic brassica species to salt stress: A review. *Biotechnol. Lett.* **2018**, *40*, 1159–1165. [CrossRef]
4. Chakraborty, S.; Newton, A.C. Climate change, plant diseases and food security: An overview. *Plant Pathol.* **2011**, *60*, 2–14. [CrossRef]

5. Soliman, A.S.; El-feky, S.A.; Darwish, E. Alleviation of salt stress on *moringa peregrina* using foliar application of nanofertilizers. *J. Hortic. For.* **2015**, *7*, 36–47.

6. Shabala, S.; Munns, R. Salinity stress: Physiological constraints and adaptive mechanisms. In *Plant Stress Physiology*; Shabala, S., Ed.; CABI: Wallingford, UK, 2017; pp. 24–63.

7. Mishra, A.; Tanna, B. Halophytes: Potential resources for salt stress tolerance genes and promoters. *Front. Plant Sci.* **2017**, *8*, 1–10. [CrossRef] [PubMed]

8. Jiang, X.-W.; Zhang, C.-R.; Wang, W.-H.; Xu, G.-H.; Zhang, H.-Y. Seed priming improves seed germination and seedling growth of *isatis indigotica* fort. Under salt stress. *HortScience* **2020**, *55*, 647–650. [CrossRef]

9. Parihar, P.; Singh, S.; Singh, R.; Singh, V.P.; Prasad, S.M. Effect of salinity stress on plants and its tolerance strategies: A review. *Environ. Sci. Pollut. Res.* **2015**, *22*, 4056–4075. [CrossRef]

10. Abdel-Haliem, M.E.F.; Hegazy, H.S.; Hassan, N.S.; Naguib, D.M. Effect of silica ions and nano silica on rice plants under salinity stress. *Ecol. Eng.* **2017**, *99*, 282–289. [CrossRef]

11. Wangsawang, T.; Chuamnakthong, S.; Kohnishi, E.; Sripichitt, P.; Sreewongchai, T.; Ueda, A. A salinity-tolerant japonica cultivar has na$^+$ exclusion mechanism at leaf sheaths through the function of a na$^+$ transporter *oshkt1;4* under salinity stress. *J. Agron. Crop Sci.* **2018**, *204*, 274–284. [CrossRef]

12. Woodrow, P.; Ciarmiello, L.F.; Annunziata, M.G.; Pacifico, S.; Iannuzzi, F.; Mirto, A.; D'Amelia, L.; Dell'Aversana, E.; Piccolella, S.; Fuggi, A.; et al. Durum wheat seedling responses to simultaneous high light and salinity involve a fine reconfiguration of amino acids and carbohydrate metabolism. *Physiol. Plant.* **2017**, *159*, 290–312. [CrossRef]

13. Gong, Z.; Xiong, L.; Shi, H.; Yang, S.; Herrera-Estrella, L.R.; Xu, G.; Chao, D.Y.; Li, J.; Wang, P.Y.; Qin, F.; et al. Plant abiotic stress response and nutrient use efficiency. *Sci. China Life Sci.* **2020**, *63*, 635–674. [CrossRef]

14. Woodrow, P.; Pontecorvo, G.; Ciarmiello, L.F.; Fuggi, A.; Carillo, P. Ttd1a promoter is involved in DNA–protein binding by salt and light stresses. *Mol. Biol. Rep.* **2011**, *38*, 3787–3794. [CrossRef]

15. Rasheed, R.; Ashraf, M.A.; Parveen, S.; Iqbal, M.; Hussain, I. Effect of salt stress on different growth and biochemical attributes in two canola (*brassica napus* L.) cultivars. *Commun. Soil Sci. Plant Anal.* **2014**, *45*, 669–679. [CrossRef]

16. Mohamed, I.A.A.; Shalby, N.; El-Badri, A.M.A.; Saleem, M.H.; Khan, M.N.; Nawaz, M.A.; Qin, M.; Agami, R.A.; Kuai, J.; Wang, B.; et al. Stomata and xylem vessels traits improved by melatonin application contribute to enhancing salt tolerance and fatty acid composition of *brassica napus* L. Plants. *Agronomy* **2020**, *10*, 1186. [CrossRef]

17. Canam, T.; Li, X.; Holowachuk, J.; Yu, M.; Xia, J.; Mandal, R.; Krishnamurthy, R.; Bouatra, S.; Sinelnikov, I.; Yu, B.; et al. Differential metabolite profiles and salinity tolerance between two genetically related brown-seeded and yellow-seeded brassica carinata lines. *Plant Sci.* **2013**, *198*, 17–26. [CrossRef]

18. Dolatabadi, N.; Toorchi, M. Rapeseed (*brassica napus* L.) genotypes response to nacl salinity. *J. Biodivers. Environ. Sci.* **2017**, *10*, 265–270.

19. HanumanthaRao, B.; Nair, R.M.; Nayyar, H. Salinity and high temperature tolerance in mungbean [*vigna radiata* (L.) wilczek] from a physiological perspective. *Front. Plant Sci.* **2016**, *7*, 957–977. [CrossRef]

20. Mohammadi, S.; Shekari, F.; Fotovat, R.; Darudi, A. Effect of laser priming on canola yield and its components under salt stress. *Int. Agrophysics* **2012**, *26*, 45–51. [CrossRef]

21. Pujari, D.S.; Chanda, S.V. Effect of salinity stress on growth, peroxidase and iaa oxidase activities in vigna seedlings. *Acta Physiol. Plant.* **2002**, *24*, 435–439. [CrossRef]

22. Kumar, V.; Khare, T. Differential growth and yield responses of salt-tolerant and susceptible rice cultivars to individual (na$^+$ and cl$^-$) and additive stress effects of nacl. *Acta Physiol. Plant.* **2016**, *38*, 170–178. [CrossRef]

23. Derakhshani, Z.; Bhave, M.; Shah, R.M. Metabolic contribution to salinity stress response in grains of two barley cultivars with contrasting salt tolerance. *Environ. Exp. Bot.* **2020**, *179*, 104229. [CrossRef]

24. Jenner, C. Effects of exposure of wheat ears to high temperature on dry matter accumulation and carbohydrate metabolism in the grain of two cultivars. I. Immediate responses. *Funct. Plant Biol.* **1991**, *18*, 165–177. [CrossRef]

25. Kumari, M.; Asthir, B. Transformation of sucrose to starch and protein in rice leaves and grains under two establishment methods. *Rice Sci.* **2016**, *23*, 255–265. [CrossRef]

26. Widodo; Patterson, J.H.; Newbigin, E.; Tester, M.; Bacic, A.; Roessner, U. Metabolic responses to salt stress of barley (*hordeum vulgare* L.) cultivars, sahara and clipper, which differ in salinity tolerance. *J. Exp. Bot.* **2009**, *60*, 4089–4103. [CrossRef]

27. Mahakham, W.; Sarmah, A.K.; Maensiri, S.; Theerakulpisut, P. Nanopriming technology for enhancing germination and starch metabolism of aged rice seeds using phytosynthesized silver nanoparticles. *Sci. Rep.* **2017**, *7*, 8263–9284. [CrossRef]

28. Gill, R.A.; Ali, B.; Islam, F.; Farooq, M.A.; Gill, M.B.; Mwamba, T.M.; Zhou, W. Physiological and molecular analyses of black and yellow seeded *brassica napus* regulated by 5-aminolivulinic acid under chromium stress. *Plant Physiol. Biochem.* **2015**, *94*, 130–143. [CrossRef]

29. Lichtenthaler, H.K. [34] chlorophylls and carotenoids: Pigments of photosynthetic biomembranes. In *Methods in Enzymology*; Academic Press: Cambridge, MA, USA, 1987; Volume 148, pp. 350–382.

30. El-Badri, A.M.A.; Batool, M.; Mohamed, I.A.A.; Khatab, A.; Sherif, A.; Wang, Z.; Salah, A.; Nishawy, E.; Ayaad, M.; Kuai, J.; et al. Modulation of salinity impact on early seedling stage via nano-priming application of zinc oxide on rapeseed (*brassica napus* L.). *Plant Physiol. Biochem.* **2021**, *166*, 376–392. [CrossRef] [PubMed]

31. Heath, R.L.; Packer, L. Photoperoxidation in isolated chloroplasts: I. *Kinetics and stoichiometry of fatty acid peroxidation Arch. Biochem. Biophys.* **1968**, *125*, 189–198. [CrossRef]

32. Abdel Latef, A.; Tran, L.S. Impacts of priming with silicon on the growth and tolerance of maize plants to alkaline stress. *Front. Plant Sci.* **2016**, *7*, 243–252. [CrossRef]

33. Ulhassan, Z.; Gill, R.A.; Ali, S.; Mwamba, T.M.; Ali, B.; Wang, J.; Huang, Q.; Aziz, R.; Zhou, W. Dual behavior of selenium: Insights into physio-biochemical, anatomical and molecular analyses of four *brassica napus* cultivars. *Chemosphere* **2019**, *225*, 329–341. [CrossRef]

34. Hashem, A.M.; Moore, S.; Chen, S.; Hu, C.; Zhao, Q.; Elesawi, I.E.; Feng, Y.; Topping, J.F.; Liu, J.; Lindsey, K.; et al. Putrescine depletion affects arabidopsis root meristem size by modulating auxin and cytokinin signaling and ros accumulation. *Int. J. Mol. Sci.* **2021**, *22*, 4094. [CrossRef] [PubMed]

35. Ahanger, M.A.; Alyemeni, M.N.; Wijaya, L.; Alamri, S.A.; Alam, P.; Ashraf, M.; Ahmad, P. Potential of exogenously sourced kinetin in protecting *solanum lycopersicum* from nacl-induced oxidative stress through up-regulation of the antioxidant system, ascorbate-glutathione cycle and glyoxalase system. *PLoS ONE* **2018**, *13*, e0202175–e0202195. [CrossRef] [PubMed]

36. Ahmad, P.; Alyemeni, M.N.; Al-Huqail, A.A.; Alqahtani, M.A.; Wijaya, L.; Ashraf, M.; Kaya, C.; Bajguz, A. Zinc oxide nanoparticles application alleviates arsenic (as) toxicity in soybean plants by restricting the uptake of as and modulating key biochemical attributes, antioxidant enzymes, ascorbate-glutathione cycle and glyoxalase system. *Plants* **2020**, *9*, 825. [CrossRef]

37. Ahmad, P.; Ahanger, M.A.; Alam, P.; Alyemeni, M.N.; Wijaya, L.; Ali, S.; Ashraf, M. Silicon (si) supplementation alleviates nacl toxicity in mung bean [*vigna radiata* (L.) wilczek] through the modifications of physio-biochemical attributes and key antioxidant enzymes. *J. Plant Growth Regul.* **2019**, *38*, 70–82. [CrossRef]

38. D'Amelia, L.; Dell'Aversana, E.; Woodrow, P.; Ciarmiello, L.F.; Carillo, P. Metabolomics for crop improvement against salinity stress. In *Salinity Responses and Tolerance in Plants*; Kumar, V., Wani, S.H., Suprasanna, P., Tran, L.-S.P., Eds.; Springer: Berlin/Heidelberg, Germany, 2018; Volume 2, pp. 267–288.

39. Chen, W.; Gao, Y.; Xie, W.; Gong, L.; Lu, K.; Wang, W.; Li, Y.; Liu, X.; Zhang, H.; Dong, H.; et al. Genome-wide association analyses provide genetic and biochemical insights into natural variation in rice metabolism. *Nat. Genet.* **2014**, *46*, 714–721. [CrossRef] [PubMed]

40. Chen, W.; Gong, L.; Guo, Z.; Wang, W.; Zhang, H.; Liu, X.; Yu, S.; Xiong, L.; Luo, J. A novel integrated method for large-scale detection, identification, and quantification of widely targeted metabolites: Application in the study of rice metabolomics. *Mol. Plant* **2013**, *6*, 1769–1780. [CrossRef]

41. Chen, J.; Hu, X.; Shi, T.; Yin, H.; Sun, D.; Hao, Y.; Xia, X.; Luo, J.; Fernie, A.R.; He, Z.; et al. Metabolite-based genome-wide association study enables dissection of the flavonoid decoration pathway of wheat kernels. *Plant Biotechnol. J.* **2020**, *18*, 1722–1735. [CrossRef]

42. Singh, A.; Singh, N.; Hussain, I.; Singh, H.; Singh, S. Plant-nanoparticle interaction: An approach to improve agricultural practices and plant productivity. *Int. J. Pharm. Sci. Invent.* **2015**, *4*, 25–40.

43. Hezaveh, T.A.; Pourakbar, L.; Rahmani, F.; Alipour, H. Interactive effects of salinity and zno nanoparticles on physiological and molecular parameters of rapeseed (*brassica napus* L.). *Commun. Soil Sci. Plant Anal.* **2019**, *50*, 698–715. [CrossRef]

44. Meot-Duros, L.; Magné, C. Effect of salinity and chemical factors on seed germination in the halophyte *crithmum maritimum* L. *Plant Soil* **2008**, *313*, 83–87. [CrossRef]

45. Ahanger, M.A.; Agarwal, R.M. Salinity stress induced alterations in antioxidant metabolism and nitrogen assimilation in wheat (*triticum aestivum* L) as influenced by potassium supplementation. *Plant Physiol. Biochem.* **2017**, *115*, 449–460. [CrossRef]

46. Turan, S.; Tripathy, B.C. Salt-stress induced modulation of chlorophyll biosynthesis during de-etiolation of rice seedlings. *Physiol. Plant.* **2015**, *153*, 477–491. [CrossRef] [PubMed]

47. Abdul Qados, A.M.S. Effect of salt stress on plant growth and metabolism of bean plant *vicia faba* (L.). *J. Saudi Soc. Agric. Sci.* **2011**, *10*, 7–15. [CrossRef]

48. Li, H.; Lei, P.; Pang, X.; Li, S.; Xu, H.; Xu, Z.; Feng, X. Enhanced tolerance to salt stress in canola (*brassica napus* L.) seedlings inoculated with the halotolerant enterobacter cloacae hsnj4. *Appl. Soil Ecol.* **2017**, *119*, 26–34. [CrossRef]

49. Zulfiqar, F.; Akram, N.A.; Ashraf, M. Osmoprotection in plants under abiotic stresses: New insights into a classical phenomenon. *Planta* **2019**, *251*, 3. [CrossRef]

50. Hasan, A.; Hafiz, H.R.; Siddiqui, N.; Khatun, M.; Islam, R.; Mamun, A.A. Evaluation of wheat genotypes for salt tolerance based on some physiological traits. *J. Crop Sci. Biotechnol.* **2015**, *18*, 333–340. [CrossRef]

51. Gupta, B.; Huang, B. Mechanism of salinity tolerance in plants: Physiological, biochemical, and molecular characterization. *Int. J. Genom.* **2014**, *2014*, 1–19. [CrossRef]

52. Alam, P.; Albalawi, T.H.; Altalayan, F.H.; Bakht, M.A.; Ahanger, M.A.; Raja, V.; Ashraf, M.; Ahmad, P. 24-epibrassinolide (ebr) confers tolerance against nacl stress in soybean plants by up-regulating antioxidant system, ascorbate-glutathione cycle, and glyoxalase system. *Biomolecules* **2019**, *9*, 640. [CrossRef] [PubMed]

53. Ahmad, I.; Maathuis, F.J.M. Cellular and tissue distribution of potassium: Physiological relevance, mechanisms and regulation. *J. Plant Physiol.* **2014**, *171*, 708–714. [CrossRef] [PubMed]

54. Mohamed, I.A.A.; Shalby, N.; Bai, C.; Qin, M.; Agami, R.A.; Jie, K.; Wang, B.; Zhou, G. Stomatal and photosynthetic traits are associated with investigating sodium chloride tolerance of *brassica napus* L. Cultivars. *Plants* **2020**, *9*, 62. [CrossRef]

55. Rajabi Dehnavi, A.; Zahedi, M.; Ludwiczak, A.; Cardenas Perez, S.; Piernik, A. Effect of salinity on seed germination and seedling development of sorghum (*sorghum bicolor* (L.) moench) genotypes. *Agronomy* **2020**, *10*, 859. [CrossRef]

56. Ahanger, M.A.; Mir, R.A.; Alyemeni, M.N.; Ahmad, P. Combined effects of brassinosteroid and kinetin mitigates salinity stress in tomato through the modulation of antioxidant and osmolyte metabolism. *Plant Physiol. Biochem.* **2020**, *147*, 31–42. [CrossRef] [PubMed]

57. Ahmad, P.; Jaleel, C.A.; Salem, M.A.; Nabi, G.; Sharma, S. Roles of enzymatic and nonenzymatic antioxidants in plants during abiotic stress. *Crit. Rev. Biotechnol.* **2010**, *30*, 161–175. [CrossRef]

58. Shah, T.; Latif, S.; Saeed, F.; Ali, I.; Ullah, S.; Abdullah Alsahli, A.; Jan, S.; Ahmad, P. Seed priming with titanium dioxide nanoparticles enhances seed vigor, leaf water status, and antioxidant enzyme activities in maize (*zea mays* L.) under salinity stress. *J. King Saud Univ. Sci.* **2021**, *33*, 101207–101214. [CrossRef]

59. Hasanuzzaman, M.; Bhuyan, M.H.M.B.; Zulfiqar, F.; Raza, A.; Mohsin, S.M.; Mahmud, J.A.; Fujita, M.; Fotopoulos, V. Reactive oxygen species and antioxidant defense in plants under abiotic stress: Revisiting the crucial role of a universal defense regulator. *Antioxidants* **2020**, *9*, 681. [CrossRef] [PubMed]

60. Zeeshan, M.; Lu, M.; Sehar, S.; Holford, P.; Wu, F. Comparison of biochemical, anatomical, morphological, and physiological responses to salinity stress in wheat and barley genotypes deferring in salinity tolerance. *Agronomy* **2020**, *10*, 127. [CrossRef]

61. Annunziata, M.G.; Ciarmiello, L.F.; Woodrow, P.; Maximova, E.; Fuggi, A.; Carillo, P. Durum wheat roots adapt to salinity remodeling the cellular content of nitrogen metabolites and sucrose. *Front. Plant Sci.* **2017**, *7*, 1–16. [CrossRef]

62. Lea, P.J.; Sodek, L.; Parry, M.A.J.; Shewry, P.R.; Halford, N.G. Asparagine in plants. *Ann. Appl. Biol.* **2007**, *150*, 1–26. [CrossRef]

63. Sanchez, D.H.; Pieckenstain, F.L.; Szymanski, J.; Erban, A.; Bromke, M.; Hannah, M.A.; Kraemer, U.; Kopka, J.; Udvardi, M.K. Comparative functional genomics of salt stress in related model and cultivated plants identifies and overcomes limitations to translational genomics. *PLoS ONE* **2011**, *6*, e17094–e17104. [CrossRef] [PubMed]

64. Kim, J.K.; Bamba, T.; Harada, K.; Fukusaki, E.; Kobayashi, A. Time-course metabolic profiling in *arabidopsis thaliana* cell cultures after salt stress treatment. *J. Exp. Bot.* **2006**, *58*, 415–424. [CrossRef]

65. Muchate, N.S.; Nikalje, G.C.; Rajurkar, N.S.; Suprasanna, P.; Nikam, T.D. Physiological responses of the halophyte sesuvium portulacastrum to salt stress and their relevance for saline soil bio-reclamation. *Flora* **2016**, *224*, 96–105. [CrossRef]

66. Benjamin, J.J.; Lucini, L.; Jothiramshekar, S.; Parida, A. Metabolomic insights into the mechanisms underlying tolerance to salinity in different halophytes. *Plant Physiol. Biochem.* **2019**, *135*, 528–545. [CrossRef]

67. López-Pérez, L.; Martínez-Ballesta, M.d.C.; Maurel, C.; Carvajal, M. Changes in plasma membrane lipids, aquaporins and proton pump of broccoli roots, as an adaptation mechanism to salinity. *Phytochemistry* **2009**, *70*, 492–500. [CrossRef]

68. Tisi, A.; Angelini, R.; Cona, A. Wound healing in plants: Cooperation of copper amine oxidase and flavin-containing polyamine oxidase. *Plant Signal. Behav.* **2008**, *3*, 204–206. [CrossRef] [PubMed]

69. Behr, J.H.; Bouchereau, A.; Berardocco, S.; Seal, C.E.; Flowers, T.J.; Zörb, C. Metabolic and physiological adjustment of *suaeda maritima* to combined salinity and hypoxia. *Ann. Bot.* **2017**, *119*, 965–976.

70. Kaur, H.; Mukherjee, S.; Baluska, F.; Bhatla, S.C. Regulatory roles of serotonin and melatonin in abiotic stress tolerance in plants. *Plant Signal. Behav.* **2015**, *10*, e1049788–e1049796. [CrossRef]

71. Dawei, L.; Ford, K.L.; Ute, R.; Siria, N.; Cassin, A.M.; Patterson, J.H.; Antony, B. Rice suspension cultured cells are evaluated as a model system to study salt responsive networks in plants using a combined proteomic and metabolomic profiling approach. *Proteomics* **2013**, *13*, 2046–2062.

72. Jorge, T.F.; Rodrigues, J.A.; Caldana, C.; Schmidt, R.; van Dongen, J.T.; Thomas-Oates, J.; António, C. Mass spectrometry-based plant metabolomics: Metabolite responses to abiotic stress. *Mass Spectrom. Rev.* **2016**, *35*, 620–649. [CrossRef] [PubMed]

73. Mechri, B.; Tekaya, M.; Cheheb, H.; Hammami, M. Determination of mannitol sorbitol and myo-inositol in olive tree roots and rhizospheric soil by gas chromatography and effect of severe drought conditions on their profiles. *J. Chromatogr. Sci.* **2015**, *53*, 1631–1638. [CrossRef]

74. Liu, J.-H.; Wang, W.; Wu, H.; Gong, X.; Moriguchi, T. Polyamines function in stress tolerance: From synthesis to regulation. *Front. Plant Sci.* **2015**, *6*, 827–836. [CrossRef] [PubMed]

75. Sandoval-Acuña, C.; Ferreira, J.; Speisky, H. Polyphenols and mitochondria: An update on their increasingly emerging ros-scavenging independent actions. *Arch. Biochem. Biophys.* **2014**, *559*, 75–90. [CrossRef] [PubMed]

76. Rouphael, Y.; Colla, G.; Bernardo, L.; Kane, D.; Trevisan, M.; Lucini, L. Zinc excess triggered polyamines accumulation in lettuce root metabolome, as compared to osmotic stress under high salinity. *Front. Plant Sci.* **2016**, *7*, 842–851. [CrossRef]

77. Thimmappa, R.; Geisler, K.; Louveau, T.; O'Maille, P.; Osbourn, A. Triterpene biosynthesis in plants. *Annu. Rev. Plant Biol.* **2014**, *65*, 225–257. [CrossRef] [PubMed]

78. Wani, S.H.; Kumar, V.; Shriram, V.; Sah, S.K. Phytohormones and their metabolic engineering for abiotic stress tolerance in crop plants. *Crop J.* **2016**, *4*, 162–176. [CrossRef]

79. Ali, B. Practical applications of brassinosteroids in horticulture—some field perspectives. *Sci. Hortic.* **2017**, *225*, 15–21. [CrossRef]

80. Ostrowski, M.; Ciarkowska, A.; Jakubowska, A. The auxin conjugate indole-3-acetyl-aspartate affects responses to cadmium and salt stress in pisum sativum L. *J. Plant. Physiol.* **2016**, *191*, 63–72. [CrossRef]

 *antioxidants*

Article

Cold Tolerance during the Reproductive Phase in Chickpea (*Cicer arietinum* L.) Is Associated with Superior Cold Acclimation Ability Involving Antioxidants and Cryoprotective Solutes in Anthers and Ovules

Anju Rani [1], Asha Kiran [2], Kamal Dev Sharma [2], P. V. Vara Prasad [3,*], Uday C. Jha [4], Kadambot H. M. Siddique [5] and Harsh Nayyar [1,*]

[1] Department of Botany, Panjab University, Chandigarh 160014, India; thakur.aanjali@gmail.com
[2] Department of Agricultural Biotechnology, CSK Himachal Pradesh Agricultural University, Palampur 176062, India; singhashakiran@gmail.com (A.K.); kamal@hillagric.ac.in (K.D.S.)
[3] Department of Agronomy, Kansas State University, Manhattan, KS 66506, USA
[4] Crop Improvement Division, Indian Institute of Pulses Research, Kanpur 208024, India; u9811981@gmail.com
[5] The UWA Institute of Agriculture, The University of Western Australia, Perth, WA 6009, Australia; kadambot.siddique@uwa.edu.au
*   Correspondence: vara@ksu.edu (P.V.V.P.); nayarbot@pu.ac.in (H.N.)

Citation: Rani, A.; Kiran, A.; Sharma, K.D.; Prasad, P.V.V.; Jha, U.C.; Siddique, K.H.M.; Nayyar, H. Cold Tolerance during the Reproductive Phase in Chickpea (*Cicer arietinum* L.) Is Associated with Superior Cold Acclimation Ability Involving Antioxidants and Cryoprotective Solutes in Anthers and Ovules. *Antioxidants* **2021**, *10*, 1693. https://doi.org/10.3390/antiox10111693

Academic Editors: Masayuki Fujita and Mirza Hasanuzzaman

Received: 14 August 2021
Accepted: 23 October 2021
Published: 26 October 2021

Publisher's Note: MDPI stays neutral with regard to jurisdictional claims in published maps and institutional affiliations.

Abstract: Chickpea is sensitive to cold stress, especially at reproductive stage, resulting in flower and pod abortion that significantly reduces seed yield. In the present study, we evaluated (a) whether cold acclimation imparts reproductive cold tolerance in chickpea; (b) how genotypes with contrasting sensitivity respond to cold acclimation; and (c) the involvement of cryoprotective solutes and antioxidants in anthers and ovules in cold acclimation. Four chickpea genotypes with contrasting cold sensitivity (cold-tolerant: ICC 17258, ICC 16349; cold-sensitive: ICC 15567, GPF 2) were grown in an outdoor environment for 40 days in November (average maximum/minimum temperature 24.9/15.9 °C) before being subjected to cold stress (13/7 °C), with or without cold acclimation in a controlled environment of walk-in-growth chambers. The 42-d cold acclimation involved 7 d exposure at each temperature beginning with 23/15 °C, 21/13 °C, 20/12 °C, 20/10 °C, 18/8 °C, 15/8 °C (12 h/12 h day/night), prior to exposing the plants to cold stress (13/7 °C, 12 h/12 h day/night; 700 μmol m^{-2} s^{-1} light intensity; 65–70% relative humidity). Cold acclimation remarkably reduced low temperature-induced leaf damage (as membrane integrity, leaf water status, stomatal conductance, photosynthetic pigments, and chlorophyll fluorescence) under cold stress in all four genotypes. It only reduced anther and ovule damage in cold-tolerant genotypes due to improved antioxidative ability, measured as enzymatic (superoxide dismutase, catalase, ascorbate peroxidase, and glutathione reductase) and non-enzymatic (ascorbate and reduced glutathione), solutes (particularly sucrose and γ-aminobutyric acid) leading to improving reproductive function and yield traits, whereas cold-sensitive genotypes were not responsive. The study concluded that cold tolerance in chickpea appears to be related to the better ability of anthers and ovules to acclimate, involving various antioxidants and cryoprotective solutes. This information will be useful in directing efforts toward increasing cold tolerance in chickpea.

Keywords: chilling; legumes; pollen; stigma; acclimatization; stress

1. Introduction

Chickpea (*Cicer arietinum* L.), the third most important grain legume in the world, is an important source of protein to human and animals in Asia and Africa. Consequently, major chickpea growing areas lie in these two continents; however, it is also cultivated in the USA, Canada, and Australia primarily for export to Asian and African countries. Chickpea evolved in the warm climates of the Mediterranean region and is thus sensitive to low temperatures [1–3]. Chickpea experiences stressful low temperatures either during vegetative or reproductive growth, depending on the cultivation region [2,4–6]. In northern India and southern Australia, chickpea experiences low temperatures (<20/10 °C) during reproductive growth wherein cold stress damages leaves and flowers, decreases pollen and ovule fertility, impairs fertilization and alters the transcription in anthers and leaves [4], leading to flower and pod abortion and reducing the yield potential [4,6–10]. The threshold temperature for chickpea is 21 °C and temperatures below are stressful to chickpea; consequently, many production regions in the world are susceptible to cold stress [2,10].

Cold-stress-induced aberrations in crops at various organizational levels, including reduced vegetative and reproductive growth, delayed phenology, enhanced leaf chlorosis and necrosis, changes in leaf hydration status, flower abnormalities, and damage to reproductive structures and yield including chickpea are well understood [1,2,4,11]. Cold stress results in fewer numbers of pods and seeds per pod leading to lower yield [6]. In cold-sensitive chickpea genotypes, cold stress at all anther development stages i.e., micro- or mega-sporogenesis, gametogenesis and at mature pollen stage results in flower abortion [2]. The flower abortion is caused either by disruption of gametogenesis or abnormal pollen/ovule development that leads to sterility [2]. Younger flowers are relatively more sensitive to cold stress compared to old flowers as younger flowers do not have developed pollen grains whereas older flowers have developed pollen grains and results in sterility [2]. In older flowers, cold stress also decreases the ability of the pollen grains to germinate and retards pollen tube growth leading to failure or lack of fertilization resulting in poor seed set and fewer seeds per pod [2,4].

Despite significant advancements in our understanding of cold stress responses of chickpea, the metabolic and molecular mechanisms affecting cold sensitivity, especially in flowers, are relatively poorly understood [4,9,10]. At the cellular level, cold stress induces damage to membranes, increases production of reactive oxygen species (ROS), denatures enzymes and proteins and causes hormonal imbalance [12]. Our recent study [4] focused on the impact of cold stress on metabolites and enzymatic antioxidants as well as expression of genes of these pathways in anthers of cold-sensitive and cold-tolerant genotypes. While starch and proline were decreased in the cold-sensitive genotypes, there was no change in the cold-tolerant genotype. This decrease in sensitive genotype resulted from down-regulation of sucrose and proline transporter genes whereas there was up-regulation of these genes in cold-tolerant genotype [4]. It was shown that pollen viability of cold-tolerant genotypes was linked to maintenance of starch, reducing sugars and proline levels [4]. Additional studies are, however, needed to elucidate complete mechanisms associated with cold-induced flower abortion.

Plants, even cold-sensitive ones, also possess the ability to acquire cold tolerance. Cold tolerance acquisition takes place when plants are exposed to gradually decreasing low non-freezing temperatures, a process known as cold acclimation [13]. In general, acclimated plants may have greater cold tolerance compared to plants those are not acclimated [14–16]. Cold acclimation has been reported in several crops such as oilseed rape (*Brassica napus*; [17]), barley (*Hordeum vulgare*; [18]), and *Arabidopsis thaliana* [19]. In oilseed rape, maximum cold tolerance was achieved by exposure to 3 d of acclimation in spring cultivars and between 6 and 9 d in the winter cultivars, and cold tolerance decreased with prolonged acclimation duration [17]. At physiological level, the cold acclimation in barley led to significant changes in tissue water content, carbohydrate content and resulted in improved tillers and growth compared to non-acclimated plants [18]. Cold acclimation also modified the photosynthetic machinery and enabled plants to survive under severe cold temperatures through manipulation of chlorophyll a fluorescence [19]. Though information is not available for chickpea, in other crops, cold acclimation encompasses several mechanisms involving membrane changes [14], osmoprotectant accumulation (e.g., carbohydrates, proline, glycine betaine), antioxidant up-regulation [13,15] coupled with changes in expression of genes of these pathways [16]. These modifications during cold acclimation prepare cells to tolerate subsequent stressful low temperatures. It appears that the differential ability of the crops or their genotypes to tolerate cold stress depends on the types of physiological or biochemical changes during the process of cold acclimation.

The impact of cold acclimation on chickpea is not well documented or understood, although a few studies showed the benefits of cold acclimation during early vegetative growth [9,20,21]. There is no information on reproductive benefits of cold acclimation in chickpea. It was hypothesized that antioxidants (enzymatic or non-enzymatic) and solutes (e.g., osmolytes and carbohydrates) accumulate in chickpea on exposure to gradually decreasing temperatures and result in cold acclimation. It is also not known whether cold-sensitive and cold-tolerant genotypes behave similarly or differently upon cold acclimation. Therefore, the objectives of this study were to evaluate (a) whether cold acclimation imparts reproductive cold tolerance in chickpea; (b) whether genotypes with contrasting cold sensitivity respond similarly or differently tc cold acclimation; and (c) the cryoprotective solutes and antioxidants are involved in cold acclimation in anthers and ovules.

2. Materials and Methods

2.1. Plant Growth Conditions and Treatments

Chickpea seeds of contrasting genotypes (cold-tolerant: ICC 17258, ICC 16349; cold-sensitive: ICC 15567, GPF 2)—selected from preliminary screening experiments involving 40 genotypes (unpublished)—were soaked for 12 h and inoculated with an appropriate culture of *Rhizobium* sp. Five inoculated seeds were sown in pots filled with sandy loam soil and farmyard manure (3:1 ratio). Tricalcium phosphate fertilizer was added (10 mg kg^{-1} soil). Fifteen days after sowing (DAS), the plants were thinned to two per pot. Sowing was undertaken in the first week of November in an outdoor natural environment in wired enclosures (to protect against birds and animals). The weather data are plotted in Figure 1 (24.9/15.9 °C mean day/night temperatures, 1300–1500 µmol m^{-2}s^{-1} light intensity, 60–70% relative humidity). At 40 DAS, plants were moved into walk-in-growth chambers for the treatments:

1. Control: 25/15 °C (12 h/12 h day/night), 700 µmol m^{-2}s^{-1} light intensity, and 65–70% relative humidity until maturity;
2. Non-acclimated, cold-stressed: 25/15 °C (12 h/12 h day/night), 700 µmol m^{-2}s^{-1} light intensity, and 65–70% relative humidity for one day; temperature then reduced to 13/7 °C (12 h/12 h day/night) over 4 days to avoid lethal shock, where it remained at this temperature until maturity; and
3. Cold-acclimated, cold-stressed: 25/18 °C (12 h/12 h day/night), 700 µmol m^{-2}s^{-1} light intensity, and 65–70% relative humidity for one day, followed by 42 d of cold acclimation, involving 7 d exposure at each decreasing temperature beginning with 23/15 °C, 21/13 °C, 20/12 °C, 20/10 °C, 18/8 °C, 15/8 °C (12 h/12 h day/night) before exposing the plants to cold stress at 13/7 °C (12 h/12 h day/night; 700 µmol m^{-2}s^{-1} light intensity, and 65–70% relative humidity). Thereafter the temperature remained at 13/7 °C until maturity.

The plants were assessed for stress injury during the reproductive stages after experiencing a minimum of 10 d exposure to normal or stressful temperatures using the procedures described below. Young leaves subtending flowers were collected from the second and third nodes. Flowers were collected at the same time. Leaf traits such as stomatal conductance and photosystem II function and biochemical traits were analyzed from 3 different randomly selected young leaves subtending flowers per plant (values were averaged), in three different plants (three replications). The data were pooled, and mean values and standard errors (SE) were estimated.

2.2. Stress Injury

2.2.1. Membrane Damage

Membrane damage was measured as electrolyte leakage (EL). Young fresh leaves located at the second/third node below flowers were collected. For analysis in anthers and ovules, flowers were collected on the day of anthesis. The tissues were washed with deionized water, dissected into smaller segments, and placed in glass vials containing 10 mL deionized water for 12 h at 25 °C. The electrical conductivity (C1) of the surrounding solution was measured after 24 h. The tissue segments were then subjected to 80 °C in a water bath for 10–15 min. The final electrical conductivity (C2) was measured after equilibration. Membrane damage was calculated as C1/C2 × 100 and expressed as a percentage [22].

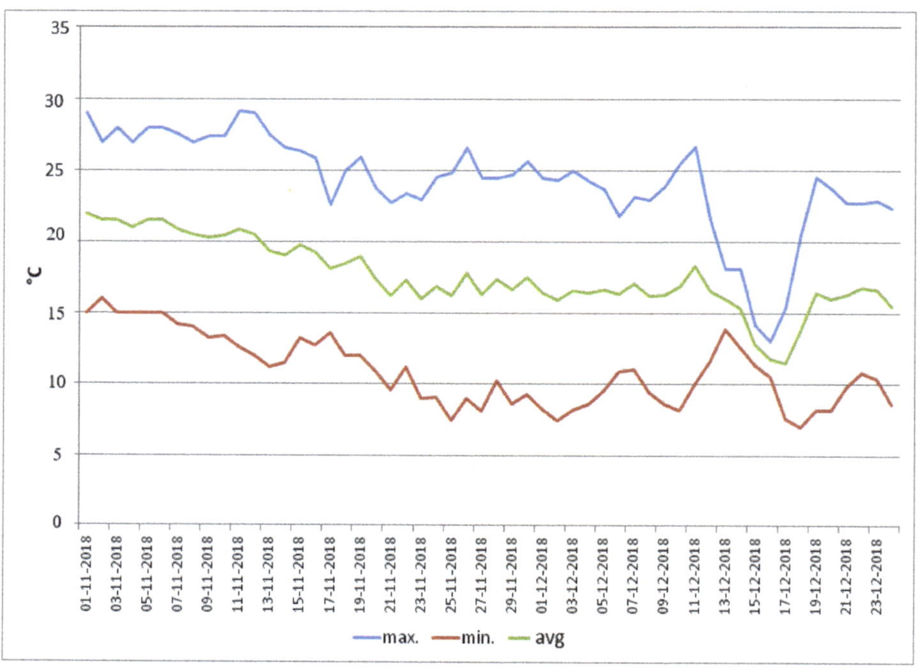

Figure 1. Weather data (maximum (max), minimum (min) and average (avg) temperature) from sowing up to 40 days, when the plants were moved to growth chamber.

2.2.2. Cellular Oxidizing Ability

Cellular oxidizing ability was assessed using 2,3,5-triphenyl tetrazolium chloride (TTC) reduction ability, involving the conversion of a colorless solution into dark red formazan due to reduction by the cells. Fresh tissue (leaves, anthers, or ovules) was immersed in an incubation solution containing 50 mM sodium phosphate (pH 7.4) and TTC (500 mg 100 mL^{-1} solutions) and kept in the dark for 1 h at 25 °C, without shaking as the reduction of TTC responds to high oxygen. The tissue samples were extracted twice (5 mL each) using 95% ethanol and combined to make a final volume of 10 mL. The developed red color was measured at 530 nm using a spectrophotometer and expressed as absorbance g^{-1} fresh weight (FW) [23].

2.2.3. Relative Leaf Water Content

Leaf water status was measured as relative leaf water content (RLWC). Fresh leaves weight (FW) (500 mg) were placed in Petri dishes containing distilled water for 2 h, removed, surface dried with filter paper, weighed initially (turgid weight, TW; weight of fully hydrated leaf), and weighted again after oven-dried at 110 °C for 24 h (dry weight, DW). RLWC was calculated as (FW − DW)/(TW − DW) × 100; expressed as a percentage [24].

2.2.4. Stomatal Conductance

Leaf stomatal conductance was measured with a portable leaf porometer (Decagon Devices, Pullman, WA, USA) and expressed as mmol^{-1} m^{-2} s^{-1} [22].

2.2.5. Photochemical Efficiency

Photochemical efficiency was assessed by recording leaf chlorophyll fluorescence (Fv/Fm ratio) with a chlorophyll fluorometer OS1-FL (Opti-Sciences, Hudson, NH, USA).

2.2.6. Chlorophyll and Carotenoids

Chlorophyll was extracted from fresh leaves (500 mg) using 80% acetone, and centrifuged at $5702 \times g$ for 15 min. The supernatant was collected, and the absorbance read at 666, 653, and 470 nm with a spectrophotometer. The pigment concentration was calculated as per the method of Lichtenthaler and Wellburn [25].

2.3. Reproductive Traits

2.3.1. Pollen Germination

Pollen grains, collected from flowers of the plants harvested for various treatments, were germinated on a growth medium containing 10% sucrose, 1640 mM boric acid, 990 mM nitrate (pH 6.5), 812 mM magnesium sulfate, and 1269 mM calcium nitrate [22,26]. The percentage germination was recorded.

2.3.2. Pollen Viability

Pollen grains were collected from flowers on the day of anthesis and examined for their viability [27]. The viability of ~200 pollen grains based on their size, shape, and color intensity was assessed in five microscopic fields using 0.5% acetocarmine and expressed as a percentage.

2.3.3. Stigma Receptivity

Stigma receptivity was measured using the esterase test, as per the method of Mattison et al. [28]. Stigmas were harvested from flowers one day prior to anthesis, kept in a solution containing α-NAA (naphthaleneacetic acid) and fast blue B (prepared in phosphate buffer) for 15 min at 37 °C. Stigma receptivity was measured based on color intensity, rated on a 1–5 scale (1-low receptivity, 5-high receptivity).

2.3.4. Ovule Viability

Ovules collected from flowers (one day before anthesis) were tested using a TTC reduction assay for their viability. The ovules were placed on a glass slide, treated with 0.5% TTC prepared in 1% solution, and then transferred to a Petri dish containing two filter papers moistened with distilled water. The ovules were incubated for 15 min at 25 °C in a growth chamber. The resulting red color was rated on a 1–5 scale (1-lowest intensity, 5-highest intensity) [22].

2.4. Oxidative Stress and Antioxidants

2.4.1. Malondialdehyde

To measure malondialdehyde (MDA) concentration, fresh tissue (anthers and ovules) was homogenized in 0.1% trichloroacetic acid (TCA) and centrifuged at $3360 \times g$ for 5 min. The supernatant (0.1 mL) was mixed with 4 mL 0.5% thiobarbituric acid (TBA), prepared in 20% TCA. The mixture was heated at 95 °C for 30 min, cooled in an ice bath, and then centrifuged at $3360 \times g$ for 10 min at 4 °C. Absorbance of the supernatant was read at 532 nm. MDA concentration was calculated using an extinction coefficient (155 mM cm^{-1}) and expressed as nmol g^{-1} DW [29].

2.4.2. Hydrogen Peroxide

Hydrogen peroxide (H_2O_2) concentration was measured from fresh tissue (anthers and ovules) extracted in chilled 80% acetone (5 mL), followed by filtration using Whatman filter paper. To this filtrate, 4 mL titanium reagent was added, followed by 5 mL ammonia solution (25%). The mixture was centrifuged at $3360 \times g$ for 10 min; the residue was dissolved in 1 M H_2SO_4. Absorbance of the resulting solution was read at 410 nm.

H_2O_2 concentration was calculated using an extinction coefficient (0.28 mmol cm^{-1}) and expressed as nmol g^{-1} DW [30].

2.4.3. Superoxide Dismutase

Superoxide dismutase (SOD) activity (E.C. 1.15.1.1) was assayed using fresh tissue extracted in a pre-cooled 50 mM phosphate buffer (pH 7.0), which was subsequently centrifuged at 3360× g for 5 min at 4 °C. SOD activity was assayed by preparing a reaction mixture comprising 0.1 mL enzyme extract, 50 mM phosphate buffer (pH 7.8), 13 mM methionine, 25 mM nitro blue tetrazolium chloride (NBT), 0.1 mM EDTA (ethylene diamine tetra acetic acid) in 3 mL total volume. Riboflavin (2 mM) was added, and the mixture was kept in fluorescent light (15 W) for 10 min. Absorbance was read at 560 nm, with the activity measured as per [31] and expressed as units mg^{-1} protein.

2.4.4. Catalase

Catalase (CAT) activity (E.C. 1.11.1.6), was assayed by adding 0.1 mL enzyme extract (as above for SOD) to a reaction mixture containing 50 mM phosphate buffer (pH 7.0) and 200 mM H_2O_2. Absorbance at 410 nm was recorded for 3 min, with the activity measured using an extinction coefficient (40 mM cm^{-1}), expressed as mmol H_2O_2 decomposed mg^{-1} protein [32].

2.4.5. Ascorbate Peroxidase

Ascorbate peroxidase (APX) activity (E.C. 1.11.1.11) was assayed by adding 0.1 mL enzyme extract (as above for SOD) to a reaction mixture containing 50 mM phosphate buffer (pH 7), 0.5 mM ascorbic acid, and 0.1 mM EDTA. H_2O_2 was added as a substrate. The activity was measured using an extinction coefficient (2.8 mM cm^{-1}) [33], expressed as mmol oxidized donor decomposed min^{-1}mg^{-1}protein.

2.4.6. Glutathione Reductase

Glutathione reductase (GR) activity (E.C. 1.6.4.2) was assayed by adding 0.1 mL enzyme extract (as above for SOD) to a reaction mixture containing 1.5 mL phosphate buffer (100 mM; pH 7.6), 0.2 mL BSA, 0.35 mL NADP (nicotinamide adenine dinucleotide phosphate), and 0.1 mL oxidized glutathione. The enzyme activity was measured as the reduction in absorbance at 340 nm for 3 min, expressed as mmol oxidized donor decomposed min^{-1} mg^{-1} protein [34].

2.4.7. Ascorbic Acid

Ascorbic acid (AsA) concentration was determined using fresh tissue extracted in 6% TCA, followed by centrifugation at 3649.15× g for 15 min. To 4 mL supernatant, 2 mL dinitrophenylhydrazine (DNPH; 2%) was added, along with one drop of 10% thiourea. The reaction mixture was boiled in a water bath for 15 min, followed by cooling at room temperature. Pre-cooled H_2SO_4 (5 mL) was added, and the absorbance recorded at 530 nm. The AsA concentration was determined from the standard curve and expressed as mg g^{-1} DW [30].

2.4.8. Glutathione

Reduced glutathione (GSH) concentration was assayed from fresh tissue homogenized in 2 mL metaphosphoric acid; the extract was centrifuged at 3650× g for 15 min. To 0.9 mL supernatant, 0.6 mL sodium citrate (10%) was added. The assay mixture comprised 100 μL extract, 100 μL distilled water, 100 μL 5,5-dithio-bis-(2)-nitrobenzoic acid (DTNB; 6 mM), and 700 μL NADPH (0.3 mM). To this mixture, 10 μL glutathione reductase (Sigma-Aldrich, Burlington, MO, USA) was added, and the absorbance was read at 412 nm. The GSH concentration was determined from a standard graph and expressed as nmol g^{-1} DW [35].

2.5. Soluble Proteins

Plant tissue was oven-dried before extraction with 0.1 M phosphate buffer (pH 7.0) and centrifuged at $514 \times g$ for 15 min. Protein concentration was measured as per [36] and explained by [37].

2.6. Solutes

2.6.1. Proline

Proline concentration was measured in plant tissue using 3% sulphosalicylic acid for extraction, centrifuged at $2150 \times g$ for 20 min at 4 °C. The supernatant was treated with acidic ninhydrin reagent, and the resulting color read at 520 nm, using toluene as a blank. The concentration was measured as nmol g^{-1} DW [38].

2.6.2. Endogenous γ-Aminobutyric Acid

Endogenous γ-aminobutyric acid (GABA) was measured in fresh tissue homogenized in TCA (8%) and centrifuged at $3360 \times g$ for 20 min at 25 °C. The supernatant was treated with 4 mL pure diethyl ether, mixed thoroughly for 10 min with a vortexer, followed by centrifugation at $3360 \times g$ for 20 min. The supernatant was left to sit to evaporate the ether (about 30 min) and tested for GABA concentration, expressed as µmol g^{-1} DW [39].

2.6.3. Trehalose

Trehalose concentration was measured using the method of [40]. The tissue was extracted in 80% hot ethanol, followed by centrifugation at $3360 \times g$ for 15 min. The supernatant (0.1 mL) was mixed with 2 mL TCA and assayed following the method of [41].

2.6.4. Sucrose

Sucrose concentration was measured in fresh tissue after extraction in 80% ethanol at 80 °C for 1.5 h (twice); the two extracts were combined and evaporated at 40 °C in an air-circulating oven. The sucrose concentration was tested as per [42].

2.6.5. In-Vitro Pollen Germination

Freshly collected pollen grains were tested for germination in a growth medium [37] at 13/7 °C; 12 h/12 h; 24 h) in the presence of 1 mM proline, GABA, sucrose, trehalose, ascorbic acid and reduced glutathione in the growth medium, along with control (not supplemented with any of these molecules).

2.7. Statistical Analysis

The experimental design was a 2 factorial randomized block design comprising four contrasting genotypes (two cold-tolerant and two cold-sensitive) and three treatments. There were 15 pots per genotype (two plants per pot) and three replications for each treatment. Five pots in triplicate (15 pots per treatment; 30 plants per treatment) were maintained separately for yield trait measurements. Analysis of variance (ANOVA) for genotype $\times$ treatment interactions was performed using Agristat software (Indian Council of Agricultural Research, Goa, India); least significant values (LSD) values were calculated ($p < 0.05$). Tukey's post hoc test was performed to compare means. In addition, principal component analysis (PCA) was conducted on non-acclimated and acclimated plants to determine the relationships among various measurements.

3. Results

3.1. Stress Injury to Leaves

3.1.1. Membrane Damage

Cold stress increased membrane damage (as electrolyte leakage; EL) in all four genotypes, more so in cold-sensitive genotypes. Cold-stressed tolerant genotypes had 18.4–20.5% EL (control: 10.5–12.5%), while cold-stressed sensitive genotypes had 26.3–28.3% EL (control: 10.1–13.4%; Figure 2A). Cold acclimation significantly reduced membrane damage in all genotypes, which decreased to 14.5–16.4% in tolerant genotypes and 20.6–21.3% in sensitive genotypes.

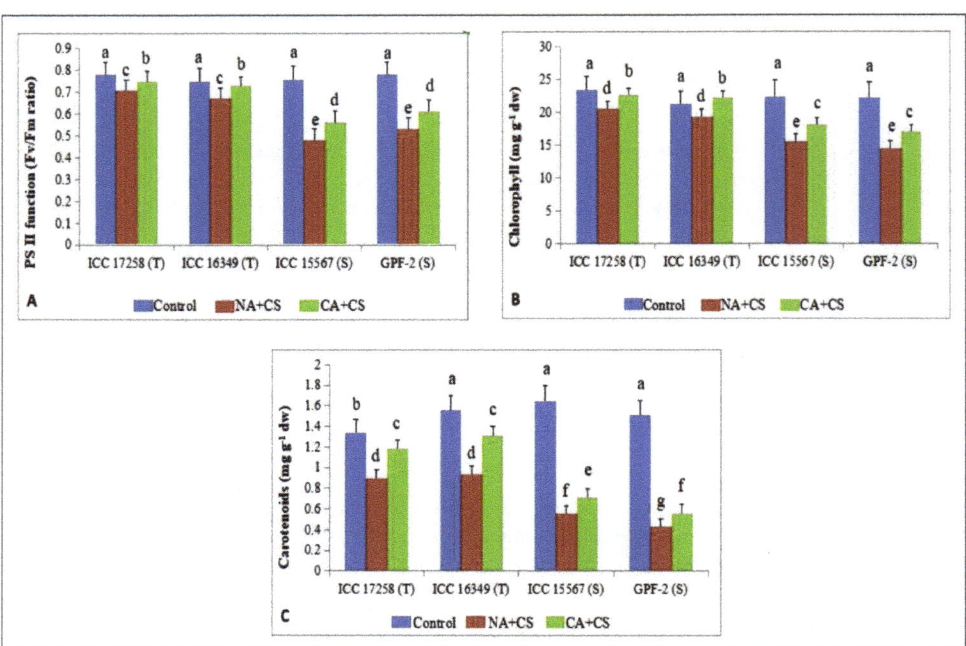

Figure 2. Membrane damage as electrolyte leakage (EL; (**A**)), relative leaf water content (RLWC; (**B**)) and stomatal conductance (*gS*; (**C**)) in leaves of control, non-acclimated, cold stressed; NA + CS) and cold-acclimated, cold stressed (CA + CS) plants of tolerant (T) and sensitive (S) genotypes. Small vertical bars represent standard errors (Mean ± S.E; *n* = 3). Different small letters on vertical bars indicate significant differences from each other (*p* < 0.05; Tukey's test). Least significant difference (LSD) for interaction (*p* < 0.05): (genotypes × treatments) EL (2.4), RLWC (3.1) and *gS* (18.5).

3.1.2. Relative Leaf Water Content

Cold stress decreased relative leaf water content (RLWC) to 69.9–70.4% (control: 81.4–82.3%) in cold-sensitive genotypes and 77.5–78.5% in cold-tolerant genotypes (control: 83.4–86.5%; Figure 2B). The RLWC is an indicator of water status of plant. Cold acclimation had a similar effect on RWLC as membrane damage, i.e., the cold-acclimated plants exposed to cold stress significantly improved their RWLC, nearly to the same extent in all genotypes.

3.1.3. Stomatal Conductance

Cold stress did not significantly affect stomatal conductance (*gS*) in non-acclimated cold-tolerant genotypes (Figure 2C), but it decreased in cold-sensitive genotypes (by 13–14%), relative to their respective controls. Cold acclimation significantly increased *gS* in cold-tolerant (8–9%) and cold-sensitive genotypes (17–18%), compared to non-acclimated plants.

3.1.4. Photosystem II Function

Photosystem II (PSII) function of control plants ranged from 0.75–0.78 Fv/Fm (variable fluorescence/maximum fluorescence) ratio with no variation between cold-tolerant and cold-sensitive genotypes (Figure 3A). Cold stress decreased PSII function in non-acclimated plants, more so in cold-sensitive genotypes (32–36%) than cold-tolerant genotypes (8–10%) as compared to controls. Cold acclimation significantly enhanced PSII function under cold stress, increasing by 6–8% in cold-tolerant genotypes and 15–16% in cold-sensitive genotypes, relative to non-acclimated plants.

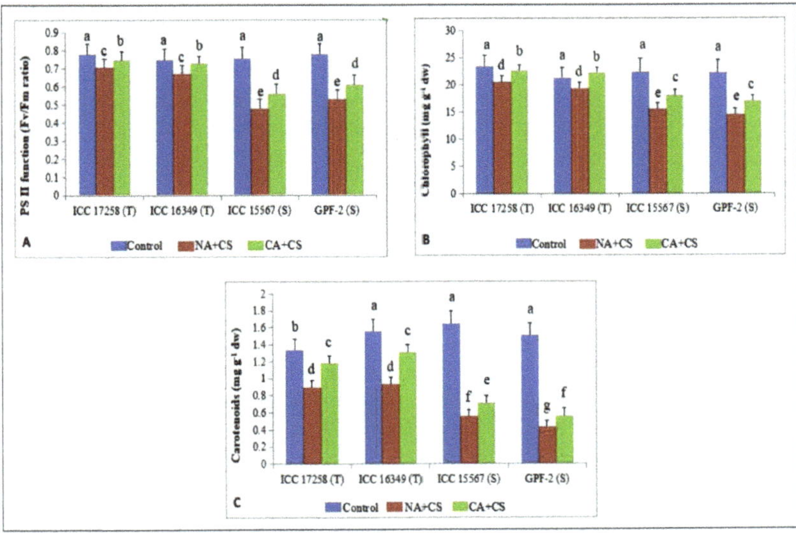

Figure 3. Photosystem II function (PS; (**A**)), chlorophyll (Chl; (**B**)) and carotenoids (Car; (**C**)) in leaves of control, non-acclimated, cold stressed; NA + CS) and cold-acclimated, cold stressed (CA + CS) plants of tolerant (T) and sensitive (S) genotypes. Small vertical bars represent standard errors (Mean ± S.E; *n* = 3). Different small letters on vertical bars indicate significant differences from each other (*p* < 0.05; Tukey's test). Least significant difference (LSD) for interaction (*p* < 0.05) (genotypes × treatments): PS (0.057), Chl (2.5), Car (0.056).

3.1.5. Photosynthetic Pigments

Cold acclimation had a similar effect on chlorophyll (Chl) content as PSII function (Figure 3B). In non-acclimated plants, cold stress significantly decreased Chl in all four genotypes, relative to their respective controls, more so in cold-sensitive genotypes (34–35%) than cold-tolerant genotypes (9–11%). Cold acclimation significantly increased leaf Chl in all four genotypes, relative to non-acclimated plants, more so in cold-sensitive genotypes (16–17%) than cold-tolerant genotypes (10–15%).

Cold had a greater impact on carotenoids than Chl (Figure 3C). In non-acclimated plants, cold stress decreased carotenoid content by 33–40% in cold-tolerant genotypes and 66–71% in cold-sensitive genotypes), compared to their respective controls. Cold acclimation increased carotenoid content in all four genotypes, but unlike PSII function and chlorophyll content, cold-tolerant genotypes increased carotenoids more (32–40%) than cold-sensitive genotypes (26–30%), relative to non-acclimated plants.

3.2. Reproductive Traits

For reproductive parameters of male or female, cold-tolerant genotypes responded better to cold acclimation than cold-sensitive genotypes. The general effects of cold stress and cold acclimation on flowers, anthers and pollen grains are shown in Figure 4.

3.2.1. Pollen Germination

Pollen germination in control plants ranged from 78.5–87.5%. Cold stress decreased pollen germination to 27.8–34.6% in non-acclimated cold-tolerant genotypes and 6.4–8.9% in non-acclimated cold-sensitive genotypes (Figure 5A). Cold-acclimated cold-tolerant genotypes had higher pollen germination (72.4–76.9%) than their non-acclimated counterparts, but cold-acclimated cold-sensitive genotypes had poor pollen germination (22.5–25.6%).

3.2.2. Stigma Receptivity

Under cold stress, the female organ in non-acclimated and cold-acclimated chickpea behaved much like the male gametophyte, suggesting that it is also highly sensitive to cold stress. Stigma receptivity was assessed on 1–5 scale using visual scoring. Under cold stress, cold-tolerant genotypes had markedly higher stigma receptivity (2.2–2.6) than cold-sensitive genotypes (1.0) (Figure 5B) but both were markedly lower than the control plants (4.2–4.8). Cold acclimation significantly improved stigma receptivity in all four genotypes, with cold-tolerant genotypes close to control values (4.2–4.3) and cold-sensitive genotypes about one-third of control values (1.6–1.8).

3.2.3. Pollen Viability

Pollen viability in cold-stressed non-acclimated plants decreased to 26.4–32.4% in cold-tolerant genotypes and 4.6–5.9% in cold-sensitive genotypes of their respective controls (81.7–88.9%) (Figure 5C). Cold-acclimated plants exposed to cold stress had much higher pollen viability than cold-stressed non-acclimated plants, being 71.8–75.6% in cold-tolerant genotypes and 25.6–27.3% in cold-sensitive genotypes.

3.2.4. Ovule Viability

Ovule viability was assessed on 1–5 scale using visual scoring. Cold stress significantly reduced ovule viability in all four genotypes, decreasing to 2.2–2.7 in cold-tolerant genotypes and 1.0 in cold-sensitive genotypes, relative to 4.3–4.8 in control plants (Figure 5D). Cold acclimation significantly improved ovule viability under cold stress, more so in cold-tolerant genotypes (4.1–4.2) than cold-sensitive genotypes (1.3–1.5).

3.2.5. Tissue Damage to Anthers and Ovules

Cold stress damaged anther and ovule tissues in chickpea, as evidenced from increased electrolyte leakage, expressed as percentage (Figure 6A). Cold-stressed non-acclimated plants of cold-sensitive genotypes had more damage (EL: 27.9–28.3% in anthers, 21.3–24.3% in ovules) than cold-tolerant genotypes (14.7–16.8% in anthers, 11.9–13.6% in ovules), with control plants ranging from 9.3–10.3% in anthers and 7.3–9.2% in ovules. Cold acclimation reduced cold-induced tissue damage in anthers and ovules, more so in cold-tolerant genotypes (11.5–12.4% in anthers, 9.3–10.8% in ovules) than cold-sensitive genotypes (22.4–23.5% in anthers, 17.6–19.8% in ovules).

Cellular viability also decreased with cold stress in anthers and ovules of non-acclimated plants (Figure 6B), more so in cold-sensitive genotypes (56–58% in anthers, 43–45% in ovules over control) than cold-tolerant genotypes (23–35% in anthers, 22–33% in ovules over control). Cold acclimation significantly increased cellular viability to 26–37% in anthers and 21–27% in ovules of cold-tolerant genotypes and 14–23% in anthers and 11–18% in ovules of cold-sensitive genotypes.

3.3. Oxidative Stress and Antioxidants

Cold acclimation decreased oxidative stress in anthers and ovules, more so in cold-tolerant genotypes than cold-sensitive genotypes.

3.3.1. Malondialdehyde

Cold stress increased malondialdehyde (MDA) concentration in the anthers and ovules of non-acclimated plants (Figure 6C), while cold acclimation reduced MDA concentration in these tissues. In anthers of cold-stressed non-acclimated plants, MDA concentrations increased more in cold-sensitive genotypes (5.5–7.8-fold) than cold-tolerant genotypes (3.7–4.5-fold), relative to their respective controls. Cold-acclimated plants had significantly lower MDA concentrations in anthers than non-acclimated plants, more so in cold-tolerant genotypes (3–3.5-fold), compared to cold-sensitive genotypes (1.26–1.32-fold).

Control plants had 1.8–2.2 nmoles g^{-1} dw of MDA in ovules, which increased with cold stress, more so in cold-sensitive genotypes (4.3–5.6-fold) than cold-tolerant genotypes

(2.4–3.2-fold), over their respective controls. Cold acclimation significantly reduced MDA concentrations in ovules, more so in cold-tolerant genotypes (1.85–1.93-fold) than cold-sensitive genotypes (1.13–1.17-fold), relative to non-acclimated plants.

3.3.2. Hydrogen Peroxide

Control plants had anther H_2O_2 concentrations of 1.6–1.9 µmol g^{-1} dw (Figure 6D), increasing under cold stress by 3.8–3.9-fold in cold-sensitive genotypes and 1.84–2.06-fold in cold-tolerant genotypes. Cold acclimation significantly decreased anther H_2O_2 concentrations, more so in cold-tolerant genotypes (1.45–1.57-fold) than cold-sensitive genotypes (1.28–1.17-fold), relative to non-acclimated plants. Control plants had ovule H_2O_2 concentrations of 1.3–1.7 µmol g^{-1} dw, increasing with cold stress by 1.93–1.84-fold in cold-tolerant genotypes and 2.28–2.41-fold in cold-sensitive genotypes. Cold acclimation decreased H_2O_2 concentrations, more so in cold-tolerant genotypes (1.6–1.5-fold) than cold-sensitive genotypes (1.14–1.20-fold), relative to non-acclimated plants.

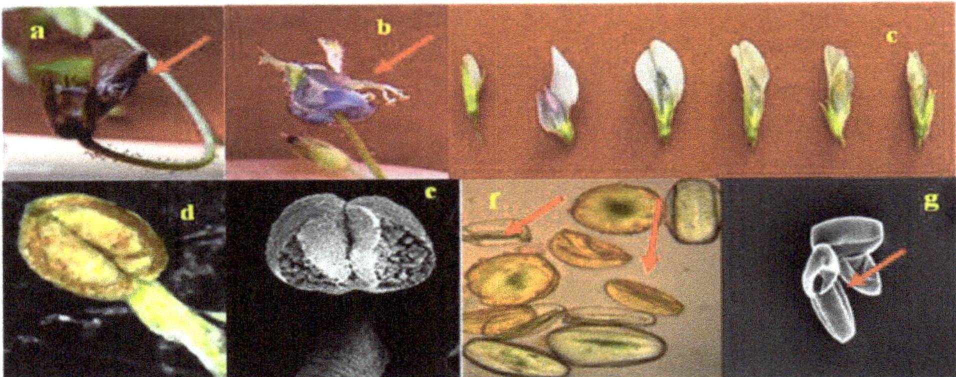

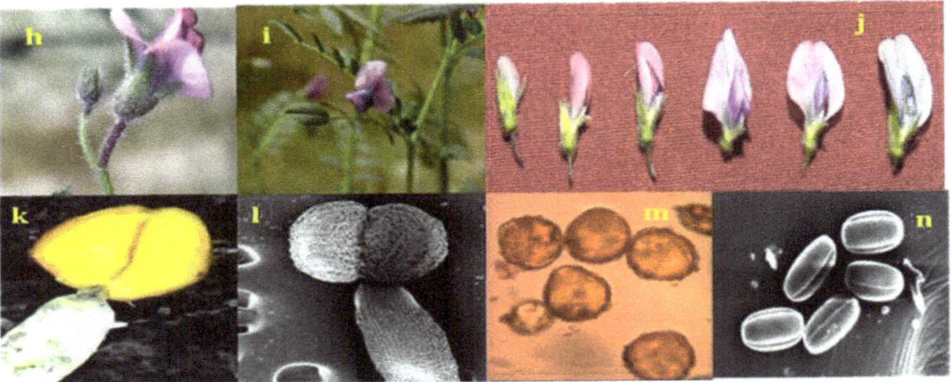

Figure 4. Images showing the effect of cold stress (above) and cold acclimation (below) on chickpea genotypes (reproductive phase). Cold stress effects from genotype ICC 15567 (cold-sensitive): aborted flower (**a**), flower with exposed anthers (**b**), developmental changes leading to abortion of flower (**c**), damaged anthers (**d,e**), and distorted and shriveled pollen grains (**f,g**). Cold acclimation effects from genotype ICC 16349 (cold-tolerant): flowers (**h,i**), developmental changes in cold-acclimated flowers (**j**), cold-acclimated anthers (**k,l**), and cold-acclimated viable pollen grains (**m,n**).

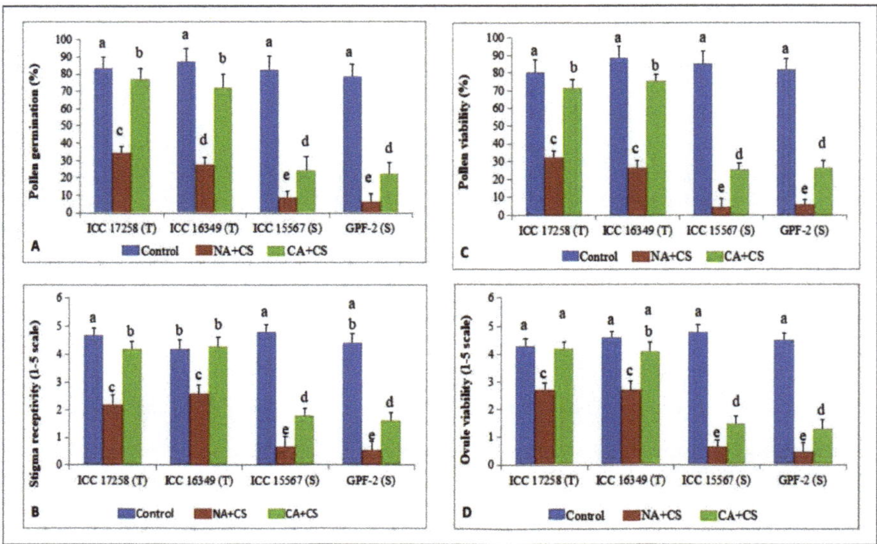

Figure 5. Pollen germination (PG; (**A**)), pollen viability (PV; (**C**)), stigma receptivity (SR; (**B**)) and ovule viability (OV; (**D**)) in control, non-acclimated, cold stressed; NA + CS) and cold-acclimated, cold stressed (CA + CS) plants of tolerant (T) and sensitive (S) genotypes. Small vertical bars represent standard errors (Mean ± S.E; n = 3). Different small letters on vertical bars indicate significant differences from each other ($p < 0.05$; Tukey's test). Least significant difference (LSD) for interaction ($p < 0.05$) (genotypes × treatments): PG (6.9), PV (7.1), SR (0.39), OV (0.36).

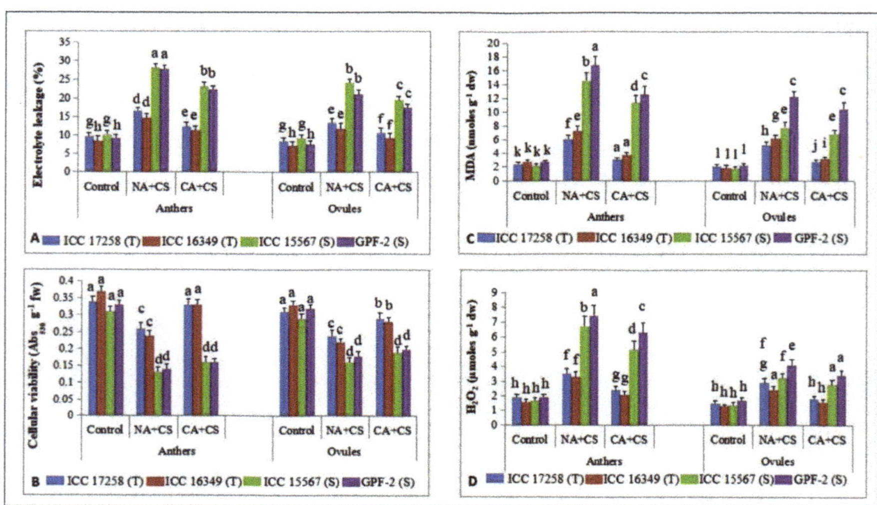

Figure 6. Electrolyte leakage (EL; (**A**)), cellular viability (CV; (**B**)), (Malondialdehyde (MDA; (**C**)) and hydrogen peroxide (H_2O_2; (**D**)) concentration in anthers and ovules of control, non-acclimated, cold stressed; NA + CS) and cold-acclimated, cold stressed (CA + CS) plants of tolerant (T) and sensitive (S) genotypes. Small vertical bars represent standard errors (Mean ± S.E; n = 3). Different small letters on vertical bars indicate significant differences from each other ($p < 0.05$; Tukey's test). Least significant difference (LSD) for interaction ($p < 0.05$) (genotypes × treatments): EL (2.4), CV (0.067), MDA (1.8), H_2O_2 (0.23).

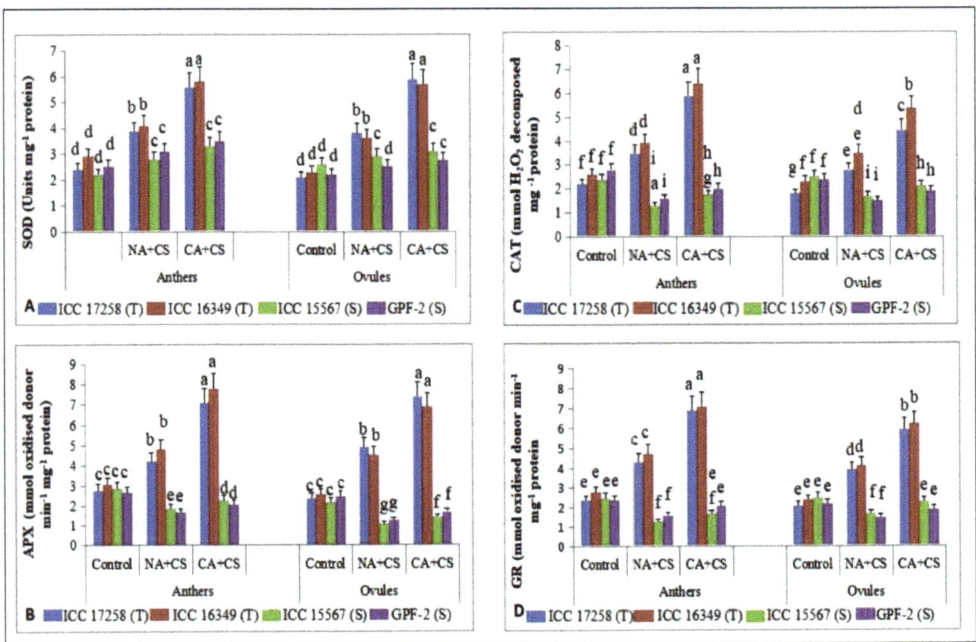

Figure 7. Superoxide dismutase (SOD; (**A**)), ascrobate peroxidase (APX; (**B**)), catalase (CAT; (**C**)) and glutathione reductase (GR; (**D**)) in anthers and ovules of control, non-acclimated, cold stressed; NA + CS) and cold-acclimated, cold stressed (CA + CS) plants of tolerant (T) and sensitive (S) genotypes. Small vertical bars represent standard errors (Mean ± S.E; *n* = 3). Different small letters on vertical bars indicate significant differences from each other (*p* < 0.05; Tukey's test). Least significant difference (LSD) for interaction (*p* < 0.05) (genotypes × treatments): SOD (0.54), CAT (0.62), APX (0.45), GR (0.49).

3.3.3. Ascorbate Peroxidase

The ascorbate peroxidase (APX) activity in anthers and ovules of control plants ranged from 2.7–3.1 and 2.2–2.6 mmol oxidized donor min^{-1} mg^{-1} protein, respectively (Figure 7B). Cold acclimation had a similar effect on APX activity as CAT activity. Cold-stressed non-acclimated plants increased APX activity in cold-tolerant genotypes (50–54% in anthers, 73–104% in ovules), but decreased APX activity in cold-sensitive genotypes (34–37% in anthers, 48–50% in ovules), relative to the controls. Cold acclimation significantly increased APX activity in all four genotypes, relative to non-acclimated plants, more so in cold-tolerant genotypes (62–69% in anthers, 51–53% in ovules) than cold-sensitive genotypes (21–23% in anthers, 30–31% in ovules).

3.3.4. Catalase

The catalase (CAT) activity in anthers and ovules of control plants ranged from 2.2–2.8 and 1.8–2.4 mmol H$_2$O$_2$ decomposed mg^{-1} protein, respectively (Figure 7C). Cold stress increased CAT activity by 50–59% in anthers and 52–58% in ovules of cold-tolerant genotypes but decreased CAT activity by 42–45% in anthers and 32–37% in ovules of cold-sensitive genotypes, relative to their respective controls. Cold acclimation increased CAT activity by 64–68% in anthers and 52–58% in ovules of cold-tolerant genotypes and 25–34% in anthers and 23–26% in ovules of cold-sensitive genotypes, relative to non-acclimated plants.

3.3.5. Glutathione Reductase

Cold stress significantly increased glutathione reductase (GR) activity (Figure 7D) in anthers (67–79%) and ovules (67–85%) of non-acclimated plants of cold-tolerant genotypes, but significantly decreased GR activity in cold-sensitive genotypes (33–48% in anthers, 31–33% in ovules), relative to their corresponding controls. Cold acclimation increased GR activity by 51–60% in anthers and 51% in ovules of cold-tolerant genotypes and 30–31% in anthers and 26–35% in ovules of cold-sensitive genotypes, relative to non-acclimated plants.

3.3.6. Ascorbate

In non-acclimated plants, cold stress significantly increased ascorbate (ASC) concentration (Figure 8A) in anthers (25–33%) and ovules (31–33%) of cold-tolerant genotypes, but significantly reduced ASC concentration in anthers (20–32%) and ovules (21–31%) of cold-sensitive genotypes, relative to control plants. Cold acclimation significantly improved ASC concentration, more so in anthers (43–50%) and ovules (46–49%) of cold-tolerant genotypes than anthers (8–10%) and ovules (14–15%) of cold-sensitive genotypes, relative to non-acclimated plants.

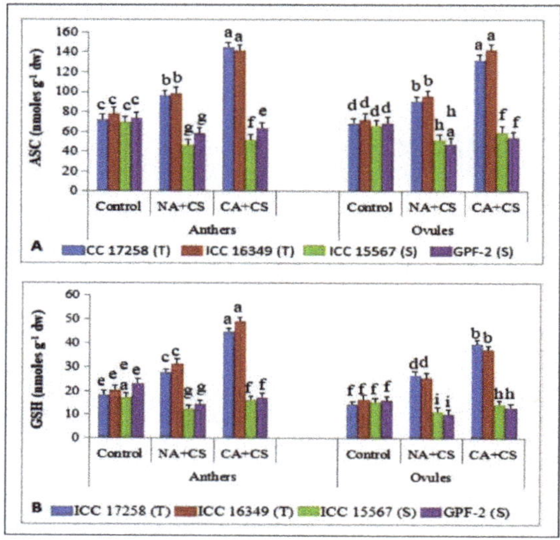

Figure 8. Ascorbate (ASC; (**A**)) and reduced glutathione (GSH; (**B**)) in anthers and ovules of control, non-acclimated, cold stressed; NA + CS) and cold-acclimated, cold stressed (CA + CS) plants of tolerant (T) and sensitive (S) genotypes. Small vertical bars represent standard errors (Mean ± S.E; $n = 3$). Different small letters on vertical bars indicate significant differences from each other ($p < 0.05$; Tukey's test). Least significant difference (LSD) for interaction ($p < 0.05$) (Genotypes × treatments): ASC (6.9), GSH (3.4).

3.3.7. Glutathione

Cold stress increased glutathione (reduced; GSH) concentrations in anthers (50–53%) and ovules (53–84%) of cold-tolerant genotypes, but decreased GSH concentrations in anthers (28–38%) and ovules (25–36%) of cold-sensitive genotypes, relative to the controls (Figure 8B). Cold acclimation significantly increased GSH levels in anthers and ovules under cold stress, relative to cold-stressed non-acclimated plants, more so in cold-tolerant genotypes (56–60% in anthers, 45–48% in ovules) than cold-sensitive genotypes (20–28% in anthers, 25–27% in ovules).

3.4. Cryoprotective Solutes

3.4.1. Proline

Cold stress increased proline (Pro) concentrations in non-acclimated plants of cold-tolerant genotypes (42–57% in anthers, 65–67% in ovules), but decreased proline concentrations in cold-sensitive genotypes (18–32% in anthers, 31–35% in ovules), relative to the controls (Figure 9A). Cold acclimation increased proline concentrations in anthers and ovules, more so in cold-tolerant genotypes (64–85% in anthers, 66–69% in ovules) than cold-sensitive genotypes (30% in anthers, 37–38% in ovules), relative to non-acclimated plants.

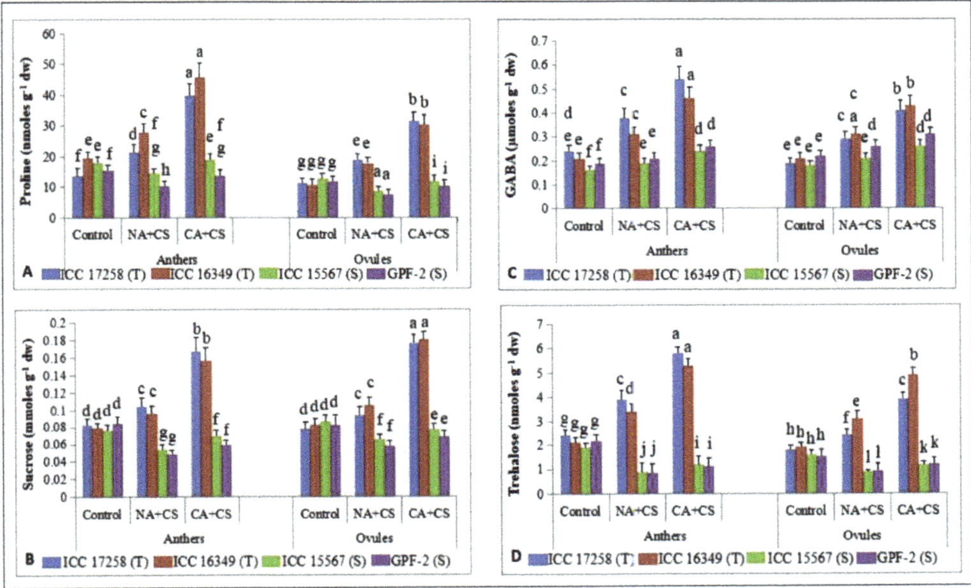

Figure 9. Proline (Pro; (**A**)), Sucrose (Suc; (**B**)), γ-amino butyric acid (GABA; (**C**)) and trehalose (Tre; (**D**)) in anthers and ovules of control, non-acclimated, cold stressed; NA + CS) and cold-acclimated, cold stressed (CA + CS) plants of tolerant (T) and sensitive (S) genotypes. Small vertical bars represent standard errors (Mean ± S.E; *n* = 3). Different small letters on vertical bars indicate significant differences from each other (*p* < 0.05; Tukey's test). Least significant difference (LSD) for interaction (*p* < 0.05) (genotypes × treatments): Pro (3.9), GABA (0.048), Suc (0.009), Tre (0.40).

3.4.2. Sucrose

Cold stress increased sucrose concentrations by 21–26% in anthers and 20–28% in ovules of cold-tolerant genotypes, but decreased sucrose concentrations by 28–42% in anthers and 24–29% in ovules of cold-sensitive genotypes, relative to their respective controls (Figure 9B). Cold acclimation increased sucrose concentrations more in cold-tolerant genotypes (60–62% in anthers, 71–87% in ovules) than cold-sensitive genotypes (23–27% in anthers, 17–18% in ovules), relative to non-acclimated plants.

3.4.3. γ-. Aminobutyric Acid

Cold stress increased γ-aminobutyric acid (GABA) concentrations more in cold-tolerant genotypes (47–58% in anthers, 46–52% in ovules) than cold-sensitive genotypes (10–18% in anthers, 16–18% in ovules), relative to their respective controls (Figure 9C). Cold acclimation further increased GABA concentrations in all four genotypes, more so in cold-tolerant genotypes (42–48% in anthers, 39–41% in ovules) than cold-sensitive genotypes (23–26% in anthers, 19–23% in ovules), relative to non-acclimated plants.

3.4.4. Trehalose

Cold stress significantly increased trehalose levels in anthers (61–62%) and ovules (33–63%) of cold-tolerant genotypes, but significantly reduced trehalose levels in anthers (53–60%) and ovules (39–45%) of cold-sensitive genotypes, relative to the controls (Figure 9D). Cold acclimation increased trehalose concentrations in all four genotypes, more so in cold-tolerant genotypes (48–55% in anthers, 58–62% in ovules) than cold-sensitive genotypes (31–35% in anthers, 31–33% in ovules), compared to non-acclimated plants.

3.5. Yield Traits

Pod set in control plants was 72.1–74.5% in cold-tolerant genotypes and 69.4–71.4% in cold-sensitive genotypes (Figure 10A). Cold stress decreased pod set to 19–22% in cold-tolerant genotypes and zero in cold-sensitive genotypes. Cold acclimation increased pod set under cold stress to 48.9–51.4% in cold-tolerant genotypes but had no effect on pod set in cold-sensitive genotypes.

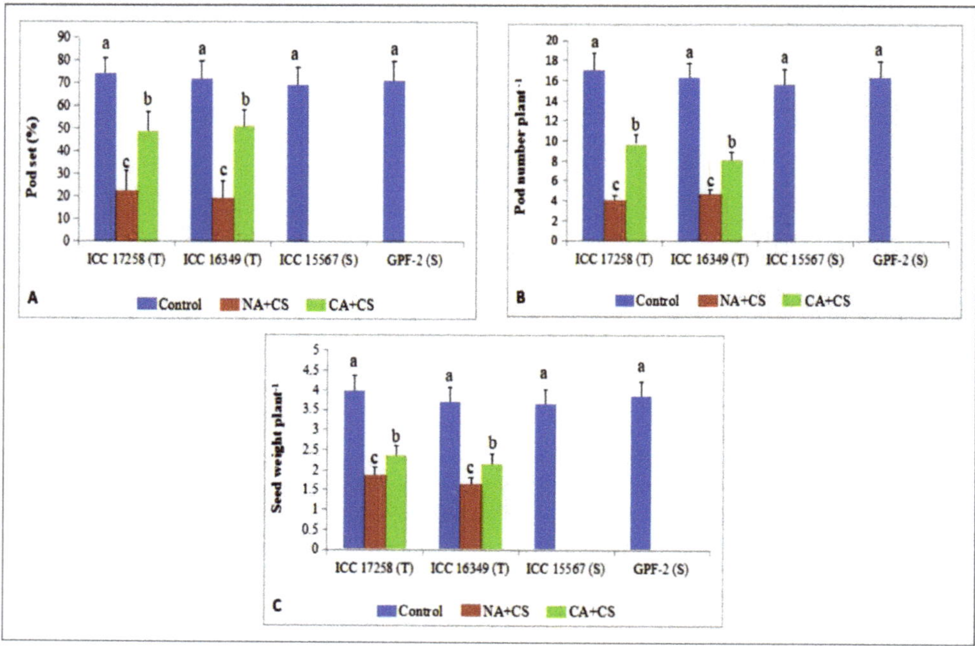

Figure 10. Pod set (**A**), pod number plant^{-1} (**B**) and seed weight plant^{-1} (**C**) in plants of control, non-acclimated, cold stressed; NA + CS) and cold-acclimated, cold stressed (CA + CS) plants of tolerant (T) and sensitive (S) genotypes. Small vertical bars represent standard errors (Mean ± S.E; *n* = 3). Different small letters on vertical bars indicate significant differences from each other (*p* < 0.05; Tukey's test). Least significant difference (LSD) for interaction (*p* < 0.05) (Genotypes × treatments): Pod set (10.3), Pod number (1.8), Seed weight (0.41).

Control plants of cold-tolerant and cold-sensitive genotypes produced 16.3–17.1 and 15.6–16.3 pods per plant, respectively (Figure 10B). Cold stress decreased pod number by 70–76% (4.1–4.7 pods plant^{-1}) in cold-tolerant genotypes, relative to the control plants, but cold-sensitive genotypes did not produce any pods. Cold acclimation improved pod set under cold stress by 70–76% (8.1–9.7 pods per plant) in cold-tolerant genotypes, relative to non-acclimated plants, while cold-sensitive genotypes did not produce any pods.

Control plants of cold-tolerant genotypes yielded 3.65–3.98 g plant^{-1}, which decreased by 52–55% under cold stress (no yield in cold-sensitive genotypes) (Figure 10C). Cold acclimation increased seed yield in cold-tolerant genotypes under cold stress, without any effect on cold-sensitive genotypes. Cold acclimation improved seed yield by 25–31% (to 2.16–2.35 g plant^{-1}) in cold-tolerant genotypes under cold stress, relative to non-acclimated plants.

3.6. Effect of Cryoprotective Solutes and Antioxidants on In-Vitro Pollen Germination

Pollen germination in control plants of cold-tolerant genotypes was 85.6–88.15% and cold-sensitive genotypes was 79.6–82.1%. Cold stress decreased pollen germination to 31.3–35.6% in cold-tolerant genotypes and 6.8–8.9% in cold-sensitive genotypes (Table 1). Exogenous supplementation of ascorbate, GSH, proline, trehalose, and sucrose improved pollen germination markedly in all four genotypes, more so with sucrose supplementation, followed by GABA and ascorbate.

Table 1. Effect of various solutes and non-enzymatic antioxidants on pollen germination under cold stress in tolerant (T) and sensitive (S) chickpea genotypes.

Treatment	ICC 17258 (T)	ICC 16348 (T)	ICC 15567 (S)	GPF2 (S)
Control	85.6 ± 5.9 a	88.1 ± 5.1 a	82.1 ± 4.8 a	79.6 ± 5.5 a
Cold-stressed (CS) (13/7 °C;12 h/12 h; 1 d)	35.6 ± 4.9 d	31.3 ± 3.6 d	8.9 ± 2.1 f	6.8 ±1.6 f
CS + Proline (1mM)	53.5 ± 4.8 cd	52.4 ± 5.4 d	21.9 ± 3.4 ef	23.4 ± 2.5 ef
CS + GABA (1 mM)	61.3 ± 4.3 bc	60.1 ± 4.7 bc	33.5 ± 2.3 e	31.5 ± 2.1 e
CS + Sucrose (1 mM)	65.6 ± 4.2 b	62.4 ± 4.6 b	31.3 ± 2.1 e	34.2 ± 3.2 e
CS + Trehalose (1 mM)	54.5 ± 3.5 d	52.4 ± 4.4 d	29.5 ± 3.1 e	31.2 ± 3.3 e
CS + Ascorbate (1 mM)	53.8 ± 3.2 d	56.1 ± 3.7 cd	28.7 ± 2.7 e	31.3 ± 2.9 e
CS + Reduced Glutathione (1 mM)	56.4 ± 3.5 cd	53.2 ± 2.9 d	26.3 ± 2.4 ef	29.6 ± 2.2 e

Different small letters indicate significant differences from each other ($p < 0.05$; Tukey's test).

3.7. Principal Component Analysis

3.7.1. Non-Acclimated (NA) Plants

Principal component analysis (PCA; Figure 11) graph for the chickpea genotypes grown under non-acclimated temperature conditions revealed a significant positive relationship among yield traits (pod set %, pod number plant^{-1} and seed weight and seed number plant^{-1}), reproductive traits (pollen germination, PG; pollen viability, PV; stigma receptivity, SR; ovule viability, OV; cellular viability, CV), leaf traits (stomatal conductance, gS; relative leaf water content, RLWC; chlorophyll, Chl; carotenoids, CAR; chlorophyll fluorescence, PSII) and biochemical traits (SOD, CAT, APX, GR, ASC, GSH and proline). All these traits were found to strongly correlate with each other except electrolyte leakage (EL) MDA, H_2O_2 that indicated negative correlation with cold tolerance.

PCA revealed that PC1 and PC2 accounted for 97.9% of the variation (PC1: 94.6% and PC2: 3.3%). PC1 showed EL in leaves, anthers and ovules, MDA and H_2O_2 in anthers and ovules. PC2 showed yield traits (pod set %, pod number plant^{-1}, seed weight and seed number plant^{-1}), reproductive traits (PG, PV, SR and OV), stress injury traits (gS, RLWC, CV, Chl, CAR, and PSII) and biochemical traits (SOD, CAT, APX, GR, ASC, GSH and proline).

The indices here formed three groups; Group 1 had six indices: proline (anthers and ovules), GR (ovules), ASC (ovules), APX (anthers), chlorophyll content (Chl, leaves), CAT (ovules and anthers). Group-2 included 13 indices: CV (anthers and ovules), PSII (leaves), ASC (anthers), SOD (anthers), RLWC, PV, PG, APX (anthers and ovules), GR (anthers and ovules), GSH (ovules), OV, pod number per plant and seed weight per plant. A strong and positive correlation was noticed in Group 1 and Group 2 with an acute angle, thus, suggesting that any of these traits may probably be used to measure the association of various traits with yield plant^{-1}. Group 3 consisted of H_2O_2, MDA (anthers and ovules), and EL (leaves, anthers and ovule), which had a negative association with yield per plant as well as with indices in Groups 1 and 2.

Narrow vector angles in the PC1-dominating variables in the arc from H_2O_2 (ovules) to EL (ovules) reveal strong correlations between these variables (H_2O_2, MDA (anthers and ovules) and EL (leaves, anther and ovule)). These traits indicate the low temperature injury to membranes and oxidative damage to the chickpea genotypes and are negatively correlated with other traits (yield and biochemical traits). The traits such as RLWC, Chl, PSII, CV, proline (anthers and ovules), SOD (anthers and ovules), CAT (ovules and anthers), APX (anthers and ovules), GR (anthers and ovules), ASC (anthers and ovules), GSH (anthers and ovules) were strongly correlated with yield traits (pod number plant^{-1} and seed eightw plant^{-1}) and reproductive traits (PV, PG, SR, and OV). Hence, it can be concluded that these traits of leaves, anthers ovules and pollen grains would be useful as indicators of yield under non-acclimated cold stress conditions in chickpea.

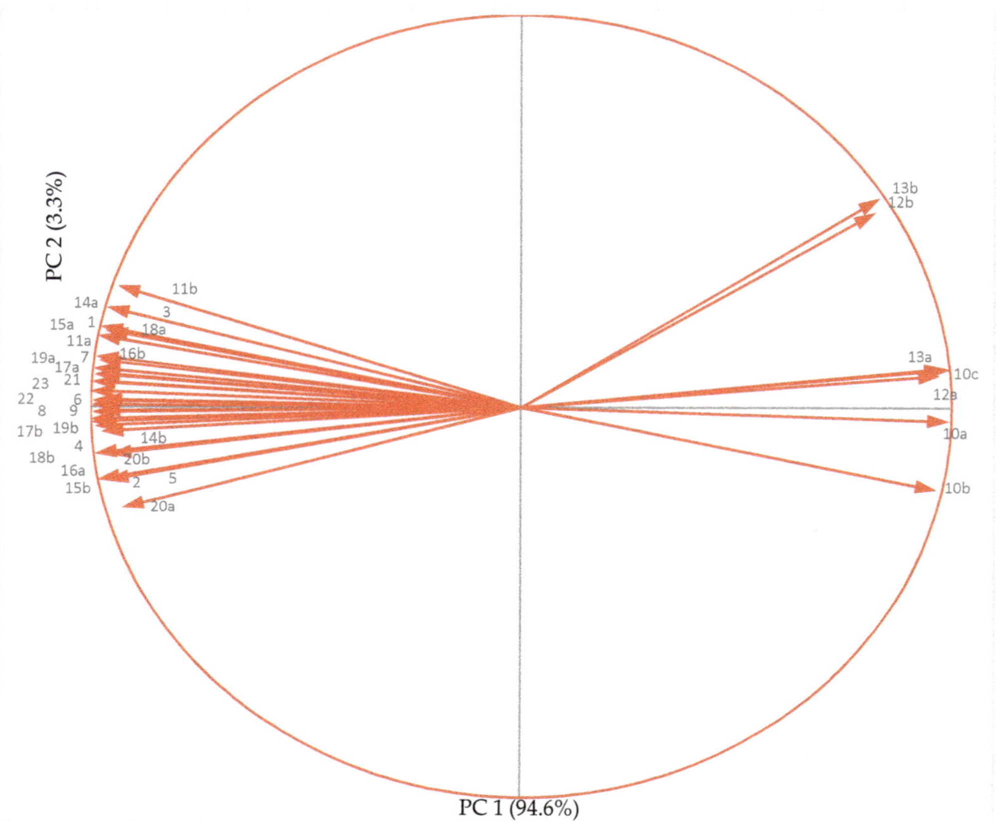

Figure 11. Principal component analysis of yield, reproductive, biochemical and leaf traits in chickpea genotypes under non acclimated conditions. Abbreviations: a-anther, b-ovule; 1-RLWC-relative leaf water content; 2-SC-stomatal conductance; 3-PSII-photosynthetic efficiency; 4-Chl-chlorophyll content; 5-CAR-carotenoids; 6-PG-pollen germination; 7-PV-pollen viability; 8-SR-stigma receptivity; 9-OV-ovule viability; 10-EL-electrolyte leakage; 11-CV-cellular viability; 12-MDA-malondialdehyde; 13-H$_2$O$_2$-hydrogen peroxide; 14-SOD-superoxide dismutase; 15-CAT- catalase; 16-APX-ascorbate peroxidase; 17-GR-glutathione reductase; 18-ASC-ascorbic acid; 19-GSH-glutathione; 20-proline; 21-pod set %; 22- pod number plant^{-1}; 23-seed weight plant^{-1}.

3.7.2. Cold-Acclimated Plants

Principal component analysis (PCA; Figure 12) for the chickpea genotypes grown under cold-acclimated conditions revealed a significant positive relationship among yield traits (pod set %, pod number plant^{-1} and seed weight plant^{-1}), reproductive traits (PG, PV, SR, OV, CV), leaf traits (*g*S, RLWC, Chl, CAR, and PSII) and biochemical traits (SOD, CAT, APX, GR, ASC, GSH and proline). All these traits showed strong correlation to each other except MDA, H$_2$O$_2$, and EL that indicated association of these traits with low temperature damage to vegetative and reproductive tissues.

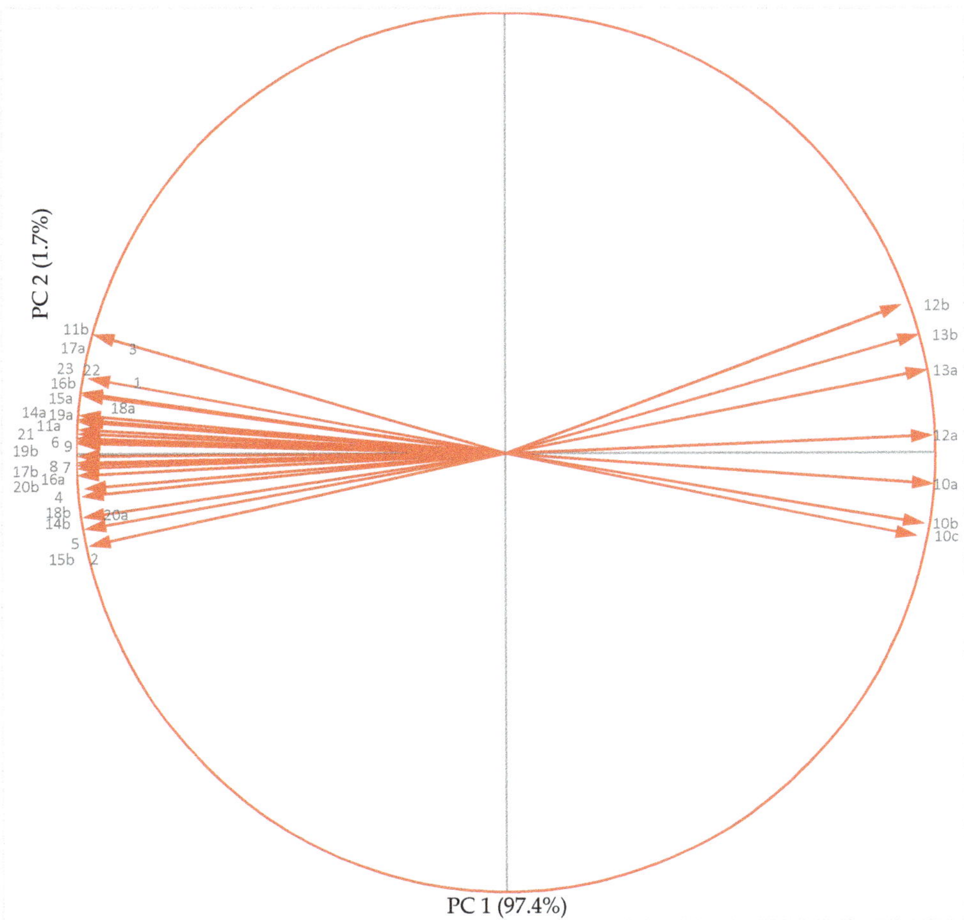

Figure 12. Principal component analysis of yield, reproductive, biochemical and leaf traits in chickpea genotypes under cold-acclimated conditions. a-anther, b-ovule; 1-RLWC-relative leaf water content; 2-SC-stomatal conductance; 3-PSII-photosynthetic efficiency; 4-Chl-chlorophyll content; 5-CAR-carotenoids; 6-PG-pollen germination; 7-PV-pollen viability; 8-SR-stigma receptivity; 9-OV-ovule viability; 10-EL-electrolyte leakage; 11-CV-cellular viability; 12-MDA-malondialdehyde; 13-H_2O_2-hydrogen peroxide; 14-SOD-superoxide dismutase; 15-CAT-catalsae; 16-APX-ascorbate peroxidase; 17-GR-glutathione reductase; 18-ASC-ascorbic acid; 19-GSH-glutathione; 20-proline; 21-pod set %; 22-pod number plant^{-1}; 23-seed weight plant^{-1}.

PCA showed that PC1 and PC2 accounted for 99.2% of the variation (PC1: 97.8% and PC2: 1.4%). PC1 described EL (leaves, anthers and ovules), MDA and H_2O_2 (anthers and ovules). Yield traits (pod set %, pod number plant^{-1} and seed weight plant^{-1}), reproductive traits (PG, PV, SR, OV), stress injury traits (gS, RLWC, CV, Chl, and PSII) and biochemical traits (SOD, CAT, APX, GR, ASC, GSH, H_2O_2 (ovules) and proline (anthers and ovules)) as indicated on PC2. The indices here formed three groups with Group-1 comprising of six indices: proline (anthers and ovules), GR, ASC (ovules), APX (anthers), Chl, CAT (ovules and anthers). Group-2 included 13 indices: CV (anthers and ovules), PSII, ASC (anthers), SOD (anthers), RLWC, PV, PG, APX (anthers and ovules), GR (anthers and ovules), GSH (anthers and ovules), OV, pod number plant^{-1} and seed weight plant^{-1}. The indices in Group-1 and Group-2 were strongly and positively correlated with an acute angle and indicated that any of these indices can probably be used to correlate antioxidative activity and yield plant^{-1}. There was third group of H_2O_2, MDA (anthers and ovules), and EL (leaves, anthers and ovule), which represent a weak link with indices in Group 1 and 2.

Narrow vector angles in the PC1-dominating variables, described in the arc from MDA (ovule) to EL (leaf) reveal strong correlations between H_2O_2, MDA (anthers and ovules) and EL (leaves, anthers and ovules). These traits indicate low temperature injury to chickpea genotypes, therefore are negatively correlated with other traits (yield and biochemical traits). Since, proline (anthers and ovules), GR (anthers and ovules), ASC (anthers and ovules), APX (anthers and ovules), chlorophyll (Chl), CAT (ovules and anthers), CV (anthers and ovules), PSII, SOD (anthers and ovules), RLWC, GR (anthers and ovule), GSH (anthers and ovules) were strongly correlated with yield traits (pod number plant^{-1} and seed weight plant^{-1}) and reproductive traits (PV, PG, SR, OV), it can be concluded that these traits of leaves, anthers, ovules and pollen grains would be useful as indicators of yield under cold-acclimated conditions.

4. Discussion

As a winter season crop in several parts of the world, chickpea suffers from cold-stress-induced damage to vegetative and reproductive tissues. Studies have reported beneficial effects of cold acclimation for chickpea during the seedling or early vegetative phase [9,21,43], there are no reports investigating the impact of cold acclimation during the reproductive stage. In the present study, cold acclimation improved leaf, anther, and ovule function under cold stress, relative to non-acclimated plants, suggesting that cold acclimation is advantageous for vegetative and reproductive organs, improving plant growth, reproduction, and yield. Our study also showed that cold acclimation improved the response of vegetative tissues (leaves) to cold stress by reducing cold-induced damage and improving cellular function, such as membrane damage or relative leaf water content, stomatal conductance, PSII function, or leaf chlorophyll and carotenoid concentrations.

Cold acclimation can improve hardiness to cold stress [17] through various mechanisms. Cold acclimation can reduce membrane damage by increasing the ratio of unsaturated to saturated fatty acids, as reported in 20-day old chickpea seedlings [20]. We observed improved leaf water status in cold-acclimated chickpea plants, as reported in barley [18], and could be due to better root hydraulic conductivity and osmolyte accumulation [44]. The observed reduction in chlorophyll loss of cold-acclimated chickpea plants might have resulted from augmented leaf water status and reduced oxidative damage [9]. The reduction in chlorophyll and PSII function agrees with previous studies on cold-acclimated chickpea seedlings [43] and *Arabidopsis thaliana* (accession C24) [19] exposed to cold stress. Carotenoids are vital for maintaining the leaf redox status, protecting them from photoinhibition under cold stress [45], in our study, cold acclimation increased leaf carotenoid concentrations in cold-stressed chickpea, which might have protected the leaves from photoinhibition by adjusting the redox status and keeping the leaves photosynthetically active.

In non-acclimated chickpea plants exposed to cold stress, the marked reductions in growth, pod set, yield-related traits (pod and seed weights), and reproductive function could be associated with increased membrane damage and decreased water status, stomatal conductance, chlorophyll concentration, and PSII function in leaves. In chickpea, low temperature stress increased membrane damage [5] and decreased leaf hydration status and stomatal conductance could be due to reduced root hydraulic conductivity [46,47], chlorophyll [48], chlorophyll fluorescence [43], pollen function, stigmatic and ovular activity [4,8,49], and pod set and yield traits [7,10] Cold-stress-induced membrane disruption results from altered lipid–protein interactions [50] or lipid peroxidation [51], chlorophyll loss in cold-stressed plants, as observed in our study, might be due to inhibited chlorophyll synthesis or increased chlorophyll degradation [52] or photooxidation-induced disorganization of chloroplasts [53], which consequently decreases chlorophyll fluorescence [43]. Leaf damage due to cold stress can disrupt photosynthetic function and sucrose synthesis and transport to developing floral organs, causing impaired reproductive function and reduced yields [54].

Unlike leaf tissues, the response of reproductive organs to cold acclimation in cold-tolerant and cold-sensitive chickpea genotypes differed. The zero pod set and zero yield in cold-acclimated cold-sensitive genotypes under cold stress indicates the lack of a cold acclimatization response. In contrast, cold-tolerant genotypes had a cold acclimation response (increased pod and seed set relative to non-acclimated plants). Interestingly, our findings and those of [9] indicate that vegetative and reproductive tissues of cold-sensitive chickpea genotypes differ in their response to cold acclimation. Indeed, reproductive organs (anthers and ovules) had significantly more tissue damage and less cellular viability in cold-sensitive genotypes than cold-tolerant genotypes; moreover, these organs were less responsive to cold acclimation in sensitive genotypes. Thus, the differential response to cold acclimation might lie in the tissue sensitivity of floral organs in cold-sensitive and cold-tolerant genotypes, as indicated by various traits related to tissue damage, but this aspect needs further study.

In chickpea, cold stress reduces pollen viability, pollen load on stigma, stigma receptivity, and ovule viability [4,49]. Chickpea plants fail to set pods at temperatures <20/10 °C due to various abnormalities related to developmental and functional factors [1,4,8,49]. In cold-tolerant chickpea genotypes, cold acclimation reduced the adverse effect of cold stress, increasing yield. Little or no cold acclimation of reproductive organs in cold-sensitive genotypes might be due to poor expression of enzymatic and non-enzymatic antioxidants and reduced accumulation of cryoprotective molecules in reproductive organs. The cold-sensitive genotypes were unable to significantly reduce cold-stress-induced oxidative stress markers, such as MDA and H_2O_2, in both male and female reproductive organs following acclimation. Consequently, these genotypes failed to detoxify ROS following the production of those by lower temperatures, impairing male and female gamete function and causing flower/pod abortion.

The role of ROS is well documented for sensitivity to abiotic stresses [55]. In chickpea, cold stress affects male and female gamete function, resulting in poor pollen germination, viability, stigmatic receptivity, and ovule viability [4,9,49,56]. The current study showed that cold stress caused tissue damage in anthers and ovules and reduced their cellular viability. The manifold increase in oxidative stress in anthers and ovules under cold stress points to its role in tissue damage and cell viability in these organs. Therefore, it cannot be ruled out that oxidative-stress-induced tissue damage disrupts developmental and functional aspects of anthers and ovules. In rice anthers, ROS accumulation has been reported under drought [57], and heat stress [58]. In cytoplasmic male sterile (CMS) rice material, the CMS line (sterile anthers) had significantly higher ROS concentrations in anthers than the corresponding maintainer line (fertile anthers) [59,60].

The cold acclimation response of cold-tolerant chickpea genotypes could be attributed to a substantial reduction in MDA and H_2C_2 levels in anthers and ovules and increased accumulation of antioxidants (enzymatic and non-enzymatic). An increase in enzymatic and non-enzymatic antioxidant levels reduced the oxidative species generated under cold stress, thus reducing the oxidative stress in anthers and ovules to levels too low to cause considerable damage to these organs. Thus, reduced oxidative damage to these organs improved anther and ovule performance under cold stress in cold-acclimated plants, relative to non-acclimated plants in the cold-tolerant genotypes. Decreased production of oxidative species and increased production of antioxidants leads to cold tolerance in crops such as rice (*Oryza sativa* L.) [61] and *Brassica* sp. [62]. Cold acclimation improved the antioxidant capacity of barley [63] and chickpea [64] leaves.

Numerous studies have demonstrated that the antioxidant enzyme system in plants can protect against ROS, but little is known about antioxidant enzymes in developing anthers [4], or the interaction between cold-induced ROS concentrations in anthers and ovules of chickpea. In some crops, antioxidant enzymes reduce ROS-induced damage and are important components of plant tolerance to environmental stresses [65,66]. In the present study, the activities of SOD (causes dismutation of peroxides), CAT (detoxifies the hydrogen peroxide), APX (detoxifies hydrogen peroxide using ascorbate as a substrate), and GR (catalyzes the reduction of glutathione disulfide to the sulfhydryl form GSH) increased in anthers and ovules of non-acclimated cold-tolerant genotypes, indicating an inherent ability of these genotypes to reduce cold-induced oxidative stress. However, the reduction in pod numbers in cold-tolerant non-acclimated genotypes exposed to cold stress suggests that the decrease in oxidative damage in anthers and ovules was not significant. In contrast, cold-sensitive genotypes had much lower antioxidant levels in anthers and ovules than cold-tolerant genotypes, causing severe oxidative damage to these organs, manifested as inhibited reproductive function and lack of pod set. The considerably greater reduction in tissue damage (as EL and cellular viability) in anthers and ovules of cold-tolerant genotypes than cold-sensitive genotypes might be due to an improvement in unsaturation of lipids [67], and reduction in oxidative stress in acclimated plants. Like anthers and ovules, cold acclimation reduced the severity of oxidative stress in chickpea seedlings [21] and barley leaves [63]. Variations, however, have been reported in the activities and the type of antioxidants in cold-acclimated plants, which might depend on the experimental conditions and plant species used [63,68]. In the present study, components of the ascorbate–glutathione pathway were greatly expressed, compared to other antioxidative enzymes, suggesting their larger role in the cold acclimation potential of cold-tolerant genotypes.

Cryoprotective molecules can maintain reproductive function in plants. Following cold acclimation, the anthers and ovules of cold-sensitive genotypes accumulated lesser amounts of cryoprotective molecules, such as proline, GABA, trehalose, and sucrose, compared to cold-tolerant genotypes. Our previous study [8] on cold-stressed chickpea revealed an association between reduced carbohydrates in ovules and floral abortion. The cold acclimation of cold-tolerant genotypes can thus be attributed to the inherent ability of these genotypes to reduce oxidative stress and enhance antioxidant levels (enzymatic and non-enzymatic) and cryoprotective solutes in reproductive organs (anthers and ovules), improving reproductive function, e.g., pollen viability, pollen load on stigma, stigma receptivity and ovule viability, and subsequently number of pods and seeds.

Cold acclimation can enhance endogenous proline (*Chrysanthemum* sp.) [69], carbohydrates (safflower, *Carthamus tincotorius*) [70], and GABA (barley and wheat; [71] levels in plants. Cryoprotective solutes, such as amino acids (proline, GABA) and carbohydrates (sucrose, trehalose), play diverse roles in plant cells [72]. Moreover, its role as an osmolyte in osmotic adjustment, proline stabilizes membranes and proteins, scavenges free radicals, and buffers cellular redox potential under stress conditions [73]. The importance of proline in cold stress mitigation can be judged because it has been used as a biomarker of cold tolerance [74]. In cold-tolerant chickpea under cold stress, higher proline levels were attributed to increased expression of the gene responsible for proline transport, *proline transporter 1* [4]. GABA is a non-protein amino acid—it has a signaling role with functions to protect from oxidative stress, maintain C and N mechanism, regulate pH in cytosol, and in osmoregulation [75] and cold tolerance [76]. Trehalose (α-D-glucopyranosyl- α-D-glucopyranoside) is a vital compatible sugar solute—it has a signaling role and stabilizes lipid membranes, dehydrated enzymes, and proteins during desiccation [77]. It has also been implicated in acquiring stress tolerance in plants, including cold stress [78]. Sucrose has been implicated in conferring cold tolerance [79] and can directly protect cell membranes by interacting with the phosphate in their lipid headgroups, thus decreasing membrane permeability [80]. Non-acclimated cold-tolerant chickpea genotypes had substantially higher levels of these solutes than non-acclimated cold-sensitive genotypes, suggesting their involvement in cold tolerance. However, their concentrations may have been inadequate to maintain reproductive competence. The depletion of proline, sucrose, and reducing sugars in flowers due to impaired mobilization and synthesis causes flower abortion due to decreased pollen viability and retarded pollen tube growth [9,56].

Sucrose, in addition to a cryoprotectant, might act as a source of carbon to developing anthers and ovules. Adequate carbohydrate supply is critical for anther function under cold stress [81] and sucrose is an important carbohydrate molecule required for proper anther function, especially under stress, e.g., in tomato (*Solanum lycopersicum*) [82] and chickpea [8]. In an earlier study, the expression of sucrose-synthesizing genes was compared in anthers of cold-stressed cold-tolerant and cold-sensitive chickpea genotypes [4]. Under cold stress, the anthers of cold-tolerant genotype, ICC 16349, had higher pollen viability than cold-sensitive, GPF2. Increased pollen viability in the cold-tolerant genotype was associated with up-regulation of sucrose-synthesizing genes, *UDP glucose pyrophosphorylase, sucrose phosphate synthase2,* and *CWIN cell wall invertase* [4].

PCA graphs of non-acclimated and cold-acclimated treatments of chickpea genotypes demonstrated strong correlation among reproductive, biochemical, anti-oxidative and yield traits. At the same time, cold-acclimated plants showed increased protective traits (CAR, Chl, CV, SC, PSII, CAT, SOD, APX, GR, ASC, proline) as compared to non-acclimated plants so that plants could achieve cold tolerance. Furthermore, cold-acclimated plants showed higher reproductive traits (PG, PV, SR, OV) than non-acclimated plants that may result in enhanced yield traits (pod set %, pod number plant^{-1}, seed weight plant^{-1}). In contrast, non-acclimated plants showed significant chilling injury traits (EL, MDA, H_2O_2) as compared to cold-acclimated plants. Moreover, there was a strongly positive correlation among various protective, reproductive and yield traits in cold-acclimated plants as compared to non-acclimated plants. Thus, cold-acclimated plants acquired substantial cold tolerance that leads to increased yield.

5. Conclusions

Vegetative and reproductive tissues respond to cold acclimation in chickpea. However, the degree of responsiveness varies between the tissues in cold-tolerant and cold-sensitive genotypes. Following cold acclimation, the leaves (vegetative) of cold-tolerant and cold-sensitive genotypes had less cold-induced membrane damage and improved cellular function (relative leaf water content, stomatal conductance, PSII function, chlorophyll and carotenoid contents) under cold stress. The degree of responsiveness of reproductive organs (anthers and ovules) to cold acclimation in cold-tolerant and cold-sensitive chickpea genotypes varied, with little to no response of cold-sensitive genotypes (zero pod set and zero yield under cold stress), while cold-tolerant genotypes improved pod set and seed yield, relative to non-acclimated plants. In cold-sensitive genotypes, the lack of cold acclimation resulted from the inability of anthers and ovules to reduce oxidative stress either through the reduced generation of oxidative molecules or enhanced production of enzymatic and non-enzymatic antioxidants in both reproductive tissues. The anthers and ovules of cold-sensitive genotypes also failed to produce enough cryoprotective solutes (proline, GABA, trehalose, and sucrose), instrumental in reducing cold-induced damage, and thus had more tissue damage, less cellular viability and lower pollen and ovule viability, pollen load on stigma, and stigma receptivity than cold-tolerant genotypes. In contrast, the responsiveness of cold-tolerant genotypes to cold acclimation resulted from their ability to produce lower amounts of oxidative molecules and increased activity/amounts of antioxidants and cryoprotective solutes in anthers and ovules, reducing damage to anthers and ovules to maintain their viability and reproductive function under cold stress, leading to improved pod set and seed yield, relative to non-acclimated plants. PCA analysis of the non-acclimated and cold-acclimated conditions cold-stressed chickpea plants revealed similarity in types of various antioxidants and cryo-protective solutes required in imparting a stable reproductive function to confer cold tolerance. However, the expression of these molecules was much stronger in cold-acclimated plants, which minimized the oxidative damage. We conclude that cold tolerance in chickpea appears to be related to the better ability of anthers and ovules to acclimate to cold stress through various antioxidants and cryoprotective solutes. This information will be useful in developing genetic, molecular, breeding and agronomic management practices toward increasing cold tolerance in chickpea.

Author Contributions: Conceptualization, H.N., K.H.M.S., P.V.V.P., K.D.S., U.C.J.; methodology, A.R., A.K.; formal analysis, H.N., K.D.S. and U.C.J.; resources, U.C.J.; writing—original draft preparation, H.N., A.R.; writing, review and editing, H.N., K.H.M.S., K.D.S., U.C.J., P.V.V.P.; supervision, H.N.; project administration, H.N.; funding acquisition, H.N., P.V.V.P. All authors have read and agreed to the published version of the manuscript.

Funding: This research received no external funding.

Institutional Review Board Statement: Not applicable.

Informed Consent Statement: Not applicable.

Data Availability Statement: The data presented in this study are available in article.

Acknowledgments: H.N. appreciates support from Panjab University. All authors are thankful to the Department of Science and Technology, Department of Biotechnology, Department of Science and Technology Purse Award from the Ministry of Agriculture, India; Consultative Group of International Agricultural Research, International Center for Agricultural Research in the Dryland Area (ICARDA—Morocco); and The University of Western Australia for funding the infrastructural facilities. A.R is thankful to Council of Scientific and Industrial Research, India for providing the fellowships during study. Contribution number 22-103-J from Kansas Agricultural Experiment Station.

Conflicts of Interest: The authors declare no conflict of interest.

References

1. Croser, J.S.; Clarke, H.J.; Siddique, K.H.M.; Khan, T.N. Low-temperature stress: Implications for chickpea (*Cicer arietinum* L.) improvement. *Crit. Rev. Plant. Sci.* **2003**, *22*, 185–219. [CrossRef]
2. Rani, A.; Devi, P.; Jha, U.C.; Sharma, K.D.; Siddique, K.H.; Nayyar, H. Developing climate-resilient chickpea involving physiological and molecular approaches with a focus on temperature and drought stresses. *Front. Plant Sci.* **2020**, *10*, 1759. [CrossRef] [PubMed]
3. Singh, K.B.; Malhotra, R.S.; Halila, M.H.; Knights, E.J.; Verma, M.M. Current status and future strategy in breeding chickpea for resistance to biotic and abiotic stresses. *Euphytica* **1993**, *73*, 137–149. [CrossRef]
4. Kiran, A.; Sharma, P.N.; Awasthi, R.; Nayyar, H.; Seth, R.; Chandel, S.S.; Sharma, K.D. Disruption of carbohydrate and proline metabolism in anthers under low temperature causes pollen sterility in chickpea. *Environ. Exp. Bot.* **2021**, *188*, 104500. [CrossRef]
5. Mir, A.H.; Bhat, M.A.; Dar, S.A.; Sofi, P.A.; Bhat, N.A.; Mir, R.R. Assessment of cold tolerance in chickpea (*Cicer* spp.) grown under cold/freezing weather conditions of North-Western Himalayas of Jammu and Kashmir, India. *Physiol. Mol. Biol. Plants* **2021**, *27*, 1105–1118. [CrossRef]
6. Srinivasan, A.; Johansen, C.; Saxena, N.P. Cold tolerance during early reproductive growth of chickpea (*Cicer arietinum* L.): Characterization of stress and genetic variation in pod set. *Field Crop. Res.* **1998**, *57*, 181–193. [CrossRef]
7. Clarke, H.J.; Siddique, K.H.M. Response of chickpea genotypes to low temperature stress during reproductive development. *Field Crop. Res.* **2004**, *90*, 323–334. [CrossRef]
8. Nayyar, H.; Bains, T.; Kumar, S. Low temperature induced floral abortion in chickpea: Relationship to abscisic acid and cryoprotectants in reproductive organs. *Environ. Exp. Bot.* **2005**, *53*, 39–47. [CrossRef]
9. Nayyar, H.; Bains, T.S.; Kumar, S. Chilling stressed chickpea seedlings: Effect of cold acclimation, calcium and abscisic acid on cryoprotective solutes and oxidative damage. *Environ. Exp. Bot.* **2005**, *54*, 275–285. [CrossRef]
10. Berger, J.D.; Kumar, S.; Nayyar, H.; Street, K.A.; Sandhu, J.S.; Henzell, J.M.; Clarke, H.C. Temperature-stratified screening of chickpea (*Cicer arietinum* L.) genetic resource collections reveals very limited reproductive chilling tolerance compared to its annual wild relatives. *Field. Crop. Res.* **2012**, *126*, 119–129. [CrossRef]
11. Thakur, P.; Kumar, S.; Malik, J.A.; Berger, J.D.; Nayyar, H. Cold stress effects on reproductive development in grain crops: An overview. *Environ. Exp. Bot.* **2010**, *67*, 429–443. [CrossRef]
12. Ritonga, F.N.; Chen, S. Physiological and molecular mechanism involved in cold stress tolerance in plants. *Plants* **2020**, *9*, 560. [CrossRef] [PubMed]
13. Thomashow, M.F. Plant cold acclimation: Freezing tolerance genes and regulatory mechanisms. *Annu. Rev. Plant Biol.* **1999**, *50*, 571–599. [CrossRef] [PubMed]
14. Orvar, B.L.; Sangwan, V.; Omann, F.; Dhindsa, R.S. Early steps in cold sensing by plant cells: The role of actin cytoskeleton and membrane fluidity. *Plant J.* **2000**, *23*, 785–794. [CrossRef] [PubMed]
15. Kang, H.M.; Saltveit, M.E. Activity of enzymatic antioxidant defense systems in chilled and heat shocked cucumber seedling radicles. *Physiol. Plant.* **2001**, *113*, 548–556. [CrossRef]
16. Pearce, R.S. Molecular analysis of acclimation to cold. *Plant Growth Regul.* **1999**, *29*, 47–76. [CrossRef]
17. Rife, C.L.; Zeinali, H. Cold tolerance in oilseed rape over varying acclimation durations. *Crop Sci.* **2003**, *43*, 96–100. [CrossRef]
18. Burchett, S.; Niven, S.; Fuller, M.P. The effect of cold-acclimation on the water relations and freezing tolerance of *Hordeum vulgare* L. *Cryo. Letters* **2006**, *27*, 295–303.

19. Mishra, K.B.; Mishra, A.; Kubasek, J.; Urban, O.; Heyer, A.G. Low temperature induced modulation of photosynthetic induction in non-acclimated and cold-acclimated Arabidopsis thaliana: Chlorophyll a fluorescence and gas-exchange measurements. *Photosyn. Res.* **2019**, *139*, 123–143. [CrossRef]

20. Kazemi-Shahandashti, S.S.; Maali-Amiri, R.; Zeinali, H.; Khazaei, M.; Talei, A.; Ramezanpour, S.S. Effect of short-term cold stress on oxidative damage and transcript accumulation of defense-related genes in chickpea seedlings. *J. Plant Physiol.* **2014**, *171*, 1106–1116.

21. Turan, O.; Ekmekçi, Y. Chilling tolerance of *Cicer arietinum* lines evaluated by photosystem II and antioxidant activities. *Turk. J. Bot.* **2014**, *38*, 499–510. [CrossRef]

22. Kaushal, N.; Awasthi, R.; Gupta, K.; Gaur, P.; Siddique, K.H.; Nayyar, H. 2013 Heat-stress-induced reproductive failures in chickpea (*Cicer arietinum* L.) are associated with impaired sucrose metabolism in leaves and anthers. *Funct. Plant Biol.* **2013**, *40*, 1334–1349. [CrossRef] [PubMed]

23. Steponkus, P.L.; Lanphear, F.O. Refinement of the triphenyl tetrazolium chloride method of determining cold injury. *Plant Physiol.* **1967**, *42*, 1423–1426. [CrossRef]

24. Barrs, H.D.; Weatherley, P.E. A re-examination of the relative turgidity technique for estimating water deficits in leaves. *Aust. J. Biol. Sci.* **1962**, *15*, 413–428. [CrossRef]

25. Lichtenthaler, H.K.; Wellburn, A.R. Determination of total carotenoids and chlorophylls a and b of leaf in different solvents. *Biochem. Soc. Trans.* **1985**, *11*, 591–592. [CrossRef]

26. Brewbaker, J.L.; Kwack, B.H. The essential role of calcium ion in pollen germination and pollen tube growth. *Am. J. Bot.* **1963**, *50*, 859–865. [CrossRef]

27. Alexander, M.P. Differential staining of aborted and non aborted pollen. *Stain. Technol.* **1969**, *44*, 117–122. [CrossRef]

28. Mattison, O.; Knox, R.B.; Heslop-Harrison, J.; Heslop-Harrison, Y. Protein pellicle of stigmatic papillae as a probable recognition site in incompatible reactions. *Nature* **1974**, *247*, 298–300. [CrossRef]

29. Heath, R.L.; Packer, L. Photoperoxidation in isolated chloroplasts: I. Kinetics and stoichiometry of fatty acid peroxidation. *Arch. Biochem. Biophys.* **1968**, *125*, 189–198. [CrossRef]

30. Mukherjee, S.P.; Choudhuri, M.A. Implications of water stress-induced changes in the levels of endogenous ascorbic acid and hydrogen peroxide in *Vigna* seedlings. *Physiol. Plant.* **1983**, *58*, 166–170. [CrossRef]

31. Dhindsa, R.S.; Matowe, W. Drought tolerance in two mosses: Correlated with enzymatic defense against lipid peroxidation. *J. Exp. Bot.* **1981**, *32*, 79–91. [CrossRef]

32. Teranishi, Y.; Tanaka, A.; Osumi, M.; Fukui, S. Catalase activities of hydrocarbon-utilizing Candida yeasts. *Agric. Biol. Chem.* **1974**, *38*, 1213–1220. [CrossRef]

33. Nakano, Y.; Asada, K. Hydrogen peroxide is scavenged by ascorbate-specific peroxidase in spinach chloroplasts. *Plant Cell Physiol.* **1981**, *22*, 867–880.

34. Mavis, R.D.; Stellwagen, E. Purification and subunit structure of glutathione reductase from bakers' yeast. *J. Biol. Chem.* **1968**, *243*, 809–814. [CrossRef]

35. Griffith, O.W. Determination of glutathione and glutathione disulfide using glutathione reductase and 2-vinylpyridine. *Anal. Biochem.* **1980**, *106*, 207–212. [CrossRef]

36. Lowry, O.H.; Rosebrough, N.J.; Farr, A.L.; Randall, R.J. Protein measurement with the Folin phenol reagent. *J. Boil. Chem.* **1951**, *193*, 265–275. [CrossRef]

37. Sita, K.; Sehgal, A.; Kumar, J.; Kumar, S.; Singh, S.; Siddique, K.H.; Nayyar, H. Identification of high-temperature tolerant lentil (*Lens culinaris* Medik.) genotypes through leaf and pollen traits. *Front. Plant Sci.* **2017**, *8*, 744. [CrossRef] [PubMed]

38. Bates, L.S.; Waldren, R.P.; Teare, I.D. Rapid determination of free proline for water-stress studies. *Plant. Soil.* **1973**, *39*, 205–207. [CrossRef]

39. Saito, T.; Matsukura, C.; Sugiyama, M.; Watahiki, A.; Ohshima, I.; Iijima, Y.; Ezura, H. Screening for γ-aminobutyric acid (GABA)-rich tomato varieties. *J. Jpn. Soc. Hort. Sci.* **2008**, *77*, 242–250. [CrossRef]

40. Trevelyan, W.E.; Harrison, J.S. Studies on yeast metabolism. 1. Fractionation and micro determination of cell carbohydrates. *Biochem. J.* **1952**, *50*, 298–303. [CrossRef] [PubMed]

41. Kumar, S.; Kaushal, N.; Nayyar, H.; Gaur, P. Abscisic acid induces heat tolerance in chickpea (*Cicer arietinum* L.) seedlings by facilitated accumulation of osmoprotectants. *Acta Physiol. Plant.* **2012**, *34*, 1651–1658. [CrossRef]

42. Jones, M.G.; Outlaw, W.H.; Lowry, O.H. Enzymic assay of 10^{-7} to 10^{-14} moles of sucrose in plant tissues. *Plant Physiol.* **1977**, *60*, 379–383. [CrossRef]

43. Turan, O.; Ekmekçi, Y. Activities of photosystem II and antioxidant enzymes in chickpea (*Cicer arietinum* L.) cultivars exposed to chilling temperatures. *Acta Physiol. Plant.* **2011**, *33*, 67–78. [CrossRef]

44. Sarkar, P.; Bosneaga, E.; Auer, M. Plant cell walls throughout evolution: Towards a molecular understanding of their design principles. *J. Exp. Bot.* **2009**, *60*, 3615–3635. [CrossRef]

45. Frank, H.A.; Brudvig, G.W. Redox functions of carotenoids in photosynthesis. *Biochemistry* **2004**, *43*, 8607–8615. [CrossRef] [PubMed]

46. Aroca, R.; Irigoyen, J.J.; Sanchez-Diaz, M. Drought enhances maize chilling tolerance. II. Photosynthetic traits and protective mechanisms against oxidative stress. *Physiol. Plant.* **2003**, *117*, 540–549. [CrossRef]

47. Lee, S.H.; Singh, A.P.; Chung, G.C.; Ahn, S.J.; Noh, E.K.; Steudle, E. Exposure of roots of cucumber (*Cucumis sativus*) to low temperature severely reduces root pressure, hydraulic conductivity and active transport of nutrients. *Physiol. Plant.* **2004**, *120*, 413–420. [CrossRef]

48. Amini, S.; Maali-Amiri, R.; Kazemi-Shahandashti, S.S.; Lopez-Gomez., M.; Sadeghzadeh, B.; Sobhani-Najafabadi, A.; Kari-man, K. Effect of cold stress on polyamine metabolism and antioxidant responses in chickpea. *J. Plant Physiol.* **2021**, *258–259*, 153387.

49. Kiran, A.; Kumar, S.; Nayyar, H.; Sharma, K.D. Low temperature-induced aberrations in male and female reproductive organ development cause flower abortion in chickpea. *Plant Cell Environ.* **2019**, *42*, 2075–2089.

50. Uemura, M.; Joseph, R.A.; Steponkus, P.L. Cold acclimation of Arabidopsis thaliana (effect on plasma membrane lipid composition and freeze-induced lesions). *Plant Physiol.* **1995**, *109*, 15–30. [CrossRef]

51. Chaki, T.; Hirata, N.; Yoshikawa, Y.; Tachibana, S.; Tokinaga, Y.; Yamakage, M. Lipid emulsion, but not propofol, induces skeletal muscle damage and lipid peroxidation. *J. Anesth.* **2019**, *33*, 628–635. [CrossRef]

52. Tewari, A.K.; Tripathy, B.C. Temperature-stress-induced impairment of chlorophyll biosynthetic reactions in cucumber and wheat. *Plant Physiol.* **1998**, *117*, 851–858. [CrossRef]

53. Camejo, D.; Jimenez, A.; Alarcon, J.J.; Torres, W.; Gomez, J.M.; Sevilla, F. Changes in photosynthetic parameters and antioxidant activities following heat-shock treatment in tomato plants. *Funct. Plant Biol.* **2006**, *33*, 177–187. [CrossRef]

54. Kaur, G.; Kumar, S.; Nayyar, H.; Upadhyaya, H.D. Cold stress injury during the pod-filling phase in chickpea (*Cicer arietinum* L.): Effects on quantitative and qualitative compo-nents of seeds. *J. Agron. Crop Sci.* **2008**, *194*, 457–464.

55. Gechev, T.; Petrov, V. Reactive Oxygen Species and Abiotic Stress in Plants. *Int. J. Mol. Sci.* **2020**, *21*, 7433. [CrossRef] [PubMed]

56. Srinivasan, A.; Saxena, N.P.; Johansen, C. Cold tolerance during early reproductive growth of chickpea (*Cicer arietinum* L.): Genetic variation in gamete development and function. *Field Crop. Res.* **1999**, *60*, 209–222. [CrossRef]

57. Fu, G.F.; Jian, S.O.N.G.; Xiong, J.; Li, Y.R.; Chen, H.Z.; Le, M.K.; Tao, L.X. Changes of oxidative stress and soluble sugar in anthers involve in rice pollen abortion under drought stress. *Agric. Sci. China* **2011**, *10*, 1016–1025. [CrossRef]

58. Zhou, R.; Yu, X.; Ottosen, C.O.; Rosenqvist, E.; Zhao, L.; Wang, Y.; Wu, Z. Drought stress had a predominant effect over heat stress on three tomato cultivars subjected to combined stress. *BMC Plant Biol.* **2017**, *17*, 24. [CrossRef] [PubMed]

59. Li, S.; Wan, C.; Kong, J.; Zhang, Z.; Li, Y.; Zhu, Y. Programmed cell death during microgenesis in a Honglian CMS line of rice is correlated with oxidative stress in mitochondria. *Funct. Plant Biol.* **2004**, *31*, 369–376. [CrossRef]

60. Wan, Z.; Jing, B.; Tu, J.; Ma, C.; Shen, J.; Yi, B.; Fu, T. Genetic characterization of a new cytoplasmic male sterility system (hau) in *Brassica juncea* and its transfer to B. napus. *Theor. Appl. Genet.* **2008**, *116*, 355–362. [CrossRef] [PubMed]

61. Nanculao, G.D.; Herrera, M.L.; Carcamo, M.P.; Velasquez, V.B. Relative expression of genes related with cold tolerance in temperate rice at the seedling stage. *Afr. J. Biotechnol.* **2014**, *13*, 2506–2512.

62. Soengas, P.; Rodriguez, V.M.; Velasco, P.; Cartea, M.E. Effect of temperature stress on antioxidant defenses in Brassica oleracea. *ACS Omega* **2018**, *3*, 5237–5243. [CrossRef]

63. Dai, K.; Peng, T.; Ke, D.; Wei, B. Photocatalytic hydrogen generation using a nanocomposite of multi-walled carbon nanotubes and TiO2 nanoparticles under visible light irradiation. *Nanotechnology* **2009**, *20*, 125603. [CrossRef] [PubMed]

64. Nazari, M.; Amiri, R.M.; Mehraban, F.H.; Khaneghah, H.Z. Change in antioxidant responses against oxidative damage in black chickpea following cold acclimation. *Russian J. Plant Physiol.* **2012**, *59*, 183–189. [CrossRef]

65. Hasanuzzaman, M.; Nahar, K.; Fujita, M. Role of tocopherol (vitamin E) in plants: Abiotic stress tolerance and beyond. In *Emerging Technologies and Management of Crop Stress Tolerance*; Academic Press: London, UK, 2014; pp. 267–289.

66. Petrov, V.; Hille, J.; Mueller-Roeber, B.; Gechev, T.S. ROS-mediated abiotic stress-induced programmed cell death in plants. *Front. Plant Sci.* **2015**, *6*, 69. [CrossRef]

67. Wodtke, E. Temperature adaptation of biological membranes. The effects of acclimation temperature on the unsaturation of the main neutral and charged phospholipids in mitochondrial membranes of the carp (*Cyprinus carpio* L.). *Biochim. Biophys. Acta Biomembr.* **1981**, *640*, 698–709. [CrossRef]

68. Pennycooke, J.C.; Cox, S.; Stushnoff, C. Relationship of cold acclimation, total phenolic content and antioxidant capacity with chilling tolerance in petunia (Petunia× hybrida). *Environ. Exp. Bot.* **2005**, *53*, 225–232. [CrossRef]

69. Chen, Y.; Jiang, J.; Chang, Q.; Gu, C.; Song, A.; Chen, S.; Chen, F. Cold acclimation induces freezing tolerance via antioxidative enzymes, proline metabolism and gene expression changes in two chrysanthemum species. *Mol. Biol. Rep.* **2014**, *41*, 815–822. [CrossRef] [PubMed]

70. Landry, E.J.; Fuchs, S.J.; Bradley, V.L.; Johnson, R.C. The effect of cold acclimation on the low molecular weight carbohydrate composition of safflower. *Heliyon* **2017**, *3*, e00402. [CrossRef] [PubMed]

71. Mazzucotell, E.; Belloni, S.; Marone, D.; De Leonardis, A.M.; Guerra, D.; Di Fonzo, N.; Cattivelli, L.; Mastrangelo, A.M. The E3 ubiquitin ligase gene family in plants: Regulation by degradation. *Curr. Genom.* **2006**, *7*, 509–522. [CrossRef]

72. Bhandari, K.; Sharma, K.D.; Rao, B.H.; Siddique, K.H.; Gaur, P.; Agrawal, S.K.; Nayyar, H. Temperature sensitivity of food legumes: A physiological insight. *Acta Physiol. Plant.* **2017**, *39*, 68. [CrossRef]

73. Hayat, S.; Hayat, Q.; Alyemeni, M.N.; Wani, A.S.; Pichtel, J.; Ahmad, A. Role of proline under changing environments: A review. *Plant Signal. Behav.* **2012**, *7*, 1456–1466. [CrossRef]

74. Vera-Hernandez, P.; Ortega Ramirez, M.A.; Martinez Nunez, M.; Ruiz-Rivas, M.; Rosas-Cárdenas, F.D.F. Proline as a probable biomarker of cold stress tolerance in sorghum (*Sorghum bicolor*). *Mex. J. Biotechnol.* **2018**, *3*, 77–86.

75. Bouche, N.; Fromm, H. GABA in plants: Just a metabolite? *Trends Plant Sci.* **2004**, *9*, 110–115. [PubMed]

76. Malekzadeh, P.; Khara, J.; Heydari, R. Alleviating effects of exogenous Gamma-aminobutyric acid on tomato seedling under chilling stress. *Physiol. Mol. Biol. Plants* **2014**, *20*, 133–137. [CrossRef]
77. Kosar, F.; Akram, N.A.; Sadiq, M.; Al-Qurainy, F.; Ashraf, M. 2019 Trehalose: A key organic osmolyte effectively involved in plant abiotic stress tolerance. *J. Plant Growth Regul.* **2019**, *38*, 606–618. [CrossRef]
78. Garg, A.K.; Kim, J.K.; Owens, T.G.; Ranwala, A.P.; Do Choi, Y.; Kochian, L.V.; Wu, R.J. Trehalose accumulation in rice plants confers high tolerance levels to different abiotic stresses. *Proc. Natl. Acad. Sci. USA* **2002**, *99*, 15898–15903. [CrossRef] [PubMed]
79. Tabaei-Aghdaei, S.R.; Pearce, R.S.; Harrison, P. Sugars regulate cold-induced gene expression and freezing-tolerance in barley cell cultures. *J. Exp. Bot.* **2003**, *54*, 1565–1575. [PubMed]
80. Strauss, G.; Hauser, H. Stabilization of lipid bilayer vesicles by sucrose during freezing. *Proc. Natl. Acad. Sci. USA* **1986**, *83*, 2422–2426. [CrossRef]
81. Parish, R.W.; Phan, H.A.; Iacuone, S.; Li, S.F. Tapetal development and abiotic stress: A centre of vulnerability. *Funct. Plant Biol.* **2012**, *39*, 553–559. [CrossRef]
82. Pressman, E.; Harel, D.; Zamski, E.; Shaked, R.; Althan, L.; Rosenfeld, K.; Firon, N. The effect of high temperatures on the expression and activity of sucrose-cleaving enzymes during tomato (*Lycopersicon esculentum*) anther development. *J. Hort. Sci. Biotechnol.* **2006**, *81*, 341–348. [CrossRef]

Article

In Silico Study of Superoxide Dismutase Gene Family in Potato and Effects of Elevated Temperature and Salicylic Acid on Gene Expression

Jelena Rudić, Milan B. Dragićević, Ivana Momčilović, Ana D. Simonović and Danijel Pantelić *

Institute for Biological Research "Siniša Stanković"—National Institute of the Republic of Serbia, University of Belgrade, Bulevar despota Stefana 142, 11060 Belgrade, Serbia; jelena.rudic@ibiss.bg.ac.rs (J.R.); mdragicevic@ibiss.bg.ac.rs (M.B.D.); ivana.momcilovic@ibiss.bg.ac.rs (I.M.); ana.simonovic@ibiss.bg.ac.rs (A.D.S.)
* Correspondence: danijel.pantelic@ibiss.bg.ac.rs

Abstract: Potato (*Solanum tuberosum* L.) is the most important vegetable crop globally and is very susceptible to high ambient temperatures. Since heat stress causes the accumulation of reactive oxygen species (ROS), investigations regarding major enzymatic components of the antioxidative system are of the essence. Superoxide dismutases (SODs) represent the first line of defense against ROS but detailed in silico analysis and characterization of the potato *SOD* gene family have not been performed thus far. We have analyzed eight functional *SOD* genes, three *StCuZnSODs*, one *StMnSOD*, and four *StFeSODs*, annotated in the updated version of potato genome (Spud DB DM v6.1). The *StSOD* genes and their respective proteins were analyzed in silico to determine the exon-intron organization, splice variants, cis-regulatory promoter elements, conserved domains, signals for subcellular targeting, 3D-structures, and phylogenetic relations. Quantitative PCR analysis revealed higher induction of *StCuZnSODs* (the major potato *SODs*) and *StFeSOD3* in thermotolerant cultivar Désirée than in thermosensitive Agria and Kennebec during long-term exposure to elevated temperature. *StMnSOD* was constitutively expressed, while expression of *StFeSODs* was cultivar-dependent. The effects of salicylic acid (10^{-5} M) on *StSODs* expression were minor. Our results provide the basis for further research on *StSODs* and their regulation in potato, particularly in response to elevated temperatures.

Keywords: superoxide dismutase; potato; AlphaFold; heat stress; salicylic acid

Citation: Rudić, J.; Dragićević, M.B.; Momčilović, I.; Simonović, A.D.; Pantelić, D. In Silico Study of Superoxide Dismutase Gene Family in Potato and Effects of Elevated Temperature and Salicylic Acid on Gene Expression. *Antioxidants* **2022**, *11*, 488. https://doi.org/10.3390/antiox11030488

Academic Editors: Masayuki Fujita, Mirza Hasanuzzaman and Juan B. Barroso

Received: 31 December 2021
Accepted: 22 February 2022
Published: 28 February 2022

Publisher's Note: MDPI stays neutral with regard to jurisdictional claims in published maps and institutional affiliations.

1. Introduction

Potato (*Solanum tuberosum* L.) is the most important vegetable crop grown worldwide, essential for global food security. It is a cool-season vegetable, very susceptible to high ambient temperatures compared to other cultivated plants. Even mildly elevated temperatures (26–30 °C) may induce biochemical, physiological, and morpho-anatomical changes that affect the growth and development of this plant species [1–3]. High temperature can accelerate stem growth, reduce leaf area and reduce or inhibit root growth in potato [2]. The most prominent effects of high temperatures relate to the reduction in tuber induction, initiation and enlargement, and decrease in partitioning of dry matter to the tubers, which results in a decline in potato yield [2,3]. At the cellular level, high temperature disrupts membranes' integrity, changes protein conformation, degrades the PSII component of the photosynthetic apparatus, and, due to impairment of electron transport chains in chloroplasts and mitochondria, promotes the production of reactive oxygen species—ROS [4,5]. Excessive production of ROS in plant cells can damage pigments, carbohydrates, lipids, proteins, nucleic acids, and in severe cases, induces cellular death [6]. On the other hand, ROS can act as signaling molecules that regulate many physiological processes during plant growth and development, and participate in various abiotic and biotic stress responses [7,8].

During evolution, an efficient defense antioxidant system developed in plants, encompassing enzymatic and non-enzymatic components that scavenge the toxic radicals and thus aid plants to cope with the large quantities of ROS. Superoxide dismutases (SOD, EC 1.15.1.1) are enzymes that catalyze the conversion or dismutation of toxic superoxide anion radical ($O_2^{\bullet-}$) into hydrogen peroxide (H_2O_2) and oxygen (O_2) and represent the first line of antioxidant defense against ROS [9]. They are metalloproteins, whose catalytic activity depends on the presence of metal prosthetic groups. In plants, SODs are classified into three groups based on their metal ions as cofactors: copper/zinc (CuZnSODs), manganese (MnSODs) and iron SODs (FeSODs). MnSODs and FeSODs share a high degree of amino acid sequence and structural homologies and are distinct from CuZnSODs [5]. Since phospholipid membranes are impermeable to charged $O_2^{\bullet-}$ radicals, SOD isoforms, although encoded by nuclear genes, are distributed in different subcellular compartments [10]. CuZnSODs are mainly distributed in cytosol, peroxisomes, chloroplasts, and/or extracellular space, FeSODs are primarily located in the plastids, and MnSOD mainly occurs in the mitochondria and peroxisomes [5]. Plant CuZnSODs can be homodimeric (cytosolic) or homotetrameric (chloroplast and extracellular), built from ~15–17 kDa subunits. Similarly, FeSODs and MnSODs are either homodimeric or homotetrameric enzymes with the subunit size of 18–27 and 18–20 kDa, respectively [5]. Many findings indicate that SODs may play a significant role in the abiotic stress tolerance of plants, which is supported by the results of studies on transgenic plants overexpressing MnSOD and/or Cu/ZnSOD [11–13].

Salicylic acid (SA) is an essential endogenous growth regulator and signaling molecule in plants, which regulates different aspects of plant physiology. SA is involved in activating plant defense responses against biotic and abiotic stresses, including drought, chilling, heavy metal toxicity, heat and salinity [14]. The application of exogenous SA could improve thermotolerance in plants. The treatment with SA at a suitable concentration generally has an acclimation-like effect, leading to enhanced heat tolerance due to promotion of heat shock factor (HSF)–DNA binding [15], enhanced accumulation of heat shock proteins (HSP) [15,16] and modulation of the antioxidant enzyme activity in plants [17,18]. When applied during heat stress, SA may also alleviate adverse effects of elevated temperatures as observed in wheat where SA caused accumulation of proline and consequently improved net photosynthesis [19]. The application of exogenous SA increased SOD activity in *Digitalis trojana* [20], rhododendron [21], and tomato [18] during heat stress. The efficiency of SA as a protective or ameliorating agent against different stresses, however, depends on the plant species, developmental stage, the applied concentration, application method and endogenous SA level [22].

Despite the importance of potato as a staple crop, detailed in silico analysis and characterization of the *S. tuberosum* SOD gene family have not been performed so far. The previous version of the potato genome was assembled using short reads and represented only 86% of the 726 Mb—large genome of the doubled monoploid potato *S. tuberosum* L. Group Phureja DM 1–3 516 R44 [23]. An updated version of the same doubled monoploid clone genome (DM v6.1), with re-estimated size of 844 Mb, is available from 2020; it is based on the Oxford Nanopore Technologies long reads coupled with proximity-by-ligation scaffolding, yielding a chromosome-scale assembly [24]. Hereby we present an in depth in silico study of potato SOD (*StSOD*) genes retrieved from the updated genome version DM v6.1 that covers exon-intron organization, splice variants, cis-regulatory promoter elements, conserved domains, encoded proteins' physicochemical properties, signals for subcellular targeting, prediction of 3D structures and phylogenetic relations with SODs from other plants. Due to global climate change, the rise in average ambient temperatures is predicted for most potato-growing regions in the 21st century [25], and investigations regarding the effects of prolonged mild heat stress on physiological, biochemical, and molecular responses of potato are gaining in importance [2]. Therefore, we analyzed *StSOD*s expression by reverse transcription quantitative PCR (qRT-PCR) after exposure of potato microplants to long-term mild heat stress (29 °C, three weeks) or slightly supraoptimal temperature treatment (26 °C, three weeks), with and without exogenous SA. Our results provide the basis for further research of *StSOD*s and can be important for a better understanding of potato antioxidant system response to elevated temperature and SA.

2. Materials and Methods

2.1. Potato Genomic Resources

Potato (*Solanum tuberosum* L.) SOD gene data, including accession numbers, chromosomal location, sequences, corresponding transcripts and proteins sequences were based on the doubled monoploid *S. tuberosum* Group Phureja DM 1-3 516 R44 and were retrieved from Spud DB Potato Genomics Resources (DM v6.1, http://spuddb.uga.edu (accessed on 15 April 2021).

2.2. StSOD Promoter Analysis

Analysis of transcription factor (TF) binding sites was performed on promoters of *StSOD*s encompassing a region 1000 nt upstream and 100 nt downstream from the transcription start site. Binding site prediction was performed using Plant Transcriptional Regulatory Map (PlantRegMap, http://plantregmap.gao-lab.org/binding_site_prediction.php, accessed on 20 June 2021) [26] with a *p*-value threshold of 10^{-5}. All binding sites with a q-value < 0.05 [27] were retained. TFs that bind identified motifs were found based on the *S. tuberosum* TF list (http://planttfdb.gao-lab.org/download.php, accessed on 20 June 2021). Gene ontology (GO) annotation of TFs was attained from Plant Transcription Factor Database v5.0 (PlantTFDB v5.0, http://planttfdb.gao-lab.org/index.php, accessed on 20 June 2021) [28]. Alternative analysis of *StSOD* promoters with PlantCARE [29] is presented only in supplementary materials.

2.3. In Silico Characterization of StSOD Protein Features

StSOD proteins features, including molecular weight, theoretical isoelectric point (pI), instability index and aliphatic index were computed using ProtParam tool (https://web.expasy.org/protparam, accessed on 22 June 2021) [30]. Subcellular localization of StSOD proteins was predicted using a combination of tools, including DeepLoc-1.0 (https://services.healthtech.dtu.dk/service.php?DeepLoc-1.0, accessed on 1 July 2021) [31], CELLO v2.5 (http://cello.life.nctu.edu.tw, accessed on 1 July 2021) [32], Light Attention (LA, https://embed.protein.properties/, accessed on 1 October 2021) [33], TargetP2 (https://services.healthtech.dtu.dk/service.php?TargetP-2.0, accessed on 1 July 2021) [34], PTS1 Predictor (https://mendel.imp.ac.at/pts1/, accessed on 2 July 2021) [35,36] and PredPlantPTS1 (http://ppp.gobics.de, accessed on 2 July 2021) [37]. Protein family (Pfam, http://pfam.xfam.org, accessed on 27 June 2021) [38] annotation was performed using hmmer 3.3.2 [39] and Pfam34 database (http://ftp.ebi.ac.uk/pub/databases/Pfam/releases/Pfam34.0/, accessed on 27 June 2021). Conserved Domain Database (CDD, https://ncbi.nlm.nih.gov/Structure/cdd/cdd.shtml, accessed on 7 September 2021) [40] annotation was performed using CD search online web server (https://ncbi.nlm.nih.gov/Structure/cdd/wrpsb.cgi, accessed on 7 September 2021) [41]. StSOD sequences processed based on TargetP2 transit peptide (or target peptide, TP) predictions were used for multiple sequence alignments and tertiary protein structure prediction. Multiple sequence alignments were generated with the DECIPHER R package [42] using 10 iterations and 10 refinements and other default options. Percent identity between aligned protein sequences was calculated as (identical positions)/(aligned positions + internal gap positions) $\times$ 100%. Tertiary protein structure was estimated using AlphaFold [43] queried via UCSF ChimeraX 1.3 [44]. The generated best structure per *StSOD* sequence was assessed using MolProbity 4.4 [45] via SWISS-MODEL Workspace (https://swissmodel.expasy.org/assess, accessed on 28 October 2021) [46] and rendered using UCSF ChimeraX 1.3. Ramachandran diagrams were created using the R package ggrama [47].

2.4. Phylogenetic Analysis of StSOD Protein Sequences

Phylogeny of StSOD proteins was assessed by comparison with homologs from 18 plant species: *Arabidopsis thaliana, Beta vulgaris, Capsicum annuum, Daucus carota, Glycine max, Helianthus annuus, Nicotiana attenuata, Phaseolus vulgaris, Solanum lycopersicum, Manihot esculenta, Oryza sativa, Saccharum spontaneum, Zea mays, Sorghum bicolor, Hordeum vulgare, Ananas comosus, Musa acuminata* and *Dioscorea rotundata*. Proteomes from these species were obtained from Ensembl release 49 (https://plants.ensembl.org/, accessed on 1 March 2021, http://ftp.ensemblgenomes.org/pub/plants/release-49/fasta/, accessed on 1 March 2021). Raw bit-score top hits per *StSOD* sequence per plant were filtered and the top seven homologs per *StSOD* sequence were used for phylogenetic analysis. Sequence alignment was performed with the DECIPHER R package [42] using 10 iterations and 10 refinements and other default options. Uninformative sites in the alignment were removed with DECIPHER R package with default options using the small-sample size correction [48]. The resulting alignment was used for fitting a maximum likelihood tree using the LG model of amino acid replacement [49]. The fitting was performed with optimization of the gamma rate parameter and proportion of invariable sites using stochastic rearrangement starting from the neighbor-joining tree using phangorn R package [50]. To assess cluster stability, non-parametric bootstrap was performed for 100 iterations. For rooting of the CuZnSOD phylogenetic tree, the CuZnSOD protein sequence from *Saccharomyces cerevisiae* was used (PDB: 1JK9). No suitable sequence was found for rooting of the Mn-FeSOD phylogenetic tree, so it was midpoint rooted.

2.5. Plant Material and Growth Conditions

Commercial *S. tuberosum* L. cultivars, Agria, Désirée and Kennebec, were used in experiments. The three unrelated potato varieties were selected to validate the presence of investigated genes and compare *StSODs* expression in tetraploid genotypes that differ in heat tolerance. Based on our unpublished data, Désirée was considered relatively heat tolerant, Kennebec as moderately sensitive, and Agria as heat-sensitive genotype.

Virus-free tubers of three potato cultivars were obtained from Solanum Komerc, Guča, Serbia. In vitro cultures were established from surface-sterilized sprouts, which were transferred on the basal medium (BM) consisting of Murashige and Skoog macro and micro-mineral salts [51], Linsmaier and Skoog vitamins [52], 0.7% agar, 3% sucrose, 100 mgL^{-1} myo-inositol and supplemented with 0.5 mgL^{-1} 6-benzylaminopurine (BAP; Sigma Aldrich, St. Louis, MO). Shoots obtained on this medium gave rise to plantlets when transferred to BM without BAP. Microplants were grown in a controlled environment (21 °C, 16 h light period, light flux 90 µmol m^{-2} s^{-1}) and were routinely subcultured every four weeks on BM using single-node stem cuttings (SNC).

2.6. SA and Temperature Treatments

SNCs (10–15 mm) from four-week-old potato microplants were transferred on BM or BM supplemented with 10^{-5} M SA in glass jars (10 SNCs per jar) with vented polypropylene caps. SA (Sigma Aldrich, St. Louis, MO, USA) was dissolved in 96% ethanol and added to the medium before the sterilization at 114 °C for 25 min, while the equivalent volume of ethanol was added in control. Based on our preliminary testing of SA concentrations in the range 10^{-4}–10^{-6} M and literature data [53], the particular SA concentration was selected as the best for alleviating adverse effects of heat treatments on microplants' growth and development. The explants were grown in the growth chamber (Aralab, Rio de Mouro, Portugal) under 16 h photoperiod, light flux 90 µmol m^{-2} s^{-1}, 70% humidity, and at three different temperatures: 21 °C, 26 °C or 29 °C. After three weeks, fully developed leaves were collected from plants grown in four jars (one biological sample), frozen in liquid nitrogen before storage at −80 °C and further used for RNA extraction. Three biological samples were used for qRT-PCR analysis.

2.7. RNA Extraction and cDNA Synthesis

Total RNA was extracted from 0.5 g of frozen potato leaves using TRIzol reagent (Invitrogen, Carlsbad, CA, USA) following the manufacturer's instructions, and stored at $-80\,^{\circ}$C until use. RNA quality and concentration were determined using NanoPhotometer N60 (Implen, Munich, BY, Germany), and all RNA samples were diluted to 1 µg µL^{-1} in RNase-free water. The purity of RNA was gauged by the absorbance ratio of A260/A280. The integrity of isolated RNA was checked electrophoretically and assessed by the ratio of 28S/18S rRNA. The genomic DNA contamination was eliminated from the total RNA by DNase I treatment (Fermentas, Hanover, MD, USA), and the first-strand cDNA was synthesized from 3 µg of RNA using Revert Aid First Strand cDNA Synthesis Kit (Fermentas, Hanover, MD, USA) with oligo(dT) primers according to manufacturer's instructions.

2.8. qRT-PCR Analysis

Oligonucleotide primers for the amplification of *StSOD* transcripts were designed using Primer-BLAST available at NCBI (https://ncbi.nlm.nih.gov/tools/primer-blast/, accessed on 10 July 2020), while the primer pair for *StCuZnSOD2* was synthesized according to [54]. For *StSOD* genes with multiple gene models (splice variants, see Table 1), the primers were designed to amplify all models of one gene (Table S1). Primers specificity was checked by PCR followed by electrophoretic sizing and by melting curve analysis. The obtained amplicons were purified from the agarose gels using Gene JET™ Gel Extraction Kit (Thermo Fisher Scientific, Waltham, MA, USA), quantified using NanoPhotometer N60 (Implen, Munich, BY, Germany), and serially diluted in a 10^9–10^2 copies µL^{-1} range to be used as standards for absolute qPCR quantification. The qPCR was run in three technical replicates for each biological replicate on MicroAmp™ Optical 96-Well Plates (Thermo Fisher Scientific, Waltham, MA, USA), in 10 µL reaction mixtures comprising of cDNA corresponding to 100 ng of total RNA, forward and reverse primers (7.5 µM each), and 5 µL Maxima SYBR Green/ROX qPCR Master Mix (2x) (Thermo Fisher Scientific, Waltham, MA, USA). The amplification was conducted using Applied Biosystems QuantStudio™ 3 Real-Time PCR system (Thermo Fisher Scientific, Waltham, MA, USA), with the program: initial denaturation at 95 °C for 10 min, followed by 40 cycles at 95 °C for 15 s and 60 °C for 1 min. Absolute expression values were normalized against the averaged expression data of the internal control genes *60SL36* and *CYC* [55].

2.9. Statistical Analysis

Statistical analysis was performed using IBM SPSS Statistics version 25 (International Business Machines Corporation, Armonk, NY, USA). Levene's test was used to verify the homogeneity of variances of the data set. One-way analysis of variance was performed with multiple comparisons analysis of means by either Tukey's HSD (for equal variances) or Dunnett's T3 (for unequal variances) post-hoc test at a significance level of 0.05. The data were shown as mean values $\pm$ standard deviation (S.D.).

Table 1. Description of the potato SOD (*StSOD*) genes and proteins analyzed in this study.

Proposed Name	Gene ID	Chromosomal Localization	Transcript ID	Transcript Length (bp)	Protein Length (aa)	Isoelectric Point (pI)	Instability Index	Aliphatic Index	Molecular Weight (kDa)
StCuZnSOD1	Soltu.DM.01G022650	Chr.01: 61,097,981-61,103,305	Soltu.DM.01G022650.1 *	459	152	5.28	32.24	80.20	15.3
StCuZnSOD2	Soltu.DM.11G020830	Chr.11: 40,784,322-40,790,930	Soltu.DM.11G020830.1 *	648	215	6.34	24.76	87.95	22.1
		Chr.11: 40,784,322-40,790,564	Soltu.DM.11G020830.2	645	214	6.34	24.25	88.36	22.1
StCuZnSOD3	Soltu.DM.03G010200	Chr.03: 28,151,770-28,156,700	Soltu.DM.03G010200.1 *	498	165	6.78	18.64	88.61	16.8
StMnSOD	Soltu.DM.06G011380	Chr.06: 34,521,324-34,524,092	Soltu.DM.06G011380.1 *	687	228	7.13	35.60	91.14	25.3
StFeSOD1	Soltu.DM.06G012180	Chr.06: 36,014,174-36,016,425	Soltu.DM.06G012180.1	747	248	6.31	33.40	70.56	27.8
		Chr.06: 36,014,151-36,016,425	Soltu.DM.06G012180.2	753	250	6.60	33.86	70.00	28.1
		Chr.06: 36,014,174-36,016,425	Soltu.DM.06G012180.3 *	759	252	6.60	34.62	70.99	28.3
		Chr.06: 36,014,174-36,016,425	Soltu.DM.06G012180.4	729	242	7.77	35.05	75.08	26.9
		Chr.06: 36,014,174-36,016,425	Soltu.DM.06G012180.5	681	226	5.96	35.28	74.38	25.1
		Chr.06: 36,014,174-36,016,425	Soltu.DM.06G012180.6	654	217	6.52	35.44	70.74	24.1
StFeSOD2	Soltu.DM.03G013800	Chr.03: 36,413,200-36,416,935	Soltu.DM.03G013800.1	915	304	5.56	41.47	73.26	34.7
		Chr.03: 36,413,200-36,416,935	Soltu.DM.03G013800.2	891	296	5.49	41.99	73.58	33.7
		Chr.03: 36,413,200-36,416,935	Soltu.DM.03G013800.3	897	298	5.83	40.95	72.11	33.8
		Chr.03: 36,413,200-36,416,935	Soltu.DM.03G013800.4 *	987	328	5.69	40.11	76.19	37.6
		Chr.03: 36,413,200-36,416,935	Soltu.DM.03G013800.5	840	279	5.20	42.01	73.87	31.9
StFeSOD3	Soltu.DM.02G001300	Chr.02: 7,378,597-7,382,073	Soltu.DM.02G001300.1 *	777	258	6.07	42.61	85.04	29.6
		Chr.02: 7,378,597-7,382,073	Soltu.DM.02G001300.2	753	250	5.96	43.46	85.80	28.5
StFeSOD4	Soltu.DM.06G012170	Chr.06: 36,008,330-36,012,608	Soltu.DM.06G012170.1	603	200	5.58	31.53	82.00	22.5
		Chr.06: 36,008,330-36,012,608	Soltu.DM.06G012170.2	540	179	5.81	33.93	78.60	20.2
		Chr.06: 36,008,330-36,012,608	Soltu.DM.06G012170.3	549	182	6.29	30.48	76.26	20.4
		Chr.06: 36,008,330-36,012,608	Soltu.DM.06G012170.4 *	684	227	7.85	39.93	73.17	25.6
		Chr.06: 36,008,330-36,012,608	Soltu.DM.06G012170.5	513	170	6.12	35.64	74.18	19.1
		Chr.06: 36,008,330-36,012,608	Soltu.DM.06G012170.6	633	210	5.89	31.93	77.71	23.8

Note: * indicates representative sequences; bp, base pair; aa, amino acid.

3. Results

3.1. StSOD Genes: Structure and Chromosomal Distribution

Ten potato genes, annotated as *SOD*, were derived from the Spud DB Potato Genomics Resources—DM v6.1. Two genes, Soltu.DM.10G011570 and Soltu.DM.06G013120, were not further investigated, since they do not encode full-length functional SOD proteins, but ~100 amino acids long polypeptides which have similarity with SOD sequences. A copper chaperone gene, Soltu.DM.08G026370, also encodes a protein that contains a copper/zinc superoxide dismutase domain (Pfam: PF00080), but it lacks crucial active site residues to be a CuZnSOD, so it was also omitted. In total, eight full-length *StSOD* genes: three *StCuZnSOD*s, one *StMnSOD*, and four *StFeSOD*s, were the object of our study (Table 1). The eight analyzed *StSOD*s are physically located on five out of 12 chromosomes: *StCuZnSOD1*, *StCuZnSOD2*, and *StFeSOD3* are on chromosomes 1, 11, and 2, respectively; two genes, *StCuZnSOD3* and *StFeSOD2*, are located on chromosome 3, while three genes, *StMnSOD*, *StFeSOD1*, and *StFeSOD4*, are on chromosome 6. *StFeSOD1* and *StFeSOD4*, separated by only 1566 bp, are considered tandemly duplicated [56].

As listed in Table 1, *StCuZnSOD2* (Soltu.DM.11G020830) and all *StFeSOD*s (Soltu.DM.06G012180, Soltu.DM.03G013800, Soltu.DM.02G001300, Soltu.DM.06G012170) have at least two and up to six transcripts. Alignment of these transcripts against their genes revealed that different gene models vary in the number of exons and introns or, in some cases, in the length of the 3′-UTR (Figure 1), suggesting that these are alternative splice variants. The differences in exon-intron organization and UTRs among *StSOD* genes and their splice variants are depicted in Figure 1, where can be seen that *StSOD* gene models possessed between 4 to 9 introns and 5 to 10 exons. *StFeSOD1* and *StFeSOD2* genes share similar exon/intron organization patterns, but with differences in exons/introns lengths, while the rest of *StSOD* genes exhibited disparate exon/intron structures.

3.2. Analysis of Promoter Regions of StSOD Genes

The potential regulation of *StSOD* genes was investigated by identification of TF binding motifs (cis-regulatory elements) in the *StSOD*s promoter regions. Based on the binding analysis using PlantRegMap [26], members of 17 TF families may bind to the eight StSOD (Figure 2A and Table S2).

The gene with the highest number of cis-regulatory elements is *StCuZnSOD1*, which contains binding motifs for 15 different TF families (Figure 2A). An especially motif dense region is located ~800 nt upstream of the transcription start site of *StCuZnSOD1*, which contains binding motifs for Dof, M-type_MADS, ERF, BBR-BPC, LBD, ARF, GRAS and MIKC_MADS TF families (Figure 2A, insert). In contrast, the *StCuZnSOD2* promoter has a single MYB binding region, while *StCuZnSOD3* contains two Dof and one M−type_MADS binding region (Figure 2A). *StMnSOD* contains three bZIP binding regions, a single bHLH and Nin-like binding sequence motif, while the remaining *StFeSOD* genes contain one to three binding regions for different TF families (Figure 2A).

To gain insight into the potential biological roles of the identified cis-regulatory elements, the biological process GO terms associated with specific potato TFs which bind the identified motifs were summarized per *StSOD* promoter and binding motif (Figure 2B and Table S3). The GO terms were obtained from PlantTFDB and non-informative terms (such as "process regulation of transcription, DNA-templated" and similar) were omitted. Based on TF-associated GO terms, *StCuZnSOD1* is regulated in response to external stressors such as cold and fungal infection, by several plant hormones, including gibberellin, ethylene, auxin and abscisic acid, and during flower development. *StCuZnSOD3* is to some extent similar in this regard to *StCuZnSOD1* (regulation by cold, gibberellins, chitin, flower development) because its promoter contains binding sites for M−type_MADS and Dof TFs, mediators of these types of regulation. *StCuZnSOD2* contains a single MYB binding cis-element associated with GO terms: response to salt stress, ethylene, auxin, jasmonic acid, chitin and cadmium concentration. The *StMnSOD* gene is regulated in response to desiccation, salt stress, abscisic and gibberellic acids, brassinosteroid and phytochrome signaling pathways.

StFeSOD1 contains binding motifs for Dof and BBR-BPC TFs and is regulated during vascular system development (Dof) and in response to ethylene (BBR-BPC). *StFeSOD2* contains a single Dof TF binding element responsive during light-mediated development. *StFeSOD3* contains an ERF and two C2H2 TFs; ERF is associated with response to chitin, cold and leaf development GO terms, while C2H2 is associated with meristem transition, unidimensional cell growth and response to brassinosteroid. *StFeSOD4* contains a BBR-BPC binding element associated with response to ethylene and regulation of developmental processes, and a MYB binding element associated with response to salt stress, chitin, ethylene, auxin and jasmonic acid.

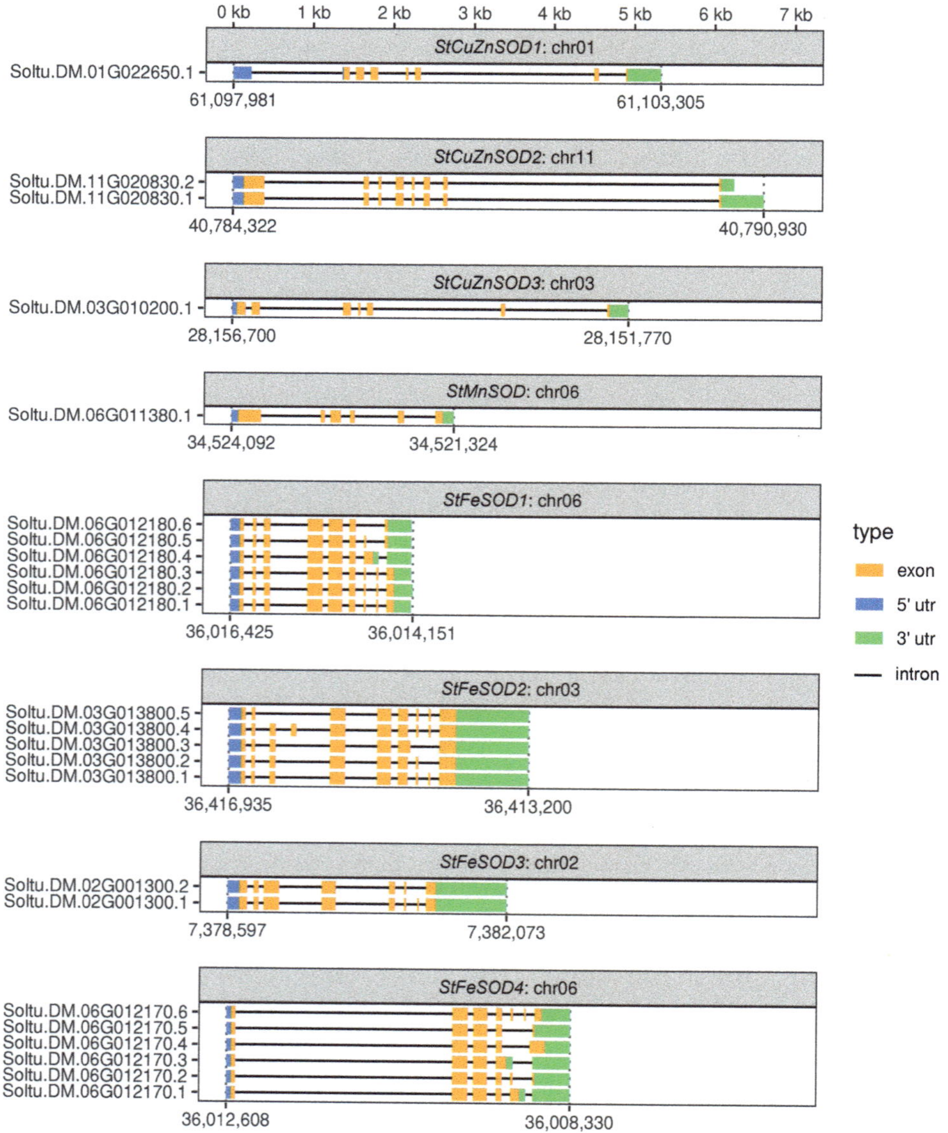

Figure 1. Exon-intron organization of *StSOD* genes. The blue and green rectangles represent UTR, the orange rectangle represent exons, and black lines represent introns.

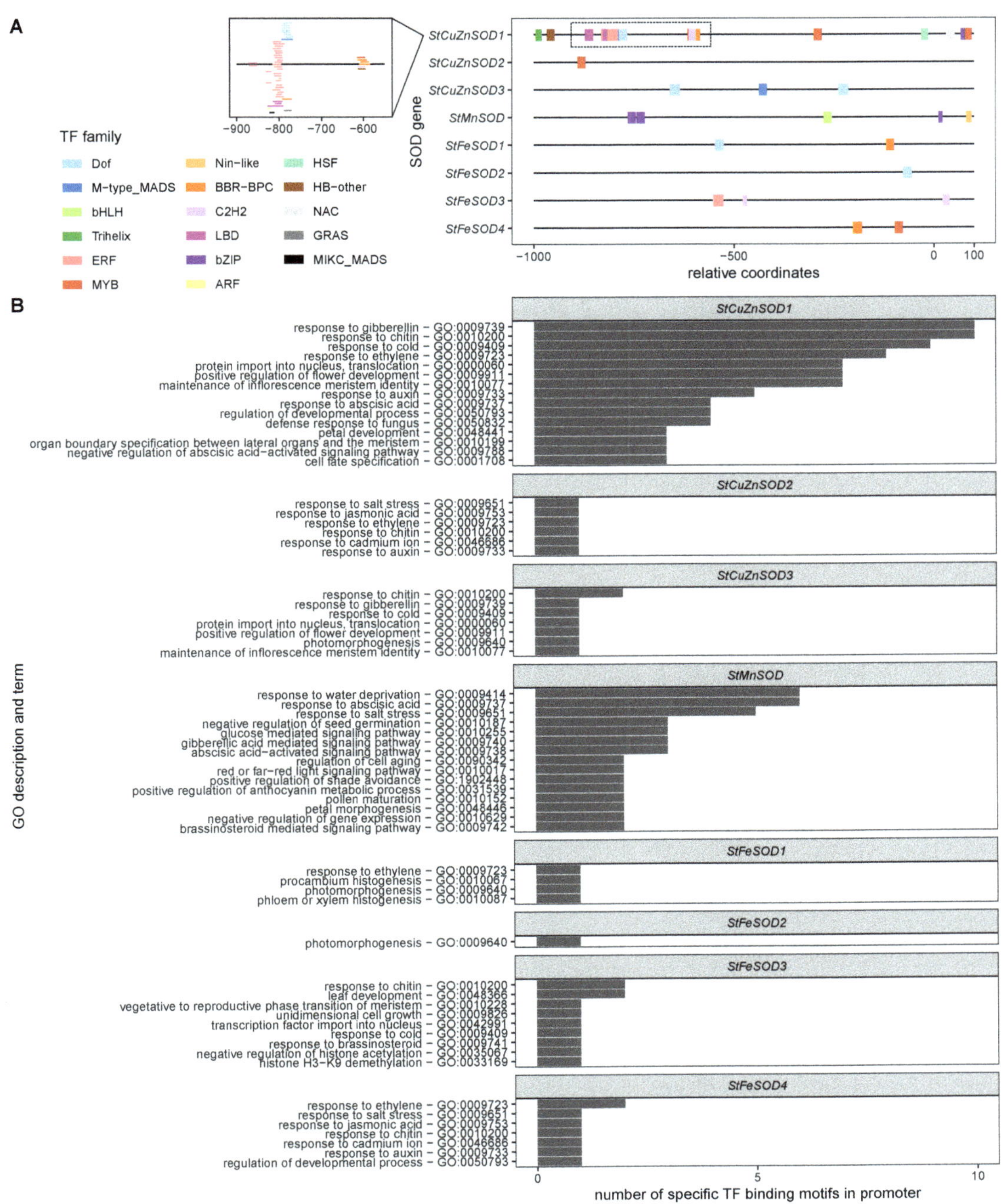

Figure 2. Transcription factor binding motifs in promoters of *StSOD* genes. (**A**) Location of binding sites for various TF families (color legend). (**B**) Biological process gene ontology terms associated with specific potato TFs which bind the identified motifs. Only unique GO terms per binding motif were counted. Non-informative terms (such as "process regulation of transcription, DNA-templated" and similar) present in all or most TF families were omitted.

3.3. Characteristics of StSOD Proteins

Physicochemical properties of StSOD proteins, including their length, isoelectric point (pI), instability index, aliphatic index, and molecular weights (MW), are shown in Table 1. Different StSODs are 152–304 amino acids long, with MW ranging from 15.3 to 37.6 kDa.

Theoretical pI values of StSODs are in a 5.20–7.85 range, indicating that almost all members of StSODs are acidic, except for StMnSOD and two StFeSODs, which are slightly basic. Based on the aliphatic index, which may be regarded as a positive factor for the enhancement of thermostability of globular proteins [57], the most thermostable protein is StMnSOD, which has the highest aliphatic index (91.14), while StCuZnSODs, with an average aliphatic index of 85.59 are generally more thermostable than StFeSODs (average aliphatic index of 76.35). The values of the instability index determine the stability of the protein in a test tube. A value of instability index above 40 predicts that the protein may be unstable. Almost all SODs were predicted to be stable (<40), except StFeSOD2 and StFeSOD3.

3.4. StSOD Protein Structure and Subcellular Localization

To gain insights into the organization of StSOD proteins structure we performed domain annotation using Pfam database, along with annotation of predicted N-terminal target peptides (Figure 3) and subcellular localization based of four algorithms: TargetP2, DeepLoc, CELLO and Light Attention (LA, Table 2). This was done for protein sequences translated in silico from all annotated StSOD gene models. The CuZnSOD protein sequences contain one Pfam domain characteristic for this class of proteins (PF00080.22), while MnSOD and FeSODs contain two Pfam domains: the N-terminal Iron/manganese SOD α-hairpin domain (PF00081.24) and the Iron/manganese SOD C-terminal domain (PF02777.20, Figure 3).

Table 2. Comparative subcellular localization prediction of potato SODs by different tools.

Gene	Protein ID	TargetP2	DeepLoc	CELLO	LA	PTS1 Predictor	PredPlant PTS1
StCuZnSOD1	Soltu.DM.01G022650.1	O	C	C	C		
StCuZnSOD2	Soltu.DM.11G020830.1	Ch	Ch	Ch	Ch		
	Soltu.DM.11G020830.2	Ch	Ch	Ch	Ch		
StCuZnSOD3	Soltu.DM.03G010200.1	O	P	C	P	Not-P	Not-P
StMnSOD	Soltu.DM.06G011380.1	Mt	Mt	Mt	Mt		
	Soltu.DM.06G012180.1	Ch	Ch	Ch	Ch		
	Soltu.DM.06G012180.2	Ch	Ch	Ch	Ch		
StFeSOD1	Soltu.DM.06G012180.3	Ch	Ch	Ch	Ch		
	Soltu.DM.06G012180.4	Ch	Ch	Ch	Ch		
	Soltu.DM.06G012180.5	Ch	Ch	Ch	Ch		
	Soltu.DM.06G012180.6	Ch	Ch	Ch	Ch		
	Soltu.DM.03G013800.1	Ch	Ch	Ch	Ch		
	Soltu.DM.03G013800.2	Ch	Ch	Ch	Ch		
StFeSOD2	Soltu.DM.03G013800.3	Ch	Ch	Ch	Ch		
	Soltu.DM.03G013800.4	Ch	Ch	Ch, C	Ch		
	Soltu.DM.03G013800.5	Ch	Ch	Ch	Mt		
StFeSOD3	Soltu.DM.02G001300.1	Mt	Ch	Ch	Mt		
	Soltu.DM.02G001300.2	Mt	Ch	Ch	Mt		
	Soltu.DM.06G012170.1	O	Mt	Ch, C	Mt		
	Soltu.DM.06G012170.2	O	Mt	Ch, C	C		
StFeSOD4	Soltu.DM.06G012170.3	O	Mt	Ch, C, Mt	C		
	Soltu.DM.06G012170.4	O	Mt	N, Mt	Mt		
	Soltu.DM.06G012170.5	O	Mt	Ch, C, Mt	Mt		
	Soltu.DM.06G012170.6	O	Mt	Ch	C		

Note: C, cytosolic; Ch, chloroplastic; P, peroxisomal; Mt, mitochondrial; N, nuclear; O, other (because TargetP2 does not assign cytosolic localization).

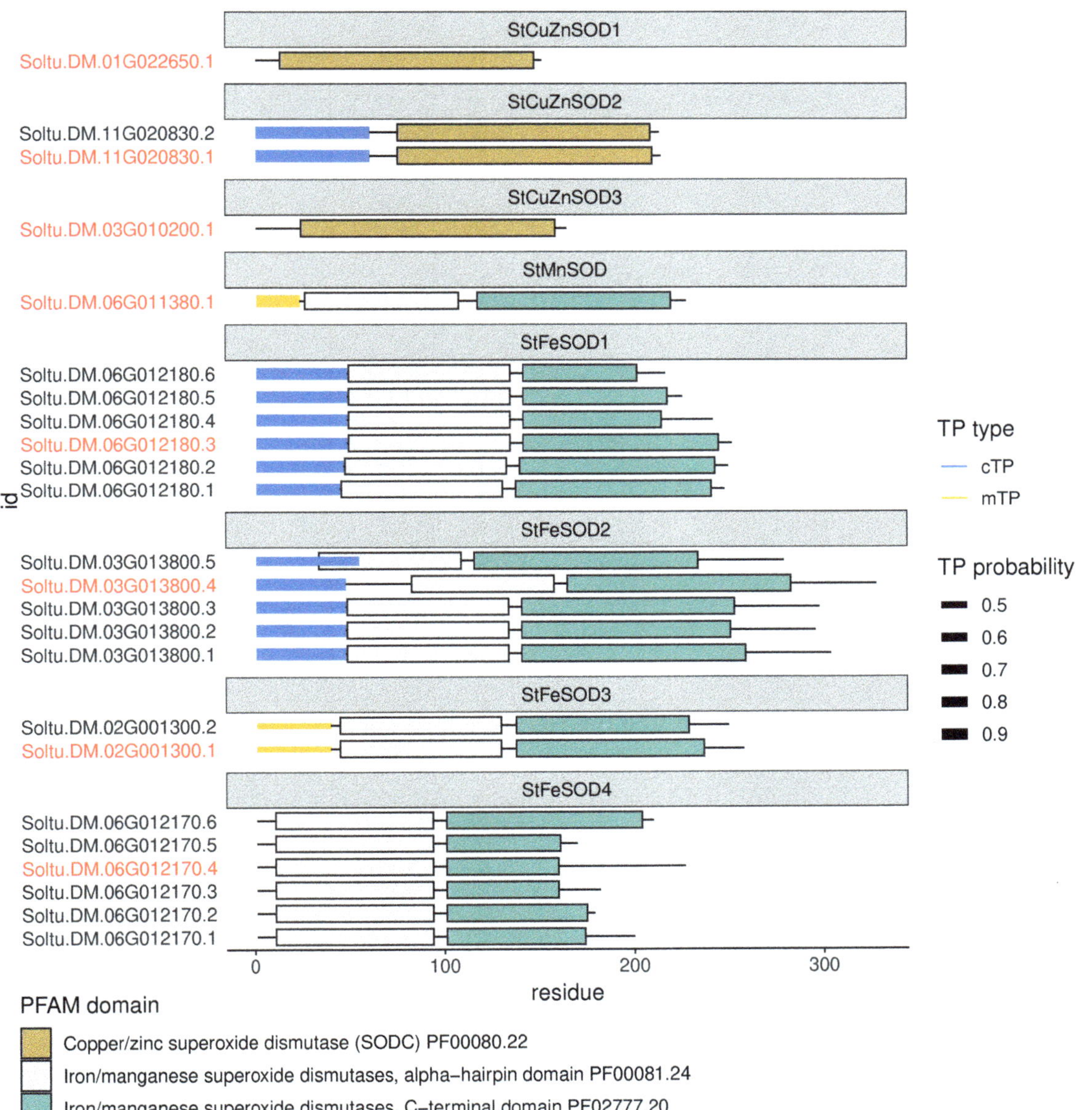

Figure 3. Annotation of protein sequences encoded by *StStSOD* genes. Protein sequences from all gene models are shown; protein products from representative (according to Spud DB DM v6.1 annotation) gene models are highlighted in red. Pfam domains are indicated by the color legend. TargetP2 annotation of target peptides (TP) is indicated by a blue (mitochondria target peptide—mTP) or yellow (chloroplast target peptide—cTP) segment on the N-terminal side. The width of the segment representing the TP corresponds to TargetP2 TP probability.

Proteins StCuZnSOD2, StFeSOD1 and StFeSOD2 are predicted to be targeted to chloroplast by all four tools used (Table 2, Figure 3). StMnSOD is predicted to be mitochondria-targeted. Analysis by TargetP2 indicates that StCuZnSOD1 and StCuZnSOD3, as well as

StFeSOD4 proteins are without predicted mitochondrial (mTP), chloroplastic (cTP) or secretory N-terminal transit peptides, so it is likely that these proteins are located in the cytosol or a cellular compartment other than mitochondria or chloroplasts (Table 2, Figure 3). The recently released subcellular localization prediction algorithm LA, based on embeddings from protein sequence language models, places the protein products of *StCuZnSOD1* and *StFeSOD4* genes in the cytosol, while the *StCuZnSOD3* product is predicted to be located in the peroxisomes. However, peroxisome-specific tools PTS1 Predictor and PredPlantPTS1 do not localize StCuZnSOD3 in the peroxisomes (Table 2). The most probable subcellular localization of isoforms StCuZnSOD3, StFeSOD3 and StFeSOD4, for which different tools gave different predictions, is discussed later in the context of literature data and phylogenetic relations.

3.5. Predicted Tertiary Structure of Potato SODs

The structural features of StSOD proteins corresponding to representative gene models (as annotated within Spud DB DM v6.1) were evaluated after tertiary structure prediction using state-of-the-art method AlphaFold. The obtained structures were compared with experimentally determined SODs from the Protein Data Bank (PDB, https://www.rcsb. org/, accessed on 16 November 2021). Proteins with predicted TP were processed prior to submission to AlphaFold. The AlphaFold models of StSOD proteins are characterized by the high percentage of Ramachandran favored residues (median: 97.8%), a low percentage of Ramachandran outliers (median: 0%), and low clash score (the number of serious clashes per 1000 atoms; median: 0.94, Table 3 and Supplementary Info S7). The structure with the highest percentage of Ramachandran outliers was StCuZnSOD3 (Soltu.DM.03G010200.1); however, the majority (3/4) of these outliers were in the first 12 N-terminal amino acids, a region predicted not to be in secondary structures by AlphaFold (Table 3).

Table 3. Evaluation of StSOD AlphaFold models.

Protein ID	Ramachandran Favored	Ramachandran Outliers	Rotamer Outliers	Clashscore
Soltu.DM.01G022650.1	98.67%	0.00%	0.00%	0.95
Soltu.DM.11G020830.1	98.03%	0.00%	0.00%	0.00
Soltu.DM.03G010200.1	92.02%	2.45%	0.00%	0.00
Soltu.DM.06G011380.1	98.51%	0.00%	0.00%	0.95
Soltu.DM.06G012180.3	98.01%	0.00%	0.00%	0.93
Soltu.DM.03G013800.1	97.24%	0.39%	0.91%	0.49
Soltu.DM.02G001300.1	95.83%	0.00%	0.00%	2.58
Soltu.DM.06G012170.6	97.60%	0.48%	0.57%	2.11

CuZnSOD protein sequences share high similarities across species belonging to different kingdoms of life. The high sequence identity is mirrored in the extraordinary similarity of tertiary structures of predicted AlphaFold models of StCuZnSODs, both among each other, as well as to experimentally determined CuZnSOD proteins from the PDB (Figure 4B,C). The eight-stranded Greek key β-barrel fold, characteristic of the eukaryotic (E-class) CuZnSODs [58], clearly describes the obtained StCuZnSOD structures (Figure 4B). The conserved residues involved in E-dimer interface (colored red in Figure 4) are located in β1 and β2 (labeled from the N-terminus) and in the coils connecting β4 to β5 and β6 to β7. The aforementioned Greek key motif is formed by β3–β6. The metal ion binding sites of CuZnSODs contains six conserved H and one conserved D. The Cu binding site consists of four H residues, while one of these H also contributes to Zn binding (Figure 4A,C). Two H residues involved in Cu ion binding are located in β4 (H46 and H48 in StCuZnSOD2, when amino acid numbering excludes the 61 amino acid-long cTP). All the residues involved in Zn ion binding (including the H involved in both Cu and Zn ion binding) are located in the loop connecting β4 and β5 (H63, H71, H80, and D83 in StCuZnSOD2), while the remaining H residue involved in Cu binding (H120 in StCuZnSOD2) is located in β7

(Figure 4B,C). The superimposition of 3D aligned structures of the StCuZnSOD2 model and the experimentally determined structure of the closely related tomato CuZnSOD (PDB: 3PU7) clearly demonstrates the conserved orientation of the active site residues (Figure 4C).

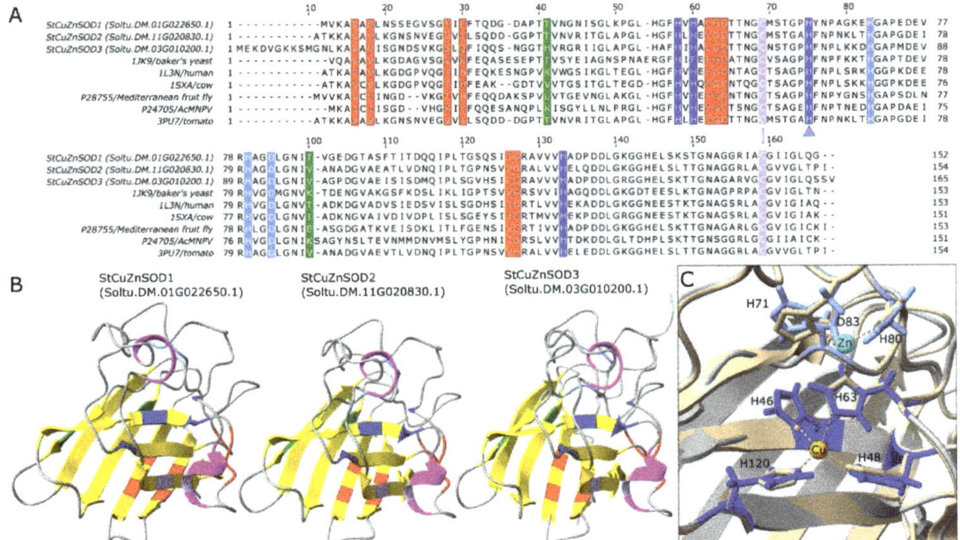

Figure 4. Structural features of StCuZnSODs. (**A**) Multiple sequence alignment of StCuZnSODs with several sequences of the same class with experimentally determined PDB structures. Processed StCuZnSOD sequences are shown based on TargetP2 transit peptide prediction. Residues are annotated according to CDD accession cd00305 (Cu/Zn_Superoxide_Dismutase): E-class dimer interface (eukaryotic polypeptide binding site) residues are colored red; P-class dimer interface (prokaryotic polypeptide binding site) residues are colored green; residues involved in Cu ion binding are colored blue, and residues involved in Zn ion binding are colored light blue. The arrow indicates the residue which is involved in binding of both Cu and Zn ions. Two cysteine residues involved in a disulfide bond as well as a bond itself (line) are colored violet. (**B**) AlphaFold models of StCuZnSODs: β-strands are colored yellow, α-helices are colored magenta, specific residues are colored as in A. (**C**) Metal binding site of StCuZnSOD2: the AlphaFold model of StCuZnSOD2 (in grey) was 3D aligned with 3PU7 (in tan) PDB structure (experimentally determined PDB structure of the tomato CuZnSOD). Metal ion position and coordinate bonds are based on 3PU7 Biological Assembly 1. StCuZnSOD2 residues involved in Cu and Zn ion binding are colored blue and light blue, respectively, and labeled according to processed peptide position. H63 residue is involved in binding of both metal ions.

The initial alignment of representative potato Mn-FeSOD protein sequences (Figure 5A) indicated that two sequences corresponding to the representative gene models Soltu.DM.03G013800.4 (*StFeSOD2* gene) and Soltu.DM.06G012170.4 (*StFeSOD4* gene) contain abnormal regions (emphasized by the red color in MSA—Figure 5A). Soltu.DM.03G013800.4 contains a 24 amino acid-long insert at the N-terminal side, which is not present in other Mn-FeSODs, while Soltu.DM.06G012170.4 contains an aberrant C-terminal region of over 60 amino acids lacking two conserved residues involved in metal ion coordination (Figure 5A). This is the reason the Iron/manganese SOD C-terminal domain is shorter in this protein sequence (Figure 3) as compared to other Mn-FeSODs. Therefore, it appears that the criterion used for establishing which sequences are representative in Spud DB DM v6.1 is sequence length; in other words, the gene model coding the longest protein sequence is used as the representative gene model by default, without taking into account sequence structural features. Based on our observations the representative sequence for

StFeSOD4 gene should be Soltu.DM.06G012170.6 instead of Soltu.DM.06G012170.4, as the longest protein sequence containing full-length PF00081 and PF02777 Pfam domains (Figure 3), and all characteristic amino acids involved in metal ion binding (Figure 5A). The representative sequence for *StFeSOD2* gene should be Soltu.DM.03G013800.1 instead of Soltu.DM.03G013800.4, because it contains the full-length PF00081 and PF02777.20 Pfam domains but lacks the non-characteristic 24 amino acid long N-terminal insert (Figure 5A). Hence, we used these sequences for tertiary structure modeling, along with StMnSOD and the default representative sequences for StFeSOD1 and StFeSOD3 (Tables 1 and 3 and Figure 5B).

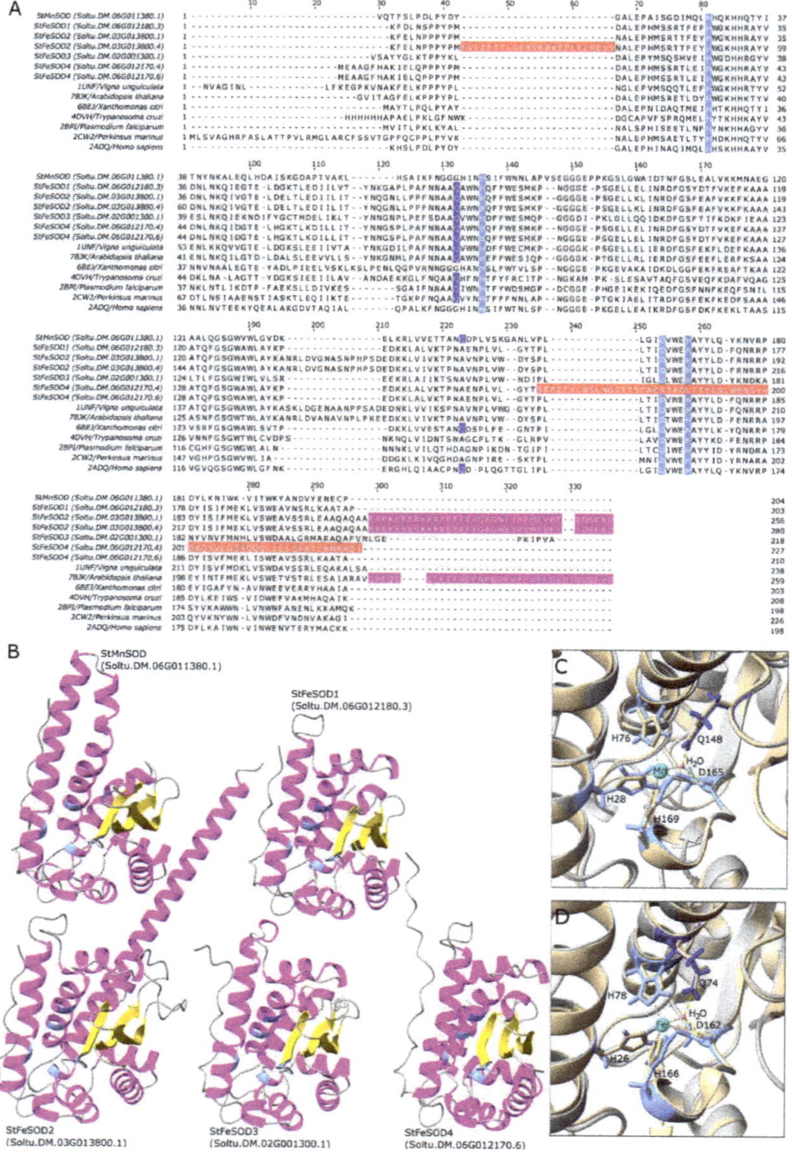

Figure 5. Structural features of StMnSOD and StFeSODs. (**A**) Multiple sequence alignment of potato

Mn-FeSODs with several sequences of the same class with experimentally determined PDB structures. Processed potato Mn-FeSOD sequences are shown based on TargetP2 transit peptide prediction. Residues involved in metal ion binding are colored light blue; residues involved in hydrogen-binding of the water molecule, which acts as the fifth metal ion ligand, are colored blue; Annotation is based on PDB structures 2ADQ, 1UNF and 7BJK. The abnormal peptide regions of two representative StFeSOD sequences, Soltu.DM.06G012170.4 (StFeSOD4) and Soltu.DM.03G013800.4 (StFeSOD2) are colored red. The long C-terminal tail of Soltu.DM.03G013800.1 and Soltu.DM.03G013800.4 is colored magenta. (**B**) AlphaFold models of StMnSOD and StFeSODs: β-strands are colored yellow, α-helices are colored magenta, specific residues involved in metal ion binding are colored as in A. (**C**) Metal binding site of StMnSOD: the AlphaFold model of StMnSOD (in grey) was 3D aligned with 2ADQ (in tan) PDB structure (experimentally determined PDB structure of the human MnSOD). Mn ion position, coordinate bonds (tan) and hydrogen-bonds (yellow) stabilizing the water molecule (magenta) which is the fifth coordination partner of the metal ion are based on 2ADQ Biological Assembly 1. StMnSOD residues involved in metal ion binding are colored light blue and labeled according to processed peptide position. The water molecule is stabilized with hydrogen-bonds (yellow) from StMnSOD Q148 (blue) and D165 (light blue). (**D**) Metal binding site of StFeSOD1: the AlphaFold model of StFeSOD1 (in grey) was 3D aligned with 7BJK (in tan) PDB structure (experimentally determined PDB structure of the chloroplastic FeSOD PAP9 from *Arabidopsis thaliana*). Metal ion position (Fe), coordinate bonds (tan) and hydrogen-bonds (yellow) stabilizing the water molecule (magenta) which is the fifth coordination partner of the metal ion are based on 7BJK Biological Assembly 1. StFeSOD1 residues involved in metal ion binding are colored light blue and labeled according to processed peptide position. The water molecule is stabilized with hydrogen-bonds from StFeSOD1 Q74 (blue) and D162 (light blue).

The obtained AlphaFold models of the five potato Mn-FeSODs display a similar fold to experimentally determined Mn-FeSODs from other organisms. The ~100 amino acid long α-hairpin domain of all StFeSODs consists of three α-helices, while the α-hairpin domain of the StMnSOD consists of two long α-helixes (Figure 5B). The C-terminal domain of all five potato Mn-FeSODs incorporates a central antiparallel β-sheet with three β-strands, which is surrounded by α-helices. In addition to the structural features common to all StFeSODs, StFeSOD2 is characterized by a long negatively charged C-terminal region rich in E/D (annotated with magenta color in Figure 5A) just after the Iron/manganese SOD C-terminal domain.

The metal binding site of Mn-FeSOD protein sequences consists of three conserved H and one conserved D residue (H28, H76, D165 and H169 in StMnSOD, where the amino acid numbering excludes the 24 amino acid-long mTP, Figure 5A,C). H28 and H76 (StMnSOD) are located in the first and last helix of the α-hairpin domain respectively while D165 and H169 (StMnSOD) are located in the C-terminal domain, specifically at the end of the third beta strand (β3) and in the short helical region just after β3 respectively. The metal ion in the catalytic center of Mn-FeSODs is penta-coordinated by these four conserved residues and a water molecule [59,60]. This ligand water molecule is stabilized by hydrogen-bonds with the D residue, also involved in metal ion coordination, and a conserved Q residue (annotated with dark blue color in Figure 5A,B). In FeSOD sequences the water-stabilizing Q is within the last helix of the α-hairpin domain (Q74 in StFeSOD1), while in MnSODs the Q (Q148 is StMnSOD) is located in the coil connecting β2 and β3 of the C-terminal domain (Figure 5A,C). Comparison of the active sites of StMnSOD with human MnSOD (PDB: 2ADQ, Figure 5C) as well as StFeSOD1 with Arabidopsis FeSOD (PAP9, Figure 5D) indicates conservation of active site residue orientation.

3.6. Phylogenetic Relations of StSOD Proteins

Phylogenetic relations of StSOD protein sequences were estimated by comparison with homologues from cultivated plant species and model organisms such as Arabidopsis. The phylogenetic trees were estimated independently for the Mn-FeSODs and CuZnSODs, because even though these two classes share some common features, these are far too sparse to produce a reliable phylogenetic model (Figure 6). The midpoint rooted maximum likelihood phylogenetic tree of potato Mn-FeSODs (Figure 6A) clearly indicates the presence of two clades, A and B, which are quite phylogenetically distant, so no suitable outgroup sequence for rooting of the tree could be found (all attempted outgroups clustered either with sequences from A or sequences from B therefore a midpoint rooted tree is presented). Cluster B is formed from MnSOD sequences homologous to StMnSOD. All of these sequences contain predicted mTP. In addition, these sequences contain the characteristic Q residue, involved in hydrogen binding with the water molecule ligand, located in the C-terminal domain (the informative sites of the Mn-FeSODs alignment used for the construction of the phylogenetic tree are provided in Figure S2, where the mentioned Q residue is colored blue). In all sequences in clade A, this Q is located in the α-hairpin domain (Figure S2). Clade A contains FeSODs, most of which are predicted to be chloroplast targeted, and is further dived into two subclades: A1 and A2. Subclade A2 consists of homologues of StFeSOD3, all of them containing a characteristic conserved sequence of 16 amino acids ([RW]A[QE][AS][FL]VNLGEPKIP[VI]A) after the Mn-Fe C-terminal domain. Interestingly, all of the sequences in A2 are predicted to be chloroplast targeted apart from the phylogenetically closest StFeSOD3 and tomato FeSOD SlSOD7 (Solyc02g021140.3.1), which are predicted by TargetP2 to be mitochondrial. However, since the subcellular localization of StFeSOD3 is ambiguous (Table 2), this issue is further considered in the Discussion section. Subclade A1 consists of all the remaining StFeSODs and their homologues including sequences with the E/D rich C-term region associated with PEP, like the Arabidopsis PAP9 (At5g51100.1-*A. thaliana*) and StFeSOD2 sequences. The phylogenetic proximity of SODs associated to PEP and SODs that are not PEP associated indicates that apart the characteristic C-term extension other sequence features are conserved in PEP SODs. All of the sequences in subclade A1 are predicted to be chloroplast targeted apart the potato FeSOD4, predicted to be localized in the cytosol (Figure 6, Table 2).

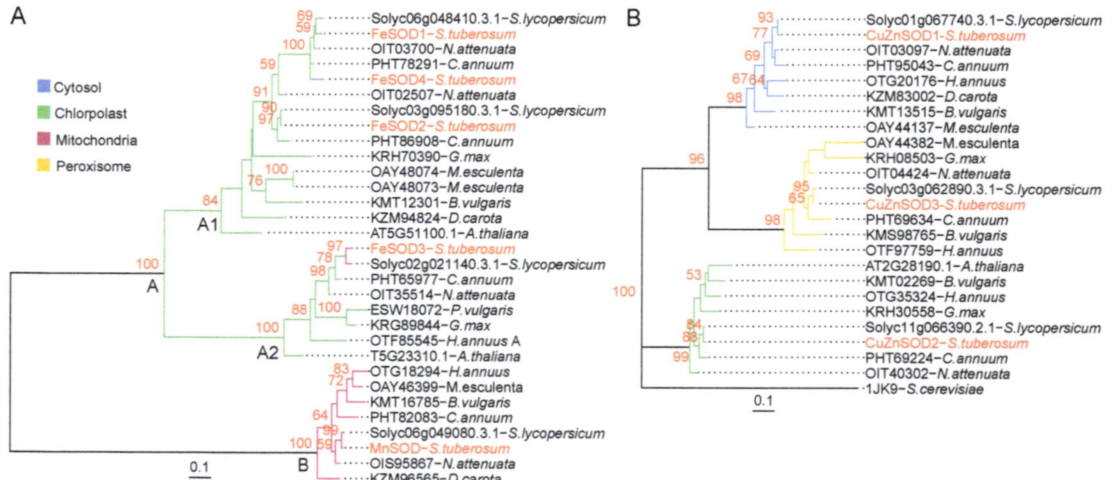

Figure 6. Maximum likelihood phylogenetic trees of StSODs. (**A**) Midpoint rooted Mn-FeSOD phylogenetic tree. (**B**) Rooted CuZnSOD phylogenetic tree. Support values were obtained using 100 iterations of non-parametric bootstrap; values over 50/100 are indicated with a red number. Clade/tip coloring is based on subcellular location (color legend) obtained using Light Attention server.

Based on the phylogenetic tree for CuZnSODs, it can be observed that the differences among the sequences belonging to this SOD class, at least sequences which are homologues to StCuZnSODs, are much less prominent compared to Mn-FeSODs. Each of the

StCuZnSODs forms a clade with closest homologues from other plants (Figure 6B). This is not surprising given the high sequence conservation of CuZnSODs among different kingdoms (Figure 5A). The clustering of sequences appears to be related to their subcellular localization—the three clusters correspond to CuZnSOD sequences from cytosol, peroxisome and chloroplast (Figure 6B and Table 2).

In general, all StSODs clustered closely to homologues from the phylogenetically closest plant species—tomato (Figure 6A,B).

3.7. Expression Profiles of the StSOD Genes in Response to Elevated Temperatures and SA Application

To evaluate the response of *StSOD* genes to elevated temperatures and SA treatment, we have analyzed the expression profiles of these genes in leaves of three potato cultivars (Agria, Désirée, and Kennebec) under three temperature treatments (21 °C, 26 °C, and 29 °C) with or without SA application. Absolute rather than relative quantification by qRT-PCR was used to allow the comparison of the expression levels between different *StSODs*. In the cases were more gene models (splice variants) were present, the primers were designed to amplify all variants. While all eight *StSOD* transcripts were detected in all three cultivars and under all treatments (Figure 7), the expression levels *StFeSOD1* and *StFeSOD4* were 3–4 orders of magnitude lower as compared to other *StSODs*, with particularly low expression in Agria and Désirée cultivars. Thus, these two isoforms can be considered minor isoforms.

Regarding the effects of elevated temperature and SA application, no general trends applicable to all *StSODs* from all cultivars were observed, but some regularities were noted (Figure 7). Mildly elevated temperature of 26 °C did not affect the expression of *StCuZnSODs* and *StMnSOD* in any of the cultivars, except for subtle induction of *StCuZnSOD3* in Agria as compared to control plants. In all examined cultivars, *StFeSOD2* and *StFeSOD3* were up-regulated under 26 °C, while minor isoforms *StFeSOD1* and *StFeSOD4* were up-regulated only in cv. Kennebec with transcript level of *StFeSOD1* increased by 2.3-fold. Treatment of 29 °C induced the expression of all *StSODs*, except minor isoforms *StFeSOD1* and *StFeSOD4*, in cv. Désirée as compared to the control treatment. The highest differences were observed for *StCuZnSOD1* and *StCuZnSOD2*, which showed 10- and 21-fold higher expression levels at 29 °C than control, respectively. Similar induction of *StCuZnSOD1*, *StCuZnSOD2*, *StFeSOD1*, *StFeSOD2* and *StFeSOD3* was seen in cv. Kennebec in response to 29 °C. In this potato cultivar, the expression level of *StCuZnSOD2* was changed as much as 33-fold when cultivated at 29 °C as compared to control plants. In cv. Agria, however, growth at 29 °C did not cause any major changes in *StSODs* expression. In this cultivar, the expression levels of *StCuZnSOD3*, *StMnSOD*, *StFeSOD1* and *StFeSOD3* slightly increased (<2.4 fold), while the expression of *StCuZnSOD1* and *StCuZnSOD2* slightly decreased (<1.5 fold) in response to 29 °C, as compared to control.

Modulation of *StSODs* expression by exogenous SA application in different potato cultivars exposed to three temperature treatments was in most cases subtle. At control temperature of 21 °C, SA had very little effect on *StSODs* expression, with the exception of a 4-fold down-regulation of *StCuZnSOD2* in cv. Agria. In plants grown at 26 °C, consistent up-regulation of all *StSODs* in response to SA treatment, as compared to plants grown at the same temperature without SA, was observed only in cv. Désirée, where greatest up-regulation of 3.6-fold was recorded for *StFeSOD4*. In response to SA treatment at 29 °C, almost all *StSODs* in cv. Agria were slightly down-regulated in comparison to plants grown without SA, with exceptions of *StCuZnSOD1* and *StFeSOD4*, which were up-regulated or unchanged, respectively. Differences in the *StSODs* expression in cv. Désirée cultivated at 29 °C with or without SA treatment were subtle, and the same is true for cv. Kennebec.

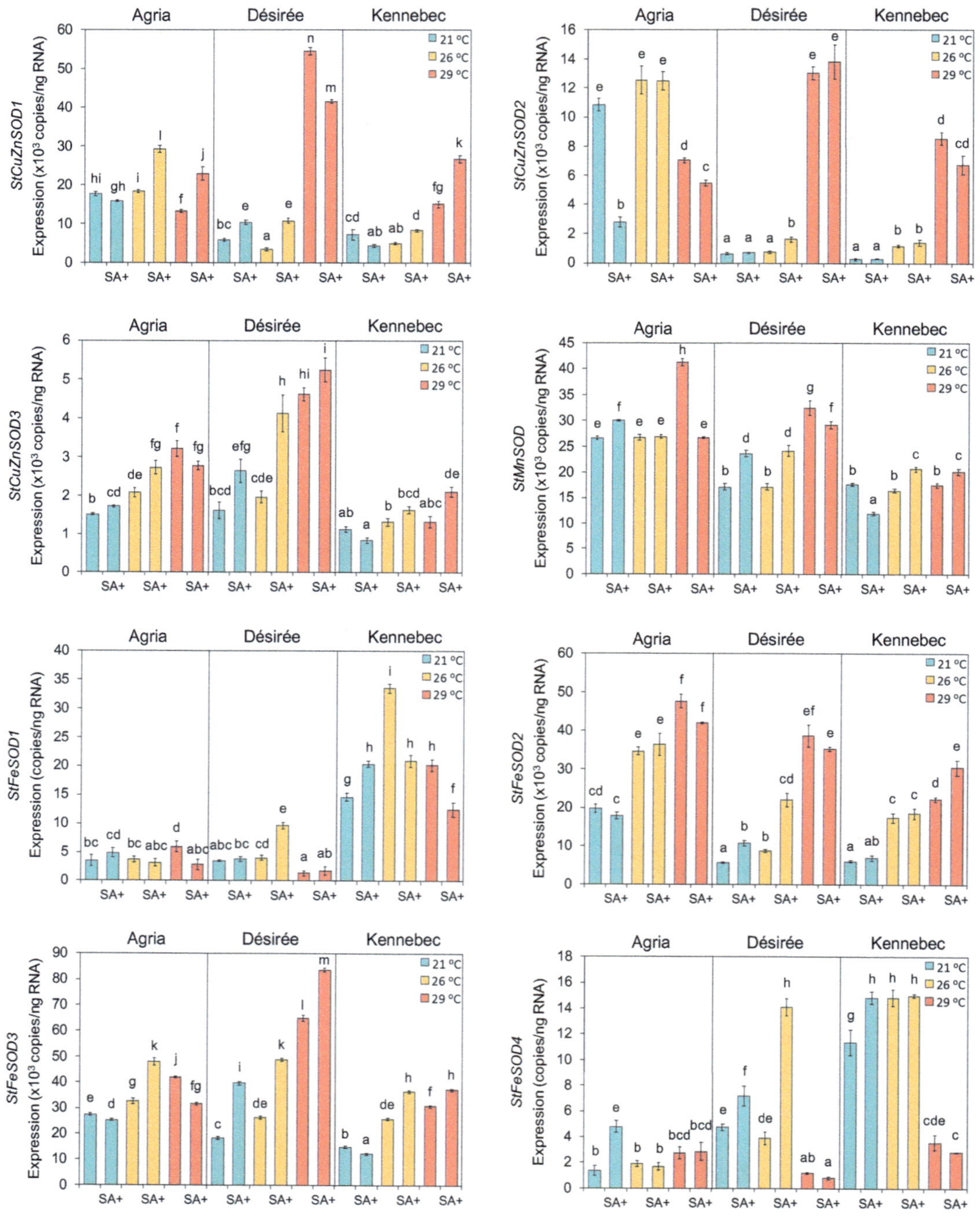

Figure 7. Expression profiles of potato *StSOD* genes under different temperature conditions and exogenous SA application. Real-time PCR was used both to validate the presence of investigated

genes and for quantitative analysis of their expression under control (21 °C) and elevated temperatures (26 and 29 °C) with SA application (10^{-5} M) or without. The analysis was conducted on three unrelated potato cultivars (Agria, Désirée, and Kennebec) grown in vitro. The scale bar represents absolute normalized expression values. The data were shown as mean values ± S.D of the three biological replicates. "SA+" indicates exogenous salicylic acid application. The different letters on bars indicate significant differences at a confidence level of $p < 0.05$.

4. Discussion

4.1. StSOD Gene Family

Comprehensive genome-wide identification and characterization of SOD family members has been conducted in almost all major crops (Table 4), but such data were not available for potato until now. Hereby we present an in depth in silico study of the three *CuZn-SODs*, named *StCuZnSOD1*, *StCuZnSOD2* and *StCuZnSOD3* in the potato genome, along with five members of the Mn-FeSOD class. Sequence Soltu.DM.06G011380 is very likely a single potato *StMnSOD*, whereas others are *FeSODs*, termed *StFeSOD1* through *StFe-SOD4*. Two SOD-like sequences annotated as SODs in DM v6.1 (Soltu.DM.10G011570 and Soltu.DM.06G013120) encode short (~100 aa) proteins, so they can be considered as pseudogenes. Cucumber gene *CsFSD3*, even though it codes for a fairly large 377 aa protein, is also considered as pseudogene because it is not expressed in any organ or under any conditions [61]. Copper chaperone CCS, which is required for the activation of CuZnSODs [62] was also identified (Soltu.DM.08G026370). Considering only papers published since 2015. (Thus probably relying on complete genomic data obtained by state-of-the-art technologies), it seems that a family of 7–9 *SOD* isoforms, with at least one *MnSOD* is typical for most species regardless of genome size (Table 4). Few exceptions with more *SODs* include banana [63], as well as *Triticum aestivum*, but in the latter case 26 *SOD* isoforms were found in three sub-genomes (A, B and D) of this hexaploid species [64].

Table 4. Overview of the *SOD* gene families in different crops.

Species	Genome Size (Mbp)	CuZnSOD	MnSOD	FeSOD	Total	Chr. num.	Introns num.	References
Solanum tuberosum	844	3	1	4	8	5	4–9	Current work
Arabidopsis thaliana	125	3	2	3	8	5	5–8	[9,65]
Oryza sativa	389	4	1	2	7	6	5–9	[9,65]
Sorghum bicolor	730	5 *	1	2	8	6	5–7	[66]
Gossypium raimondii	885	5	2	2	9	6+	4–8	[67]
Gossypium arboreum	1746	5	2	2	9	6	5–8	[67]
Cucumis sativus	367	5	1	3	9	5+	3–8	[61]
Musa acuminate	523	6	4	2	12	8	5–8	[63]
Solanum lycopersicum	828	4 *	1	4	9	6	4–8	[68]
Triticum aestivum	~17,000	17	3	6	26	3 × 3	4–7	[64]

Note: The listed number of CuZnSOD isoforms does not include Cu chaperones, but cases where it is not clear whether Cu chaperones are included are indicated with an asterisks (*). Chr. num. is the number of chromosomes where *SOD* genes are located, where "+" indicates that some *SOD* genes were found on scaffolds. In wheat, *SOD* genes are found on 3 chromosomes of each of the 3 sub-genomes.

4.2. Tandem Duplication of the FeSOD Genes Is a Characteristic of Solanum Species

StSOD genes are located on five out of 12 potato chromosomes (Table 1), with two genes on chromosome 3 and three genes on chromosome 6. In other plant species *SOD* genes are also scattered on different chromosomes (Table 4). Two physically close genes, *StFeSOD1* and *StFeSOD4*, separated by only 1566 bp on chromosome 6, are 80.77% identical at the protein level and so by definition they represent tandemly duplicated genes [56]. In addition, *StFeSOD1* is 71.91% identical to *StFeSOD2*, but since these genes are on different chromosomes, they can be considered as segmental duplications. Different types of gene duplications are major way for the expansion of gene families, which is often followed by functional divergences of the duplicated genes. However, segmental, rather than tandem duplications have been found for SODs from different plant species. For example, segmental duplications were reported for *OsCSD2* and *OsCSD3* in rice, where both genes preserved their function [65], for *SbSOD2* and *SbSOD5* in *S. bicolor* [66], *GrMSD1* and *GrMSD2* in *G. raimondii* [67], *AtMSD1* and *AtMSD2* as well as *AtFSD1* and *AtFSD2* in *A. thaliana* [67], and *SlSOD5* (Solyc06g048410.2) and *SlSOD6* (Solyc03g095180.2) in tomato [68]. None of the abovementioned species, except tomato, features tandem duplications of *SOD* genes. In tomato, just like in potato, two *FeSODs*—*SlSOD5* and *SlSOD8* (Solyc06g048420.1) are reported to be tandemly duplicated and in both genomes they are on chromosome 6 [68]. So, both in potato and in tomato one tandem duplication on chromosome 6 (*StFeSOD1/4* and *SlSOD5/8*) and one segmental duplication (*StFeSOD1/2* and *SlSOD5/6*) on chromosomes 6 and 3 were found. Relations among duplicated potato and tomato genes are clear from phylogenetic tree as well (Figure 6A). However, sequence *SlSOD8* was not included in our phylogenetic study, because different versions of this gene, Solyc06g048420.1 and Solyc06g048420.2, code for proteins of 160 and 109 residues respectively, making it unclear whether this is functional gene or a pseudogene. Based on sequence similarities which are greater between tandem pairs than between segmental duplicates in both species, and on the fact that in both species the gene on chromosome 3 does not have its tandem pair [68], it can be concluded that segmental duplication probably preceded tandem duplication event. If so, then *StFeSOD1* in potato and *SlSOD5* in tomato are ancestral genes that first gave rise to copies on chromosome 3 and then were locally duplicated. In any case, tandem duplication of *FeSOD* gene is a relatively recent event characteristic either for the genus *Solanum* or the Solenaceae family.

4.3. Gene Models of StSODs

StSOD genes have variable number of introns, ranging from 4 to 9 (Figure 1), which is comparable to the number of introns found in *SOD* genes of other species (Table 4). Even genes that are closely related, such as tandem pair *StFeSOD1* and *StFeSOD4*, have quite different exon/intron arrangements (Figure 1). The differences in exon–intron structure of duplicated genes may be accomplished by one of the three main mechanisms: exon/intron, gain/loss, exonization/pseudoexonization, and insertion/deletion [69].

As can be seen in Table 1 and Figure 1, five out of eight *StSODs* are represented with multiple gene models, where *StFeSOD1* and *StFeSOD4* have as many as six models each. Since protein-coding genes were annotated using full-length cDNAs [24], it is safe to say that these are actually different splice variants generated by alternative splicing. Alternative splicing is involved in the regulation of *SOD* gene expression [70] and has been experimentally proven in rice [71]. In addition to alternative splicing, alternative transcription start sites and alternative polyadenylation has also been reported for *SOD* genes in banana [63]. Such transcripts would share an identical ORF but have different UTRs. This is the case with *StCuZnSOD2*, where its two gene models, Soltu.DM.11G020830.1 and Soltu.DM.11G020830.2, differ in the length of 3′-UTR (Figure 1). It appears that the criterion used for establishing which sequences are representative in Spud DB DM v6.1 is sequence length; in other words, the gene model encoding the longest protein sequence is used as the representative gene model without taking into account structural features of the encoded protein. As discussed below, in some cases this default choice might not be the best choice.

Considering previously discussed tandem and segmental duplications, it is worth noting that duplicated genes, *StFeSOD1*, *StFeSOD2* and *StFeSOD4*, have significantly more splice variants in comparison to other StSOD isoforms (Table 1 and Figure 1). It is tempting to speculate that having three copies of a gene allows for splicing flexibility as a way for further divergence and molecular innovations.

It should be noted that StSOD gene models presented here are not allelic isoforms, since they are derived from a genome of a doubled monoploid clone [24], meaning that the actual molecular variety of these enzymes in potato cultivars, which are highly heterozygous autotetraploids (2n = 4x = 48), is probably even greater.

4.4. Structural Features, Subcellular Localization and Phylogenetic Relations of StCuZnSODs

Eukaryotic CuZnSODs are highly conserved proteins regarding their primary structure, position of the key residues, domain organization and tertiary and quaternary structure [72]. StCuZnSODs feature all characteristics of eukaryotic CuZnSODs, including PF00080.22 domain (Figure 3) with conserved residues involved in metal ion binding, dimerization and disulfide bridging (Figure 4A). StCuZnSODs form a typical Greek key scaffold, consisting of a β-barrel composed of eight antiparallel β-strands (Figure 4B).

CuZnSODs are commonly active as homodimers, while chloroplastic CuZnSODs are homotetrameric [5,72,73]. However, native PAGE assays with isoform-selective inhibitors or in combination with immunoblotting revealed at least 6 (and possibly 7, depending on the cultivar and growth conditions) CuZnSOD activity bands in potato [74]. As discussed later, three StCuZnSOD proteins are probably targeted to different cellular compartments, so different subunits should not combine in vivo, but various subunit combinations are possible in the leaf extracts. In the case of *StCuZnSODs*, alternative splicing as a source of protein variety is not an option, since *StCuZnSOD1* and *StCuZnSOD3* have only one gene model, while the only difference between two *StCuZnSOD2* gene models is polyadenylation site (Figure 1). Other feasible explanations as to why potato cultivars have more CuZnSOD activity bands than genes, include possible post-translational modifications [75] and allelic polymorphism of the isoforms.

Thermostability is another property of enzymes that should be considered when investigating their expression and function under elevated temperatures. Several lines of evidence suggest that CuZnSODs in general, including StCuZnSODs, are thermostable enzymes, expected to perform well during the HS. Average aliphatic index for representative StCuZnSODs is 85.59 (Table 1), which is considerably higher than that of StFeSODs (76.35) but lower than the index of StMnSOD (91.14). The aliphatic index is the relative volume of a protein occupied by aliphatic side chains and may be considered as a positive factor for the increase of thermostability of globular proteins [57]. Furthermore, CuZnSODs, unlike other SODs, feature stable Greek key scaffold which supports active site and dimer formation, and is further reinforced by a disulfide bond (Figure 4) [72]. This disulfide bond stabilizes both the subunit fold and the dimer interface and affects enzyme activity [62,72]. Cu chaperone Cu-CCS not only provides Cu for the active site of the enzyme but facilitates disulfide formation as well [62].

Plant CuZnSODs are found in different cellular compartments, including cytosol, plastids, peroxisomes and possibly extracellular space [5,73]. All four servers inquired for subcellular targeting (DeepLoc, CELLO, TargetP2 and Light Attention) indisputably located StCuZnSOD1 in the cytosol and StCuZnSOD2 in the plastids (Table 2 and Figures 3 and 8). Subcellular localization of StCuZnSOD3, however, is inconclusive. According to DeepLoc-1.0 [31], this sequence is targeted to peroxisomes (Table 2) with likelihood of 0.495, which is higher than cytoplasmic localization likelihood (0.396), whereas other possible localizations are very unlikely. Light Attention [33] also locates StCuZnSOD3 to the peroxisomes, but CELLO does not (Table 2). Programs specifically designed for peroxisomal targeting, like PTS1 Predictor [35,36] or plant-specific PredPlantPTS1 [37] do not predict that StCuZnSOD3 is targeted to peroxisomes. Since most of the peroxisomal proteins possess a peroxisome targeting signal type 1 (PTS1) consisting of a C-terminal tripeptide, the SSV> tripeptide (where ">" is C-terminal end) found in StCuZnSOD3 was compared to PTS1 signals from known peroxisomal proteins [76]. It turned out that SSV> does not belong to so-called canonical plant PTS1 tripeptides, [SA][RK][LMI]>, which confer strong peroxisome targeting efficiency, and not even to weak non-canonical PTS1 tripeptides characterized with one non-canonical amino acid residue at one of these tree positions (x[RK][LMI]>, [SA]y[LMI]> or [SA][RK]z>) [76]. However, SSV> can be found among all known 35 different PTS1 signals that are functional in plants (represented by: [SAPCFVGTLKIQ][RKNMSLHGETFPQCYDA][LMIVYF]>) [76] but its efficiency is yet to be determined. Alternative N-terminal PTS2 motif ([RK][LVIQ]x2[LVIHQ][LSGAK]x[HQ][LAF]) [76] is not found in StCuZnSOD3. Many plants have at least one SOD (whether CuZnSOD,

MnSOD, FeSOD or some combination of the isoforms) located in the matrix and/or membrane of the peroxisomes [77], so having at least one peroxisomal SOD must have some physiological advantage.

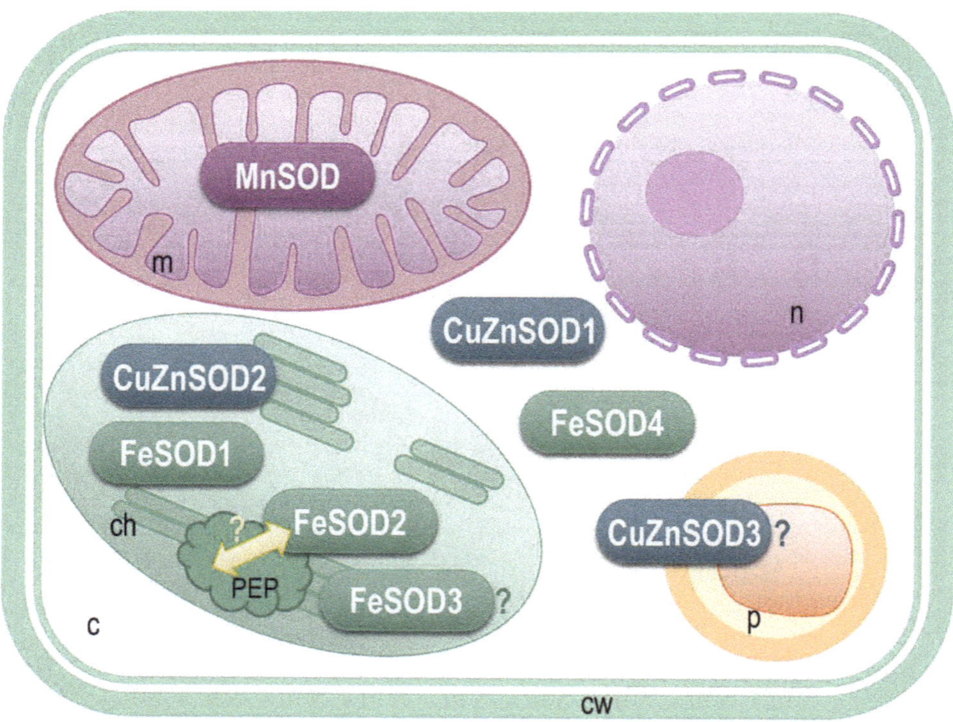

Figure 8. Proposed subcellular localization of StSODs. Prefix "St" is omitted from the potato SOD isoform names for simplicity. C—cytosol; m—mitochondrion; ch—chloroplast; p—peroxisome; n—nucleus; cw—cell wall; PEP—plastid-encoded RNA polymerase. Inconclusive localizations of certain isoforms as well as unproven interactions are indicated by a question mark.

CuZnSODs share no sequence similarity to Mn-FeSODs and they have probably evolved independently from mutually related FeSODs and MnSODs [5,73], so their relations are presented as two independent phylogenetic trees (Figure 6). In some studies, all *SOD* genes are presented by the same phylogenetic tree, but in these cases CuZnSODs are separated from Mn-FeSODs with high bootstrap values, indicating again their separate evolution [63,67]. Even though CuZnSODs, and specifically plant CuZnSODs, share high degree of homology (Figure 4A), there are some features that distinguish chloroplastic from cytosolic enzymes, which separates them into two clusters [66,68,78]. It can be seen that StCuZnSOD1 clusters with other cytosolic enzymes, whereas StCuZnSOD2 is closely related to chloroplastic CuZnSODs (Figure 6B), which supports previously discussed subcellular localizations. Peroxisomal localization of StCuZnSOD3 is supported by clear phylogenetic association with other peroxisomal SODs (Figure 6B).

4.5. StMnSOD Is a Mitochondrial Enzyme with Distinguishable Structural Features

MnSODs and *FeSODs* have apparently evolved from a common ancestral gene [5,72] and are so closely related, that in literature and sequence databases their products are often referred to as Mn-FeSODs. Both types of enzymes are characterized by the same conserved domains: iron/manganese superoxide dismutase, α-hairpin domain (PF00081.24) and iron/manganese superoxide dismutase, C-terminal domain (PF02777.20, Figure 3). Since all tested potato cultivars express one MnSOD, as confirmed by native PAGE inhibition assays and immunoblots [74], the question is which of the *StSOD* genes encodes MnSOD? We have several reasons to believe that sequence Soltu.DM.06G011380.1 encodes StMnSOD, based on subcellular targeting, structural features and phylogenetic relations. First, protein encoded by Soltu.DM.06G011380.1 is targeted to mitochondria, which is confirmed by four different servers (Table 2 and Figure 3). To our best knowledge, only MnSODs are found in plant mitochondria and it is widely accepted that MnSODs are primarily targeted to mitochondria, even though they can also be found in peroxisomes [5,72,73,77].

Second, there are some structural features that distinguish MnSODs from FeSODs analyzed in our study. In the Mn-FeSODs alignment (Figure 5A) it can be seen that most sequences, all of them being FeSODs, contain Q residue (dark blue) involved in hydrogen-bonding with water at the consensus position 132, whereas in Soltu.DM.06G011380.1 and in two other MnSODs (PDB: 2ADQ, *Homo sapiens* and 6BEJ, *Xanthomonas citri*), this Q residue is at position 223. It is important to note that this Q residue, involved in hydrogen-bonding with the water molecule, assumes a very similar position in the active center regardless of its position in the sequence (either in the loop connecting β2 and β3 strands as in Figure 5C, or in the α3 helix as in Figure 5D), but subtle differences between these two positions affecting the redox tuning of the metal ion define whether the enzyme binds Mn or Fe [79]. This also provides a reason why the amino acids interchangeable with the mentioned Q are short-chained amino acids G and A, since amino acids with longer chain would clash with the opposite Q side chain and prohibit its stabilizing effect on the water molecule ligand.

Even more interesting is the fact that, contrary to StFeSODs, the α-hairpin domain of StMnSOD consists of two long α-helices arranged as a hairpin, resembling the structure of the human mitochondrial MnSOD (PDB: 2ADQ) [60] and other eukaryotic MnSOD structures (PDB: 4X9Q, 4E4E, 4C7U to name a few). This feature is not found in all MnSODs because at least some experimentally resolved structures (for example bacterial MnSODs 6M30 from *Staphylococcus equorum* [80] or 6BEJ from *X. citri* [81]) contain a three-helix α-hairpin domain. The number of helices in the hairpin domain appears to be one of the determinants of MnSOD oligomerization state, where MnSOD variants with two α-helix hairpin domain have a preference to form tetramers. This proposition is based on interaction of the α-hairpin domains of diagonally placed SOD monomers in experimentally determined tetrameric MnSODs, where each side of the oligomer is encircled by one of the two 4-helix bundles at opposite ends of the dimer, which acts as a clamp, holding the dimers in place [82]. It should be noted not all MnSODs containing a two-helix hairpin domain form tetramers in solution, even though they crystalize in the tetrameric state [82].

Finally, these slight structural differences between MnSODs and FeSODs are reflected in the phylogenetic tree as well (Figure 6), where StMnSOD is grouped in cluster B with other MnSODs, whereas all StFeSODs are grouped in cluster A, with other FeSODs. As expected, StMnSOD is closely related to MnSODs from other Solenaceae species—tomato, wild tobacco and pepper (Figure 6).

The fact that StMnSOD has no alternative splice variants (Figure 1) probably reflects the requirement for stringent and, as discussed later, pretty constitutive and high expression, because it is the only mitochondrial SOD isoform in potato, and as such is crucial for ROS scavenging in this organelle. StMnSOD is predicted to be thermostable, having the highest aliphatic index of 91.14 of all StSODs (Table 1).

4.6. Not All StFeSOD Splice Variants Encode Functional Proteins

StFeSODs share the same conserved domains PF00081 and PF02777 typical for Mn-FeSODs (Figure 3), but some gene models have certain specificities. As explained in the Results section, sequence Soltu.DM.03G013800.4, encoded by *StFeSOD2*, contains an abnormal 24 amino acid-long insert on the N-terminal side (red highlighted in Figure 5A), while Soltu.DM.06G012170.4, encoded by *StFeSOD4*, contains an aberrant C-terminal region over 60 amino acids long which lacks two conserved amino acid residues involved in metal ion binding (red highlighted in Figure 5A). It is no wonder that these two proteins with abnormal regions, which are, as we believe, produced by abnormal splicing, are products of genes *StFeSOD2* and *StFeSOD4*—segmental and tandem duplicates, respectively, of *StFeSOD1* gene. This only confirms our previous notion that gene duplications in *StFeSOD* family provided certain splicing flexibility, which in two cases resulted in proteins with abnormal regions. While Soltu.DM.03G013800.4 could still be a fully functional enzyme, Soltu.DM.06G012170.4 probably is not. However, our conclusion that gene duplications lead to more flexible alternative splicing, based solely on the study of potato SODs, is in a complete disagreement with findings of [83]. According to comprehensive analysis of these authors, duplicated genes have fewer alternative splice forms than single-copy genes, and there is a negative correlation between the mean number of alternative splice forms and the gene family size [83].

Plant FeSODs are primarily targeted to chloroplasts, but can also be found in cytoplasm, as well as in peroxisomes [5,72,73,77] According to all queried servers, StFeSOD1 and StFeSOD2 are plastidic isoforms (Table 2 and Figures 3 and 8). StFeSOD3 is also predicted to be plastidic by DeepLoc and CELLO, but Light Attention and TargetP2 suggest its mitochondrial location, albeit with low probability (Figure 3). StFeSOD3 clusters with other chloroplastic FeSODs in clade A2 (Figure 6A), and only its closest homolog, tomato FeSOD SlSOD7 (Solyc02g021140.3.1), is also predicted to be mitochondrial by TargetP2 (Figure 6A), but ProtComp9.0 server places it in the chloroplasts [68]. Knowing that only primitive eukaryotes may contain mitochondrial FeSODs [5], StFeSOD3 is probably also targeted to chloroplasts. Finally, regarding StFeSOD4, different servers, except TargetP2, suggest different subcellular localizations for each of its gene models (Table 2). In this case we would incline with the cytosolic ("other") prediction given by TargetP2, because the N-terminal region preceding the Mn-Fe α-hairpin domain is so short, that any signal would have to overlap with the α-hairpin domain, which is highly unlikely (Figure 3). It could be argued that the chloroplast targeting signal, present in the ancestral *StFeSOD1* gene, has been preserved in segmentally duplicated *StFeSOD2*, but lost in tandemly duplicated *StFeSOD4*.

As already mentioned, StFeSOD2 (Soltu.DM.03G013800.1) is characterized by a long negatively charged C-terminal region rich in E/D (magenta highlighted in Figure 5A). This region resembles the E/D rich C-terminal region of SODs associated with the plastid-encoded RNA polymerase (PEP) multimeric enzyme, like the Arabidopsis PAP9 (also presented in Figure 5A alignment, PDB: 7BJK) [59]. PEP is essential for the proper expression of the plastid genome during chloroplast biogenesis and is composed of four plastid-encoded subunits and 12 nuclear-encoded PEP-associated proteins (PAPs) [59]. Unlike the C-term region of the Arabidopsis PAP9, which is most likely disordered because no electron density was observed for residues after G231 [59], the C-terminal region of Soltu.DM.03G013800.1 is predicted by AlphaFold to be in α-helix conformation (Figure 5B). It is worth noting that the AlphaFold model of Arabidopsis PAP9 (https://www.uniprot.org/uniprot/Q9LU64, accessed on 15 November 2021) agrees with the experimentally determined flexible C-term, so this feature differentiates Soltu.DM.03G013800.1 from PAP9. In addition, PAP9 contains a Zn ion in the active site even though it structurally falls with the Mn-FeSODs [59]. Thus, it would be interesting to experimentally characterize potato StFeSOD2.

The proposed subcellular localization of StSOD enzymes is presented in Figure 8. None of the StSODs localized to the nucleus or extracellular space. Extracellular SODs, if present, are commonly considered to be CuZnSODs [5,73], but in *A. thaliana* the only extracellular SOD is an MnSOD, ATMSD2 (AT3G56350) [9], while in *C. sativus* an FeSOD, CsFSD3, was predicted to be extracellular [61]. However, many plant species including *O. sativa* [9], *S. bicolor* [66], *G. raimondii* and *G. arboreum* [67], *S. lycopersicum* [68] and *T. aestivum* [64] do not have any SODs predicted to be apoplastic.

4.7. Expression of StSOD Genes in Response to Elevated Temperatures and Exogenous SA

The ideal temperature for the growth and development of the aerial parts of potato plants is generally 20–25 °C, while the optimal temperature range for tuber formation is 15–20 °C [84]. Even though potato is a cool-season crop, there is variation for HS tolerance across potato germplasm [85–87]. We have compared the expression of *StSODs* in three potato cultivars, Agria, Désirée and Kennebec, grown in vitro either at optimal temperature of 21 °C, or at elevated temperatures of 26 and 29 °C. Désirée is considered as relatively thermo-tolerant cultivar [85–87], Agria as thermo-sensitive [87], whereas the data for Kennebec are scarce, but it appears to belong to sensitive cultivars [86]. The described temperature treatments were combined with SA treatments (0 or 10^{-5} M SA), because SA can often enhance thermotolerance by modulating the expression and/or activity of antioxidative enzymes [88,89].

All *StSODs* identified in doubled monoploid potato genome are present and expressed in tested potato cultivars, which are highly heterozygous autotetraploids (Figure 7 and Figure S1). Even though the current expression analysis (Figure 7) cannot be directly correlated to StSOD activities [74], because different cultivars were tested (only cv. Agria is common to both studies), nevertheless the combined expression and activity data suggest that *StSODs* can be roughly categorized into: (1) *StCuZnSODs*, which have relatively high gene expression and high activity; (2) *StSODs* with relatively high gene expression but low activity, including *StMnSOD, StFeSOD2* and *StFeSOD3* and (3) minor isoforms with low expression and low activity—*StFeSOD1* and *StFeSOD4*. Since the dominance of *CuZnSODs* activities seen in potato plants grown in vitro on medium with limited Cu supply was even more pronounced in ex vitro-grown plantlets [74], the observed changes in expression of *StCuZnSODs* during different treatments should be regarded as major responses, whereas the changes of expression of other *StSODs* should be considered as fine tuning of the antioxidative defense.

StCuZnSOD1 is significantly upregulated at 29 °C in thermotolerant Désirée, and to some degree in Kennebec, but not in thermosensitive cultivar Agria (Figure 7). The expression levels of *StCuZnSOD2* and *StCuZnSOD3* are generally lower as compared to *StCuZn-SOD1*, but sharp induction of *StCuZnSOD2* in Désirée and Kennebec at 29 °C and moderate induction of *StCuZnSOD3* in Désirée and Agria at 26 and 29 °C suggest that all three StCuZn-SODs may play important roles in protecting their respective compartments—cytosol, chloroplasts and peroxisomes—form ROS generated at elevated temperatures.

StMnSOD is practically constitutively expressed with only mild induction at 29 °C in Agria and Désirée (Figure 7). The fact that *StMnSOD*, the only mitochondrial SOD in potato, is not induced by elevated temperatures, and that it has relatively low activity as compared to StCuZnSODs [74], suggests that potato susceptibility to HS may be, in part, due to inadequate antioxidative protection of mitochondria under elevated temperatures. However, when potato (cv. Désirée) was transformed with wheat mitochondrial MnSOD, it showed improvements in HS tolerance [13].

It is interesting that even moderately elevated temperature of 26 °C induced the expression of *StFeSOD2* and *StFeSOD3* in all cultivars, as well as minor *StFeSOD1* and *StFeSOD4* isoforms in cv. Kennebec. Since *StFeSOD1, StFeSOD2* and *StFeSOD3* are predicted to be chloroplastic, it seems that these isoforms, and not *StCuZnSOD2* (whose expression did not increase at 26 °C), are involved in the protection of photosynthetic apparatus during mild temperature stress. The expression of both *StFeSOD2* and *StFeSOD3* further increased with increasing temperature in all three cultivars (Figure 7), suggesting their general role in chloroplasts protection during HS. In cv. Kennebec, however, the expression of *StFe-SOD1* and *StFeSOD4* declined at 29 °C as compared to 26 °C treatment, while *StFeSOD4* expression decreased even below the control values. Overall, it seems that cv. Kennebec differs from the other two potato cultivars by generally higher expression of minor isoforms *StFeSOD1* and *StFeSOD4*, which probably have protective roles during mild stress, but somewhat lower expression of other *StSODs*. Thus, the physiological responses of cv. Kennebec to elevated temperatures should be further evaluated and compared to other cultivars. The main difference between thermosensitive Agria and thermotolerant Désirée appears to be sharp inducibility of major *StCuZnSOD1* and *StCuZnSOD2* isoforms at 29 °C in Désirée. Just like in potato, most of the *SOD* genes except mitochondrial *MnSOD* were upregulated by heat treatment in rice [65] and cucumber [61]. In banana, almost all *SOD* isoforms were induced at elevated temperatures [63], while in Arabidopsis only nuclear *CuZnSOD* (At5g18100) was induced by HS [65].

The expression of *StSODs* was generally unaffected by the application of 10^{-5} M SA, except a mild induction of all *StSODs* genes in Désirée grown at 26 °C as compared to plants grown at the same temperature on media without SA, and a downregulation of *StCuZnSOD2* in Agria grown at 21 °C (Figure 7). In banana, the expression levels of most *MaSOD* genes were up-regulated in response to the SA treatment, but in this case the plants were treated with 10^{-4} M SA [63]. Foliar application of 6×10^{-4} M SA increased both total SOD activity and *StSOD* expression of in potato [90], while in *Impatiens walleriana* grown in vitro, SA in the $1–3 \times 10^{-3}$ M range reduced SOD activity in a dose-response manner [91]. SA concentration of 10^{-5} M used in present work was selected based on unpublished results with different potato cultivars and literature data [53], but it is possible that higher concentrations of SA are required for effective regulation of *StSODs* expression in the studied cultivars. It can also be argued that in thermotolerant Désirée, SA application at 26 °C helps in stress amelioration by induction of *StSODs*, while at higher temperatures other mechanisms by which SA may enhance plants' thermotolerance are in action. Namely, SA may be directly or indirectly involved in the improvement of thermotolerance in various plant species by: stimulation of Pro synthesis to improve water uptake under HS, enhancement of different antioxidant enzymes activities and/or expression, protection of photosynthetic apparatus by increase of RUBISCO activity, inhibition of D1 protein degradation and other mechanisms, effects on other phytohormones and crosstalk among them, and induction of HSPs expression [7,92].

4.8. StCuZnSOD1 Promoter Features Many More Cis-Acting Regulatory Elements Than Other StSOD Promoters and Is the Only One Predicted to Be Regulated by Heat or SA

Gene expression is transcriptionally regulated via the change in the level or activity of TFs that bind to specific cis-acting promoter elements. We have performed in silico analysis of StSOD promoters in order to evaluate the potential for transcriptional regulation of different isoforms (Figure 2). Despite similarities in expression profiles of *StCuZnSODs*, *StCuZnSOD1* promoter substantially differs from other *StSOD* promoters in terms of the number of identified cis-regulatory elements (Figure 2A). Namely, *StCuZnSOD1* promoter is characterized by numerous, often overlapping cis-elements for binding different TFs, primarily from the ERF, Dof and LBD families, but also Heat stress TFs (HSFs)—major regulators of HS and other stress responses in plants [93] that are not found in other potato *SOD* promoters. There are two partially overlapping HSF cis-elements proximal to the transcription site, which correspond to HSFs involved in response to ROS and chitin (see Tables S2 and S3 for details). As indicated in Figure 2B, the TFs from the ERF, Dof, LBD, MYB and other families, predicted to recognize *StCuZnSOD1* promoter sequences, are implicated in responses to different stresses, phytohormones and developmental signals. Among them, an ERF TF is involved in heat acclimation, while three TFs belonging to Dof, GRAS and MYB families are involved in response to SA (Tables S2 and S3). All this suggests that cytosolic *StCuZnSOD1* is a common component of a number of signaling networks, possibly involved in controlling the level of H_2O_2, which is a well-recognized signaling molecule.

Unlike *StCuZnSOD1* promoter, *StCuZnSOD2* promoter features a single MYB binding site, which are also present in *StCuZnSOD1* and *StFeSOD4* promoters. Among other functions, some members of the MYB TF family are implicated in heat tolerance [94,95]. As a matter of fact, many other TF types with cis-elements in *StSODs* promoters (Figure 2A) can also be related to HS responses: certain members of M-type MADS, bZIP, as well as Dof family play important roles in modulating HS response in plants [95,96], while many of the ERF TFs identified in potato genome are involved in HS response [97]. However, since classification of TFs into families is generally based on their characteristic DNA-binding domains, it is not informative in terms of biological processes in which specific TFs are implicated. Thus, TFs predicted to bind *StSODs* promoters were associated with GO terms for biological processes (Figure 2B and Table S3), revealing that none of them is specifically involved in response to heat. This is in contrast with findings that majority of *SOD* promoters from other investigated species such as banana, tomato or cucumber apparently have more cis-regulatory elements as compared to potato *SODs* (with a notable exception of *StCuZnSOD1*), featuring not only more HSFs and SA-responsive elements, but different types of elements as well [61,63,68]. Common factor underlying the mentioned reports is the use of highly cited but somewhat outdated PlantCARE server for identification of plant cis-acting regulatory elements [29], whereas we have used an up to date approach based on PlantRegMap [26] and PlantTFDB v5.0 [28] which probe a more diverse motif landscape. In order to control the rate of falsely detected binding sites we used a stringent threshold for assigning a motif. Ultimately this produces quite different results compared to using PlantCARE on the same promoter sequences (see Table S4, where *StSOD* promoters were analyzed using PlantCARE, for a comparison, as well as Table S5, with comparative list of cis-elements found using both platforms). We trust that the accumulated knowledge in the almost two decades that separate PlantCARE from PlantRegMap should not be ignored.

Even though our in silico analysis of *StSOD* promoters (Figure 2 and supplementary materials) cannot fully support expressional data (Figure 7), it clearly demonstrates dependence of in silico promoter analyses on the adopted methods and a requirement for more experimental evidence on the interactions between different TFs and their cis-elements in plants. Finally, regarding the involvement of different StSODs in HS responses, it should be underlined again that StSODs, like other SODs, can probably be regulated not only at transcriptional level, but also post-transcriptionally at the level of alternative splicing [63,70,71] and by post-translational modifications [75]. CuZnSODs are additionally regulated by miRNAs, specifically miR398 [98], activated by copper chaperone CCS [62] and regulated by Cu availability [74,99].

5. Conclusions

Herby we present the first detailed insight into the SOD gene/protein family in potato. Exon-intron organization, splice variants, cis-regulatory promoter elements and chromosome localization of the eight functional *StSOD* genes has been described, along with structural features, subcellular localization, and phylogenetic relations of the StSOD proteins.

Investigation of the scope and time of the tandem *FeSOD* duplication event, characteristic for tomato and potato, would require comparative analysis of *FeSODs* in other Solenaceae species, when genomic resources become available. For determination of subcellular localization of StSODs, multiple servers were inquired and compared with phylogenetic and literature data, and yet peroxisomal targeting of StCuZnSOD3 remained inconclusive. Therefore, one of the future prospects is to experimentally determine whether there is a CuZnSOD activity in potato peroxisomes.

Higher induction of all *StCuZnSODs*, *StFeSOD3* and even *StMnSOD* in thermotolerant Désirée grown at 29 °C as compared to thermosensitive Agria and Kennebec, suggests that thermotolerance in potato might be related to induction of these isoforms. In addition, protection of chloroplasts under mild stress of 26 °C is apparently mediated by increased expression of chloroplastic *StFeSODs*. The application of low concentration of SA caused a mild induction of all *StSODs*, but only in Désirée grown at 26 °C, suggesting that ameliorating effects of SA during HS described in literature probably also include other mechanisms, as well as crosstalk among different phytohormones. Further spatio-temporal analysis of *StSODs* expression, however, should be refined to distinguish and quantify specific *StFeSODs* splice variants in different tissues and under different conditions.

We believe that our findings will aid future investigations of SODs roles and regulation in potato, particularly in relation to heat stress.

Supplementary Materials: The following supporting information can be downloaded at: https://www.mdpi.com/article/10.3390/antiox11030488/s1, Figure S1: Confirmation of amplicon size and primer specificity of studied *StSOD* genes, Figure S2: Alignments of MnFeSOD and CuZnSOD proteins used for phylogeny reconstruction, Table S1: PCR primer sequences used in this study for gene expression analysis by qRT-PCR, Table S2: Transcription factor binding sites identified in the promotor regions of *StSODs*, Table S3: Biological process GO terms associated with transcription factors binding promoters of *StSODs*, Table S4: Results of cis-element analysis in the promoter of *StSODs* using the PlantCARE web tool, Table S5: Comparison of results obtained from PlantCARE and PlantTFBD databases, Supplementary Info S6: AlphaFold models of StSOD proteins (PDB files) and Ramachandran diagrams of the respective models.

Author Contributions: Conceptualization and methodology D.P. and I.M.; investigation, J.R. and D.P.; validation J.R. and D.P.; software, M.B.D.; formal analysis M.B.D., D.P. and J.R.; writing—original draft, D.P., A.D.S. and J.R.; writing—review and editing, A.D.S., M.B.D. and I.M.; visualization D.P. and M.B.D.; supervision, I.M. and D.P. All authors have read and agreed to the published version of the manuscript.

Funding: This work was supported by the Ministry of Education, Science and Technological Development of the Republic of Serbia, Contract 451-03-9/2021-14/200007.

Institutional Review Board Statement: Not applicable.

Informed Consent Statement: Not applicable.

Data Availability Statement: Data are contained within the article and supplementary material.

Conflicts of Interest: The authors declare no conflict of interest.

References

1. Struik, P.C. Responses of the potato plant to temperature. In *Potato Biology and Biotechnology*, 1st ed.; Vreugdenhil, D., Bradshaw, J., Gebhardt, C., Govers, F., Taylor, M., MacKerron, D., Ross, H., Eds.; Elsevier B.V.: Amsterdam, The Netherlands, 2007; pp. 367–393. [CrossRef]
2. Hancock, R.D.; Morris, W.L.; Ducreux, L.J.; Morris, J.A.; Usman, M.; Verrall, S.R.; Fuller, J.; Simpson, C.G.; Zhang, R.; Hedley, P.E.; et al. Physiological, biochemical and molecular responses of the potato (*Solanum tuberosum* L.) plant to moderately elevated temperature. *Plant Cell Environ.* **2014**, *37*, 439–450. [CrossRef] [PubMed]
3. Momčilović, I.; Fu, J.; Pantelić, D.; Rudić, J.; Broćić, Z. Impact of Heat Stress on Potato: Plant Responses and Approaches to Tolerance Improvement. In *The Potato Crop: Management, Production, and Food Security*; Villa, P.M., Ed.; Nova: New York, NY, USA, 2021; pp. 91–122. [CrossRef]
4. Fu, J.; Momcilovic, I.; Prasad, P.V.V. Molecular bases and improvement of heat tolerance in crop plants. In *Heat Stress: Causes, Prevention and Treatments*; Josipovic, S., Ludwig, E., Eds.; Nova: New York, NY, USA, 2012; pp. 185–214.
5. Momčilović, I.; Pantelić, D. Plant superoxide dismutases: Important players in abiotic stress tolerance. In *Superoxide Dismutase: Structure, Synthesis and Applications*; Magliozzi, S., Ed.; Nova: New York, NY, USA, 2018; pp. 45–100.
6. Karuppanapandian, T.; Moon, J.C.; Kim, C.; Manoharan, K.; Kim, W. Reactive oxygen species in plants: Their generation, signal transduction, and scavenging mechanisms. *Aust. J. Crop Sci.* **2011**, *5*, 709–725.
7. Devireddy, A.R.; Tschaplinski, T.J.; Tuskan, G.A.; Muchero, W.; Chen, J.G. Role of reactive oxygen species and hormones in plant responses to temperature changes. *Int. J. Mol. Sci.* **2021**, *22*, 8843. [CrossRef] [PubMed]
8. Saleem, M.; Fariduddin, Q.; Castroverde, C.D.M. Salicylic acid: A key regulator of redox signalling and plant immunity. *Plant Physiol. Biochem.* **2021**, *168*, 381–397. [CrossRef]
9. Gill, S.S.; Anjum, N.A.; Gill, R.; Yadav, S.; Hasanuzzaman, M.; Fujita, M.; Mishra, P.; Sabat, S.C.; Tuteja, N. Superoxide dismutase—Mentor of abiotic stress tolerance in crop plants. *Environ. Sci. Pollut. Res.* **2015**, *22*, 10375–10394. [CrossRef] [PubMed]
10. Janků, M.; Luhová, L.; Petřivalský, M. On the origin and fate of reactive oxygen species in plant cell compartments. *Antioxidants* **2019**, *8*, 105. [CrossRef] [PubMed]
11. Kwon, S.Y.; Jeong, Y.J.; Lee, H.S.; Kim, J.S.; Cho, K.Y.; Allen, R.D.; Kwak, S.S. Enhanced tolerances of transgenic tobacco plants expressing both superoxide dismutase and ascorbate peroxidase in chloroplasts against methyl viologen-mediated oxidative stress. *Plant Cell Environ.* **2002**, *25*, 873–882. [CrossRef]
12. Kaouthar, F.; Ameny, F.K.; Yosra, K.; Walid, S.; Ali, G.; Faiçal, B. Responses of transgenic *Arabidopsis* plants and recombinant yeast cells expressing a novel durum wheat manganese superoxide dismutase *TdMnSOD* to various abiotic stresses. *J. Plant Physiol.* **2016**, *198*, 56–68. [CrossRef] [PubMed]
13. Waterer, D.; Benning, N.T.; Wu, G.; Luo, X.; Liu, X.; Gusta, M.; McHughen, A.; Gusta, L.V. Evaluation of abiotic stress tolerance of genetically modified potatoes (*Solanum tuberosum* cv. Desiree). *Mol. Breed.* **2010**, *25*, 527–540. [CrossRef]
14. Mohamed, H.I.; El-Shazly, H.H.; Badr, A. Role of salicylic acid in biotic and abiotic stress tolerance in plants. In *Plant Phenolics in Sustainable Agriculture*; Lone, R., Shuab, R., Kamili, A., Eds.; Springer: Singapore, 2020; pp. 533–554. [CrossRef]

15. Snyman, M.; Cronjé, M.J. Modulation of heat shock factors accompanies salicylic acid-mediated potentiation of Hsp70 in tomato seedlings. *J. Exp. Bot.* **2008**, *59*, 2125–2132. [CrossRef]
16. Wang, L.J.; Fan, L.; Loescher, W.; Duan, W.; Liu, G.J.; Cheng, J.S.; Luo, H.B.; Li, S.H. Salicylic acid alleviates decreases in photosynthesis under heat stress and accelerates recovery in grapevine leaves. *BMC Plant Biol.* **2010**, *10*, 34. [CrossRef]
17. Wang, L.J.; Li, S.H. Salicylic acid-induced heat or cold tolerance in relation to Ca^{2+} homeostasis and antioxidant systems in young grape plants. *Plant Sci.* **2006**, *170*, 685–694. [CrossRef]
18. Jahan, M.S.; Wang, Y.; Shu, S.; Zhong, M.; Chen, Z.; Wu, J.; Sun, J.; Guo, S. Exogenous salicylic acid increases the heat tolerance in Tomato (*Solanum lycopersicum* L) by enhancing photosynthesis efficiency and improving antioxidant defense system through scavenging of reactive oxygen species. *Sci. Hortic.* **2019**, *247*, 421–429. [CrossRef]
19. Khan, M.I.R.; Iqbal, N.; Masood, A.; Per, T.S.; Khan, N.A. Salicylic acid alleviates adverse effects of heat stress on photosynthesis through changes in proline production and ethylene formation. *Plant Signal. Behav.* **2013**, *8*, e26374. [CrossRef] [PubMed]
20. Cingoz, G.S.; Gurel, E. Effects of salicylic acid on thermotolerance and cardenolide accumulation under high temperature stress in *Digitalis trojana* Ivanina. *Plant Physiol. Biochem.* **2016**, *105*, 145–149. [CrossRef] [PubMed]
21. Shen, H.; Zhao, B.; Xu, J.; Zheng, X.; Huang, W. Effects of salicylic acid and calcium chloride on heat tolerance in *Rhododendron* 'Fen Zhen Zhu'. *J. Am. Soc. Hortic. Sci.* **2016**, *141*, 363–372. [CrossRef]
22. Miura, K.; Tada, Y. Regulation of water, salinity, and cold stress responses by salicylic acid. *Front. Plant Sci.* **2014**, *5*, 4. [CrossRef]
23. The Potato Genome Sequencing Consortium. Genome sequence and analysis of the tuber crop potato. *Nature* **2011**, *475*, 189–195. [CrossRef] [PubMed]
24. Pham, G.M.; Hamilton, J.P.; Wood, J.C.; Burke, J.T.; Zhao, H.; Vaillancourt, B.; Ou, S.; Jiang, J.; Buell, C.R. Construction of a chromosome-scale long-read reference genome assembly for potato. *Gigascience* **2020**, *9*, giaa100. [CrossRef] [PubMed]
25. Hijmans, R.J. The effect of climate change on global potato production. *Am. J. Potato Res.* **2003**, *80*, 271–279. [CrossRef]
26. Tian, F.; Yang, D.C.; Meng, Y.Q.; Jin, J.; Gao, G. PlantRegMap: Charting functional regulatory maps in plants. *Nucleic Acids Res.* **2020**, *48*, D1104–D1113. [CrossRef] [PubMed]
27. Storey, J.D.; Tibshirani, R. Statistical significance for genomewide studies. *Proc. Natl. Acad. Sci. USA* **2003**, *100*, 9440–9445. [CrossRef] [PubMed]
28. Jin, J.; Tian, F.; Yang, D.C.; Meng, Y.Q.; Kong, L.; Luo, J.; Gao, G. PlantTFDB 4.0: Toward a central hub for transcription factors and regulatory interactions in plants. *Nucleic Acids Res.* **2017**, *45*, D1040–D1045. [CrossRef] [PubMed]
29. Lescot, M.; Déhais, P.; Thijs, G.; Marchal, K.; Moreau, Y.; Van de Peer, Y.; Rouzé, P.; Rombauts, S. PlantCARE, a database of plant *cis*-acting regulatory elements and a portal to tools for in silico analysis of promoter sequences. *Nucleic Acids Res.* **2002**, *30*, 325–327. [CrossRef] [PubMed]
30. Gasteiger, E.; Hoogland, C.; Gattiker, A.; Duvaud, S.; Wilkins, M.R.; Appel, R.D.; Bairoch, A. Protein Identification and Analysis Tools on the ExPASy Server. In *The Proteomics Protocols Handbook*; Walker, J.M., Ed.; Humana Press: Totowa, NJ, USA, 2005; pp. 571–607. [CrossRef]
31. Armenteros, J.J.A.; Sønderby, C.K.; Sønderby, S.K.; Nielsen, H.; Winther, O. DeepLoc: Prediction of protein subcellular localization using deep learning. *Bioinformatics* **2017**, *33*, 3387–3395. [CrossRef] [PubMed]
32. Yu, C.S.; Chen, Y.C.; Lu, C.H.; Hwang, J.K. Prediction of protein subcellular localization. *Proteins Struct. Funct. Bioinform.* **2006**, *64*, 643–651. [CrossRef] [PubMed]
33. Stäerk, H.; Dallago, C.; Heinzinger, M.; Rost, B. Light attention predicts protein location from the language of life. *bioRxiv* **2021**. [CrossRef]
34. Armenteros, J.J.A.; Salvatore, M.; Emanuelsson, O.; Winther, O.; Von Heijne, G.; Elofsson, A.; Nielsen, H. Detecting sequence signals in targeting peptides using deep learning. *Life Sci. Alliance* **2019**, *2*, e201900429. [CrossRef] [PubMed]
35. Neuberger, G.; Maurer-Stroh, S.; Eisenhaber, B.; Hartig, A.; Eisenhaber, F. Motif refinement of the peroxisomal targeting signal 1 and evaluation of taxon-specific differences. *J. Mol. Biol.* **2003**, *328*, 567–579. [CrossRef]
36. Neuberger, G.; Maurer-Stroh, S.; Eisenhaber, B.; Hartig, A.; Eisenhaber, F. Prediction of peroxisomal targeting signal 1 containing proteins from amino acid sequence. *J. Mol. Biol.* **2003**, *328*, 581–592. [CrossRef]
37. Reumann, S.; Buchwald, D.; Lingner, T. PredPlantPTS1: A web server for the prediction of plant peroxisomal proteins. *Front. Plant Sci.* **2012**, *3*, 194. [CrossRef]
38. Mistry, J.; Chuguransky, S.; Williams, L.; Qureshi, M.; Salazar, G.A.; Sonnhammer, E.L.; Tosatto, S.C.E.; Paladin, L.; Raj, S.; Richardson, L.J.; et al. Pfam: The protein families database in 2021. *Nucleic Acids Res.* **2021**, *49*, D412–D419. [CrossRef] [PubMed]
39. Eddy, S.R. Accelerated profile HMM searches. *PLoS Comput. Biol.* **2011**, *7*, e1002195. [CrossRef]
40. Lu, S.; Wang, J.; Chitsaz, F.; Derbyshire, M.K.; Geer, R.C.; Gonzales, N.R.; Gwadz, M.; Hurwitz, D.I.; Marchler, G.H.; Song, J.S.; et al. CDD/SPARCLE: The conserved domain database in 2020. *Nucleic Acids Res.* **2020**, *48*, D265–D268. [CrossRef] [PubMed]
41. Marchler-Bauer, A.; Bryant, S.H. CD-Search: Protein domain annotations on the fly. *Nucleic Acids Res.* **2004**, *32*, W327–W331. [CrossRef]
42. Wright, E.S. DECIPHER: Harnessing local sequence context to improve protein multiple sequence alignment. *BMC Bioinform.* **2015**, *16*, 322. [CrossRef] [PubMed]
43. Jumper, J.; Evans, R.; Pritzel, A.; Green, T.; Figurnov, M.; Ronneberger, O.; Tunyasuvunakool, K.; Bates, R.; Žídek, A.; Potapenko, A.; et al. Highly accurate protein structure prediction with AlphaFold. *Nature* **2021**, *596*, 583–589. [CrossRef]

44. Pettersen, E.F.; Goddard, T.D.; Huang, C.C.; Meng, E.C.; Couch, G.S.; Croll, T.I.; Morris, J.H.; Ferrin, T.E. UCSF ChimeraX: Structure visualization for researchers, educators, and developers. *Protein Sci.* **2021**, *30*, 70–82. [CrossRef]

45. Chen, V.B.; Arendall, W.B., III; Headd, J.J.; Keedy, D.A.; Immormino, R.M.; Kapral, G.J.; Murray, L.W.; Richardson, J.S.; Richardson, D.C. MolProbity: All-atom structure validation for macromolecular crystallography. *Acta Cryst.* **2010**, *D66*, 1–21. [CrossRef] [PubMed]

46. Waterhouse, A.; Bertoni, M.; Bienert, S.; Studer, G.; Tauriello, G.; Gumienny, R.; Heer, F.T.; de Beer, T.A.P.; Rempfer, C.; Bordoli, L.; et al. SWISS-MODEL: Homology modelling of protein structures and complexes. *Nucleic Acids Res.* **2018**, *46*, W296–W303. [CrossRef] [PubMed]

47. Dragićević, M. ggrama: Ramachandran Diagrams using ggplot2 Graphics. 2021. Available online: https://github.com/missuse/ggrama (accessed on 30 November 2021). [CrossRef]

48. Yu, Y.K.; Capra, J.A.; Stojmirović, A.; Landsman, D.; Altschul, S.F. Log-odds sequence logos. *Bioinformatics* **2015**, *31*, 324–331. [CrossRef]

49. Le, S.Q.; Gascuel, O. An improved general amino acid replacement matrix. *Mol. Biol. Evol.* **2008**, *25*, 1307–1320. [CrossRef]

50. Schliep, K.P. phangorn: Phylogenetic analysis in R. *Bioinformatics* **2011**, *27*, 592–593. [CrossRef]

51. Murashige, T.; Skoog, F. A revised medium for rapid growth and bio assays with tobacco tissue cultures. *Physiol. Plant.* **1962**, *15*, 473–497. [CrossRef]

52. Linsmaier, E.M.; Skoog, F. Organic growth factor requirements of tobacco tissue cultures. *Physiol. Plant.* **1965**, *18*, 100–127. [CrossRef]

53. López-Delgado, H.; Mora-Herrera, M.E.; Zavaleta-Mancera, H.A.; Cadena-Hinojosa, M.; Scott, I.M. Salicylic acid enhances heat tolerance and potato virus X (PVX) elimination during thermotherapy of potato microplants. *Am. J. Potato Res.* **2004**, *81*, 171–176. [CrossRef]

54. Tiwari, J.K.; Buckseth, T.; Devi, S.; Varshney, S.; Sahu, S.; Patil, V.U.; Zinta, R.; Ali, N.; Moudgil, V.; Singh, R.K.; et al. Physiological and genome-wide RNA-sequencing analyses identify candidate genes in a nitrogen-use efficient potato cv. Kufri Gaurav. *Plant Physiol. Biochem.* **2020**, *154*, 171–183. [CrossRef]

55. Klatte, M.; Bauer, P. Accurate real-time reverse transcription quantitative PCR. In *Plant Signal Transduction*; Pfannschmidt, T., Ed.; Human Press: Totowa, NJ, USA, 2009; Volume 479, pp. 61–77. [CrossRef]

56. Cannon, S.B.; Mitra, A.; Baumgarten, A.; Young, N.D.; May, G. The roles of segmental and tandem gene duplication in the evolution of large gene families in *Arabidopsis thaliana*. *BMC Plant Biol.* **2004**, *4*, 10. [CrossRef] [PubMed]

57. Ikai, A. Thermostability and aliphatic index of globular proteins. *J. Biochem.* **1980**, *88*, 1895–1898. [CrossRef]

58. Getzoff, E.D.; Tainer, J.A.; Stempien, M.M.; Bell, G.I.; Hallewell, R.A. Evolution of CuZn superoxide dismutase and the Greek key β-barrel structural motif. *Proteins Struct. Funct. Genet.* **1989**, *5*, 322–336. [CrossRef]

59. Favier, A.; Gans, P.; Erba, E.B.; Signor, L.; Muthukumar, S.S.; Pfannschmidt, T.; Blanvillain, R.; Cobessi, D. The plastid-encoded RNA polymerase-associated protein PAP9 is a superoxide dismutase with unusual structural features. *Front. Plant Sci.* **2021**, *12*, 668897. [CrossRef]

60. Quint, P.; Reutzel, R.; Mikulski, R.; McKenna, R.; Silverman, D.N. Crystal structure of nitrated human manganese superoxide dismutase: Mechanism of inactivation. *Free Radic. Biol. Med.* **2006**, *40*, 453–458. [CrossRef]

61. Zhou, Y.; Hu, L.; Wu, H.; Jiang, L.; Liu, S. Genome-wide identification and transcriptional expression analysis of cucumber superoxide dismutase (SOD) family in response to various abiotic stresses. *Int. J. Genom.* **2017**, *2017*, 1–14. [CrossRef] [PubMed]

62. Furukawa, Y.; Torres, A.S.; O'Halloran, T.V. Oxygen-induced maturation of SOD1: A key role for disulfide formation by the copper chaperone CCS. *EMBO J.* **2004**, *23*, 2872–2881. [CrossRef] [PubMed]

63. Feng, X.; Lai, Z.; Lin, Y.; Lai, G.; Lian, C. Genome-wide identification and characterization of the superoxide dismutase gene family in *Musa acuminata* cv. Tianbaojiao (AAA group). *BMC Genom.* **2015**, *16*, 823. [CrossRef] [PubMed]

64. Jiang, W.; Yang, L.; He, Y.; Zhang, H.; Li, W.; Chen, H.; Ma, D.; Yin, J. Genome-wide identification and transcriptional expression analysis of superoxide dismutase (SOD) family in wheat (*Triticum aestivum*). *PeerJ* **2019**, *7*, e8062. [CrossRef]

65. Yadav, S.; Gill, S.S.; Passricha, N.; Gill, R.; Badhwar, P.; Anjum, N.A.; Francisco, J.B.J.; Tuteja, N. Genome-wide analysis and transcriptional expression pattern-assessment of superoxide dismutase (SOD) in rice and *Arabidopsis* under abiotic stresses. *Plant Gene* **2019**, *17*, 100165. [CrossRef]

66. Filiz, E.; Tombuloğlu, H. Genome-wide distribution of superoxide dismutase (SOD) gene families in *Sorghum bicolor*. *Turk. J. Biol.* **2015**, *39*, 49–59. [CrossRef]

67. Wang, W.; Xia, M.; Chen, J.; Deng, F.; Yuan, R.; Zhang, X.; Shen, F. Genome-wide analysis of superoxide dismutase gene family in *Gossypium raimondii* and *G. arboreum*. *Plant Gene* **2016**, *6*, 18–29. [CrossRef]

68. Feng, K.; Yu, J.; Cheng, Y.; Ruan, M.; Wang, R.; Ye, Q.; Zhou, G.; Li, Z.; Yao, Z.; Yang, Y.; et al. The SOD gene family in tomato: Identification, phylogenetic relationships, and expression patterns. *Front. Plant Sci.* **2016**, *7*, 1279. [CrossRef]

69. Xu, G.; Guo, C.; Shan, H.; Kong, H. Divergence of duplicate genes in exon–intron structure. *Proc. Natl. Acad. Sci. USA* **2012**, *109*, 1187–1192. [CrossRef] [PubMed]

70. Srivastava, V.; Srivastava, M.K.; Chibani, K.; Nilsson, R.; Rouhier, N.; Melzer, M.; Wingsle, G. Alternative splicing studies of the reactive oxygen species gene network in *Populus* reveal two isoforms of high-isoelectric-point superoxide dismutase. *Plant Physiol.* **2009**, *149*, 1848–1859. [CrossRef]

71. Feng, W.; Hongbin, W.; Bing, L.; Jinfa, W. Cloning and characterization of a novel splicing isoform of the iron-superoxide dismutase gene in rice (*Oryza sativa* L.). *Plant Cell Rep.* **2006**, *24*, 734–742. [CrossRef]

72. Perry, J.J.P.; Shin, D.S.; Getzoff, E.D.; Tainer, J.A. The structural biochemistry of the superoxide dismutases. *Biochim. Biophys. Acta-Proteins Proteom.* **2010**, *1804*, 245–262. [CrossRef]

73. Alscher, R.G.; Erturk, N.; Heath, L.S. Role of superoxide dismutases (SODs) in controlling oxidative stress in plants. *J. Exp. Bot.* **2002**, *53*, 1331–1341. [CrossRef] [PubMed]

74. Momčilović, I.; Pantelić, D.; Hfidan, M.; Savić, J.; Vinterhalter, D. Improved procedure for detection of superoxide dismutase isoforms in potato, *Solanum tuberosum* L. *Acta Physiol. Plant.* **2014**, *36*, 2059–2066. [CrossRef]

75. Yamakura, F.; Kawasaki, H. Post-translational modifications of superoxide dismutase. *Biochim. Biophys. Acta Proteins Proteom.* **2010**, *1804*, 318–325. [CrossRef] [PubMed]

76. Reumann, S.; Chowdhary, G. Prediction of peroxisomal matrix proteins in plants. In *Proteomics of Peroxisomes*; del Río, L., Schrader, M., Eds.; Springer: Singapore, 2018; Volume 89, pp. 125–138. [CrossRef]

77. Corpas, F.J.; González-Gordo, S.; Palma, J.M. Plant peroxisomes: A factory of reactive species. *Front. Plant Sci.* **2020**, *11*, 853. [CrossRef] [PubMed]

78. Fink, R.C.; Scandalios, J.G. Molecular evolution and structure–function relationships of the superoxide dismutase gene families in angiosperms and their relationship to other eukaryotic and prokaryotic superoxide dismutases. *Arch. Biochem. Biophys.* **2002**, *399*, 19–36. [CrossRef] [PubMed]

79. Sheng, Y.; Abreu, I.A.; Cabelli, D.E.; Maroney, M.J.; Miller, A.F.; Teixeira, M.; Valentine, J.S. Superoxide dismutases and superoxide reductases. *Chem. Rev.* **2014**, *114*, 3854–3918. [CrossRef] [PubMed]

80. Retnoningrum, D.S.; Yoshida, H.; Razani, M.D.; Meidianto, V.F.; Hartanto, A.; Artarini, A.; Ismaya, W.T. Unprecedented Role of the N73-F124 Pair in the Staphylococcus equorum MnSOD Activity. *Curr. Enzym. Inhib.* **2021**, *17*, 2–8. [CrossRef]

81. Cabrejos, D.A.L.; Alexandrino, A.V.; Pereira, C.M.; Mendonça, D.C.; Pereira, H.D.; Novo-Mansur, M.T.M.; Garratt, R.C.; Goto, L.S. Structural characterization of a pathogenicity-related superoxide dismutase codified by a probably essential gene in *Xanthomonas citri* subsp. *citri*. *PLoS ONE* **2019**, *14*, e0209988. [CrossRef]

82. Sheng, Y.; Durazo, A.; Schumacher, M.; Gralla, E.B.; Cascio, D.; Cabelli, D.E.; Valentine, J.S. Tetramerization reinforces the dimer interface of MnSOD. *PLoS ONE* **2013**, *8*, e62446. [CrossRef]

83. Su, Z.; Wang, J.; Yu, J.; Huang, X.; Gu, X. Evolution of alternative splicing after gene duplication. *Genome Res.* **2006**, *16*, 182–189. [CrossRef]

84. Van Dam, J.; Kooman, P.L.; Struik, P.C. Effects of temperature and photoperiod on early growth and final number of tubers in potato (*Solanum tuberosum* L.). *Potato Res.* **1996**, *39*, 51–62. [CrossRef]

85. Ahn, Y.J.; Claussen, K.; Zimmerman, J.L. Genotypic differences in the heat-shock response and thermotolerance in four potato cultivars. *Plant Sci.* **2004**, *166*, 901–911. [CrossRef]

86. Tang, R.; Niu, S.; Zhang, G.; Chen, G.; Haroon, M.; Yang, Q.; Rajora, O.P.; Li, X.Q. Physiological and growth responses of potato cultivars to heat stress. *Botany* **2018**, *96*, 897–912. [CrossRef]

87. Demirel, U.; Morris, W.L.; Ducreux, L.J.; Yavuz, C.; Asim, A.; Tindas, I.; Campbell, R.; Morris, J.A.; Verrall, S.R.; Hedlez, P.E.; et al. Physiological, biochemical, and transcriptional responses to single and combined abiotic stress in stress-tolerant and stress-sensitive potato genotypes. *Front. Plant Sci.* **2020**, *11*, 169. [CrossRef] [PubMed]

88. Hayat, Q.; Hayat, S.; Irfan, M.; Ahmad, A. Effect of exogenous salicylic acid under changing environment: A review. *Environ. Exp. Bot.* **2010**, *68*, 14–25. [CrossRef]

89. Herrera-Vásquez, A.; Salinas, P.; Holuigue, L. Salicylic acid and reactive oxygen species interplay in the transcriptional control of defense genes expression. *Front. Plant Sci.* **2015**, *6*, 171. [CrossRef]

90. Li, Q.; Wang, G.; Wang, Y.; Yang, D.; Guan, C.; Ji, J. Foliar application of salicylic acid alleviate the cadmium toxicity by modulation the reactive oxygen species in potato. *Ecotoxicol. Environ. Saf.* **2019**, *172*, 317–325. [CrossRef] [PubMed]

91. Antonić, D.; Milošević, S.; Cingel, A.; Lojić, M.; Trifunović-Momčilov, M.; Petrić, M.; Subotić, A.; Simonović, A. Effects of exogenous salicylic acid on *Impatiens walleriana* L. grown in vitro under polyethylene glycol-imposed drought. *S. Afr. J. Bot.* **2016**, *105*, 226–233. [CrossRef]

92. Sharma, L.; Priya, M.; Kaushal, N.; Bhandhari, K.; Chaudhary, S.; Dhankher, O.P.; Prasad, P.V.V.; Siddique, K.H.M.; Nayyar, H. Plant growth-regulating molecules as thermoprotectants: Functional relevance and prospects for improving heat tolerance in food crops. *J. Exp. Bot.* **2020**, *71*, 569–594. [CrossRef] [PubMed]

93. Guo, M.; Liu, J.H.; Ma, X.; Luo, D.X.; Gong, Z.H.; Lu, M.H. The plant heat stress transcription factors (HSFs): Structure, regulation, and function in response to abiotic stresses. *Front. Plant Sci.* **2016**, *7*, 114. [CrossRef] [PubMed]

94. Zhao, Y.; Tian, X.; Wang, F.; Zhang, L.; Xin, M.; Hu, Z.; Yao, Y.; Ni, Z.; Sun, Q.; Peng, H. Characterization of wheat MYB genes responsive to high temperatures. *BMC Plant Biol.* **2017**, *17*, 208. [CrossRef] [PubMed]

95. Islam, M.M.; Sandhu, J.; Walia, H.; Saha, R. Transcriptomic data-driven discovery of global regulatory features of rice seeds developing under heat stress. *Comput. Struct. Biotechnol. J.* **2020**, *18*, 2556–2567. [CrossRef] [PubMed]

96. Waqas, M.; Shahid, L.; Shoukat, K.; Aslam, U.; Azeem, F.; Atif, R.M. Role of DNA-binding with one finger (Dof) transcription factors for abiotic stress tolerance in plants. In *Transcription Factors for Abiotic Stress Tolerance in Plants*; Wani, S.H., Ed.; Academic Press: Cambridge, MA, USA, 2020; pp. 1–14. [CrossRef]

97. Charfeddine, M.; Saïdi, M.N.; Charfeddine, S.; Hammami, A.; Bouzid, R.G. Genome-wide analysis and expression profiling of the ERF transcription factor family in potato (*Solanum tuberosum* L.). *Mol. Biotechnol.* **2015**, *57*, 348–358. [CrossRef]
98. Sunkar, R.; Kapoor, A.; Zhu, J.K. Posttranscriptional induction of two Cu/Zn superoxide dismutase genes in *Arabidopsis* is mediated by downregulation of miR398 and important for oxidative stress tolerance. *Plant Cell* **2006**, *18*, 2051–2065. [CrossRef]
99. Cohu, C.M.; Pilon, M. Regulation of superoxide dismutase expression by copper availability. *Physiol. Plant.* **2007**, *129*, 747–755. [CrossRef]

Article

Phytohormones Producing *Acinetobacter bouvetii* P1 Mitigates Chromate Stress in Sunflower by Provoking Host Antioxidant Response

Muhammad Qadir [1], Anwar Hussain [1,*], Muhammad Hamayun [1], Mohib Shah [1], Amjad Iqbal [2], Muhammad Irshad [1], Ayaz Ahmad [3], Muhammad Arif Lodhi [3] and In-Jung Lee [4,*]

[1] Department of Botany, Garden Campus, Abdul Wali Khan University Mardan, Khyber Pakhtunkhwa, Mardan 23200, Pakistan; muhammad_qadir@awkum.edu.pk (M.Q.); hamayun@awkum.edu.pk (M.H.); mohibshah@awkum.edu.pk (M.S.); muhammad.irshad@awkum.edu.pk (M.I.)
[2] Department of Food Science & Technology, Garden Campus, Abdul Wali Khan University Mardan, Khyber Pakhtunkhwa, Mardan 23200, Pakistan; amjadiqbal@awkum.edu.pk
[3] Department of Biotechnology, Garden Campus, Abdul Wali Khan University Mardan, Khyber Pakhtunkhwa, Mardan 23200, Pakistan; ahdayazb5@awkum.edu.pk (A.A.); arif.lodhi@awkum.edu.pk (M.A.L.)
[4] Department of Applied Biosciences, Kyungpook National University, Daegu 41566, Korea
* Correspondence: drhussain@awkum.edu.pk (A.H.); ijlee@knu.ac.kr (I.-J.L.)

Citation: Qadir, M.; Hussain, A.; Hamayun, M.; Shah, M.; Iqbal, A.; Irshad, M.; Ahmad, A.; Lodhi, M.A.; Lee, I.-J. Phytohormones Producing *Acinetobacter bouvetii* P1 Mitigates Chromate Stress in Sunflower by Provoking Host Antioxidant Response. *Antioxidants* **2021**, *10*, 1868. https://doi.org/10.3390/antiox10121868

Academic Editors: Masayuki Fujita and Mirza Hasanuzzaman

Received: 26 September 2021
Accepted: 14 November 2021
Published: 24 November 2021

Abstract: Different physical and chemical techniques are used for the decontamination of Cr^{+6} contaminated sites. The techniques are expensive, laborious, and time-consuming. However, remediation of Cr^{+6} by microbes is viable, efficient, and cost-effective. In this context, plant growth-promoting rhizobacteria *Acinetobacter bouvetii* P1 isolated from the industrial zone was tested for its role in relieving Cr^{+6} induced oxidative stress in sunflower. At the elevated Cr^{+6} levels and in the absence of P1, the growth of the sunflower plants was inhibited. In contrast, the selected strain P1 restored the sunflower growth under Cr^{+6} through plant growth–promoting interactions. Specifically, P1 biotransformed the Cr^{+6} into a stable and less toxic Cr^{+3} form, thus avoiding the possibility of phytotoxicity. On the one hand, the P1 strengthened the host antioxidant system by triggering higher production of enzymatic antioxidants, including catalases, ascorbate peroxidase, superoxide dismutase, and peroxidase. Similarly, P1 also promoted higher production of nonenzymatic antioxidants, such as flavonoids, phenolics, proline, and glutathione. Apart from the bioremediation, P1 solubilized phosphate and produced indole acetic acid, gibberellic acid, and salicylic acid. The production of phytohormones not only helped the host plant growth but also mitigated the harsh condition posed by the elevated levels of Cr^{+6}. The findings mentioned above suggest that P1 may serve as an excellent phyto-stimulant and bio-remediator in a heavy metal-contaminated environment.

Keywords: Cr^{+6} bioremediation; improved antioxidant activity; plant growth promotion; rhizobacteria

1. Introduction

Heavy metals (HMs) are natural components of the earth's crust. However, their biogeochemical cycles and biochemical balances are disrupted due to rapid industrialization. Such immense industrial activities led to a higher increase and subsequent aggregation of HMs in the milieu. Additionally, the elevated levels of HMs in water, soil, and agricultural products are the result of excessive but imprudent agricultural practices, industrial and domestic wastes [1,2]. The accumulation of HMs in plants affects humans directly through the consumption of contaminated plant-based foods. Heavy metals contaminated plants can also affect humans indirectly through the consumption of meat and milk from livestock fed on such plants. Upon exposure, HMs affect cell organelles, including cell membrane, nucleus, endoplasmic reticulum, mitochondria, lysosome, enzymes, and metabolites involved in the detoxification mechanisms and injury recovery [3]. Interaction of HMs with vital cell components such as DNA and proteins leads to their damage, conformational changes, cell cycle modulation, oncogenesis, and programmed cell death [4,5]. Based on significant public health concerns, the remediation of heavy metals should take priority [6].

Amid the HMs, hexavalent chromium (Cr^{+6}) is a widespread environmental pollutant mainly from chromium-dependent industries. In undeveloped countries, industrial wastewater is released into natural water resources, most of which are not adequately treated and pose significant threats to the ecosystem. Hexavalent chromium (Cr^{+6}) is more toxic and oncogenic, and these properties are due to their higher solubility in water, high permeability through biofilms, and subsequent interaction with inter and intracellular proteins and nucleic acids [7].

In plants, metal-induced phytotoxicity and generation of reactive oxygen and nitrogen species have revealed that redox-active metals, particularly chromium (Cr), undertake redox cycling reactions. In fact, chromium owns the potential to produce reactive species, such as anions of superoxide and nitric oxide radicals in biological systems [8]. Moreover, the disturbed metals homeostasis may lead to oxidative pressure, a state where the imbalance production of reactive oxygen species (ROS) overwhelms the antioxidant defense [9]. Such conditions may cause DNA and chromosomal damage, malonaldehyde production, conformational changes in protein structure, and subsequent toxicity. The damages in the host plants due to oxidative injuries, enhanced lipid peroxidation, and altered calcium and sulfhydryl homeostasis can be translated into stunted growth. Lipid peroxides, formed by the attack of radicals on polyunsaturated fatty acid residues of phospholipids, can produce mutagenic and carcinogenic malondialdehyde, 4-hydroxynonenal, and other exocyclic DNA adducts (etheno and propano adducts) [9].

Anionic transporters in plants may take up Cr^{+6} in the form of chromate. Cr^{+6} is easily converted to Cr^{+3} under most environmental conditions. Cr in its +3 form was once thought to be a vital ingredient for animal and human existence. The role of Cr^{+3} in the glucose tolerance factor, as well as the beneficial effects of Cr^{+3} dietary supplements on diabetes and lipid metabolism, have been studied in the past [10]. Quiet recently, Cr^{+3} has been removed from the list of essential elements for humans and animals due to a lack of clear experimental evidence regarding its function in humans and animals [11].

In the current scenario, remediation of contaminated water bodies and soil is the call of the day to remove heavy metals and trace elements from the human environment in order to make it healthy. Chemical precipitation, oxidation or reduction filtration, ion exchange, reverse osmosis, membrane technology, evaporation, and electrochemical treatment are only a few methods for eliminating these heavy metals [12]. These methods, however, are usually useless when heavy metal concentrations are lower than 100 mg/L [13]. Because salts of heavy metals are readily soluble and dissolve in water, the physical separation techniques are not feasible to separate them. Furthermore, at minimal metal concentrations, physical and chemical techniques are inefficient or costly [13]. In contrast to physicochemical approaches, bioremediation (biosorption and bioaccumulation) may possibly present the optimal strategy for heavy metal elimination [14]. As a result, using microbes and plants to repair and rebuild the natural conditions of the soil is possible and sustainable. Microbial communities' responses to heavy metals depend on their concentration and availability, which is a complicated process influenced by the type of the metal, medium characteristics, and microbial species [15].

Bioremediation is a cutting-edge and promising technique for removing and recovering heavy metals from contaminated water and agrarian soil [16]. Microbes develop and utilize several detoxification processes as a result of their varied survival approaches in heavy metal-contaminated environments [17]. Biosorption, bioaccumulation, biotransformation, and biomineralization are the primary detoxification processes that could be employed for bioremediation either in ex situ or in situ conditions [18]. Polysaccharides, lipids, and proteins make up the majority of microbial cell walls, and they include various side chains that chemically bind metal ions, including carboxylate, hydroxyl, and amino acids and phosphate groups [19].

Plant growth-promoting rhizobacteria (PGPR) is an environmentally benign and cost-effective bioremediation technology for reducing heavy metal pollution in soil [20]. Metal-tolerant PGPR also affects plant survival and adaptability through nitrogen fixation, phytohormone synthesis, and physicochemical changes in polluted soil [21,22]. Plant colonization by PGPR has been driven by the release of various compounds, amino acids, proteins, and antibiotics that aid plants in the removal of heavy metal toxicity. By reducing, oxidizing, methylating, or demethylating, compartmentalizing, and converting to a less hazardous form, the PGPR have the ability to reduce the harmful influence of heavy metals [23,24]. Cr^{+6} undergoes a transition to Cr^{+3}, which is more stable, less water-soluble, and hence less harmful. Consequently, converting Cr^{+6} to Cr^{+3} is a promising approach for removing Cr^{+6} from the environment [25]. As a result, decontaminating Cr^{+6} using naturally existing microorganisms is a viable alternative. Cr^{+6}-induced phytotoxicity, on the other hand, may be reduced by using plant growth-promoting bacteria.

Furthermore, naturally existing rhizosphere microorganisms in soils contaminated with Cr^{+6} have developed resistance mechanisms to these metals' toxicity [26,27]. The Cr^{+6}-tolerant bacteria possess the substantial capacity to produce plant growth regulators, to solubilize inorganic P (salt of phosphoric acid with metals) fixation of free atmospheric N, and synthesize antimicrobial compounds and siderophores [28]. Recent studies have shown that chromate-tolerant rhizobacteria have the ability to cooperate with the host plant in many ways. Indeed, the Cr^{+6} tolerant rhizobacteria improve the host endogenous pool of phytohormones, strengthen the enzymatic and non-enzymatic arms of the antioxidant system, cut down metal uptake, and reduce the significant proportion of the accumulated metal to nontoxic or the least toxic form [29–31].

The area of study on the bioremediation of chromium from the agricultural soils by PGPR is still expanding. The present research aims to contribute to this growing body of knowledge by focusing on the premise that rhizobacteria can mitigate heavy metal stressful condition in the host plants. The objectives of the study were to establish (a) the function of PGPR in mitigation of Cr^{+6} induced oxidative stress in sunflower; (b) the effect of rhizobacterial inoculum on Cr^{+6} accumulation and translocation in the host; and (c) to explore the detoxification impact of PGPR on *Helianthus annuus* L. physiology and development under Cr^{+6}.

2. Materials and Methods

2.1. Test Plant Species (Biological Materials)

Sunflower (*H. annuus* L.), native to America and commercialized in Russia as a crop plant, is a member of the Asteraceae family. It is a significant commercial crop that ranks fourth in the world among vegetable oil seeds after soybean, oil palm, and canola [32]. In Mardan, Pakistan, HySun-33 is one of the most widely cultivated varieties of sunflower. Seeds of sunflower HySun-33 were obtained from local agriculture research Centre.

2.2. Requisition of Rhizbacterium

A Cr^{+6} resistant strain of rhizobacterium P1 (Accession No# MT478039) was acquired from the bacterial stock from the Lab of Plant Microbes Interaction, Department of Botany, Abdul Wali Khan University Mardan. The rhizobacteria were previously isolated from the rhizosphere of *Parthenium hysterophorus* L. grown on metal-polluted soil in the locality of Premier Sugar Mill Mardan [33].

2.3. Molecular Identification of Strain

The 16S rDNA marker was used for the molecular identification of the isolate. Using a Gen-Elute DNA kit (Sigma-Aldrich, Burlington, MA, USA), genomic DNA was extracted from a selected strain culture grown overnight. The GeneAmp 9700 PCR system (Thermo Fisher Scientific) was used to conduct amplification using a set of universal primers FDD2 (5′-CCG GAT CCG TCC ACA GAG TTT GAT CTT GGC GAA -3′) and RPP2 (5′-CCA AGC TTC TAG ACG GAT ACC TTG TTA CGA CTT- 3′) (Applied Biosystems Inc., Atlanta, GA, USA). The amplified fragment was subjected to sequencing through the DNA sequencing service of MACROGEN, Korea (http://dna.macrogen.com/eng accessed on 6 May 2019) [34]. Reads obtained were used to generate consensus sequence through Codon Code Aligner (version 7.2.1, Codon Technology Corporations, Hyderabad, India). The consensus sequence was used as BLAST query to find closely matching sequences in NCBI's GenBank database (https://blast.ncbi.nlm.nih.gov/Blast.cgi?PAGE_TYPE=BlastSearch accessed on 20 May 2019). Closely resembling sequences retrieved from the database were used to align with the selected sequence, and a phylogenetic tree was constructed using MEGA 7 (version 7.0.18, Pennsylvania State University, Pennsylvania, PA, USA).

2.4. Screening the Rhizobacterium

The isolate was tested for chromate tolerance, ability to release phosphate solubilizing substances, and for chromate stress alleviation in sunflower.

2.4.1. Chromate Tolerance

To check the capability of the strain to endure elevated Cr^{+6} levels, the selected rhizobacterium was subjected to 100, 300, 500, 900, and 1200 $\mu g\ mL^{-1}\ Cr^{+6}$ (in the form of K_2CrO_4 (Sigma Aldrich)) in Luria-Bertani (L.B) broth medium (Sigma Aldrich), while the nonmetal media were used as control. The cultures were kept in a shaker (Modle-Wis 20) for incubation at 130 rpm for 24 h at 28 °C. Bacterial growth in the control and metal supplemented media was recorded at 595 nm by monitoring the culture's optical density (OD).

2.4.2. Phosphate Solubilization Index

To assess the phosphate solubilization activity, the strain was grown on Pikovskaya's agar medium (HiMedia™ Laboratories). In the aseptic environment, a spot inoculation of the bacterial isolate was made on the plates and incubated at 28 °C for 7 days. Uninoculated plate with PKV agar acted as a control. Comparative calculation of the solubilization index was conducted on day 7 of incubation by calculating transparent zone and colony diameters in centimeters. The activity of the phosphate solubilization index was calculated by the method of [35] using the following formula:

$$\text{Solubilization index} = \frac{\text{Colony diameter} + \text{Halozone diameter}}{\text{Colony diameter}} \tag{1}$$

2.4.3. Potential of the Isolate to Mitigate Chromate stress in Sunflower

Helianthus annuus seeds of uniform vigor and health were surface sterilized with 0.1% $HgCl_2$ (Medical Supplier Uk) for 5 min and rinsed thrice with sterile dH_2O to remove traces of $HgCl_2$. The sterilized seeds were then placed on wet filter paper and supplemented with the following treatments in Petri plates:

Treatment 1 = 0 $\mu g\ mL^{-1}$ of Cr^{+6}
Treatment 2 = 100 $\mu g\ mL^{-1}$ of Cr^{+6}
Treatment 3 = 300 $\mu g\ mL^{-1}$ of Cr^{+6}
Treatment 4 = Bacterial inoculum.
Treatment 5 = Bacterial inoculum + 100 $\mu g\ mL^{-1}$ of Cr^{+6}
Treatment 6 = Bacterial inoculum + 300 $\mu g\ mL^{-1}$ of Cr^{+6}.

Each plate contained 10 seeds. Bacterial inoculum was prepared by harvesting overnight broth culture through centrifugation at 896 rcf for 20 min and resuspending the washed cells in 10 mM $MgCl_2$. Inoculum density was adjusted to 10^{-6} cells/mL. Each treatment was applied to a set of three Petri plates.

2.5. Plant Growth Related Metabolites in Bacterial Culture

Rhizobacteria are known to produce a number of metabolites that help in their ability to interact with the host plant species to improve their growth. In this connection, plant hormones (indole 3-acetic acid, gibberellic acid, and salicylic acid), primary metabolites (proteins), and stress-related metabolites (phenolics, flavonoids, and proline) were assessed in bacterial culture. For this purpose, the rhizobacteria were cultured in L.B broth and incubated in a shaker at 30 °C and 130 rpm. Bacterial culture supernatant (BCS) was obtained by centrifuging (Sartorius Modle: 2–16 PK) the broth culture, which was then used to determine the metabolites mentioned above as discussed below:

2.5.1. Determination of Phytohormones

Indole acetic acid was estimated in the BCS using the Salkowski reagent as described earlier [36]. Different known quantities of IAA (10–100 µg) (Sigma Aldrich, Burlington, MA, USA) were used to develop the calibration curve.

For the determination of GAs (gibberellins) in BCS, a wheat endosperm assay was performed. Wheat seeds of good health (10 embryo-less wheat seeds) were disinfected by treating them with 70% ethanol for 1 min, followed by washing with dH_2O to remove ethanol traces. The seeds were placed in plates containing 2 mL of filter-sterilized BCS and acetate buffer (10 mL, pH 4.5). After incubation in the dark for 2 days at 28 °C, the buffer was recovered from Petri plates and tested for sugars by mixing with 200 µL of Benedict's solution. The reaction mixtures were incubated for 30 min at room condition, and OD was recorded at 254 nm against a blank (all above material without BCS). The calibration curve was constructed by treating known concentrations of GA_3 (Sigma Aldrich, Burlington, MA, USA) in the same manner mentioned and presented as GA_3 equivalent [37].

To determine salicylic acid (SA) in BCS, the technique of Worrier et al. [38] was used. Different quantities (10–1000 µL) of the BCS were added to 1 mL of ethanol, and the mixture was centrifuged at 5600 rcf for 10 min. The supernatant was chilled at 4 °C in a refrigerator before estimating the SA. Approximately 100 µL of the supernatant was added to 2.9 mL of fresh $FeCl_3$ (0.1%) solution (Sigma Aldrich, Burlington, MA, USA), and the absorbance was observed at 540 nm. Different quantities (10–100 µg) of SA (Sigma Aldrich, Burlington, MA, USA) were used to construct the standard curve.

2.5.2. Flavonoids, Phenol, and Proline Determination

The aluminum chloride method was followed to estimate total flavonoid (TF) BCS [39]. Briefly, 500 µL of the supernatant was added with 100 µL of 10% $AlCl_3$ (Aluminum chloride (Sigma Aldrich, Burlington, MA, USA), 100 µL of 10% potassium acetate, and 4.8 mL of 80% methanol. The blend was vortexed and kept in an incubator for 30 min incubation. After incubation, the optical density (OD) was noted at 415 nm. Various Quercetin (10–100 µg) concentrations (TargetMol, Boston, MA, USA) were used for plotting the calibration curve.

The phenols in the BCS were estimated by minor changes in the Folin–Ciocalteu method [40]. In short, to the 0.2 mL of bacterial culture supernatant, 0.8 mL of Folin–Ciocalteu reagent (MilliporeSigma, Bangalore India) and 2 mL of 7.5% Na_2CO_3 (sodium carbonate (Sigma Aldrich, Burlington, MA, USA)) was added. The sample was diluted to 7 volumes with dH_2O and incubated for 2 h in the dark. Different concentration of catechol was used to develop a standard curve. The absorption was recorded at 765 nm against blank. Catechol (CellMark AB, Göteborg, Sweden) concentrations (1–10 mg) were used for developing a standard graph.

The technique of Bates et al. [41] was used to estimate proline concentration in the BCS. Approximately 100 µL from BCS was added to 3% sulfosalicylic acid (4 mL) (New Alliance Fine Chem Private Limited, Mumbai, India) followed by centrifugation at 504 rcf for 5 min. The supernatant was collected, added 2 mL of acidic ninhydrin, and incubated at 100 °C for 1 h. Once cooled, the reaction mixture and proline were separated through 4 mL of toluene in a separatory funnel. Optical density was documented at 520 nm using the blank (toluene). Different proline (SENOVA Technology Co., Ltd, Cambridge, UK) quantities (10–100 µg) were used for developing standard graphs.

2.6. Microbial Interaction with Host; a Strategy to Cop Excess Chromate

Healthy and good quality seeds of sunflower were surface disinfected, as mentioned earlier. Upon germination in sterilized sand, the sunflower seedlings were moved to pots with the Hoagland's (400 mL) of half strength. The selected isolate P1 was subjected to test their capacity to relieve the chromate stress in plants. The pots were stored with 3 replicates (each having 4 seedlings) in a randomized complete block design (RCBD). Treatments had two main factorial variations: chromate levels (0, 25, 50, and 100 μg mL^{-1} of Cr^{+6}) and inoculum of rhizobacterium. Plants were maintained under 28 °C, the humidity of 68%, the light intensity of 647.5 μmol m^{-2} s^{-1}, and 13 h of photoperiod in LabTech (Model; LGC-5101 G) growth chamber. The seedlings were collected after 16 days at the 4–6 leaves stage. The relative growth rate (RGR) [42], net assimilation rate (NAR) [43], P1 colonization, and various growth attributes were recorded at the end of the experiment.

$$RGR = \frac{lnW2 - lnW1}{t2 - t1} \tag{2}$$

where ln is the natural log, W1 and W2 are plant dry weights at times t1 and t2.

$$NAR = \frac{W2 - W1}{t2 - t1} \times \frac{ln L2 - ln L1}{L2 - L1} \tag{3}$$

where W2 and W1 are the plant dry weights at times t1 and t2, logeA2 and logeA1 are the natural logs of leaf areas A1 and A2 at times t1 and t2.

The experiment was repeated three times, and the data were pooled and analyzed statistically.

2.7. Determination of Plant's Metabolites

To understand plant physiological and biochemical responses under chromate stress, different metabolites were tested in the plants harvested in the experiment mentioned above. The following metabolites were estimated in seedling biomass and in exudation by their roots.

2.7.1. Estimation of Indole Acetic Acid

The plant material (leaves) used for the analysis of phytohormones were ground in liquid nitrogen. For extraction of indole acetic acid, 3–10 g of the powder was homogenized (in case of exudates 3–10 mL) in 100% methanol (2 mL/g of plant sample) and centrifuged at 5600 rcf. The supernatant was used to determine indole acetic acid using the Salkowski reagent method as mentioned above [36].

2.7.2. Estimation of Total Flavonoids, Total Phenols, and Proline

To extract total flavonoids, 0.5 g of leaves (0.5 mL in case of exudates) were crushed in 5 mL of 80% aqueous ethanol and incubated for 24 h in a shaking incubator. The extract was then centrifuged for 15 min at 5600 rcf and 25 °C. The supernatant containing flavonoids was used for the estimation of flavonoids by the AlCl$_3$ technique as described above [39].

Total phenolics were extracted by homogenizing 1 g of leaves (or 1 mL exudates) in 16 mL of ethanol. The samples were incubated at 20 °C to 80 °C for 3 h. Following incubation, we centrifuged the samples for 10 min at 5600 rcf. The samples were filtered, concentrated at 40 °C to 1 mL, and re-dissolved in 10 mL distilled water and stored at 4 °C prior to use. Total phenolics were estimated by the technique mentioned previously [44].

For the extraction of proline, 200 mg of leaves were homogenized in 1 mL of ethanol/water (40:60 *v/v*) and incubated for 24 h at 4 °C. The incubation was followed by centrifugation at 5600 rcf (5 min). For 100% extraction of proline, the process was repeated and estimated as mentioned above [41].

2.7.3. Malondialdehyde (MDA) Determination

Schmedes and Hølmer [45] method was followed to estimate MDA concentration in the host leaves. Approximately 0.2 g of leaves were crushed in 2 mL of 0.6% thiobarbituric acid (Sigma-Aldrich, Burlington, MA, USA). Centrifugation of the samples was done for 10 min at 8064 rcf followed by incubation at 100 °C for 15 min in a water bath. Following incubation, the samples were cooled down and centrifuged again at 8064 rcf. The optical density of the samples was measured at 532, 600, and 450 nm. The calibration curve was constructed in the concentration range of 0.1 to 1.0 mM of MDA (Sigma-Aldrich, Burlington, MA, USA).

2.8. Estimation of Electrolyte Leakage

To remove surface adhered electrolytes, the leaves from a 7-days stressed seedlings were washed thrice with deionized water. After rinsing, discs of equal size were scraped out of the leaves to determine electrolyte leakage. The scraped discs were then transferred to capped vails having 10 mL of dH_2O and incubated at 120 rpm and 25 °C for 24 h. The electrical conductivity (L1) was measured, and the leaves (discs) were then autoclaved for 20 min at 120 °C to estimate electrical conductivity (L2). The percent leakage of the electrolytes was measured by using the following formula [46].

$$EL = \frac{L1}{L2} \times 100 \tag{4}$$

2.9. Lignin Concentration in the Root

Root lignin was determined by following the well-established protocol [47]. For this purpose, 1 g of sunflower roots were digested in 72% H_2SO_4 at 47 °C by vigorously shaking the mixture for 7 min. The partially digested samples were then autoclaved at 121 °C for 30 min for complete digestion. The autoclaved samples were fractionated into soluble and insoluble lignin through filtration. The soluble lignin (in filtrate) was determined by taking OD at 280 nm and 215 nm against the 72% H_2SO_4 blank. Lignin was calculated by the following formula:

$$S = \frac{4.53(A215 - A280)}{300} \tag{5}$$

The equation was generated from these two formulae: A215 = 0.15 F + 70 S and A280 = 0.68 F + 18 S. where, A280 = Optical density at 280 nm, A215 = Optical density at 215 nm, F = The furfural (g), S = The soluble lignin (g).

The 0.68, 0.15, 18, and 70 represent the molar absorptivity of soluble lignin and furfural at 215 and 280 nm, respectively. Insoluble residues were burnt at 550 °C for 4 h, and ash was weighted. The difference in dry mass and ash left after burning were considered as insoluble lignin. The total lignin was determined by adding insoluble and soluble lignin (mg/g of the cell wall).

2.10. Visualization of ROS and Their Accumulation in Leaves

For visualization of ROS, the technique of O'Brien et al. [48] was used with minor modifications. Leaves from sunflower seedlings were immersed in the DAB stain solution (2 mL) and kept for incubation at 130 rpm for 4–5 h. To remove the extra DAB stain, the samples were incubated in a bleaching solution (ethanol: acetic acid: glycerol = 3:1:1). To remove chlorophyll, the samples were boiled in ethanol in a water bath, clarified the ROS strained by DAB stain, if any, and then visualized under the light microscope.

2.11. Antioxidant System of Sunflower

To assess the antioxidant system of sunflower, the following assays were performed:

2.11.1. 2,2-Diphenyl-1-picrylhydrazyl (DPPH) Radical Scavenging Activity

The DPPH-scavenging activity was determined in the leaves of the plants with minor alteration [49]. Approximately 0.1 g of plant leaves were ground in 1 mL of methanol. A 0.004%

methanolic solution of DPPH (Cayman Chemical, Ann Arbor, MI, USA) was also prepared. Initially, a DPPH solution (1 mL) was mixed with the plant sample (0.5 mL), and the mixture was left at room temperature in the dark for 30 min. A decline in color was noted at 517 nm, and the radical scavenging activity was estimated using the following equation:

$$\%\text{DPPH} = \frac{(1 - \text{AE})}{\text{AD}} \times 100 \tag{6}$$

where AE = Optical density of the mixture of DPPH and leaves extract. AD = Optical density of DPPH solution only.

2.11.2. Catalase and Ascorbate Peroxidases Activity Determination

Catalases (CAT) activity was evaluated by computing the primary rate of H_2O_2 cleavage [50]. To 0.1 mL of supernatant, 2.6 mL of 0.05 M PBS (pH 7), 0.1 mM EDTA (ChemCeed, Chippewa Falls, WI USA) and 400 µL of 3% H_2O_2 (Univar Solutions, Chicago, IL, USA) were added. Decay in H_2O_2 was recorded at 240 nm and expressed as the decay of µM H_2O_2 min^{-1}.

The protocol of Asada et al. [51] was used to estimate ascorbate peroxidases (APX) activity. The reaction mix contained 0.2 mL of leaves extract, 600 µL of PBS (50 mM, pH 7.0), 100 µL ascorbic acid (0.5 mM) (Sigma Aldrich Burlington, MA, USA), and 100 µL hydrogen peroxide (H_2O_2, 0.1 mM). Measurement of activity was noted at 290 nm and presented as unit min^{-1} mg^{-1} protein.

2.11.3. Peroxidase and Superoxide Dismutase Activity

For the measurement of peroxidase activity, guaiacol was used as a substrate for dehydrogenation [44]. Extraction of enzyme was performed in 3 mL of 100 mM PBS (pH 7.0). Fresh seedling leaves (100 mg) were crushed in 1 mL PBS and swirled for 15 min at 5 °C and 8064 rcf. To 0.1 mL of plant extract, 3 mL PBS (100 mM), 0.05 mL guaiacol (20 mM) (Parchem Fine & Specialty Chemicals, New Rochelle, NY USA) and 0.03 mL H_2O_2 (12.3 mM or 0.04%) were added and vortexed. The change in optical density by 0.1 (t) was recorded at 436 nm, and POD activity was calculated by the following formula.

$$\text{Enzyme activity} = \left(\frac{500}{\Delta t}\right) \times \left(\frac{1}{1000}\right) \times \left(\frac{\text{TV}}{\text{VU}}\right) \times \left(\frac{1}{(\text{f wt})}\right) \tag{7}$$

where; Δt = change in time; TV = total prepared volume and VU = volume of the sample used; f wt = fresh weight (g).

Extraction of SOD was performed with the buffer that consisted of 50 mM KH_2PO_4, pH 7.8 and 100 µM EDTA, 1% Triton X-100, 2% Polyvinylpyrrolidone, and a complete protease inhibitor cocktail (Roche, Mannheim, Germany). For this, 100 mg of leaves were homogenized in the buffer mentioned above, followed by sonication (2×30 s, at A = 30) (Model; LeelaSonic-50) at 4 °C and filtration (polycarbonate filters, 2.0 µm pore size; Osmonics, South Miami, FL, USA). The filtered sample was centrifuged for 20 min (10,000 rcf) at 4 °C, and the supernatant was carefully collected. The resultant supernatant was used to quantify SOD activity [52].

The activity was assessed by measuring the inhibition in the photoreduction of NBT (nitro-blue tetrazolium) by SOD [53]. The final sample composed of 50 mM PBS (pH 7.6), 100 µM EDTA, 50 mM sodium carbonate, 50 µM NBT (Sigma-Aldrich, Burlington, MA, USA), 10 µM riboflavin (BIOCHEM Bernburg GmbH, Bernburg Germany), 12 mM L methionine (Fagron Industry, Rotterdam, ZUID-HOLLAND Netherlands), and 100 µL crude extract at a final volume of 3 mL. Blank for the sample consist of all the mentioned chemical without leaves extract. Exposure to white light for 15 min initiated the reaction. Following incubation, optical density was measured at 560 nm. A unit (U) of SOD is characterized as the unit of enzyme necessary to prevent NBT from degrading photochemically by 50%.

2.11.4. Estimation of Reduced Glutathione

Reduced glutathione (GSH) (Chambio, Shanghai, China) was estimated by measuring the redox reaction of NADPH (Sigma-Aldrich, Burlington, MA, USA). The reaction mixture consisted of 0.3 mL leaves extract, 1.8 mL PBS, 0.3 mL of EDTA, 0.3 mL of NADPH, and 0.3 mL of oxidized glutathione (GSSG) (Chambio, Shanghai, China). The optical density was recorded at 340 nm [54].

2.12. Estimation of Heavy Metals

2.12.1. Colorimetric Determination as Preliminary Test

The bacterial strain and plants were grown as mentioned above in the concentration of Cr^{+6}. The isolate P1 was inoculated in the sterilized broth media and incubated at 28.2 °C for 24 h. The supernatant and pellet were isolated by centrifugation for 10 min at 5600 rcf and 4 °C. The quantity of Cr was then determined in the supernatant. Supernatant from the bacterial culture was directly subject to Cr^{+6} estimation using the di-phenyl carbazide technique; however, plants samples were exposed to wet digestion to determine Cr^{+6} following the same protocol [55].

2.12.2. BCR Sequential Extraction Method

The technique of the Community Bureau of References (BCR) was used to estimate the two different species of chromium [56]. The extraction involved three main stages:

Stage 1: Exchangeable or Acid-Soluble

To the dried plant sample (0.5 g or 0.5 mL bacterial culture/exudates), 20 mL of acetic acid (110 mM) (Sigma Aldrich, Burlington, MA, USA) was mixed, kept in a shaker for 12–24 h at 25–30 °C, and separated from the residues through centrifugation at 504 rcf at 28 °C for 3 min.

Stage 2: Cr^{+6} Extraction

The remains obtained in stage 1 were added to a freshly prepared 20 mL solution of 500 mM hydroxylamine-hydrochloride (Sigma Aldrich, Burlington, MA, USA), and pH 1.5 was adjusted with nitric acid (Sigma Aldrich, Burlington, MA, USA). The resultant solution was incubated at 25–30 °C for 16 h. The residues were separated via centrifugation, as mentioned in step1.

Stage 3: Cr^{+3} Extraction

The remains of stage 2 were treated with 30% hydrogen peroxide (5 mL) 2 times and incubated for 1 h (pH set to 2), followed by drying. To the dried samples, 1 M ammonium acetate (25 mL) was added and kept for incubation at 28 °C for 16 h.

The pellets were washed at each step of the extraction, shaken (15 min), and centrifuged (20 min) at 504 rcf. The supernatant was deliberately washed away, avoiding contamination from the residues. This procedure was repeated to eliminate any residual reactants and Cr from the preceding level.

2.12.3. Atomic Absorption Spectroscopy (AAS)

Isolated portions of the above steps were examined via atomic absorption, Perkin Elmer (AAnalyst 700, Parkin Elmer, Waltham, MA, USA). The air/acetylene flame was used to perform the measurements under the operational parameters recommended by the manufacturer.

2.13. Bioconcentration Factor

The concentrations of Cr in the seedling biomass were calculated using the following equations:

$$BCF = \frac{Cr\ in\ Plant\ biomass}{Cr\ added\ to\ media} \tag{8}$$

2.14. P1 Colonization with Host Roots

The plate count technique was followed to test the colonization of selected isolate with host roots. For this, 2 cm of host root was forcefully shaken to detach loosely attached bacteria in the sterile distilled water and then crushed into 1 mL of 10 mM $MgCl_2$ solution. The resulted mixture was filtered and serially diluted up to 10^{-6}, and 0.1 mL of each was plated on LB agar. The colonies were counted after 24 h of incubation at 28 °C.

2.15. Data Analysis

All the assays were carried out three times, and the results of the experiment were categorized into Cr^{+6} and P1 inoculations. The significance ($p < 0.05$) for *A. bouvetii* inoculation status, Cr levels, and their interaction were determined using ANOVA and Duncan's Multiple Range Test (DMRT) in IBM SPSS Statistics 21(IBM, Armonk, NY, USA). Graph Pad Prism (Version 5.03, Graph Pad Inc., San Diego, CA, USA) was used for plotting graphs.

3. Results

3.1. P1 Molecular Identification

Based on the 16S rRNA sequence, the isolate P1 showed maximum sequence homology (97.17%) with *Acinetobacter bouvetii* (Figure 1). For confirmation, the sequence was subjected to phylogenetic analysis by constructing their phylogenetic consensus by Neighbor Joining (NJ) using MEGA 7. The isolate P1 showed a close resemblance with *A. bouvetii* supported by 100% bootstrap value in the tree. In short, the strain P1 was identified as *A. bouvetii* through BLAST results and the phylogenetic analysis. The obtained sequence can be retrieved through a GenBank accession No. MT478039 on NCBI nucleotide database.

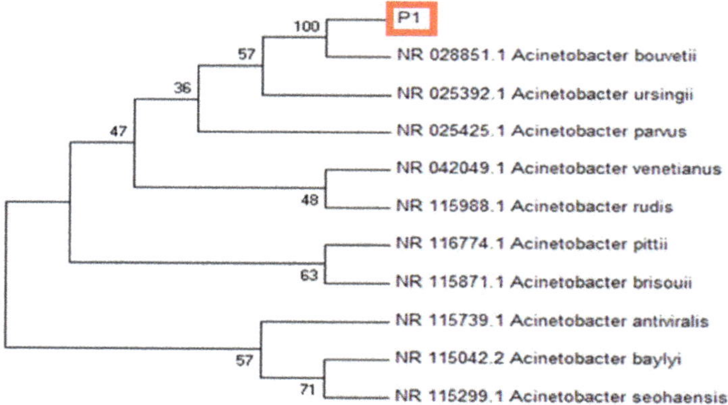

Figure 1. Molecular phylogenetic analysis of rhizobacterial strain P1 (red frame) showing evolutionary relationships of taxa using NJ (Neighbor Joining) tree method. The analysis of the evolutionary relationship was carried out in MEGA7 using 16 S rDNA sequences of our isolate and closely related sequences retrieved from NCBI GenBank. Nodes are labeled with bootstrap values.

3.2. Characterization of the Selected Strains

3.2.1. Impact of Chromate Rhizobacterial Growth

The rhizobacterial isolate P1 was allowed to grow in L.B. broth supplemented with different concentrations of Cr^{+6}. An increase in bacterial growth was recorded in an L.B. broth supplemented with 100 µg/mL of Cr^{+6}. The steady growth of rhizobacterial isolate P1 was recorded in L.B. broth supplemented with 300 µg/mL of Cr^{+6} (Figure S1). An approximately 3-fold increase was recorded in bacterial growth at 500 µg/mL supplementation of broth with Cr^{+6}. However, a decline was recorded in the growth of the rhizobacterial strain P1 as the concentration of Cr^{+6} was increased in broth. Interestingly, the recorded growth of the bacteria was higher compared with control at all supplemented levels of Cr^{+6}.

3.2.2. Production of Bioactive Compounds

The strain P1 was screened against the different concentrations of K_2CrO_4 (Potassium chromate) as a Cr^{+6} source to assess the indole acetic acid production in stressed conditions (Figure 2a). The addition of Cr^{+6} stimulated the rhizobacteria to produce elevated levels of IAA in a concentration-based manner. The release of gibberellic acid (GA_3) was also improved as the concentration of the Cr^{+6} increased. A significant concentration-dependent increase in the production of GA_3 was noted at 500 µg mL^{-1} of the Cr^{+6}, however, beyond 500 µg mL^{-1} of stress, the production of the GA_3 decreased (Figure 2b). A similar pattern was also recorded in the production of salicylic acid upon the induction of stress by the concentrations of Cr^{+6} (Figure 2c). Bacterial salicylic acid production increased with the elevation of metal levels supplemented from 100 to 1200 µg mL^{-1} of Cr^{+6} in the media.

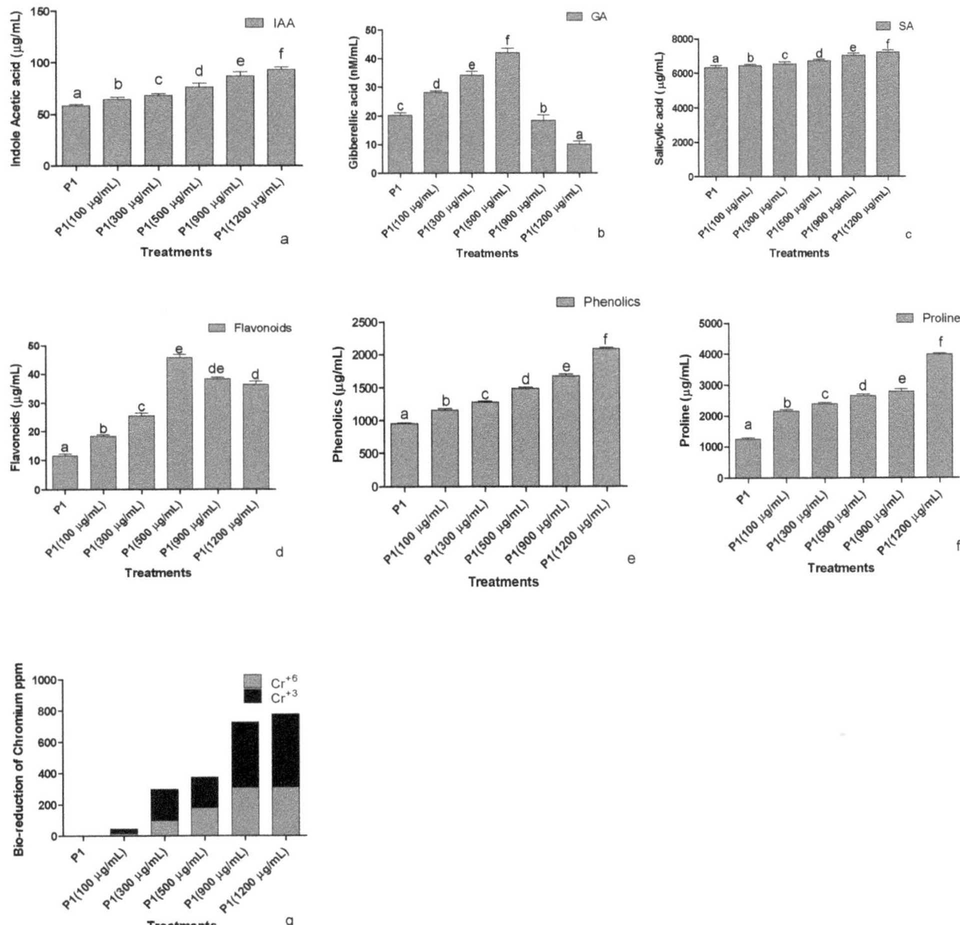

Figure 2. Effect of chromate stress on the concentration of (**a**) indole acetic acid, (**b**) gibberellic acid, (**c**) salicylic acid, (**d**) flavonoids, (**e**) phenolics, and (**f**) proline in the bacterial culture supernatant and (**g**) bioreduction capability of the rhizobacterium P1. The rhizobacterium was grown in the LB broth containing different quantities of hexavalent chromium for 24 h, and the supernatant was then assessed for the mentioned parameters. Data are means of nine replicates with ±SE of the mean. Different alphabets represent significance ($p < 0.05$).

Significant impacts were recorded on total flavonoids production by P1 at an elevated concentration of the Cr^{+6} (Figure 2d). An increasing tendency in the release of total flavonoids was noted from 0 to 500 µg mL^{-1} of the metal supplemented in media; however, a dip was noted in the flavonoid concentration of the strain upon exposure to 900 µg mL^{-1} or 1200 µg mL^{-1} of the Cr^{+6}. Higher production of phenols in Cr^{+6} spiked media was observed (Figure 2e). Further production was positively regulated with higher levels of chromate-induced stress peaking at 1200 µg mL^{-1}. Proline produced by the selected strain was determined after exposing the strain to various concentrations of the Cr^{+6} (Figure 2f). The proline concentrations showed an abrupt increase at 100 µg mL^{-1} of Cr^{+6}, which steadied till 900 µg mL^{-1} of Cr^{+6}. Elevated proline levels were recorded at 1200 µg mL^{-1}, whereas the lowest was recorded in the growth medium supplemented with 0 µg mL^{-1} of Cr^{+6}.

Beyond the production of plant regulators and stress mitigators, the selected rhizobacteria had the potential for efficient solubilization of the inorganic phosphate. The phosphate solubilization index of the strain was 3.66.

3.2.3. Reduction of Cr^{+6} to Cr^{+3}

When the selected strains were exposed to 100, 300, 500, 900, and 1200 µg mL^{-1} of the Cr^{+6}, the strains were able to reduce the valency of Cr^{+6} and biotransform it into their least toxic form Cr^{+3} (Figure 2g). After 24 h of treatment, the strain biotransformed almost half of the highly toxic Cr^{+6} to its least toxic, trivalent form.

3.3. Net Assimilation Rate (NAR) and Relative Growth Rate (RGR)

Decreasing trends were observed in the host's net assimilation rate, i.e., showing an inverse proportion to increasing metal concentration (Figure 3a). With the plant growth promoting activities of P1, the net assimilation rate of the host was positively regulated exhibiting an increase in net assimilation rate under Cr^{+6} induced stressed environment. Plant growth promoting the interaction of P1 with their host achieved the maximum net assimilation rate related to normal and stressed control plants and all other treatments.

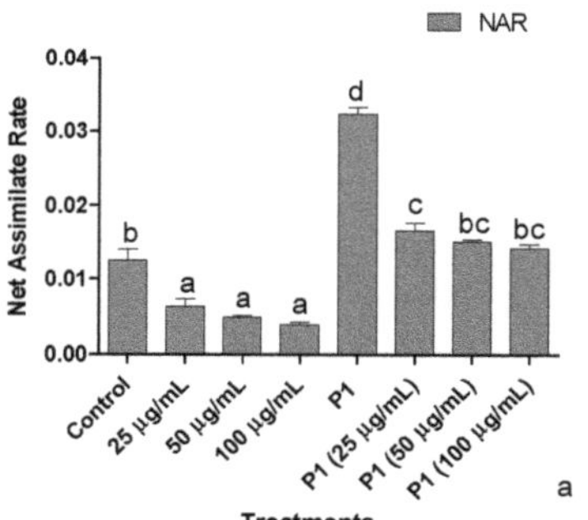
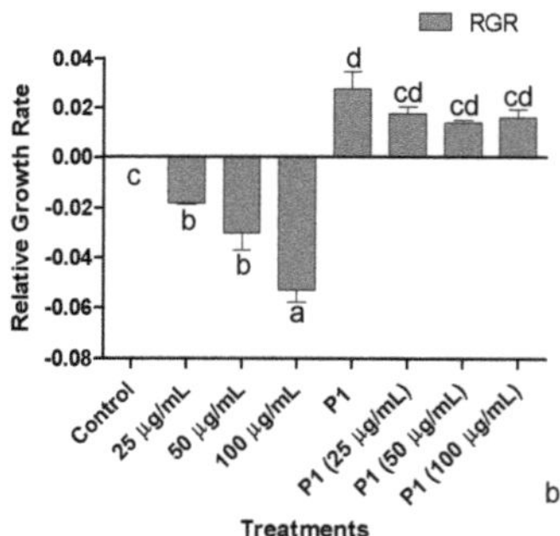

Figure 3. Effects of different concentrations of hexavalent chromium and P1 on (**a**) net assimilation rate and (**b**) relative growth rate of sunflower seedlings grown hydroponically (half-strength Hoagland's solution) for 16 days in a plant growth chamber. Data are mean of 36 replicates with SE of the mean. Different labels on mean bars represent significance (Duncan; $p < 0.05$).

The dip in relative growth rate was roughly 5-folds in the host treated with 100 µg mL^{-1} of Cr^{+6} (Figure 3b). With inoculation of P1, higher RGR was achieved by all host plants grown in Hoagland's media supplemented with the stated level of Cr^{+6}. The highest RGR was approximately 3-fold the level exhibited by the host associated with P1. Nonetheless, the decline was recorded with the metal concentration, but the RGR was still higher than normal and stressed control plants, i.e., RGR of P1 associated plants exposed to 100 µg mL^{-1} of Cr^{+6} was approximately two times that of the non-associated control plants.

3.4. Effects of P1 and Chromate Stress on Metabolites of Host Plants

Endogenous and root exudates indole acetic acid (IAA) of the chromate stressed *H. annuus* seedlings was assessed (Table S1). Inverse relation of Cr^{+6} was recorded with the endogenous and exogenous IAA levels in *H. annuus* seedlings. Inoculation of seedlings with P1 benefited the host plant by significantly improving IAA production and its exudation through the root. Different levels of chromate significantly reduced the amount of endogenous and released IAA in P1 associated seedlings, but its concentration was still higher than the seedlings grown in the absence of Pa strain.

Exposure of *H. annuus* seedlings to Cr^{+6} significantly enhanced the concentration of plant and root exuded flavonoids in a dose-dependent manner (Table S1). Inoculation of the seedlings with the strain P1 controlled the accumulation of the endogenous flavonoid's concentration keeping their levels below the noninoculated counterparts. However, P1 associated plants released significantly greater quantities of flavonoids which were synergistically enhanced upon exposure to chromate stress.

Under chromate stress, sunflower seedlings had a significantly lower concentration of total phenols in plants and in root exudates than the control seedlings (Table S1). Chromate concentration was inversely related to the phenolic concentration in the seedlings as well as in the exudates. Inoculation of *H. annuus* seedling P1 significantly enhanced the concentration of endogenous and released phenols than the control seedlings. A further increase was noted in the endogenous phenols of P1 associated sunflower seedlings exposed to chromate in a concentration-dependent manner. However, chromate-stressed P1 inoculated seedlings had a lower number of released phenols than the P1 seedlings grown in the absence of chromate stress.

Proline accumulation and release were also enhanced in the chromate-stressed seedlings than in control (Table S1). Its endogenous concentration was further enhanced in P1 inoculated seedlings, and the exposure of these seedlings to chromate caused a reduction in proline concentration. The concentration of released proline was greater in P1 inoculated seedlings than in the control seedlings. In the presence of chromate, such seedings released even higher concentrations of proline.

3.5. Root lignification, Electrolyte Leakage, and Malonaldehyde Concentration

Stressful conditions due to elevated levels of chromate upregulate the lignin biosynthesis, thereby increasing lignin deposition. An almost two times increase in lignin deposition was recorded at each increase in metal concentration (Figure 4a). Interestingly, a dip in lignin production at all treated concentrations of the metal was noted after P1 inoculation.

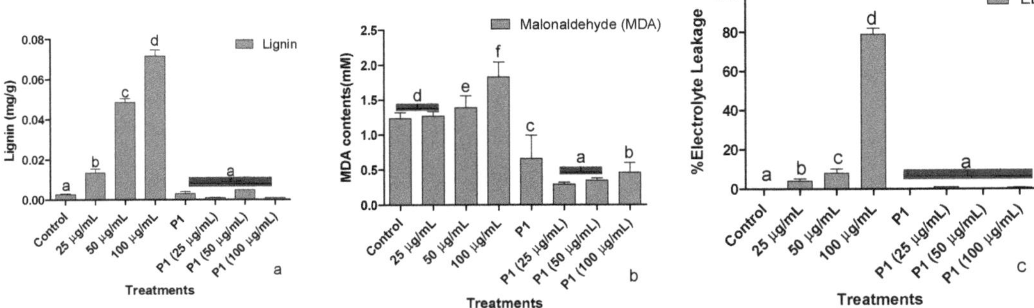

Figure 4. Effect of different concentrations of hexavalent chromium and P1 on (**a**) root lignification, (**b**) leaf malonaldehyde, and (**c**) electrolyte leakage of sunflower seedlings grown hydroponically (half-strength Hoagland's solution) for 16 days in a plant growth chamber. Data are mean of 36 replicates with SE of the mean. Different labels on mean bars represent significance (Duncan; $p < 0.05$).

Exposure of sunflower seedlings to hexavalent chromium enhanced the accumulation of malondialdehyde (MDA). An increase in the endogenous concentration of MDA was dependent on the amount of chromate present in the growth medium (Figure 4b).

Rhizobacterium P1 colonized seedlings had a significantly lower concentration of MDA, which was further reduced upon exposure to chromate.

Upon exposure to the said levels of Cr^{+6}, sunflower seedlings showed a concentration-based increase in the leakage of electrolytes. The electrolyte leakage doubled at each elevated level of Cr^{+6} (Figure 4c). The host seedlings were rescued by P1 inoculation, which reduced electrolytes leakage from sunflower leaves by several folds. Even in the presence of 100 µg/mL of chromate, electrolyte leakage from the leaves of P1 associated seedlings was lower than the control seedlings.

3.6. Antioxidant

The ability of seedling's leaf extract to scavenge DPPH free radical dropped significantly in the presence of 50 and 100 µg/mL of hexavalent chromium (Figure 5a). The lowest DPPH scavenging activity was noticed in the host plants treated with 100 µg mL^{-1} of Cr+6. However, rhizobacterium P1 Inoculated seedlings showed a stable DPPH free radical scavenging activity even in the presence of the highest concentration of chromate used in this study.

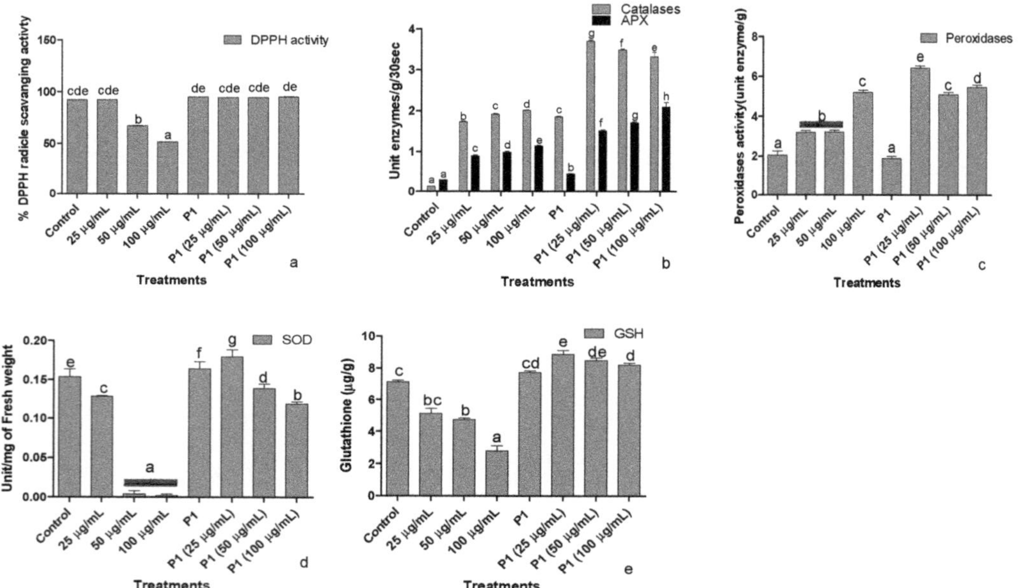

Figure 5. Effect of different concentrations of hexavalent chromium and P1 on (**a**) %DPPH radical scavenging activity, (**b**) catalases and ascorbate peroxidase, (**c**) peroxidase, (**d**) superoxide dismutase, and (**e**) reduced glutathione of sunflower seedlings grown hydroponically (half-strength Hoagland's solution) for 16 days in a plant growth chamber. Data are mean of 36 replicates with SE of the mean. Different labels on mean bars represent significance (Duncan; $p < 0.05$).

Ascorbate peroxidases and catalase activities were enhanced when sunflower seedlings were exposed to different concentrations of chromate (Figure 5b). The activity of both the enzymes increased with the increasing concentration of hexavalent chromium in the media. In P1 inoculated seedlings, activities of catalase and peroxidase were significantly greater than the control seedlings. Upon exposure to 25 µg mL^{-1} chromate, the activities of these enzymes increase by 2-fold. The rise in APX activity continued with an increase in chromate concentration. However, further increases in chromate concentration significantly reduced catalase activity in sunflower seedlings.

Exposure to chromate also induced peroxidase activity in sunflower seedlings in a concentration-dependent manner (Figure 5c). In P1 inoculated seedlings, peroxidase activity did not show any difference in comparison to the control seedlings. However, P1 associated seedlings had significantly higher peroxidase activity upon exposure to different concentrations of chromate than their respective controls.

Exposure of sunflower seedlings to hexavalent chromium was associated with reduced SOD activity than the control. Higher concentrations (50 and 100 µg mL^{-1}) more severely affected the activity of this enzyme which was reduced by several folds (Figure 5d). The rhizobacterium P1 enhanced SOD activity than its level in the noninoculated seedlings. The strain also reduced the harmful effect of chromate on SOD activity.

The concentration of reduced glutathione also declined in chromate-treated seedlings, but the effect was moderate compared with SOD activity (Figure 5e). Inoculation of seedlings with P1 significantly enhanced the concentration of reduced glutathione in the host plant compared to the control seedlings. Exposure of P1 associated seedlings either further improved the concentration of reduced glutathione or did not influence its level at all.

3.7. 3,3′-Diaminobenzidine (DAB) Stain Assay

With the exposure of sunflower seedlings, the H$_2$O$_2$ production was noted as brown spots formed by DAB strain in the leaves, and with the increase of Cr^{+6} supplementation, the size and number of brown spots increased (Figure 6). The inoculation of the P1, however, alleviated the Cr^{+6} induced stress and hence reduced the H$_2$O$_2$ production, which is recognized by spotless tissues after treatment with DAB strain.

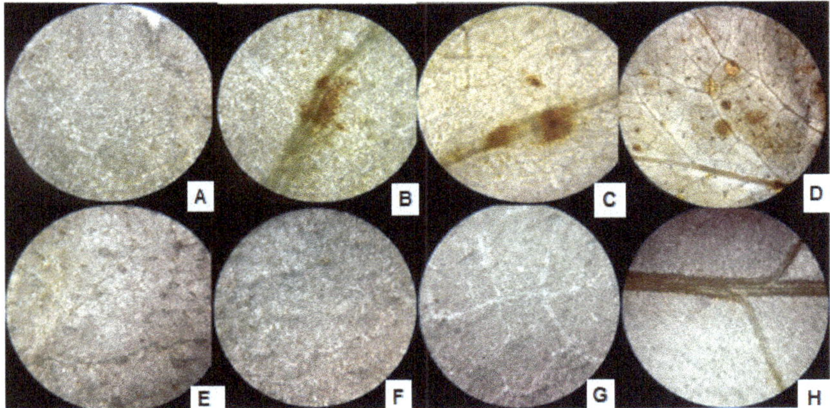

Figure 6. ROS accumulation in leaves of P1 inoculated sunflower seedlings exposed to different levels of hexavalent chromium (**A**) control (0 µg mL^{-1}), (**B**) 25 µg mL^{-1}, (**C**) 50 µg mL^{-1}, (**D**) 100 µg mL^{-1}, and (**E**) P1 strain + 0 µg mL^{-1}, (**F**) P1 + 25 µg mL^{-1}, (**G**) P1 + 50 µg mL^{-1}, and (**H**) P1 + 100 µg mL^{-1}. The seedlings were grown hydroponically (in half-strength Hoagland's solution) for 16 days in a plant growth chamber. Fully expanded leaves were detached and stained with DAB to visualize ROS.

3.8. Determination of Uptake and Accumulation of Cr^{+6} by Host Plants
3.8.1. By Colorimetric Method

The amount of Cr^{+6} accumulated by different parts of sunflower seedlings increased with an increase in the concentration of the Cr^{+6} in the medium (Figure 7a). For instance, the seedling exposed to 100 µg mL^{-1} accumulated about 10.34 µg mL^{-1} of hexavalent Cr^{+6} in their parts, which was the highest among the plants treated with various concentrations of Cr^{+6}. Accumulation of heavy metals in plant parts was significantly reduced in seedlings associated with P1.

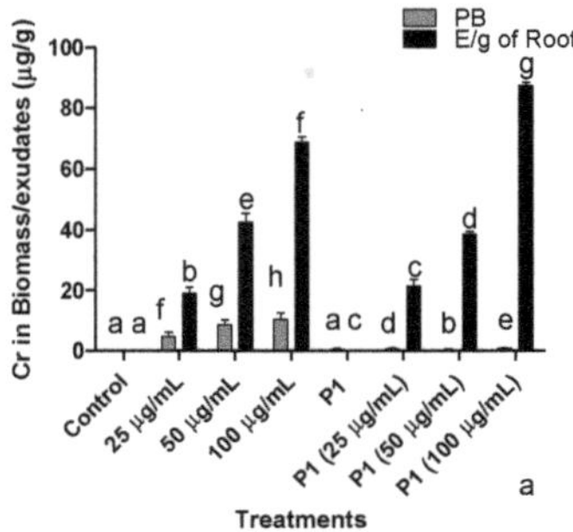
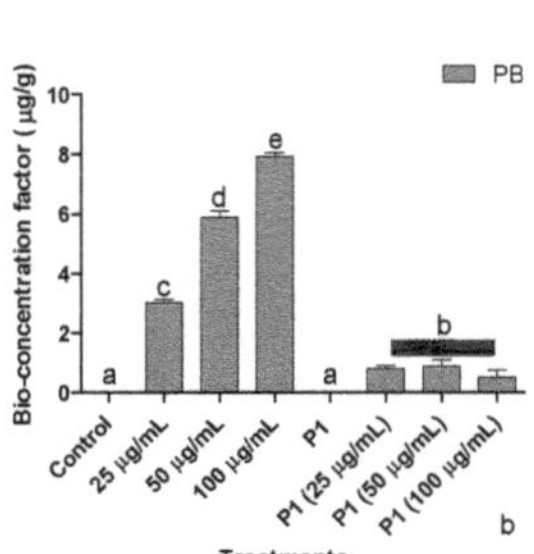
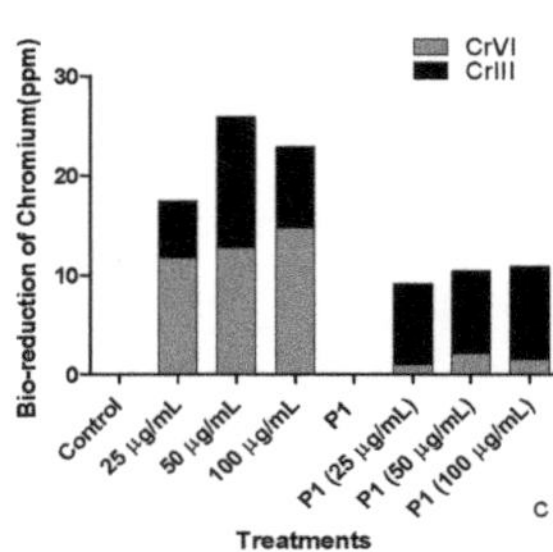

Figure 7. Influence of rhizobacterium P1 on (**a**) absorptions/accumulation, (**b**) bioconcentration, and (**c**) bio-reduction of Cr^{+6} in sunflower seedlings. The seedlings were grown hydroponically in Hoagland media (half strength) containing different concentrations of hexavalent chromium for 16 days in a plant growth chamber. Data are a mean of 36 replicates with SE and the alphabets representing significance (Duncan; $p < 0.05$).

3.8.2. Bioconcentration of Cr^{+6} in Host

Sunflower seedlings exposed to increased concentration of Cr^{+6} showed an increase in bioaccumulation of Cr^{+6} (Figure 7b). Conversely, the inoculation of *H. annuus* seedlings with rhizobacterial strain P1 inhibited the accumulation of Cr^{+6} as the concentration increased in the medium to avoid phytotoxicity. The ceased accumulation of Cr^{+6} in P1 associated *H. annuus* seedlings were noted at all treated level of Cr^{+6}.

3.8.3. Determination of Cr Species by BCR Extraction

Exposure of *H. annuus* seedlings to various concentrations of the Cr^{+6} showed higher accumulation in all parts of the plant, where a small amount has been converted to a stable Cr^{+3} form. Indeed, the bioaccumulation of Cr^{+6} increased when applied at higher concentrations (Figure 7c). With inoculating P1, a reduction in the bioaccumulation of Cr^{+6} was recorded in all parts of the host plant. Moreover, the highest percent conversion of Cr^{+6} to a nontoxic form (Cr^{+3}) was noted in the plant biomass treated with rhizobacterial strain P1.

3.8.4. Assessing Root Colonization Potential of P1

P1 shows an increase in the root colonizing capacity as the supplements of the Cr^{+6} in the medium increase. The root colonization tendency from strain control to treatments, i.e., 25 µg/mL 50 µg/mL and 100 µg/mL were 4.6×10^8, 5.0×10^8, 1.18×10^9, and 1.30×10^9 bacteria, respectively, in 2 cm of root segment.

4. Discussion

The rhizobacterial isolate P1 was able to tolerate a concentration of chromate up to 12,000 times higher than the WHO permissible levels (0.1 mg L^{-1}) [57]. A significant increase was recorded in the growth of bacterial mass at all the applied levels of Cr^{+6} in the liquid medium. The rhizobacterium was also capable of secreting substantial quantities of phytohormones, including IAA, GA, and SA, demonstrating its pro-plant nature [33,58]. We noticed that increased stress levels enabled the strain to release higher amounts of these phytochemicals. In fact, higher production of these phytohormones by P1 under elevated stress levels is an important outcome that can benefit the host plant to withstand the stress [36]. The previous findings suggest that plant species have the ability to absorb phytohormones released by the microbes in the rhizosphere [59]. Moreover, the rhizobacterium was able to colonize sunflower roots where it can directly contribute to phytohormones and improve the host fitness and growth. Sunflower seedlings colonized by the rhizobacterium P1 had comparatively higher endogenous levels of these phytohormones than the non-associated seedlings. Additionally, with the absorption of microbial contributed phytohormones, binding of salicylic acid Calcium-dependent protein Kinases

(CDPKs) activating stress-responsive downstream genes (Peroxidases, GSTs (Glutathione S-transferases), Osmotins, HSPs (Heat shot proteins)), and can be used to combat a range of abiotic and biotic stressors in the long term [60]. Increased expression of CDPKs and downstream stress-responsive genes in over-expressed lines relative to knockdown lines indicates *OsMYB-R1* mediates stress tolerance through auxin and salicylic acid-responsive signaling [61]. Among the phytohormones, gibberellin potentially enhances the agronomic attributes of plants. He et al. [62] investigated the significant effect of gibberellic acid in stress acclimation in tested plants by the upregulation of *TaMYB73* gene expression. Hyper-expression of the *Triticum TaMYB73* gene in *A. thaliana* improves the host acclimation potential to NaCl and upregulates the expression of several genes such as *AtCBF3, AtABF3, AtRD29A,* and *AtRD29B*. These are the key factors of osmotic adjustment that influence the crosstalk with microbial phytohormones in the rhizosphere, thus boosting immunity and growth of the host under a stressed environment [61].

Enhanced production of phenolic, flavonoid, and proline in the rhizosphere occurs by the stimulation of stress because they act as nonenzymatic antioxidants and enhance the capability of rhizobacteria to bear metal toxicity [63,64]. Rhizobacteria employ flavonoids to chelate soil nutrients and gain adequate nutrients when they are stressed. More crucially, flavonoids influence quorum sensing, allowing bacteria to carry out their density-dependent tasks [65]. Microbes employ phenolics and flavonoids as cross signals as part of a chemical interaction to form a plant growth-promoting relationship with the host root. Flavonoids, phenols, and proline operate as ROS scavengers inside the cell, a key step in stress tolerance [66,67]. Apart from all these characteristics, the selected rhizobacterial efficiently solubilize inorganic phosphate that helps the rhizobacteria to promote host growth normally in a stressful environment [68]. Recent studies also found that PGPR alleviate metal stress and enhance the agronomic attribute of the host. Moreover, they showed that the height of plants increased by 22–50% after inoculation with SS6, SS1, and SS3 exposed to Cr (20, 30, and 40 ppm) compared with the non-inoculated plants [27,69]. In another study, the application of *P. aeruginosa* strain OSG41, even with three times concentration of chromium, increased the dry matter accumulation, symbiotic attributes (like nodule formation), grain yield, and protein of chickpea compared with non-inoculated plants [30].

Susceptibility of sunflower seedlings was noted at elevated levels of Cr^{+6}, exhibiting serious consequences in terms of growth attributes, including low net assimilation rate and relative growth rate ($p < 0.05$) [70]. The decrease was accompanied by a sharp decline in the growth regulators, particularly IAA, GA, and SA, and a malfunctioned antioxidant system, which is necessary for the normal growth and development of the plant species [71]. Inability to produce enough phytohormones left sunflower seedlings at the mercy of chromate toxicity, leading to reduced growth and higher accumulation of ROS. Under such circumstances, the severity of chromate stress was further increased due to loss in the ability of host plant species to scavenge ROS. Recently, the role of IAA was reported to aid in stress acclimation; however, severe metal toxicity resulted in the degradation of IAA [72].

In the presence of *Acinetobacter bouvetii* P1, the seedlings had higher NAR and RGR and were more tolerant to chromate stress than the uninoculated seedlings. Sunflower seedlings received the aid of phytohormones from the associated rhizobacterial isolates, which boost the host's fitness to withstand harsh environments while growing normally. Higher IAA levels are correlated with improved growth, yield, and stress tolerance of the host plants. The same increasing pattern of total flavonoids production was noted with Cr^{+6} elevation. The flavonoids act as metal chelators, help quenching reactive oxygen species and act as a nonenzymatic antioxidant. With the inoculation of P1, a decreasing pattern in the endogenous flavonoids was observed. Nonetheless, a concentration-based increase in the exogenous flavonoids was also recorded. The lower accumulation and higher exudation of flavonoids may be a strategy of the host plant to quench the metal in the rhizosphere to avoid the possible phytotoxicity resulting from metal accumulation in the plant body. In such conditions, flavonoids chelate the toxic metal as well as ROS quenchers to detoxify the oxidative stress produced as a result of an excess of Cr^{+6} [73]. An interesting observation was the severe reduction in the total endogenous and exogenous phenolics with chromate exposure of the host, making it susceptible to various biotic and abiotic stresses [33]. Metal stress disturbs the shikimate pathway resulting in the lower production of total phenolics. The phenolics act as a part of the defense system and detoxify the ROS, working as a nonenzymatic antioxidant, and their decline leads to a susceptibility of the host [74]. Plant growth-promoting associations of P1 improve the phenolic production of the hosts and help to detoxify the metal, and subsequent oxidative impairment in the host, acclimating the host to metal toxicity and subsequent oxidative damage [75].

Higher proline production is the key response to a stressful environment acting as an osmolyte and nonenzymatic antioxidant to minimize the toxicity caused by the stressor. Their endogenous and exogenous production was further improved with the application of P1. The endogenous proline accumulation in the cell tends to protect the oxidative damage and metal toxicity whereas, the increased exogenous proline tends to chelate metal in the rhizosphere, thereby avoiding the uptake of the metal and preventing phytotoxicity and subsequent oxidative damage [76,77].

In some instances, higher lignification is a tendency of the host to develop a physical barrier to avoid the entry of such toxic substances. However, early hyper-lignification due to abiotic stressors results in stunted growth in plants [78]. A recently published review [79] suggested that hyper-lignification of the plant exposed to Cd stress sufficiently reduced the entry of Cd into the plant's cell. Another study found that, since Cd is primarily adsorbed at the root level, the anatomy and molecular structure of the cell wall in root cells is a critical parameter. Plant roots rich in suberin and lignin may be more impermeable to Cd and therefore offer resistance to Cd absorption and translocation [80]. The same tendency of root lignification was noticed in sunflower seedlings challenged with Cr^{+6}. However, the degree of lignification was not enough to completely exclude Cr^{+6} from the roots. The rhizobacterium-associated seedlings had limited access to Cr^{+6}, bringing lignification to a normal level that further promotes normal cell division and elongation [81,82]. In short, the P1 isolate helped the host plant restrict the entry of Cr^{+6}. P1 also contributed towards proline production to alleviate the stress and led to bioreduction of Cr^{+6}, strengthened antioxidant system of the host, improved host endogenous phytohormones pool, and solubilized nutrients (phosphorous).

Apart from growth-promoting activities, P1 also interfered with the seedlings' metal uptake, contributing to a 90–95% decrease in seedlings' Cr absorption. The ceased absorption of Cr^{+6} by root in Cr^{+6} supplemented Hoagland's medium may be attributable to the capacity of P1 to biotransform the highly toxic Cr^{+6} form to their least toxic Cr^{+3} form in the host's rhizosphere [83]. The application of *P. aeruginosa* strain OSG41 improved the dry matter accumulation, symbiotic features (such nodule formation), grain production, and protein contents of chickpea exposed to Cr^{+6} stress. The bioinoculant reduced Cr^{+6} absorption by 36, 38, and 40% in roots, shoots, and grains, respectively [30]. In addition to a bioreduction in the rhizosphere, chromate reduction in plants also tended to occur with isolate P1. Bioreduction of heavy metal is one of many PGPR approaches to counteract its toxic impact because Cr^{+3} is an active human oligo-element [84–87]. Also, Cr^{+3} is relatively less toxic, chiefly to the crops, and at small quantities (0.05 mg L^{-1}), it may encourage growth and production [88,89].

The capacity of P1 to enhance the vegetative attributes of sunflower seedlings stressed by Cr^{+6} implied that chromate detoxification and limited absorption are not the predominant phytostimulation process. Moreover, increased storage of enzymatic antioxidants (CAT, APX, POD, and SOD) along with reduced glutathione (nonenzymatic) may have scavenged the generated ROS in plants more efficiently to support their defensive mechanisms [90]. Evidence of improved ROS scavenging ability of rhizobacterium associated sunflower seedlings was linked to the effective DPPH quenching potential and reduced ROS accumulation. In plants, oxidative pressure can be recognized by enhanced production of ROS under stress conditions due to lower production of several antioxidant enzymes and higher malonaldehyde concentration [91,92] causing membrane disruption and hence leakage of essential electrolytes [93]. Therefore, the rhizobacterial isolate P1 in the current scenario has supported the *H. annuus* in releasing significant amounts of enzymatic and nonenzymatic antioxidants to detoxify the accumulated ROS and cease the MDA production thus, preventing electrolyte leakage in stress conditions. Within unhealthy environments, the activation of these enzymes scavenged the ROS, thereby enabling the plant to grow healthy. The bacterial strain either contributes to, or triggers, the host to produce substantial quantities of antioxidants. Liu et al. [94] unveiled the RNA-seq data that FZB42 (a PGPR) triggered the overexpression of photosynthesis-related genes, improved ROS scavenging machinery, osmoprotectants (trehalose and proline), Na1 translocation as well as jasmonic acid, auxin, and ethylene signaling in salt stress conditions. In another experiment, Bharti et al. [95] presented that *Dietzia natronolimnaea* STR1 inoculated salt-stressed wheat plants expressed the *TaWRKY10* gene. The *WRKY* TFs have indeed a significant role in stress mitigation by adjusting the osmotic balance of cells, ROS detoxification processes, and regulation of stress-related genes. The enhanced activities of defense-related enzymes contributed to the bioprotection of plants against various biotic and abiotic stress factors. This enhanced antioxidant production induces systematic resistance in the host plants.

In addition to the generation of antioxidants, colonizing roots of the host, and minimizing metal uptake, the rhizobacteria P1 often releases proline on its own to facilitate the host proline production and cope with elevated levels of metal stress [96]. Also, the strain ably converted the Cr from an extremely toxic form (Cr^{+6}) to a relatively less toxic form (Cr^{+3}), thus decreasing its entrance into the food chain and protecting the humans from the toxic effects of chromate.

5. Conclusions

From the study, it is concluded that the use of plant growth-promoting rhizobacteria P1 is a sustainable way to improve host growth under Cr^{+6} stress. The stress-induced growth reduction is accompanied by a decline in the endogenous pool of phytohormones and a compromised antioxidant system reflected by enhanced accumulation of ROS. However, the rhizobacterium *A. bouvetii* P1 association with sunflower seeds under Cr^{+6} stress enables the host to withstand up to 1200 µg/mL of Cr^{+6} without compromising its growth. Such seedlings have enough phytohormones and improved antioxidants (enzymatic and nonenzymatic) to scavenge the ROS generated during Cr^{+6} stress. Mobilization of nutrients, such as phosphate by the associated rhizobacterium, also supports the plant's better performance. Furthermore, P1 reduces Cr toxicity by making it unavailable to the plants, and in doing so, P1 removes the Cr^{+6} from the food chain, protecting humans from the toxic effects of Cr^{+6}. With all this in mind, *A. bouvetii* can be used as a potent strain in field trials instead of synthetic fertilizers that subsequently lead to soil deterioration.

Supplementary Materials: The following are available online at https://www.mdpi.com/article/10.3390/antiox10121868/s1, Figure S1: Effects of different chromium levels of bacterial growth. Data represent mean with SE ± and letter represent significance ($p < 0.05$); Table S1: Effect of Cr^{+6} and P1 on indole acetic acid contents, total flavonoids, total phenolics, and Proline contents in plant biomass (PB) and their exudation by root (E/g of Roots).

Author Contributions: M.Q.: Data curation, Formal analysis, Investigation, Methodology, Writing—original draft. A.H.: Supervision, Methodology, Writing-original draft. M.S.: Supervision, M.H.: Resources, Supervision. A.I.: Writing—review & editing, M.I.: Lab facility, A.A.: ROS determination M.A.L.: Resources for biochemical parameters, I.-J.L.: Resources. All authors have read and agreed to the published version of the manuscript.

Funding: This research was jointly supported by the Basic Science Research Program through the National Research Foundation of Korea (NRF), funded by the Ministry of Education (2017R1D1A1B04035601) and Abdul Wali Khan University Mardan.

Institutional Review Board Statement: Not applicable.

Informed Consent Statement: Not applicable.

Data Availability Statement: All the data are included in the manuscript.

Acknowledgments: This research was supported by the Abdul Wali Khan University Mardan.

Conflicts of Interest: The authors declare no conflict of interest.

References

1. Clemens, S.; Ma, J.F. Toxic heavy metal and metalloid accumulation in crop plants and foods. *Annu. Rev. Plant Biol.* **2016**, *67*, 489–512. [CrossRef]
2. Topalidis, V.; Harris, A.; Hardaway, C.J.; Benipal, G.; Douvris, C. Investigation of selected metals in soil samples exposed to agricultural and automobile activities in Macedonia, Greece using inductively coupled plasma-optical emission spectrometry. *Microchem. J.* **2017**, *130*, 213–220. [CrossRef]
3. Wang, S.; Shi, X. Molecular mechanisms of metal toxicity and carcinogenesis. *Mol. Cell. Biochem.* **2001**, *222*, 3–9. [CrossRef]
4. Beyersmann, D.; Hartwig, A. Carcinogenic metal compounds: Recent insight into molecular and cellular mechanisms. *Arch. Toxicol.* **2008**, *82*, 493. [CrossRef]
5. Chang, L.W.; Magos, L.; Suzuki, T. *Toxicology of Metals*; Taylor & Francis: Boca Raton, FL, USA, 1996.
6. Tchounwou, P.B.; Yedjou, C.G.; Patlolla, A.K.; Sutton, D.J. Heavy metal toxicity and the environment. In *Molecular, Clinical and Environmental Toxicology*; Springer: New York, NY, USA, 2012; pp. 133–164.
7. Bayramoğlu, G.; Çelik, G.; Yalçın, E.; Yılmaz, M.; Arıca, M.Y. Modification of surface properties of Lentinus sajor-caju mycelia by physical and chemical methods: Evaluation of their Cr6+ removal efficiencies from aqueous medium. *J. Hazard. Mater.* **2005**, *119*, 219–229. [CrossRef]
8. Jomova, K.; Valko, M.J.T. Advances in metal-induced oxidative stress and human disease. *Toxicology* **2011**, *283*, 65–87. [CrossRef]
9. Valko, M.; Morris, H.; Cronin, M.T.D. Metals, toxicity and oxidative stress. *Curr. Med. Chem.* **2005**, *12*, 1161–1208. [CrossRef]
10. Mertz, W. Chromium in human nutrition: A review. *J. Nutr.* **1993**, *123*, 626–633. [CrossRef]
11. Vincent, J.B. New evidence against chromium as an essential trace element. *J. Nutr.* **2017**, *147*, 2212–2219. [CrossRef]
12. Crittenden, J.C.; Trussell, R.R.; Hand, D.W.; Howe, K.; Tchobanoglous, G. *MWH's Water Treatment: Principles and Design*; John Wiley & Sons: Hoboken, NJ, USA, 2012.
13. Agoro, M.A.; Adeniji, A.O.; Adefisoye, M.A.; Okoh, O.O.J.W. Heavy metals in wastewater and sewage sludge from selected municipal treatment plants in eastern cape province, south africa. *Water* **2020**, *12*, 2746. [CrossRef]
14. Rahman, Z.; Singh, V.P.J.E.S.; Research, P. Bioremediation of toxic heavy metals (THMs) contaminated sites: Concepts, applications and challenges. *Environ. Sci. Pollut. Res.* **2020**, *27*, 27563–27581. [CrossRef]
15. Santana Martinez, J.C. The Impact of Reclamation and Vegetation Removal on Compositional and Functional Attributes of Soil Microbial Communities in the Athabasca Oil Sands Region. Master's Thesis, University of Alberta, Edmonton, AB, Canada, 2021.
16. Reyed, R.M. Biotherapeutic Approaches: Bioremediation of Industrial Heavy Metals from Ecosphere. In *Rhizobiont in Bioremediation of Hazardous Waste*; Springer: New York, NY, USA, 2021; pp. 565–592.
17. Ojuederie, O.B.; Babalola, O.O. Microbial and plant-assisted bioremediation of heavy metal polluted environments: A review. *Int. J. Environ. Res. Public Health* **2017**, *14*, 1504. [CrossRef]
18. Tekere, M. Biological Strategies for Heavy Metal Remediation. In *Methods for Bioremediation of Water and Wastewater Pollution*; Springer: New York, NY, USA, 2020; pp. 393–413.
19. Spain, O.; Plöhn, M.; Funk, C. The cell wall of green microalgae and its role in heavy metal removal. *Physiol. Plant.* **2021**, *173*, 526–535. [CrossRef]
20. Gupta, P.; Rani, R.; Chandra, A.; Varjani, S.J.; Kumar, V. Effectiveness of plant growth-promoting Rhizobacteria in phytoremediation of chromium stressed soils. In *Waste Bioremediation*; Springer: New York, NY, USA, 2018; pp. 301–312.
21. Mitra, S.; Pramanik, K.; Ghosh, P.K.; Soren, T.; Sarkar, A.; Dey, R.S.; Pandey, S.; Maiti, T.K. Characterization of Cd-resistant Klebsiella michiganensis MCC3089 and its potential for rice seedling growth promotion under Cd stress. *Microbiol. Res.* **2018**, *210*, 12–25. [CrossRef]
22. Ke, T.; Guo, G.; Liu, J.; Zhang, C.; Tao, Y.; Wang, P.; Xu, Y.; Chen, L. Improvement of the Cu and Cd phytostabilization efficiency of perennial ryegrass through the inoculation of three metal-resistant PGPR strains. *Environ. Pollut.* **2021**, *271*, 116314. [CrossRef]
23. Afsal, F.; Majumdar, A.; Kumar, J.S.; Bose, S. Microbial Inoculation to Alleviate the Metal Toxicity in Crop Plants and Subsequent Growth Promotion. In *Sustainable Solutions for Elemental Deficiency and Excess in Crop Plants*; Springer: New York, NY, USA, 2020; pp. 451–479.
24. Pandey, N.; Chandrakar, V.; Keshavkant, S. Mitigating arsenic toxicity in plants: Role of microbiota. In *Mechanisms of Arsenic Toxicity and Tolerance in Plants*; Springer: New York, NY, USA, 2018; pp. 191–218.
25. Prasad, S.; Yadav, K.K.; Kumar, S.; Gupta, N.; Cabral-Pinto, M.M.; Rezania, S.; Radwan, N.; Alam, J. Chromium contamination and effect on environmental health and its remediation: A sustainable approaches. *J. Environ. Manag.* **2021**, *285*, 112174. [CrossRef]
26. Sarma, H.; Prasad, M.N.V. Metabolic engineering of rhizobacteria associated with plants for remediation of toxic metals and metalloids. In *Transgenic Plant Technology for Remediation of Toxic Metals and Metalloids*; Elsevier: New York, NY, USA, 2019; pp. 299–318.
27. Tirry, N.; Kouchou, A.; El Omari, B.; Ferioun, M.; El Ghachtouli, N. Improved chromium tolerance of Medicago sativa by plant growth-promoting rhizobacteria (PGPR). *J. Genet. Eng. Biotechnol.* **2021**, *19*, 1–14. [CrossRef]

28. Singh, B.P. Screening and characterization of plant growth promoting rhizobacteria (PGPR): An overview. *Bull. Environ. Sci. Res.* **2015**, *4*, 1–2.
29. Qadir, M.; Hussain, A.; Shah, M.; Lee, I.J.; Iqbal, A.; Irshad, M.; Ismail; Sayyed, A.; Husna; Ahmad, A.; et al. Comparative assessment of chromate bioremediation potential of Pantoea conspicua and Aspergillus niger. *J. Hazard. Mater.* **2022**, *424*, 127314. [CrossRef]
30. Oves, M.; Khan, M.S.; Zaidi, A. Chromium reducing and plant growth promoting novel strain Pseudomonas aeruginosa OSG41 enhance chickpea growth in chromium amended soils. *Eur. J. Soil Biol.* **2013**, *56*, 72–83. [CrossRef]
31. Danish, S.; Kiran, S.; Fahad, S.; Ahmad, N.; Ali, M.A.; Tahir, F.A.; Rasheed, M.K.; Shahzad, K.; Li, X.; Wang, D.; et al. Alleviation of chromium toxicity in maize by Fe fortification and chromium tolerant ACC deaminase producing plant growth promoting rhizobacteria. *Ecotoxicol. Environ. Saf.* **2019**, *185*, 109706. [CrossRef]
32. Oshundiya, F.O.; Olowe, V.; Sowemimo, F.; Odedina, J.N. Seed yield and quality of sunflower (*Helianthus annuus* L.) as influenced by staggered sowing and organic fertilizer application in the humid tropics. *Helia* **2014**, *37*, 237–255. [CrossRef]
33. Qadir, M.; Hussain, A.; Hamayun, M.; Shah, M.; Iqbal, A.; Murad, W. Phytohormones producing rhizobacterium alleviates chromium toxicity in Helianthus annuus L. by reducing chromate uptake and strengthening antioxidant system. *Chemosphere* **2020**, *258*, 127386. [CrossRef]
34. Rokhbakhsh-Zamin, F.; Sachdev, D.; Kazemi-Pour, N.; Engineer, A.; Pardesi, K.R.; Zinjarde, S.; Dhakephalkar, P.K.; Chopade, B.A. Characterization of plant-growth-promoting traits of Acinetobacter species isolated from rhizosphere of Pennisetum glaucum. *Microbiol. Biotechnol.* **2011**, *21*, 556–566. [CrossRef]
35. Iman, M. Effect of phosphate solubilizing fungi on growth and nutrient uptake of soyabean (*Glycine max* L.) plants. *J. Appl. Sci. Res.* **2008**, *4*, 592–598.
36. Hussain, A.; Hasnain, S. Interactions of bacterial cytokinins and IAA in the rhizosphere may alter phytostimulatory efficiency of rhizobacteria. *World J. Microbiol. Biotechnol.* **2011**, *27*, 2645. [CrossRef]
37. Coombe, B.; Cohen, D.; Paleg, L.G. Barley endosperm bioassay for gibberellins. I. Parameters of the response system. *Plant Physiol.* **1967**, *42*, 105–112. [CrossRef]
38. Warrier, R.; Paul, M.; Vineetha, M. Estimation of salicylic acid in Eucalyptus leaves using spectrophotometric methods. *Genet. Plant Physiol.* **2013**, *3*, 90–97.
39. El Far, M.; Taie, H.A. Antioxidant activities, total anthocyanins, phenolics and flavonoids contents of some sweetpotato genotypes under stress of different concentrations of sucrose and sorbitol. *Aust. J. Basic Appl. Sci.* **2009**, *3*, 3609–3616.
40. Prabhavathi, R.; Prasad, M.; Jayaramu, M. Studies on qualitative and quantitative phytochemical analysis of Cissus quadrangularis. *Adv. Appl. Sci. Res.* **2016**, *7*, 11–17.
41. Bates, L.S.; Waldren, R.P.; Teare, I. Rapid determination of free proline for water-stress studies. *Plant Soil* **1973**, *39*, 205–207. [CrossRef]
42. Hoffmann, W.A.; Poorter, H. Avoiding bias in calculations of relative growth rate. *Ann. Bot.* **2002**, *90*, 37–42. [CrossRef]
43. Sudhakar, P.; Latha, P.; Reddy, P.V. Chapter 4—Photosynthetic rates. In *Phenotyping Crop Plants for Physiological and Biochemical Traits*; Sudhakar, P., Latha, P., Reddy, P.V., Eds.; Academic Press: Cambridge, MA, USA, 2016; pp. 33–39. [CrossRef]
44. Malik, C.P.; Singh, M. *Plant Enzymology and Histo-Enzymology*; Kalyani Publishers: New Delhi, India, 1980.
45. Schmedes, A.; Hølmer, G. A new thiobarbituric acid (TBA) method for determining free malondialdehyde (MDA) and hydroperoxides selectively as a measure of lipid peroxidation. *J. Am. Oil Chem. Soc.* **1989**, *66*, 813–817. [CrossRef]
46. Lutts, S.; Kinet, J.; Bouharmont, J. NaCl-induced senescence in leaves of rice (*Oryza sativa* L.) cultivars differing in salinity resistance. *Ann. Bot.* **1996**, *78*, 389–398. [CrossRef]
47. Moreira-Vilar, F.C.; de Cássia Siqueira-Soares, R.; Finger-Teixeira, A.; de Oliveira, D.M.; Ferro, A.P.; da Rocha, G.J.; Maria de Lourdes, L.F.; dos Santos, W.D.; Ferrarese-Filho, O. The acetyl bromide method is faster, simpler and presents best recovery of lignin in different herbaceous tissues than Klason and thioglycolic acid methods. *PLoS ONE* **2014**, *9*, e110000. [CrossRef] [PubMed]
48. O'Brien, J.A.; Daudi, A.; Butt, V.S.; Bolwell, G.P. Reactive oxygen species and their role in plant defence and cell wall metabolism. *Planta* **2012**, *236*, 765–779. [CrossRef]
49. Ahmad, N.; Fazal, H.; Ahmad, I.; Abbasi, B.H. Free radical scavenging (DPPH) potential in nine Mentha species. *J. Toxicol. Ind. Health* **2012**, *28*, 83–89. [CrossRef] [PubMed]
50. Radhakrishnan, R.; Lee, I.-J. Ameliorative effects of spermine against osmotic stress through antioxidants and abscisic acid changes in soybean pods and seeds. *Acta Physiol. Plant.* **2013**, *35*, 263–269. [CrossRef]
51. Asada, K. Ascorbate peroxidase—A hydrogen peroxide-scavenging enzyme in plants. *Physiol. Plant.* **1992**, *85*, 235–241. [CrossRef]
52. Janknegt, P.J.; Rijstenbil, J.W.; Van de Poll, W.H.; Gechev, T.S.; Buma, A.G.J. A comparison of quantitative and qualitative superoxide dismutase assays for application to low temperature microalgae. *J. Photochem. Photobiol. B Biol.* **2007**, *87*, 218–226. [CrossRef]
53. Mavis, R.D.; Stellwagen, E. Purification and subunit structure of glutathione reductase from bakers' yeast. *J. Biol. Chem.* **1968**, *243*, 809–814. [CrossRef]
54. Carlberg, I.; Mannervik, B. Purification and characterization of the flavoenzyme glutathione reductase from rat liver. *J. Biol. Chem.* **1975**, *250*, 5475–5480. [CrossRef]
55. Zahoor, M.; Irshad, M.; Rahman, H.; Qasim, M.; Afridi, S.G.; Qadir, M.; Hussain, A. Alleviation of heavy metal toxicity and phytostimulation of Brassica campestris L. by endophytic Mucor sp. MHR-7. *Ecotoxicol. Environ. Saf.* **2017**, *142*, 139–149. [CrossRef] [PubMed]

56. Kazi, T.; Jamali, M.; Kazi, G.; Arain, M.; Afridi, H.; Siddiqui, A.J.A. Evaluating the mobility of toxic metals in untreated industrial wastewater sludge using a BCR sequential extraction procedure and a leaching test. *Anal. Bioanal. Chem.* **2005**, *383*, 297–304. [CrossRef]

57. WHO. *Chromium in Drinking-Water*; World Health Organization: Geneva, Switzerland, 2020.

58. Ismaila, A.H.; Qadira, M.; Husnaa, M.I.; Ahmadb, A.; Hamayuna, M. Endophytic Fungi Isolated from Citrullus Colocynthesl. Leaves and Their Potential for Secretion of Indole Acetic Acid and Gibberellin. *J. Appl. Environ. Biol. Sci.* **2018**, *8*, 80–84.

59. Lugtenberg, B.J.; Malfanova, N.; Kamilova, F.; Berg, G. Plant growth promotion by microbes. *Mol. Microb. Ecol. Rhizosphere* **2013**, *2*, 561–573.

60. Pál, M.; Szalai, G.; Kovács, V.; Gondor, O.; Janda, T. Salicylic acid-mediated abiotic stress tolerance. In *Salicylic Acid*; Springer: New York, NY, USA, 2013; pp. 183–247.

61. Luo, J.; Xia, W.; Cao, P.; Xiao, Z.a.; Zhang, Y.; Liu, M.; Zhan, C.; Wang, N. Integrated transcriptome analysis reveals plant hormones jasmonic acid and salicylic acid coordinate growth and defense responses upon fungal infection in poplar. *Biomolecules* **2019**, *9*, 12. [CrossRef]

62. He, Y.; Li, W.; Lv, J.; Jia, Y.; Wang, M.; Xia, G. Ectopic expression of a wheat MYB transcription factor gene, TaMYB73, improves salinity stress tolerance in Arabidopsis thaliana. *J. Exp. Bot.* **2012**, *63*, 1511–1522.

63. Hamayun, M.; Khan, N.; Khan, M.N.; Qadir, M.; Hussain, A.; Iqbal, A.; Khan, S.A.; Rehman, K.U.; Lee, I.-J. Antimicrobial and plant growth-promoting activities of bacterial endophytes isolated from Calotropis procera (Ait.) WT Aiton. *Biocell* **2021**, *45* 363–369. [CrossRef]

64. Rolli, E.; Vergani, L.; Ghitti, E.; Patania, G.; Mapelli, F.; Borin, S. "Cry-for-help" in contaminated soil: A dialogue among plants and soil microbiome to survive in hostile conditions. *Environ. Microbiol.* **2021**, *23*, 5690–5703.

65. Hassan, S.; Mathesius, U. The role of flavonoids in root–rhizosphere signalling: Opportunities and challenges for improving plant–microbe interactions. *J. Exp. Bot.* **2012**, *63*, 3429–3444. [CrossRef]

66. Hayat, S.; Hayat, Q.; Alyemeni, M.N.; Wani, A.S.; Pichtel, J.; Ahmad, A. Role of proline under changing environments: A review. *Plant Signal. Behav.* **2012**, *7*, 1456–1466. [CrossRef]

67. Michalak, A. Phenolic compounds and their antioxidant activity in plants growing under heavy metal stress. *Pol. J. Environ. Stud.* **2006**, *15*, 523–530.

68. Mishra, J.; Fatima, T.; Arora, N.K. Role of secondary metabolites from plant growth-promoting rhizobacteria in combating salinity stress. In *Plant Microbiome: Stress Response*; Springer: New York, NY, USA, 2018; pp. 127–163.

69. Bahadur, A.; Ahmad, R.; Afzal, A.; Feng, H.; Suthar, V.; Batool, A.; Khan, A.; Mahmood-ul-Hassan, M.J.C. The influences of Cr-tolerant rhizobacteria in phytoremediation and attenuation of Cr (VI) stress in agronomic sunflower (*Helianthus annuus* L.). *Chemosphere* **2017**, *179*, 112–119. [CrossRef]

70. Zaheer, M.S.; Raza, M.A.S.; Saleem, M.F.; Khan, I.H.; Ahmad, S.; Iqbal, R.; Manevski, K. Investigating the effect of Azospirillum brasilense and Rhizobium pisi on agronomic traits of wheat (*Triticum aestivum* L.). *Arch. Agron. Soil Sci.* **2019**, *65*, 1554–1564. [CrossRef]

71. Karthik, C.; Elangovan, N.; Kumar, T.S.; Govindharaju, S.; Barathi, S.; Oves, M.; Arulselvi, P.I. Characterization of multifarious plant growth promoting traits of rhizobacterial strain AR6 under Chromium (VI) stress. *Microbiol. Res.* **2017**, *204*, 65–71. [CrossRef]

72. Bücker-Neto, L.; Paiva, A.L.S.; Machado, R.D.; Arenhart, R.A.; Margis-Pinheiro, M. Interactions between plant hormones and heavy metals responses. *Genet. Mol. Biol.* **2017**, *40*, 373–386. [CrossRef] [PubMed]

73. Baskar, V.; Venkatesh, R.; Ramalingam, S. Flavonoids (antioxidants systems) in higher plants and their response to stresses. In *Antioxidants and Antioxidant Enzymes in Higher Plants*; Springer: New York, NY, USA, 2018; pp. 253–268.

74. Mukhopadyay, M.; Bantawa, P.; Das, A.; Sarkar, B.; Bera, B.; Ghosh, P.; Mondal, T.K. Changes of growth, photosynthesis and alteration of leaf antioxidative defence system of tea [*Camellia sinensis* (L.) O. Kuntze] seedlings under aluminum stress. *Biometals* **2012**, *25*, 1141–1154. [CrossRef] [PubMed]

75. Bartwal, A.; Mall, R.; Lohani, P.; Guru, S.; Arora, S. Role of secondary metabolites and brassinosteroids in plant defense against environmental stresses. *J. Plant Growth Regul.* **2013**, *32*, 216–232. [CrossRef]

76. Liang, X.; Zhang, L.; Natarajan, S.K.; Becker, D.F. Proline mechanisms of stress survival. *Antioxid. Redox Signal.* **2013**, *19*, 998–1011. [CrossRef]

77. Sharma, S.S.; Schat, H.; Vooijs, R. In Vitro alleviation of heavy metal-induced enzyme inhibition by proline. *Phytochemistry* **1998**, *49*, 1531–1535. [CrossRef]

78. Finger-Teixeira, A.; de Lourdes Lucio Ferrarese, M.; Soares, A.R.; da Silva, D.; Ferrarese-Filho, O. Cadmium-induced lignification restricts soybean root growth. *Ecotoxicol. Environ. Saf.* **2010**, *73*, 1959–1964. [CrossRef] [PubMed]

79. Parrotta, L.; Guerriero, G.; Sergeant, K.; Cai, G.; Hausman, J.-F. Target or barrier? The cell wall of early-and later-diverging plants vs cadmium toxicity: Differences in the response mechanisms. *Front. Plant Sci.* **2015**, *6*, 133. [CrossRef] [PubMed]

80. Lux, A. Does diversity in root structure affect the diversity in cadmium uptake by plants? Opinion paper. *Agrochimica* **2010**, *54*, 342–352.

81. Ashraf, M.; Ozturk, M.; Ahmad, M.S.A. *Plant Adaptation and Phytoremediation*; Springer: New York, NY, USA, 2010.

82. Jha, Y.; Subramanian, R. Effect of root-associated bacteria on soluble sugar metabolism in plant under environmental stress. In *Plant Metabolites and Regulation Under Environmental Stress*; Elsevier: Amsterdam, The Netherlands, 2018; pp. 231–240.

83. Hussain, A.; Shah, M.; Hamayun, M.; Qadir, M.; Iqbal, A. Heavy metal tolerant endophytic fungiAspergillus welwitschiaeimproves growth, ceasing metal uptake and strengthening antioxidant system in *Glycine max* L. *Environ. Sci. Pollut. Res.* **2021**. [CrossRef]
84. Kalidhasan, S.; Ganesh, M.; Sricharan, S.; Rajesh, N. Extractive separation and determination of chromium in tannery effluents and electroplating waste water using tribenzylamine as the extractant. *J. Hazard. Mater.* **2009**, *165*, 886–892. [CrossRef]
85. Kazi, T.G.; Afridi, H.I.; Kazi, N.; Jamali, M.K.; Arain, M.B.; Jalbani, N.; Kandhro, G.A. Copper, chromium, manganese, iron, nickel, and zinc levels in biological samples of diabetes mellitus patients. *Biol. Trace Elem. Res.* **2008**, *122*, 1–18. [CrossRef]
86. Sun, P.; Liu, Z.-T.; Liu, Z.-W. Chemically modified chicken feather as sorbent for removing toxic chromium (VI) ions. *Ind. Eng. Chem. Res.* **2009**, *48*, 6882–6889. [CrossRef]
87. Uluozlu, O.D.; Tuzen, M.; Soylak, M. Speciation and separation of Cr (VI) and Cr (III) using coprecipitation with Ni2+/2-Nitroso-1-naphthol-4-sulfonic acid and determination by FAAS in water and food samples. *Food Chem. Toxicol.* **2009**, *47*, 2601–2605. [CrossRef]
88. Paiva, L.B.; de Oliveira, J.G.; Azevedo, R.A.; Ribeiro, D.R.; da Silva, M.G.; Vitória, A.P. Ecophysiological responses of water hyacinth exposed to Cr3+ and Cr6+. *Environ. Exp. Bot.* **2009**, *65*, 403–409. [CrossRef]
89. Peralta-Videa, J.R.; Lopez, M.L.; Narayan, M.; Saupe, G.; Gardea-Torresdey, J. The biochemistry of environmental heavy metal uptake by plants: Implications for the food chain. *Int. J. Biochem. Cell Biol.* **2009**, *41*, 1665–1677. [CrossRef] [PubMed]
90. Husna, H.; Hussain, A.; Shah, M.; Hamayun, M.; Iqbal, A.; Murad, W.; Irshad, M.; Qadir, M.; Kim, H.-Y. Pseudocitrobacter Anthropi Reduces Heavy Metal Uptake and Improves Phytohormones and Antioxidant System in Glycine Max L. *World J. Microbiol. Biotechnol.* **2021**, *37*, 195. [CrossRef] [PubMed]
91. Anjum, N.A.; Gill, S.S.; Duarte, A.C.; Pereira, E. Oxidative stress biomarkers and antioxidant defense in plants exposed to metallic nanoparticles. In *Nanomaterials and Plant Potential*; Springer: New York, NY, USA, 2019; pp. 427–439.
92. Panda, S.; Choudhury, S. Chromium stress in plants. *Braz. J. Plant Physiol.* **2005**, *17*, 95–102. [CrossRef]
93. Demidchik, V.; Straltsova, D.; Medvedev, S.S.; Pozhvanov, G.A.; Sokolik, A.; Yurin, V. Stress-induced electrolyte leakage: The role of K+-permeable channels and involvement in programmed cell death and metabolic adjustment. *J. Exp. Bot.* **2014**, *65*, 1259–1270. [CrossRef]
94. Liu, S.; Hao, H.; Lu, X.; Zhao, X.; Wang, Y.; Zhang, Y.; Xie, Z.; Wang, R. Transcriptome profiling of genes involved in induced systemic salt tolerance conferred by Bacillus amyloliquefaciens FZB42 in *Arabidopsis thaliana*. *Sci. Rep.* **2017**, *7*, 1–13. [CrossRef] [PubMed]
95. Bharti, N.; Pandey, S.S.; Barnawal, D.; Patel, V.K.; Kalra, A. Plant growth promoting rhizobacteria Dietzia natronolimnaea modulates the expression of stress responsive genes providing protection of wheat from salinity stress. *Sci. Rep.* **2016**, *6*, 1–16. [CrossRef]
96. Ikram, M.; Ali, N.; Jan, G.; Jan, F.G.; Rahman, I.U.; Iqbal, A.; Hamayun, M. IAA producing fungal endophyte Penicillium roqueforti Thom., enhances stress tolerance and nutrients uptake in wheat plants grown on heavy metal contaminated soils. *PLoS ONE* **2018**, *13*, e0208150. [CrossRef]

 *antioxidants*

Article

Methyl Jasmonate Protects the PS II System by Maintaining the Stability of Chloroplast D1 Protein and Accelerating Enzymatic Antioxidants in Heat-Stressed Wheat Plants

Mehar Fatma [1], Noushina Iqbal [2], Zebus Sehar [1], Mohammed Nasser Alyemeni [3], Prashant Kaushik [4], Nafees A. Khan [1,*] and Parvaiz Ahmad [3,*]

[1] Plant Physiology and Biochemistry Laboratory, Department of Botany, Aligarh Muslim University, Aligarh 202002, India; meharfatma30@gmail.com (M.F.); seharzebus5779@gmail.com (Z.S.)

[2] Department of Botany, School of Chemical and Life Sciences, Jamia Hamdard, New Delhi 110062, India; naushina.iqbal@gmail.com

[3] Botany and Microbiology Department, College of Science, King Saud University, Riyadh 11451, Saudi Arabia; mnalyemeni@gmail.com

[4] Kikugawa Research Station, Yokohama Ueki, 2265, Kamo, Kikugawa City, Shizuoka 439-0031, Japan; kaushik.prashant@yokohamaueki.co.jp

* Correspondence: na.khan.bt@amu.ac.in or naf9.amu@gmail.com (N.A.K.); Pahmad@ksu.edu.sa or parvaizbot@yahoo.com (P.A.)

Citation: Fatma, M.; Iqbal, N.; Sehar, Z.; Alyemeni, M.N.; Kaushik, P.; Khan, N.A.; Ahmad, P. Methyl Jasmonate Protects the PS II System by Maintaining the Stability of Chloroplast D1 Protein and Accelerating Enzymatic Antioxidants in Heat-Stressed Wheat Plants. *Antioxidants* **2021**, *10*, 1216. https://doi.org/10.3390/antiox10081216

Academic Editor: Maria Concetta de Pinto

Received: 14 June 2021
Accepted: 19 July 2021
Published: 28 July 2021

Abstract: The application of 10 μM methyl jasmonate (MeJA) for the protection of wheat (*Triticum aestivum* L.) photosystem II (PS II) against heat stress (HS) was studied. Heat stress was induced at 42 °C to established plants, which were then recovered at 25 °C and monitored during their growth for the study duration. Application of MeJA resulted in increased enzymatic antioxidant activity that reduced the content of hydrogen peroxide (H_2O_2) and thiobarbituric acid reactive substances (TBARS) and enhanced the photosynthetic efficiency. Exogenous MeJA had a beneficial effect on chlorophyll fluorescence under HS and enhanced the pigment system (PS) II system, as observed in a JIP-test, a new tool for chlorophyll fluorescence induction curve. Exogenous MeJA improved the quantum yield of electron transport (ET_o/CS) as well as electron transport flux for each reaction center (ET_0/RC). However, the specific energy fluxes per reaction center (RC), i.e., TR_0/RC (trapping) and DI_0/RC (dissipation), were reduced by MeJA. These results indicate that MeJA affects the efficiency of PS II by stabilizing the D1 protein, increasing its abundance, and enhancing the expression of the *psb*A and *psb*B genes under HS, which encode proteins of the PS II core RC complex. Thus, MeJA is a potential tool to protect PS II and D1 protein in wheat plants under HS and to accelerate the recovery of the photosynthetic capacity.

Keywords: antioxidant; methyl jasmonate; photosystem

1. Introduction

Wheat is the main cereal crop belonging to the Poaceae family, which contributes 30% and 50% to the world grain production and grain trade, respectively [1]. Its quality and yield are directly associated with countrywide food security [2]. However, wheat plants often suffer heat stress, which affects their quality and utilization worldwide. India and China alone are expected to show a decrease in wheat yield by 8% and 3%, respectively, due to the rise in worldwide mean temperature by 1 °C. The impact of high temperature on the growth of plants is due to the excess output of reactive oxygen species (ROS) that damage the photosynthetic apparatus of plants by decreasing the rate of photosynthetic electron transport, inactivating the pigment system (PS) II center, degrading pigments and proteins, and eventually decreasing the yield [3]. ROS produced in chloroplasts as a result of abiotic stress hinder the synthesis of D1 protein, a component of the PS II complex [4]. D1 is the main subunit of the complex which not only regulates the binding of cofactors, but also maintains the structure of the reaction center of PS II, engaged in essential charge division and electron transport [5,6]. The D1 protein encoded by the chloroplast gene *psbA* is the main target of damage under the environmental stress conditions. It was reported that quick repair of D1 protein is essential for the efficient recovery of PS II [2,7]. Plants contain established restoration mechanisms to check damage to PS II, but little is known about the mechanisms of protection of the PS II system under heat stress.

The adaptive reaction of plants to oxidative stress consists in the rise of the activity of antioxidants enzymes [8–10]. Remarkably, several studies reported the association between antioxidant activity, heat tolerance, and plant hormones [3,11–13]. Among plant hormones, jasmonates promotes protection from abiotic stresses. In particular, methyl jasmonate (MeJA) has shown promising effects in the defense of plants under diverse abiotic stresses [8,14,15]. Previous studies established that MeJA alleviated oxidative stress by improving antioxidant enzymes activities [8,16]. Exogenous application of MeJA protected *Arabidopsis thaliana* and *Citrus reticulata* × *Citrus sinensis* plants against the inhibitory effect on chlorophyll (Chl) induced by copper and cadmium and promoted photosynthetic activity [17,18]. Attaran et al. [19] reported that RNA sequencing and chlorophyll fluorescence imaging were important in studying jasmonic acid effect on photosynthetic capacity, growth, and gene expression in *Arabidopsis*. It has been reported that D1 protein turnover is more rapid than that of any other protein in the thylakoid membrane under light illumination [2]. This feature exposes PS II to photoinduced damage, causing photoinhibition and reduction in photosynthetic efficiency. Henceforth, D1 protein degradation, synthesis, and reinsertion into the PS II core complex represent an important aspect of PS II dynamics [5,20].

The regulatory mechanisms by which MeJA protects the PS II system, particularly through stabilization of D1 protein and regulation of relevant gene expression in wheat leaves under heat stress have not been investigated. Therefore, the purpose of the present study was to examine the effect of MeJA on the protection of the PS II complex, chlorophyll fluorescence, activity of enzymatic antioxidants, D1 protein abundance, and photosynthetic gene expression in wheat leaves under heat stress. Additionally, this study also investigated how MeJA affected plant growth and CO_2 assimilation in the leaves of heat-stressed wheat plants. Our findings demonstrated that MeJA application can protect the PS II complex from heat stress-induced damage by increasing the levels of enzymatic antioxidants and accelerating the stability of D1 protein in wheat leaves.

2. Materials and Methods

Wheat (*Triticum aestivum* L.) cultivar HD 2329 (winter wheat) seeds were collected from the Indian Agricultural Research Institute (IARI), New Delhi, surface-sterilized with 0.01% $HgCl_2$ followed by 3–4 frequent washings with deionized water, and then seeded in 14 cm diameter pots (14 cm diagonally to the top and 18 cm toward the bottom). The pots were filled with 4 kg sterilized acid-washed fine sand with a particle size from 125 µm to 250 µm and pH of about 7.0. The sand was first purified with a mixture of 17% w/v hydrochloric acid and 1% oxalic acid in a sand digester (electrode boiler) for 6 h. After that, the sand was washed with deionized water before being transferred to the pots. The purification of sand was performed by the following the method of Hewitt et al. [21]. A plant growth chamber (Khera KI-261, Khera Instruments Pvt. Ltd., New Delhi, India) set with day/night temperatures of $25/17 \pm 3$ °C, 12 h photoperiod (Photosynthetically Active Radiation; PAR; 350 µmol m^{-2} s^{-1}), and $65 \pm 5\%$ relative humidity was used. Three plants were kept in each pot and were given 300 mL Hoagland's solution (full strength) on alternate days. In the experimentation, the heat stress treatment was applied to 3–4 leaves at emergence stage by exposing the plants to 42 °C every day for 6 h for 10 DAS (days after sowing). After that, the plants were recovered at 25 °C (no stress; optimum temperature) and were grown for 5 more days. The experimentation continued for 30 days. Plants were supplied with Hoagland's nutrient solution alternatively in the morning. The control group of plants for each pot was kept for 30 days at 25 °C throughout the experimental duration.

The plants were sprayed with 5, 10, and 20 µM MeJA (Sigma Aldrich, St. Louis, MO, USA) in the absence (25 °C) or presence (42 °C) of heat stress at 15 DAS with a hand sprayer. Plants were sampled for the determination of oxidative stress, growth, and photosynthetic characteristics 30 DAS. A detailed examination on the effects of 10 µM MeJA (selected on the basis of the concentrations used in the experiment) in the alleviation of heat stress was performed. The MeJA solution and control distilled water were sprayed along with 0.5% surfactant Teepol. The randomized complete block design was adopted for the treatments, and the number of replicates for every treatment was four ($n = 4$). At the vegetative growth stage, the leaves were sampled and preserved (-80 °C) for physio-biochemical analyses, to examine antioxidant enzyme activities and photosynthetic pigments, Western blotting, and quantitative reverse transcription polymerase chain reaction (qRT-PCR) quantification. Harvesting was done 30 DAS, and it was assured that leaves were taken at a similar stage for the measurements.

2.1. Measurement of Reactive Oxygen Species Content and Lipid Peroxidation

Superoxide radicals (O_2^-) content was estimated with the method of Bu et al. [22] and Lang et al. [14] with slight changes. Fresh leaf tissues (200 mg) were treated with 1 mL hydroxylamine hydrochloride for 1 h, after which, 1 mL each of α-naphthylamine and p-aminobenzene sulfonic acid was added, and the solution was kept at 25 °C for 20 min. The absorbance of the solution was recorded at 530 nm, and the O_2^- content was calculated from a calibration curve using $NaNO_2$ as the standard. Determination of hydrogen peroxide (H_2O_2) was performed by the technique of Okuda et al. [23]. Fresh leaf tissues (200 mg) were ground in ice-cold 200 mM perchloric acid, followed by centrifugation at $1200 \times g$ for 10 min. After centrifugation, perchloric acid in the supernatant was neutralized with 4 M KOH, and the insoluble potassium perchlorate was eliminated by further centrifugation at $500 \times g$ for 3 min. The reaction mixture contained 1 mL of the eluate, 400 µL of 12.5 mM 3-(dimethylamino) benzoic acid in 0.375 M phosphate buffer (pH 6.5), 80 µL of 3-methyl-2-benzothiazoline hydrazone, and 20 µL of peroxidase (0.25 unit) in a final volume of 1.5 mL. The reaction was started by the addition of peroxidase at 25 °C, and the increase in absorbance was recorded at 590 nm.

Lipid peroxidation was estimated by assessing thiobarbituric reactive substances (TBARS), as per Dhindsa et al. [24]. Fresh leaf tissues were ground in 0.25% 2-thiobarbituric acid (TBA) in 10% trichloroacetic acid (TCA) using mortar and pestle. After heating at 95 °C for 30 min, the mixture was quickly cooled in an ice bath and centrifuged at $10,000 \times g$ for 10 min. To 1 mL aliquot of the supernatant, 4 mL 20% TCA containing 5% TBA was added. The absorbance of the supernatant was read at 532 nm and corrected for nonspecific turbidity by subtracting the absorbance of the same at 600 nm. The content of TBARS was calculated using the extinction coefficient (155 mM^{-1} cm^{-1}).

2.2. Assay of Antioxidant Enzymes Activities

The activity of enzymatic antioxidant catalase (CAT) was estimated following the procedure of Aebi [25] by monitoring the disappearance of H_2O_2 at 240 nm. The activity was calculated by using the extinction coefficient of 0.036 mM^{-1} cm^{-1}. One unit of enzyme is the amount necessary to decompose 1 µmol of H_2O_2 per min at 25 °C.

The activity of ascorbate peroxidase (APX) was calculated according to Foyer and Halliwell [26] by the decrease in absorbance of ascorbate at 290 nm. The assay mixture contained phosphate buffer (50 mM, pH 7.0), 0.1 mM EDTA, 0.5 mM ascorbate, 0.1 mM H_2O_2, and the enzyme extract. The activity of APX was calculated by using the extinction coefficient of 2.8 mM^{-1} cm^{-1}. One unit of the enzyme is the amount necessary to decompose 1 µmol of substrate per min at 25 °C.

The activity of glutathione reductase (GR) was determined according to Nakano and Asada [27] by monitoring the glutathione-dependent oxidation of nicotinamide adenine dinucleotide phosphate (NADPH) at 340 nm. The assay mixture contained phosphate buffer (25 mM, pH 7.8), 0.5 mM oxidized glutathione (GSSG), 0.2 mM NADPH, and the enzyme extract. The activity of GR was calculated by using the extinction coefficient of 6.2 mM^{-1} cm^{-1}. One unit of enzyme is the amount necessary to decompose 1 μmol of NADPH per min at 25 °C. The details of the method were described before by Fatma et al. [28–30].

Superoxide dismutase (SOD) activity was assayed by the method of Giannopolitis and Ries [31] also used by Beyer and Fridovich [32], with slight alterations, by monitoring the inhibition of photochemical reduction of nitro blue tetrazolium (NBT). Five mL of the reaction mixture containing 5.0 mM (4-(2-hydroxyethyl)-1-piperazineethanesulfonic acid; HEPES) having pH 7.6, 0.1 mM ethylene diaminetetraacetic acid (EDTA), 50 mM Na$_2$CO$_3$ (pH 10.0), 13 mM methionine, 0.025% (v/v) triton X-100, 63 μmol NBT, 1.3 μmol riboflavin, and the enzyme extract was illuminated for 15 min (360 μmol m^{-2} s^{-1}). A control set of experiments was also illuminated for correcting the background absorbance. A unit of SOD was defined as the amount of enzyme that inhibited NBT reduction by 50% at 560 nm.

2.3. Protein and Pigment Analysis

Fresh leaves (100 mg) were placed in 90% ammoniacal acetone for the determination of pigments and analyzed as described by Porra et al. [33]. The carotenoid content was calculated according to the method by Wellburn and Lichtenthaler [34], and Bradford [35] method was used for the estimation of total protein content, using bovine serum albumin as a standard.

2.4. Chlorophyll a Fluorescence Measurement

A Junior-PAM chlorophyll fluorometer (Heinz Walz GmbH, Eichenring, Effeltrich, Germany) was employed for the determination of chlorophyll fluorescence at room temperature. Leaves (mostly from the top of the plants) were adapted to the dark for 20 min before the fluorescence measurements [36]. Maximum (Fm) and minimal fluorescence (Fo) were studied in dark-adapted leaves with a light intensity of 131 μmol m^{-2} s^{-1} (low beam intensity). In the light-adapted state, maximum fluorescence (Fm') and minimal fluorescence (Fo') were computed in similar leaves using a saturating light intensity at 830 μmol m^{-2} s^{-1}, well balanced with steady-state fluorescence (Fs). The fluorescence values Fm-Fo and Fm'-Fo' were used for the calculation of variable fluorescence (Fv and Fv'). The intrinsic efficiency of the PS II system indicated by Fv'/Fm' and the actual PS II efficiency indicated by Fm'-Fs/Fm' were calculated. Moreover, nonphotochemical (NPQ) and photochemical quenching (qP) were evaluated by using fluorescence parameters estimated in both light- and dark-adapted conditions [37].

2.5. Analysis of OJIP Chlorophyll a Fluroscense Transient

Chlorophyll fluorescence transient was calculated in leaves dark-adapted for 20 min at room temperature by a Handy PEA (Plant Efficiency Analyzer, Hansatech Instruments, King's Lynn, Norfolk, UK). Chlorophyll fluorescence transients were computed using light excitation (at 650 nm) at high intensity (3500 μmol photons m^{-2} s^{-1}) up to 2 s, with the help of an array of 3 LEDs. The OJIP transient data were first described by Strasser et al. [38]. O (beginning) was the initial minimum fluorescence (measured), which was accompanied by an increase toward the J level (2 ms), an inflection I (30 ms), and finally the peak p (300 ms). The OJIP transient data were calculated by the JIP test as defined by Govindjee [39], Stirbet, et al. [40] and Chen et al. [41]. The parameters labelled as Fv/Fo (variable to minimal fluorescence) which reflect the water-splitting complex activity taking place on the donor side of PS II, Fv/Fm (variable to maximal fluorescence), RC/ABS (ratio of reaction centers (RCs) to PS II antenna absorption), performance index (PI), and Area, i.e., the area above the chlorophyll fluorescence curve among Fo and Fm, which estimates the plastoquinone pool size, were analyzed by Biolyzer 4 HP (Version 4.0.30.03.02, Bioenergetics Laboratory, University of Geneva, Geneva, Switzerland) and PEA plus software (Version 1.02, Hansatech Instruments, King's Lynn, Norfolk, UK). PI indicates a grouping of three self-regulating efficient phases of photosynthesis, i.e., RCs, indicating chlorophyll density, trapping of excitation energy, change of excitation energy toward the electron transport according to a particular many-parametric expression [42], and was determined as PI_{ABS}: $RC/ABS * PHIo/(1 - PHIo) * PSIo/(1 - PSIo)$, where RC/ABS is the RCs density for each PS II antenna chlorophyll, PHI is the quantity of excitons trapped per photon absorbed, and PSIo is the possibility of an electron transfer for the entire the way to PS I [38]. The fluorescence transients were normalized to Fo (at O level) for the calculation the OJIP data from different treatments. Additionally, the JIP test was performed for the parameters of the OJIP transient to quantify PS II behavior, which was established on the energy fluxes and yields. Therefore, for (A), the specific energy fluxes (per RC) as (i) Trapping (TR_0/RC); (ii) Dissipation (DI_0/RC), and (iii) Electron transport (ET_0/RC) and for (B), the yields as quantum yield of electron transport (ET_o/CS) were analyzed.

2.6. Photosynthetic and Growth Parameters

Photosynthetic efficiency determined as net photosynthesis (*Pn*), intercellular CO_2 concentration (Ci) and stomatal conductance (gs), was determined in the entirely developed uppermost plants leaves for each treatment by an Infrared Gas Analyzer (model CID-340; Bio-Science, Camas, WA, USA). The measurements were completed at 350 μmol photons m^{-2} s^{-1} with a CO_2 concentration of 370 ± 5 μmol mol^{-1} and relative humidity of $65 \pm 5\%$ at temperature 25 °C.

Estimation of leaf area was done by a leaf area meter (model LA-211; Systronics, New Delhi, India), while plant fresh weight was measured by using a digital scale (Sartorius, Göttingen, Germany). For measuring dry weight, the plants were dried in an oven at 80 °C till constant weight.

2.7. Western Blot Analysis

Fresh leaves (1 g) were crushed in liquid nitrogen to obtain a fine powder for the extraction of the thylakoid protein, as described by Zhao et al. [2] with slight modifications. For Western blot assay, Chl was measured in the thylakoid membrane extracts, and thylakoid proteins (20 μg) having equal Chl were separated by 15% sodium dodecyl sulphate–polyacrylamide gel electrophoresis (SDS-PAGE) using 6 M urea and afterward blotted on nitrocellulose membranes [43]. D1 protein immunodetection was conducted using a polyclonal antibody raised against it according to a previous report [2,44]. The membranes were incubated for 2 h with primary anti-*psb*A antibodies (PhytoAB, San Francisco, CA, USA) and then incubated again with horseradish peroxidase-conjugated anti-rabbit IgG antibodies (PhytoAB, San Francisco, CA, USA) for 2 h. The housekeeping α-tubulin protein was used as an internal loading control for normalization. Pierce ECL plus substrate (Thermo Fisher Scientific, New Delhi, India) was used for the detection of the target protein. Blots were captured on X-ray films and quantified using Image J software (Version 1.52, Wayne Rasband, National Institutes of Health, Bethesda, MD, USA).

2.8. Quantitative RT-PCR Analysis

Fresh crushed leaves (approximately 100 mg) were used for gene expression analysis. According to the instruction given in the kit, total RNA was purified by Trizol reagent (Invitrogen, Carlsbad, CA, USA) after the extraction. Nearly 0.2 μg RNA was utilized for combining the first-strand cDNA with M-MLV reverse transcriptase (Promega, Madison, WI, USA) using an oligo (dT) primer. At that time, through gel electrophoresis, the quality of cDNA was examined, and the samples were kept at −80 °C for qRT PCR. For the study of gene expression, specific primers were designed (Table 1), and the fluorescent dye SYBR Green (Toyobo, Osaka, Japan) was used. After that, according to the manual, real-time PCR was performed using the real-time PCR Master Mix (Toyobo, Osaka, Japan). The housekeeping tubulin (TUB) gene worked as an internal control. The relative amount of the target gene expression was determined by the procedure of Chen et al. [45].

Table 1. Primer sequences and data used for RT-PCR analyses.

Gene	Encoded Polypeptide	Gene ID	Forward(F) /Reverse (R)	Primer Sequences (5′–3′)	Size (bp)
*psb*A	D1 protein	7095419	F	GTATTTATTATCGCCTTCATCG	284
			R	AGGACGCATACCCAAACG	
*psb*B	CP47	7095420	F	TAGGCGTAACGGTGGA	254
			R	AACATCTCGGAACAAGG	
*psb*C	CP43	7095484	F	TAATACGGCTTATCCGAGTGAGTTT	288
			R	TCTTGCCAAGGTTGTATGTCTTT	

Zhang et al. [6]; Chen et al. [45].

2.9. Statistical Analysis

Data were evaluated statistically using analysis of variance (ANOVA) and Tukey's (post-hoc multiple comparison) test at $p < 0.05$ or $p < 0.01$ by SPSS 17.0 software (SPSS Inc., Chicago, IL, USA) for Windows. Data are presented as mean $\pm$ SE ($n = 4$).

3. Results

3.1. Screening of MeJA Concentration for Protection of Plants against Heat-Induced Oxidative Stress

The effect of different concentrations of MeJA on photosynthesis and growth was studied to assess the MeJA requirement of the crop under heat stress. Previous studies showed that MeJA plays a significant role in determining photosynthesis and growth of plants under stress [8,11]. We tested different concentrations of MeJA (5, 10, and 20 μM) to select the best concentration for maximum alleviation of heat stress. Application of 5 μM MeJA in comparison with heat stress, reduced H_2O_2 content and enhanced total protein content, Pn, plant fresh weight, but the results were statistically similar to those recorded for the control. In contrast, the application of 20 μM MeJA notably increased H_2O_2 content, declined net photosynthesis, total protein content, and plant fresh weight compared to control under no stress and with heat stress (Table 2). Moreover, the results were different with the spraying of 10 μM MeJA. Application of 10 μM MeJA more effectively minimized H_2O_2 content by 76.6% and 48.6% under no stress compared to heat stress and control, respectively. The maximum reduction in H_2O_2 content by 81.4% and 59.2% was obtained with 10 μM MeJA under heat stress compared to heat stress and control, respectively (Table 2). Application of 10 μM MeJA also increased total protein, Pn, and plant fresh weight by 24.1%, 23.8%, and 25.4% under no stress compared to control. Application of 10 μM MeJA to heat-stressed plants increased total protein content, Pn, and plant fresh weight maximally by 35.6%, 37.5%, and 32.7%, respectively, in comparison with the control; differences were significantly greater with both concentrations of 5 and 20 μM MeJA under no stress and heat stress (Table 2). This showed that the requirements of plants were met with 10 μM MeJA under heat stress.

3.2. MeJA Enhanced Antioxidant Enzymes Activity and Reduced Oxidative Damage under Heat Stress

Earlier studies have shown that MeJA decreases the accumulation of oxidative stress by increasing the activity of antioxidant enzymes under stress [14]. However, reports on the effects of MeJA on antioxidant enzymes and the content of O_2^-, H_2O_2, and TBARS under heat stress in wheat are less studied. Therefore, we tested the effect of MeJA on O_2^-, H_2O_2, and TBARS content to determine the potential of MeJA in reducing oxidative stress. We found that plants under heat stress showed a rise in the content of H_2O_2 by more than two-fold and of TBARS by about three-fold compared to control plants. Treatment with MeJA decreased oxidative stress by reducing O_2^-, H_2O_2, and TBARS by 23.2%, 12.6%, and 12.1% under no stress and by 43.6%, 42.4%, and 50% in heat-stressed plants, respectively, compared to control (Table 3). The results showed that in comparison with heat stress, the application of MeJA to heat stress-treated plants had more significantly different effects than its application under no stress, decreasing $O_2{}^-$, H_2O_2, and TBARS by 59.6%, 76.3%, and 83.7%, respectively.

Table 2. Content of H_2O_2 and total protein, net photosynthesis, and plant fresh weight of wheat leaves 30 DAS. Plants were treated with MeJA (5, 10, and 20 μM) at 42 °C (heat stress) or 25 °C (no stress). Data are presented as mean ± SE ($n = 4$). Significantly different values between control and treatments are marked with an asterisk (* $p < 0.05$, ** $p < 0.01$, *** $p < 0.001$), determined by Tukey's test. DAS, days after sowing; FW, fresh weight; HS, heat stress; H_2O_2, hydrogen peroxide; MeJA, methyl jasmonate.

Treatments	H_2O_2 Content (nmol g^{-1} Leaf FW)	Total Protein (mg g^{-1} Leaf FW)	Net Photosynthesis (μmol CO$_2$ m^{-2} s^{-1})	Plant Fresh Weight (g Plant^{-1})
Control	15.2 ± 0.89	8.70 ± 0.86	16.8 ± 0.82	3.85 ± 0.12
HS	33.4 ± 0.98 **	3.60 ± 0.42 **	10.9 ± 0.56 **	1.55 ± 0.06 **
5 μM MeJA	10.6 ± 0.76 *	9.60 ± 0.88	18.1 ± 0.93	4.07 ± 0.11
10 μM MeJA	07.8 ± 0.56 **	10.8 ± 1.21 *	20.8 ± 0.96 **	4.83 ± 0.19 *
20 μM MeJA	19.3 ± 0.84 *	5.84 ± 0.65 **	11.2 ± 0.64 **	2.83 ± 0.08 *
5 μM MeJA + HS	17.2 ± 0.73	6.65 ± 0.73 *	13.1 ± 0.71 *	3.79 ± 0.09
10 μM MeJA + HS	06.2 ± 0.66 **	11.8 ± 1.42 **	23.1 ± 0.97 ***	5.11 ± 0.22 **
20 μM MeJA+ HS	40.1 ± 1.02 **	2.73 ± 0.13 **	09.2 ± 0.54 ***	1.29 ± 0.05 **

Table 3. Production rate of superoxide radicals ($O_2{}^-$), content of H_2O_2, TBARS, activity of CAT, SOD, APX, and GR in wheat leaves at 30 DAS. Plants were treated with MeJA (10 μM) at 42 °C (heat stress) or 25 °C (no stress). Data are presented as mean ± SE ($n = 4$). Significantly different values between control and treatments are marked with an asterisk (* $p < 0.05$, ** $p < 0.01$, *** $p < 0.001$), as determined by Tukey's test. APX, ascorbate peroxidase; CAT, catalase; DAS, days after sowing; FW, fresh weight; GR, glutathione reductase; HS, heat stress; H_2O_2, hydrogen peroxide; MeJA, methyl jasmonate; SOD, superoxide dismutase; TBARS, thiobarbituric acid reactive substances.

Parameters	Control	HS	MeJA	MeJA + HS
		Treatments		
Production rate of $O_2{}^-$ (μmol g FW^{-1} min^{-1})	0.801 ± 0.05	1.118 ± 0.080 **	0.611 ± 0.060 *	0.451 ± 0.02 **
H_2O_2 content (nmol g^{-1} leaf FW)	35.60 ± 1.60	86.80 ± 02.40 ***	31.10 ± 01.1 *	20.50 ± 0.09 **
TBARS content (nmol g^{-1} leaf FW)	08.2 ± 0.12	25.3 ± 0.19 **	07.2 ± 0.09 *	04.1 ± 0.07 **
CAT activity (U mg^{-1} protein min^{-1})	119 ± 3.70	144 ± 4.00 *	210 ± 4.30 **	222 ± 5.10 **
SOD activity (U mg^{-1} protein min^{-1})	05.34 ± 0.08	07.66 ± 0.11 *	10.62 ± 0.18 **	11.5 ± 0.21 **
APX activity (U mg^{-1} protein min^{-1})	1.12 ± 0.04	1.57 ± 0.09 *	2.58 ± 0.11 **	2.72 ± 0.11 ***
GR activity (U mg^{-1} protein min^{-1})	0.197 ± 0.005	0.231 ± 0.008 *	0.288 ± 0.009 **	0.318 ± 0.01 ***

Heat stress augmented the activity of enzymatic antioxidants including CAT, SOD, APX, and GR by 21.0%, 43.4%, 40.1%, and 17.2%, respectively compared to control. Exogenous MeJA under no stress increased the activity of enzymatic antioxidants (CAT, SOD, APX, and GR) by 45.8% and 76.4%, 38.6% and 98.8%, 64.3% and 130.3%, and 24.6% and 46.1% as compared to heat stress and control, respectively. However, when heat-stressed plants were treated with MeJA, a maximum increase in the activity of CAT, SOD, APX, and GR was observed, corresponding to 54.1% and 86.5%, 50.1% and 115.3%, 73.2% and 142.8%, and 37.6% and 61.4% in comparison to heat-stress plants and control, respectively (Table 3). These results indicated that MeJA alleviation of oxidative stress was associated with increased activity of enzymatic antioxidants in wheat subjected to heat stress.

3.3. Pigments and Protein Content

Methyl jasmonate has been recognized as an important signal molecule that increases pigment and protein content in response to the different stresses [8,11]. Therefore, we studied the effects of MeJA on the pigment and protein content of plants under heat stress (Table 4). The results showed that heat stress treatment decreased Chl a, Chl b, Chl a-to-Chl b ratio, and total Chl content by 26.6%, 17.4%, 10.8%, and 25.1%, respectively, compared to control. The carotenoid content also decreased by 12.4% in heat-stressed plants. Application of MeJA under no stress enhanced the protein and pigment content significantly as compared to heat stress, but statistically the increase in Chl content was almost similar to what observed for the control under no stress. However, the application of MeJA in the presence of heat stress maximally enhanced Chl a, Chl b, Chl a-to-Chl b ratio, total Chl, and carotenoid content by 52.7% and 12.4%, 27.5% and 5.3%, 19.7% and 6.8%, 48.2% and 10.9%, and 63.3% and 43.0% compared to heat stress and control, respectively, and showed significantly more different results with respect to control, heat stress, and application of MeJA under no stress. Similarly, the protein content also increased upon MeJA application by 38.3% and 213.1% under heat stress in comparison to heat stress and control, respectively (Table 4).

Table 4. Content of Chl a, Chl b, Chl (a/b) ratio, total Chl, and carotenoids in wheat leaves at 30 DAS. Plants were treated with MeJA (10 µM) at 42 °C (heat stress) or 25 °C (no stress). Data are presented as mean ± SE ($n = 4$). Significantly different values between control and treatments are marked with an asterisk (* $p < 0.05$, ** $p < 0.01$, *** $p < 0.001$), as determined by Tukey's test. Chl, chlorophyll; DAS, days after sowing; FW, fresh weight; HS, heat stress; MeJA, methyl jasmonate.

Parameters	Treatments			
	Control	HS	MeJA	MeJA + HS
Chl a (mg g^{-1} Leaf FW)	1.71 ± 0.06	1.26 ± 0.04 **	1.75 ± 0.06 *	1.92 ± 0.08 **
Chl b (mg g^{-1} Leaf FW)	0.45 ± 0.01	0.37 ± 0.01 **	0.46 ± 0.02	0.47 ± 0.04 *
Chl (a/b)	3.79 ± 1.70	3.38 ± 1.56 **	3.80 ± 1.73	4.05 ± 1.77 **
Total chl (mg g^{-1} Leaf FW)	2.15 ± 0.07	1.62 ± 0.04 **	2.21 ± 0.07 *	2.39 ± 0.08 **
Carotenoids (mg g^{-1} Leaf FW)	0.44 ± 0.01	0.38 ± 0.01 **	0.49 ± 0.02 *	0.63 ± 0.05 ***
Total protein (mg g^{-1} Leaf FW)	08.60 ± 1.18	03.80 ± 1.09 **	10.90 ± 1.39 **	11.90 ± 1.42 **

3.4. Influence of MeJA on Chlorophyll a Fluorescence

Since previous studies have reported that MeJA repaired PS II in mustard and enhanced the photosynthetic efficiency of the system [15], we investigated the effect of MeJA on chlorophyll a fluorescence. Till date, there is no report available on the effect of MeJA on chlorophyll fluorescence under heat stress in wheat. Plants grown under heat stress exhibited reduced intrinsic PS II efficiency, actual PS II efficiency, and qP by 15.3%, 59.6%, and 33.3%, respectively, compared to control, and NPQ increased by 74.0%. Exogenous MeJA enhanced the above parameters significantly in comparison with heat-stressed plants, under no stress. However, a maximum increase in intrinsic PS II efficiency, actual PS II efficiency, and qP by 26.3% and 6.9%, 166.2% and 7.3%, and 109.4% and 39.6%, and a decline in NPQ by 70.7% and 49.0% was obtained with MeJA under heat stress, respectively, in comparison to heat stress and control (Figure 1A–D). These observations showed that MeJA increased the photosynthetic efficiency of PS II under heat stress.

It was interesting to note that MeJA was more effective under heat stress as compared to control. Possibly, efficient MeJA signaling is activated under heat stress. The chlorophyll fluorescence parameters varied under different treatments. Chlorophyll fluorescence increased from a minimum level ("O" or Fo) towards a maximum level ("*p*" or Fm). The results showed that in response to heat stress treatment, Fv/Fm and the diameter of leaves decreased to a very large degree, whereas for the MeJA-treated leaves, the results were reversed (Table 5). In heat-treated plants, both Fo and Fm decreased, and the O–J, J–I, and I–P phases showed lower amplitudes, in comparison to those in with MeJA-treated plants under control or heat stress. The OJIP curves were normalized at Fo to determine the changes in the fluorescence kinetics and show relative variable fluorescence vs. time on a logarithmic time ruler (Figure 2).

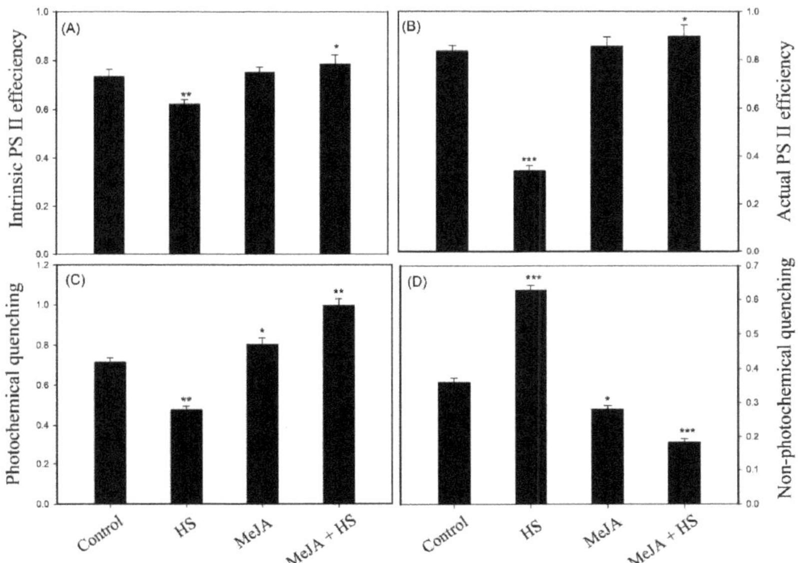

Figure 1. Intrinsic PS II efficiency (**A**), actual PS II efficiency (**B**), photochemical quenching (**C**), and non-photochemical quenching (**D**) in wheat leaves at 30 DAS. Plants were treated with MeJA (10 μM) at 42 °C (heat stress) or 25 °C (no stress). Data are presented as mean ± SE (n = 4). Significantly different values are marked with an asterisk between control and treatments (* $p < 0.05$, ** $p < 0.01$, *** $p < 0.001$), as determined by Tukey's test. DAS, days after sowing; HS, heat stress; MeJA, methyl jasmonate.

Table 5. Chl a fluorescence measurement in wheat leaves at 30 DAS. Plants were treated with MeJA (10 μM) at 42 °C (heat stress) or 25 °C (no stress). Minimal fluorescence (Fo), maximal fluorescence (Fm), maximal variable fluorescence (Fv), and Fv/Fm ratio, where Fv = (Fm−Fo), reaction center-to-PS II antenna absorption ratio (RC/ABS), performance index (PI), and Area, region above the chlorophyll fluorescence OJIP curve between Fo and Fm. Data are presented as mean ± SE (n = 4). Significantly different values between control and treatments are marked with an asterisk (* $p < 0.05$, ** $p < 0.01$, *** $p < 0.001$), as determined by Tukey's test. DAS, days after sowing; HS, heat stress; MeJA, methyl jasmonate.

Parameters	Treatments			
	Control	HS	MeJA	MeJA + HS
Fo	206 ± 03.80	187 ± 03.50 **	241 ± 04.10 **	246 ± 04.40 **
Fm	935 ± 07.30	816 ± 06.20 *	1258 ± 08.10 **	1350 ± 08.50 ***
Fv/Fo	2.863 ± 0.018	2.709 ± 0.011 *	3.445 ± 0.021 **	3.639 ± 0.026 **
Fv/Fm	0.780 ± 0.04	0.760 ± 0.04 *	0.808 ± 0.05 **	0.817 ± 0.05 **
RC/ABS	0.831 ± 0.07	0.754 ± 0.04 *	0.956 ± 0.08 **	0.992 ± 0.08 **
PI	1.675 ± 0.014	1.383 ± 0.009 *	2.423 ± 0.021 **	2.734 ± 0.023 ***
Area	21,744 ± 505	15,989 ± 421 *	23,100 ± 539	24,907 ± 611 *

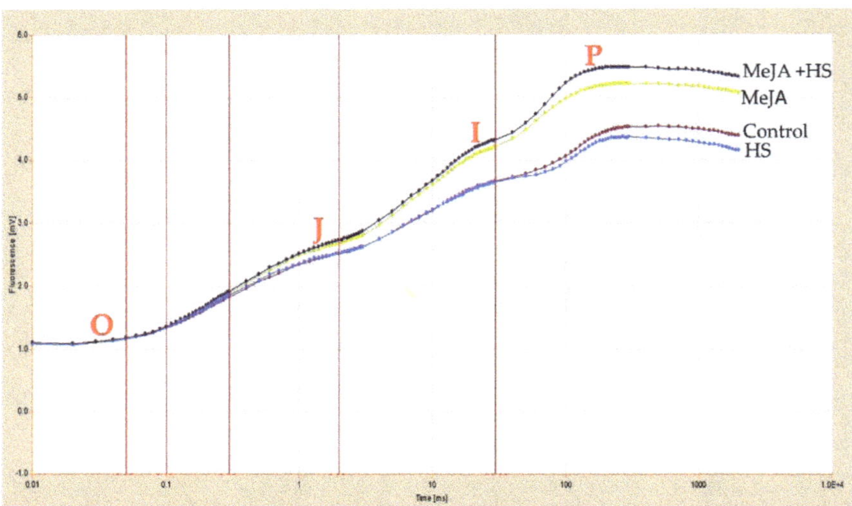

Figure 2. Chlorophyll fluorescence of OJIP curve in wheat leaves at 30 DAS. Plants were treated with MeJA (10 μM) at 42 °C (heat stress) or 25 °C (no stress). Data are presented as mean ± SE (*n* = 4). DAS, days after sowing; HS, heat stress; MeJA, methyl jasmonate.

These results suggested that heat stress influenced chlorophyll fluorescence, as presented in OJIP curves (Figure 2). The fluorescence parameters Fo, Fm, Fv/Fm, and Fv/Fo decreased by 9.2%, 12.7%, 2.6%, and 4.9%, respectively, compared to control under heat stress. Nevertheless, the fluorescence parameters were higher with the application of MeJA to heat-stressed leaves by 31.5% and 19.4%, 65.4% and 44.3%, 7.5% and 4.7%, and 34.5% and 26.8%, respectively, compared to heat stress and control. Heat treatment decreased PI by 17.4%, but in MeJA-treated plants under heat stress it increased by 97.6 and 63.2% compared to heat stress and control, respectively. These results indicate the beneficial effect of MeJA compared to control and heat-stressed plants. The ratio of RC/ABS, which reveals RCs density of the PS II antenna chlorophyll, was higher in MeJA-treated plants in heat stress by 19.3%, which decreased by 9.3% under heat stress compared to control. The leaves treated with MeJA were more resistant to heat stress and reduced-heat adversities with respect to chlorophyll fluorescence. The area over the OJIP curve, between Fo and Fm, decreased upon heat stress treatment by 26.4% but increased by 55.7% and 14.5% in MeJA-treated leaves compared to heat stress and control, respectively (Table 5; Figure 2), again confirming the benefits of the MeJA treatment in heat-stressed wheat.

Additionally, other parameters were analyzed to specify energy fluxes (per RC) and yield, as shown in Figure 3A–D. Energy fluxes (per RC), i.e., TR_0/RC (trapping) and DI_0/RC (dissipation), were higher under heat stress. However, trapping and dissipation per RC were reduced significantly after the application of MeJA under heat stress (Figure 3A,B). There were noticeable effects on ET_0/RC (electron transport flux per RC) and yield, indicated by ET_0/CS (quantum yield of electron transport), after MeJA treatment under heat stress, as MeJA increased these values by 14.7% and 9.61% and 111.6% and 63.5% compared to heat stress and control, respectively (Figure 3C,D).

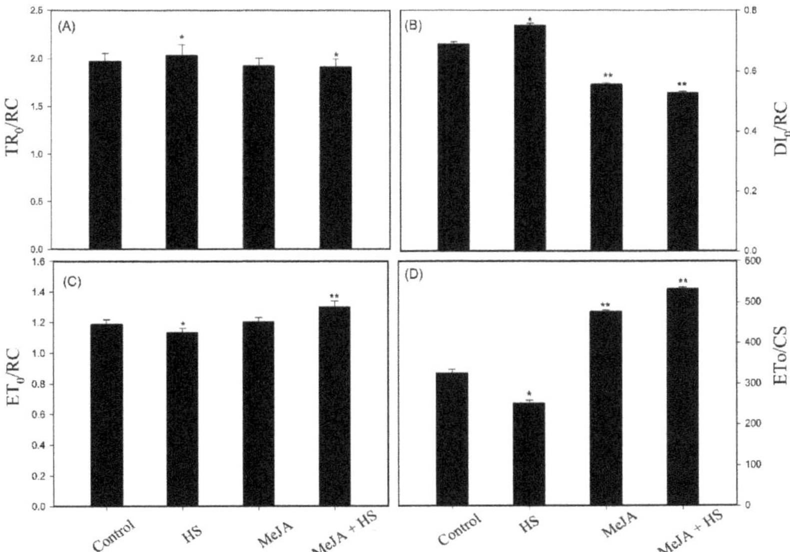

Figure 3. Photosynthetic parameters inferred from the JIP test study of chlorophyll fluorescence transients. Change of specific energy fluxes per PS II reaction center (RC) as (**A**) Trapping (TR_0/RC), (**B**) Dissipation (DI_0/RC), and (**C**) Electron transport (ET_0/RC) and for (**D**) yields as quantum yields of electron transport (ET_o/CS) in wheat leaves at 30 DAS. Plants were treated with MeJA (10 μM) at 42 °C (heat stress) or 25 °C (no stress). Data are presented as mean ± SE (n = 4). Significantly different values between control and treatments are marked with an asterisk (* $p < 0.05$, ** $p < 0.01$), as determined by Tukey's test. DAS, days after sowing; HS, heat stress; MeJA, methyl jasmonate.

3.5. Impact of MeJA on Photosynthesis and Growth under Heat Stress or without Stress

Earlier studies have shown that MeJA regulates many aspects of plant development by involving photosynthetic characteristics [15,18]. We investigated the role of MeJA in improving the photosynthetic capacity in wheat. It has been reported that MeJA exerts a positive effect in response to abiotic stress [11,14]. To test the role of MeJA on photosynthesis and growth under heat stress, we treated wheat plants with 10 μM MeJA. Our results showed that heat stress-treated plants showed a decline in Pn, Ci, and gs by 34.5%, 30.4%, and 26.4%, respectively, compared to control. Application of MeJA significantly increased these parameters (Pn, Ci, and gs) under normal conditions compared to heat stress and control. However, MeJA treatment maximally benefitted the plants under heat stress and maximally alleviated the reduction in Pn, Ci, and gs by 112.2% and 38.8%, 75.6% and 22.1%, and 65.3% and 21.6% compared to heat stress and control, respectively. These results verified that MeJA could increase the photosynthetic efficiency under heat stress. Heat stress reduced leaf area, plant fresh and dry mass by 43.3%, 47.4%, and 42.9%, respectively, compared to control. An individual dose of MeJA was effective in increasing leaf area, plant fresh and dry mass in the lack of heat stress. However, the maximal increase in the leaf area and plant fresh and dry mass by 32.6%, 34.6%, and 34.9% compared to control was noted with the application of MeJA to the heat-treated plants and proved more effective in lessening the effect of heat stress in wheat plants (Table 6).

Table 6. Net photosynthesis, intercellular CO_2 concentration, stomatal conductance, leaf area, plant fresh and dry weight of wheat leaves at 30 DAS. Plants were treated with MeJA (10 μM) at 42 °C (heat stress) or 25 °C (no stress). Data are presented as mean $\pm$ SE (n = 4). Significantly different values between control and treatments are marked with an asterisk (* $p < 0.05$, ** $p < 0.01$, *** $p < 0.001$), determined by Tukey's test. Chl, chlorophyll; DAS, days after sowing; FW, fresh weight; HS, heat stress; MeJA, methyl jasmonate.

	Treatments			
Parameters	**Control**	**HS**	**MeJA**	**MeJA + HS**
Net photosynthesis (μmol CO_2 m^{-2} s^{-1})	16.2 ± 0.91	10.6 ± 0.52 ***	20.5 ± 095 **	22.5 ± 0.99 ***
Intercellular CO_2 concentration (μmol CO_2 mol^{-1})	230 ± 9.1	160 ± 7.3 **	258 ± 10.3 **	281 ± 11.1 **
Stomatal conductance (mmol H_2O m^{-2} s^{-1})	310 ± 12.3	228 ± 8.7 **	345 ± 13.5 **	377 ± 14.1 **
Leaf area (cm^2 Plant^{-1})	106 ± 4.1	60.1 ± 2.9 ***	123 ± 4.3 **	140.6 ± 4.9 ***
Plant fresh weight (g Plant^{-1})	5.37 ± 0.09	2.82 ± 0.05 **	6.01 ± 0.10 *	7.23 ± 0.11 **
Plant dry weight (g Plant^{-1})	0.808 ± 0.04	0.461 ± 0.01 **	0.897 ± 0.06 *	1.090 ± 0.09 **

3.6. Effect of MeJA on D1 Protein Content Abundance and Gene Expression Relevant to the Photosynthetic System

The D1 protein constitutes the core of the PS II reaction center [2,5], and many studies have indicated that the PS II center in chloroplasts is the part that is the most easily damaged by environmental stresses [9]. Therefore, to identify the role of MeJA in the PS II system and in the maintenance of efficient turnover of D1 protein, the abundance of the PS II reaction center D1 protein in wheat leaves was analyzed. Our results suggest that the abundance of D1 protein in heat-treated leaves decreased significantly with respect to that in the control (Figure 4A,B). In contrast, with the application of MeJA, D1 protein abundance increased significantly compared to control and heat-stressed plants. The maximal increase in the abundance of D1 protein obtained with MeJA in heat-treated leaves was of 57.1% and 35.6% compared to heat-stressed and control leaves, respectively, and suggested that the recovery of D1 protein was effectively induced by MeJA in comparison with control. The original Immunoblot image is given in the Figure S1.

Exogenous application of MeJA modulates the photosynthetic efficiency and the expression of PS II genes [15]. Therefore, we tested changes in the expression levels of PS II genes upon exogenous application of MeJA under heat stress. The photosynthetic system was examined to obtain the expression levels of *psb*A, *psb*B, and *psb*C genes, encoding D1 protein, CP47, and CP43, to explore the protective role of MeJA in wheat leaves in heat-stressed plants. We observed that the application of MeJA enhanced the expression of *psb*A and *psb*B significantly, but it did not affect *psb*C levels under normal conditions. Heat stress decreased the expression of genes compared to control. However, MeJA under heat stress strongly stimulated *psb*A and *psb*B expression, while the effect on *psb*C expression was lower compared to that on *psb*A and *psb*B (Figure 5A–C).

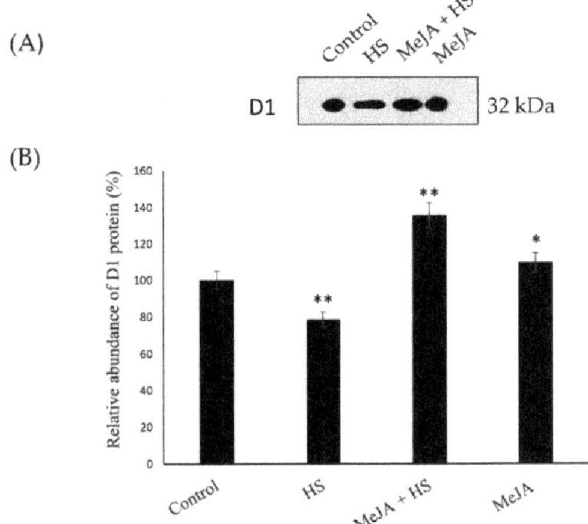

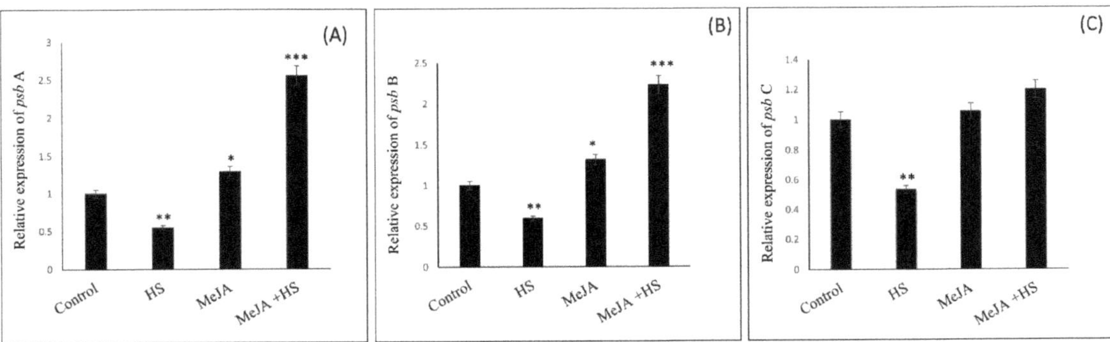

Figure 4. Immunoblot studies of thylakoid protein (D1) acquired from wheat leaves at 30 DAS. Plants were treated with MeJA (10 µM) at 42 °C (heat stress) or 25 °C (no stress). (**A**) Immunoblotting was performed with specific antibodies raised against the D1 protein. (**B**) Quantitative data for D1 protein in wheat leaves under heat stress after MeJA treatment. Results are presented relative to the respective controls (control, 100%, $n = 4$). Significantly different values are marked with an asterisk (* $p < 0.05$, ** $p < 0.01$). DAS, days after sowing; HS, heat stress; MeJA, methyl jasmonate.

Figure 5. Relative expression of the genes (**A**) *psb*A; (**B**) *psb*B, and (**C**) *psb*C in wheat leaves at 30 DAS. Plants were treated with MeJA (10 µM) at 42 °C (heat stress) or 25 °C (no stress). Results are presented relative to the respective controls (control, 1). Data are presented as mean ± SE ($n = 4$). Significantly different values between control and treatments are marked with an asterisk (* $p < 0.05$, ** $p < 0.01$, *** $p < 0.001$), as determined by Tukey's test. DAS, days after sowing; HS, heat stress; MeJA, methyl jasmonate.

4. Discussion

Heat stress causes damage in plants' photosynthetic apparatus, deterioration of leaf function, and reduction of yield. Degradation of chloroplasts and damage of PS II are important factors that affect photosynthesis. These aspects are currently the focus of research to understand the influence of stress on plants. Exogenous application of MeJA markedly regulates numerous key biochemical, physiological, and molecular processes in response to abiotic stress [14,46,47]. However, the effects of MeJA on PS II in wheat under heat stress are less known. Therefore, the present research was designed to acquire knowledge on how MeJA protects PS II and maintains the stability of D1 protein by regulating the expression of relevant genes during heat stress.

4.1. MeJA Increases Antioxidant System Activity to Mitigate the Oxidative Damage Induced by Heat Stress

Heat stress is one of the reasons for the accumulation of ROS such as H_2O_2 and superoxide ions, oxy-intermediates that cause cellular damage through the oxidation of lipids, proteins, and nucleic acids. The present study shows that exogenous MeJA markedly increased the activity of the antioxidant enzymes CAT and APX, which led to efficient detoxification of ROS and decreased lipid peroxidation in membranes under stress. Plants have developed complex defense systems as enzymatic antioxidants under heat stress; SOD changes superoxide ions into H_2O_2 and O_2; then, CAT mainly changes H_2O_2 into H_2O and O_2 that is produced by photorespiration and β-oxidation of fatty acids, while GR and APX catalyze the transformation of H_2O_2 to H_2O through dismutation [8]. Thus, we speculated that enhanced CAT, APX, GR, and SOD enzyme activities upon MeJA treatment played an effective role in the scavenging of O_2^- and H_2O_2. Lang et al. [14] reported that MeJA increased the activity of antioxidant enzymes and reversed the harmful effect of salt stress in *Glycyrrhiza uralensis*. Similarly, MeJA has been reported to increase the activity of POD and SOD in *Brassica napus* [46], of CAT, POD, and APX in *Citrus limon* [48], and of CAT, POX, and APX in *Fragaria× ananassa 'Camarosa'* [49]. Remarkably, our results indicate that MeJA improved ROS-scavenging ability in wheat plants and exerted progressive effects on the improvement of plant resistance under heat stress.

4.2. MeJA Improved the Photosynthetic Efficiency under Heat Stress

The photosynthetic pigments Chl a and Chl b are the main pigments in leaves' photosynthesis process, indicative of the physiological state of plants [32,50]. Both Chl a and Chl b absorb light energy, but only Chl a in the excited state transforms light energy into electrical energy and plays an important role in managing the stability of the light-harvesting complex related to PS II and in regulating the size of the photosynthetic antenna in plants [51]. Heat stress decreased the content of Chl a, Chl b, total Chl, carotenoid, and Chl a-to-Chl b ratio considerably (Table 4). The disturbance of Chl synthesis was linked to the decrease in Chl content in response to high temperature and cadmium stress [3,52]. Changes in the Chl a/b ratio are usually associated with changes in the size of the light-harvesting antenna of PS II. The results showed that the decrease of Chl content by heat stress was enhanced by MeJA. In addition, the Chl a/b ratio was higher under heat stress after MeJA treatment. Exogenous MeJA may act as a regulator inhibiting the disintegration of Chl molecules and protecting the photosynthetic antenna and PS II structure, thus enhancing the heat tolerance and photosynthetic efficacy of wheat through increasing the activity of antioxidant enzymes that scavenge ROS. These results are related to previous findings indicating that MeJA application protected the degradation of photosynthetic pigments under environmental stresses [8,14]. It was reported that jasmonate increased Chl a content, which is the direct photon donor to the RCs of both PS II and PS I systems [53]. Enhanced content of Chl a over Chl b helps sustain the photosynthetic capacity at a greater level, which promotes the accumulation of carbohydrates, after both jasmonate and MeJA treatments [15].

Increased Chl levels together with enhanced protein levels increased Fo in the plants (Tables 4 and 5). These results are also supported by the chlorophyll fluorescence data, which showed a decrease in PS II quantum efficiency and yield (Table 5; Figures 1–3). High temperature influences electron transport, which damages the function of PS II [41,54]. For the development of heat tolerance, wheat leaves were treated with MeJA, and then Chl fluorescence transients were examined by the JIP test (Figure 2; Table 5). Chl fluorescence kinetics transients provided additional data regarding the photochemical reaction in photosynthesis, mainly for the PS II donor side and the receptor side with RCs [41,55]. According to the JIP test, heat stress at 42 °C caused the impairment of the PS II donor side and an overreduction of the PS II acceptor side compared with those at optimal temperature (25 °C), resulting in a greater use of the excitation energy that decreased the stability of the photosystem [38] and altered the amplitudes O, J, I, and p, as observed in the OJIP curve. A significant difference upon application of MeJA under heat stress was evident when examining the OJIP curve and the data of chlorophyll a fluorescence, shown in Figures 2 and 3 and Table 5. This difference indicates that MeJA promotes heat resistance by protecting the PS II in wheat leaves through increasing quantum yield and efficiency. Heat stress lowers the stability of the PS II system [38], thereby leading to the decrease of PI. The performance index shows the photosynthetic capacity based on light absorption and is the most sensitive parameter in the OJIP curve. It comprises the maximum quantum yield of primary photochemistry, possibly related to the capability to reduce an electron acceptor at the end of PS I, and the RC/ ABS ratio or includes light absorption, trapping, and transfer excitation energy in electron transport [38,56,57]. The results showed that plants under heat stress did not adjust their light absorption and consumption, causing adverse effects on the RC. The present research shows that the application of MeJA had a beneficial impact on PI, which was higher compared to heat stress (Table 5). The application of MeJA influenced the trapped excitation and electron transport flux per RC, leading to normal levels. This proves that MeJA augmented the active RC and reduced the damage on RC in heat stress. In MeJA-treated plants under heat stress, PI and the RC/ABS ratio were higher than in the control (Table 5), suggesting improved stability of the PS II system due to comparatively higher stability of RCs, which contributed to higher PI in MeJA-treated plants exposed to heat stress. The results for PS II are consistent with those of studies on spermidine and nitric oxide application for the mitigation of PS II damage in tall fescue under heat stress [6,58].

The energy fluxes per RC, i.e., TR_0/RC and DI_0/RC, were higher, but ET_0/RC and yield (ET_o/CS) were reduced under heat stress (Figure 3). The results showed that the electron transportation efficiency and the function of Chl in PS II decreased under heat stress, but the application of MeJA remarkably improved ET_0/RC and ET_o/CS. Exogenous MeJA had a favorable effect on the acceptor and donor sides of PS II under heat stress. In addition, MeJA protected the RC of PS II and lessened the damage to the thylakoid protein complex organization by increasing Fv/Fm and qP. Exogenous MeJA protected the RCs, which were damaged under heat stress, and increased the size of light-harvesting antennas in the plants due to the increase of Chl molecules. Furthermore, the decrease in chlorophyll a fluorescence and the reduced qP are accountable for the decreased CO_2 assimilation under heat stress. MeJA worked as regulator in the stomatal response and improved *Pn*, Ci, gs, and the leaf area in plants under heat stress (Table 6), improving the efficiency of photosynthesis.

4.3. MeJA Increases D1 Protein Content and Gene Expression Relevant to the Photosynthetic System

Previous studies have shown that the PS II reaction center in chloroplasts, mainly, the D1 protein, was easily damaged by stresses [2,9]. Consistent with previous reports, the present study also showed that D1 protein, a sensitive component of the photosystem, was damaged by heat stress. It is important to maintain conformation stability of PS II RCs; therefore, D1 protein damage causes alteration of PS II RC conformation, electron transfer disruption, and destruction of PS II RCs [5,59,60]. The protection of the photosynthetic apparatus mainly depends on the stability of D1 under stress. Therefore, in the present study, wheat leaves were treated with MeJA under heat stress to examine the abundance of D1 protein and its function in PS II. It was possible that the damage of D1 protein and the repair of D1 protein under heat stress were reversible. However, in the presence of MeJA under heat stress, the level of D1 protein was much higher than in control and heat-stressed plants, indicating that MeJA treatment significantly reduced D1 protein degradation under heat stress by regulating enzymatic antioxidants. MeJA treatment not only reduced D1 protein damage or PS II function under heat stress, but also more efficiently restored D1 protein levels and PS II compared to control plants after heat stress, which was verified by the Fv/Fm results (Table 5). The application of MeJA to the heat-treated plants may increase D1 protein abundance more effectively and stimulate the assembly of PS II RCs through scavenging ROS by enhancing enzymatic antioxidants and increasing PS II activity, as PS II

RC proteins are the main targets of ROS under stress responsible for the decrease in PS II function [60,61]. The mechanisms of the recovery of D1 protein levels and PS II function are not simple, as they are not just involved in the reducing the injured proteins but, also involved in the synthesis of new proteins, in particular in relation to *psb*A gene expression in chloroplasts. Therefore, further experiments were set up to analyze gene expression and determine the mechanisms influencing the stability and repair of D1 protein in the presence of MeJA under heat stress. The damaged D1 protein RCs under heat stress disturbed energy utilization and led to a reduction in CP43 and CP47 (*psb*C- and *psb*B-encoded proteins respectively), which are main antenna protein complexes of the PS II system [62,63]. Hence, the protection of D1 under heat stress is vital for the RC of PS II. The application of MeJA stimulated CP47 transcription and increased the function of RC to some extent. The protein D1 encoded by the *psb*A gene is a crucial element of the PS II system, and an assortment of cofactors involved in electron transfer and charge separation, are collectively organized in the PS II structure [64,65]. Our results showed that in the presence of MeJA, *psb*A gene expression increased significantly, which could be responsible for PS II stability under heat stress, as the *psb*A gene may participate in the restoration and renewal of D1 protein, damaged under heat stress. As shown in Table 3, the application of MeJA stimulated antioxidant enzyme activities, helped the scavenging of oxidative stress, upregulated the expression of genes coding for proteins involved in the PS II system (Figure 5), enhanced the removal of the negative effects of heat stress, and eventually increased the photosynthetic efficiency. Thus, MeJA protection of the photosystem under heat stress was reciprocated reflected in the increased abundance of D1 protein and overexpression of the *psb*A gene. In summary, MeJA reduced PS II core proteins degradation under heat stress, thereby contributing to heat tolerance in wheat plants.

5. Conclusions

The current study revealed that MeJA is efficient in the protection of the PS II complex under heat stress. Exogenous MeJA decreased heat-induced oxidative stress through increased activity of enzymatic antioxidants that protected the photosynthetic apparatus and enhanced chlorophyll content, chlorophyll fluorescence, and CO_2 assimilation. These findings indicate that MeJA supply exerted a positive effect by maintaining the stability of chloroplast D1 protein in the PS II complex and enhancing the expression of relevant genes. This study suggests that MeJA can be utilized to enhance PS II efficiency and overall protection of photosynthesis in adverse climatic conditions.

Supplementary Materials: The following are available online at https://www.mdpi.com/article/10.3390/antiox10081216/s1, Figure S1: Immunoblot analyses of thylakoid protein (D1) obtained from wheat leaves 30 days after sowing. Plants were treated with MeJA (10 μM) at 42 °C (heat stress) or 25 °C (no stress). Immunoblotting was performed with specific antibodies raised against the D1 protein.

Author Contributions: Conceptualization, M.F and N.A.K.; methodology, M.F.; software, M.F.; formal analysis, M.F. and Z.S.; investigation, N.A.K.; resources, M.F. and P.K.; data curation, M.F.; writing—original draft preparation, M.F.; writing—review and editing, N.I., M.N.A. and P.A.; visualization, N.I.; supervision, N.A.K., funding acquisition, M.N.A., P.K. and P.A. All authors have read and agreed to the published version of the manuscript.

Funding: This research received no external funding.

Institutional Review Board Statement: Not applicable.

Informed Consent Statement: Not applicable.

Data Availability Statement: Data is contained within the article and supplementary material.

Acknowledgments: The authors would like to extend their sincere appreciation to the Researchers Supporting Project Number (RSP-2021/180), King Saud University, Riyadh, Saudi Arabia.

Conflicts of Interest: The authors declare no conflict of interest.

References

1. Poudel, P.B.; Poudel, M.R. Heat stress effects and tolerance in wheat: A Review. *J. Biol. Today World* **2020**, *9*, 1–6.
2. Zhao, H.J.; Zhao, X.J.; Ma, P.F.; Wang, Y.X.; Hu, W.W.; Li, L.H.; Zhao, Y.D. Effects of salicylic acid on protein kinase activity and chloroplast D1 protein degradation in wheat leaves subjected to heat and high light stress. *Acta Ecol. Sin.* **2011**, *31*, 259–263. [CrossRef]
3. Iqbal, N.; Umar, S.; Khan, N.A.; Corpas, F.J. Nitric oxide and hydrogen sulfide coordinately reduce glucose sensitivity and decrease oxidative stress via ascorbate-glutathione cycle in heat-stressed wheat (*Triticum aestivum* L.) plants. *Antioxidants* **2021**, *10*, 108. [CrossRef] [PubMed]
4. Gururani, M.A.; Venkatesh, J.; Tran, L.S.P. Regulation of photosynthesis during abiotic stress-induced photoinhibition. *Mol. Plant* **2015**, *8*, 1304–1320. [CrossRef] [PubMed]
5. Chen, J.H.; Chen, S.T.; He, N.Y.; Wang, Q.L.; Zhao, Y.; Gao, W.; Guo, F.Q. Nuclear-encoded synthesis of the D1 subunit of photosystem II increases photosynthetic efficiency and crop yield. *Nat. Plants* **2020**, *6*, 570–580. [CrossRef] [PubMed]
6. Zhang, L.; Hu, T.; Amombo, E.; Wang, G.; Xie, Y.; Fu, J. The alleviation of heat damage to photosystem II and enzymatic antioxidants by exogenous spermidine in tall fescue. *Front. Plant Sci.* **2017**, *8*, 1747. [CrossRef]
7. Bethmann, S.; Melzer, M.; Schwarz, N.; Jahns, P. The zeaxanthin epoxidase is degraded along with the D1 protein during photoinhibition of photosystem II. *Plant Direct* **2019**, *3*, e00185. [CrossRef]
8. Per, T.S.; Khan, N.A.; Masood, A.; Fatma, M. Methyl jasmonate alleviates cadmium-induced photosynthetic damages through increased S-assimilation and glutathione production in mustard. *Front. Plant Sci.* **2016**, *7*, 1933. [CrossRef]
9. Chen, Y.E.; Mao, H.T.; Wu, N.; Mohi Ud Din, A.; Khan, A.; Zhang, H.Y.; Yuan, S. Salicylic acid protects photosystem II by alleviating photoinhibition in *Arabidopsis thaliana* under high light. *Int. J. Mol. Sci.* **2020**, *21*, 1229. [CrossRef] [PubMed]
10. Jahan, B.; Iqbal, N.; Fatma, M.; Sehar, Z.; Masood, A.; Sofo, A.; D'Ippolito, I.; Khan, N.A. Ethylene supplementation combined with split application of nitrogen and sulfur protects salt-inhibited photosynthesis through optimization of proline metabolism and antioxidant system in mustard (*Brassica juncea* L.). *Plants* **2021**, *10*, 1303. [CrossRef] [PubMed]
11. Su, Y.; Huang, Y.; Dong, X.; Wang, R.; Tang, M.; Cai, J.; Chen, J.; Zhang, X.; Nie, G. Exogenous methyl jasmonate improves heat tolerance of perennial ryegrass through alteration of osmotic adjustment, antioxidant defense, and expression of jasmonic acid-responsive genes. *Front. Plant Sci.* **2021**, *12*, 664519. [CrossRef] [PubMed]
12. Li, Z.G.; Yi, X.Y.; Li, Y.T. Effect of pretreatment with hydrogen sulfide donor sodium hydrosulfide on heat tolerance in relation to antioxidant system in maize (*Zea mays*) seedlings. *Biologia* **2014**, *69*, 1001–1009. [CrossRef]
13. Ahammed, G.J.; Li, X.; Zhou, J.; Zhou, Y.H.; Yu, J.Q. Role of hormones in plant adaptation to heat stress. In *Plant Hormones Under Challenging Environmental Factors*, 1st ed.; Ahammed, G.J., Yu, J.Q., Eds.; Springer: Dordrecht, The Netherlands, 2016; pp. 1–21.
14. Lang, D.; Yu, X.; Jia, X.; Li, Z.; Zhang, X. Methyl jasmonate improves metabolism and growth of NaCl-stressed *Glycyrrhiza uralensis* seedlings. *Sci. Horticul.* **2020**, *266*, 109287. [CrossRef]
15. Sirhindi, G.; Mushtaq, R.; Gill, S.S.; Sharma, P.; Allah, E.F.A.; Ahmad, P. Jasmonic acid and methyl jasmonate modulate growth, photosynthetic activity and expression of photosystem II subunit genes in *Brassica oleracea* L. *Sci. Rep.* **2020**, *10*, 1–14. [CrossRef] [PubMed]
16. Faghih, S.; Zarei, A.; Ghobadi, C. Positive effects of plant growth regulators on physiology responses of *Fragaria × ananassa* cv. "Camarosa" under salt stress. *Int. J. Fruit Sci.* **2019**, *19*, 104–114. [CrossRef]
17. Maksymiec, W.; Wojcik, M.; Krupa, Z. Variation in oxidative stress and photochemical activity in *Arabidopsis thaliana* leaves subjected to cadmium and excess copper in the presence or absence of jasmonate and ascorbate. *Chemosphere* **2007**, *66*, 421–427. [CrossRef]
18. Qiu, X.; Xu, Y.; Xiong, B.; Dai, L.; Huang, S.; Dong, T.; Sun, G.; Liao, L.; Deng, Q.; Wang, X.; et al. Effects of exogenous methyl jasmonate on the synthesis of endogenous jasmonates and the regulation of photosynthesis in citrus. *Physiol. Plant.* **2020**, *170*, 398–414. [CrossRef] [PubMed]
19. Attaran, E.; Major, I.T.; Cruz, J.A.; Rosa, B.A.; Koo, A.J.; Chen, J.; Howe, G.A. Temporal dynamics of growth and photosynthesis suppression in response to jasmonate signaling. *Plant Physiol.* **2014**, *165*, 1302–1314. [CrossRef] [PubMed]
20. Demming-Adams, B.; Adams, W.; Mattoo, A. *Photoprotection, Photoinhibition, Gene Regulation and Environment*; Springer: Dordrecht, The Netherlands, 2006; pp. 1–382.
21. Hewitt, E.J. *Sand and Water Culture Methods Used in the Study of Plant Nutrition*, 2nd ed.; Commonwealth Agricultural Bureaux, Farnham Royal: Bucks, UK; Cambridge University Press: East Malling, UK, 1966; pp. 187–190.
22. Bu, R.; Xie, J.; Yu, J.; Liao, W.; Xiao, X.; Lv, J. Autotoxicity in cucumber (*Cucumis sativus* L.) seedlings is alleviated by silicon through an increase in the activity of antioxidant enzymes and by mitigating lipid peroxidation. *J. Plant Biol.* **2016**, *59*, 247–250. [CrossRef]
23. Okuda, T.; Matsuda, Y.; Yamanaka, A.; Sagisaka, S. Abrupt increase in the level of hydrogen peroxide in leaves of winter wheat is caused by cold treatment. *Plant Physiol.* **1991**, *97*, 1265–1267. [CrossRef] [PubMed]
24. Dhindsa, R.S.; Plumb-Dhindsa, P.; Thorpe, T.A. Leaf senescence: Correlated with increased levels of membrane permeability and lipid peroxidation, and decreased levels of superoxide dismutase and catalase. *J. Exp. Bot.* **1981**, *32*, 93–101. [CrossRef]
25. Aebi, H. Catalase in vitro. *Meth. Enzymol.* **1984**, *105*, 121–126.
26. Foyer, C.H.; Halliwell, B. The presence of glutathione and glutathione reductase in chloroplasts: A proposed role in ascorbic acid metabolism. *Planta* **1976**, *133*, 21–25. [CrossRef]

27. Nakano, Y.; Asada, K. Hydrogen peroxide is scavenged by ascorbate-specific peroxidase in spinach chloroplasts. *Plant Cell Physiol.* **1981**, *22*, 867–880.

28. Fatma, M.; Asgher, M.; Masood, A.; Khan, N.A. Excess sulfur supplementation improves photosynthesis and growth in mustard under salt stress through increased production of glutathione. *Environ. Exp. Bot.* **2014**, *107*, 55–63. [CrossRef]

29. Fatma, M.; Masood, A.; Per, T.S.; Khan, N.A. Nitric oxide alleviates salt stress inhibited photosynthetic performance by interacting with sulfur assimilation in mustard. *Front. Plant Sci.* **2016**, *7*, 521. [CrossRef] [PubMed]

30. Fatma, M.; Iqbal, N.; Gautam, H.; Sehar, Z.; Sofo, A.; D'Ippolito, I.; Khan, N.A. Ethylene and sulfur coordinately modulate the antioxidant system and ABA accumulation in mustard plants under salt stress. *Plants* **2021**, *10*, 180. [CrossRef] [PubMed]

31. Giannopolitis, C.N.; Ries, S.K. Superoxide dismutases: Occurrence in higher plants. *Plant Physiol.* **1977**, *59*, 309–314. [CrossRef] [PubMed]

32. Beyer, W.F., Jr.; Fridovich, I. Assaying for superoxide dismutase activity: Some large consequences of minor changes in conditions. *Anal. Biochem.* **1987**, *161*, 559–566. [CrossRef]

33. Porra, R.J.; Thompson, W.A.; Kriedemann, P.E. Determination of accurate extinction coefficients and simultaneous equations for assaying chlorophylls a and b extracted with four different solvents: Verification of the concentration of chlorophyll standards by atomic absorption spectroscopy. *Biochim. Biophys. Acta Bioenerg.* **1989**, *975*, 384–394. [CrossRef]

34. Wellburn, A.R.; Lichtenthaler, H. Formulae and program to determine total carotenoids and chlorophylls a and b of leaf extracts in different solvents. In *Advances Photosynthesis Research*, 1st ed.; Sysbesma, C., Ed.; Springer: Dordrecht, The Netherlands, 1984; Volume 2, pp. 9–12.

35. Bradford, M.M. A rapid and sensitive method for the quantitation of microgram quantities of protein utilizing the principle of protein-dye binding. *Anal. Biochem.* **1976**, *72*, 248–254. [CrossRef]

36. Demmig, B.; Björkman, O. Comparison of the effect of excessive light on chlorophyll fluorescence (77K) and photon yield of O_2 evolution in leaves of higher plants. *Planta* **1987**, *171*, 171–184. [CrossRef] [PubMed]

37. Maxwell, K.; Johnson, G.N. Chlorophyll fluorescence—A practical guide. *J. Exp. Bot.* **2000**, *51*, 659–668. [CrossRef]

38. Strasser, R.J.; Tsimilli-Michael, M.; Srivastava, A. Analysis of the chlorophyll a fluorescence transient. In *Chlorophyll a Fluorescence*, 1st ed.; Papageorgiou, G.C., Govindjee, Eds.; Springer: Dordrecht, The Netherlands, 2004; Volume 19, pp. 321–362.

39. Stirbet, A.; Govindjee. On the relation between the Kautsky effect and Photosystem II: Basics and applications of the OJIP fluorescence transient. *J. Photochem. Photobiol. B* **2011**, *104*, 236–257. [CrossRef] [PubMed]

40. Stirbet, A.; Riznichenko, G.Y.; Rubin, A.B. Modeling chlorophyll a fluorescence transient: Relation to photosynthesis. *Biochemist* **2014**, *79*, 291–323. [CrossRef] [PubMed]

41. Chen, K.; Sun, X.; Amombo, E.; Zhu, Q.; Zhao, Z.; Chen, L.; Fu, J. High correlation between thermotolerance and photosystem II activity in tall fescue. *Photosynth. Res.* **2014**, *122*, 305–314. [CrossRef] [PubMed]

42. Tsimilli-Michael, M.; Eggenberg, P.; Biro, B.; Köves-Pechy, K.; Vörös, I.; Strasser, R.J. Synergistic and antagonistic effects of arbuscular mycorrhizal fungi and *Azospirillum* and *Rhizobium* nitrogen-fixers on the photosynthetic activity of alfalfa, probed by the polyphasic chlorophyll a fluorescence transient OJIP. *Appl. Soil Ecol.* **2000**, *15*, 169–182. [CrossRef]

43. Towbin, H.; Staehelin, T.; Gordon, J. Electrophoretic transfer of proteins from polyacrylamide gels to nitrocellulose sheets: Procedure and some applications. *Proc. Natl. Acad. Sci. USA* **1979**, *76*, 4350–4354. [CrossRef] [PubMed]

44. Callahan, F.E.; Ghirardi, M.L.; Sopory, S.K.; Mehta, A.M.; Edelman, M.; Mattoo, A.K. A novel metabolic form of the 32kDa-D1 protein in the grana localized reaction center of photosystem II. *J. Biol. Chem.* **1990**, *265*, 15357–15360. [CrossRef]

45. Chen, Y.; Gelfond, J.A.; McManus, L.M.; Shireman, P.K. Reproducibility of quantitative RT-PCR array in miRNA expression profiling and comparison with microarray analysis. *BMC Genom.* **2009**, *10*, 1–10. [CrossRef]

46. Farooq, M.A.; Gill, R.A.; Islam, F.; Ali, B.; Liu, H.; Xu, J.; Zhou, W. Methyl jasmonate regulates antioxidant defense and suppresses arsenic uptake in *Brassica napus* L. *Front. Plant Sci.* **2016**, *7*, 468. [CrossRef]

47. Verma, G.; Srivastava, D.; Narayan, S.; Shirke, P.A.; Chakrabarty, D. Exogenous application of methyl jasmonate alleviates arsenic toxicity by modulating its uptake and translocation in rice. *Ecotoxicol. Environ. Saf.* **2020**, *201*, 110735. [CrossRef] [PubMed]

48. Serna-Escolano, V.; Martínez-Romero, D.; Giménez, M.J.; Serrano, M.; García-Martínez, S.; Valero, D.; Valverde, J.M.; Zapata, P.J. Enhancing antioxidant systems by preharvest treatments with methyl jasmonate and salicylic acid leads to maintain lemon quality during cold storage. *Food Chem.* **2021**, *338*, 128044. [CrossRef] [PubMed]

49. Zuñiga, P.E.; Castañeda, Y.; Arrey-Salas, O.; Fuentes, L.; Aburto, F.; Figueroa, C.R. Methyl jasmonate applications from flowering to ripe fruit stages of strawberry (*Fragaria* × *ananassa* "Camarosa") reinforce the fruit antioxidant response at post-harvest. *Front. Plant Sci.* **2020**, *11*, 538. [CrossRef]

50. Singh, A.K.; Rana, H.K.; Pandey, A.K. Analysis of chlorophylls. In *Recent Advances in Natural Products Analysis*, 1st ed.; Silva, A.S., Nabavi, S.F., Saeedi, M., Nabavi, S.M., Eds.; Elsevier: Tehran, Iran, 2020; pp. 635–650.

51. Yamasato, A.; Nagata, N.; Tanaka, R.; Tanaka, A. The N-terminal domain of chlorophyllide a oxygenase confers protein instability in response to chlorophyll b accumulation in *Arabidopsis*. *Plant Cell* **2005**, *17*, 1585–1597. [CrossRef] [PubMed]

52. Parmar, P.; Kumari, N.; Sharma, V. Structural and functional alterations in photosynthetic apparatus of plants under cadmium stress. *Bot. Stud.* **2013**, *54*, 1–6. [CrossRef]

53. Cotado, A.; Müller, M.; Morales, M.; Munné-Bosch, S. Linking jasmonates with pigment accumulation and photoprotection in a high-mountain endemic plant, *Saxifraga longifolia*. *Environ. Exp. Bot.* **2018**, *154*, 56–65. [CrossRef]

54. Bi, A.; Fan, J.; Hu, Z.; Wang, G.; Amombo, E.; Fu, J.; Hu, T. Differential acclimation of enzymatic antioxidant metabolism and photosystem II photochemistry in tall fescue under drought and heat and the combined stresses. *Front. Plant Sci.* **2016**, *7*, 453. [CrossRef]

55. Giorio, P.; Sellami, M.H. Polyphasic OKJIP Chlorophyll a fluorescence transient in a landrace and a commercial cultivar of sweet pepper (*Capsicum annuum, L.*) under long-term salt stress. *Plants* **2021**, *10*, 887. [CrossRef] [PubMed]

56. Yunus, M.; Pathre, U.; Mohanty, P. The fluorescence transient as a tool to characterize and screen photosynthetic samples. In *Probing Photosynthesis: Mechanisms, Regulation and Adaptation*; Strasser, A., Srivastava, A., Tsimilli-Michael, M., Eds.; Taylor and Francis: London, UK, 2000; pp. 445–483.

57. Fan, Y.; Liu, Z.; Zhang, F.; Zhao, Q.; Wei, Z.; Fu, Q.; Li, H. Tunable mid-infrared coherent perfect absorption in a graphene meta-surface. *Sci. Rep.* **2015**, *5*, 1–8. [CrossRef] [PubMed]

58. Chen, K.; Chen, L.; Fan, J.; Fu, J. Alleviation of heat damage to photosystem II by nitric oxide in tall fescue. *Photosynth. Res.* **2013**, *116*, 21–31. [CrossRef] [PubMed]

59. Yamamoto, Y.; Aminaka, R.; Yoshioka, M.; Khatoon, M.; Komayama, K.; Takenaka, D.; Yamamoto, Y. Quality control of photosystem II: Impact of light and heat stresses. *Photosynth. Res.* **2008**, *98*, 589–608. [CrossRef]

60. Yamamoto, Y.; Hori, H.; Kai, S.; Ishikawa, T.; Ohnishi, A.; Tsumura, N.; Morita, N. Quality control of Photosystem II: Reversible and irreversible protein aggregation decides the fate of Photosystem II under excessive illumination. *Front. Plant Sci.* **2013**, *4*, 433. [CrossRef] [PubMed]

61. Kale, R.; Hebert, A.E.; Frankel, L.K.; Sallans, L.; Bricker, T.M.; Pospíšil, P. Amino acid oxidation of the D1 and D2 proteins by oxygen radicals during photoinhibition of Photosystem II. *Proc. Natl. Acad. Sci. USA* **2017**, *114*, 2988–2993. [CrossRef]

62. Vani, B.; Saradhi, P.P.; Mohanty, P. Alteration in chloroplast structure and thylakoid membrane composition due to in vivo heat treatment of rice seedlings: Correlation with the functional changes. *J. Plant Physiol.* **2000**, *158*, 583–592. [CrossRef]

63. Sehar, Z.; Iqbal, N.; Khan, M.I.R.; Masood, A.; Rehman, M.T.; Hussain, A.; Alajmi, M.F.; Ahmad, A.; Khan, N.A. Ethylene reduces glucose sensitivity and reverses photosynthetic repression through optimization of glutathione production in salt-stressed wheat (*Triticum aestivum* L.). *Sci. Rep.* **2021**, *11*, 1–12. [CrossRef] [PubMed]

64. Bredenkamp, G.J.; Baker, N.R. Temperature-sensitivity of D1 protein metabolism in isolated *Zea mays* chloroplasts. *Plant Cell Environ.* **1994**, *17*, 205–210. [CrossRef]

65. Adamiec, M.; Misztal, L.; Kosicka, E.; Paluch-Lubawa, E.; Luciński, R. *Arabidopsis thaliana* egy2 mutants display altered expression level of genes encoding crucial photosystem II proteins. *J. Plant Physiol.* **2018**, *231*, 155–167. [CrossRef] [PubMed]

Article

Nitric Oxide Regulates Plant Growth, Physiology, Antioxidant Defense, and Ion Homeostasis to Confer Salt Tolerance in the Mangrove Species, *Kandelia obovata*

Mirza Hasanuzzaman [1,2,*], Masashi Inafuku [1], Kamrun Nahar [3], Masayuki Fujita [4] and Hirosuke Oku [1,*]

1   Molecular Biotechnology Group, Center of Molecular Biosciences (COMB), Tropical Biosphere Research Center, University of the Ryukyus, 1 Senbaru, Nishihara, Okinawa 903-0213, Japan; h098648@eve.u-ryukyu.ac.jp

2   Department of Agronomy, Faculty of Agriculture, Sher-e-Bangla Agricultural University, Dhaka 1207, Bangladesh

3   Department of Agricultural Botany, Faculty of Agriculture, Sher-e-Bangla Agricultural University, Dhaka 1207, Bangladesh; knahar84@yahoo.com

4   Laboratory of Plant Stress Responses, Department of Applied Biological Science, Faculty of Agriculture, Kagawa University, 2393 Ikenobe, Miki-cho, Kita-gun, Kagawa 761-0795, Japan; fujita@ag.kagawa-u.ac.jp

*   Correspondence: mhzsauag@yahoo.com (M.H.); okuhiros@comb.u-ryukyu.ac.jp (H.O.)

Citation: Hasanuzzaman, M.; Inafuku, M.; Nahar, K.; Fujita, M.; Oku, H. Nitric Oxide Regulates Plant Growth, Physiology, Antioxidant Defense, and Ion Homeostasis to Confer Salt Tolerance in the Mangrove Species, *Kandelia obovata*. *Antioxidants* **2021**, *10*, 611. https://doi.org/10.3390/antiox10040611

Academic Editor: Luca Sebastiani

Received: 3 March 2021
Accepted: 12 April 2021
Published: 16 April 2021

Abstract: Facultative halophyte *Kandelia obovata* plants were exposed to mild (1.5% NaCl) and severe (3% NaCl) salt stress with or without sodium nitroprusside (SNP; 100 µM; a NO donor), hemoglobin (Hb, 100 µM; a NO scavenger), or Nω-nitro-L-arginine methyl ester (L-NAME, 100 µM; a NO synthase inhibitor). The plants were significantly affected by severe salt stress. They showed decreases in seedling growth, stomatal conductance, intercellular CO_2 concentration, SPAD value, photosynthetic rate, transpiration rate, water use efficiency, and disrupted antioxidant defense systems, overproduction of reactive oxygen species, and visible oxidative damage. Salt stress also induced ion toxicity and disrupted nutrient homeostasis, as indicated by elevated leaf and root Na^+ contents, decreased K^+ contents, lower K^+/Na^+ ratios, and decreased Ca contents while increasing osmolyte (proline) levels. Treatment of salt-stressed plants with SNP increased endogenous NO levels, reduced ion toxicity, and improved nutrient homeostasis while further increasing Pro levels to maintain osmotic balance. SNP treatment also improved gas exchange parameters and enhanced antioxidant enzymes' activities (catalase, ascorbate peroxidase, monodehydroascorbate reductase, and dehydroascorbate reductase). Treatment with Hb and L-NAME reversed these beneficial SNP effects and exacerbated salt damage, confirming that SNP promoted stress recovery and improved plant growth under salt stress.

Keywords: halophytes; antioxidants; reactive oxygen species; soil salinity; signaling molecules; abiotic stress

1. Introduction

Among the plethora of abiotic stresses experienced by plants, salt stress has attracted the attention of plant scientists because of its complexity and its widespread effects over different regions of the globe. The global rise of temperature is gradually increasing the sea level and threatening coastal regions with salinity problems [1,2]. Salinity causing significant challenges in the way of crop cultivation and food production for a growing global population. Salt stress induces osmotic/dehydration stress and ion toxicity that disrupts the structure and function of biomolecules, including lipids, proteins, and DNA, thereby destroying the structure of biomembranes and causing damage to cell organelles [3]. These changes adversely affect vital phenological processes and disrupt critical physiological functions, including water uptake, water use efficiency, nutrient uptake and mobilization, transpiration, respiration, photosynthesis, and assimilation of photosynthate [4–6].

Current strategies to address salt stress have included incorporating modern cultivation practices and the introduction of plant species that can adapt to or tolerate salt stress.

Kandelia obovata, a mangrove species widely distributed in extremely saline environments in eastern Asia [7]. Learning from such halophytes and tailoring the traits associated with high salinity tolerance may build a foundation for salt tolerance in glycophytes.

Reducing photosynthesis is one of the major consequences of salt stress. Salt stress can damage chloroplast membranes and destroy the structure of these organelles. It also reduces stomatal conductance, thereby reducing carbon dioxide uptake, decreasing the carboxylation reaction of RuBisCO, and depressing photosystem II (PSII) activity, electron transport, and photophosphorylation activity [8–10]. This salt-induced stomatal closure and inhibition of photosynthesis exposes chloroplasts to excessive excitation energy that leads to the aggravated generation of reactive oxygen species (ROS), including superoxide ($O_2^{\bullet-}$), hydrogen peroxide (H_2O_2), hydroxyl radical ($OH^{\bullet}$), and singlet oxygen ($^1O_2$) that cause serious damage to plant cells [11–13]. Plants have internal antioxidant defense systems that scavenge ROS and reduce this oxidative stress to some extent, and some plant species with highly active antioxidant systems are more tolerant of ROS and, therefore, of salt stress. Nonenzymatic antioxidants (ascorbate (AsA), glutathione (GSH), α-tocopherol, phenolic compounds, alkaloids, and nonprotein amino acids) and antioxidant enzymes [(superoxide dismutase (SOD), catalase (CAT), ascorbate peroxidase (APX), monodehydroascorbate reductase (MDHAR), dehydroascorbate reductase (DHAR), glutathione reductase (GR), glutathione peroxidase (GPX), glutathione *S*-transferase (GST), and peroxidases (POD)] function to scavenge ROS and decrease their levels under stress conditions [13–15].

Nitric oxide (NO) has recently been recognized as a potential signaling molecule and is often termed a plant hormone [16]. Enhancement of endogenous NO levels, as well as the exogenous application of NO, has been reported to improve salt tolerance via several different mechanisms. NO can stimulate the activity of proton pumps and the Na^+/H^+ antiport in the tonoplast, thereby improving the K^+/Na^+ ratio [17,18]. NO-induced improvement in the functioning of antioxidant systems also alleviates oxidative stress [19–21]. NO regulates the activity of phosphoenolpyruvate carboxylase kinase to mediate responses of plants to salinity [22]. NO also modulates stomatal conductance, promotes quenching of excess energy, and raises the quantum yield of PSII to improve photosynthesis [23–26].

Based on previous research, we hypothesized that NO could have a potential role in the tolerance of mangrove plants to high salinity. For this reason, we exposed mangrove plants to very high salt concentrations to understand the response and physiology behind the adaptative behavior of *K. obovata* growing in highly saline conditions. Our overall goal is to exploit this knowledge to instill salt tolerance in other cultivated plants that today face severe damage and crop losses due to salinity. Here, we examined mangrove plants' response to high salt stress and determined the effects of exogenously applied NO on salt tolerance in this plant.

2. Materials and Methods

2.1. Growth Condition and Treatments

Healthy propagules were collected from the Iriomote mangrove forest, Okinawa, Japan. The propagules were washed and planted in Wagner pots containing salt-free sand. The propagules were allowed to grow in a glasshouse at the University of the Ryukyus, Okinawa, Japan, under controlled environmental conditions (relative humidity 60–70%, temperature $25 \pm 2\ ^{\circ}C$, light 600 μmol m^{-1} s^{-1} and photoperiod from 12 to 14 h) in stagnant water. The experiment was conducted following a completely randomized design (CRD) with three replicates. After two months, the seedlings were exposed to freshwater or to water containing two salt levels (1.5 and 3% NaCl). Each set of seedlings was treated with 100 μM sodium nitroprusside (SNP, a NO donor), 100 μM hemoglobin (Hb, a NO scavenger), and 100 μm Nω-nitro-L-arginine methyl ester (L-NAME, a NO synthase inhibitor) and allowed to grow for two more months. The solutions were renewed every 15 days. At the end of the experiment, the plants were harvested and separated into shoots and roots. After washing well with distilled water, the plant parts were ground to a fine powder in liquid nitrogen and stored at $-80\,^{\circ}$C until further use. The experiment was replicated three times.

2.2. Quantification of Nitric Oxide

Nitric oxide was measured indirectly by quantifying nitrite. To determine the leaves' NO contents, 0.5 g of freshly plucked leaves were ground in an ice-cold mortar with acetic acid buffer (50 mM, pH 3.6) and then centrifuged ($10,000\times g$, 15 min) to remove the residue. The supernatants were decolorized with charcoal and mixed with Griess reagent. Griess reagent was prepared with N-1-naphthylethylenediamine dihydrochloride, sulfanilamide and phosphoric acid. The absorbance was taken at 540 nm, and the amount of NO was calculated using a standard curve [21].

2.3. Measurement of Gas Exchange and Photosynthetic Parameters

Data of stomatal conductance, intercellular CO_2, photosynthetic rate, transpiration rate and water use efficiency were determined using a LiCOR 6400 open system portable infrared red gas analyzer (IRGA) (LI-COR Biosciences, Lincoln, NE, USA) from fully expanded leaves at 10 a.m. to 2.00 p.m. The conditions, which were used for the equipment/leaf chamber were as follows: ambient pressure 99.2 kPa, atmospheric CO_2 concentration (Cref) 400 μmol mol^{-1}, leaf surface area 6 cm^2, PAR (Qleaf) was maximum up to 900–1000 μmol m^{-2} s^{-1} and the chamber water vapor pressure varied from 4.0 to 5.8 mbar [27]. SPAD values were recorded using a SPAD-502 m (Konica-Minolta, Tokyo, Japan).

2.4. Estimation of Proline Content

Proline (Pro) content was estimated by the method of Bates et al. [28]. Leaf was extracted by sulfosalicylic acid (3%) and glacial acetic acid, acid ninhydrin, and then the mixture was heated at the boiling water bath at 100 $^{\circ}$C for 60 min. The reaction was terminated by placing the tube in an ice bath for 15 min. Toluene was then used as reaction reagents with the previous mixture. Upper toluene chromophore was used for spectrophotometric observation at 520 nm. The content of proline was calculated in the test sample using a standard curve.

2.5. Determination of Electrolyte Leakage

Electrolyte leakage (EL) was determined following the method of Dionisio-Sese and Tobita [29]. Electrical conductivity (EC) was measured by using a portable EC meter (Eutech Instruments, Singapore). Initial (EC$_1$) and final (EC$_2$) EC were recorded, and EL was determined by the following formula: EL(%) = (EC$_1$/EC$_2$) $\times$ 100.

2.6. Measurement of Lipid Peroxidation

The level of lipid peroxidation was measured by estimating malondialdehyde (MDA), a decomposition product of the per-oxidized polyunsaturated fatty acid composition of the membrane lipid, using 2-thiobarbituric acid (TBA) as the reactive material following the method of Heath and Packer [30]. The tissue was homogenized with ice-cold trichloroacetic acid (TCA) and then was centrifuged at $11,500\times g$ at 4 $^{\circ}$C. The supernatant was mixed with a reaction mixture of thiobarbituric acid (TBA) and TCA and heated in a water bath. Then the cooled supernatant mixture was read at 532 nm. The concentration of MDA was calculated by using the extinction coefficient of 155 mM^{-1} cm^{-1} and was expressed as nmol of MDA g^{-1} FW.

2.7. Measurement of Hydrogen Peroxide Generation

The amount of H_2O_2 generated was recorded using the method described by Yu et al. [31]. Fresh leaf samples of 0.5 g were extracted with 3 mL of potassium phosphate (K–P) buffer (50 mM, pH 6.5) and centrifuged at $11,500\times g$ for 15 min to obtain a clear supernatant. A total of 2 mL of this supernatant was mixed with 666.4 µL of the reaction mixture (5.5 mM $TiCl_4$ in 20% H_2SO_4), incubated for 10 min at room temperature and centrifuged again at $11,500\times g$ for 12 min. To determine the H_2O_2 content, the absorption of the final supernatant was read at 410 nm with a spectrophotometer using 0.28 μM^{-1} cm^{-1} as the extinction coefficient, and the results are expressed as nmol g^{-1} FW.

2.8. Measurement of Antioxidant Enzyme Activities

Leaf samples were homogenized using potassium-phosphate (K-P) buffer (pH 7.0), AsA, β-mercaptoethanol, KCl, and glycerol as an extraction buffer. After centrifuging, leaf extracts were used to determine protein and enzyme activity. Soluble protein concentration was measured using Coomassie brilliant blue dye following the technique of Bradford [32]. The optical absorbance was then recorded at 595 nm, and a standard curve prepared by using bovine serum albumin was used for calculation.

Ascorbate peroxidase (EC: 1.11.1.11) activity was assayed following Nakano and Asada [33] by using K-P buffer (pH 7.0), AsA, ethylenediaminetetraacetic acid (EDTA), and H_2O_2 and absorbance were recorded at 290 nm. APX activity was calculated with an extinction co-efficient 2.8 mM^{-1} cm^{-1}.

The activity of MDHAR (EC: 1.6.5.4) was measured by using a buffer solution containing ascorbate oxidase (AO), Tris-HCl buffer (pH 7.5), AsA, and nicotinamide adenine dinucleotide phosphate (NADPH) according to the method described by Parvin et al. [34]. Optical absorbance was observed at 340 nm. An extinction coefficient of 6.2 mM^{-1} cm^{-1} was used for calculating its activity.

Dehydroascorbate reductase (EC: 1.8.5.1) activity was assayed according to Nakano and Asada [33] by using dehydroascorbate (DHA), K-P buffer (pH 7.0), GSH, and EDTA and absorbance were recorded at 265 nm and extinction coefficient 14 mM^{-1} cm^{-1} was considered for calculating DHAR activity.

Catalase (EC: 1.11.1.6) activity was determined by using K-P buffer (pH 7.0) and H_2O_2 according to the method described by Hasanuzzaman et al. [35]. Optical absorbance was recorded at 240 nm and calculated using 39.4 M^{-1} cm^{-1} as the extinction coefficient.

Peroxidase (EC: 1.11.1.7) activity was estimated by using K-P buffer (pH 7), catechol, and H_2O_2 according to the method described by Gong et al. [36]. Absorbance was recorded at 470 nm and expressed as $Umin^{-1}$ mg^{-1} protein.

Glutathione *S*-transferase (EC: 2.5.1.18) activity was assayed by using 1-chloro-2,4-dinitrobenzene (CDNB), Tris-HCl buffer (pH 6.5), and GSH according to the technique described by Rahman et al. [28] and absorbance was measured at 340 nm.

2.9. Measurement of Ion Contents

The contents of Na^+, K^+ and Ca^{2+} were measured from the dry shoot and root tissues by following Rahman et al. [37]. From ground homogenous dry plant samples, 100 mg was digested with the acid mixture (HNO_3: $HClO_4$; 5:1). Digestion of the sample was done at 70 °C for 48 h. The Na^+, K^+ and Ca^{2+} contents were determined from the digested solution by using the different lamps for the different minerals of atomic absorption spectrophotometer (AA-7000, Shimadzu, Japan), and the content was calculated using the standard curve of the respective mineral.

2.10. Statistical Analysis

Data accumulated from different parameters were subjected to analysis using CoStat v.6.400 [38] and one-way analysis of variance (ANOVA). For finding out mean differences among the replications, Tukey's honest significant difference (HSD) test at the 5% level of significance was applied.

3. Results

3.1. Root and Shoot Length

Both root and shoot growth were significantly inhibited with the increase in NaCl concentration from 1.5% to 3%. Root length decreased by 8% of the control in response to 1.5% salt but was decreased by 25% by exposure to 3.0% NaCl (Figure 1A). Shoot length showed a similar pattern of reduction to increasing salt stress (Figure 1B). The inclusion of SNP in the salt medium somewhat alleviated the negative effect of salt stress, as indicated by an increase in root length of 14 and 22%, and an increase in shoot length of 8 and 11%, in plants growing in 1.5% and 3% NaCl, respectively (Figure 1A,B). The beneficial role of SNP was further established by treatment with Hb and L-NAME, as scavenging NO or inhibiting its production in plants under salt stress further exacerbated the decreases in root and shoot length, compared to the plants treated with salt alone (Figure 1A,B).

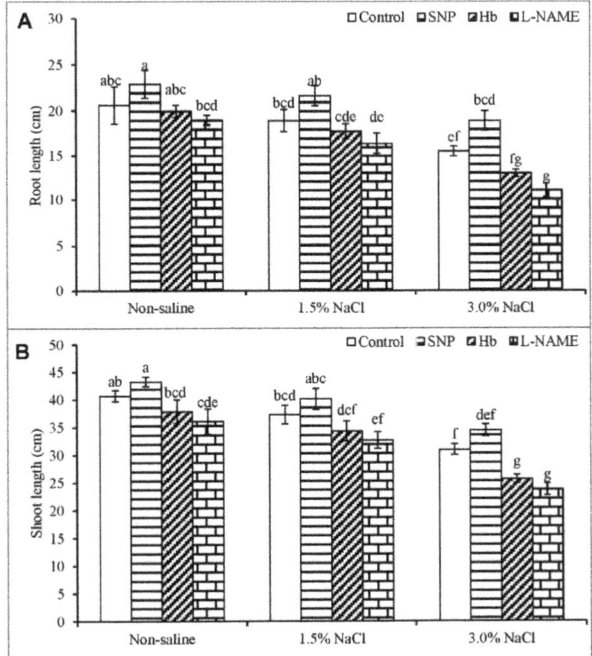

Figure 1. Root length (**A**) and shoot length (**B**) of *Kandelia obovata* treated with different salt concentrations and with 100 μM sodium nitroprusside (SNP, a NO donor), hemoglobin (Hb, a NO scavenger), or Nω-nitro- L-arginine methyl ester (L-NAME, a NO synthase inhibitor). Mean ($\pm$ SD) was calculated from three replicates for each treatment. Bars with different letters are significantly different at $p \leq 0.05$, applying Tukey's HSD test.

3.2. Endogenous Nitric Oxide Level

Treatment with SNP increased the NO content in nonstressed plants compared to the untreated control. Salt stress at 1.5% and 3% increased the NO content by 48 and 61%, respectively, in the roots (Figure 2A) and by 8 and 12%, respectively, in the leaf (Figure 2B), compared to the control treatment. Treatment of plants with SNP increased the NO content by 15% and 25% in the root (Figure 2A) and by 39% and 37% in the leaf (Figure 2B) in plants growing in 1.5 and 3% NaCl respectively, compared to salt stress alone. Treatment of salt-stressed plants with either Hb or L-NAME decreased the NO level drastically, both the root and the shoot, confirming the inhibitory effect of Hb or L-NAME on NO levels.

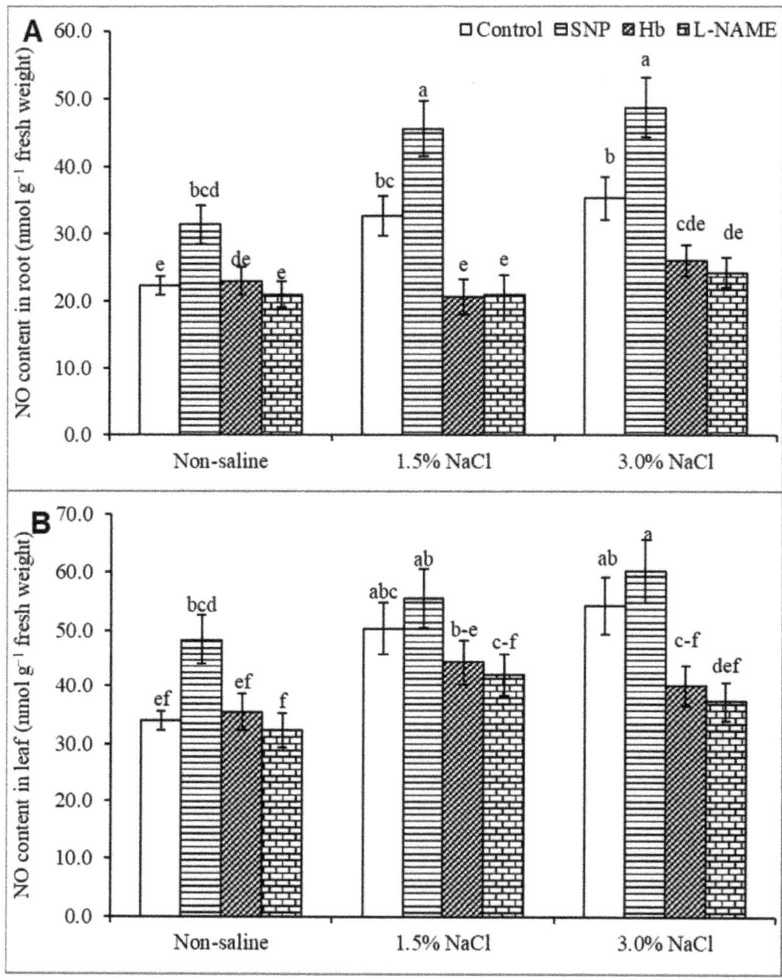

Figure 2. Endogenous NO content of *Kandelia obovata* root (**A**) and leaf (**B**) treated with different salt concentrations and with 100 μM sodium nitroprusside (SNP, a NO donor), hemoglobin (Hb, a NO scavenger), or Nω-nitro- L-arginine methyl ester (L-NAME, a NO synthase inhibitor). Mean (± SD) was calculated from three replicates for each treatment. Bars with different letters are significantly different at $p \leq 0.05$, applying Tukey's HSD test.

3.3. Gas Exchange and Photosynthetic Parameters

Various photosynthetic parameters were affected by salt exposure. Stomatal conductance (g_s) did not change under the mild 1.5% salt stress but was substantially reduced by 3% NaCl stress than the unstressed control (Figure 3A). Treatment of the salt-stressed plants with SNP increased g_s, even in the plants treated with 1.5% NaCl (Figure 3A). Salt stress did not alter the C_i value at 1.5% NaCl, but the C_i value was decreased severely, by 50%, in the 3% salt treatment compared to the unstressed control. Treatment of the salt-stressed plants with SNP had no significant effect on the C_i level than salt stress alone. However, treatment with Hb or L-NAME decreased the C_i value drastically (Figure 3B). The SPAD value also decreased in response to 3% NaCl compared to the unstressed control. The SPAD value increased when salt-treated plants were treated with SNP, compared to plants exposed to salt stress alone. Treatment of salt-stressed plants with Hb or L-NAME substantially decreased the SPAD value (Figure 3C). As shown in Figure 3D, the P_n was not affected significantly at 1.5% NaCl, but it was decreased by 24% under 3% NaCl stress compared to the unstressed control. Treatment of the salt-stressed plants with SNP somewhat alleviated the detrimental effects of salt stress, resulting in increased P_n compared to the untreated salt-stressed plants. Treatment with Hb or L-NAME further decreased the P_n value compared to salt stress alone, supporting a positive effect of NO on photosynthesis (Figure 3D).

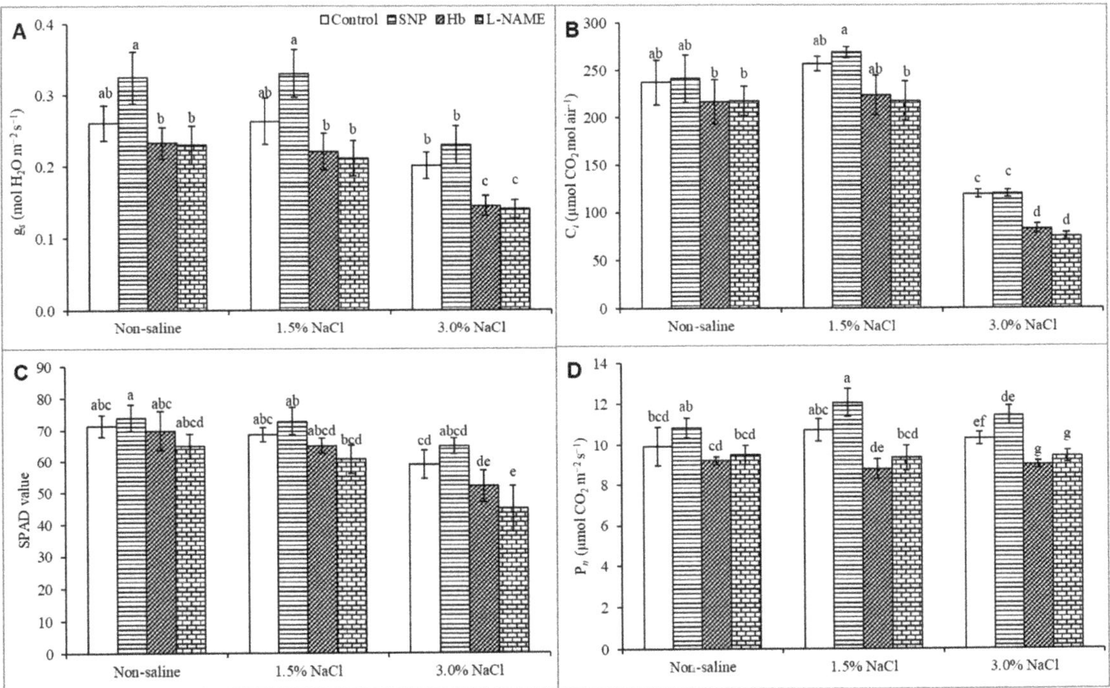

Figure 3. Stomatal conductance, g_s (**A**); internal $[CO_2]$, C_i (**B**); SPAD value (**C**) and net photosynthesis, P_n (**D**) of *Kandelia obovata* treated with different salt concentrations and with 100 μM sodium nitroprusside (SNP, a NO donor), hemoglobin (Hb, a NO scavenger), or Nω-nitro- L-arginine methyl ester (L-NAME, a NO synthase inhibitor). Mean ($\pm$ SD) was calculated from three replicates for each treatment. Bars with different letters are significantly different at $p \leq 0.05$, applying Tukey's HSD test.

3.4. Transpiration Rate and Water Use Efficiency

The T_r value was unaffected by treatment with 1.5% NaCl but was decreased by 20% in response to 3% NaCl stress compared to the unstressed control (Figure 4A). Treatment of salt-stressed plants with SNP did not significantly increase the T_r value above that seen with salt stress alone. Treatment of salt-stressed plants with Hb or L-NAME further decreased T_r (Figure 4A). The WUE value did not change under salt stress than the unstressed control, and treatment with SNP did not significantly increase WUE (Figure 4B). However, treatment with Hb or L-NAME substantially decreased WUE, indicating a role for NO (Figure 4B).

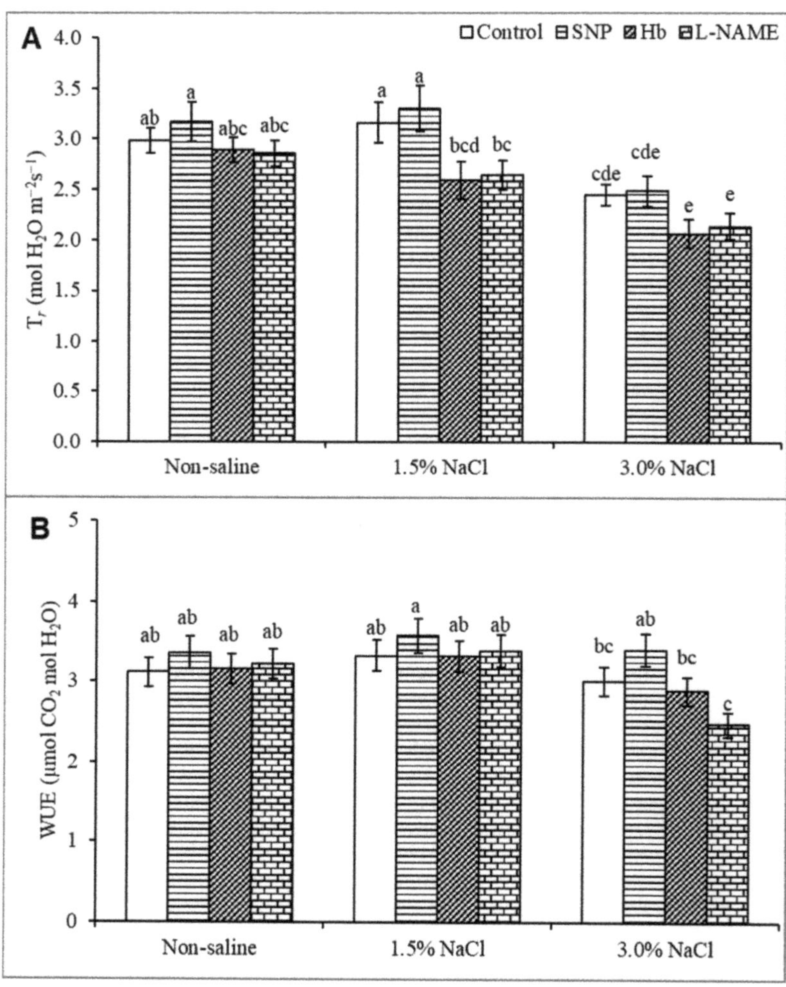

Figure 4. Transpiration rate (**A**) and water use efficiency, WUE (**B**) of *Kandelia obovata* treated with different salt concentrations and with 100 µM sodium nitroprusside (SNP, a NO donor), hemoglobin (Hb, a NO scavenger), or Nω-nitro- L-arginine methyl ester (L-NAME, a NO synthase inhibitor). Mean (± SD) was calculated from three replicates for each treatment. Bars with different letters are significantly different at $p \leq 0.05$, applying Tukey's HSD test.

3.5. Proline Content and Electrolyte Leakage

The plant Pro content increased progressively with the increase in salt concentration. Compared to the control treatment, the Pro level increased by 66 and 197% in the leaf and by 104% and 217% in the root in response to 1.5 and 3% NaCl, respectively. Treatment of salt-stressed plants with SNP further increased the Pro content, whereas treatment with Hb or L-NAME decreased the Pro content in both the root and shoot (Figure 5A,B). Salt stress caused electrolyte leakage (EL) from the root and leaf, with EL levels increasing by 49% and 111% from the leaf and 33% and 60% from the root in response to 1.5% and 3% NaCl, respectively compared to the unstressed control. Treatment with SNP decreased EL from the salt-stressed plants., whereas treatment with Hb or L-NAME increased EL (Figure 5C,D).

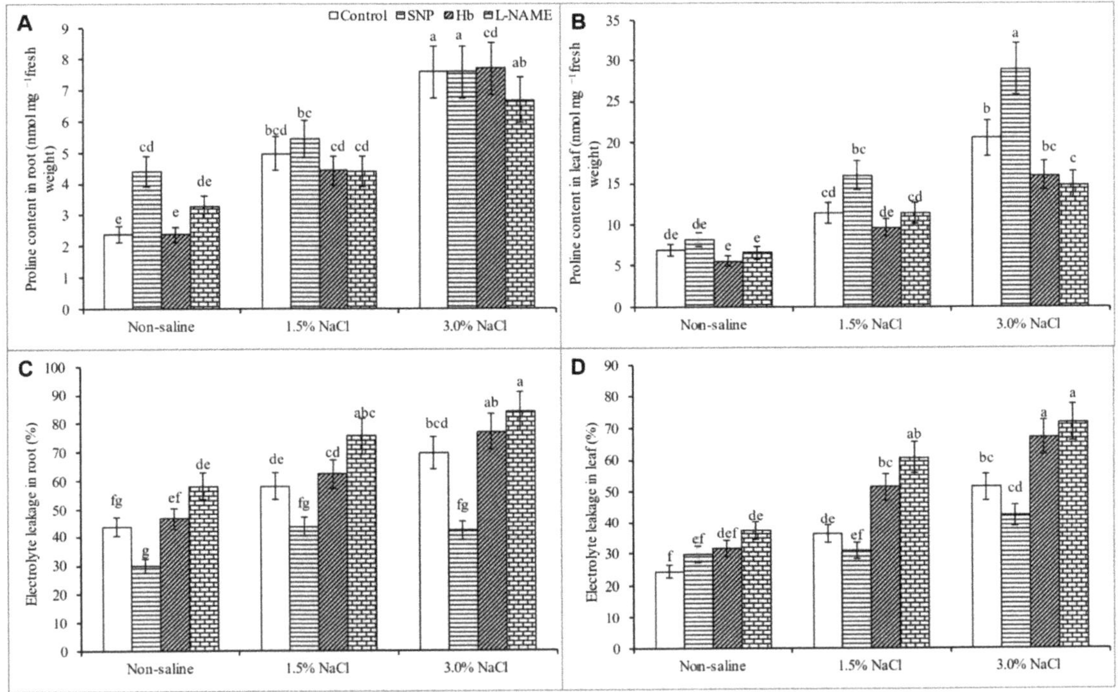

Figure 5. Proline content (**A,B**) and electrolyte leakage (**C,D**) in roots and leaves of *Kandelia obovata* treated with different salt concentrations and with 100 μM sodium nitroprusside (SNP, a NO donor), hemoglobin (Hb, a NO scavenger), or Nω-nitro-L-arginine methyl ester (L-NAME, a NO synthase inhibitor). Mean (± SD) was calculated from three replicates for each treatment. Bars with different letters are significantly different at $p \leq 0.05$, applying Tukey's HSD test.

3.6. Oxidative Stress Markers

The higher 3% NaCl level increased the MDA levels in both the leaf and root by 97% and 60%, respectively, compared to the unstressed control, whereas the lower 1.5% salt stress caused no significant change in MDA levels compared to the untreated control. Treatment with SNP decreased the MDA level in seedlings growing in 3% NaCl, whereas treatment with Hb or L-NAME caused increases in MDA levels in the salt-treated plants (Figure 6A,B). The hydrogen peroxide (H_2O_2) content was not significantly increased in the leaf or root in response to 1.5% NaCl stress and only showed a slight increase in response to 3% salt. Treatment of salt-stressed plants with SNP decreased the H_2O_2 levels compared to salt stress alone. Treatment of salt-stressed plants with Hb or L-NAME further increased the H_2O_2 levels above those observed in the SNP-treated salt-stressed plants (Figure 6C,D).

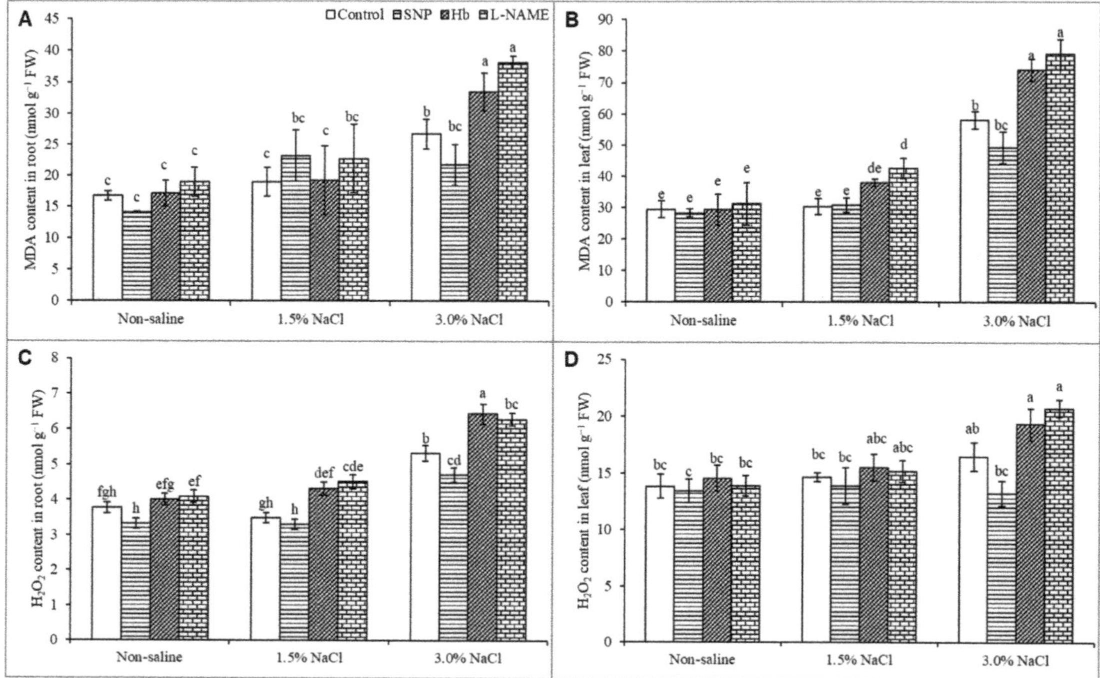

Figure 6. MDA content (**A,B**) and H₂O₂ content (**C,D**) in roots and leaves of *Kandelia obovata* treated with different salt concentrations and with 100 µM sodium nitroprusside (SNP, a NO donor), hemoglobin (Hb, a NO scavenger), or Nω-nitro-L-arginine methyl ester (L-NAME, a NO synthase inhibitor). Mean (± SD) was calculated from three replicates for each treatment. Bars with different letters are significantly different at $p \leq 0.05$, applying Tukey's HSD test.

3.7. Activities of Antioxidant Enzymes

In general, antioxidant enzyme activities (CAT, APX, MDHAR, and DHAR) increased in response to salt stress compared to the unstressed control. The exceptions were leaf and root DHAR activity in response to 1.5% NaCl stress and leaf MDHAR in response to both levels of salt stress. Treatment with SNP did not cause significant increases in enzyme activity in salt-treated plants, except for the leaf DHAR activity in the 3% NaCl treatment (Figure 7A–F).

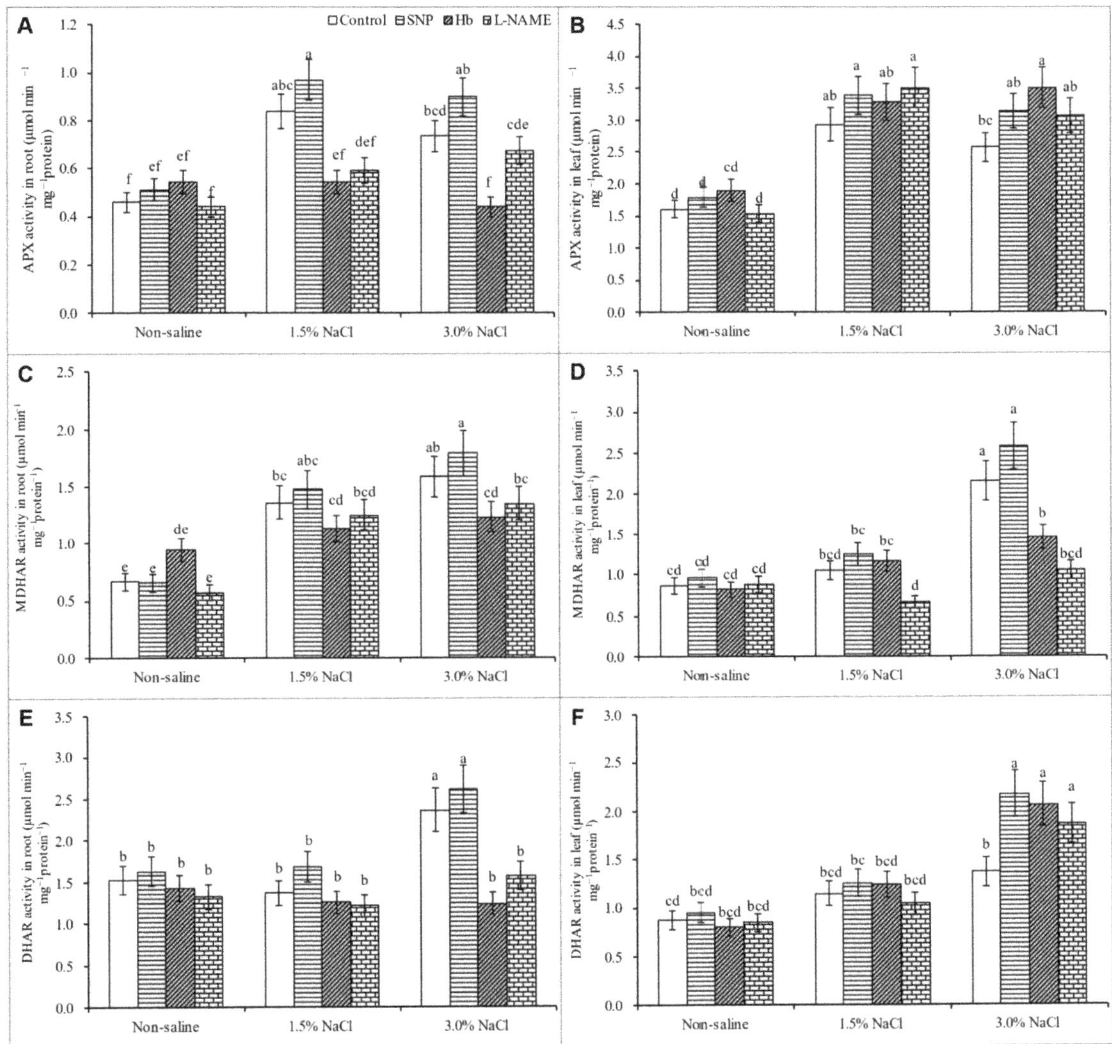

Figure 7. Activity of APX (**A,B**), MDHAR (**C,D**) and DHAR (**E,F**) in roots and leaves of *Kandelia obovata* treated with different salt concentrations and with 100 µM sodium nitroprusside (SNP, a NO donor), hemoglobin (Hb, a NO scavenger), or Nω-nitro- L-arginine methyl ester (L-NAME, a NO synthase inhibitor). Mean ($\pm$ SD) was calculated from three replicates for each treatment. Bars with different letters are significantly different at $p \leq 0.05$, applying Tukey's HSD test.

The POD activity increased by 118 and 74% in the root and by 117% and 34% in the leaf in response to 1.5 and 3% NaCl treatment, respectively, compared to the unstressed control (Figure 8A,B). The activity of CAT increased only in response to 1.5% NaCl stress, compared to the unstressed control (Figure 8B,C). No changes were noted in root and leaf GST activity under salt stress compared to the unstressed control (Figure 8E,F). Treatment with SNP increased the CAT activity in the leaf and root above that observed in plants exposed to salt stress alone (Figure 8A–F).

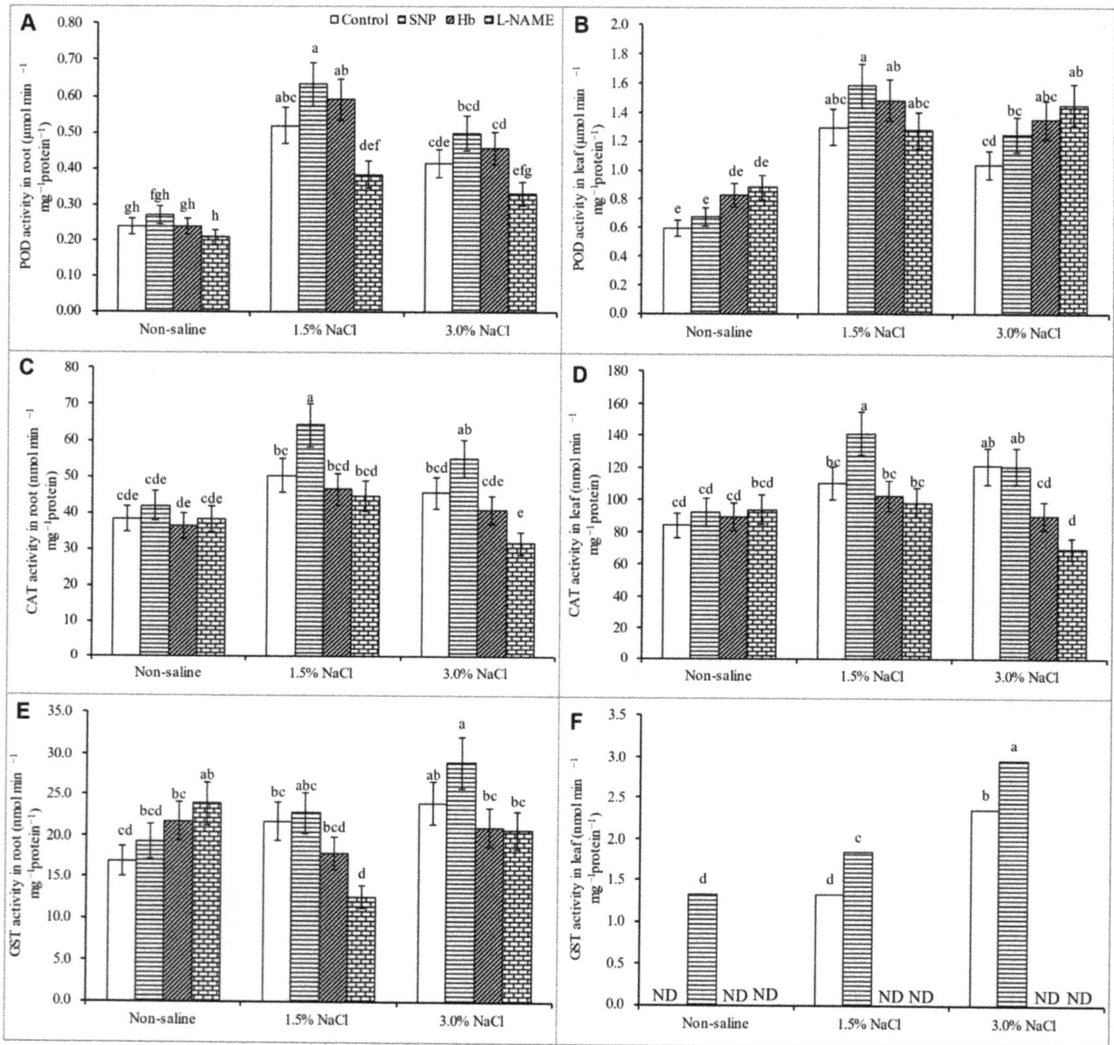

Figure 8. Activity of POD (**A**,**B**), CAT (**C**,**D**) and GST (**E**,**F**) in roots and leaves of *Kandelia obovata* treated with different salt concentrations and with 100 μM sodium nitroprusside (SNP, a NO donor), hemoglobin (Hb, a NO scavenger), or Nω-nitro-L-arginine methyl ester (L-NAME, a NO synthase inhibitor). Mean ($\pm$ SD) was calculated from three replicates for each treatment. Bars with different letters are significantly different at $p \leq 0.05$, applying Tukey's HSD test.

3.8. Ion Contents

Salt stress created ion toxicity by increasing the plant Na^+ content, decreasing the K^+ content, and disrupting the Na^+/K^+ balance; however, it also disrupted nutrient homeostasis. Treatment of plants with 1.5% NaCl increased the Na^+ levels in the leaf and root by 317% and 738%, respectively, while treatment with 3% NaCl increased these levels by 725% and 465%, respectively compared to the unstressed control (Figure 9A,B). By contrast, the K^+ content in the leaf and root decreased by 23% and 14%, respectively, at 1.5% NaCl stress, and by 45% and 41%, respectively, at 3% NaCl compared to the unstressed control (Figure 9C,D). Consequently, the K^+/Na^+ ratio decreased considerably in both the leaf and the root (Figure 9E,F).

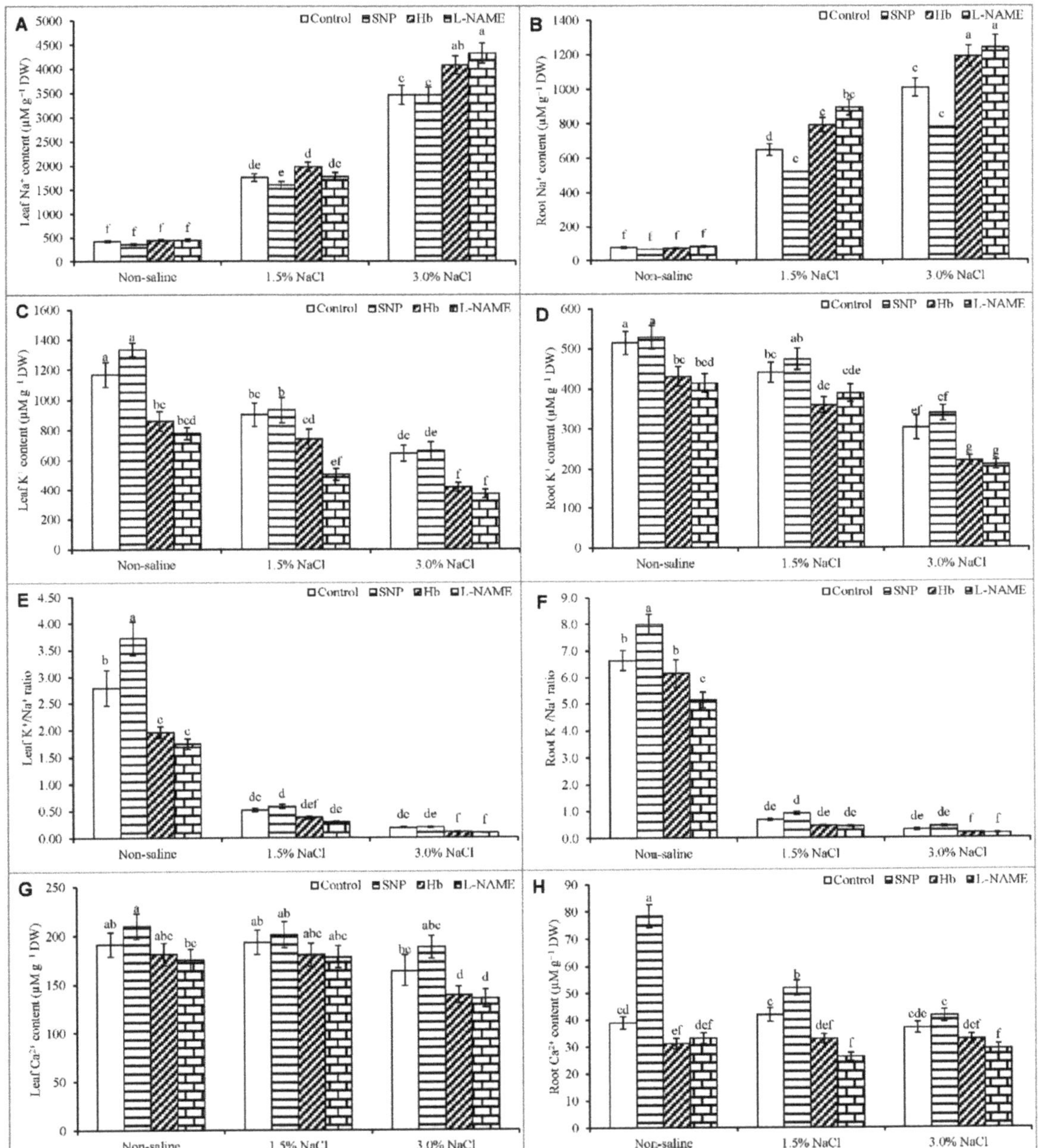

Figure 9. Na$^+$ (**A,B**), K$^+$ (**C,D**), K$^+$/Na$^+$ ratio (**E,F**) and Ca^{2+} content in roots (**G**) and leaves (**H**) of *Kandelia obovata* treated with different salt concentrations and with 100 μM sodium nitroprusside (SNP, a NO donor), hemoglobin (Hb, a NO scavenger), or Nω-nitro- L-arginine methyl ester (L-NAME, a NO synthase inhibitor). Mean (± SD) was calculated from three replicates for each treatment. Bars with different letters are significantly different at $p \leq 0.05$, applying Tukey's HSD test.

The content of Ca^{2+} decreased in response to 3% NaCl stress, whereas the root Ca^{2+} was not significantly affected compared to the unstressed control (Figure 9G,H). Treatment of salt-stressed plants with SNP decreased the Na^+ content, increased the K^+ content, and improved the K^+/Na^+ ratio in both the root and the leaf, compared to the plants treated with salt stress alone. The treatment with SNP also improved the Ca^{2+} level under salt stress, compared to salt treatment alone. Treatment with Hb or L-NAME caused further decreases in the Ca^{2+} level than the salt-stressed plants treated with SNP (Figure 9A–H).

4. Discussion

In this study, we investigated how the supplementation and inhibition of NO regulates the physiology of *K. obovate*. In salt-affected *K. obovata* plants, the endogenous NO level also increased, as reported in previous studies that have examined plants growing in stressful environments [39,40]. This rise in NO is suggested to play a signaling function in response to stress [16]. Treatment of salt-stressed *K. obovata* seedlings with SNP further increased the endogenous NO level, indicating an efficient uptake and accumulation of NO donated from SNP, in agreement with a similar role documented for SNP in previous studies [41,42]. By contrast, treatment with Hb and L-NAME decreased the NO levels both under normal and salt stress conditions, indicating the efficacy of these agents in scavenging NO or inhibiting its biosynthesis.

Kandelia obovata plants subjected to salt stress showed a significant decrease in shoot and root growth, which is due to the disruption of plant water relations that occurred from osmotic stress imposed by the salinity. Moreover, ion toxicity aggravates the stress response upon entry of the salt into the plant cells, which triggers secondary damage to the cell [43]. In the present study, NO supplementation via SNP treatment appeared to reinstate the plant growth suppressed by salt stress, while NO removal by scavenging with Hb or suppression of NO production with L-NAME prevented this restoration and worsened the effects of salt stress in *K. obovata* plants. A similar effect of NO on the restoration of plant growth under salt stress also has been demonstrated by other researchers [25,40,44–47].

In this study, *K. obovata* showed substantial changes in several gas exchange and photosynthetic parameters, namely P_n, g_s, T_r, C_i, and WUE, especially when exposed to 3% NaCl. Salt stress creates a physiological drought through the generation of osmotic stress. This physiological drought can cause stomatal closure, thereby reducing photosynthetic CO_2 assimilation. In the halophyte *K. obovata*, a high salt stress level caused a decrease in g_s and a subsequent marked decrease in C_i. As facultative halophyte, *K. obovata* plant was able to tolerate 1.5% salt and did not show any changes in P_n under that level of stress; however, its P_n value decreased in response to growth in 3% salt compared to the unstressed control. Several reasons may explain this response. In the present study, g_s, T_r, C_i, and WUE decreased at the higher level of salt stress. Reductions in g_s and C_i under high salt stress are the direct reasons for the observed decrease in P_n. The salinity-induced decline in P_n was positively related to the reduction in g_s and C_i, as reported previously in soybean plants [48].

Treatment with Hb and L-NAME reversed the positive effects of SNP in the absence of salt stress. In *K. obovata* plants, the negative effects on the SPAD value, P_n, g_s, T_r, C_i, and WUE were more prominent following treatment with the NO inhibitors. Treatment of the salt-treated plants with Hb and L-NAME caused a further decrease in SPAD value, P_n, g_s, T_r, C_i, and WUE than was observed with salt stress alone. These inhibitors hindered NO accumulation in two ways, by scavenging and by suppressing the NO biosynthesis that would be produced by plants under salt stress. Inhibition of uptake and accumulation of NO from exogenous sources like SNP would also be suppressed.

Plants facing osmotic stress accumulate osmolytes inside the cells to correct their water balance. The large increase in Pro, which increased gradually with the rise in salt level, indicates that *K. obovata* can cope with salt-induced osmotic stress. This capacity was further increased by SNP treatment, as indicated by the further increase in Pro content. Proline is widely known for its osmoprotective effects, as well as its proficiency at scavenging hydroxyl radicals and stabilizing the structure and function of macromolecules, including DNA, protein, and membranes [48–50]. Proline protects the photosynthetic machinery and thus enhances photosynthesis. It acts as an energy storage material during salt stress, thereby improving survival and adaptation [45,51]. Treatment of salt-stressed *K. obovata* plants with SNP increased the Pro level, thereby improving osmoregulation, ROS scavenging, and the stabilization of biomembranes and biomolecules. In the present study, exogenous SNP promoted an increase in the Pro level in salt-affected plants, as previously reported for Pro levels in cadmium-stressed mung bean [52]. The decrease in Pro following Hb and L-NAME treatments that decreased NO availability also abruptly decreased the Pro level, indicating a potential role for SNP in regulating Pro levels.

Salinity is similar to all other abiotic stresses in that it causes oxidative stress, and salt-induced oxidative stress has been reported in different plant species [4,53]. The resulting oxidative damage is a consequence of altered photosystem activity and stomatal movement, ion toxicity, disrupted nutrient homeostasis and disturbed antioxidant defense mechanisms in the salt-affected plants [44,54,55]. In our study, salt-induced oxidative damage was observed in *K. obovata* plants, with the more severe damage noted at the higher salt level. The high levels of H_2O_2 measured in this plant confirmed a salt-induced ROS overproduction. The high MDA levels are the result of membrane lipid peroxidation due to salt-induced oxidative damage. Some reports support similar findings in other plants under salt stress [44,54].

In the present study, SNP decreased the levels of H_2O_2 and MDA in the salt-affected plants. Previous studies on mung bean have shown that SNP treatment can induce inhibition of lipoxygenase (LOX) activity and reduce oxidative stress [52]. Nitric oxide can react with $O_2^{\bullet-}$ to form peroxynitrite ($ONOO^-$), which can then be detoxified by peroxiredoxins [56]. Treatment with SNP reduced the production of H_2O_2 and $O_2^{\bullet-}$ and led to a reduction in MDA accumulation in ryegrass under copper toxicity stress [19]. Nitric oxide lessens oxidative damage in several ways, through reduction of $O_2^{\bullet-}$ and subsequent oxidative stress, through inhibition of LOX activity, and through NO-induced regulation of molecules like Pro, which also scavenge ROS and stabilize biomembranes and biomolecules.

The results from Hb and L-NAME treatments in the present study confirm that oxidative damage was exacerbated to an even greater extent by these NO modifiers to even higher levels than were observed by salt stress alone. Treatment of salt-stressed plants with Hb and L-NAME promoted the production of H_2O_2, raised MDA levels, and increased electrolyte leakage above the levels seen in the salt-stressed, SNP-treated plants. This result confirmed that Hb and L-NAME reduced NO levels, thereby promoting greater oxidative damage.

In the present study, both leaf Na^+ and root Na^+ contents were strongly increased to levels expected to create ion toxicity [18,57]. The increased Na^+ levels decreased K^+ levels, and decreased K^+/Na^+ ratios in both the leaf and root indicate a condition of disrupted and imbalanced ion homeostasis. These cations directly compete with each other to enter the plants under salt stress and represents a common sign of salt-induced ion toxicity, as demonstrated in several previous studies [44,46,58]. The leaf and root Ca^{2+} contents also decreased similarly to the K^+ contents, in agreement with the findings of other studies [44,46]. Treatment of salt-stressed plants with NO in the form of SNP resulted in a decreased root and leaf Na^+ content, an increased K^+ content, an increased K^+/Na^+ ratio, and an increased Ca^{2+} content, compared to salt stress alone. The NO provided by SNP promoted cell membrane reconstruction and restored the ability to exclude toxic ions and to take up nutrient elements necessary for plant development [44]. Treatment of cotton plants with SNP upregulated the expression of vital salt-tolerant genes, including the plasma membrane Na^+/H^+ antiporter (*SOS1*) and the vacuolar Na^+/H^+ antiporter (*NHX1*), allowing salt-stressed plants to decrease Na^+ uptake from the salt solution, while simultaneously promoting an increased K^+ uptake [46]. Several other studies have also shown an SNP-induced reduction of Na^+ uptake and augmentation of K^+ uptake, with an eventual increase in the K^+/Na^+ ratio [44,58].

Some evidence also supports our findings regarding SNP-induced improvement in nutrient uptake. For example, the concentrations of minerals in the leaves and roots of salt-affected wheat plants were enhanced by treatment with SNP [18]. Improved K^+ levels in salt-affected cotton were also demonstrated in response to SNP treatment [44,58–60]. In the present study, *K. obovata* plants treated with Hb and L-NAME showed reduced endogenous NO levels under salt stress but higher levels of Na^+, lower levels of K^+ and Ca^{2+} and a reduced K^+/Na^+ ratio in both roots and leaves. These results indicate that Hb and L-NAME removed NO and inhibited its accumulation, thereby promoting the development of ion toxicity and disruption of mineral homeostasis that was even more severe than were observed under the salt-stress condition. Hb and L-NAME inhibited the endogenous production of NO in salt-affected plants and also eliminated the beneficial effect of exogenous SNP in salt-stressed *K. obovata* plants.

Nitric oxide enhances the transformation of $O_2^{\bullet-}$ to H_2O_2 and O_2 through enhanced activities of SOD; the H_2O_2 is then detoxified by the H_2O_2 scavenging enzymes APX, GPX, and GST [59]. The activities of the enzymes MDHAR and DHAR, which are AsA recycling enzymes, were increased by treatment with SNP. Therefore, SNP treatment most likely increased the levels of AsA, allowing greater scavenging of ROS. Treatment with SNP was reported to increase GR's activity, which increased the GSH content by enhancing its recycling process, and this increased GSH also participated in ROS detoxification processes [21,59]. Treatment with SNP improved the activities of SOD, POD, and CAT in wheat plants subjected to salt stress [44], in agreement with the present findings regarding antioxidant defense systems, ROS detoxification, and oxidative stress alleviation by SNP. Treatment of salt-stressed plants with Hb or L-NAME reversed these responses, leading to a reduction in antioxidant enzyme activities and an enhancement of oxidative stress. These findings further confirm a role for NO in enhancing the antioxidant defense system and oxidative stress tolerance.

5. Conclusions

The facultative mangrove plant, *K. obovata*, showed the least damage under low-level salt stress (1.5%). This low stress did not change the studied physiological attributes significantly, thereby confirming the salt tolerance of this plant species. At the higher level of salt stress (3%), the plants showed oxidative damage. When the salt-affected plants were treated with SNP, the endogenous NO level increased, and the plants showed significant improvements in most of the studied physiological attributes compared to salt-treated control plants. The effect of NO was further confirmed by reducing the NO levels in salt-stressed plants with the NO scavenger Hb and the NO biosynthesis inhibitor L-NAME. The use of these NO modulators exacerbated the adverse effects of salt stress, providing strong evidence for a role for NO in improving salt adaptation and tolerance traits in the *K. obovata* plant.

Author Contributions: Conceptualization, M.H., M.F. and H.O.; methodology, M.H., M.I., M.F. and K.N.; formal analysis, M.H. and K.N.; investigation, M.H. and M.I.; resources, H.O. and M.F.; data curation, M.H.; writing—original draft preparation, M.H. and K.N.; writing—review and editing, M.H. and M.I.; visualization, M.H.; supervision, H.O.; All authors have read and agreed to the published version of the manuscript.

Funding: This research was funded by the Japan Society for the Promotion of Science (JSPS).

Institutional Review Board Statement: Not applicable.

Informed Consent Statement: Not applicable.

Data Availability Statement: Not applicable.

Acknowledgments: We acknowledge Shin Watanabe for providing the propagules and the portable photosynthesis meter. We also thank Yoshinobu Kawamitsu for helping in measuring gas exchange parameters and photosynthesis.

Conflicts of Interest: The authors declare no conflict of interest.

References

1. Pitman, M.G.; Läuchli, A. Global impact of salinity and agricultural ecosystems. In *Salinity: Environment–Plants–Molecules*; Läuchli, A., Lüttge, U., Eds.; Kluwer: Dordrecht, The Netherlands, 2002; pp. 3–20.
2. Chen, Y.P.; Ye, Y. Growth and physiological responses of saplings of two mangrove species to intertidal elevation. *Mar. Ecol. Prog. Ser.* **2013**, *482*, 107–118. [CrossRef]
3. Hasanuzzaman, M.; Nahar, K.; Fujita, M. Plant response to salt stress and role of exogenous protectants to mitigate salt-induced damages. In *Ecophysiology and Responses of Plants under Salt Stress*; Ahmad, P., Azooz, M., Prasad, M., Eds.; Springer: New York, NY, USA, 2013; pp. 25–87.
4. Kawana, Y.; Sasamoto, H.; Ashihara, H. Mechanism of salt tolerance in mangrove plants. *Bull. Soc. Sea Water Sci.* **2008**, *62*, 207–214.
5. Ali, A.; Yun, D.J. Salt stress tolerance; what do we learn from halophytes? *J. Plant Biol.* **2017**, *60*, 431–439. [CrossRef]
6. Jiang, G.-F.; Goodale, U.M.; Liu, Y.-Y.; Hao, G.-Y.; Cao, K.-F. Salt management strategy defines the stem and leaf hydraulic characteristics of six mangrove tree species. *Tree Physiol.* **2017**, *37*, 389–401. [CrossRef]
7. Zheng, C.; Tang, J.; Chen, J.; Liu, W.; Qiu, J.; Peng, X.; Ye, Y. Mechanisms on inhibition of photosynthesis in *Kandelia obovata* due to extreme cold events under climate change. *Ecol. Process.* **2016**, *5*, 20. [CrossRef]
8. Liu, J.; Shi, D.C. Photosynthesis, chlorophyll fluorescence, inorganic ion and organic acid accumulations of sunflower in responses to salt and salt-alkaline mixed stress. *Photosynthetica* **2010**, *48*, 127–134. [CrossRef]
9. Naeem, M.S.; Jin, Z.L.; Wan, G.L.; Liu, D.; Liu, H.B.; Yoneyama, K.; Zhou, W.J. 5-Aminolevulinic acid improves photosynthetic gas exchange capacity and ion uptake under salinity stress in oilseed rape (*Brassica napus* L.). *Plant Soil.* **2010**, *332*, 405–415. [CrossRef]
10. Wu, X.X.; Ding, H.D.; Zhu, Z.W.; Yang, S.J.; Zha, D.S. Effects of 24-epibrassinolide on photosynthesis of eggplant (*Solanum melongena* L.) seedlings under salt stress. *Afr. J. Biotechnol.* **2012**, *11*, 8665–8671.
11. Ahmad, P.; Sharma, S. Salt stress and phyto-biochemical responses of plants. *Plant Soil Environ.* **2008**, *54*, 89–99.
12. Ahmad, P.; Nabi, G.; Ashraf, M. Cadmium-induced oxidative damage in mustard. [*Brassica juncea* (L.) Czern. & Coss.] plants can be alleviated by salicylic acid. *S. Afr. J. Bot.* **2011**, *77*, 36–44.
13. Hasanuzzaman, M.; Bhuyan, M.H.M.B.; Zulfiqar, F.; Raza, A.; Mohsin, S.M.; Mahmud, J.A.; Fujita, M.; Fotopoulos, V. Reactive oxygen species and antioxidant defense in plants under abiotic stress: Revisiting the crucial role of a universal defense regulator. *Antioxidants* **2020**, *9*, 681. [CrossRef]
14. Hasanuzzaman, M.; Bhuyan, M.H.M.B.; Parvin, K.; Bhuiyan, T.F.; Anee, T.I.; Nahar, K.; Hossen, M.S.; Zulfiqar, F.; Alam, M.M.; Fujita, M. Regulation of ROS metabolism in plants under environmental stress: A review of recent experimental evidence. *Int. J. Mol. Sci.* **2020**, *21*, 8695. [CrossRef] [PubMed]
15. Sachdev, S.; Ansari, S.A.; Ansari, M.I.; Fujita, M.; Hasanuzzaman, M. Abiotic stress and reactive oxygen species: Generation, signaling and defense mechanisms. *Antioxidants* **2021**, *10*, 277. [CrossRef] [PubMed]
16. Molassiotis, A.; Tanou, G.; Diamantidis, G. NO says more than 'YES' to salt tolerance Salt priming and systemic nitric oxide signaling in plants. *Plant Signal. Behav.* **2010**, *5*, 209–212. [CrossRef] [PubMed]
17. Beligni, M.V.; Lamattina, L. Nitric oxide interferes with plant photooxidative stress by detoxifying reactive oxygen species. *Plant Cell Environ.* **2002**, *25*, 737–740. [CrossRef]
18. Chen, J.; Xiong, D.-Y.; Wang, W.-H.; Hu, W.-J.; Simon, M.; Xiao, Q.; Chen, J.; Liu, T.-W.; Liu, X.; Zheng, H.L. Nitric oxide mediates root K^+/Na^+ balance in a mangrove plant, *Kandelia obovata*, by enhancing the expression of AKT1-Type K^+ channel and Na^+/H^+ antiporter under high salinity. *PLoS ONE* **2013**, *8*, e71543. [CrossRef]
19. Dong, Y.; Xu, L.; Wang, Q.; Fan, Z.; Kong, J.; Bai, X. Effects of exogenous nitric oxide on photosynthesis, antioxidative ability, and mineral element contents of perennial ryegrass under copper stress. *J. Plant Interact.* **2014**, *9*, 402–411. [CrossRef]
20. Fatma, M.; Khan, N.A. Nitric oxide protects photosynthetic capacity inhibition by salinity in Indian mustard. *J. Funct. Environ. Bot.* **2014**, *4*, 106–116. [CrossRef]
21. Nahar, K.; Hasanuzzaman, M.; Alam, M.M.; Rahman, A.; Suzuki, T.; Fujita, M. Polyamine and nitric oxide crosstalk: Antagonistic effects on cadmium toxicity in mung bean plants through upregulating the metal detoxification, antioxidant defense and methylglyoxal detoxification systems. *Ecotoxicol. Environ. Saf.* **2016**, *126*, 245–255. [CrossRef]
22. Monreal, J.A.; Arias-Baldrich, C.; Pérez-Montaño, F.; Gandullo, J.; Echevarría, C.; García-Mauriño, S. Factors involved in the rise of phosphoenolpyruvate carboxylase-kinase activity caused by salinity in sorghum leaves. *Planta* **2013**, *237*, 1401–1413. [CrossRef] [PubMed]
23. Tan, J.; Zhao, H.; Hong, J.; Han, Y.; Li, H.; Zhao, W. Effects of exogenous nitric oxide on photosynthesis, antioxidant capacity and proline accumulation in wheat seedlings subjected to osmotic stress. *World J. Agric. Sci.* **2008**, *4*, 307–313.
24. Wu, X.X.; Zhu, X.H.; Chen, J.L.; Yang, S.J.; Ding, H.D.; Zha, D.S. Nitric oxide alleviates adverse salt-induced effects by improving the photosynthetic performance and increasing the antioxidant capacity of eggplant (*Solanum melongena* L.). *J. Hortic. Sci. Biotechnol.* **2013**, *88*, 352–360. [CrossRef]

25. Fatma, M.; Masood, A.; Per, T.S.; Khan, N.A. Nitric oxide alleviates salt stress inhibited photosynthetic performance by interacting with sulfur assimilation in mustard. *Front. Plant Sci.* **2016**, *7*, 521. [CrossRef]

26. Fatma, M.; Masood, A.; Per, T.S.; Rasheed, F.; Khan, N.A. Interplay between nitric oxide and sulfur assimilation in salt tolerance in plants. *Crop J.* **2016**, *4*, 153–161. [CrossRef]

27. Ali, Q.M.; Ashraf, M.; Humera, H. Ameliorating effect of foliar applied proline on nutrient uptake in water stressed maize (*Zea mays* L.) plants. *Pak. J. Bot.* **2008**, *40*, 211–219.

28. Bates, L.S.; Waldren, R.P.; Teare, I.D. Rapid determination of free proline for water stress studies. *Plant Soil* **1973**, *39*, 205–207. [CrossRef]

29. Dionisio-Sese, M.L.; Tobita, S. Antioxidant responses of rice seedlings to salinity stress. *Plant Sci.* **1998**, *135*, 1–9. [CrossRef]

30. Heath, R.L.; Packer, L. Photoperoxidation in isolated chloroplast. I. Kinetics and stoichiometry of fatty acid peroxidation. *Arch. Biochem. Biophy.* **1968**, *125*, 189–198. [CrossRef]

31. Yu, C.W.; Murphy, T.M.; Lin, C.H. Hydrogen peroxide induced chilling tolerance in mung beans mediated through ABA-independent glutathione accumulation. *Funct. Plant Biol.* **2003**, *30*, 955–963. [CrossRef] [PubMed]

32. Bradford, M.M. A rapid and sensitive method for the quantitation of microgram quantities of protein utilizing the principle of protein-dye binding. *Anal. Biochem.* **1976**, *72*, 248–254. [CrossRef]

33. Nakano, Y.; Asada, K. Hydrogen peroxide is scavenged by ascorbate-specific peroxidase in spinach chloroplasts. *Plant Cell Physiol.* **1981**, *22*, 867–880.

34. Parvin, K.; Nahar, K.; Hasanuzzaman, M.; Bhuyan, M.H.M.B.; Mohsin, S.M.; Fujita, M. Exogenous vanillic acid enhances salt tolerance of tomato: Insight into plant antioxidant defense and glyoxalase systems. *Plant Physiol. Biochem.* **2020**, *150*, 109–120. [CrossRef] [PubMed]

35. Hasanuzzaman, M.; Alam, M.M.; Nahar, K.; Mohsin, S.M.; Bhuyan, M.H.M.B.; Parvin, K.; Hawrylak-Nowak, B.; Fujita, M. Silicon-induced antioxidant defense and methylglyoxal detoxification works coordinately in alleviating nickel toxicity in *Oryza sativa* L. *Ecotoxicology* **2019**, *28*, 261–276. [CrossRef]

36. Gong, H.; Zhu, X.; Chen, l.; Wang, S.; Zhang, C. Silicon alleviates oxidative damage of wheat plants in pots under drought. *Plant Sci.* **2005**, *169*, 313–321. [CrossRef]

37. Rahman, A.; Nahar, K.; Hasanuzzaman, M.; Fujita, M. Calcium supplementation improves Na^+/K^+ ratio, antioxidant defense and glyoxalase systems in salt-stressed rice seedlings. *Front. Plant Sci.* **2016**, *7*, 609. [CrossRef]

38. CoStat. *CoStat-Statistics Software Version 6.400*; CoHort Software: Monterey, CA, USA, 2008.

39. Hasanuzzaman, M.; Hossain, M.A.; Fujita, M. Nitric oxide modulates antioxidant defense and methylglyoxal detoxification system and reduces salinity-induced damage in wheat seedling. *Plant Biotechnol. Rep.* **2011**, *5*, 353–365. [CrossRef]

40. Huang, J.; Zhu, C.; Hussain, S.; Huang, J.; Liang, Q.; Zhu, L.; Cao, X.; Kong, Y.; Li, Y.; Wang, L.; et al. Effects of nitric oxide on nitrogen metabolism and the salt resistance of rice (*Oryza sativa* L.) seedlings with different salt tolerances. *Plant Physiol. Biochem.* **2020**, *155*, 374–383. [CrossRef]

41. Groppa, M.D.; Rosales, E.P.; Iannone, M.F.; Benavides, M.P. Nitricoxide, polyamines and Cd-induced phytotoxicity in wheat roots. *Phytochemistry* **2008**, *69*, 2609–2615. [CrossRef] [PubMed]

42. Li, X.; Gong, B.; Xu, K. Interaction of nitric oxide and polyamines involves antioxidants and physiological strategies against chilling-induced oxidative damage in Zingiber officinale Roscoe. *Sci. Hort.* **2014**, *170*, 237–248. [CrossRef]

43. Deinlein, U.; Stephan, A.B.; Horie, T.; Luo, W.; Xu, G.; Schroeder, J.I. Plant salt-tolerance mechanisms. *Trends Plant Sci.* **2014**, *19*, 371–379. [CrossRef]

44. Tian, X.; He, M.; Wang, Z.; Zhang, J.; Song, Y.; He, Z.; Dong, Y. Application of nitric oxide and calcium nitrate enhances tolerance of wheat seedlings to salt stress. *Plant Growth Regul.* **2015**, *77*, 343–356. [CrossRef]

45. Reddy, P.S.; Jogeswar, G.; Rasineni, G.K.; Maheswari, M.; Reddy, A.R.; Varshney, R.K.; Kishor, P.B.K. Proline overaccumulation alleviates salt stress and protects photosynthetic and antioxidant enzyme activities in transgenic sorghum (*Sorghum bicolor* (L.) Moench). *Plant Physiol. Biochem.* **2015**, *94*, 104–113. [CrossRef]

46. Kong, X.; Wang, T.; Li, W.; Tang, W.; Zhang, D.; Dong, H. Exogenous nitric oxide delays salt-induced leaf senescence in cotton (*Gossypium hirsutum* L.). *Acta Physiol. Plant.* **2016**, *38*, 61. [CrossRef]

47. Li, X.; Pan, Y.; Chang, B.; Wang, Y.; Tang, Z. NO promotes seed germination and seedling growth under high salt may depend on EIN3 protein in *Arabidopsis*. *Front. Plant Sci.* **2016**. [CrossRef]

48. Ahmad, P.; Jaleel, C.A.; Salem, M.A.; Nabi, G.; Sharma, S. Roles of enzymatic and non-enzymatic antioxidants in plants during abiotic stress. *Crit. Rev. Biotechnol.* **2010**, *30*, 161–175. [CrossRef] [PubMed]

49. Lu, K.X.; Cao, B.H.; Feng, X.P.; He, Y.; Jiang, D.A. Photosynthetic response of salt-tolerant and sensitive soybean varieties. *Photosynthetica* **2009**, *47*, 381–387. [CrossRef]

50. Khan, M.N.; Siddiqui, M.H.; Mohammad, F.; Naeem, M.; Khan, M.M.A. Calcium chloride and gibberellic acid protect linseed (*Linum usitatissimum* L.) from NaCl stress by inducing antioxidative defence system and osmoprotect ant accumulation. *Acta Physiol. Plant.* **2010**, *32*, 121–132. [CrossRef]

51. Jogaiah, S.; Govind, S.R.; Tran, L.S.P. Systems biology-based approaches toward understanding drought tolerance in food crops. *Crit. Rev. Biotechnol.* **2013**, *33*, 23–39. [CrossRef] [PubMed]

52. Nahar, K.; Hasanuzzaman, M.; Rahman, A.; Alam, M.M.; Mahmud, J.A.; Suzuki, T.; Fujita, M. Polyamines confer salt tolerance in mung bean (*Vigna radiata* L.) by reducing sodium uptake, improving nutrient homeostasis, antioxidant defense, and methylglyoxal detoxification systems. *Front. Plant Sci.* **2016**, *7*, 1104. [CrossRef]

53. Hasanuzzaman, M.; Oku, H.; Nahar, K.; Bhuyan, M.H.M.B.; Mahmud, J.A.; Baluska, F.; Fujita, M. Nitric oxide-induced salt stress tolerance in plants: ROS metabolism, signaling, and molecular interactions. *Plant Biotechnol. Rep.* **2018**, *12*, 77–92. [CrossRef]

54. Shabala, S.; Munns, R. Salinity stress: Physiological constraints and adaptive mechanisms. In *Plant Stress Physiology*; Shabala, S., Ed.; CABI: Wallingford, UK, 2012; pp. 59–93.

55. Khan, M.I.R.; Iqbal, N.; Masood, A.; Per, T.S.; Khan, N.A. Salicylic acid alleviates adverse effects of heat stress on photosynthesis through changes in proline production and ethylene formation. *Plant Signal. Behav.* **2013**, *8*, e26374. [CrossRef]

56. Leitner, M.; Vandelle, E.; Gaupels, F.; Bellin, D.; Delledonne, M. NO signals in the haze: Nitric oxide signaling in plant defense. *Curr. Opin. Plant Biol.* **2009**, *12*, 451–458. [CrossRef] [PubMed]

57. Shabala, S. Learning from halophytes: Physiological basis and strategies to improve abiotic stress tolerance in crops. *Ann. Bot.* **2013**, *112*, 1209–1221. [CrossRef] [PubMed]

58. Munns, R.; Tester, M. Mechanisms of salinity tolerance. *Annu. Rev. Plant Biol.* **2008**, *59*, 651–681. [CrossRef] [PubMed]

59. Sheldon, A.R.; Dalal, R.C.; Kirchhof, G.; Kopittke, P.M.; Menzies, N.W. The effect of salinity on plant-available water. *Plant Soil.* **2017**, *418*, 477–491. [CrossRef]

60. Khan, M.N.; Siddiqui, M.H.; Mohammad, F.; Naeem, M. Interactive role of nitric oxide and calcium chloride in enhancing tolerance to salt stress. *Nitric Oxide* **2012**, *27*, 210–218. [CrossRef] [PubMed]

Article

Nitric Oxide Prevents Fe Deficiency-Induced Photosynthetic Disturbance, and Oxidative Stress in Alfalfa by Regulating Fe Acquisition and Antioxidant Defense

Md Atikur Rahman [1], Ahmad Humayan Kabir [2,3], Yowook Song [1], Sang-Hoon Lee [1], Mirza Hasanuzzaman [4] and Ki-Won Lee [1,*]

[1]   Grassland and Forage Division, National Institute of Animal Science, Rural Development Administration, Cheonan 31000, Korea; atikbt@korea.kr (M.A.R.); songs0806@korea.kr (Y.S.); sanghoon@korea.kr (S.-H.L.)
[2]   Molecular Plant Physiology Laboratory, Department of Botany, University of Rajshahi, Rajshahi 6205, Bangladesh; ahmad.kabir@ru.ac.bd
[3]   Department of Genetics, University of Georgia, Athens, GA 30602, USA
[4]   Department of Agronomy, Faculty of Agriculture, Sher-e-Bangla Agricultural University, Sher-e-Bangla Nagar, Dhaka 1207, Bangladesh; mhzsauag@yahoo.com
*   Correspondence: kiwon@korea.kr; Tel.: +82-41-580-6757

Abstract: Iron (Fe) deficiency impairs photosynthetic efficiency, plant growth and biomass yield. This study aimed to reveal the role of nitric oxide (NO) in restoring Fe-homeostasis and oxidative status in Fe-deficient alfalfa. In alfalfa, a shortage of Fe negatively affected the efficiency of root and shoot length, leaf greenness, maximum quantum yield PSII (Fv/Fm), Fe, S, and Zn accumulation, as well as an increase in H_2O_2 accumulation. In contrast, in the presence of sodium nitroprusside (SNP), a NO donor, these negative effects of Fe deficiency were largely reversed. In response to the SNP, the expression of Fe transporters (*IRT1*, *NRAMP1*) and S transporter (*SULTR1;2*) genes increased in alfalfa. Additionally, the detection of NO generation using fluorescence microscope revealed that SNP treatment increased the level of NO signal, indicating that NO may act as regulatory signal in response to SNP in plants. Interestingly, the increase of antioxidant genes and their related enzymes (Fe-SOD, APX) in response to SNP treatment suggests that Fe-SOD and APX are key contributors to reducing ROS (H_2O_2) accumulation and oxidative stress in alfalfa. Furthermore, the elevation of Ascorbate-glutathione (AsA-GSH) pathway-related genes (*GR* and *MDAR*) Fe-deficiency with SNP implies that the presence of NO relates to enhanced antioxidant defense against Fe-deficiency stress.

Keywords: nitric oxide; iron-deficiency; chlorosis; antioxidant; alfalfa

Citation: Rahman, M.A.; Kabir, A.H.; Song, Y.; Lee, S.-H.; Hasanuzzaman, M.; Lee, K.-W. Nitric Oxide Prevents Fe Deficiency-Induced Photosynthetic Disturbance, and Oxidative Stress in Alfalfa by Regulating Fe Acquisition and Antioxidant Defense. *Antioxidants* **2021**, *10*, 1556. https://doi.org/10.3390/antiox10101556

Academic Editor: Luca Sebastiani

Received: 31 August 2021
Accepted: 27 September 2021
Published: 29 September 2021

Publisher's Note: MDPI stays neutral with regard to jurisdictional claims in published maps and institutional affiliations.

1. Introduction

Iron (Fe) is an essential micro-nutrient for plants, as it participates in numerous physiological processes such as photosynthesis, respiration, and nitrogen assimilation [1]. Therefore, any restriction in Fe acquisition hampers plant growth, development, and productivity [2]. In calcareous soils or at high pH levels, Fe is readily oxidized which forms insoluble ferric oxide (Fe^{3+}), resulting in Fe (Fe^{2+}) deficiency-induced growth inhibition and leaf chlorosis [1]. To deal with this problem, the plant evolved two strategies to deal with soluble Fe (Fe^{2+}) shortage and recover Fe from soils. First, there are strategy-I plants (all dicots and non-graminaceous monocots), in which Fe^{3+} is reduced into Fe^{2+} by a plasma membrane ferric reductase enzyme encoded by the *FRO* (ferric reduction oxidase) gene, before being transported across the rhizodermis cell by a Fe^{2+} transporter, encoded by an *IRT* (iron-regulated transporter) gene [1,3].

On the other hand, strategy-II plants produce phytosiderophores (PS) capable of chelating Fe^{3+}, which are then taken up by specific epidermal root cell plasma membrane transporters [4]. However, alfalfa is strategy-I species. Under Fe-deficient conditions, strategy-I species undergo several morphological and physiological changes to aid in nutrient transportation and acquisition. For example, high expression of *IRT1* gene enhances Fe^{2+} acquisition in plants [3]. Similarly, the up-regulation of Fe responsive genes *IRT1* and *FRO1* in *Brassica* at an early stage of S deficiency has been reported [5]. Furthermore, Fe deficiency in plants can cause oxidative stress at the cellular level by impairing photosystem II efficiency, increasing H_2O_2 levels, cellular injury, and disrupting redox homeostasis, all of which can lead to programmed cell death (PCD) [6]. In contrast, Fe deficiency was found to regulate antioxidant mechanisms in *Prunus* rootstocks, where superoxide dismutase (SOD), peroxidase (POD), and catalase (CAT) activities were differentially induced [7].

Nitric oxide (NO) is a small gaseous signaling molecule, involved in the alleviation of oxidative stress, induction of antioxidant activity, and plant sustenance against environmental stimuli [8]. An increasing number of studies have revealed the protective role of NO in plant tolerance to different abiotic stresses. In response to salt stress, NO supplementation regulates SOD, CAT, and osmolyte activities in chickpea [9]. It was also reported that NO donor sodium nitroprusside (SNP) could act as an antioxidant in barley, leading to prevention of PCD [10]. In response to cadmium stress, exogenous NO increases chlorophyll and the Chl a/b ratio in *Brassica* [11]. Despite these significant advances in NO response in multiple stress tolerance in plants, little is known about NO response in nutrient-deficient/nutrient sufficient conditions. Therefore, it is imperative to investigate NO-mediated Fe-homeostasis, alleviation of Fe-induced oxidative stress, and balancing of redox state in plants under Fe-deficiency.

Alfalfa (*Medicago sativa* L.) is a perennial legume crop that serves as nutritious fodder for livestock. Nutrient deficiency negatively impacts on forage growth, biomass yield, quality parameters, digestibility, and ultimate animal performance. It is therefore critical to investigate long-term strategies for maintaining nutrient content in forage in changing climates. Alfalfa is also a good source of protein; it is well known that Fe acts as a cofactor in various enzymes and proteins [12]. Unfortunately, Fe deficiency causes chlorosis, a decrease in protein and chlorophyll content, and reduces in forage yield and quality, especially in alkaline soils [13]. Therefore, considering the above facts the aim of the study was designed to explore NO-mediated mechanisms associated with plant protection from Fe-induced chlorosis, photosynthetic disruption, transcriptional regulation of Fe-responsive genes, and antioxidant defense.

2. Materials and Methods

2.1. Plant Cultivation and Treatment

Viable seeds of alfalfa (*Medicago sativa* L.) were treated with 70% ethanol for 1 min, washed properly using deionized water, and then placed for germination up to 2–3 days. Five-day-old seedlings were transferred into plastic boxes supplemented with micro and macro-elements [14]. There were four treatments: control (25 μM FeNaEDTA); −Fe (0.1 μM FeNaEDTA); −Fe (0.1 μM FeNaEDTA) and sodium nitroprusside (SNP; 100 μM) as nitric oxide (NO) donor; and SNP (100 μM). The plants were maintained at 25 °C under 60–65% relative humidity, 200 $\mu molm^{-2} s^{-1}$ light intensity and light/dark cycle (14 h/10 h dark). The plants were harvested 2 weeks following treatments.

2.2. Measurement of Morphological Features and Photosynthetic Parameters

The root and shoot lengths were measured in centimeters (cm) scale. The fresh weight (FW) of the plants was determined using a digital balance. Alfalfa plants were kept at dark for 1 h before physiological indices were measured. The leaf greenness of young alfalfa leaves was determined using a SPAD meter (Minolta, Japan). The maximum yield of photosystem II (PSII; Fv/Fm) was determined using a portable fluorometer after plants were dark-adapted for 20 min at room temperature (PAM 200, Effeltrich, Germany).

2.3. Nitric Oxide (NO) Localization Using Fluorescent Histochemical Staining

Endogenous NO formation in root tips was measured using 4,5-diaminofluore scein diacetate (DAF-2DA) (Sigma-Aldrich, Burlington, MA, USA) as a NO-specific fluorescent dye [15]. In brief, dimethyl sulfoxide (DMSO) was used to prepare 10 µM DAF-2DA. Root tips of alfalfa were soaked in 10 mM Tris−HCl buffer (pH 6.5) containing 10 µM DAF-2DA for 30 min at dark conditions. The incubated root tips were washed with diethyl pyrocarbonate (DEPC) treated water and observed using a fluorescence microscope (Logos Biosystems, Anyang, South Korea) with 495 nm excitation and 515 nm emissions.

2.4. Estimation of Elemental Concentration

Following treatments, alfalfa roots were washed with deionized water to remove nutrient components from the surface area, and excess water was blotted with tissue paper. Root and shoot were dried for 72 h at 70 °C. Equivalent amounts of plant samples were weighed and digested with a solution ($HClO_4/HNO_3$; 1:3 v/v), and elements were measured using inductively coupled plasma mass spectroscopy (ICP-MS, Agilent 7700, Santa Clara, CA, USA). To prepare with a standard curve, a multi-element ICP-standard-solution (ROTI®STAR, Roth, Germany) was considered Elements were analyzed from samples of the three biological replications.

2.5. Analysis of Soluble Protein Content

The amount of soluble protein was determined according to the protocol described previously [16]. Shortly, 100 mg of plant tissue was homogenized with Tris-HCl (50 mM, pH 7.5), EDTA (2 mM), and 0.04% (v/v) β-mercaptoethanol (β-ME). The mixture was centrifuged at 10,000 rpm for 15 min. Following that, 1 mL supernatant was mixed with 1 mL Coomassie Brilliant Blue (CBB), and the absorbance was measured at 595 nm.

2.6. Hydrogen Peroxide Accumulation

The accumulation of hydrogen peroxide (H_2O_2) was detected spectrophotometrically using the previously described protocol [17]. Shortly, 100 mg of ground sample was homogenized with KP-buffer (50 mM, pH adjusted to 7.0) containing catalase inhibitor hydroxylamine (1 mM). The mixture was centrifuged for 20 min at 12,000 rpm then supernatant (0.7 mL) was transferred to a new tube and 0.7 mL 20% H_2SO_4 containing titanium chloride (TiCl) was added. The mixture was centrifuged at 12,000 rpm for 15 min. Finally, 1 mL of supernatant was taken and measured with absorbance at 410nm using a spectrophotometer (UV-1650PC, Shimadzu, Japan).

2.7. Measurement of Cell Death

Cell death percentages (%) were measured according to the method used earlier [18]. Shortly, 200 mg of plant tissue was homogenized with Evan's blue solution (2 mL) for 15 min. The mixture was treated with 80% ethanol for 8–10 min. The solution was incubated at 50 °C for 20 min in a water bath (Vision Scientific, Daejeon, Korea) system. The mixture was centrifuged at 12,000 rpm for 10 min. The supernatant (1 mL) was then exposed to a wavelength of at 600 nm. The percentage of cell death of tissue was calculated according to the fresh weight basis.

2.8. Gene Expression Analysis by Real-Time PCR

The RNeasy plant mini kit was used to isolate total RNA from plant tissue (QIAGEN, Hilden, Germany). Shortly, 0.1 g of ground tissue was mixed with RNA extraction buffer containing 2M DDT and 1% (v/v) β-ME. The mixture was vortex thoroughly before being centrifuged at 13,000 rpm for 2 min. Following multiple centrifugation and washing steps, total RNA yield was obtained. The RNA concentration in the sample was determined using a nano-drop UV/Vis spectrophotometer (UVISDrop-99, Taipei, Taiwan). For further molecular analyses, RNA concentrations of more than 300 ng/µL were considered. The cDNA synthesis kit (Bio-Rad, USA) was used for the synthesis of cDNA. The CFX96

Real-Time system (Bio-Rad, USA) was used to analyze gene expression. The gene-specific primers were used for qPCR (Supplementary Table S1) analysis. The reaction mixture (20 µL) contained 10 µL of SYBR Green, 1 µL of cDNA, 1 µL of each forward and reverse primer (10 µM), and rest of DEPC treated water. The qPCR system was set to 95 °C for 3 min, 40 amplification cycles of 5 s at 95 °C, 30 sec of annealing 60 °C, and 5 min of extension at 60 °C. The expression of the target genes was analyzed using the dd$^{-\Delta Ct}$ method [19], where *actin* was considered as an internal control.

2.9. Antioxidant Enzyme Activity

Antioxidant enzyme activities of plant tissue were measured following the protocol used previously [20]. Briefly, 100 mg tissue was homogenized in 0.5 mL of 100 mM (KP-buffer, pH 7.0), vortex well. Then the solution was centrifuged at 10,000 rpm for 15 min, and this supernatant was used for further enzymatic analysis. In order to assess SOD, 100 µL extract was added to EDTA (0.1 mM), $NaHCO_3$ (50 mM, pH 9.8) and epinephrine (0.6 mM). The adrenochrome was confirmed by exposing the solution at 475 nm. The activity of APX was determined according to the method used previously [21]. The reaction buffer consisted of100 µL of sample extract, EDTA (0.1 mM), KP-buffer (50 mM, pH 7.0), hydrogen peroxide (0.1 mM), and ascorbic acid (0.5 mM). The 1 mL supernatant was taken and the absorbance was measured at 290 nm and the activity was calculated at extinction co-efficient (2.8 mM^{-1} cm^{-1}). The CAT activity was measured using a mixture containing KP-buffer (100 mM, pH 7.0), hydrogen peroxide (6%), and 100 µL sample extract, and the mixture was read at 240 nm (extinction co-efficient 0.036 mM^{-1} cm^{-1}) considering between 30s-60s. For GR activity, 100 µL plant extract was added to KP-buffer (100 mM), EDTA (1 mM), GSSG (20 mM) and NADPH (0.2 mM). The reaction was triggered with GSSG, which was reduced in absorbance at 340 nm in response to NADPH oxidation. The GR accumulation was ascertained using the extinction co-efficient of 6.12 mM^{-1} cm^{-1} [22].

2.10. Statistical Analysis

All experiments were conducted with the three biological replications for each sampling. The significance level ($p \leq 0.05$) was considered by one-way analysis of variance (ANOVA) followed by Tukey honestly significant test, which was followed by the software SPSS Statistics 20.0. The software GraphPad Prism (version 6.0) was used for graphical presentation.

3. Results

3.1. Alteration of Morphological Features

Iron deficiency-significantly altered the morphological features in alfalfa following 14 days of plant cultivation in the medium. Fe deficiency inhibited the growth of alfalfa seedlings compared with those growth was sufficiently stimulated with SNP (Figure 1). Root length, root dry weight, shoot length, and shoot dry weight of SNP-supplied plants (−Fe+SNP) were significantly higher than those of Fe-deficient plants (Figure 2a–d). However, the addition of SNP to the control plants (+SNP) resulted in an increase in biomass production.

Figure 1. Morphological changes in alfalfa cultivated with different growth conditions: control (25 µM FeNaEDTA); −Fe (0.1 µM FeNaEDTA); −Fe (0.1 µM FeNaEDTA) and sodium nitroprusside (SNP, 100 µM) as nitric oxide (NO) donor; and SNP (100 µM).

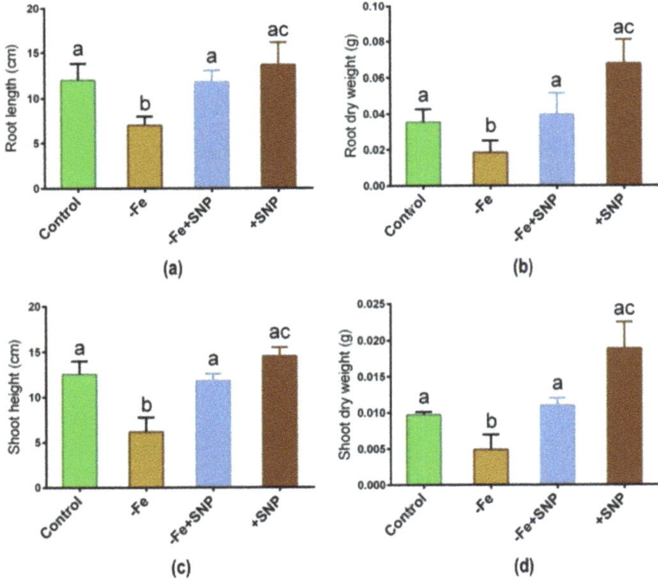

Figure 2. Root length (**a**), root dry weight (**b**), shoot height (**c**), and shoot dry weight (**d**) in alfalfa cultivated with different growth conditions: control (25 µM FeNaEDTA); −Fe (0.1 µM FeNaEDTA); −Fe (0.1 µM FeNaEDTA) and sodium nitroprusside (SNP, 100 µM) as nitric oxide (NO) donor; and SNP (100 µM). Different letters above the error bar indicate significant differences ($p < 0.05$) among means ± SD of treatments ($n = 3$).

3.2. Fe Deficiency-Induced Chlorosis and Regulation of Photosynthetic Parameters

Fe-deficient alfalfa plants exhibited chlorosis and photosynthetic disturbance, and reduced chlorophyll levels. As shown in Figure 3a,b, the leaf greenness and maximum quantum yield of PSII were significantly reduced in response to Fe deficiency, whereas these parameters were improved following exogenous SNP supplementation to the Fe deficient condition (−Fe+SNP). These parameters remained unchanged while the control plants were treated with SNP (+SNP) and untreated control.

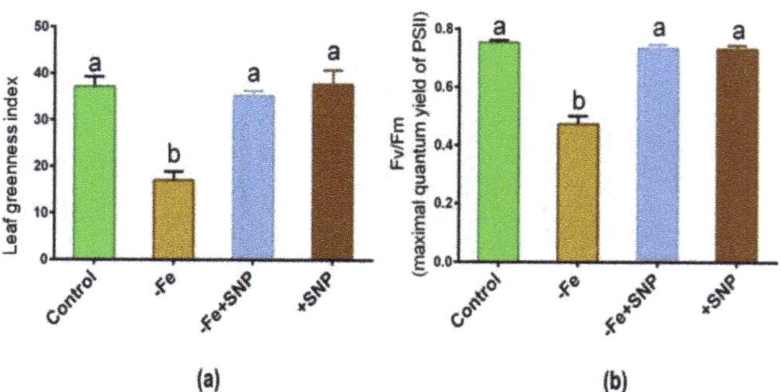

(a)   (b)

Figure 3. Leaf greenness (**a**) and maximum quantum yield of PSII (**b**) in alfalfa cultivated with different growth conditions: control (25 μM FeNaEDTA); −Fe (0.1 μM FeNaEDTA); −Fe (0.1 μM FeNaEDTA) and sodium nitroprusside (SNP, 100 μM) as nitric oxide (NO) donor; and SNP (100 μM). Different letters above the error bar indicate significant differences ($p < 0.05$) among means ± SD of treatments ($n = 3$).

3.3. Changes of Cellular Stress Indicators

Fe deficiency caused a significant increase in cell death (%) compared to the plants treated with SNP along with non-treated control (Figure 4a). However, cell death (%) was found to be reduced after adding SNP with Fe-deficiency. Total soluble protein was significantly reduced in Fe deficient plants while it was increased in SNP-treated plants (Figure 4b). Furthermore, Fe deficiency increased the rate of H_2O_2 production in alfalfa while it showed a significant reduction in SNP-treated with Fe-deficiency plants (Figure 4c).

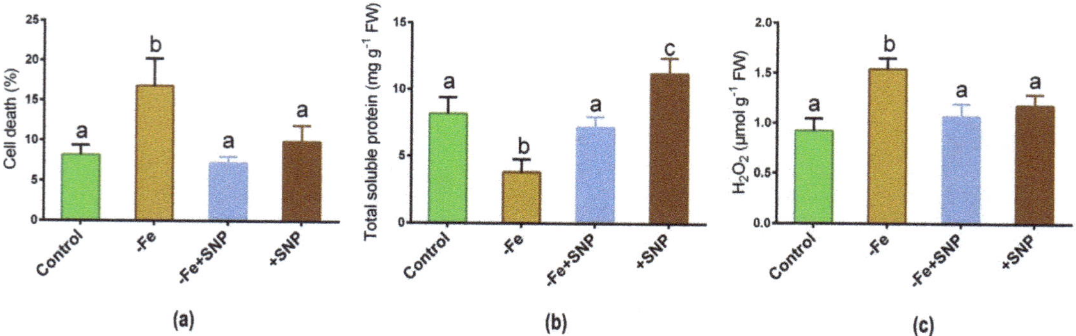

(a)   (b)   (c)

Figure 4. Cell death (**a**), total soluble protein (**b**), and hydrogen peroxide (H_2O_2) (**c**) in alfalfa cultivated with different growth conditions: control (25 μM FeNaEDTA); −Fe (0.1 μM FeNaEDTA); −Fe (0.1 μM FeNaEDTA) and sodium nitroprusside (SNP, 100 μM) as nitric oxide (NO) donor; and SNP (100 μM). Different letters above the error bar indicate significant differences ($p < 0.05$) among means ± SD of treatments ($n = 3$).

3.4. Regulation of Endogenous NO Level

Endogenous NO level in alfalfa root was slightly induced in Fe deficient condition that was lower compared to SNP treated plants. However, the addition of SNP in combination with or without Fe deficient plants showed a strong green fluorescent signal compared to control. The non-treated control plant exhibited very weak fluorescence intensity in alfalfa root (Figure 5).

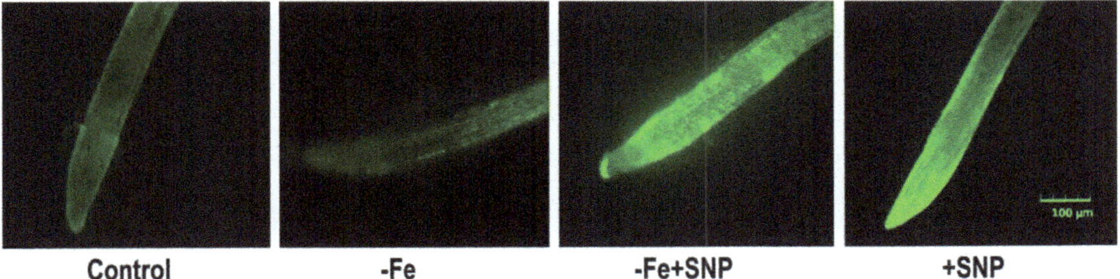

Figure 5. Endogenous accumulation of nitric oxide (NO) in alfalfa cultivated with different growth conditions: control (25 µM FeNaEDTA); −Fe (0.1 µM FeNaEDTA); −Fe (0.1 µM FeNaEDTA) and sodium nitroprusside (SNP, 100 µM) as nitric oxide (NO) donor; and SNP (100 µM). The pictures of the stained roots were taken at 10× magnification. Scale bar = 100 µm.

3.5. Regulation of Mineral Nutrition and Transporter Expression

Fe deficiency-regulated the concentrations of Fe, Zn, and S in alfalfa roots and shoots. Fe level was significantly decreased in Fe-deficient condition in root and shoot while it was increased after SNP supplementation (Figure 6a). Zn concentration was unchanged in roots under Fe-deficient and control condition but it was elevated in response to SNP-treated roots and shoots (Figure 6b). The S level was significantly reduced both in roots and shoots under Fe-deficiency (Figure 6c). However, neither plants treated with SNP nor untreated control plants had significantly different S concentrations in their roots or shoots.

As a consequence, several candidate genes involved in transporting metallic ions expressed differently in alfalfa roots and shoots. The Fe regulating gene *IRT1* exhibited subtaintial downregulation in response to Fe deficiency, and demonstrated significant upregulation in response to −Fe+SNP or +SNP treatments (Figure 6d). Fe-deficiency induced-stress significantly declined the expression of *NRAMP1* while the expression of the gene enhanced after SNP supplementation (Figure 6e). Sulfate transporter gene *SULT1;2* showed a significant downregulation in Fe-deficiency stress. However, the expression of *SULT1;2* was not significantly different among the −Fe+SNP, +SNP and non-treated control plants (Figure 6f).

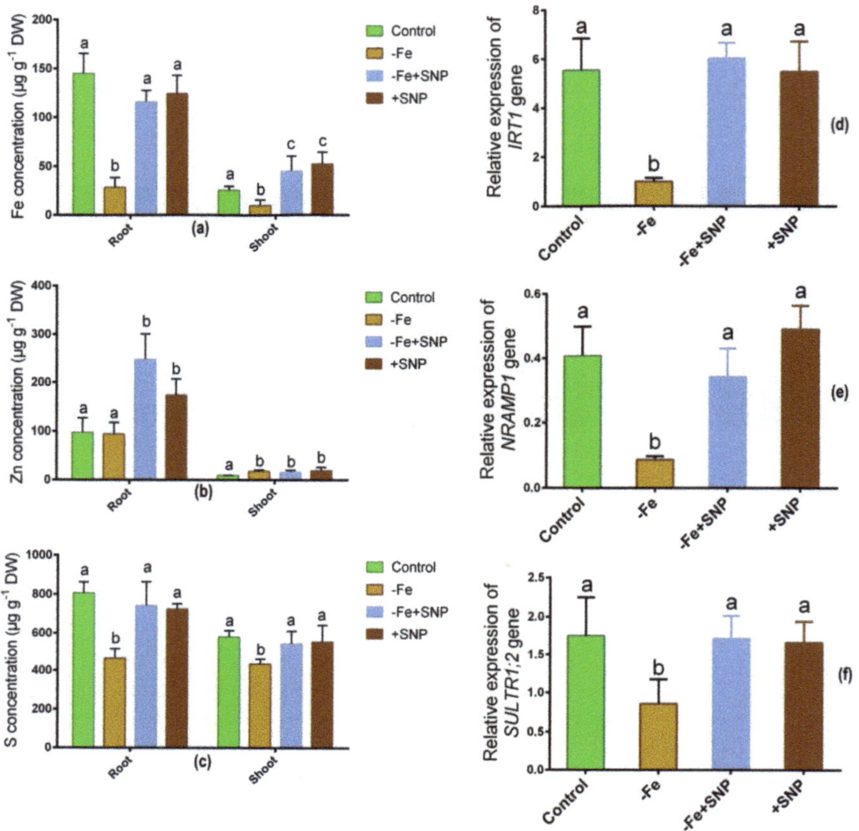

Figure 6. Fe (**a**), Zn (**b**) and S (**c**) concentration, along with *IRT1* (**d**), *NRAMP1* (**e**), and *SULT1;2* (**f**) candidate gene expression in alfalfa cultivated with different growth conditions: control (25 μM FeNaEDTA); −Fe (0.1 μM FeNaEDTA); −Fe (0.1 μM FeNaEDTA) and sodium nitroprusside (SNP, 100 μM) as nitric oxide (NO) donor; and SNP (100 μM). Different letters above the error bar indicate significant differences ($p < 0.05$) among means ± SD of treatments ($n = 3$).

3.6. Antioxidant Enzyme Activity

Fe deficiency-induced stress along with SNP supplementation significantly regulated the activity of key antioxidant enzymes in alfalfa. The addition of SNP to Fe deficient conditions led to a significant increase in SOD activity, although this increase was not consistent across the different treatments (Figure 7a). CAT activity was slightly lifted in SNP-treated plants, but this enzyme did not exhibit significant changes due to Fe deficiency or SNP treatments (Figure 7b). Fe deficiency showed a significant decrease in APX activity while it increased and it was consistent in SNP treated (+SNP) plants (Figure 7c). GR activity was greatly increased under Fe deficiency with or without SNP supplementation compared to non-treated control (Figure 7d). DHAR showed a significant increase in its activity in Fe-deficient condition, along with Fe deficiency with SNP (−Fe+SNP) treatment (Figure 7e). Furthermore, significant MDAR activity was observed only in Fe deficiency with SNP of all assayed treatments (Figure 7f).

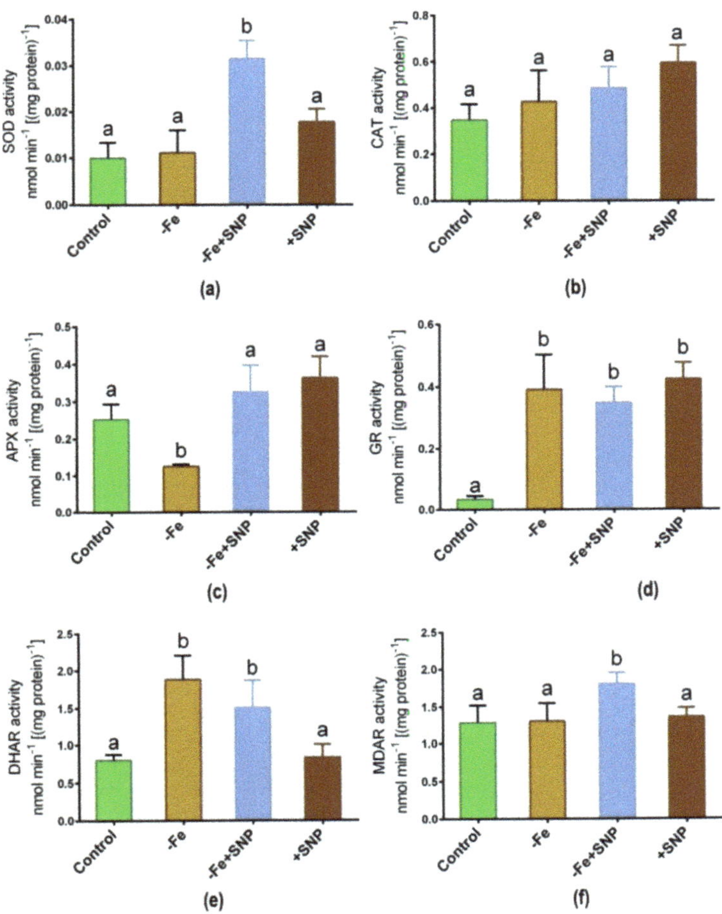

Figure 7. SOD (**a**), CAT (**b**), APX (**c**), GR (**d**), DHAR (**e**) and MDAR (**f**) enzyme activity in alfalfa cultivated with different growth conditions: control (25 μM FeNaEDTA); −Fe (0.1 μM FeNaEDTA); −Fe (0.1 μM FeNaEDTA) and sodium nitroprusside (SNP, 100 μM) as nitric oxide (NO) donor; and SNP (100 μM). Different letters above the error bar indicate significant differences ($p < 0.05$) among means ± SD of treatments ($n = 3$).

3.7. Expression of Key Genes Involved in ROS Homeostasis

The expression of major antioxidative enzyme genes is regulated differentially in response to Fe-deficiency with or without SNP supplementation. As shown in Figure 8, the transcript of the *Fe-SOD* gene significantly was increased due to the addition of SNP to the Fe deficient (−Fe+SNP) condition (Figure 8a). *CAT* gene showed the expression with nearly the same pattern among the treatments (Figure 8b). Fe-deficiency significantly declined the expression of the *APX* gene, while the pattern remained unchanged in the rest of the treatment groups (Figure 8c). Interestingly, *GR* gene showed a significant up-regulation under Fe-deficiency along with SNP-treated plants compared to control (Figure 8d). Ascorbate-glutathione (AsA-GSH) cycle gene *DHAR* is significantly upregulated in response to Fe deficiency (Figure 8e). However, the combination of SNP with Fe deficiency highly induced the transcript of *MDAR* gene, while the rest of the treatments showed nearly the same pattern (Figure 8f).

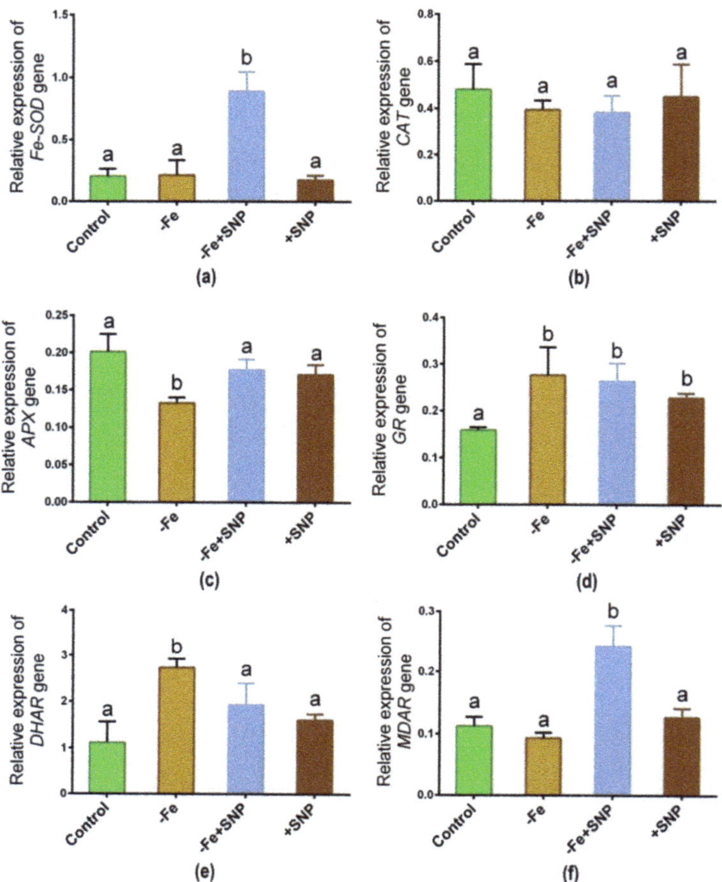

Figure 8. *Fe-SOD* (**a**), *CAT* (**b**), *APX* (**c**), *GR* (**d**), *DHAR* (**e**) and *MDAR* (**f**) candidate gene expression in alfalfa cultivated with different growth conditions: control (25 μM FeNaEDTA); −Fe (0.1 μM FeNaEDTA); −Fe (0.1 μM FeNaEDTA) and sodium nitroprusside (SNP, 100 μM) as nitric oxide (NO) donor; and SNP (100 μM). Different letters above the error bar indicate significant differences ($p < 0.05$) among means ± SD of treatments ($n = 3$).

4. Discussion

This study showed a mechanistic basis related to NO-mediated protection of chlorosis, photosynthetic disturbance, and oxidative stress in alfalfa. The role of NO and its signaling response in alfalfa roots reveals novel features along with antioxidant enzymes and corresponding candidate genes linked to the protection of plants from Fe deficiency- induced damages. These findings can be utilized by the farmer to improve alfalfa production in Fe-deficient soils.

4.1. NO Mitigated Chlorosis, Photosynthetic Disruption, and Plant Growth Reduction

In Fe-deficient condition, a reduction of root-shoot biomass, chlorosis, and photo-synthetic disturbance was observed, which confirming the inhibition of morph physiological features is the adverse effects of Fe-deficiency in plants. Reduction of chlorophyll synthesis in chlorosis leaves, fresh weight, and photosynthetic rate is associated with Fe-deficiency [23]. However, chlorosis is also involved in reducing Fe accumulation and other nutritional imbalances that are related to growth retardation in alfalfa. Another report suggests that nutrient deficiencies caused by Fe-deficiency have an impact onplant growth and development [24]. Our study indicates that NO is capable of coping with Fe-deficiency as well as growth retardation, which demonstrates the role of NO in Fe homeostasis particularly in Fe deficient plants [25]. NO also protects the photosynthetic disturbance by regulating the leaf greenness in alfalfa. However, it is well documented that NO protects against photosynthetic disturbance in plants under stressful conditions [26]. In this study, the ratio of Fv/Fm manifested a remarkable reduction due to Fe deficiency. It indicates that the PSII photochemical reaction considerably affected growth attributes, which impaired the initial growth phase of Fe-deficiency in alfalfa. This is supported by the evidence in plants where reduction of PSII is often associated with Fe-deficiency in leaves [27]. A study documented that the redox state of PSII acceptors was negatively influenced by Fe-deficiency [28]. In contrast, the quantum yield of PSII was well maintained following supplementation of NO in alfalfa. It implies that Fe-deficiency inhibits the Fe uptake but Fe acquisition can be well maintained by NO. Thus, it suggests that NO may play an important role in maintaining photosynthesis capacity and it is required for the mechanism involved in Fe-deficiency tolerance in alfalfa.

4.2. Endogenous NO Level Reduced ROS-Induced Cellular Damages

In plants, Fe deficiency causes reactive oxygen species (ROS), cellular injury, and non-autolytic programmed cell death (PCD) [29]. In this present study, Fe-deficiency induced ROS (H_2O_2) level, increased cell death percentage (%). At the same time, these parameters significantly recovered after addition of SNP. Thus, it suggests that Fe-deficiency stress linked to ROS generation and oxidative stress, which involved in cellular injury in alfalfa. In contrast, alleviation of Fe deficiency and reduction of ROS induced cellular damage by NO suggests that generation of endogenous NO (detected using DAF-2 DA fluorescence probe) in root cells is associated with Fe-homeostasis. A recent study evidenced that NO is involved in Fe-homeostasis especially in Fe deficient plants [25]. Hence, low intensity of temporal accumulation of NO is also visualized in Fe-deficient alfalfa roots.It is not surprising that NO signal can be induced in plant due to Fe-deficiency [29].

4.3. NO Involved in Regulating Nutrients Accumulation and Transporter Gene Expression

The retardation of alfalfa growth with chlorotic symptoms as a consequence of low Fe accumulation may be associated with lower uptake and translocation of Fe in Fe- deficient plants. However, this occurrence did not happen in the case of SNP treated plants. It is evidenced that *IRT1* gene is involved in transporting Fe in the cytosol of cells [30]. In Arabidopsis, it is also documented that *IRT1* transporter expressed in the root, and it is responsible for Fe uptake from the soil for plant growth [31]. In this study, *NRAMP1* gene significantly downregulated in Fe-deficiency but sharply induced after the addition of SNP with Fe deficiency, indicating that *NRAMP1* is associated with the low Fe uptake and transport. In contrast, the induction of *NRAMP1* due to SNP supplementation suggests that it may be involved in Fe homeostasis, as it is evidenced that *NRAMP1* is a crucial metal transporter involved in Fe transport as well as homeostasis in plants [32]. In addition, *SULTR1;2* gene associated with sulfate transport showed increased expression due to SNP supplementation and declined to Fe deficiency. However, it has yet to disclose *SULTR* gene function in *Medicago* species, *SULTR* study in legume plant *Lotus japonicas* suggested that *LjSULTR1:2* is involved in sulfate uptake [33].Consequently, knockout of *SULTR1;2* was reported to decline sulfate uptake and growth in Arabidopsis [34]. This evidence suggests that S accumulation is linked to the response of *SULTR1;2* and activity depend on Fe availability in plants.

4.4. Antioxidant Genes Expressions Was Tightly Related to the Changes of Corresponding Enzymes Activities

An intimate relationship between antioxidant enzyme activities and Fe deficiency-induced stress was reported in soybean plants [35]. In this study, we found that ROS scavenging-related genes expressions were tightly related to the alterations of corresponding enzymes activities in alfalfa. Changes in leaf greenness and gene expression indicate the disturbance of chloroplast integrity under Fe deficiency. Fe can interact with *SOD* as a co-factor, so it is expected that a possible interaction exists between the response of *Fe-SOD* and Fe availability/efficiency in plants. Fe-SOD enzyme involved in detoxification of superoxide formed during photosynthetic electron transport and function in ROS metabolism [36]. Fe-availability is a vital determinant of Fe-SOD expression. As analysis of alfalfa phenotype provides support for the role of this enzyme in ROS scavenging. In this study, Fe-deficiency declined the expression of *Fe-SOD* transcript but was significantly induced after supplementation of SNP in alfalfa. Our findings supported the study in the *Arabidopsis* plant where *Fe-SOD* transcript was downregulated in response to Fe deficiency [37]. *CAT* expression and its corresponding enzyme activity were not significantly increased, though slightly induced in response to SNP, indicating that CAT may not be actively involved in particularly Fe-deficiency induced stress alleviation due to overproduction of ROS in plants [38].

The activity of the Fe-containing enzyme APX was lower in Fe-deficiency stress but it tended to be induced by SNP. This finding indicates that a low level of Fe in tissue possibly influenced the activity of the APX enzyme along with the expression of APX transcript. However, the high *Fe-SOD* and *APX* activity presence of SNP indicates that *Fe-SOD* and *APX* improve alfalfa to reduce ROS-induced oxidative injury in plants. The report suggests that Fe deficiency leads to chlorosis, ROS generation, and oxidative stress, which induce cellular injury and non-autolytic PCD in plants [6]. In this study, the supplementation of SNP stimulated the transcripts of *GR*, *DHAR*, and *MDAR* as well as their corresponding enzyme activities, part of the ASC-GSH pathway. These comprehensive insights suggest a mechanistic basis of SNP-mediated protection of Fe deficiency-induced chlorosis, photosynthetic disturbance, and oxidative stress in alfalfa (Figure 9).

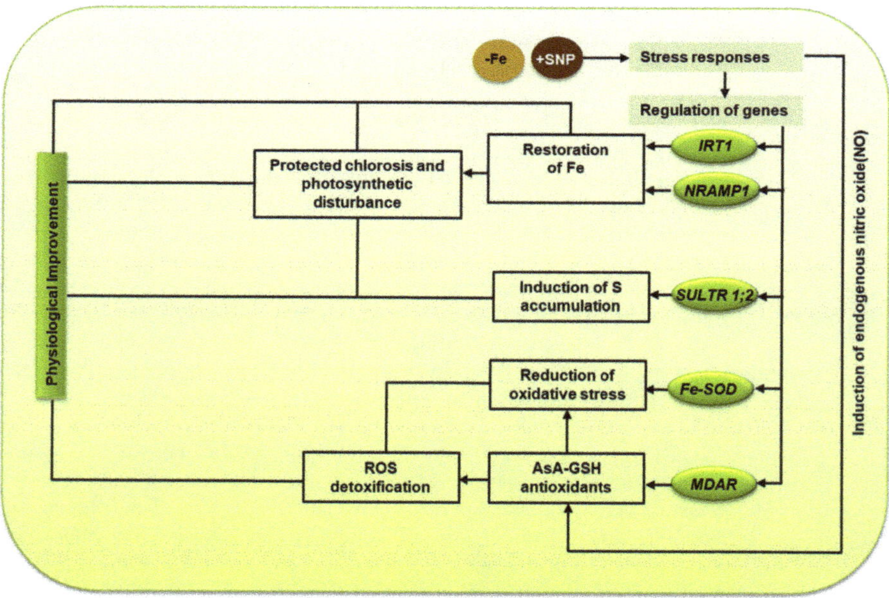

Figure 9. The mechanistic basis of nitric oxide-mediated protection of Fe deficiency-induced chlorosis, photosynthetic disturbance, and oxidative stress in alfalfa.

5. Conclusions

The results of this study shed light on the mechanisms that underlie SNP-mediated alleviation of Fe deficiency-induced growth retardation, chlorosis, photosynthetic disturbance, and oxidative stress in plants. Fe deficiency-induced chlorosis is one of the consequences of Fe deficiency when plants are grown in such a condition. In this study, Fe-deficiency induced stress significantly impacted root-shoot length, leaf greenness, maximum quantum yield PSII (Fv/Fm), Fe, S, and Zn accumulation, increased H_2O_2 content in alfalfa. Surprisingly, these negative impacts of Fe-deficiency stress were largely restored due to SNP supplementation. The response of Fe and S transports in under Fe-deficiency suggested that Fe shortage declined the Fe, Zn and S accumulation both in root and shoots as well as significant decreased of *IRT1*, *NRAMP1* and *SULTR 1;2* gene transcripts in alfalfa. Furthermore, SNP-induced antioxidant candidate genes along with their corresponding enzyme activities indicate that SNP-induced antioxidant enzymes are involved in preventing of Fe deficiency stress-induced chlorosis, photosynthetic disturbance, and ROS accumulation as well as oxidative stress in alfalfa.

Supplementary Materials: The following are available online at https://www.mdpi.com/article/10.3390/antiox10101556/s1, Table S1: Primer sequences used in qRT-PCR.

Author Contributions: Conceptualization, M.A.R., A.H.K. and K.-W.L.; Data curation, M.A.R.; Formal analysis, M.A.R. and A.H.K.; Funding acquisition, K.-W.L.; Investigation; S.-H.L.; Methodology, M.A.R., M.H. and Y.S.; Supervision, K.-W.L.; Writing—original draft, M.A.R.; Writing—review and editing, S.-H.L., M.H. and K.-W.L. All authors have read and agreed to the published version of the manuscript.

Funding: This research was funded by the Cooperative Research Program for Agriculture Science & Technology Development (Project No. PJ01592501). This study was also supported by the Postdoctoral Fellowship Program of the National Institute of Animal Science, Rural Development Administration, Korea.

Institutional Review Board Statement: Not applicable.

Informed Consent Statement: Not applicable.

Data Availability Statement: Data are contained within the article and Supplementary Materials.

Conflicts of Interest: There is no conflict of interest related to this research.

References

1. Angulo, M.; García, M.J.; Alcántara, E.; Pérez-Vicente, R.; Romera, F.J. Comparative Study of Several Fe Deficiency Responses in the *Arabidopsis thaliana* Ethylene Insensitive Mutants *ein2-1* and *ein2-5*. *Plants* **2021**, *10*, 262. [CrossRef]
2. Briat, J.-F.; Dubos, C.; Gaymard, F. Iron nutrition, biomass production, and plant product quality. *Trends Plant. Sci.* **2015**, *20*, 33–40. [CrossRef]
3. Brumbarova, T.; Bauer, P.; Ivanov, R. Molecular mechanisms governing Arabidopsis iron uptake. *Trends Plant. Sci.* **2015**, *20*, 124–133. [CrossRef] [PubMed]
4. Kobayashi, T.; Nishizawa, N.K. Iron uptake, translocation, and regulation in higher plants. *Annu. Rev. Plant. Biol.* **2012**, *63*, 131–152. [CrossRef]
5. Muneer, S.; Lee, B.-R.; Kim, K.-Y.; Park, S.-H.; Zhang, Q.; Kim, T.-H. Involvement of sulphur nutrition in modulating iron deficiency responses in photosynthetic organelles of oilseed rape (*Brassica napus* L.). *Photosyn. Res.* **2014**, *119*, 319–329. [CrossRef] [PubMed]
6. Tewari, R.K.; Hadacek, F.; Sassmann, S.; Lang, I. Iron deprivation-induced reactive oxygen species generation leads to non-autolytic PCD in *Brassica napus* leaves. *Environ. Exp. Bot.* **2013**, *91*, 74–83. [CrossRef] [PubMed]
7. Molassiotis, A.N.; Diamantidis, G.C.; Therios, I.N.; Tsirakoglou, V.; Dimassi, K.N. Oxidative stress, antioxidant activity and Fe(III)-chelate reductase activity of five *Prunus* rootstocks explants in response to Fe deficiency. *Plant Growth Regul.* **2005**, *46*, 69–78. [CrossRef]
8. Sami, F.; Faizan, M.; Faraz, A.; Siddiqui, H.; Yusuf, M.; Hayat, S. Nitric oxide-mediated integrative alterations in plant metabolism to confer abiotic stress tolerance, NO crosstalk with phytohormones and NO-mediated post translational modifications in modulating diverse plant stress. *Nitric Oxide* **2018**, *73*, 22–38. [CrossRef]
9. Ahmad, P.; Abdel Latef, A.A.; Hashem, A.; Abd_Allah, E.F.; Gucel, S.; Tran, L.-S.P. Nitric oxide mitigates salt stress by regulating levels of osmolytes and antioxidant enzymes in Chickpea. *Front. Plant Sci.* **2016**, *7*, 347. [CrossRef] [PubMed]

10. Beligni, M.V.; Fath, A.; Bethke, P.C.; Lamattina, L.; Jones, R.L. Nitric oxide acts as an antioxidant and delays programmed cell death in barley aleurone layers. *Plant Physiol.* **2002**, *129*, 1642–1650. [CrossRef]

11. Jhanji, S.; Setia, R.C.; Kaur, N.; Kaur, P.; Setia, N. Role of nitric oxide in cadmium-induced stress on growth, photosynthetic components and yield of *Brassica napus* L. *J. Environ. Biol.* **2012**, *33*, 1027–1032.

12. Balk, J.; Schaedler, T.A. Iron cofactor assembly in plants. *Annu. Rev.* **2014**, *65*, 125–153. [CrossRef]

13. López-Millán, A.-F.; Grusak, M.; Abadia, A.; Abadía, J. Iron deficiency in plants: An insight from proteomic approaches. *Front. Plant Sci.* **2013**, *4*, 254. [CrossRef]

14. Hoagland, D.R.; Arnon, D.I. The water-culture method for growing plants without soil. *Circ. Calif. Agric. Exp. Stn.* **1950**, *347*, 32.

15. Singh, S.; Prasad, S.M. Management of chromium (VI) toxicity by calcium and sulfur in tomato and brinjal: Implication of nitric oxide. *J. Hazard. Mater.* **2019**, *373*, 212–223. [CrossRef] [PubMed]

16. Guy, C.; Haskell, D.; Neven, L.; Klein, P.; Smelser, C. Hydration-state-responsive proteins link cold and drought stress in spinach. *Planta* **1992**, *188*, 265–270. [CrossRef]

17. Rahman, M.A.; Kim, Y.-G.; Alam, I.; Gongshe, L.; Lee, H.; Lee, J.J.; Lee, B.-H. Proteome analysis of alfalfa roots in response to water deficit stress. *J. Integr. Agric.* **2016**, *15*, 1275–1285. [CrossRef]

18. Zhao, J.; Fujita, K.; Sakai, K. Oxidative stress in plant cell culture: A role in production of β-thujaplicin by *Cupresssus lusitanica* suspension culture. *Biotechnol. Bioeng.* **2005**, *90*, 621–631. [CrossRef]

19. Livak, K.J.; Schmittgen, T.D. Analysis of relative gene expression data using real-time quantitative PCR and the $2^{-\Delta\Delta CT}$ method. *Methods* **2001**, *25*, 402–408. [CrossRef] [PubMed]

20. Haque, A.F.M.M.; Tasnim, J.; El-Shehawi, A.M.; Rahman, M.A.; Parvez, M.S.; Ahmed, M.B.; Kabir, A.H. The Cd-induced morphological and photosynthetic disruption is related to the reduced Fe status and increased oxidative injuries in sugar beet. *Plant Physiol. Biochem.* **2021**, *166*, 448–458. [CrossRef] [PubMed]

21. Hasanuzzaman, M.; Nahar, K.; Gill, S.S.; Alharby, H.F.; Razafindrabe, B.H.N.; Fujita, M. Hydrogen peroxide pretreatment mitigates cadmium-induced oxidative stress in *Brassica napus* L.: An intrinsic study on antioxidant defense and glyoxalase systems. *Front. Plant Sci.* **2017**, *8*, 115. [CrossRef] [PubMed]

22. Halliwell, B.; Foyer, C. Properties and physiological function of a glutathione reductase purified from spinach leaves by affinity chromatography. *Planta* **1978**, *139*, 9–17. [CrossRef]

23. Li, J.; Cao, X.; Jia, X.; Liu, L.; Cao, H.; Qin, W.; Li, M. Iron Deficiency Leads to Chlorosis Through Impacting Chlorophyll Synthesis and Nitrogen Metabolism in *Areca catechu* L. *Front. Plant Sci.* **2021**, *12*, 710093. [CrossRef] [PubMed]

24. Oberschelp, G.P.J.; Gonçalves, A.N. Analysis of nutrient deficiencies affecting in vitro growth and development of Eucalyptus dunnii Maiden. *Physiol. Mol. Biol. Plants* **2018**, *24*, 693–702. [CrossRef]

25. Tewari, R.K.; Horemans, N.; Watanabe, M. Evidence for a role of nitric oxide in iron homeostasis in plants. *J. Exp. Bot.* **2020**, *72*, 990–1006. [CrossRef]

26. Song, L.; Yue, L.; Zhao, H.; Hou, M. Protection effect of nitric oxide on photosynthesis in rice under heat stress. *Acta Physiol. Plant* **2013**, *35*, 3323–3333. [CrossRef]

27. Basa, B.; Lattanzio, G.; Solti, Á.; Tóth, B.; Abadía, J.; Fodor, F.; Sárvári, É. Changes induced by cadmium stress and iron deficiency in the composition and organization of thylakoid complexes in sugar beet (*Beta vulgaris* L.). *Environ. Exp. Bot.* **2014**, *101*, 1–11. [CrossRef]

28. Devadasu, E.R.; Madireddi, S.K.; Nama, S.; Subramanyam, R. Iron deficiency cause changes in photochemistry, thylakoid organization, and accumulation of photosystem II proteins in *Chlamydomonas reinhardtii*. *Photosyn. Res.* **2016**, *130*, 469–478. [CrossRef]

29. Zhai, L.; Xiao, D.; Sun, C.; Wu, T.; Han, Z.; Zhang, X.; Xu, X.; Wang, Y. Nitric oxide signaling is involved in the response to iron deficiency in the woody plant *Malus xiaojinensis*. *Plant Physiol. Biochem.* **2016**, *109*, 515–524. [CrossRef]

30. Palmer, C.M.; Guerinot, M.L. Facing the challenges of Cu, Fe and Zn homeostasis in plants. *Nat. Chem. Biol.* **2009**, *5*, 333–340. [CrossRef]

31. Vert, G.; Grotz, N.; Dédaldéchamp, F.; Gaymard, F.; Guerinot, M.L.; Briat, J.-F.; Curie, C. *IRT1*, an Arabidopsis transporter essential for iron uptake from the soil and for plant growth. *Plant Cell* **2002**, *14*, 1223–1233. [CrossRef]

32. Curie, C.; Alonso, J.M.; Le Jean, M.; Ecker, J.R.; Briat, J.F. Involvement of *NRAMP1* from *Arabidopsis thaliana* in iron transport. *Biochem. J.* **2000**, *347*, 749–755. [CrossRef]

33. Giovannetti, M.; Tolosano, M.; Volpe, V.; Kopriva, S.; Bonfante, P. Identification and functional characterization of a sulfate transporter induced by both sulfur starvation and mycorrhiza formation in *Lotus japonicus*. *New Phytol.* **2014**, *204*, 609–619. [CrossRef] [PubMed]

34. Shibagaki, N.; Rose, A.; McDermott, J.P.; Fujiwara, T.; Hayashi, H.; Yoneyama, T.; Davies, J.P. Selenate-resistant mutants of Arabidopsis thaliana identify Sultr1;2, a sulfate transporter required for efficient transport of sulfate into roots. *Plant J.* **2002**, *29*, 475–486. [CrossRef]

35. Santos, C.S.; Ozgur, R.; Uzilday, B.; Turkan, I.; Roriz, M.; Rangel, A.O.S.S.; Carvalho, S.M.P.; Vasconcelos, M.W. Understanding the Role of the antioxidant system and the tetrapyrrole cycle in iron deficiency chlorosis. *Plants* **2019**, *8*, 348. [CrossRef]

36. Pilon, M.; Ravet, K.; Tapken, W. The biogenesis and physiological function of chloroplast superoxide dismutases. *Biochim. Biophys. Acta Bioenerg.* **2011**, *1807*, 989–998. [CrossRef]

37. Waters, B.M.; McInturf, S.A.; Stein, R.J. Rosette iron deficiency transcript and microRNA profiling reveals links between copper and iron homeostasis in *Arabidopsis thaliana*. *J. Exp. Bot.* **2012**, *63*, 5903–5918. [CrossRef] [PubMed]
38. Jelali, N.; Donnini, S.; Dell'Orto, M.; Abdelly, C.; Gharsalli, M.; Zocchi, G. Root antioxidant responses of two *Pisum sativum* cultivars to direct and induced Fe deficiency. *Plant Biol.* **2014**, *16*, 607–614. [CrossRef] [PubMed]

 *antioxidants*

Review

Melatonin-Induced Water Stress Tolerance in Plants: Recent Advances

Mohamed Moustafa-Farag [1,2,*,†], Ahmed Mahmoud [2,3,†], Marino B. Arnao [4],
Mohamed S. Sheteiwy [5], Mohamed Dafea [2], Mahmoud Soltan [6,7], Amr Elkelish [8],
Mirza Hasanuzzaman [9] and Shaoying Ai [1,*]

1 Institute of Agricultural Resources and Environment, Guangdong Academy of Agricultural Sciences, Guangzhou 510640, China
2 Horticulture Research Institute, Agriculture Research Center, 9 Gmaa St, Giza 12619, Egypt; 11716103@zju.edu.cn (A.M.); mohameddafea@yahoo.com (M.D.)
3 Laboratory of Germplasm Innovation and Molecular Breeding, Institute of Vegetable Science, Zhejiang University, Hangzhou 310058, China
4 Department of Plant Physiology, Faculty of Biology, University of Murcia, 30100 Murcia, Spain; marino@um.es
5 Department of Agronomy, Faculty of Agriculture, Mansoura University, Mansoura 35516, Egypt; salahco_2010@mans.edu.eg
6 Horticulture and Crop Science Department, Ohio Agricultural Research and Development Center, Columbus, The Ohio State University, Columbus, OH 43210, USA; dsoltan2012@gmail.com
7 Vegetable Production under Modified Environment Department, Horticulture Research Institute, Agriculture Research Center, Cairo 11865, Egypt
8 Botany Department, Faculty of Science, Suez Canal University, Ismailia 41522, Egypt; amr.elkelish@science.suez.edu.eg
9 Department of Agronomy, Faculty of Agriculture, Sher-e-Bangla Agricultural University, Dhaka 1207, Bangladesh; mhzsauag@yahoo.com
* Correspondence: m_m_kamel2005@gdaas.cn (M.M.-F.); aishaoying@gdaas.cn (S.A.); Tel.: +86-020-3288-5970 (S.A.)
† These authors contributed equally to this work.

Received: 1 July 2020; Accepted: 18 August 2020; Published: 1 September 2020

Abstract: Water stress (drought and waterlogging) is severe abiotic stress to plant growth and development. Melatonin, a bioactive plant hormone, has been widely tested in drought situations in diverse plant species, while few studies on the role of melatonin in waterlogging stress conditions have been published. In the current review, we analyze the biostimulatory functions of melatonin on plants under both drought and waterlogging stresses. Melatonin controls the levels of reactive oxygen and nitrogen species and positively changes the molecular defense to improve plant tolerance against water stress. Moreover, the crosstalk of melatonin and other phytohormones is a key element of plant survival under drought stress, while this relationship needs further investigation under waterlogging stress. In this review, we draw the complete story of water stress on both sides—drought and waterlogging—through discussing the previous critical studies under both conditions. Moreover, we suggest several research directions, especially for waterlogging, which remains a big and vague piece of the melatonin and water stress puzzle.

Keywords: melatonin; water stress; drought; waterlogging; abiotic stress; antioxidants; stress signaling; phytohormones

1. Introduction

With the notable increase in global warming, rainfall disparity, and poor drainage, water stress (drought and waterlogging) is becoming one of the fiercest environmental challenges in the agriculture sector, mainly in the arid and semiarid regions for drought stress [1,2], and in the areas of heavy rainfall, inadequate draining, and flooding for waterlogging stress [3], which could seriously threaten food security by 2050, whenthe world's population is predicted to reach ten billion [4]. The key impact of water stress is the massive generation of reactive oxygen species (ROS) and malondialdehyde (MDA) over the cell tolerance ability [5], therefore, directly and/or indirectly damaging the cell membrane, nucleic acids, and proteins (Figure 1). This adversely affects gas exchange and photosynthesis and decreases plant growth, as well as yield quality and quantity [6–8]. Practically, a global-scale analysis of published studies over the last four decades on maize and wheat revealed that 20–40% of yield reductions were due to water scarcity [9].Meanwhile, the destructive effect of waterlogging on crop yield has been estimated at a 40–80% loss in an area of more than 1.7 billion hectares [10–12].

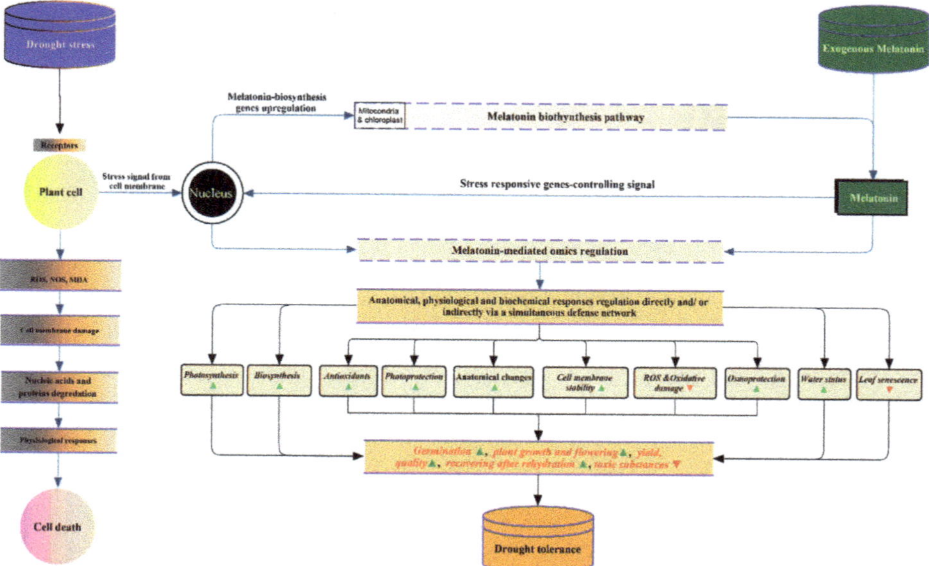

Figure 1. A schematic model explaining the mechanism underlying the melatonin-mediated drought stress response. At the cellular level, a stress signal from the cell membrane is received by the nucleus, which starts to activate the melatonin biosynthesis pathway from its precursor, tryptophan, in mitochondria and chloroplasts by upregulating the melatonin-biosynthesis genes. Melatonin sends its feedback on such stress to the nucleus to activate omics regulation. Consequently, the genes encoding the proteins related to plant anatomical, physiological, and biochemical responses are regulated directly and/or indirectly via a simultaneous defense network. The omics-mediated responses include photosynthesis, biosynthesis, enzymatic and nonenzymatic antioxidants, photoprotection, cell membrane stability, ROS and oxidative damage, osmoprotection, water status, and leaf senescence, in addition to the anatomical changes, which lead to drought tolerance. Consequently, the whole plant status is enhanced, including growth and development, flowering, yield, quality, and survival rate, while the toxic substances are decreased.

Indeed, plants have developed several strategies to cope with water stress. In drought, plants avoid the drastic effects of stress through the induction of stomatal closure, accumulation of compatible solutes, and biosynthesis of wax [4]. Moreover, plants increase their tolerance by the activation of antioxidative abilities and the induction of some molecular chaperones to alleviate oxidative damage [8,13]. In waterlogging, plants avert stress by altering plant metabolism toward anaerobic, glycolytic, and fermentative metabolism. In response to anoxia, the plant activates the antioxidant machinery, expression of heat shock transcript, and accumulation of osmolytes [14]. Previous publications have stated that the various plant responses to water stress are mediated by essential regulators such as phytohormones [15]. Among them, melatonin is a unique antioxidant and plant master regulator that protects plants from oxidative stress and regulates various plant responses to environmental disorders, especially water stress [16–18]. Although accumulating reviews about the beneficial effects of melatonin have been published over the last decade, it still needs more discussion in order to update and discover melatonin functions, especially under biotic and abiotic stresses [19–21]. Herein, we will discuss the most recent and relevant studies of the protective roles of melatonin-induced water stress tolerance, including anatomical changes, and physiological and molecular mechanisms, as well as its central role in the hormonal system. Moreover, we will address the potential triple relationship, melatonin–nitric oxide–hydrogen sulfide, an emerging research point, in the light of previous water stress research. A grasp of the current situation and consideration of the future perspectives of the roles of melatonin in water stress tolerance will also be deeply discussed.

2. Melatonin-Induced Drought Stress Tolerance

2.1. An Overview

Among plant growth substances, melatonin (N-acetyl-5-methoxytryptamine) is an amazing and powerful naturally occurring antioxidant that effectively copes with the drastic effects of water deficit in plants [16,22]. Thus, melatonin is strongly recommended to mitigate drought stress in several plant species, including model plants [23,24], field crops [25,26], fruit crops [27,28], vegetable crops [29,30], as well as ornamental and medicinal plants [31,32] (Table 1). Melatonin treatment ranges from a very low concentration (50 nM) in grape [33] to a high dosage (1 mM) in maize [34] (Table 1). Moreover, melatonin can be applied in different forms to alleviate drought stress, including seed priming [35], seed coating [36], direct soil treatment [37], foliar application [32], in nutrient solutions and hydroponic systems [38], supplemented with irrigation [27], and roots pretreatment [39] (Table 1).

Table 1. Roles of melatonin in drought stress tolerance.

Common Name	Scientific Name	Drought Treatment	Melatonin Treatment		Effects	Reference
			Concentration *	Application Form		
Model Plants						
Arabidopsis	*Arabidopsis thaliana*	Water withholding (21 d)	50 µM	Supplemented with nutrient solution	Stress-responsive genes ▲, soluble sugars ▲	[40]
Field Crops						
Rice	*Oryza sativa*	Water drainage from vessels (5 d)	100 µM	Pretreatment in growing distilled water	Plant growth ▲, osmoprotectants proline ▲, stress-responsive genes ▲, mitochondrial structure ▲, ROS ▼, electroleakage ▼	[41]
Maize	*Zea mays*	Water withholding (8 d), melatonin application during recovery, followed by withholding (8 d).	1 mM	Supplemented with irrigation	Photoprotection (PSII efficiency) ▲	[34]
Maize	*Z. mays*	30–60% SWC (8d)	100 µM	Foliar application	Recovering after rehydration ▲, photosynthesis ▲, stomatal conductance ▲, transpiration rates ▲, cell turgor and water holding capacity ▲, enzymatic and nonenzymatic antioxidants ▲, osmotic potential ▼, ROS ▼	[42]
Maize	*Z. mays*	20% PEG6000(3 d)	10–100 µM	Foliar application pre-treatment	Photosynthesis ▲, antioxidant enzymes ▲, carbon fixation ▲, amino acids and secondary metabolites biosynthesis ▲, ROS ▼	[26]
Maize	*Z. mays*	Water withholding (7 d)	100 µM	Two methods (root-irrigation and foliar application)	Photosynthesis ▲, ROS ▼	[43]
Maize	*Z. mays*	40–45% field capacity (50 d)	50 µM (foliar spray) and 100 µM (soil drench)	Foliar application or soil treatment	Photosynthesis ▲, antioxidant enzymes ▲, ROS ▼	[44]

Table 1. *Cont.*

Common Name	Scientific Name	Drought Treatment	Melatonin Treatment		Effects	Reference
			Concentration *	Application Form		
Wheat	*Triticum aestivum*	40% and 60% field capacity (7 d)	500 µM	Soil application	Chloroplast structure▲, photosynthesis ▲, cell turgor and water holding capacity ▲, GSH and AsA contents ▲, antioxidant enzymes▲, GSH–AsA cycle-related genes ▲, ROS ▼, membrane damage ▼	[45]
Wheat	*T. aestivum*	30% pot holding capacity (8 d)	100 µM	Soil application	Recovering after rehydration ▲, biomass and root/shoot ratio ▲, water holding capacity ▲, chlorophyll ▲, photosynthesis ▲, ROS ▼, MDA ▼	[46]
Wheat	*T. aestivum*	20% PEG 6000 (7 d)	10 and 100 µM (variety-dependent)	Seeds treatment	Germination percentage ▲, germination index ▲, germination potential ▲, radicle length and number ▲, plumule length ▲, lysine (germination-related amino acid) ▲	[47]
Tartary Buckwheat	*Fagopyrum tataricum*	20% field capacity (15 d)	100 µM	Foliar application	Water status ▲, osmoprotection ▲, secondary metabolites▲, antioxidant enzymes▲, photosynthesis ▲, ROS ▼	[48]
Barley	*Hordeum vulgare*	(Combined drought and cold)	1 mM	Foliar or soil application	Endogenous melatonin▲, ABA ▲, water status ▲, antioxidants ▲, photosynthesis ▲, PSII efficiency ▲	[49]
Soybean	*Glycine max*	20% field capacity (10 d)	50 µM	Seed coating	Seedlings growth ▲, biomass ▲, electrolyte leakage ▼	[36]
Soybean	*G. max*	15% PEG 6000 (7 d)	100 µM	Supplemented with nutrient solution	Seedlings growth ▲, photosynthesis ▲	[38]
Soybean	*G. max*	45% RSWC (15 d)	100 µM	Foliar application	Antioxidant enzymes ▲, osmolytes ▲, MDA ▼	[25]
Soybean	*G. max*	15% PEG6000 (3 d)	100 µM	Foliar and root application	Plant growth and flowering ▲, seed yield ▲, gas exchange▲, PSII efficiency ▲, antioxidant enzymes ▲, MDA ▼	[50]

Table 1. *Cont.*

Common Name	Scientific Name	Drought Treatment	Melatonin Treatment		Effects	Reference
			Concentration *	Application Form		
Cassava	*Manihot esculenta*	20% PEG 6000 (11 d)	100 μM	Soil application	POD activity ▲, ROS ▼	[37]
Cotton	*Gossypium hirsutum*	10% PEG 6000 (7 d)	100 μM	Seeds pre-soaking	Number and opening of stomata in cotton testa ▲, germination parameters▲, antioxidant enzymes ▲, osmoprotection ▲, GA3 ▲, ABA ▼, ROS ▼, MDA ▼	[51]
Alfalfa	*Medicago sativa*	Water withholding (7 d)	10 μM	Soil application	Chlorophyll ▲, stomatal conductance ▲, osmoprotection ▲, Nitro-oxidative homeostasis ▲, cellular redox disruption ▼,MDA ▼, ROS ▼	[52]
Fruits						
Apple	*Malus* spp.	Water withholding (6 d)	100 μM	Soil application	Water holding capacity ▲, chlorophyll ▲, photosynthesis ▲, antioxidants ▲, stomatal opening regulation ▲, melatonin biosynthesis genes ▲, electrolyte leakage ▼, ROS ▼, ABA ▼ through ABA synthesis gene▼ and catabolic genes ▲	[53]
Apple	*M. domestica*	50% field capacity (3 months with sampling every month)	100 μM	Soil application	Plant growth ▲, nutrients uptake fluxes ▲, N metabolism ▲, endogenous melatonin ▲, chlorophyll ▲, photosynthesis ▲, relative water content ▲, stomatal status ▲, electrolyte leakage ▼, ROS ▼	[54]
Apple	*M. domestica*	50% field capacity (3 months with sampling every month)	100 μM	Soil application	Chlorophyll ▲, photosynthesis ▲, photoprotection ▲, antioxidant enzymes ▲, GSH and AsA contents ▲, oxidative damage ▼, leaf senescence ▼, senescence-associated gene 12 ▼, pheophorbide a oxygenase-related gene ▼, ROS ▼	[55]

Table 1. *Cont.*

Common Name	Scientific Name	Drought Treatment	Melatonin Treatment		Effects	Reference
			Concentration *	Application Form		
Grape	*Vitis vinifer*	10% PEG 6000 (12 d)	50, 100 and 200 nM	Roots pretreatment	Photoprotection ▲, leaf thickness ▲, spongy tissue ▲, stoma size ▲, chloroplast structure ▲, enzymatic and nonenzymatic antioxidants ▲, osmoprotectants (free proline) ▲, ultrastructural damage ▼, oxidative injury ▼	[33]
Grapevine	*V. amurensis V. vinifera* and *V. labruscana*	10% PEG 6000 (4 d)	Endophyte colonization of secreted-melatonin bacteria	*Bacillus amyloliquefaciens* SB-9 colonization	Melatonin synthesis and its intermediates ▲, plant growth ▲, ROS ▼, MDA ▼	[56]
Grape	*V. vinifer*	Water withholding (18 d)	100 μM	Supplemented with irrigation	MDA ▼, relative conductivity ▼	[57]
Grape	*V. vinifer*	Water withholding (18 d)	100 μM	Supplemented with irrigation	Chlorophyll ▲, SOD activity ▲	[28]
Kiwifruit	*Actinidia. chinensis var. deliciosa*	Water withholding (9 d) (RWC below 35% field capacity)	100 μM	Supplemented with irrigation	Root vigor ▲, osmoprotectants ▲, proteins biosynthesis ▲, chlorophyll ▲, photosynthesis ▲, light energy absorption ▲, photoprotection ▲, CO_2 fixation-associated genes ▲, MDA ▼, cell membranes damage ▼, stomatal closure ▼	[58]
Kiwifruit	*A. chinesis*	water withholding (9 days)	100 μM	Irrigation pretreatment	Water holding capacity ▲, antioxidant enzymes-related genes▲, GSH–AsA cycle-related genes ▲, ROS ▼, MDA ▼	[27]
Chinese hickory	*Carya cathayensis*	30% PEG 6000 (10–40 d)	100 μM	Foliar application pretreatment	Recovering after rehydration ▲, photosynthesis ▲, antioxidants ▲, osmoprotectants ▲, metabolic pathways-related genes ▲, antioxidant enzymes-related genes▲, ROS ▼	[59]

Table 1. *Cont.*

Common Name	Scientific Name	Drought Treatment	Melatonin Treatment		Effects	Reference
			Concentration *	Application Form		
Vegetables						
Tomato	*Solanum lycopersicum*	Water withholding for (5–20 d after moderate drought)	0.1 mM	Supplemented with irrigation	Photosynthesis ▲, root vigor ▲, PSII efficiency ▲, antioxidants ▲, toxic substances ▼	[60]
Tomato	*S. lycopersicum*	10% PEG (7 d)	200 μM	Foliar application	Chlorophyll ▲, p-coumaric acid content ▲, antioxidant enzymes ▲, MDA ▼	[29]
Pepper	*Capsicum annuum*	10% PEG (8 d)	50 μM	Seed pretreatment	Water holding capacity ▲, endogenous melatonin▲, GSH content ▲, chlorophyll ▲, carotenoids ▲, proline ▲, antioxidant enzymes ▲, MDA ▼	[30]
Watermelon	*Citrullus lanatus*	Water withholding (4 d)	150 μM	Root pretreatment	Wax accumulation ▲, melatonin–ABA crosstalk ▲	[39]
Cucumber	*Cucumis sativus*	18% PEG 6000 (days)	100 μM	Seeds priming and nutrient solution	Seed germination ▲, root growth ▲, root/shoot ratio ▲, roots vigor ▲, chlorophyll ▲, photosynthesis ▲, chloroplasts ultrastructure ▲, antioxidant enzymes ▲, ROS ▼	[61]
Rapeseed	*Brassica napus*	4% PEG 6000 (7 d)	0.05 mM	In PEG solution	Plant growth ▲, antioxidants ▲, osmoprotectants ▲, ROS ▼	[62]
Rapeseed	*B. napus*	−0.3 and −0.4 Mpa PEG 6000 (7 d)	500 μM	Seed priming	Chlorophyll ▲, stomatal regulation ▲, chloroplast structure ▲, cell expansion and cell wall ▲, antioxidant enzymes ▲, osmoprotectants ▲, oxidative injury ▼	[35]
Ornamental and Medicinal Plants						
Jinyu Chuju	*Dendranthma morifolium*	40% field capacity (6 d)	100 μM	Foliar application	Chlorophyll ▲, photosynthesis ▲, biomass ▲, osmoprotectants (TSS and proline) ▲, cell membrane damage ▼, relative conductivity ▼, MDA ▼	[63]

Table 1. *Cont.*

Common Name	Scientific Name	Drought Treatment	Melatonin Treatment		Effects	Reference
			Concentration *	Application Form		
Moldavian balm (Dragon head)	*Dracocephalum moldavica*	40–60% field capacity	100 μM	Foliar application	Plant growth and flowering ▲, antioxidants ▲, chlorophyll ▲, water holding capacity ▲, ROS ▼, MDA ▼	[64]
Creeping bentgrass	*Agrostisstolonifera*	Water withholding (14 d)	20 μM	Foliar application	Visual quality ▲, PSII efficiency ▲, chlorophyll ▲, water holding capacity ▲, melatonin biosynthesis genes ▲, dehydration responsive genes ▲, Chlorophyll-degradation genes ▼, leaf senescence ▼, ROS ▼, MDA ▼	[65]
Tall fescue	*Festuca arundinacea*	Water withholding (10 d)	20 μM	Irrigation pretreatment	Plant growth ▲, chlorophyll ▲, antioxidant enzymes ▲, ROS ▼, MDA ▼	[66]
Bermudagrass	*Cynodon dactylon*	Withholding water (21 d)	20 and 100 μM	Irrigation pretreatment	Plant growth ▲, chlorophyll ▲, survival rate ▲, antioxidant enzymes ▲, stress-responsive genes ▲, metabolic regulation ▲, hormonal signaling-related genes regulation ▲, ROS ▼	[67]
Fenugreek	*Trigonella foenum-graecum*	19.5% PEG 6000(21 d)	100 and 300 μM	Foliar application pre-treatment	Endogenous melatonin and secondary metabolites ▲, chlorophyll ▲, antioxidant enzymes ▲, ROS ▼	[68]
Coffee	*Coffea arabica*	40% of max moisture retention capacity (21 d)	300 μM	Soil application	Root vigor ▲, photoprotection ▲, gas exchange ▲, carboxylation efficiency ▲, chlorophyll ▲, antioxidants ▲, MDA ▼	[31]
Tea	*Camellia sinensis*	20% PEG 6000 (2 d)	100 μM	Foliar application pre-treatment	Photosynthesis ▲, GSH and AsA contents ▲, antioxidant enzymes ▲, antioxidant enzymes-related genes ▲, ROS ▼, MDA ▼	[32]
Other Crops						
Tobacco, Tomato and Cucumber	*Nicotiana benthamiana, S. lycopersicum* and *C. sativus*	Water withholding (6 d)	10 μM	Foliar application	MDA ▼, drought tolerance ▲	[23]

▲ or ▼, enhanced or decreased compared to control. ROS, reactive oxygen species; PSII, photosystem II; GSH, glutathione; AsA, ascorbate; MDA, malondialdehyde; ABA, abscisic acid; GA3, gibberellic acid; SOD, superoxide dismutase; POD, peroxidase; TSS, total soluble sugar. * Only those doses of exogenous melatonin that had a superior positive impact on plant tolerance against drought stress have been selected.

2.2. Melatonin is Involved in Drought Stress Tolerance

Given the wide use of melatonin in drought stress alleviation, it has been of interest for the scientific community to investigate the direct evidence of melatonin involvement in drought tolerance. This takes place through melatonin biosynthesis genes such as tryptophan decarboxylase (*TDC*), N-acetylserotonin methyltransferase (*ASMT*), serotonin N-acetyltransferase (*SNAT*), and caffeic acid O-methyltransferase (COMT). In this respect, the endogenous melatonin levels change with the alteration of the environmental conditions of plant growth. The melatonin level is increased, with a protective role, in response to different abiotic stressors such as cold, heat, heavy metals, UV radiation, water deficit, and waterlogging [18,22]. Thus, the expressions of the biosynthesis enzyme transcripts (*TDC*, *SNAT*, *ASMT*, and *COMT* genes) occur in stressful situations, producing a burst in the levels of endogenous melatonin. The global influence of environmental factors on the melatonin levels of plant organs was demonstrated in barley, tomato, and lupin plants by Arnao and coworkers [69–71]. Some representative examples of melatonin induction by drought can be consulted in studies on *Arabidopsis* [24], barley [49], bermudagrass [67], apple [53], grapevine [56], and rice [72]. In these cases, an increase in the melatonin level, between 2- and 6-fold, in one or more transcripts of melatonin biosynthesis enzymes due to stress conditions have been described [72,73].

2.3. Mechanisms of Melatonin-Induced Drought Stress Tolerance

2.3.1. Anatomical Changes and Physiological Mechanisms

In the last few years, the role of melatonin as a multifunctional regulator of plant status under drought conditions, including (i) anatomical and (ii) physiological aspects, have been progressively studied and, notably, reached more than 42 studies within seven years (Table 1). (i) The anatomical changes are induced by melatonin within the different plant organs, including less cell membrane damage [63], more intact grana lamella of the chloroplast [45], alleviation of chloroplast ultrastructural damage and preservation of its system [33,61], safeguarding of the mitochondrial structure [67], maintenance of cell expansion [35], better leaf thickness, spongy tissue, and stomata size [33,35], cuticle formation [74], and wax accumulation [39]. (ii) By increasing drought severity, melatonin, which is biosynthesized in mitochondria and chloroplasts [75,76], exhibits more defense strategies. It promotes the physiological aspects, including the antioxidant system [27,59], to alleviate the oxidative damage, leading to less accumulation of reactive oxygen and nitrogen species (ROS and RNS) [25,52], less electrolyte leakage [41], lower lipid peroxidation (malondialdehyde reduction) [27,65], lower relative conductivity [57], the easing of toxic substances content [60], cellular redox disruption limitation [52], better nitro-oxidative homeostasis [52], and enhanced ascorbate (AsA)–glutathione(GSH) cycle capacity (higher GSH and AsA contents) [54]. Such beneficial effects are carried out by regulating enzymatic activity involving peroxidase (POD), ascorbate peroxidase (APX), catalase (CAT), and superoxide dismutase (SOD), as well as nonenzymatic antioxidants and osmoprotectants (proline and others) [37,44,64], and also secondary metabolites such as flavonoids, phenolics, and phenylalanine ammonialyase [48]. Simultaneously, melatonin improves the plant photosystem, as indicated by higher chlorophyll content [58], greater photosynthetic rates [43], and higher transpiration rates [31]. Moreover, melatonin has been proven to enhance photoprotection via improving photosystem II efficiency [34]. As a multifunctional substance, melatonin also regulates the osmotic potential of the cell [42] via the accumulation of soluble sugars and proline [62]. Moreover, water status is one of the most important priorities of melatonin to control under drought conditions. In this respect, melatonin enhances plant resistance via higher stomatal conductance [42], higher cell turgor and water holding capacity [65], and stomatal opening regulation [77]. Consequently, the whole plant status is enhanced, including seed germination efficiency [47], root generation vitality and strength [61], growth and flowering [36,50], visual quality [65], seed yield [38], leaf senescence alleviation [54], and quick recovery after rehydration [59].

2.3.2. Molecular Mechanisms

Omics of Redox Hemostasis and Plant Built-In Processes

The protective mechanisms of melatonin have also been studied, and the ability of melatonin to protect plant cells against redox homeostasis disruption in response to drought stress has been focused on. Melatonin regulates ROS/reactive nitrogen species (ROS/RNS) levels and antioxidant-related genes, including SOD, POD, CAT, APX, glutathione S-transferase (*GSTP*), monodehydroascorbate reductase (*MDHAR*), dehydroascorbate reductase (*DHAR*), and glutathione reductase (*GR*) [27,30,32,37,43,45,52,56,59,78], as well as osmoprotective elements via the regulation of proline biosynthesis genes [52]. Melatonin not only alleviates oxidative damage, but also regulates plant built-in-associated genes, including carbohydrate/fatty and amino acids metabolism [26,36,37], the carbon metabolic pathway [67], nitrogen metabolism and transport [37,54], plant secondary metabolism [59], energy production [37,78], carotenoid metabolism and photosynthesis [27,36,37,59], and cuticle wax biosynthesis [74]. In this regard, the metabolism of carbohydrate/fatty acids has been reported to be upregulated via the seed-coating of soybean with a melatonin solution as a means to improve its tolerance to drought stress [36]. Melatonin is also a key regulator of nitrogen (N) metabolism and transport, as indicated by the higher expression levels of N uptake genes (*AMT2-1, AMT1-2, AMT1-6, AMT1-5, NRT1-1, NRT2-5, NRT2,* and *7NRT2-4*) and metabolic genes (*NADH-GOGAT NR, Fd-GOGAT, NiR,* and *GS*) in the leaves of apple trees [54].

Omics of Energy Production, Photosynthesis, and Wax Biosynthesis

Melatonin promotes energy production under water scarcity through regulating glycolytic protein expression and electron transport in the respiratory chain [78]. Moreover, melatonin governs the photosynthesis process via the regulation of molecular elements involved in the enzymatic activities of carbon dioxide (CO_2)fixation (*PGK, TKT, FBA, RPI, FBP, GAPA, TIM, RPK, Rubisco, SEBP,* and *RPE*) [58], protein expression for carbon fixation [26], light reaction of photosynthesis (cytochrome P450) [37], and tetrapyrrole pigment biosynthesis [37,56,65]. Photosynthesis has also been reported to be upregulated via the seed-coating of soybean with a melatonin solution as a means to improve its tolerance to drought and salinity stress [36]. Among the interesting genes upregulated by melatonin, there are two subunits of photosystem I (*PS I; PsaG* and *PsaK*) and two elements (*PsbO* and *PsbP*) related to the oxygen-evolving complex of PS II (oxygen-evolving enhancer proteins) [36]. Moreover, melatonin upregulates the relative expression of the *PetF* ferredoxin gene(which controls the amount of reduced ascorbate and protects chlorophyll from degradation) and the *VTC4* gene, encoding the L-galactose 1-P-phosphatase for ascorbate biosynthesis [36]. In another study, Ma et al. [65] reported that melatonin inhibited the gene expression and enzyme activities of chlorophyll-degradation genes, including *chlase, Chl-PRX,* and *PPH,* in melatonin-treated plants during drought stress, which directly affects photosynthesis performance. On the other hand, Ding et al., [74] tested the relative expression of four wax biosynthesis-related genes, including *KCS1*(responsible for fatty acid elongation), *CER3*(involved in alkane synthesis), *TTS1*(associated with triterpenoids synthesis), and *LTP1*(accountable for lipids transport). It was remarked that the transcripts of the four genes were triggered by drought stress and were further induced as a result of melatonin treatment, demonstrating the role of melatonin in enhancing wax biosynthesis [74].

Omics of Stomatal Movement, Autophagy, and Others

Melatonin-mediated stomatal closure mechanism has also been investigated, suggesting that melatonin is a phytohormone that triggers stomatal closure via the signaling pathway ofPMTR1, which controls hydrogen peroxide (H_2O_2) production and the Ca^{2+} signalingtransduction cascade [77]. PMTR1 is a phytomelatonin receptor that has a receptor-like topology and interacts with the subunit of G-protein A (*GPA1*) in the plasma membrane [77]. The phytomelatonin–receptor binding drives the dissociation of Gγβ and Gα (heterotrimeric G-proteins), which triggers NADPH oxidase-dependent

H_2O_2 release and activates Ca^{2+} as well as K^+ efflux, leading to stomatal closure [77]. In addition, NAPDH oxidase, as a respiratory burst oxidase, generates superoxide radicals, which then undergoes dismutation to hydrogen peroxide either enzymatic or nonenzymatically. Under excessive drought, plants resort to getting rid of dysfunctional or unnecessary cellular components in order to facilitate the orderly degradation and recycling of cellular components through the autophagy mechanism. The regulatory role of melatonin in autophagy is elucidated in wheat seedlings via the enhancement of the metabolic process associated with autophagy, represented by the upregulation of the fused signal recognition particle receptor, Rab-related protein, serine protease, and aspartyl protease at the protein or mRNA level [78]. Moreover, melatonin regulates the action of key transcription factors such as *Myb4*, *AP37*, and zinc finger [41,67] in parallel with some transporter proteins, including proton transporter (*UCP1*), potassium transporter (*HKT1*), and water channel protein (*PIP2;1*) [41], which are all essential elements in stress tolerance. Moreover, melatonin application orchestrates some stress-signaling genes such as calcium and protein kinases-related genes, implying that kinase signaling could prove to have essential roles in drought tolerance [67].

All in all, as shown in Figure 1, it can be concluded that once the plants feel water scarcity under drought conditions, the protective and regulatory role of melatonin, in parallel with other anti-stress strategies, will start to prevent, alleviate, or stop the harmful effects of the stress [18,79]. At the cellular level, stress signals from the cell membrane inform the nucleus that "cell life is under threat" to cope with the drastic effects of the drought [77,80]. Quickly, the nucleus starts to activate the melatonin biosynthesis pathway from its precursor, tryptophan, in mitochondria and chloroplasts [75,76,81] through the upregulation of the melatonin-biosynthesis genes [53,65]. Remarkably, melatonin starts by sending its feedback on such stress to the nucleus to trigger the appropriate stress response through omics regulation [40,45,54,65]. As a result, the genes involved in the anatomical, physiological, and biochemical aspects are regulated directly and/or indirectly via a simultaneous defense network. The omics-mediated responses include photosynthesis, biosynthesis, antioxidants, photoprotection, cell membrane stability, osmoprotection, water status, leaf senescence, and oxidative damage alleviation, in addition to the anatomical changes. Consequently, the whole plant status is enhanced, including growth and development, flowering, yield, quality, and survival rate (recovering after rehydration), while the toxic substances are decreased, which collectively lead to drought tolerance.

2.3.3. Melatonin Orchestrates other Phytohormones in the Regulatory–Defense Network

Melatonin is a central molecule in the hormonal system and, thus, increases plant tolerance to drought stress through the regulation of phytohormone levels such as abscisic acid (ABA), auxins (Auxs), cytokinins (CKs), gibberellins (GAs), brassinosteroids (BRs), jasmonic acid (JA), and salicylic acid (SA). The key physiological aspects that are much regulated by phytohormones in response to drought stress include antioxidant metabolism, carbohydrate production (carbon metabolism), stomatal movement, and leaf senescence [82]. Drought stress upregulates ABA, BRs, and JA [59,82] and downregulates CKs and GAs [51,59], while melatonin enhances the levels of BRs, GAs, JA, and CKs [59] and decreases the ABA level [59] (Figure 2).

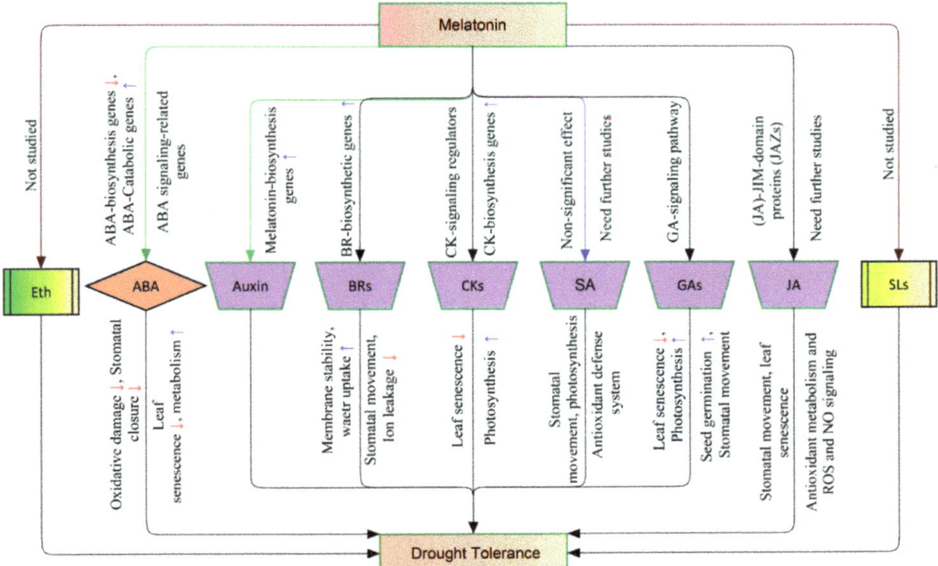

Figure 2. A schematic model explaining the effect of melatonin on other phytohormones under drought stress: Under drought, melatonin enhances the levels of brassinosteroids (BRs), cytokinins (CKs), gibberellins (GAs), and jasmonates (JAs) and decreases the abscisic acid (ABA) level and auxins. Eth, ethylene; ABA, abscisic acid; BRs, brassinosteroids; CKs, cytokinins; SA, salicylic acid; GAs, gibberellins; JA, jasmonic acid; SLs, strigolactones. Red connectors, not studied; green connectors, reduced; black connectors, enhanced; blue connectors, nonsignificant effect. ↑, upregulated; ↓, downregulated.

Water scarcity stimulates abscisic acid (ABA) biosynthesis [13,83], which in turn downregulates the main metabolic pathways [59], induces stomatal closure [82], and contributes to leaf senescence [82]. Moreover, the overaccumulation of abscisic acid upregulates the ROS generation pathways and causes oxidative damage [13]. However, melatonin maintains the abscisic acid homeostasis (low to moderate concentrations) by positive regulation of its biosynthetic genes and negative regulation of the catabolic genes [49,51,53,59]. Li et al. [53] clarified that melatonin effectively downregulates *MdNCED3*, an abscisic acid synthesis gene, and upregulates its catabolic genes, *MdCYP707A2* and *MdCYP707A1*, causing abscisic acid reduction. Moreover, melatonin regulates abscisic acid signaling-related genes such as *SnRK2* (SNF1-related protein kinases 2), *RCAR/PYR/PYL*, and *NCED* (nine-cis-epoxycarotenoid dioxygenase) [67]. Cytokinins (CKs) are an essential group of phytohormones in the inhibition of leaf senescence and chlorophyll degradation under water stress, which in turn suppresses cytokinin biosynthesis and transport, causing cytokinin reduction and faster leaf senescence [84–86]. Melatonin treatment upregulates cytokininlevels and some related signaling factors, leading to better photosynthesis efficiency and drought-induced tolerance [59,65,87]. The first demonstration that melatonin inhibits leaf senescence was made in barley [88]. Melatonin-induced alleviation of leaf senescence in creeping bentgrass is associated with the downregulation of chlorophyll catabolism and synergistic interaction with cytokinins-biosynthesis genes and signaling pathways in melatonin-treated ipt-transgenic plants [65].

Brassinosteroids (BRs) possess an apparent ability as drought stress-protective molecules in plants [89]. Melatonin regulates the biosynthesis of brassinosteroidsvia the stimulation of various brassinosteroid–biosynthetic genes like DWARF4, D11, and RAVL1 [90], which control stomatal movement [91], enhance cell membrane constancy and water uptake, and decrease membrane damage-induced ion leakage in the case of water limitation [59,92]. Jasmonic acid (JA) is a crucial plant hormone in the regulation of drought responses such as stomatal movement, leaf senescence, antioxidant metabolism, and ROS and nitro-oxide signaling [93–98]. Jasmonic acid levels are increased in response to drought stress and are highly stimulated as a result of melatonin application, which induces drought tolerance [59]. The melatonin–jasmonic acid crosstalk is stated by regulating molecular transcripts such as JA–JIM-domain proteins (JAZs) in jasmonic acid signaling [67].

Moreover, melatonin interacts with gibberellins (GAs) via GA-signaling, which further controls the biosynthesis of auxins [24,59]. Gibberellins are regulators of stomatal movement [99,100], photosynthesis [101], seed germination [102], and leaf senescence [4]. Drought stress inhibits gibberellin biosynthesis [51,103], which is much enhanced in response to melatonin treatment, causing drought tolerance [59]. Salicylic acid (SA) accumulation plays a vital role in stomatal movement, photosynthesis, and the antioxidant defense system [4]. In maize plants, under drought conditions, an increase (but nonsignificant) in the defense hormonesalicylic acid has been described in melatonin-treated plants [34]. Enhanced drought tolerance was achieved using mainly transgenic plants through the overexpression of melatonin-biosynthesis genes under drought conditions [24,104,105], which led to a decrease of indole-3-acetic acid (IAA) that may be due to the competition for the same precursor, tryptophan. The plant root is the first plant organ to touch the environment, and it represents a priority for plant breeders to improve its efficiency under abiotic stresses, including drought. Interestingly, melatonin targets plant roots, showing an auxin-like action [106]. In this regard, Pelagio-Flores et al. [106] provided direct evidence supporting the mechanism of this action in *Arabidopsis thaliana* via inspiring lateral and adventitious root formation, conferring a widespread root system. The auxin-like effect of melatonin in roots was elucidated using auxin-responsive marker constructs. It was suggested that melatonin neither activates auxin-inducible gene expression nor induces the degradation of *HS:AXR3NT-GUS*, indicating that root developmental changes elicited by melatonin are independent of auxin signaling [106]. To date, under drought situations, there has been no comprehensive study revealing the interaction between melatonin and ethylene or strigolactones; thus, further investigations are needed. All the above details confirmed that melatonin acts as a relevant regulator of many plant hormone elements, a so-called plant master regulator [107,108], making the plants more tolerant when irrigation water is limited (Figure 2).

2.3.4. The Crosstalk of Melatonin, Nitric Oxide, and Hydrogen Sulfide in Melatonin–Water Stress Research

Melatonin, nitric oxide (NO), and hydrogen sulfide (H$_2$S) are essential small molecules in the plant defense network [109]. Melatonin controls various plant responses under water stress, as described throughout the text. Nitric oxide is a fundamental signaling molecule working as a pro-oxidant and antioxidant element against adverse environments, which is determinant by its endogenous concentration and locational production status [110]. Hydrogen sulfide is a master metabolic regulator in plants, which alleviates the destructive effects of environmental stresses such as drought and waterlogging through the regulation of enzymatic antioxidants [111,112]. The relationship of melatonin, nitric oxide, and hydrogen sulfide has been studied in fruit ripening regulation [113], as well as under biotic [114] and abiotic stresses [110] such as salinity [115] and drought [52]. For instance, nitric oxide and ethylene crosstalk is mediated by hydrogen sulfide and melatonin activity, which regulate various metabolic pathways associated with fruit ripening [113]. Moreover, salt stress alone or combined with iron deficiency expands endogenous hydrogen sulfide and nitric oxide, which are much enhanced due to melatonin treatment [116]. To date, there has only been one published report addressing the relationship between melatonin and nitric oxide under water scarcity [52], while the melatonin–hydrogen sulfide relationship and the triple crosstalk of melatonin–nitric oxide–hydrogen sulfide under water stress remain unknown. In that report, the authors suggested that melatonin mitigates drought damage in alfalfa plants by modulating nitro-oxidative homeostasis through the regulation of reactive oxygen and nitrogen species metabolic enzymes at the enzymatic and/or transcript level [52]. However, how endogenous melatonin interacts with nitric oxide under water scarcity is still a research point [110]. The question that still needs to be answered is whether the crosstalk of melatonin, nitric oxide, and hydrogen sulfide under water stress is similar to the situation under other environmental stresses or if they have a unique interaction in each situation.

3. Melatonin-Induced Waterlogging Stress Tolerance

3.1. An Overview

Despite the importance of melatonin in mitigating the harmful effects of abiotic stresses, the research on melatonin-induced waterlogging tolerance has only recently started to emerge (Table 2). The first report was registered as a patent in 2015 by Chen et al. [117]. In this report, the authors indicated that melatonin has a great ability to eliminate ROS, alleviate oxidative damage, resist waterlogging, and, consequently, revert losses in yield and quality [117]. After this ground-breaking work, Zheng et al. [118] elucidated that melatonin is an effective phytohormone to protect apple plants under waterlogging stress. Melatonin application improved endogenous melatonin levels, antioxidant enzyme activities, chlorophyll content and photosynthesis, and aerobic respiration, while it suppressed chlorosis, wilting, ROS, malondialdehyde, and anaerobic respiration [118]. Moreover, melatonin-biosynthesis enzymes (*MbT5H1*, *MbAANAT3*, and *MbASMT9*) were upregulated due to melatonin treatment [118]. In recent work, Zhang et al. [119] investigated the impact of melatonin pretreatment on alfalfa under waterlogging stress and indicated that melatonin could alleviate the stress damage and enhance plant growth, chlorophyll content, and PSII efficiency. Moreover, melatonin treatment increased polyamine (putrescine, spermidine, and spermine) levels and decreased ethylene under stress, which are controlled via changes in gene expression [119].

Table 2. Roles of melatonin in waterlogging stress tolerance.

Species	Scientific Name	Waterlogging Treatment	Melatonin Treatment		Functions	References
			Concentration *	Application Form		
Apple	*Malus baccata*	Waterlogging stress (9 d)	200 µM (foliar spraying) 600 µM (root irrigation)	Foliar spraying or root irrigation	Endogenous melatonin ▲, antioxidant enzymes ▲, chlorophyll ▲, photosynthesis ▲, aerobic respiration ▲, synthetic enzymes ▲ ROS ▼, MDA ▼, anaerobic respiration ▼, chlorosis and wilting ▼	[118]
Alfalfa	*Medicago sativa*	Waterlogging stress (10 d)	100 µM	Foliar spraying pretreatment	Endogenous melatonin ▲, gene expression regulation ▲, photosynthesis ▲, electroleakage ▼, MDA ▼, leaf senescence ▼, polyamine and ethylene metabolism reprogramming	[119]

▲ or ▼, enhanced or decreased compared to control. ROS, reactive oxygen species; MDA, malondialdehyde. * Only those doses of exogenous melatonin that had a superior positive impact on plant tolerance against waterlogging stress have been selected.

3.2. Mechanisms of Melatonin-Mediated Waterlogging Stress Tolerance

Melatonin application is a practical approach to suppress the drastic effects of waterlogging (Figure 3). To date, there are two published mechanisms induced by melatonin to enforce waterlogging tolerance [118,119]. Zheng et al. [118] proposed the first mechanism of melatonin-mediated waterlogging tolerance in apple seedlings, which keeps aerobic respiration and preserves photosynthesis by efficient repression of the ROS burst and consequent mitochondrial degradation. Zhang et al. [119] suggested another model in alfalfa through interaction with or direct regulation of the metabolic pathways of ethylene and polyamines (PAs). Waterlogging stress induced an increase of endogenous melatonin levels of 2- to 5-fold compared with unstressed plants. Melatonin starts by reducing ethylene production via the downregulation of ethylene synthesis-associated genes and alleviation of waterlogging-caused growth inhibition, chlorosis, and premature senescence [119]. Then, melatonin enhances polyamines levels by promoting the gene expression of the involved enzymes in polyamine metabolism [119]. The authors proposed that melatonin increases waterlogging tolerance, at least partially, by regulating polyamines and ethylene biosynthesis due to ethylene suppression and polyamine promotion, leading to more stable cell membranes, better photosynthesis, and less ethylene-responsive senescence [119].

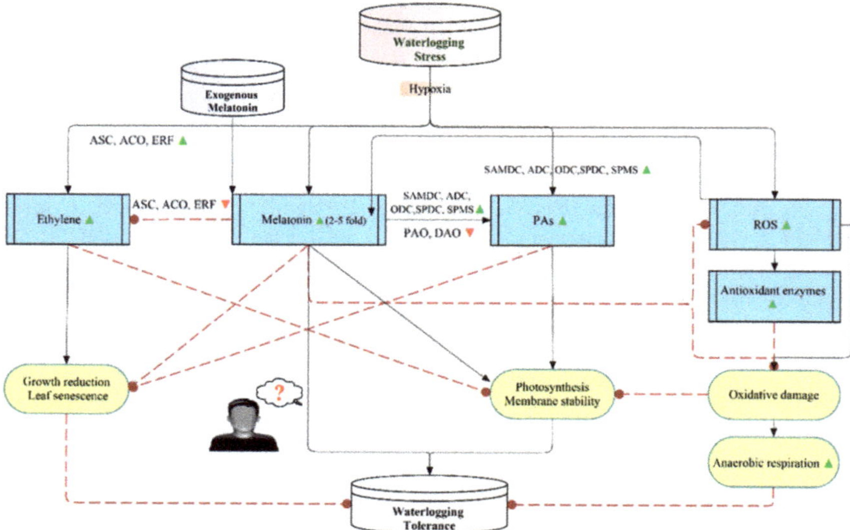

Figure 3. Schematic model explaining the protective mechanisms of melatonin in waterlogging tolerance. The solid arrows indicate stimulation, while the dashes indicate inhibition. ▲ and ▼ shapes indicate enhanced or decreased levels, respectively. Waterlogging induces ethylene, melatonin (2- to 5-fold), polyamines (PAs), and reactive oxygen species (ROS). Melatonin is also induced in response to ROS generation and exogenous melatonin. Melatonin stimulates PA biosynthesis, photosynthesis, and membrane stability, while it inhibits ethylene biosynthesis, growth reduction, leaf senescence, ROS, and oxidative damage. Excessive ROS causes oxidative damage leading to anaerobic respiration, which is scavenged by antioxidant enzymes. Additionally, growth reduction and leaf senescence are increased by ethylene, while they are decreased by PAs. Moreover, photosynthesis and membrane stability is enhanced by PAs, while they are reduced by ethylene induction and oxidative damage. The role of melatonin in waterlogging tolerance still needs further study. This figure is a combination of the two published mechanisms of Zheng et al. [118] and Zhang et al. [119], with some modifications.

Collectively, waterlogging induces ethylene, melatonin (2- to 5-fold), polyamines (PAs), and ROS. Melatonin is also produced in response to ROS generation and exogenous melatonin. Melatonin stimulates polyamines biosynthesis, photosynthesis, and membrane stability, while it inhibits ethylene biosynthesis, growth reduction, leaf senescence, ROS, and oxidative damage. Excessive ROS causes oxidative damage leading to anaerobic respiration, which is scavenged by antioxidant enzymes. Additionally, growth reduction and leaf senescence are increased by ethylene, while they are decreased by polyamines. Moreover, photosynthesis and membrane stability are enhanced by polyamines, while they are reduced by ethylene induction and oxidative damage (Figure 3). The role of melatonin in waterlogging tolerance, especially molecular evidence, still needs further study.

4. Conclusions

Water stress tolerance (drought stress and waterlogging) may be attributed to structural and functional adaptations at the cellular and whole-plant levels, including root enhancement, growth promotion, oxidative damage alleviation, osmotic potential modulation, leaf water potential, cell wall elasticity control, stomatal closure, and the accumulation of osmolytes, thereby easing the harmful impacts of such destructive stresses [4,14]. Melatonin may be considered a core part of the regulatory network controlling all of these mechanisms, and it represents a promising material for future studies and practical use [18,107,108]. Melatonin research has been experiencing hypergrowth in the last two decades; however, its roles in water stress tolerance need further investigation. The regulation of melatonin and its metabolism pathway under water stress is still unclear. Understanding the role of melatonin in nutrient uptake will give us an excellent opportunity to take advantage of such a useful molecule for strengthening plant tolerance and adaptability to water stress. Furthermore, in-depth studies to clarify the molecular mechanisms using microarray, transcriptomic, and proteomic analyses will help to figure out the genes regulating plant anatomical, physiological, and biochemical aspects in response to exogenous melatonin applications under water stress Exploring new receptor-mediated phytomelatonin signaling plays a role in such physiological processes in future works. Additionally, the best-known information on the relationship of melatonin with other small signaling molecules, such as NO and H_2S, can be relevant. In recent decades, significant advancement in the knowledge of the mechanism of NO and H_2S signaling and their crosstalk with melatonin has been made [113,115]. Therefore, the distribution of melatonin in plant organs and their interrelations with NO and H_2S should be further studied [115]. The molecular mechanisms revealing the crosstalk between melatonin and other phytohormones such as strigolactones and ethylene in promoting water stress tolerance are worth further studies on mutagenesis or genetic modulation and aquatic model plants. The relationship between melatonin and multiple stressor combinations is a topic to be taken into account in future research due to the complexity of the interaction of plants with diverse environmental agents. Lastly, the use of synthetic melatonin, a relatively cheap compound, or phytomelatonin-rich extracts should be an interesting approach to improving plant tolerance [120,121].

Author Contributions: M.M.-F. and A.M. contributed to the writing of the first draft of the manuscript. M.B.A., M.S., M.D., M.S.S., A.E., and M.H. contributed to the planning of the main ideas, visualization, and the revision of the manuscript. S.A. was the team leader and mentor. All authors have read and agreed to the published version of the manuscript.

Funding: This work was financially supported by the Dean Funding Project Scheme of the Guangdong Academy of Agricultural Sciences, Guangzhou, China (No. 201811B). Science and Technology Planning Project of Guangdong Province (No. 2016B020240009).

Acknowledgments: M.M.-F. would like to thank his group team at Guangdong Academy of Agriculture Sciences, Yichun Li, Linfeng Li, Mingdeng Tang, and Yanhong Wang, for their support and cooperation during the writing of this review. Also, thanks to Osama ElSawah for his help in improving quality of the figures.

Conflicts of Interest: The authors declare that there is no conflict of interest.

References

1. Xoconostle-Cázares, B.; Ramirez-Ortega, F.A.; Flores-Elenes, L.; Ruiz-Medrano, R. Drought tolerance in crop plants. *Am. J. Plant Physiol.* **2010**, *5*, 1–16.

2. Stuart, M.E.; Gooddy, D.C.; Bloomfield, J.P.; Williams, A.T. A review of the impact of climate change on future nitrate concentrations in groundwater of the UK. *Sci. Total Environ.* **2011**, *409*, 2859–2873. [CrossRef] [PubMed]

3. Jackson, M.B.; Colmer, T.D. Response and adaptation by plants to flooding stress. *Ann. Bot.* **2005**, *96*, 501–505. [CrossRef] [PubMed]

4. Hasanuzzaman, M.; Islam, M.T.; Nahar, K.; Anee, T.I. Drought stress tolerance in wheat: Omics approaches in enhancing antioxidant defense. In *Abiotic Stress-Mediated Sensing and Signaling in Plants: An Omics Perspective*; Zargar, S.M., Ed.; Springer: New York, NY, USA, 2016; pp. 267–307.

5. Arbona, V.; Hossain, Z.; Lopez-Climent, M.F.; Perez-Clemente, R.M.; Gomez-Cadenas, A. Antioxidant enzymatic activity is linked to waterlogging stress tolerance in citrus. *Physiol. Plant.* **2008**, *132*, 452–466. [CrossRef] [PubMed]

6. Hossain, Z.; Lopez-Climent, M.F.; Arbona, V.; Perez-Clemente, R.M.; Gomez-Cadenas, A. Modulation of the antioxidant system in Citrus under waterlogging and subsequent drainage. *J. Plant Physiol.* **2009**, *166*, 1391–1404. [CrossRef] [PubMed]

7. Boru, G.; Vantoai, T.; Alves, J.; Hua, D.; Knee, M. Responses of Soybean to Oxygen Deficiency and Elevated Root-zone Carbon Dioxide Concentration. *Ann. Bot.* **2003**, *91*, 447–453. [CrossRef]

8. Laxa, M.; Liebthal, M.; Telman, W.; Chibani, K.; Dietz, K.J. The Role of the Plant Antioxidant System in Drought Tolerance. *Antioxidants* **2019**, *8*, 94. [CrossRef]

9. Daryanto, S.; Wang, L.; Jacinthe, P.-A. Global synthesis of drought effects on maize and wheat production. *PLoS ONE* **2016**, *11*, e0156362. [CrossRef]

10. Voesenek, L.A.; Sasidharan, R. Ethylene—and oxygen signalling—drive plant survival during flooding. *Plant Biol.* **2013**, *15*, 426–435. [CrossRef]

11. Shabala, S.; Shabala, L.; Barcelo, J.; Poschenrieder, C. Membrane transporters mediating root signalling and adaptive responses to oxygen deprivation and soil flooding. *Plant Cell Environ.* **2014**, *37*, 2216–2233. [CrossRef]

12. Dennis, E.S.; Dolferus, R.; Ellis, M.; Rahman, M.; Wu, Y.; Hoeren, F.U.; Grover, A.; Ismond, K.P.; Good, A.G.; Peacock, W.J. Molecular strategies for improving waterlogging tolerance in plants. *J. Exp. Bot.* **2000**, *51*, 89–97. [CrossRef] [PubMed]

13. Elkeilsh, A.; Awad, Y.M.; Soliman, M.H.; Abu-Elsaoud, A.; Abdelhamid, M.T.; El-Metwally, I.M. Exogenous application of β-sitosterol mediated growth and yield improvement in water-stressed wheat (*Triticum aestivum*) involves up-regulated antioxidant system. *J. Plant Res.* **2019**, *132*, 881–901. [CrossRef] [PubMed]

14. Tewari, S.; Mishra, A. Flooding Stress in Plants and Approaches to Overcome. In *Plant Metabolites and Regulation under Environmental Stress*; Elsevier: Amsterdam, The Netherlands, 2018; pp. 355–366.

15. Verma, V.; Ravindran, P.; Kumar, P.P. Plant hormone-mediated regulation of stress responses. *BMC Plant Biol.* **2016**, *16*, 86. [CrossRef] [PubMed]

16. Arnao, M.B.; Hernández-Ruiz, J. Melatonin: A New Plant Hormone and/or a Plant Master Regulator? *Trends Plant Sci.* **2019**, *24*, 38–48. [CrossRef] [PubMed]

17. Sharma, A.; Zheng, B. Melatonin Mediated Regulation of Drought Stress: Physiological and Molecular Aspects. *Plants* **2019**, *8*, 190. [CrossRef] [PubMed]

18. Ahmad, S.; Cui, W.; Kamran, M.; Ahmad, I.; Meng, X.; Wu, X.; Su, W.; Javed, T.; El-Serehy, H.A.; Jia, Z.; et al. Exogenous Application of Melatonin Induces Tolerance to Salt Stress by Improving the Photosynthetic Efficiency and Antioxidant Defense System of Maize Seedling. *J. Plant Growth Regul.* **2020**, *2020*, 1–14. [CrossRef]

19. Sharif, R.; Xie, C.; Zhang, H.; Arnao, M.B.; Ali, M.; Ali, Q.; Muhammad, I.; Shalmani, A.; Nawaz, M.A.; Chen, P.; et al. Melatonin and Its Effects on Plant Systems. *Molecules* **2018**, *23*, 2352. [CrossRef]

20. Moustafa-Farag, M.; Almoneafy, A.; Mahmoud, A.; Elkelish, A.; Arnao, M.B.; Li, L.; Ai, S. Melatonin and Its Protective Role against Biotic Stress Impacts on Plants. *Biomolecules* **2020**, *10*, 54. [CrossRef]

21. Bose, S.K.; Howlader, P. Melatonin plays multifunctional role in horticultural crops against environmental stresses: A review. *Environ. Exp. Bot.* **2020**, *176*, 104063. [CrossRef]

22. Arnao, M.B.; Hernandez-Ruiz, J. Functions of melatonin in plants: A review. *J. Pineal Res.* **2015**, *59*, 133–150. [CrossRef]

23. Back, K.; Lee, H.-J. 2-Hydroxymelatonin confers tolerance against combined cold and drought stress in tobacco, tomato, and cucumber as a potent anti-stress compound in the evolution of land plants. *Melatonin Res.* **2019**, *2*, 35–46.

24. Yang, W.J.; Du, Y.T.; Zhou, Y.B.; Chen, J.; Xu, Z.S.; Ma, Y.Z.; Chen, M.; Min, D.H. Overexpression of *TaCOMT* Improves Melatonin Production and Enhances Drought Tolerance in Transgenic *Arabidopsis*. *Int. J. Mol. Sci.* **2019**, *20*, 652. [CrossRef] [PubMed]

25. Cao, L.; Jin, X.J.; Zhang, Y.X. Melatonin confers drought stress tolerance in soybean (*Glycine max* L.) by modulating photosynthesis, osmolytes, and reactive oxygen metabolism. *Photosynthetica* **2019**, *57*, 812–819. [CrossRef]

26. Su, X.; Fan, X.; Shao, R.; Guo, J.; Wang, Y.; Yang, J.; Yang, Q.; Guo, L. Physiological and iTRAQ-based proteomic analyses reveal that melatonin alleviates oxidative damage in maize leaves exposed to drought stress. *Plant Physiol. Biochem.* **2019**, *142*, 263–274. [CrossRef] [PubMed]

27. Xia, H.; Ni, Z.; Hu, R.; Lin, L.; Deng, H.; Wang, J.; Tang, Y.; Sun, G.; Wang, X.; Li, H.; et al. Melatonin Alleviates Drought Stress by a Non-Enzymatic and Enzymatic Antioxidative System in Kiwifruit Seedlings. *Int. J. Mol. Sci.* **2020**, *21*, 852. [CrossRef]

28. Lin, Z.; Wang, Y.; Xia, H.; Liang, D. Effects of exogenous melatonin and abscisic acid on the antioxidant enzyme activities and photosynthetic pigment in 'Summer Black' grape under drought stress. *IOP Conf. Ser. Earth Environ. Sci.* **2019**, *295*, 012013. [CrossRef]

29. Karaca, P.; Cekic, F.Ö. Exogenous melatonin-stimulated defense responses in tomato plants treated with polyethylene glycol. *Int. J. Veg. Sci.* **2019**, *25*, 601–609. [CrossRef]

30. Kaya, A.; Doganlar, Z.B. Melatonin improves the multiple stress tolerance in pepper (*Capsicum annuum*). *Sci. Hortic.* **2019**, *256*, 108509. [CrossRef]

31. Campos, C.N.; Avila, R.G.; Dazio de Souza, K.R.; Azevedo, L.M.; Alves, J.D. Melatonin reduces oxidative stress and promotes drought tolerance in young *Coffea arabica* L. plants. *Agric. Water Manag.* **2019**, *211*, 37–47. [CrossRef]

32. Li, J.; Yang, Y.; Sun, K.; Chen, Y.; Chen, X.; Li, X. Exogenous Melatonin Enhances Cold, Salt and Drought Stress Tolerance by Improving Antioxidant Defense in Tea Plant (*Camellia sinensis* (L.) O. Kuntze). *Molecules* **2019**, *24*, 1826. [CrossRef]

33. Meng, J.F.; Xu, T.F.; Wang, Z.Z.; Fang, Y.L.; Xi, Z.M.; Zhang, Z.W. The ameliorative effects of exogenous melatonin on grape cuttings under water-deficient stress: Antioxidant metabolites, leaf anatomy, and chloroplast morphology. *J. Pineal Res.* **2014**, *57*, 200–212. [CrossRef] [PubMed]

34. Fleta-Soriano, E.; Díaz, L.; Bonet, E.; Munné-Bosch, S. Melatonin may exert a protective role against drought stress in maize. *J. Agron. Crop Sci.* **2017**, *203*, 286–294. [CrossRef]

35. Khan, M.N.; Zhang, J.; Luo, T.; Liu, J.; Rizwan, M.; Fahad, S.; Xu, Z.; Hu, L. Seed priming with melatonin coping drought stress in rapeseed by regulating reactive oxygen species detoxification: Antioxidant defense system, osmotic adjustment, stomatal traits and chloroplast ultrastructure perseveration. *Ind. Crop. Prod.* **2019**, *140*, 111597. [CrossRef]

36. Wei, W.; Li, Q.T.; Chu, Y.N.; Reiter, R.J.; Yu, X.M.; Zhu, D.H.; Zhang, W.K.; Ma, B.; Lin, Q.; Zhang, J.S.; et al. Melatonin enhances plant growth and abiotic stress tolerance in soybean plants. *J. Exp. Bot.* **2015**, *66*, 695–707. [CrossRef] [PubMed]

37. Ding, Z.; Wu, C.; Tie, W.; Yan, Y.; He, G.; Hu, W. Strand-specific RNA-seq based identification and functional prediction of lncRNAs in response to melatonin and simulated drought stresses in cassava. *Plant Physiol. Biochem.* **2019**, *140*, 96–104. [CrossRef]

38. Zou, J.N.; Jin, X.J.; Zhang, Y.X.; Ren, C.Y.; Zhang, M.C.; Wang, M.X. Effects of melatonin on photosynthesis and soybean seed growth during grain filling under drought stress. *Photosynthetica* **2019**, *57*, 512–520. [CrossRef]

39. Li, H.; Mo, Y.; Cui, Q.; Yang, X.; Guo, Y.; Wei, C.; Yang, J.; Zhang, Y.; Ma, J.; Zhang, X. Transcriptomic and physiological analyses reveal drought adaptation strategies in drought-tolerant and -susceptible watermelon genotypes. *Plant Sci.* **2019**, *278*, 32–43. [CrossRef]

40. Shi, H.; Qian, Y.; Tan, D.X.; Reiter, R.J.; He, C. Melatonin induces the transcripts of *CBF/DREB1s* and their involvement in both abiotic and biotic stresses in Arabidopsis. *J. Pineal Res.* **2015**, *59*, 334–342. [CrossRef]

41. Lee, H.-J.; Back, K. 2-Hydroxymelatonin promotes the resistance of rice plant to multiple simultaneous abiotic stresses (combined cold and drought). *J. Pineal Res.* **2016**, *61*, 303–316. [CrossRef]

42. Ye, J.; Wang, S.; Deng, X.; Yin, L.; Xiong, B.; Wang, X. Melatonin increased maize (*Zea mays* L.) seedling drought tolerance by alleviating drought-induced photosynthetic inhibition and oxidative damage. *Acta Physiol. Plant.* **2016**, *38*, 48. [CrossRef]

43. Huang, B.; Chen, Y.E.; Zhao, Y.Q.; Ding, C.B.; Liao, J.Q.; Hu, C.; Zhou, L.J.; Zhang, Z.W.; Yuan, S.; Yuan, M. Exogenous Melatonin Alleviates Oxidative Damages and Protects Photosystem II in Maize Seedlings Under Drought Stress. *Front. Plant Sci.* **2019**, *10*, 677. [CrossRef] [PubMed]

44. Ahmad, S.; Kamran, M.; Ding, R.; Meng, X.; Wang, H.; Ahmad, I.; Fahad, S.; Han, Q. Exogenous melatonin confers drought stress by promoting plant growth, photosynthetic capacity and antioxidant defense system of maize seedlings. *PeerJ* **2019**, *7*, e7793. [CrossRef] [PubMed]

45. Cui, G.; Zhao, X.; Liu, S.; Sun, F.; Zhang, C.; Xi, Y. Beneficial effects of melatonin in overcoming drought stress in wheat seedlings. *Plant Physiol. Biochem.* **2017**, *118*, 138–149. [CrossRef] [PubMed]

46. Ye, J.; Deng, X.P.; Wang, S.W.; Yin, L.N.; Chen, D.Q.; Xiong, B.L.; Wang, X.Y. Effects of melatonin on growth, photosynthetic characteristics and antioxidant system in seedling of wheat under drought stress. *J. Triticeae Crop.* **2015**, *35*, 1275–1283.

47. Li, D.; Zhang, D.; Wang, H.; Li, H.; Song, S.; Li, H.; Li, R. Effects of melatonin on germination and amino acid content in different wheat varieties seeds under polyethylene glycol stress. *BioRxiv* **2019**, *2019*, 710954.

48. Hossain, M.S.; Li, J.; Sikdar, A.; Hasanuzzaman, M.; Uzizerimana, F.; Muhammad, I.; Yuan, Y.; Zhang, C.; Wang, C.; Feng, B. Exogenous Melatonin Modulates the Physiological and Biochemical Mechanisms of Drought Tolerance in Tartary Buckwheat (*Fagopyrum tataricum* (L.) Gaertn). *Molecules* **2020**, *25*, 2828. [CrossRef]

49. Li, X.; Tan, D.X.; Jiang, D.; Liu, F. Melatonin enhances cold tolerance in drought-primed wild-type and abscisic acid-deficient mutant barley. *J. Pineal Res.* **2016**, *61*, 328–339. [CrossRef]

50. Zhang, M.; He, S.; Zhan, Y.; Qin, B.; Jin, X.; Wang, M.; Zhang, Y.; Hu, G.; Teng, Z.; Wu, Y. Exogenous melatonin reduces the inhibitory effect of osmotic stress on photosynthesis in soybean. *PLoS ONE* **2019**, *14*, e0226542. [CrossRef]

51. Bai, Y.; Xiao, S.; Zhang, Z.; Zhang, Y.; Sun, H.; Zhang, K.; Wang, X.; Bai, Z.; Li, C.; Liu, L. Melatonin improves the germination rate of cotton seeds under drought stress by opening pores in the seed coat. *PeerJ* **2020**, *8*, e9450. [CrossRef]

52. Antoniou, C.; Chatzimichail, G.; Xenofontos, R.; Pavlou, J.J.; Panagiotou, E.; Christou, A.; Fotopoulos, V. Melatonin systemically ameliorates drought stress-induced damage in Medicago sativa plants by modulating nitro-oxidative homeostasis and proline metabolism. *J. Pineal Res.* **2017**, *62*, e12401. [CrossRef]

53. Li, C.; Tan, D.X.; Liang, D.; Chang, C.; Jia, D.; Ma, F. Melatonin mediates the regulation of ABA metabolism, free-radical scavenging, and stomatal behaviour in two *Malus* species under drought stress. *J. Exp. Bot.* **2015**, *66*, 669–680. [CrossRef] [PubMed]

54. Liang, B.; Ma, C.; Zhang, Z.; Wei, Z.; Gao, T.; Zhao, Q.; Ma, F.; Li, C. Long-term exogenous application of melatonin improves nutrient uptake fluxes in apple plants under moderate drought stress. *Environ. Exp. Bot.* **2018**, *155*, 650–661. [CrossRef]

55. Wang, P.; Sun, X.; Li, C.; Wei, Z.; Liang, D.; Ma, F. Long-term exogenous application of melatonin delays drought-induced leaf senescence in apple. *J. Pineal Res.* **2013**, *54*, 292–302. [CrossRef] [PubMed]

56. Jiao, J.; Ma, Y.; Chen, S.; Liu, C.; Song, Y.; Qin, Y.; Yuan, C.; Liu, Y. Melatonin-producing endophytic bacteria from grapevine roots promote the abiotic stress-induced production of endogenous melatonin in their hosts. *Front. Plant Sci.* **2016**, *7*, 1387. [CrossRef] [PubMed]

57. Niu, X.; Deqing, C.; Liang, D. Effects of exogenous melatonin and abscisic acid on osmotic adjustment substances of 'Summer Black' grape under drought stress. *IOP Conf. Ser. Earth Environ. Sci.* **2019**, *295*, 012012. [CrossRef]

58. Liang, D.; Ni, Z.; Xia, H.; Xie, Y.; Lv, X.; Wang, J.; Lin, L.; Deng, Q.; Luo, X. Exogenous melatonin promotes biomass accumulation and photosynthesis of kiwifruit seedlings under drought stress. *Sci. Hortic.* **2019**, *246*, 34–43. [CrossRef]

59. Sharma, A.; Wang, J.; Xu, D.; Tao, S.; Chong, S.; Yan, D.; Li, Z.; Yuan, H.; Zheng, B. Melatonin regulates the functional components of photosynthesis, antioxidant system, gene expression, and metabolic pathways to induce drought resistance in grafted *Carya cathayensis* plants. *Sci. Total Environ.* **2020**, *713*, 136675. [CrossRef]

60. Liu, J.; Wang, W.; Wang, L.; Sun, Y. Exogenous melatonin improves seedling health index and drought tolerance in tomato. *Plant Growth Regul.* **2015**, *77*, 317–326. [CrossRef]

61. Zhang, N.; Zhao, B.; Zhang, H.-J.; Weeda, S.; Yang, C.; Yang, Z.-C.; Ren, S.; Guo, Y.-D. Melatonin promotes water-stress tolerance, lateral root formation, and seed germination in cucumber (Cucumis sativus L.). *J. Pineal Res.* **2013**, *54*, 15–23. [CrossRef]

62. Li, J.; Zeng, L.; Cheng, Y.; Lu, G.; Fu, G.; Ma, H.; Liu, Q.; Zhang, X.; Zou, X.; Li, C. Exogenous melatonin alleviates damage from drought stress in *Brassica napus* L. (rapeseed) seedlings. *Acta Physiol. Plant.* **2018**, *40*, 43. [CrossRef]

63. Yan, W.; Hongyan, L.; Xuejiao, M.; Xuejuan, W.; Yuanbing, Z. Effect of Foliar Spraying Exogenous Melatonin on Physiological and Biochemical Characteristics of *Dendranthema morifolium*. *Acta Bot. Boreali-Occident. Sin.* **2016**, *36*, 2241–2246.

64. Kabiri, R.; Hatami, A.; Oloumi, H.; Naghizadeh, M.; Nasibi, F.; Tahmasebi, Z. Foliar application of melatonin induces tolerance to drought stress in Moldavian balm plants (*Dracocephalum moldavica*) through regulating the antioxidant system. *Folia Hortic.* **2018**, *30*, 155–167. [CrossRef]

65. Ma, X.; Zhang, J.; Burgess, P.; Rossi, S.; Huang, B. Interactive effects of melatonin and cytokinin on alleviating drought-induced leaf senescence in creeping bentgrass (*Agrostis stolonifera*). *Environ. Exp. Bot.* **2018**, *145*, 1–11. [CrossRef]

66. Alam, M.N.; Wang, Y.; Chan, Z. Physiological and biochemical analyses reveal drought tolerance in cool-season tall fescue (*Festuca arundinacea*) turf grass with the application of melatonin. *Crop Past. Sci.* **2018**, *69*, 1041–1049. [CrossRef]

67. Shi, H.; Jiang, C.; Ye, T.; Tan, D.-X.; Reiter, R.J.; Zhang, H.; Liu, R.; Chan, Z. Comparative physiological, metabolomic, and transcriptomic analyses reveal mechanisms of improved abiotic stress resistance in bermudagrass [*Cynodon dactylon* (L). Pers.] by exogenous melatonin. *J. Exp. Bot.* **2015**, *66*, 681–694. [CrossRef] [PubMed]

68. Zamani, Z.; Amiri, H.; Ismaili, A. Improving drought stress tolerance in fenugreek (*Trigonella foenum-graecum*) by exogenous melatonin. *Plant Biosyst.* **2019**, *2019*, 1–13. [CrossRef]

69. Arnao, M.B.; Hernández-Ruiz, J. Growth conditions determine different melatonin levels in *Lupinus albus* L. *J. Pineal Res.* **2013**, *55*, 149–155. [CrossRef]

70. Arnao, M.B.; Hernández-Ruiz, J. Chemical stress by different agents affects the melatonin content of barley roots. *J. Pineal Res.* **2009**, *46*, 295–299. [CrossRef]

71. Arnao, M.B.; Hernández-Ruiz, J. Growth conditions influence the melatonin content of tomato plants. *Food Chem.* **2013**, *138*, 1212–1214. [CrossRef]

72. Wei, Y.; Zeng, H.; Hu, W.; Chen, L.; He, C.; Shi, H. Comparative Transcriptional Profiling of Melatonin Synthesis and Catabolic Genes Indicates the Possible Role of Melatonin in Developmental and Stress Responses in Rice. *Front. Plant Sci* **2016**, *7*, 676. [CrossRef]

73. Arnao, M.B.; Hernández-Ruiz, J. Role of Melatonin to Enhance Phytoremediation Capacity. *Appl. Sci.* **2019**, *9*, 5293. [CrossRef]

74. Ding, F.; Wang, G.; Wang, M.; Zhang, S. Exogenous Melatonin Improves Tolerance to Water Deficit by Promoting Cuticle Formation in Tomato Plants. *Molecules* **2018**, *23*, 1605. [CrossRef] [PubMed]

75. Tan, D.X.; Manchester, L.C.; Liu, X.; Rosales-Corral, S.A.; Acuna-Castroviejo, D.; Reiter, R.J. Mitochondria and chloroplasts as the original sites of melatonin synthesis: A hypothesis related to melatonin's primary function and evolution in eukaryotes. *J. Pineal Res.* **2013**, *54*, 127–138. [CrossRef] [PubMed]

76. Back, K.; Tan, D.X.; Reiter, R.J. Melatonin biosynthesis in plants: Multiple pathways catalyze tryptophan to melatonin in the cytoplasm or chloroplasts. *J. Pineal Res.* **2016**, *61*, 426–437. [CrossRef] [PubMed]

77. Wei, J.; Li, D.X.; Zhang, J.R.; Shan, C.; Rengel, Z.; Song, Z.B.; Chen, Q. Phytomelatonin receptor PMTR1-mediated signaling regulates stomatal closure in *Arabidopsis thaliana*. *J. Pineal Res.* **2018**, *65*, e12500. [CrossRef] [PubMed]

78. Cui, G.; Sun, F.; Gao, X.; Xie, K.; Zhang, C.; Liu, S.; Xi, Y. Proteomic analysis of melatonin-mediated osmotic tolerance by improving energy metabolism and autophagy in wheat (*Triticum aestivum* L.). *Planta* **2018**, *248*, 69–87. [CrossRef]

79. Arnao, M.B.; Hernandez-Ruiz, J. Melatonin and its relationship to plant hormones. *Ann. Bot.* **2018**, *121*, 195–207. [CrossRef]

80. Zhou, L.; Zhou, J.; Xiong, Y.; Liu, C.; Wang, J.; Wang, G.; Cai, Y. Overexpression of a maize plasma membrane intrinsic protein ZmPIP1;1 confers drought and salt tolerance in *Arabidopsis*. *PLoS ONE* **2018**, *13*, e0198639. [CrossRef]

81. Wang, L.; Feng, C.; Zheng, X.; Guo, Y.; Zhou, F.; Shan, D.; Liu, X.; Kong, J. Plant mitochondria synthesize melatonin and enhance the tolerance of plants to drought stress. *J. Pineal Res.* **2017**, *63*, e12429. [CrossRef]

82. Burgess, P.; Huang, B. Mechanisms of Hormone Regulation for Drought Tolerance in Plants. In *Drought Stress Tolerance in Plants, Vol 1: Physiology and Biochemistry*; Hossain, M.A., Wani, S.H., Bhattacharjee, S., Burritt, D.J., Tran, L.-S.P., Eds.; Springer International Publishing: Cham, Swizerland, 2016; pp. 45–75.

83. Behnam, B.; Iuchi, S.; Fujita, M.; Fujita, Y.; Takasaki, H.; Osakabe, Y.; Yamaguchi-Shinozaki, K.; Kobayashi, M.; Shinozaki, K. Characterization of the promoter region of an *Arabidopsis* gene for 9-cis-epoxycarotenoid dioxygenase involved in dehydration-inducible transcription. *DNA Res.* **2013**, *20*, 315–324. [CrossRef]

84. Kudoyarova, G.R.; Vysotskaya, L.B.; Cherkozyanova, A.; Dodd, C. Effect of partial rootzone drying on the concentration of zeatin-type cytokinins in tomato (*Solanum lycopersicum* L.) xylem sap and leaves. *J. Exp. Bot.* **2007**, *58*, 161–168. [CrossRef] [PubMed]

85. Choi, J.; Hwang, I. Cytokinin: Perception, signal transduction, and role in plant growth and development. *J. Plant Biol.* **2007**, *50*, 98–108. [CrossRef]

86. Lim, P.O.; Kim, H.J.; Nam, H.G. Leaf senescence. *Annu. Rev. Plant Biol.* **2007**, *58*, 115–136. [CrossRef] [PubMed]

87. Xu, Y.; Burgess, P.; Huang, B. Transcriptional regulation of hormone-synthesis and signaling pathways by overexpressing cytokinin-synthesis contributes to improved drought tolerance in creeping bentgrass. *Physiol. Plant.* **2017**, *161*, 235–256. [CrossRef] [PubMed]

88. Arnao, M.B.; Hernández-Ruiz, J. Protective effect of melatonin against chlorophyll degradation during the senescence of barley leaves. *J. Pineal Res.* **2009**, *46*, 58–63. [CrossRef] [PubMed]

89. Tanveer, M.; Shahzad, B.; Sharma, A.; Khan, E.A. 24-Epibrassinolide application in plants: An implication for improving drought stress tolerance in plants. *Plant Physiol. Biochem.* **2019**, *135*, 295–303. [CrossRef]

90. Hwang, O.J.; Back, K. Melatonin is involved in skotomorphogenesis by regulating brassinosteroid biosynthesis in rice plants. *J. Pineal Res.* **2018**, *65*, e12495. [CrossRef]

91. Xia, X.-J.; Gao, C.-J.; Song, L.-X.; Zhou, Y.-H.; Shi, K.A.I.; Yu, J.-Q. Role of H_2O_2 dynamics in brassinosteroid-induced stomatal closure and opening in *Solanum lycopersicum*. *Plant Cell Environ.* **2014**, *37*, 2036–2050. [CrossRef]

92. Bhargava, S.; Sawant, K.; Tuberosa, R. Drought stress adaptation: Metabolic adjustment and regulation of gene expression. *Plant Breed.* **2013**, *132*, 21–32. [CrossRef]

93. Anjum, S.A.; Wang, L.; Farooq, M.; Khan, I.; Xue, L. Methyl Jasmonate-Induced Alteration in Lipid Peroxidation, Antioxidative Defence System and Yield in Soybean Under Drought. *J. Agron. Crop Sci.* **2011**, *197*, 296–301. [CrossRef]

94. Shan, C.; Zhou, Y.; Liu, M. Nitric oxide participates in the regulation of the ascorbate-glutathione cycle by exogenous jasmonic acid in the leaves of wheat seedlings under drought stress. *Protoplasma* **2015**, *252*, 1397–1405. [CrossRef] [PubMed]

95. Anjum, S.A.; Tanveer, M.; Hussain, S.; Tung, S.A.; Samad, R.A.; Wang, L.; Khan, I.; Rehman, N.u.; Shah, A.N.; Shahzad, B. Exogenously applied methyl jasmonate improves the drought tolerance in wheat imposed at early and late developmental stages. *Acta Physiol. Plant* **2015**, *38*, 25. [CrossRef]

96. Wasternack, C. Jasmonates: An update on biosynthesis, signal transduction and action in plant stress response, growth and development. *Ann. Bot.* **2007**, *100*, 681–697. [CrossRef] [PubMed]

97. Murata, Y.; Mori, I.C. Stomatal regulation of plant water status. In *Plant Abiotic Stress*; Jenks, M.A., Hasegawa, P.M., Eds.; Wiley: New York, NY, USA, 2014; pp. 47–67.

98. Balbi, V.; Devoto, A. Jasmonate signalling network in *Arabidopsis thaliana*: Crucial regulatory nodes and new physiological scenarios. *New Phytol.* **2007**, *177*, 301–318. [CrossRef] [PubMed]

99. Goh, C.-H.; Lee, D.J.; Bae, H.-J. Gibberellic acid of *Arabidopsis* regulates the abscisic acid-induced inhibition of stomatal opening in response to light. *Plant Sci.* **2009**, *176*, 136–142. [CrossRef]

100. Teszlák, P.; Kocsis, M.; Gaál, K.; Nikfardjam, M.P. Regulatory effects of exogenous gibberellic acid (GA3) on water relations and CO_2 assimilation among grapevine (*Vitis vinifera* L.) cultivars. *Sci. Hortic.* **2013**, *159*, 41–51. [CrossRef]

101. Kang, S.M.; Radhakrishnan, R.; Khan, A.L.; Kim, M.J.; Park, J.M.; Kim, B.R.; Shin, D.H.; Lee, I.J. Gibberellin secreting rhizobacterium, *Pseudomonas putida* H-2-3 modulates the hormonal and stress physiology of soybean to improve the plant growth under saline and drought conditions. *Plant Physiol. Biochem.* **2014**, *84*, 115–124. [CrossRef]

102. Li, Z.; Lu, G.Y.; Zhang, X.K.; Zou, C.S.; Cheng, Y.; Zheng, P.Y. Improving drought tolerance of germinating seeds by exogenous application of gibberellic acid (GA3) in rapeseed (*Brassica napus* L.). *Seed Sci. Technol.* **2010**, *38*, 432–440. [CrossRef]

103. Krugman, T.; Peleg, Z.; Quansah, L.; Chague, V.; Korol, A.B.; Nevo, E.; Saranga, Y.; Fait, A.; Chalhoub, B.; Fahima, T. Alteration in expression of hormone-related genes in wild emmer wheat roots associated with drought adaptation mechanisms. *Funct. Integr. Genom.* **2011**, *11*, 565–583. [CrossRef]

104. Wang, L.; Zhao, Y.; Reiter, R.J.; He, C.; Liu, G.; Lei, Q.; Zuo, B.; Zheng, X.D.; Li, Q.; Kong, J. Changes in melatonin levels in transgenic 'Micro-Tom' tomato overexpressing ovine *AANAT* and ovine *HIOMT* genes. *J. Pineal Res.* **2014**, *56*, 134–142. [CrossRef]

105. Zuo, B.; Zheng, X.; He, P.; Wang, L.; Lei, Q.; Feng, C.; Zhou, J.; Li, Q.; Han, Z.; Kong, J. Overexpression of *MzASMT* improves melatonin production and enhances drought tolerance in transgenic Arabidopsis thaliana plants. *J. Pineal Res.* **2014**, *57*, 408–417. [CrossRef] [PubMed]

106. Pelagio-Flores, R.; Muñoz-Parra, E.; Ortiz-Castro, R.; López-Bucio, J. Melatonin regulates Arabidopsis root system architecture likely acting independently of auxin signaling. *J. Pineal Res.* **2012**, *53*, 279–288. [CrossRef] [PubMed]

107. Arnao, M.B.; Hernández-Ruiz, J. Melatonin and reactive oxygen and nitrogen species: A model for the plant redox network. *Melatonin Res.* **2019**, *2*, 152–168. [CrossRef]

108. Arnao, M.B.; Hernández-Ruiz, J. Is Phytomelatonin a New Plant Hormone? *Agronomy* **2020**, *10*, 95. [CrossRef]

109. Bhuyan, M.H.M.B.; Hasanuzzaman, M.; Parvin, K.; Mohsin, S.M.; Al Mahmud, J.; Nahar, K.; Fujita, M. Nitric oxide and hydrogen sulfide: Two intimate collaborators regulating plant defense against abiotic stress. *Plant Growth Regul.* **2020**, *90*, 409–424. [CrossRef]

110. Zhu, Y.; Gao, H.; Lu, M.; Hao, C.; Pu, Z.; Guo, M.; Hou, D.; Chen, L.Y.; Huang, X. Melatonin-Nitric Oxide Crosstalk and Their Roles in the Redox Network in Plants. *Int. J. Mol. Sci.* **2019**, *20*, 6200. [CrossRef]

111. Jin, Z.; Shen, J.; Qiao, Z.; Yang, G.; Wang, R.; Pei, Y. Hydrogen sulfide improves drought resistance in *Arabidopsis thaliana*. *Biochem. Biophys. Res. Commun.* **2011**, *414*, 481–486. [CrossRef]

112. Cheng, W.; Zhang, L.; Jiao, C.; Su, M.; Yang, T.; Zhou, L.; Peng, R.; Wang, R.; Wang, C. Hydrogen sulfide alleviates hypoxia-induced root tip death in *Pisum sativum*. *Plant Physiol. Biochem.* **2013**, *70*, 278–286. [CrossRef]

113. Mukherjee, S. Recent advancements in the mechanism of nitric oxide signaling associated with hydrogen sulfide and melatonin crosstalk during ethylene-induced fruit ripening in plants. *Nitric Oxide* **2019**, *82*, 25–34. [CrossRef]

114. Shi, H.; Chen, Y.; Tan, D.-X.; Reiter, R.J.; Chan, Z.; He, C. Melatonin induces nitric oxide and the potential mechanisms relate to innate immunity against bacterial pathogen infection in *Arabidopsis*. *J. Pineal Res.* **2015**, *59*, 102–108. [CrossRef]

115. Mukherjee, S.; David, A.; Yadav, S.; Baluška, F.; Bhatla, S.C. Salt stress-induced seedling growth inhibition coincides with differential distribution of serotonin and melatonin in sunflower seedling roots and cotyledons. *Physiol. Plant.* **2014**, *152*, 714–728. [CrossRef] [PubMed]

116. Kaya, C.; Higgs, D.; Ashraf, M.; Alyemeni, M.N.; Ahmad, P. Integrative roles of nitric oxide and hydrogen sulfide in melatonin-induced tolerance of pepper (*Capsicum annuum* L.) plants to iron deficiency and salt stress alone or in combination. *Physiol. Plant.* **2020**, *168*, 256–277. [CrossRef] [PubMed]

117. Chen, H.; Feng, C.; Kong, J.; Wang, L.; Wang, N.; Zheng, X.; Zhou, Y.; Chan, D. Use of Product Containing Melatonin as Effective Component for Improving Waterlogging Stress Resistance in Plants. CN105076136-A, CN105076136-B, CN105076136-A, 25 November 2015. A01N-043/38 201612.

118. Zheng, X.; Zhou, J.; Tan, D.X.; Wang, N.; Wang, L.; Shan, D.; Kong, J. Melatonin Improves Waterlogging Tolerance of *Malus baccata* (Linn.) Borkh. Seedlings by maintaining aerobic respiration, photosynthesis and ROS Migration. *Front. Plant Sci.* **2017**, *8*, 483. [CrossRef] [PubMed]

119. Zhang, Q.; Liu, X.; Zhang, Z.; Liu, N.; Li, D.; Hu, L. Melatonin improved waterlogging tolerance in alfalfa (*Medicago sativa*) by reprogramming polyamine and ethylene metabolism. *Front. Plant Sci.* **2019**, *10*, 44. [CrossRef] [PubMed]

120. Arnao, M.B.; Hernández-Ruiz, J. Melatonin as a chemical substance or as phytomelatonin rich-extracts for use as plant protector and/or biostimulant in accordance with EC legislation. *Agronomy* **2019**, *9*, 570. [CrossRef]

121. Pérez-Llamas, F.; Hernández-Ruiz, J.; Cuesta, A.; Zamora, S.; Arnao, M.B. Development of a phytomelatonin-rich extract from cultured plants with excellent biochemical and functional properties as an alternative to synthetic melatonin. *Antioxidants* **2020**, *9*, 158. [CrossRef]

 *antioxidants*

Article

Alterations of Endogenous Hormones, Antioxidant Metabolism, and Aquaporin Gene Expression in Relation to γ-Aminobutyric Acid-Regulated Thermotolerance in White Clover

Hongyin Qi [†], Dingfan Kang [†], Weihang Zeng [†], Muhammad Jawad Hassan, Yan Peng, Xinquan Zhang, Yan Zhang, Guangyan Feng and Zhou Li *

College of Grassland Science and Technology, Sichuan Agricultural University, Chengdu 611130, China; qihongyin@stu.sicau.edu.cn (H.Q.); kangdingfan24@163.com (D.K.); zengwh0123@163.com (W.Z.); jawadhassan3146@gmail.com (M.J.H.); pengyanlee@163.com (Y.P.); zhangxq@sicau.edu.cn (X.Z.); zhangyan1111zy@126.com (Y.Z.); feng0201@sicau.edu.cn (G.F.)

* Correspondence: lizhou1986814@163.com

† These authors contributed equally to this work.

Citation: Qi, H.; Kang, D.; Zeng, W.; Jawad Hassan, M.; Peng, Y.; Zhang, X.; Zhang, Y.; Feng, G.; Li, Z. Alterations of Endogenous Hormones, Antioxidant Metabolism, and Aquaporin Gene Expression in Relation to γ-Aminobutyric Acid-Regulated Thermotolerance in White Clover. *Antioxidants* **2021**, *10*, 1099. https://doi.org/10.3390/antiox10071099

Academic Editors: Masayuki Fujita and Mirza Hasanuzzaman

Received: 17 June 2021
Accepted: 2 July 2021
Published: 8 July 2021

Abstract: Persistent high temperature decreases the yield and quality of crops, including many important herbs. White clover (*Trifolium repens*) is a perennial herb with high feeding and medicinal value, but is sensitive to temperatures above 30 °C. The present study was conducted to elucidate the impact of changes in endogenous γ-aminobutyric acid (GABA) level by exogenous GABA pretreatment on heat tolerance of white clover, associated with alterations in endogenous hormones, antioxidant metabolism, and aquaporin-related gene expression in root and leaf of white clover plants under high-temperature stress. Our results reveal that improvement in endogenous GABA level in leaf and root by GABA pretreatment could significantly alleviate the damage to white clover during high-temperature stress, as demonstrated by enhancements in cell membrane stability, photosynthetic capacity, and osmotic adjustment ability, as well as lower oxidative damage and chlorophyll loss. The GABA significantly enhanced gene expression and enzyme activities involved in antioxidant defense, including superoxide dismutase, catalase, peroxidase, and key enzymes of the ascorbic acid–glutathione cycle, thus reducing the accumulation of reactive oxygen species and the oxidative injury to membrane lipids and proteins. The GABA also increased endogenous indole-3-acetic acid content in roots and leaves and cytokinin content in leaves, associated with growth maintenance and reduced leaf senescence under heat stress. The GABA significantly upregulated the expression of *PIP1-1* and *PIP2-7* in leaves and the *TIP2-1* expression in leaves and roots under high temperature, and also alleviated the heat-induced inhibition of *PIP1-1*, *PIP2-2*, *TIP2-2*, and *NIP1-2* expression in roots, which could help to improve the water transportation and homeostasis from roots to leaves. In addition, the GABA-induced aquaporins expression and decline in endogenous abscisic acid level could improve the heat dissipation capacity through maintaining higher stomatal opening and transpiration in white clovers under high-temperature stress.

Keywords: adaptability; heat stress; water homeostasis; hormone; oxidative damage; stoma; transpiration

1. Introduction

Nowadays, the major impact of global warming is already discernible in animal and plant populations [1]. In terms of plants, a global high temperature has resulted in the increase in mortality of forest trees [2], substantial reduction in crop yield [3], and changes in the spatial distribution of herbs [4]. Plant hormones and other plant growth regulators (PGRs) perform pivotal functions in regulating plant adaptability to different abiotic stresses. γ-aminobutyric acid (GABA), a non-protein amino acid comprising four carbons, exists abundantly in living organisms including microbes, plants, and vertebrates. Over the past 50 years, GABA has been regarded as a very important PGR that affects plant growth and the adaptability to high temperature or other environmental stresses [5]. A recent study revealed that the improvement in GABA content or metabolism is an important defense mechanism triggered by the exogenous spermidine (Spd) in white clover (*Trifolium repens*) under heat stress [6]. GABA affected glutathione, carbon, and amino acid metabolism, associated with better heat tolerance in creeping bentgrass (*Agrostis stolonifera*) [7]. Nayyar et al. found that GABA could safeguard rice (*Oryza sativa*) seedlings during high temperature by improving osmotic protection and antioxidant defense [8]. However, study of the regulation mechanisms and physiological function of GABA is still in its infancy as compared to other endogenous hormones (auxin (IAA), gibberellin (GA), cytokinin (CTK), and abscisic acid (ABA)) crucial for plants growth and development [9,10]. As an important signal molecule, ABA participates in the regulation of stomatal movement and multiple gene expression in response to abiotic stress [11]. GA or IAA could effectively alleviate heat-induced growth inhibition and the decline in yield of lemon (*Citrus limon*) trees or rice [12,13]. The CTK-regulated heat tolerance is associated with enhancement of antioxidant defense and delayed leaf senescence in creeping bentgrass [14]. However, it remains unclear so far whether the GABA-regulated heat tolerance is linked with changes in endogenous hormones in plants.

Oxidative damage is one of the most imperative stress indicators when plants are suffering from heat stress. High temperature leads toward massive production of reactive oxygen species (OH^-, $O_2^{\bullet-}$, and H_2O_2) in plants, resulting in increased oxidative stress to various cell organelles [15]. Plants have evolved an efficient antioxidant defense system (enzymatic and non-enzymatic) to scavenge reactive oxygen species (ROS) under unfavorable environmental conditions. For the enzymatic mechanism, superoxide dismutases (SODs) including Cu/ZnSOD, MnSOD, and FeSOD are the first line of defense against ROS and are involved in the dismutation of superoxide anion to produce hydrogen peroxide and molecular oxygen. The Cu/ZnSOD dominates the main function in higher plants, while the lower plants are dominated by MnSOD and FeSOD [16]. Catalase (CAT), peroxidase (POD), and the ascorbic acid (ASA)–glutathione (GSH) cycle can convert H_2O_2 into H_2O and O_2. For the ASA–GSH cycle, ascorbate peroxidase (APX), glutathione peroxidase (GPX), monodehydroascorbate reductase (MR), dehydroascorbate reductase (DR), and glutathione reductase (GR) remove H_2O_2 through catalyzing the oxidation–reduction of ASA and GSH. In this process, ASA and GSH act as non-enzymatic antioxidants to scavenge ROS [17]. Previous studies showed that the GABA-regulated tolerance under unfavorable conditions was closely related to enhanced antioxidant defense in different plant species. For instance, pepper (*Capsicum annuum*) could resist low-light stress by increasing antioxidant defense after GABA treatment [18], and GABA could slow down the oxidative injury induced by salinity in white clover [19]. However, how GABA interacts with endogenous hormones to affect antioxidant defense systems (enzymatic and non-enzymatic) in leaf and root remains unclear when plants undergo a prolonged period of heat stress.

Under high-temperature stress, even if water supply is sufficient, plants will suffer from physiological water shortage due to the aggravation of transpiration loss and the decrease in water absorbing capacity in the roots [20]. Water transport in plants is regulated mainly by three ways, including the apoplastic pathway, symplastic pathway, and transcellular pathway. Among them, the transcellular pathway is mainly carried out by aquaporins (AQPs). Four kinds of AQPs have been widely recognized in plants: plasma membrane endogenous proteins (PIPS), vacuolar endogenous proteins (TIPS), nodule protein 26 (NIPS), and small basic endogenous proteins (SIPS) [21]. PIPS are divided into PIP1 and PIP2, that sense the change in water inside and outside of cell membranes, and PIP2 shows a greater water channel efficiency than PIP1. TIPS are located in the vacuolar membrane for sensing the change in water content in the cytoplasm, which performs a vital role in maintaining cell permeability. NIPS is responsible for transporting a wide range of substances, such as water, urea, glycerol, and some metal ions. There are only a few studies on SIPS that are mainly located in endoplasmic reticulum [22–25]. It was reported that AQPs perform crucial functions in plant water regulation under osmotic or extreme temperature stress. The study of Peng et al. showed that the overexpression of an AQP gene *pgtip1* enhanced the stress tolerance in *Arabidopsis thaliana* under low-temperature, salinity, and water stress [26]. Alexandersson et al. found that *atpip2-2* affected root water transport under drought stress [27]. In soybean (*Glycine max*), *TIP2-6* responds to high-temperature stress through hormone regulation [28]. The expression of *CsT1P1-3*, *CsT1P2-3*, and *CsPIP2-4* was enhanced by high temperature in citrus (*Citrus reticulata Blanco*) [29]. Up to now, it is not well documented whether GABA-regulated thermotolerance is related to AQP expression in plants.

White clover is a perennial herb with high feeding and medicinal value. It is cultivated worldwide as an important forage due to high crude protein content and nutritional value, and extract productions derived from white clover such as flavonoids have multiple effects, such as reducing blood fat, cancer prevention, immunity enhancement, and anti-aging [30,31]. However, white clover is sensitive to high temperature and its growth is inhibited significantly at temperatures above 30 °C [32]. With global warming, the heat-induced damage to white clover is becoming increasingly prominent in summer. The objective of the present study was to determine the impact of GABA on improvement in thermotolerance associated with alterations of endogenous hormones, antioxidant metabolism, water regulation, and AQP-related gene transcript level in leaf and root of white clover. This will help to reveal the potential regulatory mechanism of GABA in herbs under heat stress and improve the cultivation and utilization of white clover or other herbs in more regions.

2. Materials and Methods

2.1. Plant Materials and Treatments

Seeds of white clover (cultivar "Ladino") were surface sterilized with 0.1% mercuric chloride ($HgCl_2$) solution for 4 min and rinsed thrice with deionized water. Seeds (20 g/m^2) were placed in a container (24 cm length, 15 cm width, and 8 cm height) comprising quartz sand. The containers were kept in controlled laboratory conditions (23 °C/19 °C day/night, 12 h photoperiod, 700 µmol m^{-2} s^{-1} photosynthetically active radiation, and 75% relative humidity). Seeds were firstly germinated in ddH$_2$O for 7 days and then cultured in Hoagland nutrient solution for 24 days [33]. Before the beginning of heat stress, plant materials were grown on Hoagland's solution with or without 2 mM GABA for 3 days. The half of non-pretreated or GABA-pretreated plants was kept in the controlled growth chamber (conditions as illustrated above) as the control (C) or the control + GABA (C+GABA) for 30 days, and the other half was transferred into a high-temperature chamber (38/33 °C day/night and other conditions the same as the normal growth chamber) as heat stress (H) and heat+GABA (H+GABA) for 30 days. Fresh Hoagland's solution was applied to the plants every day. The plastic containers were laid out in completely randomized design with four biological replicates. Leaf samples were taken on 0, 5, 10, 15, 20, and 30 d after heat stress, respectively. Root samples were taken at 15 d for gene expression and at 30 d for other parameters.

2.2. Determination of Endogenous GABA and Phytohormones Content

The GABA content was estimated by using the Test Kit obtained from Suzhou Comin Biotechnology Co. Ltd, Suzhou, China. following the manufacturer's guidelines. For the determination of ABA, GA, and IAA, fresh samples (0.4 g) were crushed with 3 mL methanol:isopropanol (1:4, *v/v*) and 1% glacial acetic acid. The mixture was centrifuged for 1 h at 4 °C. A total of 2 mL of supernatant was collected, dried, and mixed in CH$_3$OH (300 µL). The reaction solution was then filtered by passing through a 0.22 µm PTFE filter [34]. The concentrations of endogenous IAA, GA, and ABA were determined by Waters Acquity UPLCSCIEX Se-lex ION Triple Quad 5500 mass spectrometer (waters, Milford, MA, USA). A 5 µL sample was loaded into the Acquity UPLC beh C18 column (1.7 µm, 50 × 2.1 mm; waters, Waxford, Ireland) at 40 °C. The mobile phase was composed of 40% acetic acid solution and 60% CH$_3$OH, and the flow rate was 1 mL min^{-1}. The

CTK content was estimated by using enzyme-linked immunosorbent assay (ELISA) in accordance with the manufacturer's guidelines. The Test Kit was obtained from Beijing Fang Cheng Biotechnology Co., Ltd, Beijing, China.

2.3. Measurement of Water Status and Photosynthetic Parameters

For leaf relative water content (RWC), the fresh weight (FW) was weighed instantly when leaves were cut off from plants. The leaf samples were drenched in deionized water for 24 h to obtain the saturated weight (SW) and later kept in an oven at 85 °C for 72 h to obtain the dry weight (DW). The RWC was formulated as RWC (%) = [(FW − DW)/ (SW − DW)] × 100% [35]. Root viability was estimated by following the procedure of McMichael and Burke [36]. The osmotic potential (OP) was detected by using the protocols of Blum. Fresh leaves or roots were immerged in distilled water for 12 h and quickly frozen in liquid nitrogen for 10 min after the surface moisture being absorbed, and then thawed at 4 °C for 30 min. The cell fluids were squeezed from leaves or roots, and the osmolality of cell sap was determined by using a vapor pressure osmometer (Wescor, Logan, UT, USA), and the OP was converted based on $- c \times 2.58 \times 10^{-3}$ [37]. For electrolyte leakage (EL), the procedure of Blum and Ebercon was utilized [38]. Chlorophyll (Chl) was measured by Arnon's method [39]. The photochemical efficiency (Fv/Fm) and performance index absorption basis (PIABS) were measured by using a Chl fluorescence system (Pocket PEA, Hansatech, UK). Prior to analysis, fresh leaf samples were kept in dark conditions for 30 min with attached leaf clips. Net photosynthetic efficiency (Pn), water use efficiency (WUE), stomatal conductance (Gs), intercellular CO_2 concentration (Ci), and transpiration rate (Tr) were recorded with a movable photosynthetic system (CIRAS-3, PP Systems, Amesbury, MA, USA) that supplied CO_2 (400 μL^{-1}) and red and blue light (800 μmol photon m^{-2}), respectively.

2.4. Determination of Antioxidant Metabolism

The protein carbonyl content and the total antioxidant capacity (TAC) was measured by using the Test Kit purchased from Suzhou Kangmin Biotechnology Co, Suzhou, China. Superoxide anion ($O_2^{\bullet -}$) or hydrogen peroxide (H_2O_2) were measured following the method of Elstner and Heupel [40] or Velikova et al. [41], respectively. As for malondialdehyde (MDA), SOD, CAT, POD, APX, DR, GR, and MR activities, fresh tissues (0.15 g) were taken and mechanically ground in 1.5 mL of precooled phosphoric acid buffer solution (150 mM and pH 7.0), and then transferred into a centrifugal tube. The homogenate was centrifuged at 15,000× *g* at 4 °C for 20 min to obtain the supernatant. For MDA content, supernatant (0.5 mL) was added in reaction solution (1 mL) having trichloroacetic acid (20%) and thiobarbituric acid (0.5%). The mixture was kept in a boiling water bath (95 °C) for 15 min and later hastily cooled in an ice bath. The solution was centrifuged at 8000× *g* for 10 min at 4 °C. The absorbance value was noticed spectrometrically at 532 and 600 nm [42]. The activities of SOD, CAT, POD, or APX were estimated by noting absorbance values at 560, 240, 470, or 290 mm, respectively [43–45]. The GR, MR, or DR activity was detected by using the protocols of Cakmak et al. [46]. Bradford's method was used to detect soluble protein content [47]. ASA and GSH content were measured using the method of Gossett et al. [48].

2.5. Total RNA Extraction and qRT-PCR Analysis

Real time quantitative polymerase chain reaction (qRT-PCR) was used to determine the gene expression. The Rneasy Mini Kit (Qiagen, Duesseldorf, Germany) was used for extracting total RNA in fresh leaves or roots. The RNA samples were treated with DNAsa to remove possible DNA and then reverse-transcribed to cDNA using a revert Aid First Stand cDNA Synthesis Kit (Fermentas, Lithuania). Primers of antioxidant enzyme genes (*Cu/ZnSOD, MnSOD, FeSOD, CAT, POD, APX, DR, GR, MR*) and aquaporin genes (*PIP1-1, PIP2-2, PIP2-7, SIP1-1, TIP1-1, TIP2-1, TIP2-2, NIP1-2, NIP2-1*) are recorded in Table S1 (*β*-actin as an internal control). For all genes, PCR conditions (iCycler iQ qRT-PCR detection system with SYBR Green Supermix, Bio-Rad, Hercules, CA, USA) were as follows: 5 min at

94 °C and 30 s at 95 °C (45 repeats of denaturation), annealing and extending for 45 s at 58 to 66 °C (Table S1), and amplicon from 60 to 95 °C to obtain the melting curve.

2.6. Statistical Analysis

Statistical analysis of all the data was conducted using SPSS 23 (IBM, Armonk, NY, USA). Significant differences among various treatments were estimated by one-way ANOVA together with LSD test at 5% probability level ($p \leq 0.05$).

3. Results

3.1. Changes in Endogenous GABA and Hormones Content

The endogenous GABA content significantly increased after exogenous GABA pretreatment (Figure 1A). On the 15th day, the GABA content increased significantly in the GABA-pretreated and non-pretreated plants under the high-temperature condition, and the increase was more significant in the GABA-pretreated plants (Figure 1A). On the 30th day, the GABA content in the roots and leaves of two heat-stressed treatments ("H" and "H + GABA") was still significantly higher in contrast to normal treatments ("C" and "C + GABA"), and GABA pretreatment could further increase heat-induced GABA content in roots and leaves (Figure 1A,B). The GABA pretreatment had no significant effect on ABA, GA, IAA, and CTK contents in leaves and roots under normal conditions (Figure 2A–D). The ABA content increased significantly, but GA, IAA, and CTK contents decreased significantly in leaves and roots after 30 d of heat stress (Figure 2A–D). Under the high-temperature condition, the GABA application significantly inhibited heat-induced increases in ABA content in roots and leaves (Figure 2A), but did not affect the GA content (Figure 2B). The IAA content in roots and leaves as well as CTK content in leaves were increased significantly by the GABA pretreatment (Figure 2C,D).

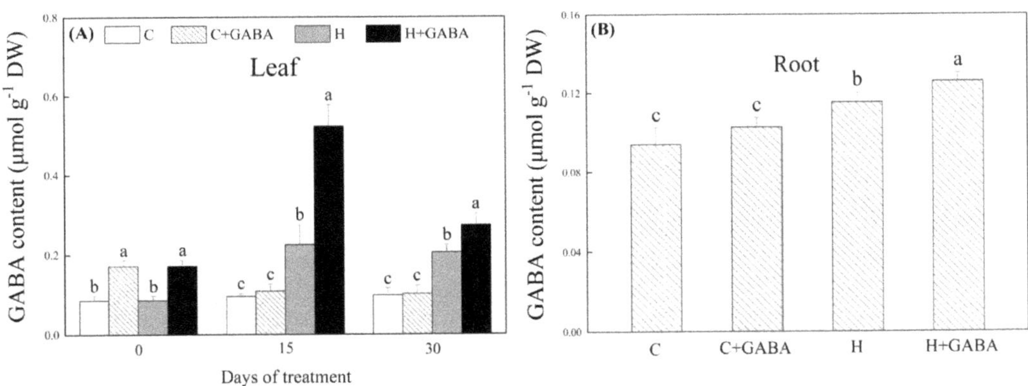

Figure 1. Effects of GABA application on the content of GABA in plant leaves (**A**) and roots (30th day) (**B**) under normal-temperature and high-temperature stress. The vertical bar above the column in (**A**) or (**B**) represents + SE of the mean ($n = 4$) and the different letters in the columns indicate significant differences at a given day of treatment based on LSD ($p \leq 0.05$). C, control (normal condition); C + GABA, control plants pretreated with GABA (normal condition); H, heat stress; H + GABA, heat-stressed plants pretreated with GABA.

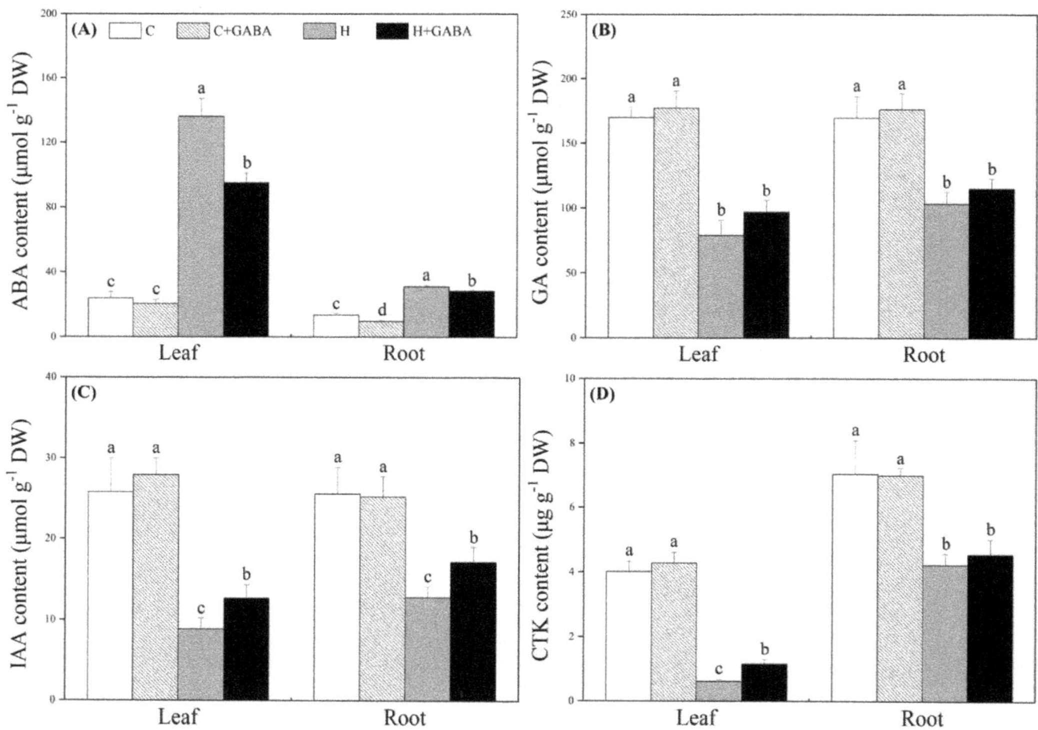

Figure 2. Effects of GABA application on ABA content (**A**), GA content (**B**), IAA content (**C**), and CTK content (**D**) of leaves (30th day) and roots (30th day) under normal-temperature and high-temperature stress. The vertical bar above the column in (**A**), (**B**), (**C**), or (**D**) represents + SE of the mean (*n* = 4) and the different letters in the column indicate significant differences in leaf or root based on LSD ($p \leq 0.05$). C, control (normal condition); C + GABA, control plants pretreated with GABA (normal condition); H, heat stress; H + GABA, heat-stressed plants pretreated with GABA.

3.2. Effects of GABA on Cell Membrane Stability and Oxidative Damage

The MDA, H_2O_2, and $O_2^{\bullet-}$ content significantly increased in leaves and roots of the GABA-treated and untreated plants due to high-temperature stress and the GABA application could significantly alleviate these effects (Figure 3A–C). However, no significant differences in the accumulation of $O_2^{\bullet-}$, H_2O_2, and MDA content between GABA- treated and untreated plants were observed under normal conditions (Figure 3A–C). EL in leaves gradually increased with the extension of heat stress time. The GABA-pretreated plants maintained a 12.86, 22.48, or 20.27% decrease in EL in leaves as compared to the non-pretreated plants on 15, 20, or 30 d of high-temperature stress, respectively (Figure 4A). Similarly, exogenous GABA application also effectively alleviated the heat-induced increase in EL in roots after 30 d (Figure 4B). With the passage of stress time, the carbonyl content increased significantly under high temperature, but the application of GABA showed a significant inhibitory effect on the increase in carbonyl content in leaves and roots in response to heat stress (Figure 4C,D). The GABA-treated white clover plants showed a significant reduction in carbonyl content when compared to untreated plants on the 15th, 20th, and 30th days of heat stress (Figure 4C). Under heat stress, the GABA-pretreated plants maintained a 20.86% greater decrease in the carbonyl content than non-pretreated plants in roots (Figure 4D).

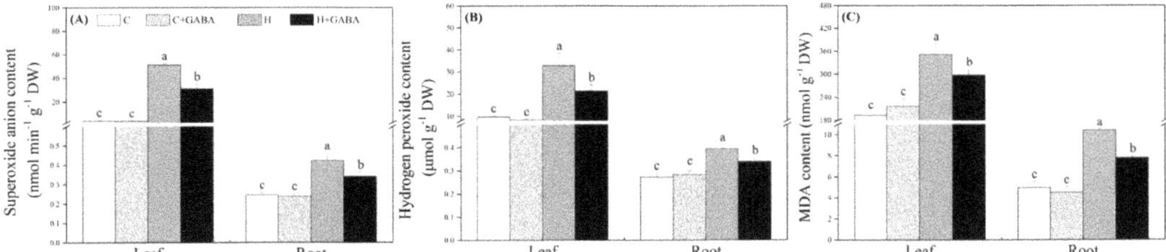

Figure 3. Effects of GABA application on superoxide anion content (**A**), hydrogen peroxide content (**B**), and malondialde-hyde (MDA) content (**C**) of plants leaves (30th day) and roots (30th day) under normal-temperature and high-temperature stress. The vertical bar above the column in (**A**), (**B**), or (**C**) represents + SE of the mean (n = 4) and the different letters in the column indicate significant differences in leaf or root based on LSD ($p \leq 0.05$). C, control (normal condition); C + GABA, control plants pretreated with GABA (normal condition); H, heat stress; H + GABA, heat-stressed plants pretreated with GABA.

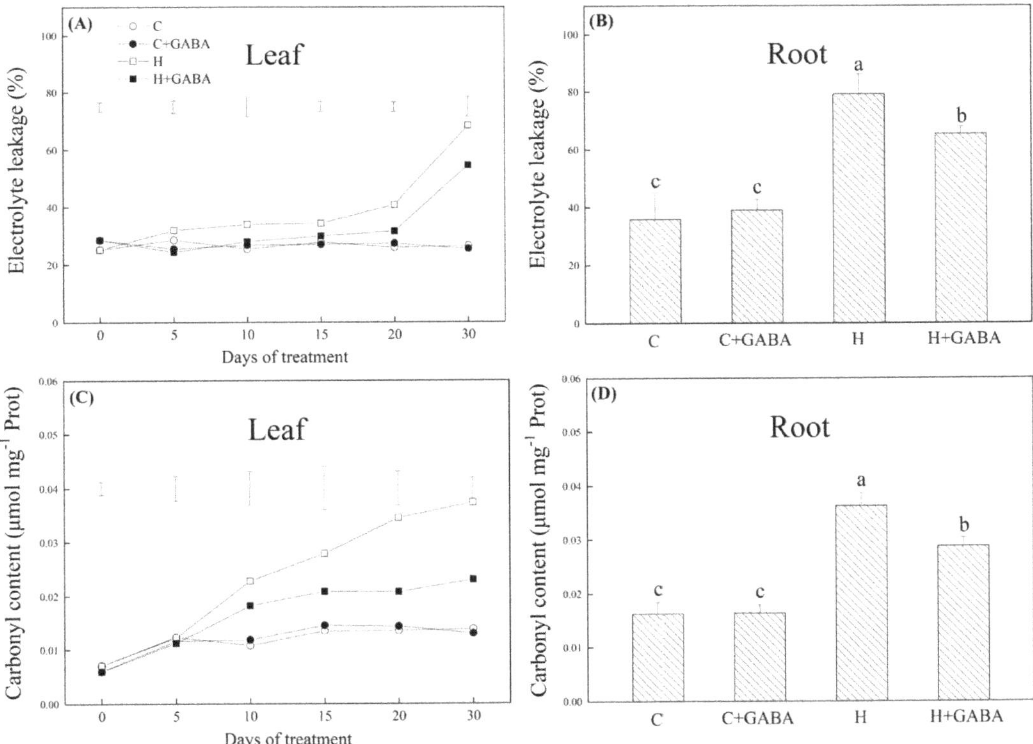

Figure 4. Effects of GABA application on electrolyte leakage of leaves (**A**) or roots (30th day) (**B**), carbonyl content of leaves (**C**) or roots (30th day) (**D**) of plants under normal-temperature and high-temperature stress. Vertical bars above curves in (**A**) or (**C**) represent the least significant difference (LSD) values at a particular day (n = 4; $p \leq 0.05$). The vertical bar above the column in (**B**) or (**D**) represents + SE of the mean (n = 4) and the different letters in the columns indicate significant differences based on LSD ($p \leq 0.05$). C, control (normal); C + GABA, control plants pretreated with GABA; H, heat stress; H + GABA, heat-stressed plants pretreated with GABA.

3.3. Effect of GABA on Antioxidant Metabolism

The TAC in leaves of the GABA-pretreated and non-pretreated plants gradually increased from 10 to 20 d of heat stress and then declined on the 30th day of heat stress (Figure 5A). The GABA pretreatment did not greatly influence the TAC in leaves under normal condition, but the GABA-pretreated plants exhibited 20.03, 18.39, or 37.75% higher TAC compared to the non-pretreated plants on the 15th, 20th, or 30th day of heat stress, respectively (Figure 5A). In the root system, the TAC in the heat-treated plants was significantly lower in contrast to the control group "C", but with the application of GABA, the TAC in roots increased significantly under heat stress (Figure 5B). Under normal conditions, the GABA pretreatment also significantly improved the TAC in roots (Figure 5B). For changes in antioxidant enzyme activities, heat stress significantly improved SOD and POD activities in leaves and roots, and the GABA-pretreated plants showed significantly higher SOD and POD activities in leaves and roots compared with the non-pretreated plants under high temperature (Figure 6A,C). Under heat stress, the GABA-pretreated plants maintained an 86.81 or 90.14% increase in CAT activity in leaves or roots than non-pretreated plants, respectively (Figure 6B). Heat stress significantly improved APX activity in leaves and DR activity in roots, but the GABA pretreatment did not significantly affect APX activity in leaves and DR activity in roots under normal as well as stressed conditions (Figure 6D,E). However, the GABA pretreatment effectively alleviated heat-induced declines in AXP activity in roots, DR activity in leaves, and GR activities in leaves and roots (Figure 6D–F). The GABA-pretreated plants showed a 217.05% increase in MR activity in leaves compared to non-pretreated white clover plants subjected to heat stress (Figure 6G). In leaves, ASA content was substantially ameliorated by the exogenous application of GABA under normal as well as stressed conditions. In roots, the GABA application could also stop the heat-caused decline in ASA content (Figure 6H). GSH content significantly declined in leaves and was enhanced in roots under heat stress; however, the GABA application did not exert its significant effects on GSH content under any of the above-mentioned treatments (Figure 6I).

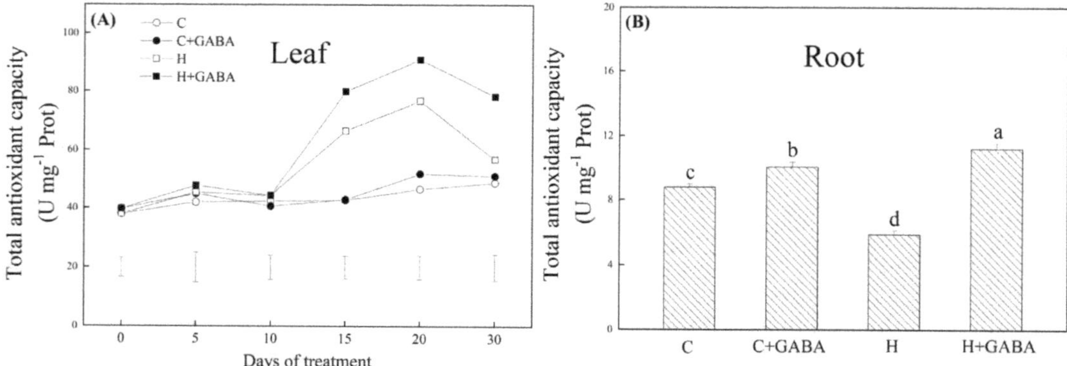

Figure 5. Effects of GABA application on total antioxidant capacity of leaves (**A**) or roots (30th day) (**B**). Vertical bars above curves in (**A**) represent the least significant difference (LSD) values at a particular day ($n = 4$; $p \leq 0.05$). The vertical bar above the column in (**B**) represents + SE of the mean ($n = 4$) and the different letters in the columns indicate significant differences based on LSD ($p \leq 0.05$). C, control (normal condition); C + GABA, control plants pretreated with GABA (normal condition); H, heat stress; H + GABA, heat-stressed plants pretreated with GABA.

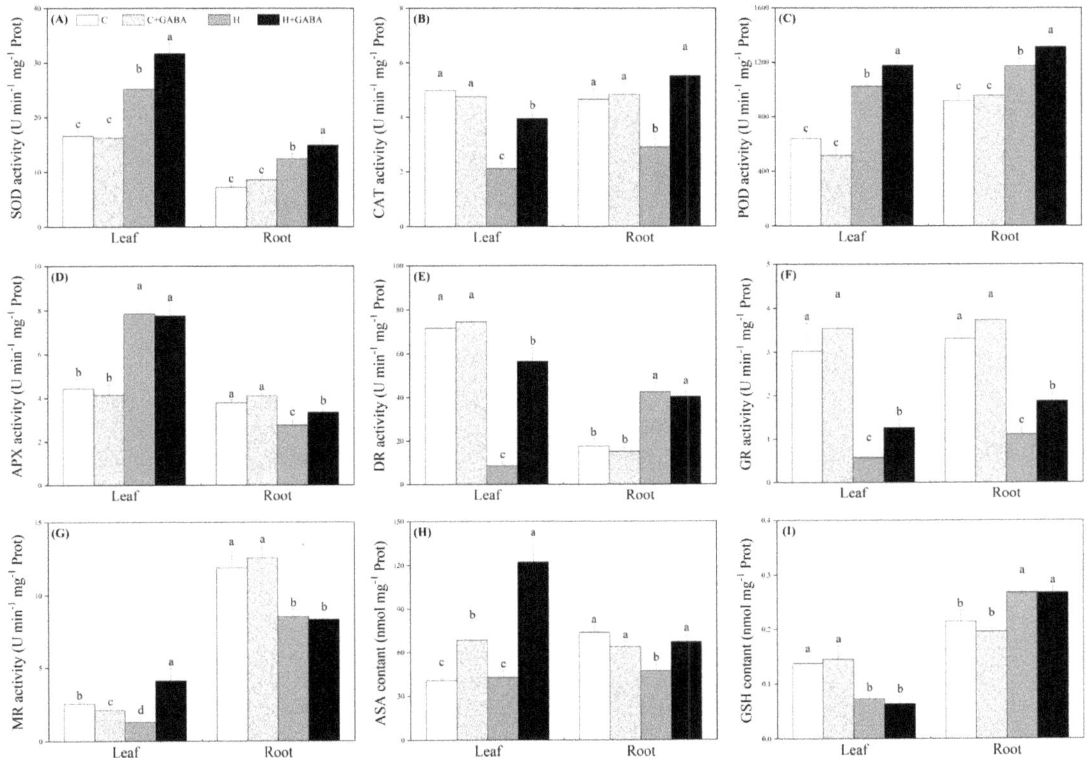

Figure 6. Effects of GABA application on SOD content (**A**), CAT content (**B**), POD content (**C**), APX content (**D**), DR content (**E**), GR content (**F**), MR content (**G**), ASA content (**H**), and GSH content (**I**) of plant leaves (30th day) and roots (30th day) under normal-temperature and high-temperature stress. The vertical bar above the column in (**A–I**) represents + SE of the mean (n = 4) and the different letters in the column indicate significant differences in leaf or root based on LSD ($p \leq 0.05$). C, control (normal condition); C + GABA, control plants pretreated with GABA (normal condition); H, heat stress; H + GABA, heat-stressed plants pretreated with GABA.

Heat stress significantly upregulated the expression level of *Cu/ZnSOD*, *POD*, and *DR* in leaves treated with or without GABA (Figure 7A). A 30.34, 32.50, or 81.14% increase in *Cu/ZnSOD*, *POD*, or *DR* expression was observed in leaves of the "H + GABA" as compared to that of the "H". The *MnSOD* expression in leaves was only significantly improved in the GABA-pretreated plants by heat stress. The *FeSOD* and *APX* expression significantly decreased in leaves under heat stress, and there were no significant differences in *FeSOD* expression between the GABA-treated and untreated plants under control as well as stressed conditions (Figure 7A). *CAT* and *MR* expression in leaves showed similar results in response to the GABA pretreatment and heat stress, as reflected by a significant decline in the non-pretreated plants and a significant increase in the GABA-pretreated plants under heat stress. The GABA pretreatment significantly enhanced *GR* expression in leaves under control and stressed conditions (Figure 7A). For gene expression in roots, the GABA-treated plants exhibited significantly higher *Cu/ZnSOD*, *MnSOD*, *CAT*, *APX*, *DR*, and *MR* than the non-pretreated plants under heat stress (Figure 7B). In addition, heat stress significantly inhibited the *FeSOD* and *GR* expression and improved *POD* expression in roots of the GABA-pretreated and non-pretreated plants, but the GABA pretreatment did not show any significant impact on the expression of these genes under control as well as heat-stress conditions (Figure 7B).

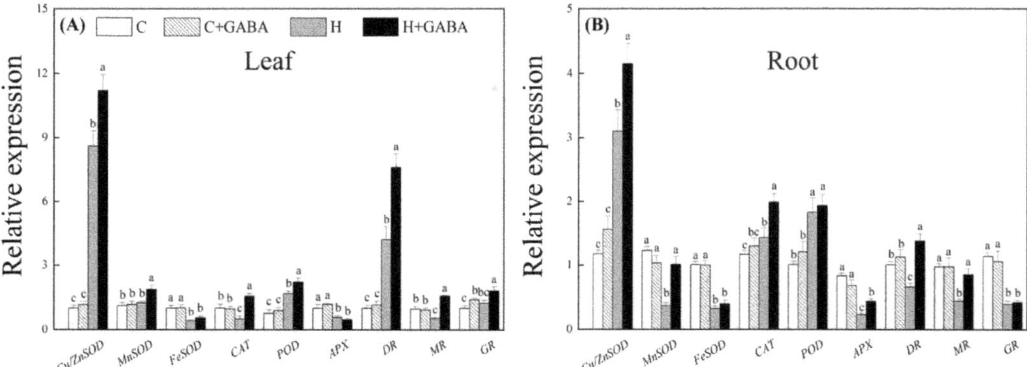

Figure 7. Effect of GABA on relative expression of *Cu/ZnSOD, MnSOD, FeSOD, CAT, POD, APX, DR, MR,* and *GR* genes in leaves (15th day) (**A**) and roots (15th day) (**B**). The vertical bar above the column in (**A**) and (**B**) represents + SE of the mean (*n* = 4) and the different letters in the columns indicate significant differences based on LSD (*p* ≤ 0.05). C, control (normal condition); C + GABA, control plants pretreated with GABA (normal condition); H, heat stress; H + GABA, heat-stressed plants pretreated with GABA.

3.4. Effects of GABA on Photosynthesis and Water Status

Chl a, Chl b, total Chl content, Fv/Fm, PIABS, Pn, and WUE significantly decreased under heat stress, but the application of GABA significantly alleviated their declines under heat stress (Figure 8A–C,G,H). Upon exposure to high temperature, Gs and Tr increased significantly on the 15th day, but decreased on the 30th day. GABA induced significant increases in Gs on the 15th and 30th days of heat stress and Tr on the 15th day of heat stress (Figure 8D,F). The Ci was not changed significantly by heat stress and the GABA application (Figure 8E). WUE and Tr showed a significant downward trend during heat stress, but the downward trend was significantly alleviated by GABA application (Figure 8G,H). Leaf RWC, root activity, and OP in leaf and root did not show any significant difference by foliar application of GABA under controlled conditions (Figure 9A–D). Under heat stress, the leaf RWC and root activity declined markedly in GABA treated and untreated white clover plants, but the GABA-pretreated plants exhibited a 17.23% or 22.73% greater increase in leaf RWC or root activity than non-pretreated plants on the 30th day of heat stress, respectively (Figure 9A,B). High-temperature stress significantly reduced OP in leaves, while it increased OP in roots (Figure 9C,D). The GABA pretreatment further decreased heat-induced decline in OP in leaves on the 20th and 30th days of heat stress (Figure 9C). In addition, the GABA-pretreated plants also demonstrated a 15.32% or 15.22% greater decline in OP in roots than the non-pretreated plants on the 20th and 30th days of heat stress, respectively (Figure 9D).

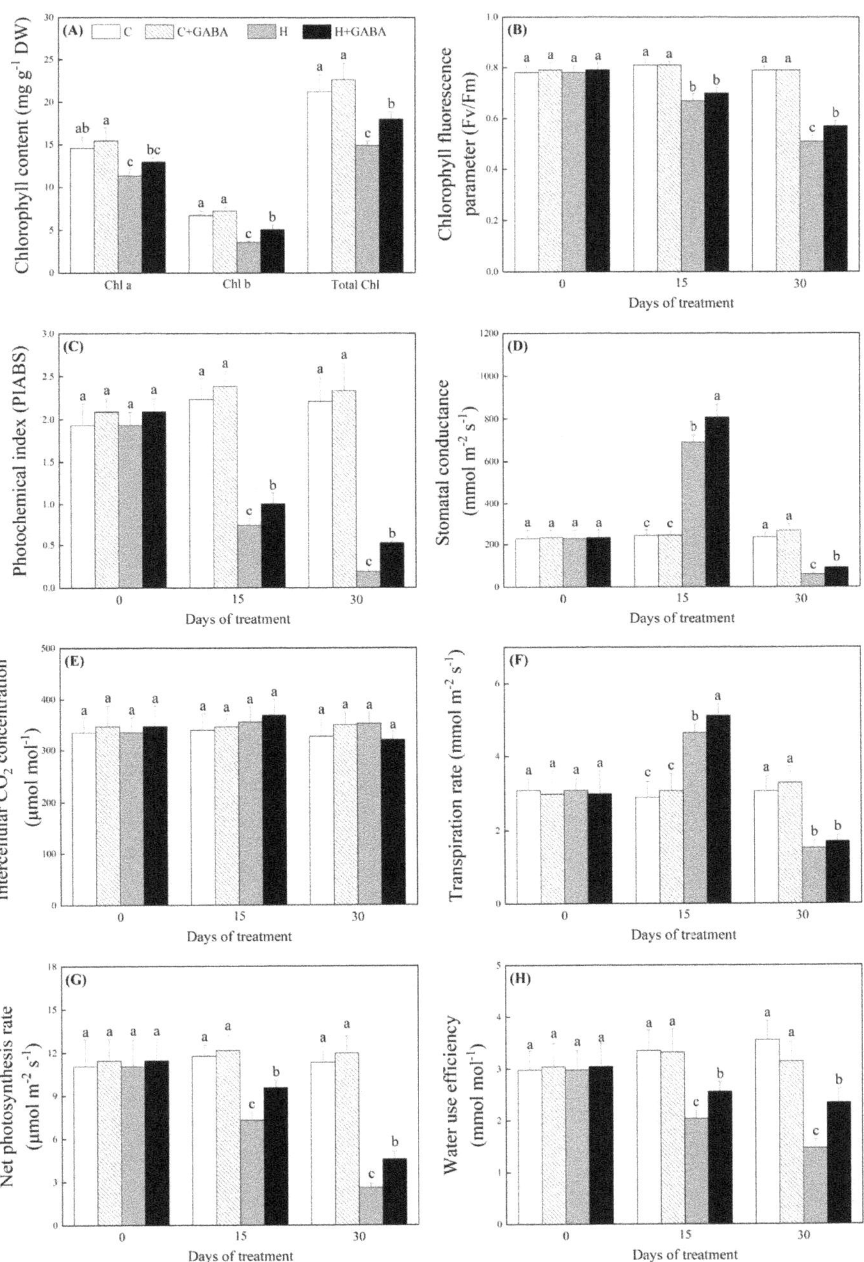

Figure 8. Effects of GABA application on chlorophyll a, b, or total content (**A**), chorophyll fluorescence parameter (**B**), photochemical index (**C**), stomatal conductance (**D**), intercellular CO$_2$ concentration (**E**), transpiration rate (**F**), net photosynthesis rate (**G**), or water use efficiency (**H**) of plants under normal-temperature and high-temperature stress. The vertical bar above the column in (**A**) to (**H**) represents + SE of the mean (*n* = 4) and the different letters in the columns indicate significant differences at a given day of treatment based on LSD (*p* ≤ 0.05). C, control (normal condition); C + GABA, control plants pretreated with GABA (normal condition); H, heat stress; H + GABA, heat-stressed plants pretreated with GABA.

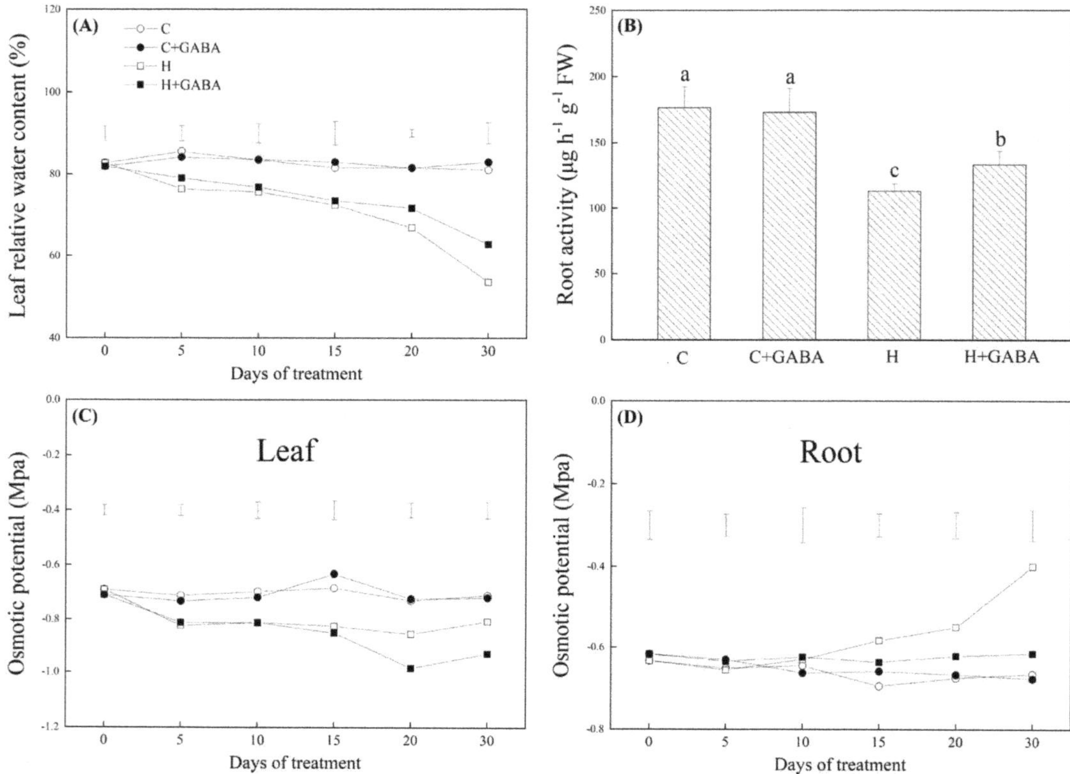

Figure 9. Effects of GABA application on leaf relative water content (**A**), root activity (30th day) (**B**), and osmotic potential of leaves (**C**) or of roots (**D**) of plants under normal-temperature and high-temperature stress. Vertical bars above curves in (**A**), (**C**), or (**D**) represent the least significant difference (LSD) values at a particular day ($n = 4$; $p \leq 0.05$). The vertical bar above the column in (**B**) represents + SE of the mean ($n = 4$) and the different letters in the columns indicate significant differences based on LSD ($p \leq 0.05$). C, control (normal condition); C + GABA, control plants pretreated with GABA; H, heat stress (normal condition); H + GABA, heat-stressed plants pretreated with GABA.

3.5. Effect of GABA on AQP Gene Expression

For the transcription level of genes encoding AQPs, heat stress significantly upregulated the transcript level of *PIP1-1*, *PIP2-7*, *TIP1-1*, and *TIP2-1* in leaves of the GABA-pretreated and untreated plants (Figure 10A). As compared with the "H" treatment, the transcript level of *PIP1-1*, *PIP2-7*, or *TIP2-1* in the "H + GABA" treatment increased by 10.81, 74.28, or 35.48%, respectively. Under heat stress, the expression of *NIP1-2* and *NIP2-1* in leaves decreased significantly, and no significant difference between the GABA-treated and untreated plants was observed. As compared to the "C" treatment, the expression of *SIP1-1* did not change significantly in the "H" treatment under heat stress, but increased by 52.41% in the "H + GABA" treatment (Figure 10A). For AQP gene expression in roots, heat stress significantly inhibited the expression of *PIP1-1*, *PIP2-2*, *SIP1-1*, *TIP2-2*, *NIP1-2*, and *NIP2-1* in plants without the GABA pretreatment. However, the GABA-pretreated plants maintained a significantly higher expression of *PIP1-1*, *PIP1-2*, *PIP2-1*, *SIP1-1*, *TIP2-1*, *TIP2-2*, and *NIP2-1* when compared with non-pretreated plants under high-temperature stress (Figure 10B). The GABA application further upregulated the heat-induced increase in *TIP2-1* expression in roots (Figure 10B).

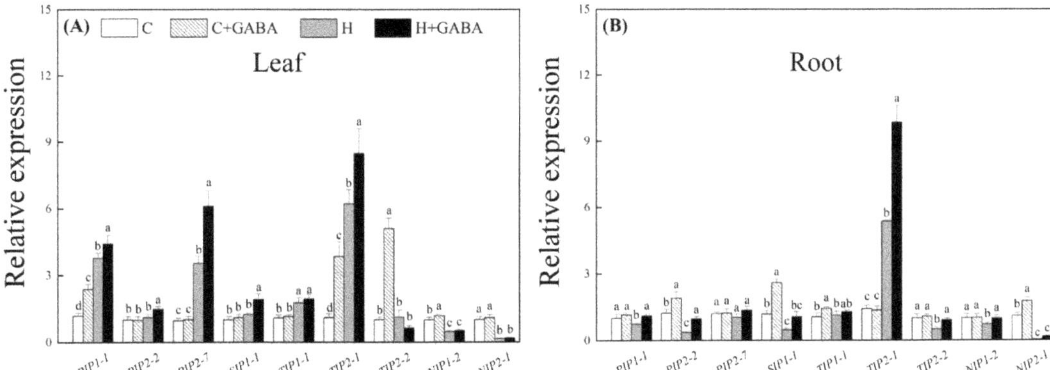

Figure 10. Effect of GABA on relative expression of *PIP1-1, PIP2-2, PIP2-7, SIP1-1, TIP1-1, TIP2-1, TIP2-2, NIP1-2,* and *NIP2-1* genes in leaves (15th day) (**A**) and roots (15th day) (**B**). The vertical bar above the column in (**A,B**) represents + SE of the mean (n = 4) and the different letters in the column indicate significant differences based on LSD ($p \leq 0.05$). C, control (normal condition); C + GABA, control plants pretreated with GABA (normal condition); H, heat stress; H + GABA, heat-stressed plants pretreated with GABA.

4. Discussion

When plants are exposed to high temperatures, GABA as a metabolite or signal molecule regulates intracellular pH environment, carbon (C)–nitrogen (N) nutrient metabolism, oxidative and osmotic balance, and signal transmission [49]. In this study, the GABA pretreatment not only further enhanced the heat-induced GABA content in roots and leaves of white clover, but also significantly alleviated Chl loss and declines in Pn and Fv/Fm under high temperature. A previous study proved that the increase in endogenous GABA could improve the heat resistance of plants such as creeping bentgrass, perennial ryegrass (*Lolium perenne*), and vegetable soybean (*Glycine max*) [50–52]. Under unfavorable environmental conditions, changes in endogenous hormones are critical modulatory factors of stress tolerance in plants [53]. The study of Lancien and Roberts indicated that potential interactions between GABA, ABA, and ethylene modulated key genes involved in C–N metabolism in *Arabidopsis thaliana* under normal conditions [54]. Enhanced GABA accumulation induced by exogenous GABA application in roots of maize (*Zea mays*) were accompanied by a substantial reduction in ABA content and increment in GA, IAA, and CTK content during saline conditions [55]. Our present findings demonstrate that exogenous GABA significantly inhibited ABA accumulation in leaves and roots, and improved IAA content in leaves and roots and CTK content in leaves of white clover, which implied that the GABA-mediated tolerance under extreme temperature was linked with changes in ABA, GA, and IAA. Stress-induced increase in ABA content activates defensive responses, including stomatal closure, defensive gene expression, and protein accumulation [56,57]. However, stomatal closure is unfavorable for transpiration and thermolysis. The IAA accumulation in roots promotes the development of lateral roots, thus enhancing the water absorption performance of *Arabidopsis thaliana* [58], and can also improve the heat tolerance of rice [12]. Among the most pivotal roles of CTK in leaves is to delay leaf senescence by inhibiting Chl loss, membrane deterioration, and protein degradation [59]. Maintenance of higher CTK levels was proved to be of importance for heat tolerance [60].

High temperature caused a large amount of ROS accumulation which caused damage to proteins and membrane lipids [61]. Beneficial effects of GABA in antioxidant enzyme activities and ROS scavenging have been exhibited in leaves of different plant species under water-deficient conditions, salt, and high-temperature stress [62–64], but relatively little research has been conducted to discuss the effects of GABA in antioxidant metabolism in roots under abiotic stress. The current findings show that high-temperature stress results in ROS ($O_2{}^{\bullet-}$ and H_2O_2) accumulation and oxidative damage to membrane lipids or proteins, as shown by the considerable upsurge in MDA content or protein carbonyl content in roots and leaves of white clover. For comparing the difference in oxidative damage between leaves and roots, it can be seen that the leaves suffered more severe oxidative damage than roots after the same duration of heat stress, which is consistent with the earlier research on bermudagrass (*Cynodon dactylon*) [65]. In order to minimize ROS damage, the antioxidant defense was comprehensively elevated during heat stress [66]. After 30 d of heat stress, white clover plants pretreated with GABA exhibited significantly higher TAC, SOD, CAT, and POD activities in roots and leaves than untreated plants. For the ASA–GSH cycle, exogenous GABA promoted APX and GR activities as well as ASA content in roots and leaves during high-temperature stress. The ASA is one of the most crucial non-enzymatic antioxidants for ROS scavenging and the maintenance of its content also plays a key role in the ASA–GSH cycle [67,68]. The study of Li et al. reported that GABA ameliorated the ASA–GSH cycle in favor of thermotolerance in creeping bentgrass, which is a heat-sensitive perennial grass [62]. Interestingly, DR and MR activities were only significantly improved by the GABA in leaves, but not in roots of white clover under heat stress. This could indicate that the GABA-regulated differential antioxidant enzymes might be dependent on the severity of stress, since the leaves suffered from more severe oxidative damage. In terms of genes involved in antioxidant metabolism, the *Cu/ZnSOD* expression in leaves and roots was extremely enhanced by heat stress and the GABA application, reflecting its more imperative function in white clover under heat stress as compared to *MnSOD* and *FeSOD*. It was found that *Cu/ZnSOD* dominates the main function in higher plants [69]. The GABA-regulated genes encoding antioxidant enzymes were consistent with the enzyme activities under heat stress. These findings indicate that GABA helped to maintain antioxidant metabolism in leaves and roots of white clover through regulating enzyme activities and gene expression under heat stress.

In addition to oxidative damage and photosynthetic damage, plants also suffer physiological drought, which often appears in the later period of heat stress due to increased transpiration and root death, leading to a decline in water uptake, even if the water supply is adequate in soil [20]. This could explain why the leaves RWC in white clover slightly decreased with the development of heat stress and then declined sharply after 20 d of heat stress. Different from drought stress, ABA was not significantly induced in the early stage of heat stress [70], which means that stomata will remain open to increase heat dissipation in creeping bentgrass. At the later stage of heat stress, the plants synthesized ABA to regulate stomatal closure in order to decrease the transpiration rate [71]. Our results demonstrate that white clover plants pretreated with GABA exhibited significantly higher stomatal conductance and transpiration rate, which could be contributed to better thermolysis of white clover under heat stress. This might be linked with inhibition of ABA accumulation induced by the GABA in white clover. In addition, the maintenance of higher root activity, Pn, and leaf WUE was observed in the GABA-pretreated white clover. The WUE reflects the homeostasis between transpiration and Pn, which is good for adaptation under water-deficient conditions. Kumar et al. reported that the GABA-promoted WUE helped to keep water homeostasis in plants under abiotic stress [72]. Higher root activity is propitious to water and nutrient transport in plants upon exposure to harsh environments [73]. Heat stress also induced continual decline in OP in leaves, but an increase in OP in roots in both GABA-pretreated and non-pretreated white clover, demonstrating different responses to heat stress between leaves and roots. However, the GABA-pretreated white clover exhibited significantly lower OP in leaves and roots than untreated plants during the late stage of heat stress (20 and 30 d) associated with higher RWC in leaves of the GABA-pretreated white clover. Our earlier findings in creeping bentgrass demonstrate that GABA enhanced organic metabolites accumulation (amino acids and sugars) leading to the decline in OP, thereby improving water balance in leaves under elevated temperatures, salinity, and water-deficient conditions [74].

The GABA application further upregulated heat-induced increases in *PIP1-1* and *PIP2-7* expression in leaves and *TIP2-1* expression in leaves and roots of white clover under high-temperature stress. According to previous studies, the main function of PIP and TIP was to transport water molecules through cell or vacuole membranes, which is very important to maintain water uptake and homeostasis of cells [75–77]. Forrest and Bhave also proved that PIP and TIP as key AQPs had a strong function of water regulation against osmotic stress in wheat [76]. The spermine increased the *PIP2-7* expression and PIP2-7 accumulation was associated with better water transportation and balance in white clover and *Arabidopsis* during water shortage [77]. *TIP2-1* is considered to be an important pressure-gated water channel in grapevine (*Vitis vinifera*) subjected to various abiotic stresses [78]. Moreover, heat stress did not significantly affect *PIP2-2* and *SIP1-1* expression in leaves of white clover, but the GABA transcripts for these two genes under heat stress. Interestingly, the GABA effectively alleviated heat-induced inhibition of *PIP1-1*, *PIP2-2*, *TIP2-2*, and *NIP1-2* expression in roots of white clover. In the previous study of Grondin et al., the expression of *PIP1-1* and *PIP2-2* was correlated with water flux in roots [79]. The study of Liu et al. also proved that the overexpression of *PIP2-2* in roots could increase the osmotic adjustment ability of plants under salt stress [80]. Na^+ induced PIP1 and TIP1 accumulation and gene expression in roots and leaves of white clover, which could enhance water transport under drought stress [81]. These findings indicated that the GABA-regulated tolerance in white clover might be related with increases in water transport and homeostasis through activating AQPs expression in leaves and roots under heat stress.

Apart from the main function of AQPs for transporting H_2O or various molecules, including CO_2, H_2O_2, and $(NH_2)_2CO$, across cells, AQPs also play a vital role in the regulation of stomatal conductance and transpiration in plants [82–84]. The study of Lin et al. found that *Arabidopsis* plants overexpressing *PgTIP1* showed significantly higher stomatal conductance and transpiration rate in contrast to the wild type [85]. Plant vigor and transpiration rate were also significantly improved due to the overexpression of a *PIP1b* in transgenic tobacco (*Nicotiana tabacum*) under normal conditions [86]. In addition, constitutive expression of a *TIP2-2* in transgenic tomato (*Lycopersicon esculentum*) plants increased water mobility and transpiration under well-watered as well as drastic environmental conditions [87]. On the contrary, the stomatal conductance and transpiration could be inhibited significantly due to the downregulation of AQP expression in various plant species [88–90]. Our current findings show that white clover increased transpiration through increasing stomatal conductance in favor of the improvement in heat dissipation on the 15th day of high-temperature stress. Interestingly, stomatal conductance and transpiration were further increased in leaves of white clover because of the further increase in endogenous GABA level by the exogenous GABA application. The variation tendency of stomatal conductance and transpiration was consistent with the AQP gene expression regulated by GABA in white clover under heat stress. Our findings could indicate that the GABA-upregulated AQP gene expression could contribute to better heat dissipation performance related to stomatal opening and transpiration under high-temperature stress.

5. Conclusions

The improvement in endogenous GABA in leaf and root by GABA pretreatment could significantly alleviate the damage to white clover during high-temperature stress, as demonstrated by enhancements in cell membrane stability, photosynthetic capacity, and osmotic adjustment ability, as well as lower oxidative damage and Chl loss. The GABA significantly upregulated the transcript levels of antioxidant enzyme genes, and improved antioxidant enzyme activity (SOD, CAT, and POD) and key enzymes involved in the ASA–GSH cycle, thus reducing the oxidative damage to membrane lipids and proteins. In terms of plant hormones, GABA increased IAA content in roots and leaves and CTK content in leaves, associated with growth maintenance and reduced leaf senescence under heat stress. GABA upregulated the expression of *PIP1-1* and *PIP2-7* in leaves and *TIP2-1* expression in leaves and roots under high temperatures, and also alleviated the heat-induced inhibition of the expression of *PIP1-1*, *PIP2-2*, *TIP2-2*, and *NIP1-2* in roots. This could help to improve the water transportation and homeostasis under heat stress. In addition, GABA induced AQP expression and a decline in endogenous ABA levels, which could enhance the heat dissipation capacity through maintaining higher stomatal opening and transpiration in white clovers under high-temperature stress.

Supplementary Materials: The following are available online at https://www.mdpi.com/article/10.3390/antiox10071099/s1, Table S1: Primer sequences used in qRT-PCR.

Author Contributions: Conceptualization, Z.L.; methodology, Z.L. and Y.P.; formal analysis, H.Q. and D.K.; investigation, H.Q., D.K., and W.Z.; resources, Z.L.; data curation, H.Q.; writing—original draft preparation, H.Q.; writing—review and editing, Z.L., M.J.H., Y.P., Y.Z., G.F., and X.Z; supervision, Z.L.; project administration, Z.L., Y.P., and X.Z.; funding acquisition, Z.L., Y.P., and X.Z. All authors have read and agreed to the published version of the manuscript.

Funding: This research was funded by the China Agriculture Research System of MOF and MARA, and the Sichuan Forage Innovation Team Project of the Industrial System Construction of Modern Agriculture of China, grant number sccxtd-2020-16.

Institutional Review Board Statement: Not applicable.

Informed Consent Statement: Not applicable.

Data Availability Statement: Data are contained within the article and Supplementary Materials.

Conflicts of Interest: The authors declare no conflict of interest.

References

1. Root, T.L.; Price, J.T.; Hall, K.R.; Schneider, S.H.; Rosenzweigk, C.; Pounds, J.A. Fingerprints of global warming on wild animals and plants. *Nat. Cell Biol.* **2003**, *421*, 57–60. [CrossRef]
2. Allen, C.D.; Macalady, A.K.; Chenchouni, H.; Bachelet, D.; McDowell, N.; Vennetier, M.; Kitzberger, T.; Rigling, A.; Breshears, D.D.; (Ted) Hogg, E.H.; et al. A global overview of drought and heat-induced tree mortality reveals emerging climate change risks for forests. *For. Ecol. Manag.* **2010**, *259*, 660–684. [CrossRef]
3. Akter, N.; Islam, M.R. Heat stress effects and management in wheat. A review. *Agron. Sustain. Dev.* **2017**, *37*, 37. [CrossRef]
4. Lenoir, J.; Gégout, J.-C.; Marquet, P.A.; De Ruffray, P.; Brisse, H. A significant upward shift in plant species optimum elevation during the 20th century. *Science* **2008**, *320*, 1768–1771. [CrossRef] [PubMed]
5. Li, Z.; Peng, Y.; Huang, B. Alteration of transcripts of stress-protective genes and transcriptional factors by γ-aminobutyric acid (GABA) associated with improved heat and drought tolerance in creeping bentgrass (*Agrostis stolonifera*). *Int. J. Mol. Sci.* **2018**, *19*, 1623. [CrossRef]
6. Luo, L.; Li, Z.; Tang, M.Y.; Cheng, B.Z.; Zeng, W.H.; Peng, Y.; Nie, G.; Zhang, X.Q. Metabolic regulation of polyamines and γ-aminobutyric acid in relation to spermidine-induced heat tolerance in white clover. *Plant Biol.* **2020**, *22*, 794–804. [CrossRef] [PubMed]
7. Li, Z.; Zeng, W.; Cheng, B.; Huang, T.; Peng, Y.; Zhang, X. γ-Aminobutyric acid enhances heat tolerance associated with the change of proteomic profiling in creeping bentgrass. *Molecules* **2020**, *25*, 4270. [CrossRef] [PubMed]
8. Nayyar, H.; Kaur, R.; Kaur, S.; Singh, R.S. γ-Aminobutyric acid (GABA) imparts partial protection from heat stress injury to rice seedlings by improving leaf turgor and upregulating osmoprotectants and antioxidants. *J. Plant Growth Regul.* **2014**, *33*, 408–419. [CrossRef]
9. Li, L.; Dou, N.; Zhang, H.; Wu, C. The versatile GABA in plants. *Plant Signal. Behav.* **2021**, *16*, 1862565. [CrossRef]
10. Bown, A.W.; Shelp, B.J. Plant GABA: Not just a metabolite. *Trends Plant Sci.* **2016**, *21*, 811–813. [CrossRef]
11. Gong, Z.; Xiong, L.; Shi, H.; Yang, S.; Herrera-Estrella, L.R.; Xu, G.; Chao, D.-Y.; Li, J.; Wang, P.-Y.; Qin, F.; et al. Plant abiotic stress response and nutrient use efficiency. *Sci. China Life Sci.* **2020**, *63*, 635–674. [CrossRef] [PubMed]
12. Sarwar, N.; Atique-ur-Rehman; Farooq, O.; Wasaya, A.; Saliq, S.; Mubeen, K. Improved auxin level at panicle initiation stage enhance the heat stress tolerance in rice plants. In Proceedings of the Agronomy Australia Conference, Wagga Wagga, NSW, Australia, 25–29 August 2019.
13. Valero, D.; Martínez-Romero, D.; Serrano, M.; Riquelme, F. Postharvest gibberellin and heat treatment effects on polyamines, abscisic acid and firmness in lemons. *J. Food Sci.* **2006**, *63*, 611–615. [CrossRef]
14. Liu, X.; Huang, B. Cytokinin effects on creeping bentgrass response to heat stress. *Crop. Sci.* **2002**, *42*, 466–472. [CrossRef]
15. Bi, A.; Fan, J.; Hu, Z.; Wang, G.; Amombo, E.; Fu, J.; Hu, T. Differential acclimation of enzymatic antioxidant metabolism and photosystem ii photochemistry in tall fescue under drought and heat and the combined stresses. *Front. Plant Sci.* **2016**, *7*, 453. [CrossRef]

16. Ahmad, P.; Jaleel, C.A.; Salem, M.A.; Nabi, G.; Sharma, S. Roles of enzymatic and nonenzymatic antioxidants in plants during abiotic stress. *Crit. Rev. Biotechnol.* **2010**, *30*, 161–175. [CrossRef]

17. Mittler, R. Oxidative stress, antioxidants and stress tolerance. *Trends Plant Sci.* **2002**, *7*, 405–410. [CrossRef]

18. Li, Y.; Fan, Y.; Ma, Y.; Zhang, Z.; Yue, H.; Wang, L.; Li, J.; Jiao, Y. Effects of exogenous γ-aminobutyric acid (GABA) on photosynthesis and antioxidant system in pepper (*Capsicum annuum* L.) seedlings under low light stress. *J. Plant Growth Regul.* **2017**, *36*, 436–449. [CrossRef]

19. Cheng, B.; Li, Z.; Liang, L.; Cao, Y.; Zeng, W.; Zhang, X.; Ma, X.; Huang, L.; Nie, G.; Liu, W.; et al. The γ-aminobutyric acid (GABA) alleviates salt stress damage during seeds germination of white clover associated with Na_+/K_+ transportation, dehydrins accumulation, and stress-related genes expression in white clover. *Int. J. Mol. Sci.* **2018**, *19*, 2520. [CrossRef]

20. Morales, D.; Rodríguez, P.; Dell'Amico, J.; Nicolás, E.; Torrecillas, A.; Sánchez-Blanco, M. High-temperature preconditioning and thermal shock imposition affects water relations, gas exchange and root hydraulic conductivity in tomato. *Biol. Plant.* **2003**, *46*, 203–208. [CrossRef]

21. Singh, R.K.; Deshmukh, R.; Muthamilarasan, M.; Rani, R.; Prasad, M. Versatile roles of aquaporin in physiological processes and stress tolerance in plants. *Plant Physiol. Biochem.* **2020**, *149*, 178–189. [CrossRef]

22. Chaumont, F.; Barrieu, F.; Jung, R.; Chrispeels, M.J. Plasma membrane intrinsic proteins from maize cluster in two sequence subgroups with differential aquaporin activity. *Plant Physiol.* **2000**, *122*, 1025–1034. [CrossRef]

23. Johanson, U.; Karlsson, M.; Johansson, I.; Gustavsson, S.; Sjö¨vall, S.; Fraysse, L.; Weig, A.R.; Kjellbom, P. The complete set of genes encoding major intrinsic proteins in Arabidopsis provides a framework for a new nomenclature for major intrinsic proteins in plants. *Plant Physiol.* **2001**, *126*, 1358–1369. [CrossRef]

24. Mitani-Ueno, N.; Yamaji, N.; Zhao, F.-J.; Ma, J.F. The aromatic/arginine selectivity filter of NIP aquaporins plays a critical role in substrate selectivity for silicon, boron, and arsenic. *J. Exp. Bot.* **2011**, *62*, 4391–4398. [CrossRef]

25. Ishikawa, F.; Suga, S.; Uemura, T.; Sato, M.H.; Maeshima, M. Novel type aquaporin SIPs are mainly localized to the ER membrane and show cell-specific expression in *Arabidopsis thaliana*. *FEBS Lett.* **2005**, *579*, 5814–5820. [CrossRef] [PubMed]

26. Peng, Y.; Lin, W.; Cai, W.; Arora, R. Overexpression of a *Panax ginseng* tonoplast aquaporin alters salt tolerance, drought tolerance and cold acclimation ability in transgenic *Arabidopsis plants*. *Planta* **2007**, *226*, 729–740. [CrossRef] [PubMed]

27. Alexandersson, E.; Fraysse, L.; Sjövall-Larsen, S.; Gustavsson, S.; Fellert, M.; Karlsson, M.; Johanson, U.; Kjellbom, P. Whole gene family expression and drought stress regulation of aquaporins. *Plant Mol. Biol.* **2005**, *59*, 469–484. [CrossRef]

28. Feng, Z.J.; Liu, N.; Zhang, G.-W.; Niu, F.-G.; Xu, S.-C.; Gong, Y.-M. Investigation of the AQP family in soybean and the promoter activity of tip2;6 in heat stress and hormone responses. *Int. J. Mol. Sci.* **2019**, *20*, 262. [CrossRef] [PubMed]

29. Shafqat, W.; Jaskani, M.J.; Maqbool, R.; Chattha, W.S.; Ali, Z.; Naqvi, S.A.; Haider, M.S.; Khan, I.A.; Vincent, C.I. Heat shock protein and aquaporin expression enhance water conserving behavior of citrus under water deficits and high temperature conditions. *Environ. Exp. Bot.* **2021**, *181*, 104270. [CrossRef]

30. Marshall, A.H.; Williams, T.A.; Abberton, M.T.; Michaelson-Yeates, T.P.T.; Olyctt, P.; Powell, H.G. Forage quality of white clover (*Trifolium repens* L.) × Caucasian clover (*T. ambiguum M. Bieb.*) hybrids and their grass companion when grown over three harvest years. *Grass Forage Sci.* **2004**, *59*, 91–99. [CrossRef]

31. Li, S.; Liu, C.; Huang, Y.; Li, S.; Zhang, Y.; Wu, P. Optimization of extraction technology of flavonoids from white clover by Ionic liquids. *Liaoning J. Tradit. Chin. Med.* **2018**, *45*, 1445–1448. [CrossRef]

32. He, J.; He, S.; Fu, Q.; Lei, H.; Yang, Y. Effects on relative water content, membrane permeability, praline content and chlorophyll content of white clover under high temperature stress. *Guizhou Anim. Vet. Med.* **2007**, *126*, 1358–1369. [CrossRef]

33. Hoagland, D.R.; Arnon, D.I. The water culture method for growing plants without soil. *Calif. Agric. Exp. Stn.* **1950**, *347*, 357–359. [CrossRef]

34. Müller, M.; Munné-Bosch, S. Rapid and sensitive hormonal profiling of complex plant samples by liquid chromatography coupled to electrospray ionization tandem mass spectrometry. *Plant Methods* **2011**, *7*, 37. [CrossRef] [PubMed]

35. Barrs, H.; Weatherley, P. A Re-examination of the relative turgidity technique for estimating water deficits in leaves. *Aust. J. Biol. Sci.* **1962**, *15*, 413–428. [CrossRef]

36. McMichael, B.; Burke, J. Metabolic activity of cotton roots in response to temperature. *Environ. Exp. Bot.* **1994**, *34*, 201–206. [CrossRef]

37. Blum, A. Osmotic adjustment and growth of barley genotypes under drought stress. *Crop. Sci.* **1989**, *29*, 230–233. [CrossRef]

38. Blum, A.; Ebercon, A. Cell membrane stability as a measure of drought and heat tolerance in wheat1. *Crop. Sci.* **1981**, *21*, 43–47. [CrossRef]

39. Arnon, D.I. Copper enzymes in isolated chloroplasts. polyphenoloxidase in beta vulgaris. *Plant Physiol.* **1949**, *24*, 1–15. [CrossRef]

40. Elstner, E.F.; Heupel, A. Inhibition of nitrite formation from hydroxylammoniumchloride: A simple assay for superoxide dismutase. *Anal. Biochem.* **1976**, *70*, 616–620. [CrossRef]

41. Velikova, V.; Yordanov, I.; Edreva, A. Oxidative stress and some antioxidant systems in acid rain-treated bean plants. *Plant Sci.* **2000**, *151*, 59–66. [CrossRef]

42. Dhindsa, R.S.; Dhindsa, P.P.; Thorpe, T.A. Leaf senescence: Correlated with increased leaves of membrane permeability and lipid peroxidation, and decreased levels of superoxide dismutase and catalase. *J. Exp. Bot.* **1981**, *32*, 93–101. [CrossRef]

43. Nakano, Y.; Asada, K. Hydrogen peroxide is scavenged by ascorbate-specific peroxidase in spinach chloroplasts. *Plant Cell Physiol.* **1981**, *22*, 867–880. [CrossRef]

44. Giannopolitis, C.N.; Ries, S.K. Superoxide dismutase. I. Occurrence in higher plants. *Plant physiol.* **1997**, *59*, 309–314. [CrossRef]
45. Chance, B.; Maehly, A. Assay of catalases and peroxidases. *Odorant Bind. Chemosens. Proteins* **1955**, *2*, 764–775. [CrossRef]
46. Cakmak, I.; Strbac, D.; Marschner, H. Activities of hydrogen peroxide-scavenging enzymes in germinating wheat seeds. *J. Exp. Bot.* **1993**, *44*, 127–132. [CrossRef]
47. Bradford, M.M. A rapid and sensitive method for the quantitation of microgram quantities of protein utilizing the principle of protein-Dye binding. *Anal. Biochem.* **1976**, *72*, 248–254. [CrossRef]
48. Gossett, D.R.; Millhollon, E.P.; Lucas, M.C. Antioxidant response to NaCl stress in salt-tolerant to and salt-sensitivity cultivars of cotton. *Crop Sci.* **1994**, *34*, 706–714. [CrossRef]
49. Fait, A.; Fromm, H.; Walter, D.; Galili, G.; Fernie, A.R. Highway or byway: The metabolic role of the GABA shunt in plants. *Trends Plant Sci.* **2008**, *13*, 14–19. [CrossRef] [PubMed]
50. Li, Z.; Cheng, B.; Zeng, W.; Liu, Z.; Peng, Y. The transcriptional and post-transcriptional regulation in perennial creeping bentgrass in response to γ-aminobutyric acid (GABA) and heat stress. *Environ. Exp. Bot.* **2019**, *162*, 515–524. [CrossRef]
51. Wang, R.; Wang, Z.; Xiang, Z. Effect of γ-aminobutyric acid on photosynthetic characteristics and carbohydrate metabolism under high temperature stress in perennial ryegrass. *Acta Prataculturae Sin.* **2019**, *28*, 168–178. [CrossRef]
52. Takahashi, Y.; Sasanuma, T.; Abe, T. Accumulation of gamma-aminobutyrate (GABA) caused by heat-drying and expression of related genes in immature vegetable soybean (edamame). *Breed. Sci.* **2013**, *63*, 205–210. [CrossRef]
53. Dreher, K.; Callis, J. Ubiquitin, Hormones and biotic stress in plants. *Ann. Bot.* **2007**, *99*, 787–822. [CrossRef]
54. Lancien, M.; Roberts, M.R. Regulation of arabidopsis thaliana 14-3-3 gene expression by gamma-aminobutyric acid. *Plant Cell Environ.* **2006**, *29*, 1430–1436. [CrossRef]
55. Wang, Y.; Zhang, B.; Gu, W.; Li, Z.; Mao, J.; Guo, J.; Shao, R.; Yang, Q. γ-Aminobutyric acid on oxidative damage and endogenous hormones in maize seedling roots under salt stress. *Chin. J. Pestic. Sci.* **2018**, *20*, 607–617. [CrossRef]
56. Fujita, M.; Fujita, Y.; Maruyama, K.; Seki, M.; Hiratsu, K.; Ohme-Takagi, M.; Tran, L.-S.P.; Yamaguchi-Shinozaki, K.; Shinozaki, K. A dehydration-induced NAC protein, RD26, is involved in a novel ABA-dependent stress-signaling pathway. *Plant J.* **2004**, *39*, 863–876. [CrossRef] [PubMed]
57. Tardieu, F.; Zhang, J.; Katerji, N.; Bethenod, O.; Palmer, S.; Davies, W.J. Xylem ABA controls the stomatal conductance of field-grown maize subjected to soil compaction or soil drying. *Plant Cell Environ.* **1992**, *15*, 193–197. [CrossRef]
58. Zeng, J.; Wang, Q.; Lin, J.; Deng, K.; Zhao, X.; Tang, N.; Liu, X. Arabidopsis cryptochrome-1 restrains lateral roots growth by inhibiting auxin transport. *J. Plant Physiol.* **2010**, *167*, 670–673. [CrossRef] [PubMed]
59. Liu, L.; Li, H.; Zeng, H.; Cai, Q.; Zhou, X.; Yin, C. Exogenous jasmonic acid and cytokinin antagonistically regulate rice flag leaf senescence by mediating chlorophyll degradation, membrane deterioration, and senescence-associated genes expression. *J. Plant Growth Regul.* **2016**, *35*, 366–376. [CrossRef]
60. Wu, C.; Cui, K.; Wang, W.; Li, Q.; Fahad, S.; Hu, Q.; Huang, J.; Nie, L.; Mohapatra, P.K.; Peng, S. Heat-induced cytokinin transportation and degradation are associated with reduced panicle cytokinin expression and fewer spikelets per panicle in rice. *Front. Plant Sci.* **2017**, *8*, 371. [CrossRef] [PubMed]
61. Yamashita, A.; Nijo, N.; Pospíšil, P.; Morita, N.; Takenaka, D.; Aminaka, R.; Yamamoto, Y.; Yamamoto, Y. Quality control of photosystem II: Reactive oxygen species are responsible for the damage to photosystem II under moderate heat stress. *J. Biol. Chem.* **2008**, *283*, 28380–28391. [CrossRef]
62. Li, Z.; Yu, J.; Peng, Y.; Huang, B. Metabolic pathways regulated by γ-aminobutyric acid (GABA) contributing to heat tolerance in creeping bentgrass (*Agrostis stolonifera*). *Sci. Rep.* **2016**, *6*, 30338. [CrossRef]
63. Li, Z.; Cheng, B.; Zeng, W.; Zhang, X.; Peng, Y. Proteomic and metabolomic profilings reveal crucial functions of γ-aminobutyric acid in regulating ionic, water, and metabolic homeostasis in creeping bentgrass under salt stress. *J. Proteome Res.* **2020**, *19*, 769–780. [CrossRef]
64. Tang, M.; Li, Z.; Luo, L.; Cheng, B.; Zhang, Y.; Zeng, W.; Peng, Y. Nitric oxide signal, nitrogen metabolism, and water balance affected by γ-aminobutyric acid (GABA) in relation to enhanced tolerance to water stress in creeping bentgrass. *Int. J. Mol. Sci.* **2020**, *21*, 7460. [CrossRef] [PubMed]
65. Peng, H.; Du, H. Effectsof heat stress on differential expression of proteins in leaves and roots of hybrid bermudagrass (*Cynodon transvaalensis* × *Cynodon dactylon*). *J. Shanghai Jiaotong Univ. (Agric.)* **2016**, *34*, 62–68. [CrossRef]
66. Jaleel, C.A.; Riadh, K.; Gopi, R.; Manivannan, P.; Inès, J.; Al-Juburi, H.J.; Chang-Xing, Z.; Hong-Bo, S.; Panneerselvam, R. Antioxidant defense responses: Physiological plasticity in higher plants under abiotic constraints. *Acta Physiol. Plant.* **2009**, *31*, 427–436. [CrossRef]
67. Kaya, C.; Ashraf, M.; Al-Huqail, A.A.; Alqahtani, M.A.; Ahmad, P. Silicon is dependent on hydrogen sulphide to improve boron toxicity tolerance in pepper plants by regulating the ASA-GSH cycle and glyoxalase system. *Chemosphere* **2020**, *257*, 127241. [CrossRef] [PubMed]
68. Han, Y.; Wang, X.; Xu, X.; Gao, Y.; Wen, G.; Zhang, R.; Wang, Y. Responses of anti-oxidant enzymes and the ascorbate-glutathione cycle to heat, drought, and synergistic stress in Phyllostachys edulis seedlings. *J. Zhejiang AF Univ.* **2018**, *35*, 268–276. [CrossRef]
69. Dong, L.; He, Y.; Wang, Y.; Dong, Z. Research progress on application of superoxide dismutase (SOD). *J. Agric. Sci. Technol.* **2013**, *15*, 53–58. [CrossRef]
70. Larkindale, J.; Huang, B. Effects of abscisic acid, salicylic acid, ethylene and hydrogen peroxide in thermotolerance and recovery for creeping bentgrass. *Plant Growth Regul.* **2005**, *47*, 17–28. [CrossRef]

71. Wahid, A.; Gelani, S.; Ashraf, M.; Foolad, M.R. Heat tolerance in plants: An overview. *Environ. Exp. Bot.* **2007**, *61*, 199–223. [CrossRef]
72. Kumar, N.; Dubey, A.K.; Upadhyay, A.K.; Gautam, A.; Ranjan, R.; Srikishna, S.; Sahu, N.; Behera, S.K.; Mallick, S. GABA accretion reduces Lsi-1 and Lsi-2 gene expressions and modulates physiological responses in Oryza sativa to provide tolerance towards arsenic. *Sci. Rep.* **2017**, *7*, 1–11. [CrossRef] [PubMed]
73. Liu, H.; Zheng, G.; Guan, J.; Li, G. Changes of root activity and membrane permeability under drought stress in maize. *Acta Agric. Boreall—Sin.* **2002**, *2*, 21–23. [CrossRef]
74. Li, Z.; Cheng, B.; Peng, Y.; Zhang, Y. Adaptability to abiotic stress regulated by γ-aminobutyric acid in relation to alterations of endogenous polyamines and organic metabolites in creeping bentgrass. *Plant Physiol. Biochem.* **2020**, *157*, 185–194. [CrossRef]
75. Regon, P.; Panda, P.; Kshetrimayum, E.; Panda, S.K. Genome-wide comparative analysis of tonoplast intrinsic protein (TIP) genes in plants. *Funct. Integr. Genom.* **2014**, *14*, 617–629. [CrossRef] [PubMed]
76. Forrest, K.L.; Bhave, M. The PIP and TIP aquaporins in wheat form a large and diverse family with unique gene structures and functionally important features. *Funct. Integr. Genom.* **2007**, *8*, 115–133. [CrossRef] [PubMed]
77. Li, Z.; Hou, J.; Zhang, Y.; Zeng, W.; Cheng, B.; Hassan, M.J.; Zhang, Y.; Pu, Q.; Peng, Y. Spermine regulates water balance associated with Ca^{2+}-dependent aquaporin (TrTIP2-1, TrTIP2-2 and TrPIP2-7) expression in plants under water stress. *Plant Cell Physiol.* **2020**, *61*, 1576–1589. [CrossRef] [PubMed]
78. Vandeleur, R.; Mayo, G.; Shelden, M.C.; Gilliham, M.; Kaiser, B.; Tyerman, S.D. The role of plasma membrane intrinsic protein aquaporins in water transport through roots: Diurnal and drought stress responses reveal different strategies between isohydric and anisohydric cultivars of grapevine. *Plant Physiol.* **2009**, *149*, 445–460. [CrossRef]
79. Grondin, A.; Mauleon, R.; Vadez, V.; Henry, A. Root aquaporins contribute to whole plant water fluxes under drought stress in rice (*Oryza sativa* L.). *Plant Cell Environ.* **2016**, *39*, 347–365. [CrossRef]
80. Liu, P.; Yin, L.; Wang, S.; Zhang, M.; Deng, X.; Zhang, S.; Tanaka, K. Enhanced root hydraulic conductance by aquaporin regulation accounts for silicon alleviated salt-induced osmotic stress in *Sorghum bicolor* L. *Environ. Exp. Bot.* **2015**, *111*, 42–51. [CrossRef]
81. Li, Z.; Peng, D.; Zhang, X.; Peng, Y.; Chen, M.; Ma, X.; Huang, L.; Yan, Y. Na$^+$ induces the tolerance to water stress in white clover associated with osmotic adjustment and aquaporins-mediated water transport and balance in root and leaf. *Environ. Exp. Bot.* **2017**, *144*, 11–24. [CrossRef]
82. Maurel, C.; Verdoucq, L.; Luu, D.-T.; Santoni, V. Plant Aquaporins: Membrane channels with multiple integrated functions. *Annu. Rev. Plant Biol.* **2008**, *59*, 595–624. [CrossRef] [PubMed]
83. Zwiazek, J.J.; Xu, H.; Tan, X.; Rodenas, A.N.; Morte, A. Significance of oxygen transport through aquaporins. *Sci. Rep.* **2017**, *7*, 40411. [CrossRef] [PubMed]
84. Li, G.; Santoni, V.; Maurel, C. Plant aquaporins: Roles in plant physiology. *Biochim. Biophys. Acta (BBA) Gen. Subj.* **2014**, *1840*, 1574–1582. [CrossRef]
85. Lin, W.; Peng, Y.; Li, G.; Arora, R.; Tang, Z.; Su, W.; Cai, W. Isolation and functional characterization of PgTIP1, a hormone-autotrophic cells-specific tonoplast aquaporin in ginseng. *J. Exp. Bot.* **2007**, *58*, 947–956. [CrossRef]
86. Aharon, R.; Shahak, Y.; Wininger, S.; Bendov, R.; Kapulnik, Y.; Galili, G. Overexpression of a plasma membrane aquaporin in transgenic tobacco improves plant vigor under favorable growth conditions but not under drought or salt Stress. *Plant Cell* **2003**, *15*, 439–447. [CrossRef]
87. Sade, N.; Vinocur, B.J.; Diber, A.; Shatil, A.; Ronen, G.; Nissan, H.; Wallach, R.; Karchi, H.; Moshelion, M. Improving plant stress tolerance and yield production: Is the tonoplast aquaporin SlTIP2;2 a key to isohydric to anisohydric conversion? *New Phytol.* **2009**, *181*, 651–661. [CrossRef]
88. Siefritz, F.; Tyree, M.T.; Lovisolo, C.; Schubert, A.; Kaldenhoff, R. PIP1 plasma membrane aquaporins in tobacco: From cellular effects to function in plants. *Plant Cell* **2002**, *14*, 869–876. [CrossRef]
89. Flexas, J.; Ribas-Carbo, M.; Hanson, D.T.; Bota, J.; Otto, B.; Cifre, J.; McDowell, N.G.; Medrano, H.; Kaldenhoff, R. Tobacco aquaporin NtAQP1 is involved in mesophyll conductance to CO_2 in vivo. *Plant J.* **2006**, *48*, 427–439. [CrossRef]
90. Sade, N.; Gallé, A.; Flexas, J.; Lerner, S.; Peleg, G.; Yaaran, A.; Moshelion, M. Differential tissue-specific expression of NtAQP1 in Arabidopsis thaliana reveals a role for this protein in stomatal and mesophyll conductance of CO_2 under standard and salt-stress conditions. *Planta* **2014**, *239*, 357–366. [CrossRef] [PubMed]

 *antioxidants*

Article

Molybdenum-Induced Regulation of Antioxidant Defense-Mitigated Cadmium Stress in Aromatic Rice and Improved Crop Growth, Yield, and Quality Traits

Muhammad Imran [1,2,3], Saddam Hussain [4], Longxin He [1,2,3], Muhammad Furqan Ashraf [5], Muhammad Ihtisham [6], Ejaz Ahmad Warraich [4] and Xiangru Tang [1,2,3,*]

1. State Key Laboratory for Conservation and Utilization of Subtropical Agro-Bioresources, College of Agriculture, South China Agricultural University, Guangzhou 510642, China; muhammadimran@scau.edu.cn (M.I.); helx@stu.scau.edu.cn (L.H.)
2. Scientific Observing and Experimental Station of Crop Cultivation in South China, Ministry of Agriculture and Rural Affairs, Guangzhou 510642, China
3. Guangzhou Key Laboratory for Science and Technology of Aromatic Rice, Guangzhou 510642, China
4. Department of Agronomy, University of Agriculture Faisalabad, Punjab 38040, Pakistan; shussain@uaf.edu.pk (S.H.); ewarraich@uaf.edu.pk (E.A.W.)
5. College of Life Sciences, South China Agricultural University, Guangzhou 510642, China; furqan2210uaf@scau.edu.cn
6. College of Landscape Architecture, Chengdu Campus, Sichuan Agricultural University, Wenjiang 611100, China; ihtisham@sicau.edu.cn
* Correspondence: author: tangxr@scau.edu.cn

Citation: Imran, M.; Hussain, S.; He, L.; Ashraf, M.F.; Ihtisham, M.; Warraich, E.A.; Tang, X. Molybdenum-Induced Regulation of Antioxidant Defense-Mitigated Cadmium Stress in Aromatic Rice and Improved Crop Growth, Yield, and Quality Traits. *Antioxidants* **2021**, *10*, 838. https://doi.org/10.3390/antiox10060838

Academic Editor: Luca Sebastiani

Received: 12 April 2021
Accepted: 20 May 2021
Published: 24 May 2021

Publisher's Note: MDPI stays neutral with regard to jurisdictional claims in published maps and institutional affiliations.

Abstract: Cadmium (Cd) stress causes serious disruptions in plant metabolism, physio-biochemical responses, crop yield, and grain quality characteristics. A pot experiment was conducted to investigate the role of molybdenum (Mo) in mitigating Cd-induced adversities on plant growth, yield attributes, and grain quality characteristics of a popular aromatic rice cultivar 'Xiangyaxiangzhan'. The Mo was applied at 0.15 mg kg^{-1} soil in both control (no Cd) and Cd-contaminated (100 mg kg^{-1}) soils. A treatment with Mo-free ($-$Mo) soil was also maintained for comparison. The results showed that Cd toxicity significantly ($p < 0.05$) reduced plant dry biomass, grain yield, photosynthetic efficiency, and pigment contents, and impaired chloroplast ultra-structural configuration and simultaneously destabilized the plant metabolism owing to higher accumulation of hydrogen peroxide, electrolyte leakage, and malondialdehyde contents. However, Mo supply improved grain yield and 2-acetyl-1-pyrroline content by 64.75% and 77.09%, respectively, under Cd stress, suggesting that Mo supply mitigated Cd-provoked negative effects on yield attributes and grain quality of aromatic rice. Moreover, Mo supply enhanced photosynthesis, proline, and soluble protein content, and also strengthened plant metabolism and antioxidant defense through maintaining higher activities and transcript abundance of ROS-detoxifying enzymes at the vegetative, reproductive, and maturity stages of aromatic rice plants under Cd toxicity. Collectively, our findings indicated that Mo supply strengthened plant metabolism at prominent growth stages through an improved enzymatic and non-enzymatic antioxidant defense system, thereby increasing grain yield and quality characteristics of aromatic rice under Cd toxicity.

Keywords: molybdenum; cadmium stress; photosynthetic pigments; oxidative damage; quality characters; fragrant rice

1. Introduction

Aromatic rice, the finest quality rice, is recognized around the world for its unique fragrance and taste [1]. The rice milling standards could be measured by head rice rate, milling recovery percentage, and milled rice rate, which are directly connected to market values, whereas the rice chalkiness percentage and grain protein contents belong to cooking characteristics of aromatic rice [2].

Cadmium (Cd) toxicity is one of the third major contaminants in agricultural soils across the globe and is considered to be the only metal that poses health hazards to humans and animals at a plant tissue concentration that is usually non-phytotoxic [3,4]. In paddy soils, Cd is taken up and moved to above ground components by plant roots and affects natural plant metabolism, morpho-physiological characteristics, crop growth, and productivity [5]. Specifically, Cd toxicity leads to stunted plant growth, photosynthesis disturbance, reductions in chlorophyll biosynthesis, deformation of cellular structures, and higher production of reactive oxygen species (ROS), disturbance in cellular functions, and ultimately plant death [6–8].

Molybdenum (Mo), an indispensable microelement, is involved in multiple metabolic and cellular processes in higher plants [9,10]. It plays a key role in different plant physio-biochemical processes, such as photosynthesis, chloroplast configuration and ultra-structural integrity, and chlorophyll biosynthesis and also acts as a stress-resilient element to enhance the oxidative stress tolerance under salinity [11], drought [12], low temperature [13], and ammonium toxicity [14]. Several recent studies have documented the beneficial functions of Mo in alleviating Cd toxicity in *Brassica napus* L. [15] and *Cannabis sativa* L. [16]. The Mo-induced mitigation of Cd toxicities has been ascribed to following four mechanisms: (i) obstructing the absorption and translocation of Cd and manipulating its fractions in plant cells; (ii) repairing damaged cell membranes, ultra-structures of chloroplasts, and reinforcing the photosynthetic system; (iii) regulating the uptake of essential elements; and (iv) activating the antioxidant defense systems (both enzymatic and non-enzymatic) in order to detoxify ROS [4].

However, little is known regarding the effects of Mo on plant physio-biochemical processes, grain yield, and grain quality traits of aromatic rice plants under Cd toxicity. Therefore, the present study was conducted to investigate the protective effects of Mo supply against Cd toxicity on plant growth, physio-biochemical processes, and antioxidant defense responses of aromatic rice plants at three prominent growth stages (vegetative, reproductive, and maturity stages), and consequently examine the Mo-induced enhancement in rice yield and grain quality characteristics under Cd toxicity.

2. Materials and Methods

2.1. Experimental Location, Treatments, and Crop Husbandry

A pot experiment was performed during July–November 2020 at the Experimental Research Farm (113°21′ E, 23°14′ N), College of Agriculture, South China Agricultural University (SCAU), Guangzhou, China, in a rain-sheltered wire house under open-air conditions. The soil used in this experiment was under paddy cultivation for many years. The above-soil layer (0–20 cm) was used to fill the pots (10 kg air-dried soil in each pot with 25 cm in height and 32 cm in diameter). The essential chemical properties of the soil used in this experiment are shown in Table S1. The experimental treatments consisted of Mo (0 and 0.15 mg kg^{-1} soil) in the form of ammonium molybdate [$(NH_4)_6Mo_7O_{24} \cdot 4H_2O$]) and Cd (0 and 100 mg kg^{-1} soil) supplied as $CdCl_2 \cdot 2.5\ H_2O$. About two weeks before transplantation, the treatment concentrations were carefully mixed in the soil of respective pots and a water layer (2–3 cm) was maintained over the soil surface to create puddled-like and anaerobic conditions.

The seeds of aromatic rice cultivar "Xiangyaxiangzhan" were collected from South China Agricultural University, Guangdong, China. This aromatic rice cultivar is locally renowned for its special aroma-producing quality and therefore widely cultivated by local farmers of South China on a commercial scale. However, previous studies have reported that heavy metals (e.g., Cd, Pb) significantly affect the grain yield, aroma, and other quality traits of this aromatic rice cultivar [2,17]. The uniform and well-grown rice seedlings (22-days-old) were transplanted in soil-filled pots (3–4 seedlings per hill and 5 hills per pot) by following a cross mark (+) type of planting pattern (four seedlings at a side each and the fifth in the middle of the pot). All the experimental pots were supplied with 2.25, 3.33, and 1.35 g N-P-K in the form of urea, phosphorus pentoxide, and potassium oxide, respectively.

2.2. Sampling

At three prominent growth stages (vegetative, reproductive, and maturity stages), fresh plant leaves were sampled and well-preserved ($-80\ ^\circ$C) for the assessment of photosynthetic pigments, various physio-biochemical analysis, antioxidant enzyme activities, and qRT-PCR quantification. At maturity, plants were harvested to measure dry biomass, grain yield, and 2AP content to analyze the qualitative characteristics of aromatic rice grains.

2.3. Estimation of Photosynthetic Efficiency, Pigment Contents, and Transmission Electron Microscopy

For the photosynthetic measurement, one fully expanded mature leaf from the top was used per plant and five plants were selected in each replication of a particular treatment and the results were averaged. The photosynthetic rate was recorded with a portable photosynthetic system (LI- 6400xt, Li-Cor, Inc., Lincoln, NE, USA). The photosynthetic pigments were determined in accordance with the method described in previous studies [18,19]. Fresh leaf samples were randomly collected from different aromatic rice plants for TEM analysis. Small sections of leaves 1–3 mm in length were fixed in 4% glutaraldehyde (*v/v*) in 0.2 mol/L PBS (sodium phosphate buffer, pH 7.2) for 6–8 h and post-fixed in 1% OsO_4 (Osmium (VIII) oxide) for 1 h, then in 0.2 mol/L PBS (pH 7.2) for 1–2 h. Dehydration was performed in a graded ethanol series (50%, 60%, 70%, 80%, 90%, 95%, and 100%) followed by acetone, then samples were filtrated and embedded in Spurr's resin. Ultra-thin sections (80 nm) were prepared and mounted on copper grids for viewing under a transmission electron microscope (TALOS L120C) [20].

2.4. Measurement of Soluble Protein and Proline Contents

The protein contents in leaves were estimated according to the method of [21] using G-250. The fresh leaf samples (0.2 g) were homogenized in 50 mM sodium phosphate buffer (pH 7.0) containing 2% polyvinylpyrrolidine-40 (PVP-40) and 1 mM EDTA-Na2. The reaction mixture was centrifuged at $10,000\times g$ for 15 min at 4 $^\circ$C. The absorbance of the reaction mixture was read at 595 nm in triplicate and final protein contents were expressed as mg g^{-1} FW. The proline contents were estimated according to the method of [22] using ninhydrin. The reaction mixture was extracted with 5 mL toluene and the absorbance of the red chromophore in the toluene fraction was measured at 520 nm. The amount of proline was estimated by comparing with a standard curve and expressed as µg g^{-1} FW.

2.5. Measurement of Reactive Oxygen Species and Lipid Peroxidation

Fresh leaf samples (0.2 g) at different growth stages were crushed in liquid nitrogen and homogenized with 1 mL of 0.1% trichloroacetic acid (TCA) and centrifuged at $12,000\times g$ for 20 min (4 $^\circ$C) for the measurement of H_2O_2 content in aromatic rice plants. The reaction mixture consisted of 0.5 mL of potassium phosphate buffer (pH 6.8, 100 mM), 1 mL potassium iodide (1M), and 0.5 mL supernatant. The H_2O_2 contents were measured by spectrophotometer (UV-VIS 2550, Shimadzu, Japan) at 390 nm [23]. The malondialdehyde (MDA) contents in leaves at different growth stages of aromatic rice were estimated by a prior method [24]. Fresh leaf samples (0.2 g) were homogenized in 2 mL 0.5% thiobarbituric acid (TBA) solution in 10% trichloroacetic acid (TCA) and boiled in the water bath at 100 $^\circ$C for 30 min. The boiled samples were then cooled down in an ice bath and centrifuged at $4000\times g$ for 15 min. The absorbance of the reaction mixtures was read at 532, 600, and 450 nm in triplicate. The MDA content in the reaction solution was calculated as follows: MDA content = 6.45 (ΔOD532-600) − (0.56 OD450) and the final contents were expressed as µmol g^{-1} FW. To measure the electrolyte leakage percentage at respective growth stages of aromatic rice plants, the fresh leaf samples were thoroughly washed with deionized water and cut into small pieces. Leaf discs (0.3 g) were then placed in 10 mL deionized water and incubated for 6 h at 25 $^\circ$C and electrical conductivity (EC) EC1 was recorded using an EC meter (SX-650, Sanxin, China). The leaf samples were again incubated for 2 h at 90 °C to record EC2. The electrolyte leakage (EL) percentage in leaf tissues of aromatic rice plants was estimated as: EL (%) = (EC1/EC2 × 100) [23].

2.6. Measurement of Enzymatic and Non-Enzymatic Antioxidant

In order to estimate antioxidant enzyme assay, fresh leaf samples (0.3 g) were crushed in liquid nitrogen and homogenized in 6 mL of 50 mM sodium phosphate buffer (pH 7.8) with a mortar and pestle in an ice bath. The homogenate was centrifuged (12,000× g, 10 min) to separate the supernatant from crude fibers and the aliquot of the supernatant was used to measure the activities of catalase, superoxide dismutase, peroxidase, and ascorbate peroxidase in aromatic rice leaves according to the methods described in a previous study [24]. The contents of non-enzymatic antioxidants (GSH and GSSG) at different growth stages of aromatic rice plants were determined by 'A006-2-1' and 'A061-2-1' kits, respectively, bought from Nanjing Jiancheng Bioengineering Institute (www.njjcbio.com (accessed on 3 September 2020)), China.

2.7. Total RNA Extraction and qRT-PCR Analysis

Fresh aromatic rice leaves were sampled at three prominent growth stages for total RNA extraction. Total RNA was extracted using TRIzol reagent (Invitrogen, Carlsbad, CA, USA). However, subsequent procedures for assessing the quality and quantity of total RNA, cDNA synthesizing, and Real Time PCR setting up were carefully adopted as described in our previous experiment [25]. Data on nucleotides sequence and specific annealing temperature can be found in Table S2. Three biological replicates of each sample were used, and by normalizing the Ct value relative to the ACTIN Ct value, the expression levels for each gene were calculated according to the 2-ΔΔCt method for quantification [26].

2.8. Estimation of Molybdenum and Cadmium Concentration in Different Plant Parts

The aromatic rice plant leaves were sampled at three prominent growth stages and ears (at reproductive stage) and grains (at maturity stage) and oven dried for estimating Mo and Cd concentrations. The plant samples (0.3 g) were digested in 5:1 (v/v) HNO$_3$:HClO$_4$ (5 mL) in a microwave oven (MLS 1200, Milestone, FKV, Italy). Mo and Cd concentrations were measured by ICP-mass spectrometry (ICP-MS) (ELAN DRC-e, PerkinElmer SCIEX, DE) and Atomic Absorption Spectrophotometer (AAS) (AA6300C, Shimadzu, Japan) [2,27].

2.9. Estimation of 2-acetyl-1-pyrroline (2AP) Content

Freshly collected grains were homogenized with dichloromethane and sodium sulfate for 2AP measurement. The 2AP content in aromatic rice grains was measured in combination with the Gas Chromatograph Mass Spectrometer (GCMS-QP 2010 Plus, Shimadzu Corporation, Japan) by the synchronization distillation and extraction (SDE) method. The measuring conditions were as follows: the gas chromatograph fitted with silica capillary column (30 m × 0.32 mm × 0.25 μm) of RTX-5MS (Shimadzu, Japan) and highly pure helium gas (carrier gas) with a flow rate of 2.0 mL min^{-1} (99.99%, Guangzhou Gases Co., Ltd., Guangzhou, China) [28,29].

2.10. Measurement of Grain Yield, Yield Attributes, and Grain Quality Characters of Aromatic Rice

The grain yield and yield contributing attributes of aromatic rice cultivar 'Xiangyaxiangzhan' were measured at the harvesting time. The total numbers of tillers pot^{-1} and numbers of productive tillers pot^{-1} were counted manually. One thousand grains were counted and weighed on the digital electric balance to achieve a 1000-grain weight (g), while the overall paddy weight obtained from each pot was calculated as grain yield pot^{-1}. The grain quality features of aromatic rice 'Xiangyaxiangzhan' cultivar were determined in the harvested grains. The brown rice was obtained by a rice huller (Jiangsu, China) and the percentage of brown rice was determined as: (weight of brown rice/paddy weight × 100). For measuring brown rice milling percentage, a Jingmi testing rice miller (Zhejiang, China) was used and the percentage was calculated as: (weight of milled rice/original weight × 100). The milling degree percentage was determined as: (weight of milled rice/weight of brown rice × 100). The percentages of chalkiness degree and chalkiness rate were calculated via an SDE-A light box (Guangzhou, China). The head rice rate percentage was determined by separating the whole milled grains out of 100 grains. Infratec 1241 (FOSS-TECATOR) was used to determine the protein and moisture content percentage in aromatic rice grains.

2.11. Statistical Analysis

The data collected were statistically analyzed through two-way ANOVA within a particular growth stage using 'Statistix 8.1' (Analytical Software, Tallahassee, FL, USA). Mean variances were separated by an LSD test at $p < 0.05$. SigmaPlot 10.0 was used for graphical representations. The heat-map hierarchical analysis between treatments and different studied parameters was conducted by R 3.5.1.

3. Results

3.1. Effect of Mo Supply on Photosynthetic Pigments and Photosynthetic Efficiency and Chloroplast Configuration of Aromatic Rice Plants under Cd Toxicity

In the absence of Mo, Cd stress significantly ($p < 0.05$) decreased the photosynthetic pigments in leaves of aromatic rice plants at different growth stages; however, Mo supply enhanced the chlorophyll contents under − / + Cd toxicity (Figure 1A,B). Under Cd stress, Mo supply enhanced chlorophyll a content by 80.88%, 94.37%, and 88.33%, and chlorophyll b content by 89.12%, 99.32%, and 92.27% at vegetative, reproductive, and maturity stages, respectively, as compared to −Mo treatments (Figure 1). Similarly, Cd toxicity significantly reduced the photosynthetic efficiency at different growth stages while Mo application enhanced the photosynthetic rate of aromatic rice plants. Under Cd stress, Mo application enhanced the photosynthetic rate by 61.67%, 72.79%, and 79.66% at the vegetative, reproductive, and maturity stages, respectively (Figure 1D).

The current study revealed that Cd toxicity severely affected the cell ultra-structural configurations and the integrity of chloroplasts in aromatic rice plants (Figure 2). However, under Mo-supplied treatments, the chloroplasts were observed to be comparatively regularly oval shaped, more consistent, and with well-organized arrangement and accumulated greater starch grains (Figure 2).

3.2. Influence of Mo Supply on Osmo-Regulation under Cd Toxicity

Our findings showed that Cd stress significantly decreased the soluble protein content in aromatic rice at different crop growth stages; however, such reductions were more severe at maturity stage (Figure 3A). However, Mo supply enhanced the soluble protein content with/without Cd toxicity compared to −Mo treatment. Under Cd stress, Mo supply enhanced the soluble protein content by 89.30%, 84.15%, and 115.37% at the vegetative, reproductive, and maturity stages, respectively, compared with −Mo treatment (Figure 3A). In contrast to soluble protein, proline content was significantly enhanced under Cd stress at different growth stages compared to without Cd toxicity (Figure 3B). Moreover, under Cd stress, Mo supply further enhanced the proline contents by 41.96%, 51.95%, and 72.54% at the vegetative, reproductive, and maturity stages, respectively, relative to −Mo treatments (Figure 3B).

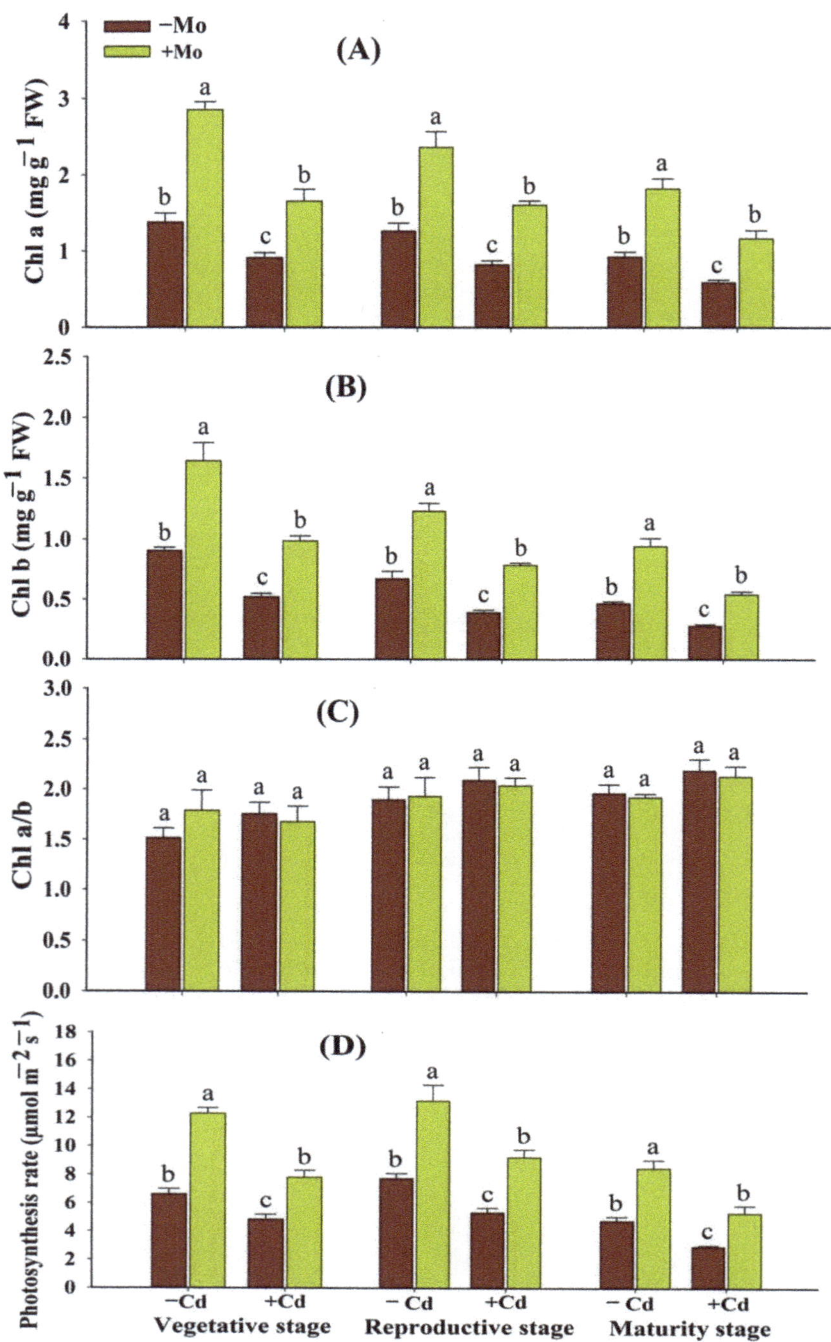

Figure 1. Effect of Mo supply and Cd toxicity on Chl a (**A**), Chl b (**B**), ratio of Chl a/b (**C**), and photosynthetic rate (P_n) (**D**), in leaves of aromatic rice 'Xiangyaxiangzhan' cultivar, at the vegetative, reproductive, and maturity stages. Bars above means indicate ±S.E. of four independent replicates ($n = 4$) and different small alphabetical letters above means reveal significant differences among treatments within a particular growth stage according to the LSD test ($p < 0.05$).

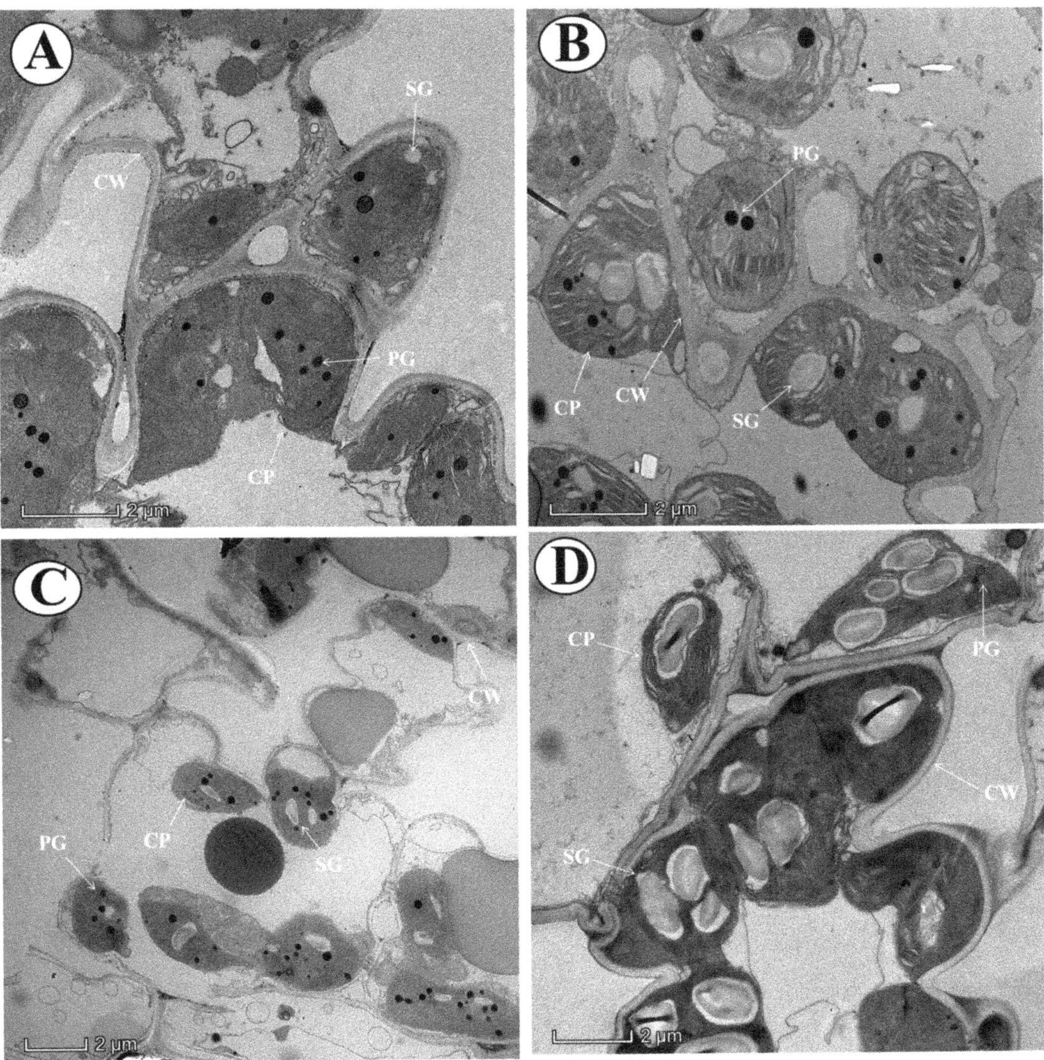

Figure 2. Effect of Mo supply and Cd toxicity on the transmission electron microscopy (TEM) images of aromatic rice 'Xiangyaxiangzhan' cultivar. Aromatic rice plants were treated with (**A**) (0 and 0 mg), (**B**) (0.15 and 0 mg), (**C**) (0 and 100 mg), and (**D**) (0.15 and 100 mg) of Mo and Cd, respectively. CP, chloroplasts; PG, plastoglobuli; CW, cell wall; SG, starch grains.

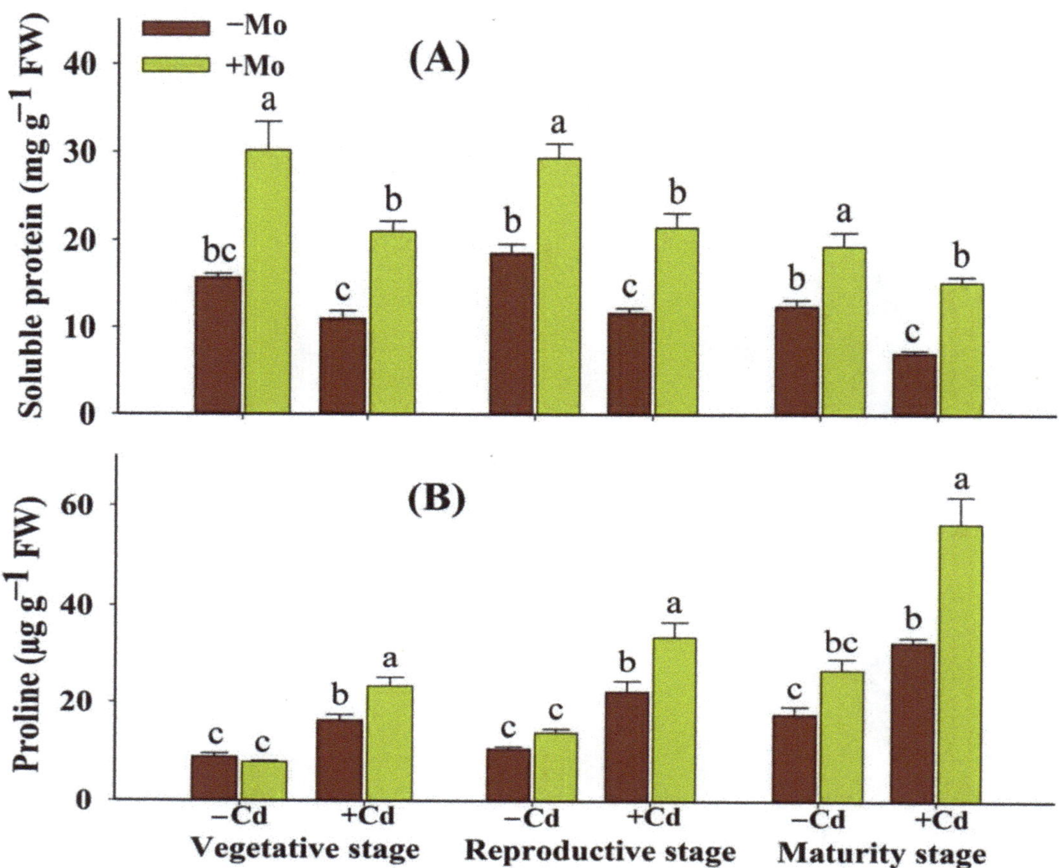

Figure 3. Effect of Mo supply and Cd toxicity on soluble protein (**A**) and proline contents (**B**) in leaves of aromatic rice 'Xiangyaxiangzhan' cultivar, at the vegetative, reproductive, and maturity stages. Bars above means indicate ±S.E. of four independent replicates (*n* = 4) and different small alphabetical letters above means reveal significant differences among treatments within a particular growth stage according to the LSD test (*p* < 0.05).

3.3. Effect of Mo Supply on Membrane Integrity in Aromatic Rice under Cd Stress

The lipid peroxidation (MDA production), electrolyte leakage (EL), and H_2O_2 accumulations are associated with the membrane integrity. The results indicated that Cd toxicity considerably increased the levels of H_2O_2, MDA, and EL in leaves of aromatic rice plants (Figure 4). However, Mo application reduced H_2O_2, MDA, and EL in leaf tissues of aromatic rice plants. Under Cd stress, Mo supply reduced the oxidative damage by lowering the levels of H_2O_2 (43.81%, 40.55%, and 29.95%), MDA (45.43%, 37.48%, and 31.24%), and EL (47.15%, 32.71%, and 25.14%) at the vegetative, reproductive, and maturity stages, respectively, compared with −Mo treatment (Figure 4).

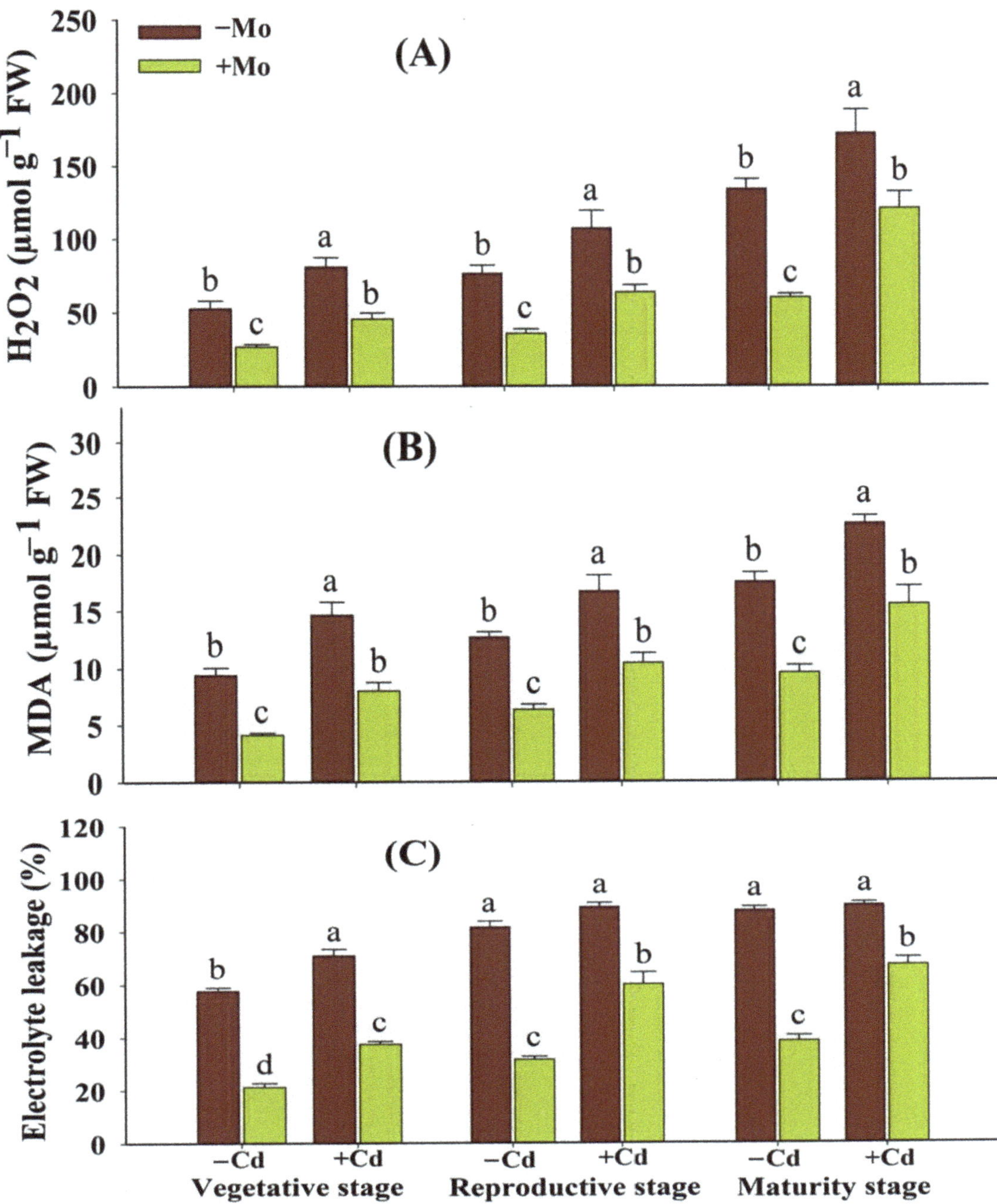

Figure 4. Effect of Mo supply and Cd toxicity on hydrogen peroxide (H_2O_2) content (**A**), malondialdehyde (MDA) (B), and electrolyte leakage (EL) (**C**), in leaves of aromatic rice 'Xiangyaxiangzhan' cultivar, at the vegetative, reproductive, and maturity stages. Bars above means indicate ±S.E. of four independent replicates ($n = 4$) and different small alphabetical letters above means reveal significant differences among treatments within a particular growth stage according to the LSD test ($p < 0.05$).

3.4. Effect of Mo Supply on the Activities and Transcript Abundance of Enzymatic Antioxidants under Cd Toxicity in Aromatic Rice

Under Cd stress, the antioxidant enzyme activities and their respective transcript levels were higher at the vegetative stage while being decreased at the maturity stage. The Mo supply significantly enhanced the enzymatic antioxidant activities and transcript abundance at different growth stages (vegetative, reproductive, and maturity) (Figure 5). Mo application regulated the activities of SOD by 90.29%, 109.99%, and 233.30%; POD by 80.96%, 135.73%, and 204.01%; CAT by 91.69%, 162.42%, and 225.60%; and APX by 70.09%, 147.42%, and 153.48% under Cd toxicity, at the vegetative, reproductive, and maturity stages, respectively, compared with −Mo treatment (Figure 5). Similarly, Mo application increased the expression levels of SOD, POD, CAT, and APX genes in the leaf tissues of aromatic rice plants with/without Cd stress at different growth stages (Figure 6), indicating that Mo supply strengthened the antioxidant defense system of aromatic rice plants through increasing antioxidant enzyme activities and expressions under stress conditions.

3.5. Influence of Mo Supply on Non-Enzymatic Antioxidants in Aromatic Rice under Cd-Stress

The Cd stress reduced the GSH and total glutathione contents while significantly increasing GSSG concentration at different growth stages of aromatic rice (Figure 7). However, Mo supply enhanced the GSH (165.82%, 193.56%, and 99.11%) and GSH + GSSG (104.12%, 117.27%, and 62.77%) levels while decreasing GSSG (25.39%, 35.43%, and 15.29%) contents under Cd stress, at the vegetative, reproductive, and maturity stages, respectively (Figure 7A–C). Moreover, Mo supply increased the GSH/GSSG in leaves at prominent growth stages of aromatic rice plants (Figure 7D).

3.6. Mo and Cd Concentrations in Different Plant Parts of Aromatic Rice Plants

The results revealed that Mo concentrations in leaves, ears, and grains of aromatic rice plants were significantly increased with Mo supply; however, non-significant effects were observed in Mo concentration under −/+ Cd stress in various plant parts of aromatic rice (Table S3). The Cd concentrations were increased in leaves, ears, and grains of aromatic rice plants under Cd stress. However, Mo application decreased the Cd concentration in leaves by 32.51%, 36.40%, and 24.82% at the vegetative, reproductive, and maturity stages, respectively, and 25.35% in ears (at reproductive stage) and 29.75% in grains (at maturity stage) under Cd toxicity (Table S3), suggesting that Mo supply inhibited the Cd accumulation and consequently mitigated the Cd toxicity through the least accretion in the edible part (grains) of aromatic rice.

3.7. Influence of Mo Supply on Yield Traits and Grain Quality Characteristics of Aromatic Rice under Cd Stress

Pronounced variations in the growth of aromatic rice plants were observed under the influence of Mo supply and Cd toxicity treatments (Figure S1). The Cd stress caused considerable reductions in grain yield and yield contributing traits and also deteriorated the grain quality characteristics of aromatic rice. However, Mo supply mitigated Cd-provoked adversities in aromatic rice and significantly improved grain yield and associated attributes (except 1000-grain weight) under Cd toxicity (Table 1). Compared with −Mo treatment, Mo supply improved the total number of tillers pot^{-1} (61.04%), productive tillers pot^{-1} (69.86%), filled grain percentage (25.07%), 1000-grain weight (10.15%), grains panicle^{-1} (40.07%), and grain yield pot^{-1} (64.75%) of aromatic rice under Cd stress (Table 1).

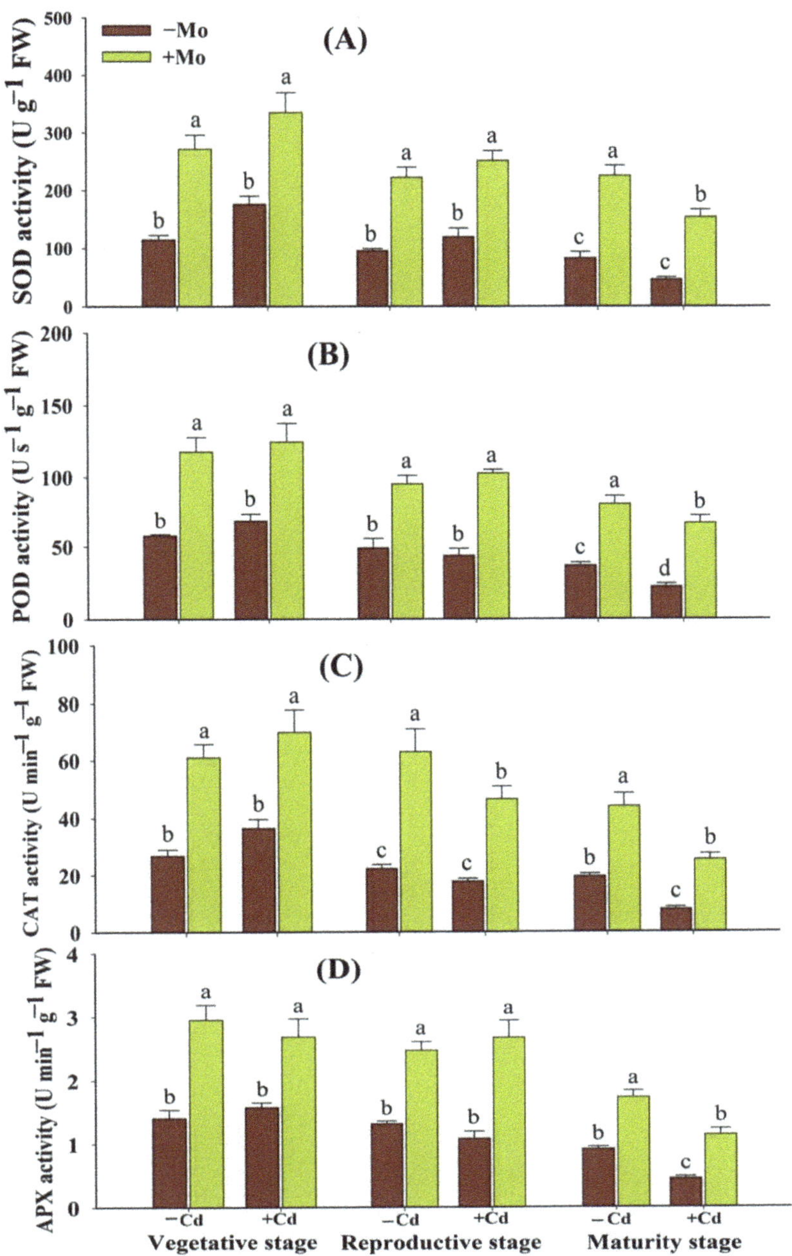

Figure 5. Effect of Mo supply and Cd toxicity on the activities of enzymatic antioxidants: superoxide dismutase (SOD) (**A**); peroxidase (POD) (**B**); catalase (CAT) (**C**); and ascorbate peroxidase (APX) (**D**), in the leaves of aromatic rice 'Xiangyaxiangzhan' cultivar, at the vegetative, reproductive, and maturity stages. Bars above means indicate ±S.E. of four independent replicates ($n = 4$) and different small alphabetical letters above means reveal significant differences among treatments within a particular growth stage according to the LSD test ($p < 0.05$).

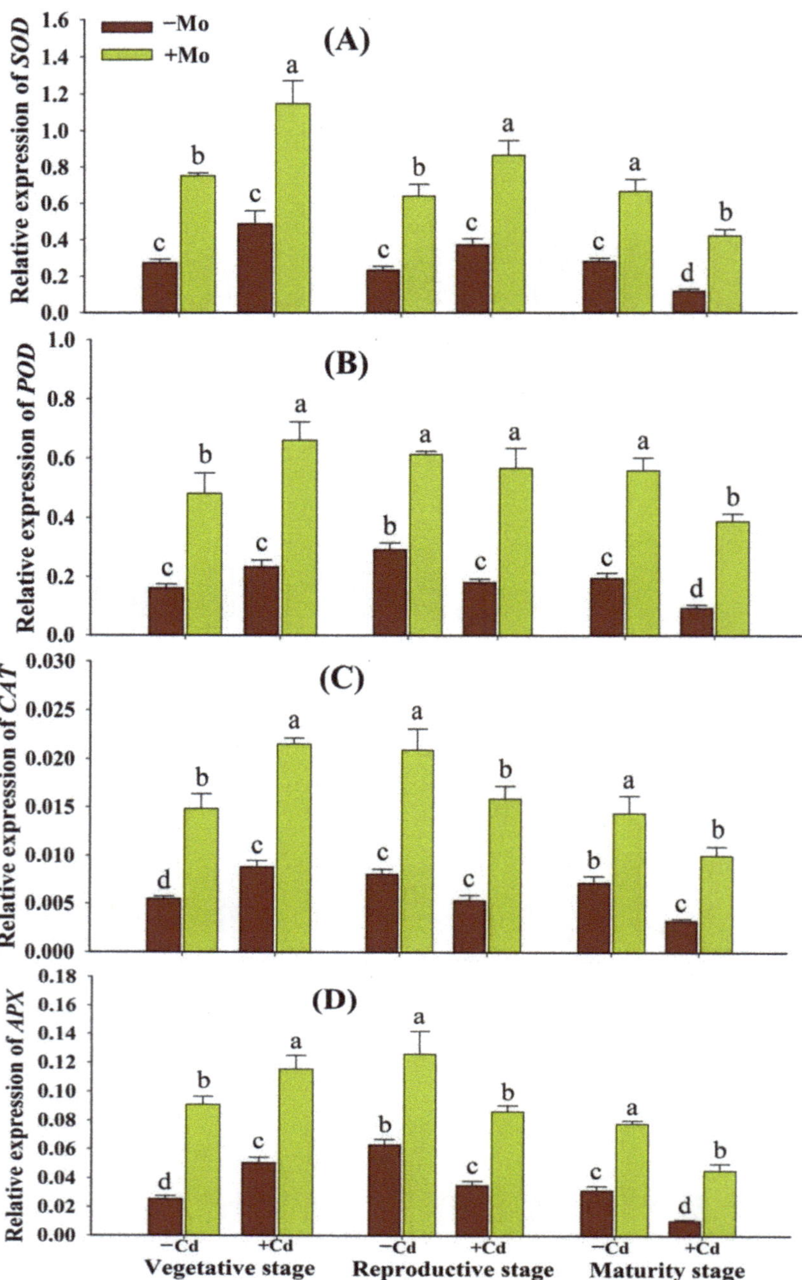

Figure 6. Effect of Mo supply and Cd toxicity on qRT-PCR analysis of antioxidant enzyme-related transcripts of superoxide dismutase (SOD) (**A**); peroxidase (POD) (**B**); catalase (CAT) (**C**); and ascorbate peroxidase (APX) (**D**), in the leaves of aromatic rice 'Xiangyaxiangzhan' cultivar, at the vegetative, reproductive, and maturity stages. Bars above means indicate ±S.E. of four independent replicates ($n = 4$) and different small alphabetical letters above means reveal significant differences among treatments within a particular growth stage according to the LSD test ($p < 0.05$).

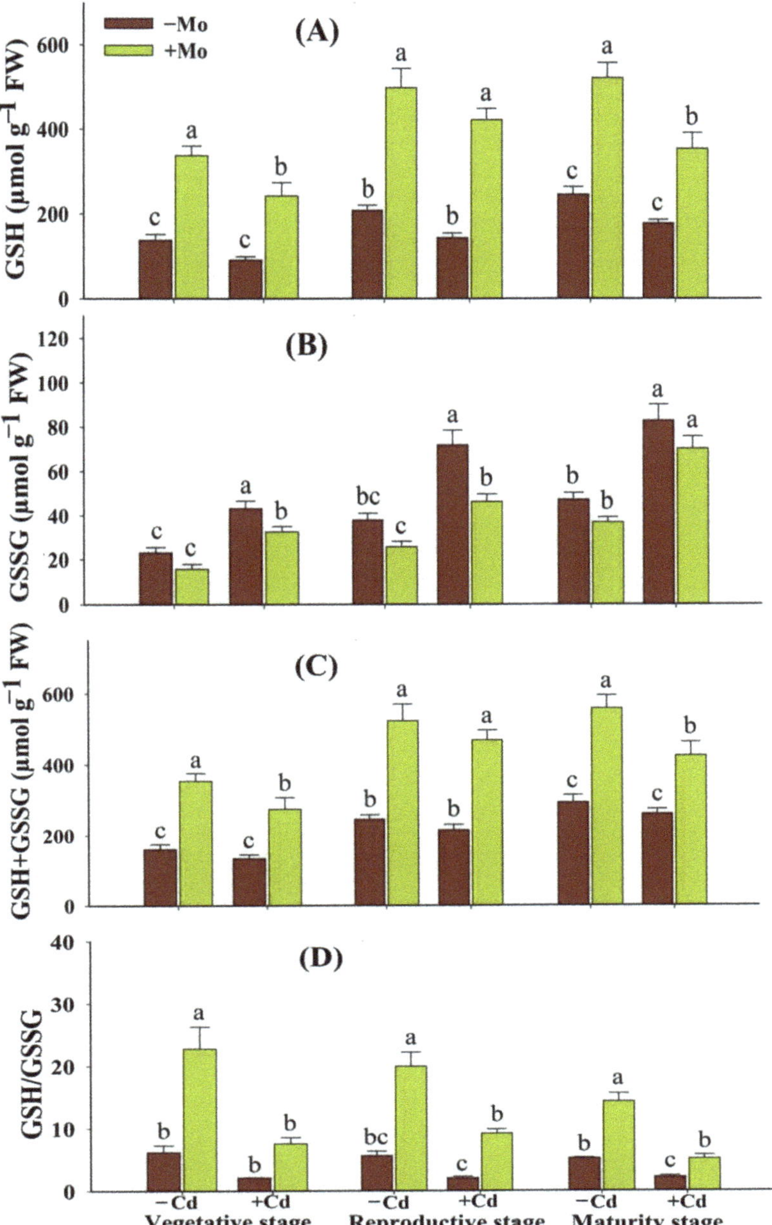

Figure 7. Effect of Mo supply and Cd toxicity on reduced glutathione (GSH) (**A**); oxidized glutathione (GSSG) (**B**); total glutathione (GSSG + GSH) (**C**); and ratio of GSH/GSSG (**D**), in the leaves of aromatic rice 'Xiangyaxiangzhan' cultivar, at the vegetative, reproductive, and maturity stages. Bars above means indicate ±S.E. of four independent replicates ($n = 4$) and different small alphabetical letters above means reveal significant differences among treatments within a particular growth stage according to the LSD test ($p < 0.05$).

Rice quality attributes including 2AP contents, brown rice rate, milled rice rate, head rice rate, and protein and moisture contents were considerably reduced while chalkiness rate and chalkiness degree increased under Cd toxicity, indicating that Cd stress deteriorated aromatic rice quality. However, Mo supply restored the grain quality characters under Cd toxicity (Table 2) and improved aromatic rice quality under Cd stress. Compared with −Mo treatment, Mo supply increased the 2AP contents (77.09%), milling degree (14.24%), milled rice rate (12.35%), head rice rate (25.74%), protein contents (29.35%), brown rice rate (5.59%), and moisture content (2.67%), while reducing the chalkiness degree (26.08%) and chalkiness rate (22.71%) in aromatic rice grains under Cd stress (Table 2).

3.8. Relationships

The hierarchical analysis revealed that Cd toxicity negatively affected the growth parameters, yield attributes, and grain quality characteristics of aromatic rice plants and concomitantly these inhibitory effects are correlated with the reduced photosynthetic pigments and weakened enzymatic and non-enzymatic antioxidant defense mechanism while triggering oxidative damage. However, Mo supply established strongly positive correlations with the grain yield attributes, quality characters, strengthening photosynthetic apparatus, and antioxidant defense system, and reduced the oxidative damage. In essence, this hierarchical correlation analysis reveals that Mo supply played an efficient role in mitigating the Cd-induced toxicity effects on the plant growth, grain yield, and yield-contributing attributes, and quality characteristics of aromatic rice (Figure 8).

Table 1. Influence of Mo supply and Cd toxicity on aromatic rice yield and associated attributes.

Cd Toxicity	Mo Application	Tillers Pot^{-1}	Productive Tillers Pot^{-1}	Grains Panicle^{-1}	Filled Grain Percentage	1000-Grain Weight (g)	Grain Yield Pot^{-1} (g)
−Cd	−Mo	26.97 ± 1.36 [b]	21.31 ± 1.48 [c]	128.91 ± 4.59 [b]	68.24 ± 2.23 [b]	18.11 ± 0.42 [b]	46.11 ± 2.47 [c]
	+Mo	40.08 ± 2.70 [a]	32.46 ± 1.47 [a]	165.21 ± 8.61 [a]	83.35 ± 5.15 [a]	20.30 ± 0.85 [a]	73.36 ± 4.32 [a]
+Cd	−Mo	19.89 ± 0.61 [c]	16.02 ± 1.40 [d]	103.72 ± 3.48 [c]	57.76 ± 2.81 [c]	16.97 ± 0.67 [b]	35.26 ± 1.43 [d]
	+Mo	32.05 ± 1.59 [b]	27.20 ± 1.18 [b]	145.28 ± 5.67 [b]	72.25 ± 1.88 [b]	18.70 ± 0.64 [ab]	58.08 ± 4.19 [b]
LSD $(p < 0.05)$		5.35	4.27	18.20	10.09	2.05	10.26

−Mo and +Mo denote 0 and 0.15 mg of molybdenum supplied as ammonium molybdate [(NH$_4$)$_6$Mo$_7$O$_{24}$·4H$_2$O]), whilst −Cd and +Cd denote 0 and 100 mg of cadmium supplied as cadmium chloride (CdCl$_2$·2.5H$_2$O) kg^{-1} soil, respectively. The numeric values reflect means from four independent replicates (±S.E.) with different treatments. Data presented in individual columns indicated with dissimilar letters differ significantly by LSD test at $p < 0.05$.

Table 2. Influence of Mo supply and Cd toxicity on aromatic rice grain quality traits.

Cd Toxicity	Mo Application	Brown Rice Rate (%)	Milled Rice Rate (%)	Milling Degree (%)	Head Rice Rate (%)	Chalkiness Rate (%)	Chalkiness Degree (%)	Moisture Content (%)	Protein Content (%)	2AP Content (ng g^{-1} FW)
−Cd	−Mo	73.89 ± 3.78 [a]	60.14 ± 1.52 [bc]	77.91 ± 1.59 [b]	56.89 ± 3.82 [bc]	33.66 ± 2.44 [b]	16.94 ± 0.53 [b]	12.27 ± 0.08 [b]	6.47 ± 0.14 [c]	144.35 ± 7.40 [b]
	+Mo	77.81 ± 1.87 [a]	66.42 ± 1.70 [a]	85.83 ± 1.46 [a]	68.53 ± 5.26 [a]	24.06 ± 1.28 [c]	11.75 ± 0.58 [c]	12.68 ± 0.18 [a]	8.49 ± 0.22 [a]	297.08 ± 23.91 [a]
+Cd	−Mo	71.24 ± 1.78 [a]	55.20 ± 1.71 [c]	71.55 ± 1.82 [c]	49.57 ± 2.04 [c]	39.78 ± 1.70 [a]	20.62 ± 0.94 [a]	12.12 ± 0.13 [b]	5.50 ± 0.09 [d]	97.01 ± 6.39 [c]
	+Mo	75.22 ± 2.31 [a]	62.02 ± 2.35 [ab]	81.74 ± 2.05 [ab]	62.34 ± 2.94 [ab]	30.74 ± 1.71 [b]	15.25 ± 0.56 [b]	12.45 ± 0.07 [ab]	7.12 ± 0.12 [b]	171.79 ± 14.58 [b]
LSD $(p < 0.05)$		7.89	5.69	5.38	11.42	5.50	2.07	0.38	0.47	45.69

−Mo and +Mo denote 0 and 0.15 mg of molybdenum supplied as ammonium molybdate [(NH$_4$)$_6$Mo$_7$O$_{24}$·4H$_2$O]), whilst −Cd and +Cd denote 0 and 100 mg of cadmium supplied as cadmium chloride (CdCl$_2$·2.5H$_2$O) kg^{-1} soil, respectively. The numeric values reflect means from four independent replicates (±S.E.) with different treatments. Data presented in individual columns indicated with dissimilar letters differ significantly by LSD test at $p < 0.05$.

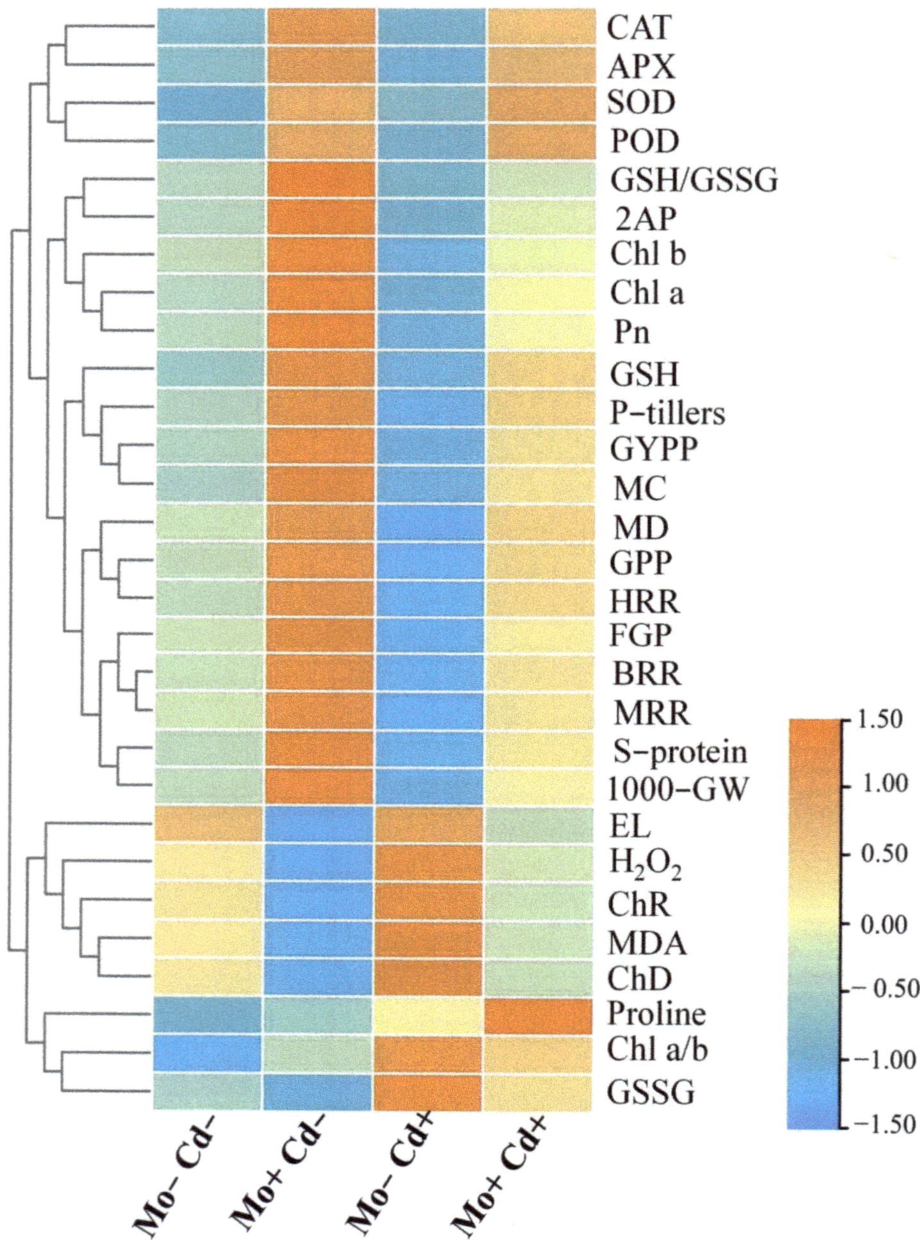

Figure 8. Heat-map reveals a hierarchical clustering analysis between different treatments and studied parameters of aromatic rice plants. Abbreviated names are as follows: 2AP (2-acetyl-1-pyrroline), SOD (superoxide dismutase activity), POD (peroxide activity), CAT (catalase activity), APX (ascorbate peroxidase activity), Chl a (chlorophyll a), Chl b (chlorophyll b), Pn (photosynthetic rate), GSH (reduced glutathione), GSSG (oxidized glutathione), P-tillers (productive tillers), HRR (head rice rate), MRR (milled rice rate), FGP (filled grain percentage), BRR (brown rice rate), S-protein (soluble protein), 1000-GW (1000-grain weight), EL (electrolyte leakage), H_2O_2 (hydrogen peroxide), ChR (chalkiness rate), ChD (chalkiness degree), MDA (malondialdehyde contents), MC (moisture content), MD (milling degree), GPP (grains per panicle), GYPP (grain yield per pot).

4. Discussion

All heavy metals, particularly Cd toxicity, inevitably provoke perturbations in various physio-biochemical processes, photosynthetic apparatus, plant metabolism, and antioxidant protection mechanisms, thereby resulting in significant yield reduction and quality deterioration in crop plants [4,30–32]. Thus, Cd stress-alleviating approaches in plant growth and developmental cycles, yield reductions, and quality deteriorations through strengthened plant metabolism continue to be challenging goals for plant scientists. Molybdenum (Mo), an essential and anti-stress micronutrient, has garnered substantial consideration due to its primary involvement in multiple plant growth and production processes and enhancing oxidative stress resistance under cold, drought, and heavy metal toxicities [13,15,24]. The current experiment established a number of mitigating roles of Mo supply on photosynthetic apparatus, plant metabolism, antioxidant defense system, yield attributes, and quality characteristics of aromatic rice cultivar 'Xiangyaxiangzhan' under Cd toxicity.

Our results revealed that Cd toxicity significantly reduced the photosynthetic pigments and photosynthetic efficiency at the vegetative, reproductive, and maturity stages of aromatic rice plants (Figure 1), which is in line with the results of several previous reports, who documented the Cd-induced reduction in chlorophyll contents and photosynthetic efficiency [33,34]. The Mo application was found to increase the chlorophyll contents and photosynthetic efficiency in Cd-stressed aromatic rice plants (Figure 1), which might be ascribed to Mo-induced reductions in oxidative damage and maintenance of chloroplast ultra-structure (Figures 2 and 4) [35]. Similarly, previous studies reported that Mo supply increased the leaf gaseous exchange attributes, photosynthetic pigments, and chloroplast integrity and configurations in Triticum aestivum [36] and Fragaria ananassa [37].

Biosynthesis and relative accumulation of compatible solutes/osmolytes are well-recognized due to a wide range of their functions, i.e., protection and stabilization of cellular membranes, protection of several enzymes, acting as osmoticum for turgor maintenance, and scavenging roles against ROS (e.g., proline) in plants [38]. In our study, Cd toxicity reduced soluble protein contents, which might be due to greater oxidative damage, and these results are concomitant with previous studies that Cd toxicity stimulated soluble protein degradation through higher protease activity [39] and unnecessary ROS generation [40]. The Mo supply was found to enhance the levels of both soluble protein and proline contents at different growth stages of aromatic rice plants under Cd stress (Figure 3), suggesting the ameliorative role of Mo in improving and maintaining higher plant osmotic balance under stress conditions.

The Cd toxicity causes overproduction of free radicals and ROS, causing ultra-structural and functional alterations in cell nuclei, DNA, lipids, and proteins. The present study revealed that Cd stress stimulated the oxidative stress as evident from higher production of H_2O_2, MDA, and loss of membrane integrity (greater EL) in aromatic rice plants (Figure 4). However, Mo supply significantly reduced MDA, H_2O_2, and EL in leaf tissues at prominent growth stages of aromatic rice plants (Figure 4), suggesting that Mo supply mitigated Cd-provoked intracellular membrane disruptions throughout the growth period of aromatic rice plants. These Mo-induced mitigating approaches to preserve and retain bilayer membranes and safeguard cell membranes from oxidative stress destructions have also been reported in strawberry [41].

Under stress conditions, both enzymatic and non-enzymatic antioxidants play key roles to scavenge the ROS and counteract the oxidative damage in plants. In the present study, Cd toxicity aggravated oxidative damage in aromatic rice plants, while Mo application alleviated oxidative stress, which could be ascribed to the enhancement of antioxidant enzyme activities and expressions (Figures 5 and 6). The protection of plants by antioxidant defense systems against oxidative damage was triggered by antioxidant enzymes including SOD, CAT, APX, and POD [42–44]. Generally, ROS is scavenged by antioxidant enzymes. For example, SOD catalyzes O_2^- into H_2O_2, and then H_2O_2 is disintegrated by CAT and APX [42,45,46]. Previous studies also reported that Mo supply relieved oxidative damage by improving the antioxidant defense ability of plants under cold, drought, salt stress, and heavy metals [11,13,47]. However, the mechanisms by which Mo improves antioxidant defense ability remain to be investigated. The non-enzymatic antioxidants (GSH and GSSG) also serve as a redox buffer under various heavy metals [48]. GSH shields guard cells from oxidative destruction and plays a major role in reducing the bulk of ROS. However, under adverse conditions as cells undergo greater oxidative stress damages, GSSG accumulates and the GSH to GSSG ratio decreases. Thus, estimating the GSH to GSSG ratio (GSH/GSSG) is a valuable indicator for assessing the rate of oxidative stress in cells and tissues [49]. In this study, under Cd toxicity, Mo supply maintained a higher GSH/GSSG ratio in leaves of aromatic rice plants at the vegetative, reproductive, and maturity stages (Figure 7D), demonstrating that Mo supply mitigated the Cd-provoked stress adversities in aromatic rice. These results thus indicate that Mo supply has improved both enzymatic and non-enzymatic antioxidants during the growth cycle of aromatic rice plants, i.e., the vegetative, reproductive, and maturity stages and mitigated Cd-induced toxicity effects on plant physio-biochemical processes as is evident from significant reductions in ROS production, MDA contents, and electrolyte leakage under Cd-stressed treatments.

In rice, grain yield depends upon productive tillers, sterility percentage, number of grains panicle-1, and 1000-grain weight, while 2AP content, milling recovery, and chalkiness rate are important indices for quality estimation of aromatic rice [2,28]. In the present study, Cd toxicity reduced the yield-contributing attributes, i.e., productive tillers per pot, number of grains per panicle, and filled grain percentage (Table 1), and deteriorated the rice quality traits through reduced 2AP contents, brown rice rate, milled rice rate, milling degree, head rice rate, protein contents, while increasing chalkiness rate and chalkiness degree (Table 2), indicating that Cd toxicity reduced aromatic rice yield and deteriorated quality traits through all the contributing traits being affected. Previous studies have also reported significant reductions in yield and quality traits under Cd toxicity in different aromatic rice cultivars [8,17]. Similarly, our findings agree with previous studies reporting that Cd toxicity resulted in yield reductions and quality deterioration through a brutally impeded plant metabolism and photosynthetic system, and higher Cd concentration in plant parts and stimulated oxidative stress [4,15]. However, Mo supply substantially increased grain yield and quality traits of aromatic rice under Cd stress. The probable explanation is that Mo supply mitigated Cd-induced inhibitions on plant metabolism through strengthening the enzymatic and non-enzymatic antioxidant defense system, fortifying photosynthetic apparatus, and inhibiting the absorption and translocation of Cd in aromatic rice plants.

Taken together, our findings infer that Cd toxicity hampered aromatic rice plant growth, yield attributes, and grain quality traits through destabilizing plant metabolism, increasing oxidative damage, undermining the plant protection system, and distressing the photosynthetic apparatus. However, Mo supply alleviated the Cd-induced inhibitions on plant metabolism and improved aromatic rice yield and grain quality traits by strengthening the photosynthetic system and antioxidant protection mechanisms. Future studies can, however, be meditated to examine the possible role/s of Mo supply during 2AP biosynthesis pathways in different aromatic rice cultivars under heavy metal-polluted soils.

5. Conclusions

The current experiment revealed that Cd stress hampered the plant growth, grain yield, and quality traits of aromatic rice. The Cd toxicity triggered the production of H_2O_2 and electrolyte leakage, presumably by desynchronizing the ROS-scavenging mechanism. Nevertheless, Mo supply proficiently relieved Cd-provoked inhibitory effects on the plant metabolism, grain yield, and quality characteristics of aromatic rice, which could mainly be ascribed to reduced Cd uptake, reinforced photosynthetic apparatus, consistent chloroplast ultra-structure, and a higher scavenging ROS amount. Thus, our findings concluded that Mo supply mitigated the Cd-provoked inhibitory effects on plant growth, physio-biochemical processes, and antioxidant defense system during the plant growth cycle (vegetative, reproductive, and maturity stages), thereby enhancing grain yield and quality traits of aromatic rice.

Supplementary Materials: The following are available online at https://www.mdpi.com/article/10.3390/antiox10060838/s1, Figure S1: Pictorial view of aromatic rice plants grown under molybdenum (Mo) and cadmium (Cd) treatments, Table S1: The experimental soil's chemical properties, Table S2: Primer sequences used for qRT-PCR amplification, Table S3: Influence of molybdenum supplementation and cadmium toxicity on Mo and Cd concentration at different growth stages in leaves, ears and grains of aromatic rice.

Author Contributions: Conceptualization, methodology, software, validation, writing—original draft, writing—review and editing, M.I. (Muhammad Imran), S.H., and E.A.W.; data curation, formal analysis, funding acquisition, investigation, resources, visualization, L.H., M.F.A., M.I. (Muhammad Ihtisham), and X.T.; project administration, supervision, X.T. All authors have read and agreed to the published version of the manuscript.

Funding: This study was supported by National Natural Science Foundation of China (31971843), The Technology System of Modern Agricultural Industry in Guangdong (2020KJ105), and Guangzhou Science and Technology Project (202103000075).

Institutional Review Board Statement: Ethical review and approval were waived for this study.

Informed Consent Statement: Not applicable.

Data Availability Statement: Data is contained within the article and Supplementary Materials.

Conflicts of Interest: The authors declare that they have no known competing financial interests or personal relationships that could have appeared to influence the work reported in this paper.

References

1. Bryant, R.; McClung, A. Volatile profiles of aromatic and non-aromatic rice cultivars using SPME/GC–MS. *Food Chem.* **2011**, *124*, 501–513. [CrossRef]
2. Ashraf, U.; Kanu, A.S.; Deng, Q.; Mo, Z.; Pan, S.; Tian, H.; Tang, X. Lead (Pb) toxicity; physio-biochemical mechanisms, grain yield, quality, and Pb distribution proportions in scented rice. *Front. Plant Sci.* **2017**, *8*, 259. [CrossRef]
3. Grant, C.; Clarke, J.; Duguid, S.; Chaney, R. Selection and breeding of plant cultivars to minimize cadmium accumulation. *Sci. Total Environ.* **2008**, *390*, 301–310. [CrossRef]
4. Ismael, M.A.; Elyamine, A.M.; Moussa, M.G.; Cai, M.; Zhao, X.; Hu, C. Cadmium in plants: Uptake, toxicity, and its interactions with selenium fertilizers. *Metallomics* **2019**, *11*, 255–277. [CrossRef]
5. Uraguchi, S.; Mori, S.; Kuramata, M.; Kawasaki, A.; Arao, T.; Ishikawa, S. Root-to-shoot Cd translocation via the xylem is the major process determining shoot and grain cadmium accumulation in rice. *J. Exp. Bot.* **2009**, *60*, 2677–2688. [CrossRef]
6. Hussain, S.; Khaliq, A.; Noor, M.A.; Tanveer, M.; Hussain, H.A.; Hussain, S.; Shah, T.; Mehmood, T. Metal Toxicity and Nitrogen Metabolism in Plants: An Overview. In *Carbon and Nitrogen Cycling in Soil*; Springer: Berlin/Heidelberg, Germany, 2020; pp. 221–248.
7. Cao, F.; Wang, R.; Cheng, W.; Zeng, F.; Ahmed, I.M.; Hu, X.; Zhang, G.; Wu, F. Genotypic and environmental variation in cadmium, chromium, lead and copper in rice and approaches for reducing the accumulation. *Sci. Total Environ.* **2014**, *496*, 275–281. [CrossRef]
8. Imran, M.; Hussain, S.; Rana, M.S.; Saleem, M.H.; Rasul, F.; Ali, K.H.; Potcho, M.P.; Pan, S.; Duan, M.; Tang, X. Molybdenum improves 2-acetyl-1-pyrroline, grain quality traits and yield attributes in fragrant rice through efficient nitrogen assimilation under cadmium toxicity. *Ecotoxicol. Environ. Saf.* **2021**, *211*, 111911. [CrossRef] [PubMed]
9. Rana, M.S.; Hu, C.X.; Shaaban, M.; Imran, M.; Afzal, J.; Moussa, M.G.; Elyamine, A.M.; Bhantana, P.; Saleem, M.H.; Syaifudin, M. Soil phosphorus transformation characteristics in response to molybdenum supply in leguminous crops. *J. Environ. Manag.* **2020**, *268*, 110610. [CrossRef] [PubMed]
10. Rana, M.S.; Sun, X.; Imran, M.; Ali, S.; Shaaban, M.; Moussa, M.G.; Khan, Z.; Afzal, J.; Binyamin, R.; Bhantana, P. Molybdenum-induced effects on leaf ultra-structure and rhizosphere phosphorus transformation in *Triticum aestivum* L. *Plant Physiol. Biochem.* **2020**, *153*, 20–29. [CrossRef] [PubMed]
11. Zhang, M.; Hu, C.; Zhao, X.; Tan, Q.; Sun, X.; Cao, A.; Cui, M.; Zhang, Y. Molybdenum improves antioxidant and osmotic-adjustment ability against salt stress in Chinese cabbage (*Brassica campestris* L. ssp. Pekinensis). *Plant Soil* **2012**, *355*, 375–383. [CrossRef]
12. Wu, S.; Hu, C.; Tan, Q.; Xu, S.; Sun, X. Nitric oxide mediates molybdenum-induced antioxidant defense in wheat under drought stress. *Front. Plant Sci.* **2017**, *8*, 1085. [CrossRef] [PubMed]

13. Sun, X.; Hu, C.; Tan, Q.; Liu, J.; Liu, H. Effects of molybdenum on expression of cold-responsive genes in abscisic acid (ABA)-dependent and ABA-independent pathways in winter wheat under low-temperature stress. *Ann. Bot.* **2009**, *104*, 345–356. [CrossRef] [PubMed]

14. Imran, M.; Sun, X.; Hussain, S.; Ali, U.; Rana, M.S.; Rasul, F.; Saleem, M.H.; Moussa, M.G.; Bhantana, P.; Afzal, J. Molybdenum-Induced Effects on Nitrogen Metabolism Enzymes and Elemental Profile of Winter Wheat (Triticum aestivum L.) Under Different Nitrogen Sources. *Int. J. Mol. Sci.* **2019**, *20*, 3009. [CrossRef] [PubMed]

15. Ismael, M.A.; Elyamine, A.M.; Zhao, Y.Y.; Moussa, M.G.; Rana, M.S.; Afzal, J.; Imran, M.; Zhao, X.H.; Hu, C.X. Can selenium and molybdenum restrain cadmium toxicity to pollen grains in Brassica napus? *Int. J. Mol. Sci.* **2018**, *19*, 2163. [CrossRef]

16. Ali, N.; Hadi, F.; Ali, M. Growth stage and molybdenum treatment affect cadmium accumulation, antioxidant defence and chlorophyll contents in Cannabis sativa plant. *Chemosphere* **2019**, *236*, 124360. [CrossRef]

17. Kanu, A.S.; Ashraf, U.; Mo, Z.; Fuseini, I.; Mansaray, L.R.; Duan, M.; Pan, S.; Tang, X. Cadmium uptake and distribution in fragrant rice genotypes and related consequences on yield and grain quality traits. *J. Chem.* **2017**, *2017*, 1405878. [CrossRef]

18. Ihtisham, M.; Liu, S.; Shahid, M.O.; Khan, N.; Lv, B.; Sarraf, M.; Ali, S.; Chen, L.; Liu, Y.; Chen, Q. The Optimized N, P, and K Fertilization for Bermudagrass Integrated Turf Performance during the Establishment and Its Importance for the Sustainable Management of Urban Green Spaces. *Sustainability* **2020**, *12*, 294. [CrossRef]

19. Ihtisham, M.; Fahad, S.; Luo, T.; Larkin, R.M.; Yin, S.; Chen, L. Optimization of nitrogen, phosphorus, and potassium fertilization rates for overseeded perennial ryegrass turf on dormant bermudagrass in a transitional climate. *Front. Plant Sci.* **2018**, *9*, 487. [CrossRef]

20. Saleem, M.H.; Ali, S.; Rehman, M.; Rana, M.S.; Rizwan, M.; Kamran, M.; Imran, M.; Riaz, M.; Soliman, M.H.; Elkelish, A. Influence of phosphorus on copper phytoextraction via modulating cellular organelles in two jute (Corchorus capsularis L.) varieties grown in a copper mining soil of Hubei Province, China. *Chemosphere* **2020**, *248*, 126032. [CrossRef]

21. Bradford, M.M. A rapid and sensitive method for the quantitation of microgram quantities of protein utilizing the principle of protein-dye binding. *Anal. Biochem.* **1976**, *72*, 248–254. [CrossRef]

22. Bates, L.S.; Waldren, R.P.; Teare, I. Rapid determination of free proline for water-stress studies. *Plant Soil* **1973**, *39*, 205–207. [CrossRef]

23. Valentovic, P.; Luxova, M.; Kolarovic, L.; Gasparikova, O. Effect of osmotic stress on compatible solutes content, membrane stability and water relations in two maize cultivars. *Plant Soil Environ.* **2006**, *52*, 184.

24. Imran, M.; Sun, X.; Hussain, S.; Ali, U.; Rana, M.S.; Rasul, F.; Shaukat, S.; Hu, C. Molybdenum application regulates oxidative stress tolerance in winter wheat under different nitrogen sources. *J. Soil Sci. Plant Nutr.* **2020**, *20*, 1827–1837. [CrossRef]

25. Imran, M.; Sun, X.; Hussain, S.; Rana, M.S.; Saleem, M.H.; Riaz, M.; Tang, X.; Khan, I.; Hu, C. Molybdenum supply increases root system growth of winter wheat by enhancing nitric oxide accumulation and expression of NRT genes. *Plant Soil* **2021**, *459*, 235–248. [CrossRef]

26. Pfaffl, M.W. A new mathematical model for relative quantification in real-time RT–PCR. *Nucleic Acids Res.* **2001**, *29*, e45. [CrossRef] [PubMed]

27. Filipiak-Szok, A.; Kurzawa, M.; Szłyk, E. Determination of toxic metals by ICP-MS in Asiatic and European medicinal plants and dietary supplements. *J. Trace Elem. Med. Biol.* **2015**, *30*, 54–58. [CrossRef]

28. Mo, Z.; Li, Y.; Nie, J.; He, L.; Pan, S.; Duan, M.; Tian, H.; Xiao, L.; Zhong, K.; Tang, X. Nitrogen application and different water regimes at booting stage improved yield and 2-acetyl-1-pyrroline (2AP) formation in fragrant rice. *Rice* **2019**, *12*, 74. [CrossRef]

29. Mo, Z.; Ashraf, U.; Tang, Y.; Li, W.; Pan, S.; Duan, M.; Tian, H.; Tang, X. Nitrogen application at the booting stage affects 2-acetyl-1-pyrroline, proline, and total nitrogen contents in aromatic rice. *Chil. J. Agric. Res.* **2018**, *78*, 165–172. [CrossRef]

30. Cao, F.; Cai, Y.; Liu, L.; Zhang, M.; He, X.; Zhang, G.; Wu, F. Differences in photosynthesis, yield and grain cadmium accumulation as affected by exogenous cadmium and glutathione in the two rice genotypes. *Plant Growth Regul.* **2015**, *75*, 715–723. [CrossRef]

31. Fahad, S.; Rehman, A.; Shahzad, B.; Tanveer, M.; Saud, S.; Kamran, M.; Ihtisham, M.; Khan, S.U.; Turan, V.; ur Rahman, M.H. Rice responses and tolerance to metal/metalloid toxicity. In *Advances in Rice Research for Abiotic Stress Tolerance*; Elsevier: Amsterdam, The Netherlands, 2019; pp. 299–312.

32. Saleem, M.H.; Rehman, M.; Zahid, M.; Imran, M.; Xiang, W.; Liu, L. Morphological changes and antioxidative capacity of jute (Corchorus capsularis, Malvaceae) under different color light-emitting diodes. *Braz. J. Bot.* **2019**, *42*, 581–590. [CrossRef]

33. Najeeb, U.; Jilani, G.; Ali, S.; Sarwar, M.; Xu, L.; Zhou, W. Insights into cadmium induced physiological and ultra-structural disorders in Juncus effusus L. and its remediation through exogenous citric acid. *J. Hazard. Mater.* **2011**, *186*, 565–574. [CrossRef] [PubMed]

34. Padmaja, K.; Prasad, D.; Prasad, A. Inhibition of chlorophyll synthesis in Phaseolus vulgaris L. seedlings by cadmium acetate. *Photosynthetica* **1990**, *24*, 399–405.

35. Liu, Z.; Gao, J.; Gao, F.; Liu, P.; Zhao, B.; Zhang, J. Photosynthetic characteristics and chloroplast ultrastructure of summer maize response to different nitrogen supplies. *Front. Plant Sci.* **2018**, *9*, 576. [CrossRef] [PubMed]

36. Imran, M.; Hu, C.; Hussain, S.; Rana, M.S.; Riaz, M.; Afzal, J.; Aziz, O.; Elyamine, A.M.; Ismael, M.A.F.; Sun, X. Molybdenum-induced effects on photosynthetic efficacy of winter wheat (Triticum aestivum L.) under different nitrogen sources are associated with nitrogen assimilation. *Plant Physiol. Biochem.* **2019**, *141*, 154–163. [CrossRef] [PubMed]

37. Liu, L.; Xiao, W.; Li, L.; Li, D.-M.; Gao, D.-S.; Zhu, C.-y.; Fu, X.-L. Effect of exogenously applied molybdenum on its absorption and nitrate metabolism in strawberry seedlings. *Plant Physiol. Biochem.* **2017**, *115*, 200–211. [CrossRef]

38. Hasanuzzaman, M.; Inafuku, M.; Nahar, K.; Fujita, M.; Oku, H. Nitric Oxide Regulates Plant Growth, Physiology, Antioxidant Defense, and Ion Homeostasis to Confer Salt Tolerance in the Mangrove Species, Kandelia obovata. *Antioxidants* **2021**, *10*, 611. [CrossRef]
39. Palma, J.M.; Sandalio, L.M.; Corpas, F.J.; Romero-Puertas, M.C.; McCarthy, I.; Luis, A. Plant proteases, protein degradation, and oxidative stress: Role of peroxisomes. *Plant Physiol. Biochem.* **2002**, *40*, 521–530. [CrossRef]
40. Romero-Puertas, M.; Palma, J.; Gómez, M.; Del Rio, L.; Sandalio, L. Cadmium causes the oxidative modification of proteins in pea plants. *Plant Cell Environ.* **2002**, *25*, 677–686. [CrossRef]
41. Li, L.; Wei, X.; Ji, M.-L.; Chao, Y.; Ling, L.; Gao, D.-S.; Fu, X.-L. Effects of molybdenum on nutrition, quality, and flavour compounds of strawberry (Fragaria× ananassa Duch. cv. Akihime) fruit. *J. Integr. Agric.* **2017**, *16*, 1502–1512. [CrossRef]
42. Wu, Z.; Zhao, X.; Sun, X.; Tan, Q.; Tang, Y.; Nie, Z.; Qu, C.; Chen, Z.; Hu, C. Antioxidant enzyme systems and the ascorbate–glutathione cycle as contributing factors to cadmium accumulation and tolerance in two oilseed rape cultivars (Brassica napus L.) under moderate cadmium stress. *Chemosphere* **2015**, *138*, 526–536. [CrossRef]
43. Moustafa-Farag, M.; Mahmoud, A.; Arnao, M.B.; Sheteiwy, M.S.; Dafea, M.; Soltan, M.; Elkelish, A.; Hasanuzzaman, M.; Ai, S. Melatonin-induced water stress tolerance in plants: Recent advances. *Antioxidants* **2020**, *9*, 809. [CrossRef] [PubMed]
44. Hasanuzzaman, M.; Bhuyan, M.; Zulfiqar, F.; Raza, A.; Mohsin, S.M.; Mahmud, J.A.; Fujita, M.; Fotopoulos, V. Reactive oxygen species and antioxidant defense in plants under abiotic stress: Revisiting the crucial role of a universal defense regulator. *Antioxidants* **2020**, *9*, 681. [CrossRef] [PubMed]
45. Sachdev, S.; Ansari, S.A.; Ansari, M.I.; Fujita, M.; Hasanuzzaman, M. Abiotic stress and reactive oxygen species: Generation, signaling, and defense mechanisms. *Antioxidants* **2021**, *10*, 277. [CrossRef] [PubMed]
46. Hussain, S.; Khan, F.; Cao, W.; Wu, L.; Geng, M. Seed priming alters the production and detoxification of reactive oxygen intermediates in rice seedlings grown under sub-optimal temperature and nutrient supply. *Front. Plant Sci.* **2016**, *7*, 439. [CrossRef] [PubMed]
47. Imran, M.; Hussain, S.; El-Esawi, M.A.; Rana, M.S.; Saleem, M.H.; Riaz, M.; Ashraf, U.; Potcho, M.P.; Duan, M.; Rajput, I.A. Molybdenum Supply Alleviates the Cadmium Toxicity in Fragrant Rice by Modulating Oxidative Stress and Antioxidant Gene Expression. *Biomolecules* **2020**, *10*, 1582. [CrossRef]
48. Sharma, S.S.; Dietz, K.-J. The relationship between metal toxicity and cellular redox imbalance. *Trends Plant Sci.* **2009**, *14*, 43–50. [CrossRef]
49. Monostori, P.; Wittmann, G.; Karg, E.; Túri, S. Determination of glutathione and glutathione disulfide in biological samples: An in-depth review. *J. Chromatogr. B* **2009**, *877*, 3331–3346. [CrossRef]

 *antioxidants*

MDPI

Article

Seed Priming with Brassinosteroids Alleviates Chromium Stress in Rice Cultivars via Improving ROS Metabolism and Antioxidant Defense Response at Biochemical and Molecular Levels

Farwa Basit [1,†], Min Chen [1,†], Temoor Ahmed [2], Muhammad Shahid [3,*], Muhammad Noman [2], Jiaxin Liu [1], Jianyu An [1], Abeer Hashem [4], Al-Bandari Fahad Al-Arjani [4], Abdulaziz A. Alqarawi [5], Mashail Fahad S. Alsayed [4], Elsayed Fathi Abd_Allah [5], Jin Hu [1] and Yajing Guan [1,*]

[1] Institute of Crop Sciences, College of Agriculture and Biotechnology, Zhejiang University, Hangzhou 310058, China; 11916138@zju.edu.cn (F.B.); 11616036@zju.edu.cn (M.C.); liujiaxin@zju.edu.cn (J.L.); anjianyu@live.cn (J.A.); jhu@zju.edu.cn (J.H.)
[2] State Key Laboratory of Rice Biology, Institute of Biotechnology, Zhejiang University, Hangzhou 310058, China; temoorahmed@zju.edu.cn (T.A.); nomansiddique834@gmail.com (M.N.)
[3] Department of Bioinformatics and Biotechnology, Government College University, Faisalabad 38000, Pakistan
[4] Botany and Microbiology Department, College of Science, King Saud University, P.O. Box 2460, Riyadh 11451, Saudi Arabia; habeer@ksu.edu.sa (A.H.); aarjani@ksu.edu.sa (A.-B.F.A.-A.); malsayed@ksu.edu.sa (M.F.S.A.)
[5] Plant Production Department, College of Food and Agricultural Sciences, King Saud University, P.O. Box 2460, Riyadh 11451, Saudi Arabia; alqarawi@ksu.edu.sa (A.A.A.); eabdallah@ksu.edu.sa (E.F.A.)
* Correspondence: mshahid@gcuf.edu.pk (M.S.); vcguan@zju.edu.cn (Y.G.)
† These authors contributed equally to this work.

Citation: Basit, F.; Chen, M.; Ahmed, T.; Shahid, M.; Noman, M.; Liu, J.; An, J.; Hashem, A.; Fahad Al-Arjani, A.-B.; Alqarawi, A.A.; et al. Seed Priming with Brassinosteroids Alleviates Chromium Stress in Rice Cultivars via Improving ROS Metabolism and Antioxidant Defense Response at Biochemical and Molecular Levels. *Antioxidants* 2021, 10, 1089. https://doi.org/10.3390/antiox10071089

Academic Editors: Masayuki Fujita and Mirza Hasanuzzaman

Received: 18 April 2021
Accepted: 28 June 2021
Published: 7 July 2021

Publisher's Note: MDPI stays neutral with regard to jurisdictional claims in published maps and institutional affiliations.

Abstract: This research was performed to explore the vital role of seed priming with a 0.01 µM concentration of brassinosteroids (EBL) to alleviate the adverse effects of Cr (100 µM) in two different rice cultivars. Seed priming with EBL significantly enhanced the germination attributes (germination percentage, germination energy, germination index, and vigor index, etc.), photosynthetic rate as well as plant growth (shoot and root length including the fresh and dry weight) under Cr toxicity as compared to the plants primed with water. Cr toxicity induced antioxidant enzyme activities (SOD, POD, CAT, and APX) and ROS level (MDA and H_2O_2 contents) in both rice cultivars; however, a larger increment was observed in YLY-689 (tolerant) than CY-927 (sensitive) cultivar. EBL application stimulatingly increased antioxidant enzyme activities to scavenge ROS production under Cr stress. The gene expression of SOD and POD in EBL-primed rice plants followed a similar increasing trend as observed in the case of enzymatic activities of SOD and POD compared to water-primed rice plants. Simultaneously, Cr uptake was observed to be significantly higher in the water-primed control compared to plants primed with EBL. Moreover, Cr uptake was significant in YLY-689 compared to CY-927. In ultra-structure studies, it was observed that EBL priming relieved the rice plants from sub-cellular damage. Conclusively, our research indicated that seed priming with EBL could be adopted as a promising strategy to enhance rice growth by copping the venomous effect of Cr.

Keywords: antioxidants; brassinosteroids; chromium; heavy metals; rice

1. Introduction

Soil contamination is turning into an alarming situation because of its negative influences on crop productivity. Abiotic stresses can cause >50% reduction in crop yield worldwide [1]. Besides, heavy metal toxicity is more dangerous towards crop production in this era. Chromium (Cr) is the 7th utmost copious heavy metal in the Earth's crust and it spreads in soil by different industries such as paints, leather, and fertilizer [2]. According to the US Environmental Protection Agency, Cr contamination is a major cause of human carcinoma [2]. Heavy metals, including Cr, become part of the soil in various ways, cause hazardous effects on plant growth and development, and become a major source of human health complications via entrance into the food chain [3]. Moreover, Cr pollution in the soil is extending day-by-day in numerous parts of the world. Almost 30, 896, and 142 metric tons of Cr are released into the air, water, and soil every year, respectively. Cr has various valance states, i.e., 0 to 6, but trivalent chromite (Cr^{+3}) and hexavalent chromate (Cr^{+6}) are prevalent in the environment. The Cr^{+6} is considered more unstable, extremely peripatetic, and hazardous, especially with a higher pH than Cr^{+3}. Latently, the issue of Cr contamination in agriculture is increasing day by day [4].

Rice is a vital food source worldwide due to a huge global population's dependence to fulfill their nutrition requirements with this crop. China is the leading rice producer, consumer, and importer country. Hence, Cr contamination is a more concerning topic in rice-growing areas of China. It is approximated that almost 10% of soil is contaminated with heavy metals in China [5]. Apart from rice, some other plants such as *Hibiscus esculentus* L. [6], *Pisum sativum* L. [7], *Triticum aestivum* L. [8], *Glycine max* L. [9], *Zea mays* L. [10], and *Lycopersicon esculentum* [11] are also being contaminated with Cr. Once, Cr is taken up by plants, nutrient transportation as well as metabolic activities are disturbed, and consequently, crop yield is compromised [5]. For example, uptake of Cr^{+6} causes accumulation and disturbs the uptake of micronutrients such as Mn, Zn, Cu, Fe, etc.

Moreover, Cr toxicity also induces cellular oxidative stress by triggering the accumulation of reactive oxygen species (ROS), which leads to the necrosis of plants. Several studies have reported that the accumulation of Cr in plants causes severe damage to crop production by increasing toxicity and inhibiting plant growth [12–14]. This growth retardation could be linked to the huge disturbance in cellular homeostasis and sub-cellular organelle damage [15,16]. To reduce the ROS activity, plants initiate an efficient mechanism of detoxification by organizing the autoxidation defense mechanism, which consists of superoxide dismutase (SOD), catalase (CAT), peroxidase (POD), and ascorbate peroxidase (APX) as an enzymatic compound as well as glutathione, ascorbic acid, proline, etc. as a non-enzymatic compound. However, plants can build multiple approaches to cope with heavy metal stresses such as Cr, with an increased antioxidant enzyme activity and decreased cellular ROS concentration [13,17,18].

Brassinosteroids (BRs) are considered as the sixth class of plant steroid hormones, which have pleiotropic properties in plants. They can protect plants under various biotic and abiotic stress conditions. Exogenous solicitation of BRs increases stress tolerance in plants [19]. Brassinolide, 28-homobrassinolide (28-HomoBL), and 24-epibrassinolide (EBL) are the three important BRs investigated in various aspects. 24-Epibrassinolide plays an important role in mitigating various plant stresses, including biotic and abiotic stresses. The 24-epibrassinolide is the most biologically active BR compound involved in developmental processes, cell division, elongation, gene expression, and vascular differentiation in plants [20]. It is known to improve plant growth by enhancing the chlorophyll contents, antioxidant enzymes, and up-regulate stress-response gene expression [21].

Latent investigations indicate that EBL plays an essential role in mitigating the Cr^{+6} toxicity by stimulating the antioxidant enzyme activities. The present study was further deepened to understand the Cr^{+6} induced sub-cellular damage in rice plants, followed by relieving this stress damage at the physiological and molecular level by applying EBL in a sensitive and tolerant rice cultivar. To the best of our knowledge, this is the first report investigating the comparative physiological and molecular responses of rice plants to the application of EBL in stress-sensitive and tolerant rice cultivars.

2. Materials and Methods

2.1. Seed Materials and Brassinosteroids (EBL) Preparation

The two rice cultivars, Chunyou 927 (CY-927, sensitive), Yliangyou 689 (YLY-689 tolerant), were used in this experiment which provided by the Zhejiang Nongke Seeds CO., LTD. Hangzhou, Zhejiang Province, China. 24-Epibrassinolide (EBL) was obtained from the Shanghai Aladdin Biochemical Technology Co., Ltd., China. The BR liquefication was made in an appropriate amount of ethanol and a standard solution (10^{-5} M) concentration was prepared by adding ddH$_2$O and 0.05% Tween-20 as a surfactant.

2.2. Seed Priming and Germination Test

For seed priming, rice seeds were firstly immersed in 5% (*w/v*) sodium hypochlorite (NaOCl) solution for 15 min sterilization and gently rinsed with distilled water to remove residual chloride. The sterilized seeds were then primed with 0.01 µM EBL at 30 °C in darkness for 24 h, and were dried at room temperature to the seed original moisture con-tent. The seeds primed with water (H2O) were considered as a control.

The seed germination test was carried out after priming and, for this purpose, 50 seeds per box (12 × 18 cm) were germinated with three replications. The germination experiment was carried out in a growth chamber at 25 °C with an alternation cycle of 8 h illumination and 16 h dark conditions for 14 days [22]. Incubated seeds were exposed to different concentrations (0, 50, 100, 200, 300, and 400 µM) of Cr. The selection of Cr concentration for further experimentation was based on this primary experiment, which concluded that 100 µM Cr caused significant damage to plant growth without killing the plants.

Total germinated seeds (seed radicle visibly protrudes through the seed coat and the radicle reaches to the length of the whole seed) were counted on the 5th day of germination and deliberated as germination energy (G.E). Moreover, the germination percentage (G.P) was recorded on the 14th day. Germination Index (G.I), Mean Germination Time (MGT), as well as Vigor Index (V.I), was measured by following formulas [22].

$$G.I = \Sigma(Gt/Tt) \tag{1}$$

$$MGT = \Sigma(Gt \times Tt)/\Sigma Gt \tag{2}$$

$$V.I = Germination\% \times [Shoot\ length + Root\ length] \tag{3}$$

Gt is the total calculated germinated seeds on day t, and Tt is the time conforming to Gt in days [22].

2.3. Plant Growth Conditions

The incubated seeds were treated with a 100 µM concentration of K$_2$Cr$_2$O$_7$ with a nutrient media solution. The composition of the nutrient solution was 0.5 µM potassium nitrate (KNO$_3$), 0.5 µM calcium nitrate (Ca(NO$_3$)$_2$), 0.5 µM magnesium sulfate (MgSO$_4$), 2.5 µM monopotassium phosphate (KH$_2$PO$_4$), 2.5 µM ammonium chloride (NH$_4$Cl), 100 µM ferric EDTA (Fe–K–EDTA), 30 µM boric acid (H$_3$BO$_3$), 5 µM manganese monosulfate (MnSO$_4$), 1 µM copper sulfate (CuSO$_4$), 1 µM zinc sulfate (ZnSO$_4$), and 1 µM ammonium heptamolybdate ((NH$_4$)6Mo$_7$O$_{24}$) per liter. The pH of the nutrient solution was adjusted to 5.0 with HCl and NaOH. The concentration was based on findings from a preliminary experiment with various Cr^{6+} (K$_2$Cr$_2$O$_7$) concentrations solutions (0, 50, 100, 200, 300, and 400 µM). The Cr concentrations 50 µM exhibited slight damage to plant growth. Although, a Cr concentration of 100 µM exhibited substantial damage to plant growth; however, concentrations greater than 100 µM were excessively toxic for the growth and killed the plants.

2.4. Experimental Design and Treatment Pattern

The experiment was conducted in hydroponic conditions. Two-week-old seedlings (primed with water and with 0.01 µM EBL) were treated with 100 µM Cr concentration and without Cr treatment considered as control (CK). The experiment was conducted through a completely randomized design (CRD) and the boxes were repositioned every day inside the growth chamber. Sampling was carried out at 21 days to perform numerous observations and measurements.

2.5. Plant Growth Investigation

The plants were harvested and immersed in a bucket containing ddH$_2$O to remove the remnants of the disinfectant and inspect the safety of the roots. The plants were uprooted, and the lengths of shoots and roots were calculated, including the measurement of fresh mass. To calculate their dry masses, roots and shoots were dried in an oven at 80 °C for 24 h.

2.6. Measurement of Chlorophyll Pigments

The photosynthetic pigments such as chlorophyll a, b, and total chlorophyll were determined by following the referenced method [23]. In short, fresh leaves (0.2 g) were standardized inside 3 mL ethanol (95%, *v/v*). The centrifugation of homogenate was done at $5000 \times g$ for 10 min and, subsequently, the supernatant was obtained. Next, 9 mL ethanol (95%, *v/v*) was supplemented with a 1 mL aliquot of the supernatant. Afterward, the measurements were made at the wavelengths of 665 and 649 nm through using an ultraviolet-visible spectrophotometer (UV-1900, Shimadzu, Japan). The following equations were utilized for calculating the pigment contents.

$$\text{Chlorophyll a (Chla)} = 13.95\,_{A_{665}} - 6.88\,A_{649} \tag{4}$$

$$\text{Chlorophyll b (Chlb)} = 24.96\,A_{649} - 7.32\,_{A_{665}} \tag{5}$$

$$\text{Total chlorophyll content} = \text{Chla} + \text{Chlb} \tag{6}$$

The quantities of pigments were expressed as milligrams per liter of plant material.

2.7. Measurement of Cr Contents

Elemental analysis to deliberate Cr contents was accomplished on dried roots and shoots. Dry plant samples (0.2 g) for each treatment were assimilated through 5 mL concentrated HNO_3 and $HCLO_4$ (5:1, *v/v*) in a furnace at 70 °C for nearly 5 h. The dilution of the processed samples was done with 2% HNO_3 to make an ultimate quantity of 10 mL for Cr content investigation. The filtrate was used to investigate Cr and microelements such as Mn, Cu, and Zn through the use of an atomic absorption spectrophotometer (iCAT-6000-6300, Thermo-Fisher Scientific, Waltham, MA, USA) [24].

2.8. Transmission Electron Microscopy Analysis

Shoot sections deprived of veins (8–10) per treatment after 14 days of treatment were obtained from indiscriminately collected seedlings and put into 2.5% (*v/v*) glutaraldehyde in 0.1 M PBS (sodium phosphate buffer with pH 7.4) as well as washed thrice with the alike PBS. Further, the leaves were postfixed in 1% OsO_4 (osmium (VIII) oxide) for 1 h. After that, it was washed three times in 0.1 M PBS, with 10 min gaps between each wash. After 15–20 min, the leaves were dehydrated within the classified categorization of ethanol (50%, 60%, 70%, 80%, 90%, 95%, and 100%, respectively) and splashed with absolute acetone 20 min. Then, the leaves were soaked in Spurr's resin for overnight. Consequently, ultrathin sections (80 nm) of samples were cut and placed in copper nets to observe through a transmission electron microscope (JEOLTEM-1230EX) at a hastening voltage of 60.0 kV.

2.9. Investigation of Malondialdehyde (MDA) Contents and H_2O_2 Production

The measurement of MDA concentration was carried out with 2-thiobarbituric acid (TBA). Approximately, 1.5 mL of extract was homogenized in 2.5 mL of 5% TBA diluted in 5% trichloroacetic acid (TCA). The homogenate sample was heated at 95 °C for 15 min, quickly chilled in ice, and centrifuged at $5000 \times g$ for 10 min. The absorbance of the supernatant was determined at two wavelengths, 532 and 600 nm, using a UV–vis spectrophotometer (Hitachi U-2910, Tokyo, Japan) [25]. MDA values were expressed as nmol mg^{-1} protein. After absorbance determination, calculations were made according to the following formula.

$$\text{MDA (nmol g}^{-1}\text{ FW)} = [(\text{OD532} - \text{OD600}) \times A \times V] \div (a \times E \times W)$$

$$A = \text{Total Reaction Solution} + \text{Enzyme extract}$$

$$V = \text{Total volume of buffer used for enzyme extract}$$

$$a = \text{Volume of the enzyme extract used}$$

$$W = \text{Fresh weight of the sample}$$

$$E = \text{Constant for MDA } (1.55 \times 10^{-1})$$

The reaction mixture without enzyme extract was used as a control and its measurements were subtracted from treatments for accuracy.

To measure hydrogen peroxide (H_2O_2), the plant tissues were homogenized in phosphate buffer followed by centrifugation at $6000 \times g$. The supernatant was mixed with 0.1% titanium sulfate containing 20% (v/v) H_2SO_4, followed by centrifugation. The intensity of yellow color was estimated colorimetrically at 410 nm using the above-mentioned unit of the UV–vis spectrophotometer [26]. H_2O_2 concentration was calculated by using a standard curve constructed with the known concentrations of H_2O_2. A control reaction mixture without the plant tissues was used and its measurements were subtracted from treatments for accuracy. The H_2O_2 concentration was calculated in terms of (μmol g^{-1} FW) at $25 \pm 2\,^{\circ}$C.

2.10. Measurement of Antioxidant Enzyme Activity

Vigorous samples (0.5 g) of both shoots and roots were standardized inside 8 mL of 50 mM potassium phosphate buffer (i.e., pH 7.0, comprising 1 mM EDTANa$_2$ in addition to 0.5% PVP, w/v) on ice. Accordingly, centrifugation of the homogenate was conducted for 20 min at $12,000 \times g$ at 4 $^{\circ}$C. The supernatant was used for the measurement of POD, SOD, APX, and CAT.

SOD activity was determined by measuring its aptitude to inhibit the photochemical reduction of nitroblue tetrazolium chloride (NBT). NBT reaction solution consisted of 50 mmol L^{-1} phosphate buffer (pH 7.8), 13 mmol L^{-1} methionine, 75 μmol L^{-1} NBT, 2 μmol L^{-1} riboflavin, and 0.1 mmol L^{-1} EDTA. The reaction began after adding 2 μmol L^{-1} riboflavin and placing the reaction tubes under 15 W fluorescent lamps for 15 min. The reaction mixture without enzyme extract was used as a control. One unit of SOD activity was elucidated as the quantity of enzyme required to cause 50% inhibition of the NBT reduction and photoreduction was observed at 560 nm [27].

The CAT activity was observed with an extermination constant of 39.4 mM^{-1} cm^{-1} at 240 nm absorbance due to the reduction of extinction H_2O_2. The 3 mL reaction mixture comprised of 2.8 mL phosphate buffer (25 mM, pH 7.0), 0.1 mL H_2O_2 (0.4%), and 0.1 mL enzyme extract. The reaction began by adding H_2O_2 [28] and enzyme activity was calculated in terms of 1 M of H_2O_2 g^{-1} FW min^{-1} at $25 \pm 2\,^{\circ}$C.

The POD activity was assessed using the reaction mixture consisting of 10 mM H_2O_2 in 50 mM Tris buffer (pH7.0) and 25 μL of enzyme extract in a total volume of 750 μL [29]. The total of 750 μL reaction solution contained 25 μL enzyme extract mixed with 1% (v/v) guaiacol and 0.4% (v/v) H_2O_2 assembled in 50 mM Tris buffer (pH 7.0). Changes in absorbance related to the oxidation of guaiacol ($\varepsilon = 25.5$ mM^{-1} cm^{-1}) were calculated at 470 nm. The enzyme activity was calculated in terms of 1 M of guaiacol oxidized g^{-1} FW min^{-1} at $25 \pm 2\,^{\circ}$C.

The APX activity was calculated depending on the decrease in absorbance at 290 nm as ascorbate was oxidized. The 3 mL reaction mixture contained 2.7 mL phosphate buffer (25 mM, pH 7.0), 0.1 mL ascorbate (7.5 mM), 0.1 mL H_2O_2 (0.4%), and 0.1 mL of enzyme extract. The reaction began by adding H_2O_2 and the enzyme activity was calculated as μmol min^{-1} mg^{-1} protein at $25 \pm 2\,^{\circ}$C [30].

2.11. RNA Extraction and Gene Expression Analysis

Gene expression of antioxidant enzymes was measured through quantitative real-time PCR (qRT-PCR). Frozen shoot and root samples were ground comprehensively in liquid nitrogen using a mortar and pestle. Total RNA was extracted from both shoots and roots by Trizol mixture by following an already described method [31]. The RNA concentration and its purity were determined through NanoDrop 2000/2000 c (Thermo-Fisher Scientific, Waltham, MA, USA). For the synthesis of cDNA, the PrimeScript™ RT reagent kit was utilized. The primers used to quantify the expression of *SOD-Cu-Zn*, *SOD-Fe$_2$*, *APX02*, *APX08*, *CATa*, *CATb*, *POX1*, and *POX2* genes are listed in Supplementary Table S1. The 20-µL reaction mixture was made with 2XSYBR Green Master Mix reagent (10 µL volume) (Applied Biosystems, Foster City, CA, USA), cDNA samples (6 µL volume), and 200 nM gene-specific primers. The thermocycler was set as follows: 95 °C for 3 min; 40 cycles of 95 °C for 30 s, 60 °C for 30 s, and 72 °C for 1 min. The relative change in the expression of genes was determined according to Livak et al. [32]. The *OsActin* was used for internal calibration as a control gene to normalize the other genes.

2.12. Statistical Analysis

Different treatments for one-way analysis of variance through the least significant differences (LSD) were pragmatic as a posthoc test at 95% assurance interlude amongst frequent data set by using SPSS v16.0 (SPSS, Inc., Chicago, IL, USA). The analysis of variance [33] was conducted via Duncan's multiple range test amongst various treatment means to accomplish the significant difference at $p < 0.05$ and 0.01 level among mean values. The principal component analysis (PCA) and agglomerative hierarchical clustering (AHC) were accomplished to examine the classification and grouping of two different cultivars of rice for their vulnerability to Cr by using Minitab software version 18.1.

3. Results

3.1. Determination of the Significant Effect of Cr (VI) on Seed Vigor and Plant Development

Based on the preliminary experiment results, Cr toxicity caused clear phenotypic changes in both the cultivars at various concentrations (0, 50, 100, 200, 300, and 400 µM). Cr imposed higher toxicity at 200 µM and above concentrations, and hence, a 100 µM Cr concentration was selected for further experimentation based on phenotypical alterations. Brassinosteroids (0.01 µM) showed a positive effect on stress mitigation under Cr exposure on all concentrations (Figure 1).

Figure 1. Physiological effect of Cr toxicity on two different rice cultivars. (**a**) Physiological effect of Cr on the YLY-689 cultivar and stress alleviation effect of 0.01 µM EBL under 100 µM Cr concentration. (**b**) Physiological effect of Cr on cultivar CY-927 and stress alleviation effect of 0.01 µM EBL under 100 µM Cr concentration.

Germination energy and the germination percentage were significantly reduced by the exposure of Cr with a concentration of 100 μM compared to control in both cultivars (Table 1). Seed germination was observed to be better in plants treated with 0.01 μM brassinosteroids (EBL) compared to seeds primed with water (H_2O). A clear reduction in germination energy and germination percentage was observed in cultivar CY-927 under Cr stress. Germination index and vigor index were decreased in seeds primed with water in both cultivars but the mean germination time was decreased in seeds primed with EBL under Cr stress (Table 1).

Table 1. Effect of Cr toxicity on the seed germination parameter in seeds primed with EBL compared to seeds primed with water in two different rice cultivars.

Varieties Name	Treatment	G.E	G.P	G.I	MGT	V.I
CY-927	H_2O	90.00 ± 2.00 [b]	94.67 ± 1.15 [b]	20.13 ± 1.17 [b]	2.91 ± 0.17 [b]	2.04 ± 0.11 [b]
	EBL	96.00 ± 2.00 [a]	99.33 ± 1.15 [a]	28.01 ± 1.47 [a]	2.16 ± 0.11 [c]	2.87 ± 0.13 [a]
	H_2O+Cr	38.67 ± 3.06 [d]	46.67 ± 3.06 [d]	7.15 ± 0.74 [d]	4.06 ± 0.07 [a]	0.31 ± 0.07 [d]
	EBL+Cr	72.67 ± 3.06 [c]	82.67 ± 3.06 [c]	12.90 ± 0.35 [c]	3.21 ± 0.14 [b]	0.77 ± 0.03 [c]
YLY-689	H_2O	93.33 ± 2.31 [a]	99.33 ± 1.15 [a]	27.29 ± 0.76 [b]	2.44 ± 0.13 [c]	2.63 ± 0.09 [b]
	EBL	98.00 ± 2.00 [a]	100.00 ± 0.00 [a]	32.98 ± 0.59 [a]	2.07 ± 0.08 [d]	3.24 ± 0.05 [a]
	H_2O+Cr	54.00 ± 2.00 [c]	66.00 ± 2.00 [c]	12.26 ± 0.88 [d]	3.66 ± 0.19 [a]	0.90 ± 0.07 [d]
	EBL+Cr	86.67 ± 1.15 [b]	89.33 ± 1.15 [b]	21.50 ± 0.30 [c]	2.74 ± 0.14 [b]	2.00 ± 0.03 [c]

Each treatment value represents the mean of three replicates ± standard deviation. Same letters are representing no significant differentiation at 95% probability level ($p < 0.05$). The data presented here represent the selected Cr concentration (100 μM). Here, germination energy (G.E); germination percentage (G.P); germination index (G.I); mean germination time (MGT); vigor index (V.I); brassinosteroids (EBL).

Shoot length, root length, fresh weight, and dry weight were also affected by Cr toxicity in both cultivars. Shoot length and root length were significantly decreased in both cultivars under stress conditions compared to control, but more lessening was noted in cultivar CY-927. Fresh weight and dry weight were also lower under Cr stress in plants primed with water. Priming with EBL showed mitigation behavior toward Cr toxicity in both cultivars (Table 2).

Table 2. Improvement in shoot length, root length, and fresh and dry weight by seed priming with EBL compared to the control under Cr toxicity.

Varieties Name	Treatment	S.L	R.L	F/W	D/W
CY-927	H_2O	15.51 ± 0.02 [b]	13.04 ± 0.02 [b]	0.90 ± 0.01 [b]	0.10 ± 0.001 [b]
	EBL	17.19 ± 0.04 [a]	15.03 ± 0.16 [a]	0.96 ± 0.01 [a]	0.10 ± 0.001 [a]
	H_2O+Cr	8.30 ± 0.03 [d]	7.56 ± 0.03 [d]	0.51 ± 0.01 [d]	0.05 ± 0.001 [d]
	EBL+Cr	11.59 ± 0.02 [c]	10.96 ± 0.02 [c]	0.68 ± 0.01 [c]	0.06 ± 0.001 [c]
YLY-689	H_2O	15.16 ± 0.08 [b]	13.31 ± 0.02 [b]	0.91 ± 0.01 [b]	0.10 ± 0.003 [a]
	EBL	18.39 ± 0.06 [a]	15.28 ± 0.04 [a]	0.96 ± 0.01 [a]	0.10 ± 0.001 [a]
	H_2O+Cr	9.34 ± 0.04 [d]	8.43 ± 0.02 [d]	0.59 ± 0.01 [d]	0.07 ± 0.002 [c]
	EBL+Cr	12.22 ± 0.04 [c]	11.22 ± 0.03 [c]	0.68 ± 0.01 [c]	0.09 ± 0.001 [b]

Each treatment value represents the mean of three replicates ± standard deviation. Same letters represent no significant difference at the 95% probability level ($p < 0.05$). Here, shoot length (S.L); root length (R.L); fresh weight (F/W); dry weight (D/W); brassinosteroids (EBL).

3.2. Seed Priming with EBL Significantly Enhanced Photosynthetic Pigments under Cr Stress

Results represented that Cr toxicity caused a clear reduction in chlorophyll a (Chla), chlorophyll b (Chlb), and total chlorophyll contents compared to the control (Figure 2). In cultivar CY-927, decreased Chla, Chlb, and total chlorophyll content was noted to be more pronounced than YLY-689 under Cr toxicity. Seed priming with EBL exhibited a significant increase in Chla, Chlb, and total chlorophyll at 100 μM Cr stress compared to seeds primed with water (Figure 2).

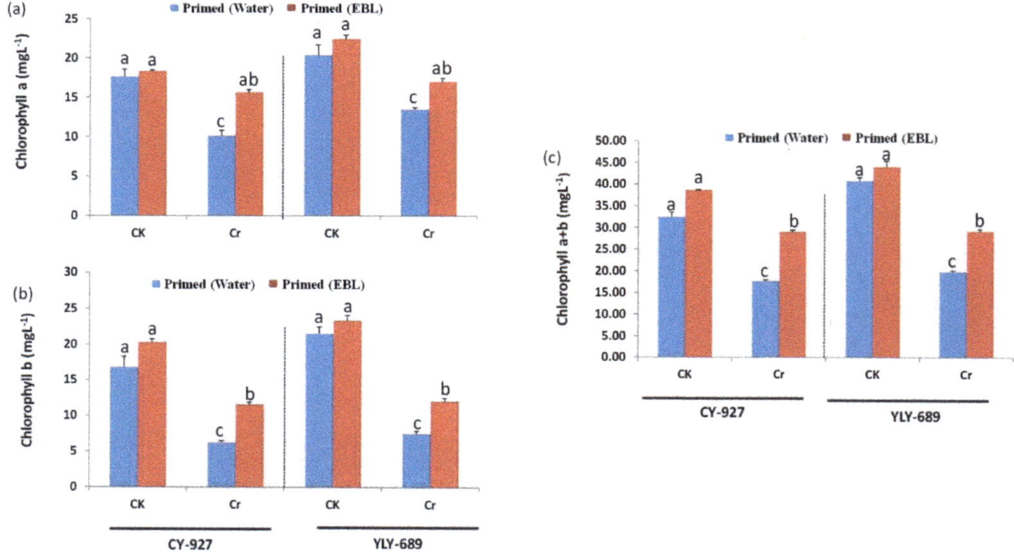

Figure 2. Effect of brassinosteroids on chlorophyll contents in two different rice cultivars under Cr toxicity. (**a**) Chlorophyll a contents in both rice cultivars; (**b**) Chlorophyll b contents in both rice cultivars; (**c**) Total Chlorophyll contents (Chlorophyll a+b) in both rice cultivars. The values presented are means ± SDs (*n* = 3). Different letters (a–c) above bars show a significant difference at *p* < 0.05 among treatments.

3.3. Accumulation of Cr Contents Was Reduced Significantly by EBL Seed Priming

Results demonstrated that Cr accumulation was more pronounced inside roots compared to shoots and uptake was higher in cultivar YLY-689 than CY-927 (Supplementary Tables S2 and S3). Cr uptake caused macronutrient imbalance as well. Under Cr toxicity, Mn, Cu, and Zn uptake were decreased in both cultivars; however, this decrease was more obvious in cultivar CY-927 than YLY-689 in both roots (Supplementary Table S3) as well as shoots (Supplementary Table S2).

Results demonstrated that the effect of brassinosteroids diminished the uptake and accumulation of Cr and maintained the nutrient balance in both cultivars of rice. Mn, Zn, and Cu contents were increased by EBL treatment under Cr stress in shoots and the same trend of nutrient balance was observed in roots of both cultivars (Supplementary Tables S1 and S2).

3.4. Significant Reduction of MDA Contents and H_2O_2 Production by Seed Priming with EBL

Lipid peroxidation was estimated in both cultivars of rice plants. Cr stress enhanced MDA contents in both roots and shoots compared to the control. MDA content was higher in YLY-689, estimated as 72.50% and 38.41% in shoots and roots, respectively, compared to CY-927 (64% and 34.60% in shoots and roots, respectively) (Figure 3). MDA contents remained pronounced in shoots of both plants. The application of EBL represented the significant decline of MDA contents in shoots and roots of both varieties (e.g., 36.90 and 15.40% for YLY-689 and 38.40 and 20% for CY-927, respectively) compared to the plants primed with water.

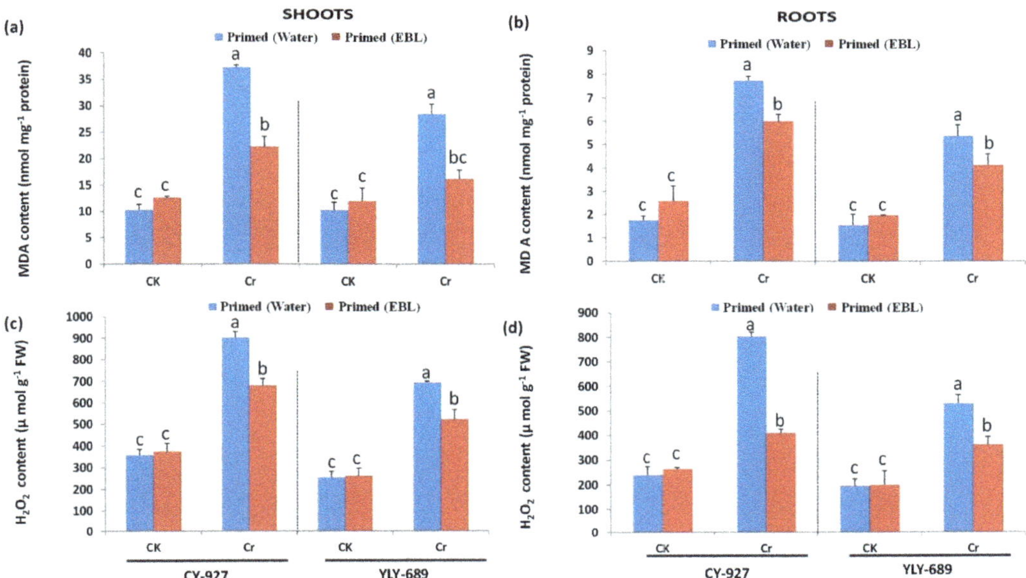

Figure 3. Effect of Cr toxicity on MDA contents as well as H_2O_2 production ard alleviation behavior of EBL toward MDA contents and H_2O_2 production under Cr stress in both varieties of rice. (**a**) MDA contents in shoots of both rice cultivars; (**b**) MDA contents in roots of both rice cultivars; (**c**) H_2O_2 contents in shoots of both rice cultivars; (**d**) H_2O_2 contents in roots of both rice cultivars. The values presented are means $\pm$ SDs ($n = 3$). Different letters (a–c) above bars show a significant difference at $p < 0.05$ among treatments.

Recent outcomes revealed that H_2O_2 production was enhanced under 100 μM Cr toxicity in both cultivars as compared to the control (Figure 3). More production was recorded in cultivar YLY-689 with 58.1% and 65.2% in both shoots and roots, respectively, CY-927 with 57.3 and 62.2% individually in both shoots and roots. Seeds primed with brassinosteroids showed lower MDA contents in shoots and roots of both varieties YLY-689 and CY-927 at 45.6, 35.3, 44.8, and 46.6%, respectively, compared to control.

3.5. Regulation of Antioxidant Activities via Seed Priming with EBL

Under Cr stress alone, antioxidant activities were modulated differently. The results validated that 100 μM Cr stress significantly enhanced the antioxidants (SOD, CAT, POD, and APX) compared to the control in both cultivars. A larger increment was observed in shoots and the same trend was noticed in both cultivars. The change was more pronounced in YLY-689 as compared to the CY-927 cultivar (Figure 4). Seed priming with 0.01 μM EBL showed very interesting behavior towards antioxidant enzyme activities and enhanced more SOD, POD, CAT, and APX under 100 μM Cr stress compared to plants primed with water. The results revealed that SOD activity was more prominent than other antioxidants in shoots of YLY-689 cultivar compared to the non-stressed plants. SOD, POD, CAT, and APX were enhanced by Cr treatment 75.7%, 64.80%, 57%, and 60.7%, respectively in shoots and 38.5%, 45.8%, 43.8%, and 48.4% in roots, respectively in YLY-689. Further, 45.3%, 56.4%, 41.3%, and 58.4% in shoots correspondingly alongside 37.8%, 20.6%, 33.1%, and 38.5% individually was enhanced in roots of CY-927 cultivars and it further enhanced by EBL together in roots and shoots of both cultivars.

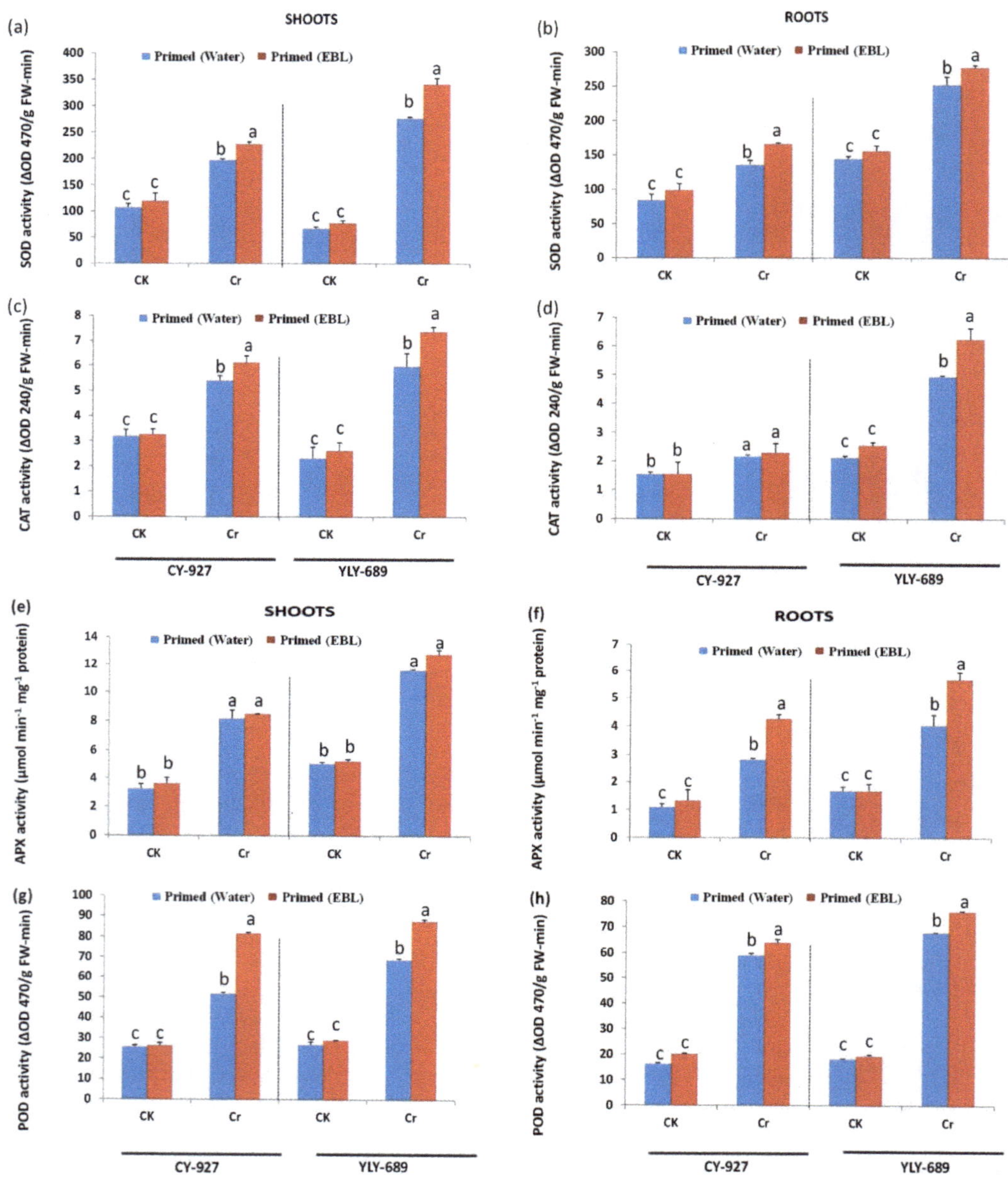

Figure 4. Upregulation of antioxidant enzyme activities by seed priming with EBL under Cr stress in both cultivars. (**a**) and (**b**) superoxide dismutase (SOD) in shoots and roots of both rice cultivars, respectively; (**c,d**) catalase in shoots and roots of both rice cultivars, respectively; (**e,f**) ascorbate peroxidase (APX) in shoots and roots of both rice cultivars, respectively; (**g,h**) peroxidase (POD) in shoots and roots of both rice cultivars, respectively. The values presented are means ± SDs ($n = 3$). Different letters (a–c) above bars show a significant difference at $p < 0.05$ among treatments.

3.6. Determination of Ultrastructure Analysis

The current study demonstrated the ultrastructural changing inside mesophyll cells of rice cultivars primed with water and with 0.01 μM EBL growing under control conditions as well as under the exposure of 100 μM Cr stress. The TEM micrographs of leaf mesophyll cells of CY-927 and YLY-689 primed with water and with 0.01 µM EBL represented normal structure with the hygienic and thin cell wall, normal organelles, as well as healthy chloroplast and granule thylakoids (Figure 5). Although, mesophyll cells of CY-927 primed with water under 100 µM Cr stress showed a damaged structure of the nucleolus as well as a double-layered nuclear membrane expansion (Figure 5). Comparatively, mesophyll cells of CY-927 primed with 0.01 µM EBL showed slight damage in nucleolus development; however, the chloroplast was normal compared to the mesophyll cells of CY-927 primed with water (Figure 5). A nuclear membrane was also observed normally in mesophyll cells of CY-927 primed with 0.01 µM EBL.

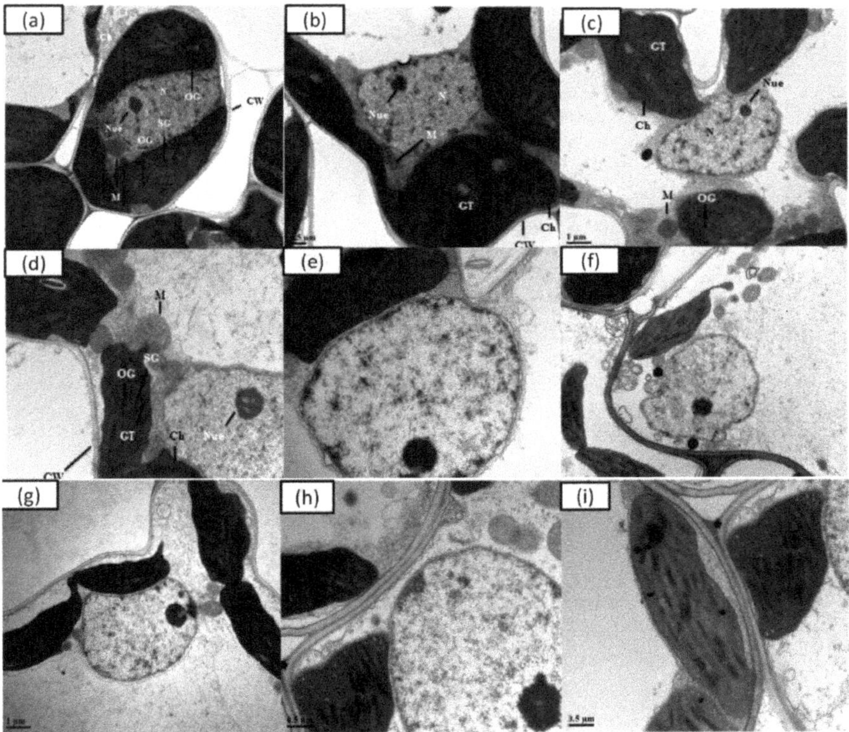

Figure 5. Electron micrographs of leaf mesophyll cells of two various rice cultivars (CY-927, YLY-689) primed with water as well as primed with 0.01 µM EBL grow under control and exposure of 100 µM Cr stress. (**a**) Leaf mesophyll cell of CY-927 (primed with water) at the control level. (**b**) Leaf mesophyll cell of CY-927 (primed with 0.01 µM EBL) at the control level. (**c**) Leaf mesophyll cell of YLY-689 (primed with water) at the control level. (**d**) Leaf mesophyll cell of YLY-689 (primed with 0.01 µM EBL) at the control level. (**e**) Leaf mesophyll cell of CY-927 (primed with water) underexposure of 100 µM Cr toxicity. (**f**) Leaf mesophyll cell of CY-927 (primed with 0.01 µM EBL) under 100 µM Cr toxicity. (**g**) Leaf mesophyll cell of YLY-689 (primed with water) under the disclosure of 100 µM Cr stress. (**h**,**i**) Leaf mesophyll cell of YLY-689 (primed with 0.01 µM EBL) under 100 µM Cr toxicity, N (nucleus); CW (cell wall); Ch (chloroplast); GT (granule thylakoids); M (mitochondria); Nue (nucleolus); NM (nuclear membrane).

The TEM micrographs of leaf mesophyll cells of YLY-689 primed with water as well as primed with 0.01 µM EBL in control plants showed normal growth with a healthy structure of organelles (Figure 5). Whereas, mesophyll cells of YLY-689 primed with water under 100 µM Cr stress showed damage in chloroplast development and granule thylakoids, which were thinner compared to the control. In addition, the nucleolus was also observed to be abnormal (developed double nucleolus) (Figure 5). On the other hand, the mesophyll cells of YLY-689 primed with 0.01 µM EBL demonstrated better chloroplast development with little damage to thylakoid development. Furthermore, it comprised normal cell arrangements as well as the normal development of the nucleolus besides the nuclear membrane (Figure 5).

3.7. Determination of Gene Expression Analysis

The expression of *SOD-Cu-Zn* and *SOD-Fe2* was significantly higher in seeds primed with water under Cr toxicity compared with control. Moreover, expression was increased in seeds primed with 0.01 μM EBL compared to seeds primed with water in both cultivars (Supplementary Figure S1). The transcription level was higher in the YLY-689 cultivar compared to CY-927 ($p < 0.01$). Gene expression was noted to be higher in shoots of both cultivars than roots.

Moreover, the transcription level of genes *APX02* and *APX08* were observed to be high in both shoots and roots of the CY-927 cultivar under Cr stress. Nevertheless, significant upregulation was observed in shoots of CY-927 compared to roots (Supplementary Figure S2). Furthermore, an upregulation in the expression of both genes was noted in seedlings of plants primed with 0.01 μM EBL compared to seeds primed with water under Cr toxicity. Interestingly, the behavior of cultivar YLY-689 was quite different from CY-927 in terms of gene expression. A significant downregulation of genes *APX02* and *APX08* was calculated in roots and shoots under Cr stress. However, expression was higher in plants primed with 0.01 μM EBL compared to plants primed with water but this increment was non-significant (Supplementary Figure S2).

The same trend in transcription level of both *CATa, CATb* genes was observed. A significant increase was observed in gene expression of *CATa* and *CATb* in CY-927 for both shoots and roots, but shoots represented a higher transcription level compared to roots (Supplementary Figure S3). Expression was more prominent in seeds primed with 0.01 μM EBL compared to seeds primed with water. Downregulation was observed in both shoots and roots of cultivar YLY-689 under Cr stress. Seeds primed with 0.01 μM EBL showed a significantly enhanced transcription level compared to seeds primed with water under Cr stress conditions (Supplementary Figure S3).

A significant upregulation of genes *POX1* and *POX2* were measured in both roots and shoots under 100 μM Cr concentration in both rice cultivars. Significant upregulation of *POX2* was noted in shoots of the CY-927 cultivar under Cr stress. In both cultivars, data represented that the transcription level of *POX1* and *POX2* was significantly enhanced in plants primed with 0.01 μM EBL, and this trend was found to be similar to the results of antioxidant enzymatic activity under Cr stress. Furthermore, upregulation was more obvious in shoots of both cultivars than roots (Supplementary Figure S4).

3.8. Determination of Interaction among Growth and Physiological Parameters through Principal Component Analysis, Clustering, and Correlation Analysis

Principle component analysis presented the interaction of measured growth and physiological parameters under the influence of different treatments in both varieties (Figure 6). In the PCA analysis, it was found that there was a close interrelation among MDA, MGT, and H_2O_2. Although, MGT, H_2O_2, and MDA showed a significantly negative correlation with V.I, F/W, D/W, S.L, R.L, G.E, and G.P demonstrated a negative correlation with SOD, POD, CAT, and APX as well (Figure 6). It demonstrated the maximum contribution of F1 (87.6%) followed by F2 (10.80%) with a total contribution of 97.80% in CY927 and for YLY689 the maximum contribution of F1 (88.8%) followed by F2 (9.2%), with a total contribution of 98.0% was noticed (Figure 6). ACH outcomes also confirmed both varieties' responses under distinct treatments (Figure 6). The dendrogram represented three groups that characterized the close correlation between both cultivars (CY927 and YLY689) primed with EBL as well as primed with water under Cr stress. Cultivars primed with EBL under Cr stress showed a close correlation with both controls (primed with water and EBL) compared to plants primed with water under Cr toxicity (Figure 6). The dendrogram showed the same pattern of treatments as demonstrated by PCA (Figure 6).

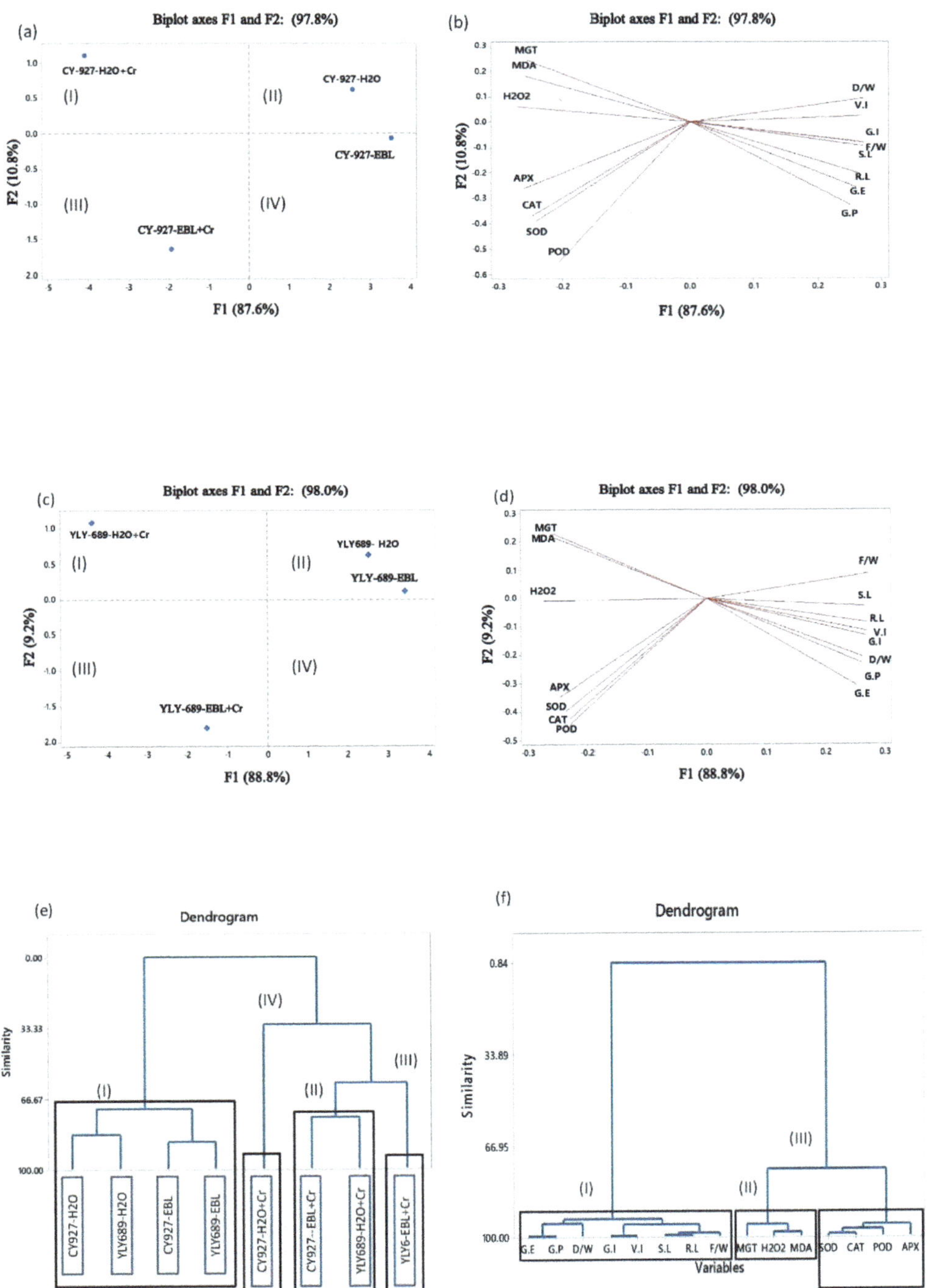

Figure 6. Bioplot of principle components 1 and 2 of the PCA extracted from the results obtained from physiological data of two various rice cultivars (CY927, YLY689) under different treatments such as control primed with water (CY927-H$_2$O,

YLY689-H$_2$O), control primed with EBL (CY927-EBL, YLY689-EBL), seed primed with EBL, and treatment under Cr stress (CY927-EBL + Cr, YLY689-EBL + Cr), seed primed with H$_2$O under Cr stress (CY92-Cr + H$_2$O, YLY689-Cr + H$_2$O). A sharp angle represented a positive correlation, an obtuse angle showed a negative correlation, and a right angle demonstrated a correlation between parameters. (**a**) Representation of the correlation between different treatments in rice cultivar CY927. (**b**) Physiological parameters of rice variety CY927 are represented through Pearson's correlation coefficients under different treatments. (**c**) Illustration of correlation between various treatments in rice cultivar YLY689. (**d**) Physiological parameters of rice variety YLY689 representation via Pearson's correlation coefficients under various treatments. Distance between each line represented the strength of the correlation. (I) contains MDA, MGT, and H$_2$O$_2$; (II) shows G.I, F/W, D/W, G.E, G.P, and V.I; (III) illustrates POD, CAT, APX, and SOD; (IV) represents R.L and S.L. (**e**) Dendrogram of two different rice cultivars under various treatments obtained through agglomerative hierarchical clustering using Ward's method based on physiological traits. (**f**) The dendrogram demonstrated a correlation between various physiological parameters of two different rice cultivars under numerous treatments obtained through agglomerative hierarchical clustering using Ward's method.

4. Discussion

4.1. EBL Improve Physio-Biochemical Effects Caused by Cr Toxicity in Rice Plants

Recently, increasing industrialization has become a serious threat to soil contamination which is now the main source of plant growth inhibition by disturbing various mechanisms of plants at the physiological and molecular level [34]. This study examined that chromium caused plant growth inhibition and induced negative physio-biochemical processes in two rice varieties. Reportedly, plants develop various mechanisms to overcome stress by inducing antioxidant activities, biochemical mechanisms, phytochelatins, and various hormones [35].

In the current research, the role of brassinosteroids (EBL) has been elaborated to overcome Cr toxicity. Cr has reduced the plant biomass, altered the structure of roots, and affected seed germination energy, percentage, vigor index, and mean germination time. Seed germination was affected because of heavy metal toxicity through the access of rice plants' embryonic tissues and due to the structure of seed coats. When a seed radicle comes into contact with Cr; it lowers the germination rate by lowering the α and β amylase activities, reducing the sugar supply to the seeds, and constrain the seed germination rate [36]. Root and shoot length also reduce under Cr stress and it may be caused by nutrient uptake disturbance inside rice plants and may decrease the cell division and elongation.

The root is the primary point of contact with Cr; however, the toxicity and accumulation of Cr hinder root elongation in plants. Cr uptake might be caused by the imbalance of nutrients such as Zn, Mn, and Cu, which is necessary to regulate various plant growth mechanisms. The current study revealed that the application of EBL caused a positive effect on plant growth and development under Cr stress by reducing the uptake of Cr and by enhancing nutrient contents (Zn, Cu, and Mn) significantly in both rice cultivars. This has been translated to improve shoot length, root length, and fresh and dry weight of plants under Cr stress. It also maintained the germination rate in rice cultivars by enhancing cell division and elongation processes [37]. Similar studies have been focused on various plant species such as *Raphanus sativus* [38], *Hordeum vulgaris* [39], and *Lycopersicon esculentum* [40].

4.2. EBL Prevented Degradation of Chlorophyll Pigments under Cr Stress

This research showed that Cr caused a significant reduction in Chla, Chlb, and total chlorophyll contents (Figure 2) in both rice varieties. Degradation of chlorophyll is the primary sign of Cr toxicity, which is the leading indicator of phytotoxicity in plants [41]. The reduction of photosynthetic pigments might result from ROS activity enhancement under Cr toxicity [42]. In this study, seed priming with EBL showed a significant increment in photosynthetic pigments under Cr toxicity as compared to plants primed with water. Based on previous studies, CO$_2$ acclimatization was augmented by EBL, and it enhanced some specific genes' expression to increase the antioxidant enzyme activities and to scavenge ROS activity [36].

4.3. Seed Priming with EBL Reduced MDA Contents as Well as H_2O_2 Production

The results revealed that there is a higher production of H_2O_2 and MDA contents under Cr stress compared to the control plants (Figure 3). ROS activity was relatively higher in shoots compared to roots. Previous studies also indicated that *Arabidopsis Thaliana* exposed to heavy metals and rice caused an increase in ROS activity [43,44]. The higher ROS concentration causes oxidative damage in biomolecules, such as DNA, RNA, proteins, and pigments in addition to lipid peroxidation [45]. In this study, seeds primed with EBL reduced ROS activity under Cr stress by protecting membrane damage as compared to seeds primed with water. Moreover, EBL also protected mung bean plants from membrane damage by reducing ROS activity. EBL also proved this protective behavior towards the green bell pepper under chilling stress by reducing MDA contents [46]. Besides, EBL foliar spray in tomato plants decreased oxidative membrane damage and lipid peroxidation [47]. EBL enhanced the resistance capability of various plants by reducing ROS activity because of enhancing antioxidant activity to scavenge ROS accumulation in plants such as the mung bean, soybean, pea epicotyls, bean, sunflower, and cucumber hypocotyls, *Arabidopsis* peduncles, and *Hordeum vulgare* [48–50]. Similarly, a recent study demonstrated the protective behavior of EBL towards Zn-induced oxidative damage in *Solanum nigrum* L. plants [51]. In another study, Jan et al. [12] reported that EBL significantly reduced the Cr toxicity in tomato plants and also improved the growth, physio-biochemical attributes, and antioxidant activity. It also has been proven that EBL mitigated the lead toxicity in *Brassica juncea* L. by scavenging ROS activity [52].

4.4. EBL Enhanced Antioxidant Activity to Mitigate Cr Toxicity

An increase in MDA contents and H_2O_2 production causes oxidative and lipid peroxidation damage inside plants and disturbs the metabolic processes, function, and structure of membranes [53]. As a result, physiological processes besides growth inhibition also occur in rice seedlings. To scavenge the higher production of MDA and H_2O_2, the antioxidant defense mechanism (SOD, POD, CAT, and APX) is stimulated in plants [54,55]. In a recent study, Khan et al. [56] revealed that EBL significantly increased plant growth and triggered the antioxidant defense system in wheat plants against drought stress. Similarly, Rattan et al. [57] reported that EBL alleviated salt stress in maize seedlings by regulating the antioxidant enzyme activities. Another study revealed that EBL detoxified the combined toxicity of salinity and potassium deficiency in barley plants by modulating the antioxidant defense mechanism [58], and enhanced manganese tolerance in *Arabidopsis thaliana* L. by regulating the antioxidant defense mechanism [59].

4.5. Rice Cultivar YLY-689 Was More Resistant to Cr Stress Than CY-927

The biomass (fresh and dry weight, and shoot and root length), germination energy, germination percentage, germination index, and vigor index were reduced while mean germination time was increased in CY-927 YLY-689 cultivar under Cr toxicity (Tables 1 and 2). These results further indicated the EBL role in mitigating Cr toxicity in rice plants in both sensitive and tolerant varieties. Cr caused a decrease in the growth of plant roots and shoots in addition to the height [55,60,61]. Likewise, Cr instigated a reduction in biomass and some other seed germination parameters (Tables 1 and 2). Nonetheless, YLY-689 showed more resistance toward Cr stress compared to CY-927.

4.6. Ultrastructural Changing Induced by Cr in Rice Plants

The influence of 100 µM Cr toxicity on the ultrastructure of leaf mesophyll cells was observed in both cultivars. Outcomes demonstrated that both rice cultivars (CY-927 and YLY-689) primed with water and with 0.01 µM EBL had well-developed chloroplast, granule thylakoids, nucleolus, nuclear membrane, cell wall, mitochondria as well as thylakoid membranes (Figure 5). The plants under exposure of Cr (100 µM) showed ruptured chloroplasts with damaged granule thylakoids (Figure 5) besides having an abnormal nucleolus with nuclear membrane aberrations (Figure 5). Similar outcomes were observed in rice and *Lolium perenne* L. [62,63]. The reduction in chlorophyll contents occurred because of the inhibition of specific enzyme biosynthesis involved in chlorophyll content productions. Cr accrued inside cells and caused damage inside chloroplasts by producing abnormalities in granule thylakoids through swollen chloroplasts. It directly affected the photosynthesis rate under a stressed condition. The thylakoid membrane ruptured because of Cr toxicity in *Brassica napus* L. and inevitably changed chloroplast structure [64].

There is a close relationship between the structure and function of rice seedlings. However, the Cr impact on chloroplasts has an essential role in physiological alternations in plants. Chloroplast aberrations also have a role in the reduction of chlorophyll contents. The decrease in chlorophyll contents and photosynthetic pigments mainly caused a reduction in photosynthetic rate [65].

In the current study, both cultivars primed with 0.01 μM EBL showed resistance towards Cr toxicity. In comparison with the control, less damage was observed in chloroplasts and granule thylakoids compared to the control under Cr toxicity. Nucleolus and nuclear membrane aberrations were also lower in plants primed with EBL than plants primed with water and exposed to Cr toxicity. Hence, it was found that priming with EBL has positively affected both cultivars and played its role in maintaining chlorophyll contents in rice plants.

4.7. Gene Expression Level

The expression level of genes was investigated at the mRNA level in both shoots and roots of the rice seedlings of both cultivars. The estimation of antioxidant gene expression together with antioxidant enzyme activities presented a better evaluation of these measurements after priming with EBL. It was investigated that Cr toxicity is mitigated in rice plants because of the upregulation of SOD, APX, CAT, and POD activities [65,66]. In our study, SOD and POD gene expression was observed to be upregulated in both cultivars. Whereas upregulation of the transcription level of CAT and APX genes was observed in cultivar CY-927. The same trend of gene expression level was observed in various plants such as *Raphanus Sativa* [38], *Cicer arietinum* [67], and Tomato [12]. Additionally, SOD, POD, CAT, and APX activities increased in plants treated with EBL under Cr toxicity in both cultivars, and the upregulation of their genes was noted with the same pattern. The increase in antioxidant activities and transcription level of antioxidant activity-related genes was due to the accretion of salicylic acid contents through BR signaling pathways. Furthermore, it mitigates the Cr toxicity and enhances the tolerance mechanism inside plants against heavy metal toxicity. Interestingly, in our findings, APX and CAT transcriptional levels in both roots and shoots of the YLY-689 cultivar were quite the opposite of cultivar CY-927. It also varied from the results of antioxidant enzyme activities. Although, gene expression of SOD and POD showed the same trend with antioxidant enzyme activities in both rice cultivars. The distinction between APX and CAT gene expression with its enzymatic activities was also investigated in *Brassica napus* [64] and cotton [55] plants as well under Cr stress. The discrepancy between antioxidant enzyme activities and gene expressions under Cr toxicity demonstrated that it might have some vital role inside the plant defense system against Cr stress mitigation.

4.8. Clustering and Correlation Analysis

We used principal component analysis (PCA) to recognize and categorize the large dataset in terms of growth and physiological parameters into a small number of dynamically interrelated variables in our study [68,69]. It was found that EBL priming of rice seedlings in both cultivars was distinctly separate in PCA compared to Cr stressed plants. This placement of EBL-primed plants in separate coordinates compared with Cr-stressed plants was more prominent in the case of variety CY-927 (Figure 6) than variety YLY-689 (Figure 6). This disclosed the interaction between different genotypes of rice based on distinct treatments (Figure 6). On the basis of physiological characteristics, different treatments were exploited to discriminate the sensitive and tolerant genotype besides representing the correlation among various traits by using the amalgamation of both PCA and ACH (Figure 4). V.I, F/W, D/W, S.L, R.L, G.E, and G.P showed to be a group with a significantly positive correlation between each other but instantaneously negative relation with MGT, H_2O_2, and MDA.

5. Conclusions

Chromium toxicity caused adverse effects on rice plants' physiological, biochemical, and molecular mechanisms, which negatively affected seed germination parameters and further reduced the plants' growth and development. Seed priming with brassinosteroids (EBL) improved seed germination attributes and capped the worst effect of Cr on both cultivars of rice plants by triggering its physio-biochemical processes such as through maintaining chlorophyll contents, increasing mineral uptakes via reducing Cr uptake and accumulation, by enhancing antioxidant enzyme activities, as well as lessening ROS production under Cr stress. Furthermore, EBL improved the ultrastructure of both cultivars of rice under Cr toxicity. Our studies ensured the competency of EBL to cope and detoxify the nastiest effects of Cr in rice plants. Moreover, CY-927 was more affected by Cr stress rather than the YLY-689 cultivar. Thus, the YLY-689 cultivar is more tolerant toward Cr stress.

Supplementary Materials: The following are available online at https://www.mdpi.com/article/10.3390/antiox10071089/s1.

Author Contributions: F.B., Conceptualization, Investigation, Writing—Original Draft; Y.G. and M.S., Conceptualization, Investigation; J.H., Investigation; T.A., M.N., J.A., and J.L., Investigation, Formal Analysis; A.H., A.-B.F.A.-A., M.F.S.A., A.A.A., and E.F.A., Funding Acquisition; T.A., M.C., A.H., E.F.A., and J.A., Writing—Revision and Editing; F.B., Y.G., and M.S., Conceptualization, Supervision, Validation, Writing—Review and Editing, and Funding Acquisition. All authors have read and agreed to the published version of the manuscript.

Funding: This research was funded by the National Natural Science Foundation of China (grant number 32072127), the Key Research and Development Program of Zhejiang Province (grant number 2019C02011), Dabeinong Funds for Discipline Development and Talent Training in Zhejiang University, and Jiangsu Collaborative Innovation Center for Modern Crop Production, China. The authors would like to extend their sincere appreciation to the Researchers Supporting Project Number (RSP-2021/356), King Saud University, Riyadh, Saudi Arabia.

Institutional Review Board Statement: Not applicable.

Informed Consent Statement: Not applicable.

Data Availability Statement: The data presented in this study are available within the article.

Acknowledgments: The authors would like to extend their sincere appreciation to the Researchers Supporting Project Number (RSP-2021/356), King Saud University, Riyadh, Saudi Arabia.

Conflicts of Interest: The authors declare no conflict of interest.

Abbreviations

Cr	Chromium
CK	Control
Chla	Chlorophyll a
Chlb	Chlorophyll b
Chl a+b	Chlorophyll a+b
MDA	Malondialdehyde
H_2O_2	Hydrogen peroxide
SOD	Superoxide dismutase
CAT	Catalase
APX	Ascorbate peroxidase

POD	Peroxidase
G.E	Germination energy
G.P	Germination percentage
G.I	Germination index
V.I	Vigor index
MGT	Mean germination time
S.L	Shoot length
R.L	Root length
F/W	Fresh weight
D/W	Dry weight

References

1. Ali, S.; Bai, P.; Zeng, F.; Cai, S.; Shamsi, I.H.; Qiu, B.; Wu, F.; Zhang, G. The ecotoxicological and interactive effects of chromium and aluminum on growth, oxidative damage and antioxidant enzymes on two barley genotypes differing in Al tolerance. *Environ. Exp. Bot.* **2011**, *70*, 185–191. [CrossRef]
2. Noman, M.; Shahid, M.; Ahmed, T.; Tahir, M.; Naqqash, T.; Muhammad, S.; Song, F.; Abid, H.M.A.; Aslam, Z. Green copper nanoparticles from a native Klebsiella pneumoniae strain alleviated oxidative stress impairment of wheat plants by reducing the chromium bioavailability and increasing the growth. *Ecotoxicol. Environ. Saf.* **2020**, *192*, 110303. [CrossRef]
3. Singh, J.; Kalamdhad, A.S. Effects of heavy metals on soil, plants, human health and aquatic life. *Int. J. Res. Chem. Environ.* **2011**, *1*, 15–21.
4. Farid, M.; Ali, S.; Rizwan, M.; Ali, Q.; Abbas, F.; Bukhari, S.A.H.; Saeed, R.; Wu, L. Citric acid assisted phytoextraction of chromium by sunflower; morpho-physiological and biochemical alterations in plants. *Ecotoxicol. Environ. Saf.* **2017**, *145*, 90–102. [CrossRef]
5. Shanker, A.K.; Cervantes, C.; Loza-Tavera, H.; Avudainayagam, S. Chromium toxicity in plants. *Environ. Int.* **2005**, *31*, 739–753. [CrossRef] [PubMed]
6. Amin, H.; Arain, B.A.; Amin, F.; Surhio, M.A. Phytotoxicity of chromium on germination, growth and biochemical attributes of *Hibiscus esculentus* L. *Am. J. Plant Sci.* **2013**, *2013*.
7. Tiwari, K.; Dwivedi, S.; Singh, N.; Rai, U.; Tripathi, R. Chromium (VI) induced phytotoxicity and oxidative stress in pea (*Pisum sativum* L.): biochemical changes and translocation of essential nutrients. *J. Environ. Biol.* **2009**, *30*, 389–394. [PubMed]
8. Ali, S.; Chaudhary, A.; Rizwan, M.; Anwar, H.T.; Adrees, M.; Farid, M.; Irshad, M.K.; Hayat, T.; Anjum, S.A. Alleviation of chromium toxicity by glycinebetaine is related to elevated antioxidant enzymes and suppressed chromium uptake and oxidative stress in wheat (*Triticum aestivum* L.). *Environ. Sci. Pollut. Res.* **2015**, *22*, 10669–10678. [CrossRef] [PubMed]
9. Amin, H.; Arain, B.A.; Amin, F.; Surhio, M.A. Analysis of growth response and tolerance index of *Glycine max* (L.) Merr. under hexavalent chromium stress. *Adv. Life Sci.* **2014**, *1*, 231–241.
10. Singh, S.; Srivastava, P.K.; Kumar, D.; Tripathi, D.K.; Chauhan, D.K.; Prasad, S.M. Morpho-anatomical and biochemical adapting strategies of maize (*Zea mays* L.) seedlings against lead and chromium stresses. *Biocatal. Agric. Biotechnol.* **2015**, *4*, 286–295. [CrossRef]
11. Karthik, C.; Kadirvelu, K.; Bruno, B.; Maharajan, K.; Rajkumar, M.; Manoj, S.R.; Arulselvi, P.I. Cellulosimicrobium funkei strain AR6 alleviate Cr (VI) toxicity in Lycopersicon esculentum by regulating the expression of growth responsible, stress tolerant and metal transporter genes. *Rhizosphere* **2021**, *18*, 100351. [CrossRef]
12. Jan, S.; Noman, A.; Kaya, C.; Ashraf, M.; Alyemeni, M.N.; Ahmad, P. 24-Epibrassinolide alleviates the injurious effects of Cr (VI) toxicity in tomato plants: Insights into growth, physio-biochemical attributes, antioxidant activity and regulation of Ascorbate–glutathione and Glyoxalase cycles. *J. Plant. Growth Regul.* **2020**, *39*, 1587–1604. [CrossRef]
13. Ahmad, R.; Ali, S.; Rizwan, M.; Dawood, M.; Farid, M.; Hussain, A.; Wijaya, L.; Alyemeni, M.N.; Ahmad, P. Hydrogen sulfide alleviates chromium stress on cauliflower by restricting its uptake and enhancing antioxidative system. *Physiol. Plant.* **2020**, *168*, 289–300. [CrossRef] [PubMed]
14. Lei, K.; Sun, S.; Zhong, K.; Li, S.; Hu, H.; Sun, C.; Zheng, Q.; Tian, Z.; Dai, T.; Sun, J. Seed soaking with melatonin promotes seed germination under chromium stress via enhancing reserve mobilization and antioxidant metabolism in wheat. *Ecotoxicol. Environ. Saf.* **2021**, *220*, 112241. [CrossRef] [PubMed]
15. Wakeel, A.; Xu, M.; Gan, Y. Chromium-induced reactive oxygen species accumulation by altering the enzymatic antioxidant system and associated cytotoxic, genotoxic, ultrastructural, and photosynthetic changes in plants. *Int. J. Mol. Sci.* **2020**, *21*, 728. [CrossRef]
16. Fan, W.-J.; Feng, Y.-X.; Li, Y.-H.; Lin, Y.-J.; Yu, X.-Z. Unraveling genes promoting ROS metabolism in subcellular organelles of *Oryza sativa* in response to trivalent and hexavalent chromium. *Sci. Total Environ.* **2020**, *744*, 140951. [CrossRef]
17. Wang, M.; Zhang, S.; Ding, F. Melatonin mitigates chilling-induced oxidative stress and photosynthesis inhibition in tomato plants. *Antioxidants* **2020**, *9*, 218. [CrossRef] [PubMed]
18. Dreyer, A.; Dietz, K.-J. Reactive oxygen species and the redox-regulatory network in cold stress acclimation. *Antioxidants* **2018**, *7*, 169. [CrossRef]

19. Sirhindi, G. Brassinosteroids: biosynthesis and role in growth, development, and thermotolerance responses. In *Molecular Stress Physiology of Plants*; Springer: Berlin, Germany, 2013; pp. 309–329.

20. Tanveer, M.; Shahzad, B.; Sharma, A.; Khan, E.A. 24-Epibrassinolide application in plants: An implication for improving drought stress tolerance in plants. *Plant Physiol. Biochem.* **2019**, *135*, 295–303. [CrossRef] [PubMed]

21. Shahzad, B.; Tanveer, M.; Che, Z.; Rehman, A.; Cheema, S.A.; Sharma, A.; Song, H.; ur Rehman, S.; Zhaorong, D. Role of 24-epibrassinolide (EBL) in mediating heavy metal and pesticide induced oxidative stress in plants: a review. *Ecotoxicol. Environ. Saf.* **2018**, *147*, 935–944. [CrossRef]

22. Zheng, Y.; Hu, J.; Zhang, S.; Gao, C.; Song, W. Identification of chilling-tolerance in maize inbred lines at germination and seedling growth stages. *J. Zhejiang Univer. Agric. Life Sci.* **2006**, *32*, 41–45.

23. Lichtenthaler, H.K.; Wellburn, A.R. *Determinations of total carotenoids and chlorophylls a and b of leaf extracts in different solvents*; Portland Press: London, UK, 1983.

24. Shentu, J.; He, Z.; Yang, X.-E.; Li, T. Accumulation properties of cadmium in a selected vegetable-rotation system of southeastern China. *J. Agric. Food Chem.* **2008**, *56*, 6382–6388. [CrossRef] [PubMed]

25. Heath, R.L.; Packer, L. Photoperoxidation in isolated chloroplasts: I. Kinetics and stoichiometry of fatty acid peroxidation. *Arch. Biochem. Biophys.* **1968**, *125*, 189–198. [CrossRef]

26. Kwasniewski, M.; Chwialkowska, K.; Kwasniewska, J.; Kusak, J.; Siwinski, K.; Szarejko, I. Accumulation of peroxidase-related reactive oxygen species in trichoblasts correlates with root hair initiation in barley. *J. Plant Physiol.* **2013**, *170*, 185–195. [CrossRef] [PubMed]

27. Giannopolitis, C.N.; Ries, S.K. Superoxide dismutases: I. Occurrence in higher plants. *Plant Physiol.* **1977**, *59*, 309–314. [CrossRef] [PubMed]

28. Aebi, H. Catalase in vitro. *Meth. Enzymol.* **1984**, *105*, 121–126.

29. Chance, B.; Maehly, A. *136, Assay of Catalases and Peroxidases*; Elsevier: Amsteradam, The Netherlands, 1955.

30. Nakano, Y.; Asada, K. Hydrogen peroxide is scavenged by ascorbate-specific peroxidase in spinach chloroplasts. *Plant Cell Physiol.* **1981**, *22*, 867–880.

31. Sah, S.K.; Kaur, G.; Kaur, A. Rapid and reliable method of high-quality RNA extraction from diverse plants. *Am. J. Plant Sci.* **2014**, *5*, 3129. [CrossRef]

32. Livak, K.J.; Schmittgen, T.D. Analysis of relative gene expression data using real-time quantitative PCR and the $2^{-\Delta\Delta CT}$ method. *Methods* **2001**, *25*, 402–408. [CrossRef]

33. Shakirova, F.; Allagulova, C.; Maslennikova, D.; Fedorova, K.; Yuldashev, R.; Lubyanova, A.; Bezrukova, M.; Avalbaev, A. Involvement of dehydrins in 24-epibrassinolide-induced protection of wheat plants against drought stress. *Plant Physiol. Biochem.* **2016**, *108*, 539–548. [CrossRef]

34. Hayat, S.; Khalique, G.; Irfan, M.; Wani, A.S.; Tripathi, B.N.; Ahmad, A. Physiological changes induced by chromium stress in plants: an overview. *Protoplasma* **2012**, *249*, 599–611. [CrossRef] [PubMed]

35. Patel, M.; Surti, M.; Ashraf, S.A.; Adnan, M. Physiological and Molecular Responses to Heavy Metal Stresses in Plants. *Harsh Environ. Plant Resil. Mol. Funct. Asp.* **2021**, *171*. [CrossRef]

36. Bewley, J.D.; Black, M. *Physiology and Biochemistry of Seeds in Relation to Germination: Volume 2: Viability, Dormancy, and Environmental Control*; Springer: Berlin, Germany, 2012.

37. Shinwari, K.I.; Jan, M.; Shah, G.; Khattak, S.R.; Urehman, S.; Daud, M.; Naeem, R.; Jamil, M. Seed priming with salicylic acid induces tolerance against chromium (VI) toxicity in rice (*Oryza sativa* L.). *Pak. J. Bot.* **2015**, *47*, 161–170.

38. Sharma, I.; Pati, P.K.; Bhardwaj, R. Effect of 28-homobrassinolide on antioxidant defence system in *Raphanus sativus* L. under chromium toxicity. *Ecotoxicology* **2011**, *20*, 862–874. [CrossRef]

39. Ali, A.A.; Abdel-Fattah, R.I. Osmolytes-antioxidant behaviour in *Phaseolus vulgaris* and *Hordeum vulgare* with brassinosteroid under salt stress. *J. Agron.* **2006**.

40. Hayat, S.; Hasan, S.A.; Hayat, Q.; Ahmad, A. Brassinosteroids protect *Lycopersicon esculentum* from cadmium toxicity applied as shotgun approach. *Protoplasma* **2010**, *239*, 3–14. [CrossRef]

41. Vajpayee, P.; Tripathi, R.; Rai, U.; Ali, M.; Singh, S. Chromium (VI) accumulation reduces chlorophyll biosynthesis, nitrate reductase activity and protein content in *Nymphaea alba* L. *Chemosphere* **2000**, *41*, 1075–1082. [CrossRef]

42. Yu, X.-Z.; Lin, Y.-J.; Zhang, Q. Metallothioneins enhance chromium detoxification through scavenging ROS and stimulating metal chelation in *Oryza sativa*. *Chemosphere* **2019**, *220*, 300–313. [CrossRef]

43. Wakeel, A.; Ali, I.; Wu, M.; Kkan, A.R.; Jan, M.; Ali, A.; Liu, Y.; Ge, S.; Wu, J.; Gan, Y. Ethylene mediates dichromate-induced oxidative stress and regulation of the enzymatic antioxidant system-related transcriptome in *Arabidopsis thaliana*. *Environ. Exp. Bot.* **2019**, *161*, 166–179. [CrossRef]

44. Hussain, A.; Ali, S.; Rizwan, M.; ur Rehman, M.Z.; Hameed, A.; Hafeez, F.; Alamri, S.A.; Alyemeni, M.N.; Wijaya, L. Role of zinc–lysine on growth and chromium uptake in rice plants under Cr stress. *Plant Growth Regul.* **2018**, *37*, 1413–1422. [CrossRef]

45. Sharma, A.; Kapoor, D.; Wang, J.; Shahzad, B.; Kumar, V.; Bali, A.S.; Jasrotia, S.; Zheng, B.; Yuan, H.; Yan, D. Chromium bioaccumulation and its impacts on plants: an overview. *Plants* **2020**, *9*, 100. [CrossRef] [PubMed]

46. Yang, P.; Wang, Y.; Li, J.; Bian, Z. Effects of brassinosteroids on photosynthetic performance and nitrogen metabolism in pepper seedlings under chilling stress. *Agronomy* **2019**, *9*, 839. [CrossRef]

47. Ahanger, M.A.; Mir, R.A.; Alyemeni, M.N.; Ahmad, P. Combined effects of brassinosteroid and kinetin mitigates salinity stress in tomato through the modulation of antioxidant and osmolyte metabolism. *Plant Physiol. Biochem.* **2020**, *147*, 31–42. [CrossRef] [PubMed]

48. Ali, B.; Hasan, S.; Hayat, S.; Hayat, Q.; Yadav, S.; Fariduddin, Q.; Ahmad, A. A role for brassinosteroids in the amelioration of aluminium stress through antioxidant system in mung bean (*Vigna radiata* L. Wilczek). *Environ. Exp. Bot.* **2008**, *62*, 153–159. [CrossRef]

49. Somssich, M.; Vandenbussche, F.; Ivakov, A.; Funke, N.; Ruprecht, C.; Vissenberg, K.; Van Der Straeten, D.; Persson, S.; Suslov, D. *Brassinosteroids Influence Arabidopsis Hypocotyl Graviresponses Through Changes in Mannans and Cellulose*; University Ghent: Ghent, Belguim, 2019.

50. Chen, L.; Hu, W.-f.; Long, C.; Wang, D. Exogenous plant growth regulator alleviate the adverse effects of U and Cd stress in sunflower (*Helianthus annuus* L.) and improve the efficacy of U and Cd remediation. *Chemosphere* **2021**, *262*, 127809. [CrossRef] [PubMed]

51. Sousa, B.; Soares, C.; Oliveira, F.; Martins, M.; Branco-Neves, S.; Barbosa, B.; Ataíde, I.; Teixeira, J.; Azenha, M.; Azevedo, R.A. Foliar application of 24-epibrassinolide improves *Solanum nigrum* L. tolerance to high levels of Zn without affecting its remediation potential. *Chemosphere* **2020**, *244*, 125579. [CrossRef]

52. Soares, T.F.S.N.; dos Santos Dias, D.C.F.; Oliveira, A.M.S.; Ribeiro, D.M.; dos Santos Dias, L.A. Exogenous brassinosteroids increase lead stress tolerance in seed germination and seedling growth of *Brassica juncea* L. *Ecotoxicol. Environ. Saf.* **2020**, *193*, 110296. [CrossRef]

53. Horton, A.; Fairhurst, S.; Bus, J.S. Lipid peroxidation and mechanisms of toxicity. *Crit. Rev. Toxicol.* **1987**, *18*, 27–79. [CrossRef]

54. Khan, M.; Samrana, S.; Zhang, Y.; Malik, Z.; Khan, M.D.; Zhu, S. Reduced glutathione protects subcellular compartments from Pb-induced ROS injury in leaves and roots of upland cotton (*Gossypium hirsutum* L.). *Front. Plant Sci.* **2020**, *11*, 412. [CrossRef]

55. Samrana, S.; Ali, A.; Muhammad, U.; Azizullah, A.; Ali, H.; Khan, M.; Naz, S.; Khan, M.D.; Zhu, S.; Chen, J. Physiological, ultrastructural, biochemical, and molecular responses of glandless cotton to hexavalent chromium (Cr^{6+}) exposure. *Environ. Pollut.* **2020**, *266*, 115394. [CrossRef]

56. Khan, I.; Awan, S.A.; Ikram, R.; Rizwan, M.; Akhtar, N.; Yasmin, H.; Sayyed, R.Z.; Ali, S.; Ilyas, N. Effects of 24-epibrassinolide on plant growth, antioxidants defense system, and endogenous hormones in two wheat varieties under drought stress. *Physiol Plant.* **2020**. [CrossRef]

57. Rattan, A.; Kapoor, D.; Kapoor, N.; Bhardwaj, R.; Sharma, A. Brassinosteroids regulate functional components of antioxidative defense system in salt stressed maize seedlings. *Plant Growth Regul.* **2020**, *39*, 1465–1475. [CrossRef]

58. Liaqat, S.; Umar, S.; Saffeullah, P.; Iqbal, N.; Siddiqi, T.O.; Khan, M.I.R. Protective Effect of 24-Epibrassinolide on Barley Plants Growing Under Combined Stress of Salinity and Potassium Deficiency. *Plant Growth Regul.* **2020**, *39*, 1543–1558. [CrossRef]

59. Surgun-Acar, Y.; Zemheri-Navruz, F. Exogenous application of 24-epibrassinolide improves manganese tolerance in *Arabidopsis thaliana* L. via the modulation of antioxidant system. *Plant Growth Regul.* **2021**, 1–12.

60. Kabir, A. Biochemical and molecular changes in rice seedlings (*Oryza sativa* L.) to cope with chromium stress. *Plant Biol.* **2016**, *18*, 710–719. [CrossRef]

61. Ma, J.; Lv, C.; Xu, M.; Chen, G.; Lv, C.; Gao, Z. Photosynthesis performance, antioxidant enzymes, and ultrastructural analyses of rice seedlings under chromium stress. *Environ. Sci. Pollut. Res.* **2016**, *23*, 1768–1778. [CrossRef]

62. Qiu, B.; Zeng, F.; Cai, S.; Wu, X.; Haider, S.I.; Wu, F.; Zhang, G. Alleviation of chromium toxicity in rice seedlings by applying exogenous glutathione. *J. Plant Physiol.* **2013**, *170*, 772–779. [CrossRef]

63. Vernay, P.; Gauthier-Moussard, C.; Hitmi, A. Interaction of bioaccumulation of heavy metal chromium with water relation, mineral nutrition and photosynthesis in developed leaves of *Lolium perenne* L. *Chemosphere* **2007**, *68*, 1563–1575. [CrossRef]

64. Li, L.; Long, M.; Islam, F.; Farooq, M.A.; Wang, J.; Mwamba, T.M.; Shou, J.; Zhou, W. Synergistic effects of chromium and copper on photosynthetic inhibition, subcellular distribution, and related gene expression in *Brassica napus* cultivars. *Environ. Sci. Pollut. Res.* **2019**, *26*, 11827–11845. [CrossRef]

65. dos Santos, R.W.; Schmidt, É.C.; Martins, R.d.P.; Latini, A.; Maraschin, M.; Horta, P.A.; Bouzon, Z.L. Effects of cadmium on growth, photosynthetic pigments, photosynthetic performance, biochemical parameters and structure of chloroplasts in the agarophyte *Gracilaria domingensis* (Rhodophyta, Gracilariales). *Am. J. Plant Sci.* **2012**, *3*, 1077–1084. [CrossRef]

66. Choudhury, S.; Panda, P.; Sahoo, L.; Panda, S.K. Reactive oxygen species signaling in plants under abiotic stress. *Plant Signal Behav.* **2013**, *8*, e23681. [CrossRef] [PubMed]

67. Saif, S.; Khan, M.S. Assessment of toxic impact of metals on proline, antioxidant enzymes, and biological characteristics of *Pseudomonas aeruginosa* inoculated *Cicer arietinum* grown in chromium and nickel-stressed sandy clay loam soils. *Environ. Monit. Assess.* **2018**, *190*, 1–18. [CrossRef] [PubMed]

68. Aziz, A.; Mahmood, T.; Mahmood, Z.; Shazadi, K.; Mujeeb-Kazi, A.; Rasheed, A. Genotypic Variation and Genotype× Environment Interaction for Yield-Related Traits in Synthetic Hexaploid Wheats under a Range of Optimal and Heat-Stressed Environments. *Crop Sci.* **2018**, *58*, 295–303. [CrossRef]

69. Ali, Q.; Perveen, R.; El-Esawi, M.A.; Ali, S.; Hussain, S.M.; Amber, M.; Iqbal, N.; Rizwan, M.; Alyemeni, M.N.; El-Serehy, H.A. Low doses of *Cuscuta reflexa* extract act as natural biostimulants to improve the germination vigor, growth, and grain yield of wheat grown under water stress: photosynthetic pigments, antioxidative defense mechanisms, and nutrient acquisition. *Biomolecules* **2020**, *10*, 1212. [CrossRef]

Article

Genome-Wide Analysis and Expression Profile of Superoxide Dismutase (SOD) Gene Family in Rapeseed (*Brassica napus* L.) under Different Hormones and Abiotic Stress Conditions

Wei Su [†], Ali Raza [†], Ang Gao, Ziqi Jiao, Yi Zhang, Muhammad Azhar Hussain, Sundas Saher Mehmood, Yong Cheng, Yan Lv * and Xiling Zou *

Key Lab of Biology and Genetic Improvement of Oil Crops, Oil Crops Research Institute, Chinese Academy of Agricultural Sciences (CAAS), Wuhan 430062, China; 82101181063@caas.cn (W.S.); alirazamughal143@gmail.com (A.R.); 1696555087@163.com (A.G.); jiaziqi0507@163.com (Z.J.); zhangyi20210615@163.com (Y.Z.); azharhussain301@gmail.com (M.A.H.); snookas.saher90@gmail.com (S.S.M.); chengyong@caas.cn (Y.C.)
* Correspondence: lvyan01@caas.cn (Y.L.); zouxiling01@caas.cn (X.Z.)
† These authors equally contributed to this work.

Citation: Su, W.; Raza, A.; Gao, A.; Jia, Z.; Zhang, Y.; Hussain, M.A.; Mehmood, S.S.; Cheng, Y.; Lv, Y.; Zou, X. Genome-Wide Analysis and Expression Profile of Superoxide Dismutase (SOD) Gene Family in Rapeseed (*Brassica napus* L.) under Different Hormones and Abiotic Stress Conditions. *Antioxidants* **2021**, *10*, 1182. https://doi.org/10.3390/antiox10081182

Academic Editor: Juan B. Barroso

Received: 8 June 2021
Accepted: 22 July 2021
Published: 25 July 2021

Publisher's Note: MDPI stays neutral with regard to jurisdictional claims in published maps and institutional affiliations.

Abstract: Superoxide dismutase (SOD) is an important enzyme that acts as the first line of protection in the plant antioxidant defense system, involved in eliminating reactive oxygen species (ROS) under harsh environmental conditions. Nevertheless, the *SOD* gene family was yet to be reported in rapeseed (*Brassica napus* L.). Thus, a genome-wide investigation was carried out to identify the rapeseed *SOD* genes. The present study recognized 31 *BnSOD* genes in the rapeseed genome, including 14 *BnCSDs*, 11 *BnFSDs*, and six *BnMSDs*. Phylogenetic analysis revealed that *SOD* genes from rapeseed and other closely related plant species were clustered into three groups based on the binding domain with high bootstrap values. The systemic analysis exposed that *BnSODs* experienced segmental duplications. Gene structure and motif analysis specified that most of the *BnSOD* genes displayed a relatively well-maintained exon–intron and motif configuration within the same group. Moreover, we identified five hormones and four stress- and several light-responsive *cis*-elements in the promoters of *BnSODs*. Thirty putative bna-miRNAs from seven families were also predicted, targeting 13 *BnSODs*. Gene ontology annotation outcomes confirm the *BnSODs* role under different stress stimuli, cellular oxidant detoxification processes, metal ion binding activities, SOD activity, and different cellular components. Twelve *BnSOD* genes exhibited higher expression profiles in numerous developmental tissues, i.e., root, leaf, stem, and silique. The qRT-PCR based expression profiling showed that eight genes (*BnCSD1*, *BnCSD3*, *BnCSD14*, *BnFSD4*, *BnFSD5*, *BnFSD6*, *BnMSD2*, and *BnMSD10*) were significantly up-regulated under different hormones (ABA, GA, IAA, and KT) and abiotic stress (salinity, cold, waterlogging, and drought) treatments. The predicted 3D structures discovered comparable conserved BnSOD protein structures. In short, our findings deliver a foundation for additional functional investigations on the *BnSOD* genes in rapeseed breeding programs.

Keywords: abiotic stress; antioxidant defense systems; gene ontology; miRNA; phytohormones; 3D structures

1. Introduction

Several environmental cues, including abiotic and biotic traumas, are considered key influences affecting the plants' productivity [1,2]. Under stressful conditions, the plant amends its homeostatic apparatus by developing an increased reactive oxygen species (ROS) in plant cells. Usually, ROS overproduction results in several molecular and cellular damages, and ultimately programmed cell death [3,4]. ROS such as superoxide anion, hydrogen peroxide, hydroxyl radical, perhydroxyl radicals, alkoxy radicals, peroxy radicals, singlet oxygen, and organic hydroperoxide, are considered to be major signaling molecules, regulating abiotic and biotic stress responses, and also participate in plant productivity [3,4]. Mainly, ROS are formed in the apoplast, mitochondria, plasma membrane, chloroplast, peroxisomes, endoplasmic reticulum, and cell walls [3,4]. Therefore, to manage ROS noxiousness, plants have established well-organized and composite antioxidant defense systems, including numerous non-enzymatic and enzymatic antioxidants [3,4].

Among numerous antioxidant enzymes, superoxide dismutase (SOD; EC 1.15.1.1), a set of metalloenzymes, largely exists in alive organisms. SOD acts as the first line of ROS scavenging and plays a crucial role in plants' physio-biochemical procedures to manage environmental cues [3]. Findings revealed that SOD catalyzes the superoxide radicals' dismutation into oxygen and hydrogen peroxide via disproportionation, and protects the plant cells from oxidative injury [5,6]. According to metal cofactors, plant SODs are largely characterized into four major groups such as copper-zinc SOD (Cu/ZnSOD), iron SOD (FeSOD), manganese SOD (MnSOD), and nickel SOD (NiSOD) [7,8]. Among them, NiSOD mainly exists in Streptomyces, cyanobacteria, and marine life, and is yet to be described in plants [9]. However, FeSODs and MnSODs are primarily extant in lower plants, while Cu/ZnSODs exist in higher plants [10,11]. These SODs are widely distributed in various cell organs. For instance, Cu/ZnSODs are amply present, and mostly distributed in chloroplasts, cytoplasm, and peroxisomes [12]. FeSOD is distributed in the chloroplasts, and MnSODs are distributed in mitochondria and peroxisomes [8,13].

Recently, numerous investigations have revealed that the transcript level of the plant *SODs* gene responds to several environmental cues and helps plants cope with harsh environmental conditions. For instance, the increased SOD activity helps plants to show resistance to salinity and drought stress in *Brassica juncea* plants [14], temperature-induced oxidative damage in *Acutodesmus dimorphus* [15], cold-induced oxidative damage in tomato (*Solanum lycopersicum*) [16], etc. Moreover, the expression of the *Cu/ZnSOD* gene was persuaded by the copper-nanoparticle application in cucumber (*Cucumis sativus*) [17]. Under cold stress, trehalose application modulated the expression profile of the *Cu/ZnSOD* gene in tomato plants [16]. In a recent study, the overexpression of *SikCuZnSOD3* enhances tolerance to cold, drought, and salinity stresses in cotton (*Gossypium hirsutum*) [18]. Under water deficit conditions, wheat (*Triticum aestivum*) plants showed a higher expression of different antioxidant encoding genes, including *MnSOD* [19]. Likewise, under 1-methylcyclopropene (1-MCP) supplementation, apple (*Malus × domestica* Borkh.) fruits showed a higher expression of *Cu/ZnSODs*, *MnSODs*, and *FeSODs* [20]. In short, these findings showed that improved SOD activity and higher expression of SOD-encoding genes can contribute to plant tolerance to multiple stresses.

Additionally, earlier investigations recommended that miRNA-mediated regulation of ROS-accompanying genes is crucial for plant productivity [21,22] and stress resistance [23–26]. For example, miR398 targets two *Cu/ZnSOD* genes in *Arabidopsis thaliana* [27]. In another study, 20 miRNAs are found to be targeting 14 cotton *SOD* genes at 33 predicted sites [28]. Moreover, ghr-miR414c, ghr-miR7267, m0081, m0166, and m0362 are found to play a vital role in cotton fiber variation and progress [29,30], and ghr-miR3 controls the transcript level of targeted genes throughout cotton somatic embryogenesis [31]. In *A. thaliana*, two *Cu/ZnSOD* genes (*CSD1* and *CSD2*) are induced via the down-regulation of miR398, which helps plants to enhance tolerance to oxidative injury [32]. These discoveries stated that miRNAs might play active roles against environmental cues and plant development through modifying the *SOD* genes.

Rapeseed (*Brassica napus* L.) is the second imperative oilseed crop and possesses a complex genome. Several abiotic stresses significantly limit rapeseed productivity [33–36]. To date, *SOD* family genes have not been reported in rapeseed. Thus, in the present study, we performed a genome-wide analysis to identify *SOD* genes in the rapeseed genome. Additionally, their phylogenetic relationships, synteny analysis, gene structures, conserved motifs, *cis*-elements, miRNA predictions, functional annotations, and 3D structures have been characterized. Moreover, the expression profile in numerous tissues/organs and under numerous hormone and abiotic stress conditions have been extensively appraised, which deeply boosted our understanding of the *SOD* genes in rapeseed.

2. Materials and Methods

2.1. Identification and Characterization of SOD Genes in Rapeseed

According to our recent study [37], we used two methods to identify *SOD* genes in the *B. napus* genome, i.e., BLASTP (protein blast) and the Hidden Markov Model (HMM) [37]. The *B. napus* genome sequence was downloaded from the BnPIR database (Available online: http://cbi.hzau.edu.cn/bnapus/index.php, accessed on 1 April 2021) [38]. For BLASTP, we used eight *A. thaliana* SODs (AT1G08830.1/*AtCSD1*, AT2G28190.1/*AtCSD2*, AT5G18100.1/*AtCSD3*, AT4G25100.1/*AtFSD1*, AT5G51100.1/*AtFSD2*, AT5G23310.1/*AtFSD3*, AT3G10920.1/*AtMSD1*, and AT3G56350.1/*AtMSD2*) amino acid sequences as a query with an e-value set to $1e^{-5}$. The amino acid sequences of eight *AtSODs* were obtained from the TAIR *Arabidopsis* genome database (Available online: http://www.arabidopsis.org/, accessed on 1 April 2021). [39]. Further, the local HMMER 3.1 web server (Available online: http://www.hmmer.org/, accessed on 1 April 2021) [40] was used to search the *SOD* genes with default parameters. Then, the HMM file of the Sod_Cu (PF00080.21) and Sod_Fe_C (PF02777.19) having *SOD* genes were downloaded from the Pfam protein domain database (Available online: http://pfam.xfam.org/, accessed on 1 April 2021) [41]. Finally, 31 *BnSOD* genes were identified by combining the two methods in the rapeseed genome. Moreover, we identified *SOD* genes in different plant species, such as *Brassica rapa*, and *Brassica oleracea*, with the genome downloaded from the JGI Phytozome 12.0 database (Available online: https://phytozome.jgi.doe.gov/pz/portal.html, accessed on 1 April 2021) [42] via the same method.

The physico-chemical properties of molecular weight, and isoelectric points, were analyzed by the online ProtParam tool (Available online: http://web.expasy.org/protparam/, accessed on 1 April 2021) [43]. The subcellular localization of BnSOD proteins was predicted from the WoLF PSORT server (Available online: https://wolfpsort.hgc.jp/, accessed on 1 April 2021) [44]. *BnSOD* gene structures were constructed via TBtools software (V 1.068; https://github.com/CJ-Chen/TBtools, accessed on 1 April 2021) [45]. The conserved motifs about BnSOD protein sequences were identified using the MEME website (Available online: https://meme-suite.org/meme/db/motifs, accessed on 1 April 2021) [46].

2.2. Phylogenetic Tree and Synteny Analysis of BnSOD Proteins

To observe the evolutionary relationship of the *BnSOD* gene family, we constructed a phylogenetic tree about *B. napus*, *B. oleracea*, *B. rapa*, and *A. thaliana* protein sequences. The sequence alignment was performed by MEGA 7 software (Available online: https://megasoftware.net/home, accessed on 1 April 2021) [47]. The neighbor-joining (NJ) method was performed to construct a phylogenetic tree with 1000 bootstrap replicates using the Evolview v3 website (Available online: https://www.evolgenius.info/evolview, accessed on 1 April 2021) [48] to display the phylogenetic tree. Synteny relationships of *SOD* genes were developed by the python-package, JCVI (Available online: https://github.com/tanghaibao/jcvi, accessed on 1 April 2021) [49] from *B. napus*, *B. oleracea*, *B. rapa*, and *A. thaliana*. We calculated the Ka/Ks ratios of all *SODs* using KaKs_Calculator 2.0 (Available online: https://sourceforge.net/projects/kakscalculator2/, accessed on 1 April 2021) [50].

2.3. Analysis of Cis-Acting Regulatory Elements in the BnSODs Promoters

To analyze the putative *cis*-elements in the *BnSODs* promoters, we extracted the 2Kb sequence upstream of start codons in the *B. napus* genome. Then, the promoter sequence of each gene was analyzed using the PlantCARE website (Available online: http://bioinformatics.psb.ugent.be/webtools/plantcare/html/, accessed on 1 April 2021) [51] and figures drawn using TBtools (V 1.068) [45].

2.4. Prediction of Putative miRNA Targeting BnSOD Genes and GO Annotation Analysis

The coding sequence (CDS) of *BnSODs* was used to identify possible target miRNAs in the psRNATarget database (Available online: http://plantgrn.noble.org/psRNATarget/, accessed on 01 April 2021) [52] with default parameters. We drew the interaction network figure between the miRNAs and *BnSOD* genes by Cytoscape software (V3.8.2; Available online: https://cytoscape.org/download.html, accessed on 1 April 2021). Gene ontology (GO) annotation analysis was performed by uploading all BnSODs protein sequences to the eggNOG website (Available online: http://eggnog-mapper.embl.de/, 1 April 2021) [53]. TBtools was used to perform GO enrichment analysis.

2.5. Expression Profiling of BnSOD Genes in Different Tissues

For tissue-specific expression profiling, we downloaded RNA-seq data (BioProject ID: PRJCA001495) of rapeseed from the National Genomics Data Center. The complete method has been described in our recent work [37]. Briefly, Cuffquant and Cuffnorm were used to produce normalized counts in transcripts per million (TPM) values. Based on TPM values, the expression heat map was created using GraphPad Prism 8 software (https://www.graphpad.com/, accessed on 1 April 2021). [54].

2.6. Plant Material and Stress Conditions

In this study, a typical cultivated variety, "ZS11," was used for stress treatments. The seeds of the "ZS11" genotype were provided by the OCRI-CAAS, Wuhan, China. The stress treatments were carried out as described in our recent work [37]. Briefly, the vigor seeds were cultivated on water-saturated filter paper in a chamber (25 °C day/night and 16 h/8 h light/dark cycle) till the radicle's extent extended around 5 mm. For stress treatment, germinated seeds were subjected to 150 mM NaCl solution for salinity stress, 15% PEG6000 solution for drought stress, and 4 °C for cold stress on water-saturated filter paper. For waterlogging stress, the seeds were flooded with water in the Eppendorf tube. To evaluate the impact of diverse hormones, the germinated seeds were grown in Murashige and Skoog (MS) medium supplied with 100 μM abscisic acid (ABA), 100 μM gibberellic acid (GA), 100 μM indole-acetic acid/auxin (IAA), and 100 μM kinetin (KT). The samples were collected at 0 (CK), 2, 4, 6, and 8 h after the treatments. All of the treatments were performed with three biological replications. All of the samples were immediately frozen in liquid nitrogen and were stored at −80 °C for further experiments.

2.7. RNA Extraction and qRT-PCR Analysis

Total RNA extraction and cDNA synthesis were performed using a TransZol Up Plus RNA Kit (TransGen Biotech, Beijing, China) and cDNA Synthesis SuperMix (TransGen Biotech, Beijing, China) according to manufacturer instructions. The detailed information of qRT-PCR reactions has been described in our recently published work [37]. Initially, the expression data were analyzed using the $2^{-\Delta\Delta CT}$ method. Due to the large difference in the expression levels, we used the log2 fold change method to calculate the expression results for better visualization of differently expressed genes under stress treatments. All of the primers used in this experiment are listed in Table S1. The heatmap was created using GraphPad Prism 8 software [54].

2.8. Prediction of the 3D Structure of BnSOD Proteins

The predicted 3D structures of BnSODs were created with the 3D LigandSite website (https://www.wass-michaelislab.org/3dlig/index.html, accessed on 1 April 2021) [55]. The probability score of the predicted sites shows the possibility of apiece residue to be elaborated in binding. During the 3D model predictions, we choose cluster 1 with a higher Z-score. The higher Z-score value indicates the reliability and trueness of the cluster/model [55].

3. Results

3.1. Identification of SOD Gene Family in B. napus

The current study identified 31 *BnSOD* genes in the rapeseed genome using eight *A. thaliana* (*AtSODs*) protein sequences as queries (Table 1; Table S2). According to the domain scrutiny, 14 proteins were found to have a Cu/Zn-SODs domain (Pfam: 00080), 11 proteins were found to have a Fe-SODs domain (Pfam: 00081), and six proteins were found to have an Mn-SODs domain (Pfam: 02777); hereafter, these genes were named as *BnCSD1-BnCSD14*, *BnFSD1-BnFSD11*, and *BnMSD1-BnMSD6*, respectively (Table 1). Comprehensive statistics of 31 *BnSOD* genes were documented in Table 1. Out of 31 *BnSODs*, 15 genes were located on the A subgenome, and 16 genes were located on the C subgenome (Table 1). The gene length extended from 826 bp (*BnCSD4*) to 6898 bp (*BnFSD5*) with 3–9 exons in the individual gene sequences. The CDS length, protein length, and molecular weight extended from 441–1173 bp, 146–390 amino acids, and 14.55–42.34 kDa (*BnCSD9-BnCSD2*), respectively. The isoelectric points extended from 4.88 (*BnFSD7*) to 9.56 (*BnMSD3*) (Table 1). The subcellular localization results prophesied that 15 proteins were located on the chloroplast, nine proteins were located on the cytoskeleton, five proteins were located on the mitochondrion, and the remaining two proteins (*BnFSD9* and *BnMSD6*) were located on the endoplasmic reticulum (Table 1).

Additionally, 14 SOD genes were also identified from *Brassica oleracea* (*BolCSD1-BolCSD6*, *BolFSD1-BolFSD5*, and *BolMSD1-BolMSD3*), and *Brassica rapa* (*BraCSD1-BraCSD7*, *BraFSD1-BraFSD4*, and *BraMSD1-BraMSD3*) genomes (Table S2).

3.2. Phylogenetic Relationships of SOD Genes

In the current study, the evolutionary relationships were explored between *BnSODs*, *BolSODs*, *BraSODs*, and *AtSODs* genes. Based on domains (Cu/Zn-SODs, Fe-SODs, and Mn-SODs) and a phylogenetic tree, 67 *SODs* were clustered into three major groups (Figure 1). Results presented that the Cu/Zn-SODs group consists of 30 *SODs* members (14 *BnSODs*, 7 *BraSODs*, 6 *BolSODs*, and 3 *AtSODs*), the Mn-SODs group consists of 14 *SODs* members (6 *BnSODs*, 3 *BraSODs*, 3 *BolSODs*, and 2 *AtSODs*), and the Fe-SODs group consists of 23 *SODs* members (11 *BnSODs*, 4 *BraSODs*, 5 *BolSODs*, and 3 *AtSODs*) (Figure 1). Interestingly, Cu/Zn-SODs and Fe-SODs groups possessed a greater number of *SODs* than the Mn-SOD group. It was also found that the *BnSODs* exhibited a more closely phylogenetic relationship with *BolSODs* and *BraSODs* in each group.

3.3. Chromosomal Locations and Synteny Analysis of SOD Genes

Gene duplications (tandem and segmental) are considered the main driving forces in promoting new genomic evolution [56]. Thus, gene duplication events were evaluated between *BnSODs*, *BolSODs*, *BraSODs*, and *AtSODs* (Table S3). The chromosol location of 10 *BnSODs* gene pairs was examined (Figure 2). Twelve out of 19 chromosomes harbored *BnSODs*. Particularly, chromosomes A01, A04, A05, A08, A10, C04, C05, and C07 possessed one gene, chromosome C01 contained two genes, chromosomes C08 and C09 possessed three genes, and chromosome A09 contained four genes (Figure 2). Surprisingly, the residual chromosomes did not comprehend *BnSOD* genes. Our results show that segmental duplications have played vital parts in developing *BnSOD* genes in the rapeseed genome (Table S3). Moreover, no tandem duplication events were detected. Notably, a gene pair (*BnMSD6* and *BnMSD6*) was found to be dispersed, while the remaining gene pairs experienced segmental duplications (Table S3).

Table 1. The data of 31 *SOD* genes identified in rapeseed genome.

Gene ID	Gene Name	Genomic Position (bp)	Gene Length (bp)	CDS Length (bp)	Exons	Protein Length (Amino Acids)	Molecular Weight (kDa)	Isoelectric Point (pI)	Subcellular Localization
BnaA04T0182200ZS	*BnCSD1*	A04-18599782:18600942 (+)	1160	639	6	212	22.15	7.16	Chloroplast
BnaA06T0053200ZS	*BnCSD2*	A06-3364923:3368393 (+)	3470	1173	9	390	42.34	8.71	Cytoskeleton
BnaA08T0278700ZS	*BnCSD3*	A08-26336887:26338407 (+)	1520	954	6	317	33.76	5.19	Chloroplast
BnaA09T0129500ZS	*BnCSD4*	A09-7757584:7758410 (−)	826	660	3	219	23.60	5.95	Chloroplast
BnaA09T0647100ZS	*BnCSD5*	A09-61965958:61967448 (+)	1490	954	6	317	33.45	5.26	Chloroplast
BnaA09T0664700ZS	*BnCSD6*	A09-62938095:62939938 (−)	1843	459	8	152	15.17	5.64	Cytoskeleton
BnaA10T0190600ZS	*BnCSD7*	A01-21329437:21330797 (−)	1360	465	6	154	15.94	7.12	Cytoskeleton
BnaC04T0482300ZS	*BnCSD8*	C04-60854831:60856275 (+)	1444	627	6	208	21.60	7.84	Chloroplast
BnaC05T0066000ZS	*BnCSD9*	C05-3767908:3769181 (+)	1273	441	7	146	14.55	5.44	Cytoskeleton
BnaC08T0217500ZS	*BnCSD10*	C08-32001876:32003366 (−)	1490	951	5	316	33.71	6.91	Chloroplast
BnaC08T0505400ZS	*BnCSD11*	C08-51947641:51949188 (+)	1547	957	6	318	33.58	4.97	Chloroplast
BnaC08T0529200ZS	*BnCSD12*	C08-53323394:53325251 (−)	1857	672	7	223	23.42	6.66	Cytoskeleton
BnaC09T0137500ZS	*BnCSD13*	C09-10172656:10173492 (−)	836	660	3	219	23.72	6.83	Chloroplast
BnaC09T0484500ZS	*BnCSD14*	C09-59599637:59601046 (−)	1409	510	6	169	17.63	6.82	Cytoskeleton
BnaA01T0146300ZS	*BnFSD1*	A01-8654405:8657035 (−)	2630	561	6	186	21.12	6.06	Cytoskeleton
BnaA03T0141400ZS	*BnFSD2*	A03-7188761:7190553 (+)	1792	906	8	301	34.51	4.97	Chloroplast
BnaA06T0320600ZS.1	*BnFSD3*	A06-40740605:40744473 (+)	3868	879	9	292	33.60	7.58	Chloroplast
BnaA09T0072700ZS	*BnFSD4*	A09-4315462:4316968 (−)	1506	792	8	263	30.16	7.76	Chloroplast
BnaA10T0083700ZS	*BnFSD5*	A10-13303579:13310477 (−)	6898	738	6	245	28.02	9.44	Mitochondrion
BnaC01T0186100ZS	*BnFSD6*	C01-13777631:13779146 (−)	1515	618	5	205	22.99	6.3	Cytoskeleton
BnaC03T0164000ZS	*BnFSD7*	C03-9109698:9111720 (+)	2022	903	9	300	34.39	4.88	Chloroplast
BnaC07T0373300ZS	*BnFSD8*	C07-49894566:49896148 (−)	1582	780	6	259	29.58	8.58	Chloroplast
BnaC07T0462000ZS	*BnFSD9*	C07-55868729:55870093 (−)	1364	648	5	215	24.17	6.65	Endoplasmic reticulum
BnaC09T0062700ZS	*BnFSD10*	C09-4091989:4094054 (−)	2065	783	8	260	29.72	6.66	Chloroplast

Table 1. *Cont.*

Gene ID	Gene Name	Genomic Position (bp)	Gene Length (bp)	CDS Length (bp)	Exons	Protein Length (Amino Acids)	Molecular Weight (kDa)	Isoelectric Point (pI)	Subcellular Localization
BnaC09T0329400ZS	*BnFSD11*	C09-41111988:41116227 (−)	4239	1077	7	358	40.40	8.8	Chloroplast
BnaA01T0376200ZS	*BnMSD1*	A01-33774266:33775498 (−)	1232	699	6	232	25.50	8.38	Mitochondrion
BnaA05T0446100ZS	*BnMSD2*	A05-41549634:41550932 (−)	1298	696	6	231	25.41	8.47	Mitochondrion
BnaA09T0519400ZS	*BnMSD3*	A09-55258451:55262461 (−)	4010	966	9	321	35.60	9.56	Cytoskeleton
BnaC01T0471300ZS	*BnMSD4*	C01-53548352:53549620 (−)	1268	693	6	230	25.36	8.94	Mitochondrion
BnaC05T0492700ZS	*BnMSD5*	C05-53531802:53533097 (−)	1295	696	6	231	25.43	7.83	Mitochondrion
BnaC08T0362400ZS	*BnMSD6*	C08-43431847:43433140 (−)	1293	729	6	242	27.11	6.14	Endoplasmic reticulum

In the genomic position, the positive (+) and negative (−) sign shows the presence of a gene on the positive and negative strand of that specific marker correspondingly. CSD means Cu/Zn-SOD; FSD means Fe-SOD; and MSD means Mn-SOD.

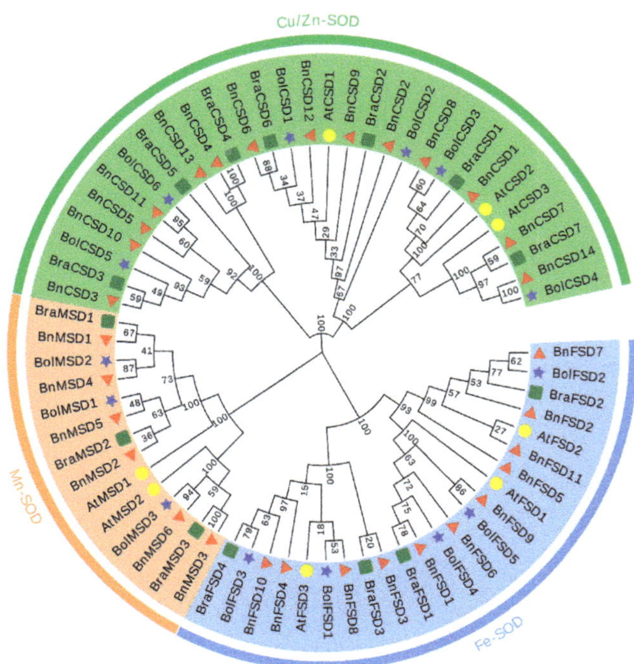

Figure 1. A neighbor-joining phylogenetic tree of 53 SOD proteins from *B. napus*, *B. oleracea*, *B. rapa*, and *A. thaliana*. Overall, 31 *BnSODs* (red triangle), 12 *BolSODs* (blue star), 12 *BraSODs* (green box), and 8 *AtSODs* (yellow circles) were grouped into three groups based on domain and 1000 bootstrap values. All of the notes indicate the percentage of bootstrap values.

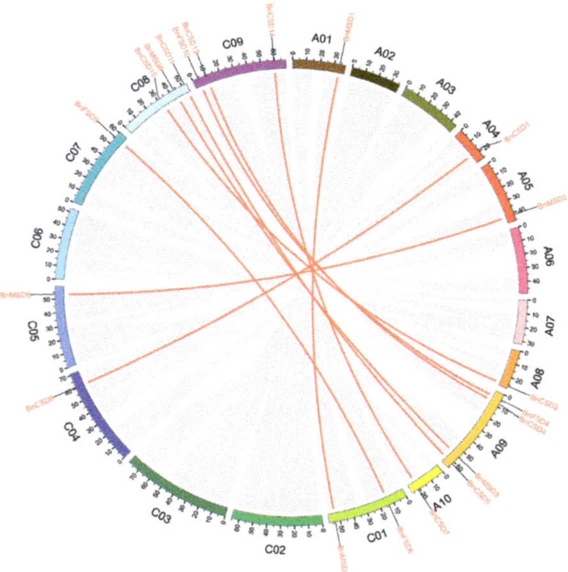

Figure 2. Chromosomal locations and inter-chromosomal associations of *BnSOD* genes. Grey lines in the background display all syntenic blocks in the *B. napus* genome, and the red lines display syntenic *BnSODs* gene pairs.

Collinearity investigation revealed strong orthologs of *SOD* genes among *B. napus* and three closely related species (*B. rapa*, *B. oleracea*, and *A. thaliana*) (Figure 3). In summary, in the A subgenome, seven and five *B. napus* genes demonstrated syntenic relations with *BraSODs* and *AtSODs*, respectively. Three genes exhibited syntenic relations with *BolSODs* and *AtSODs* (Figure 3). In contrast, in the C subgenome, six and three *B. napus* genes displayed syntenic relations with *BolSODs* and *AtSODs*. One gene revealed syntenic relations with *BnSODs* and *AtSODs* (Figure 3). Our results revealed numerous *A. thaliana*, *B. rapa*, and *B. oleracea* homologous sustained syntenic relations with *BnSODs*, suggesting that whole-genome duplication or segmental duplications played a significant part in *BnSODs* gene family progression (Table S3).

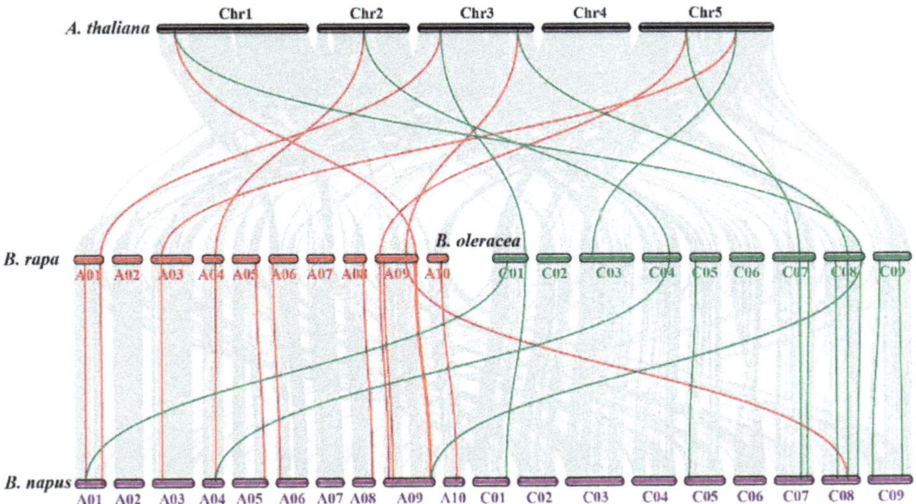

Figure 3. Collinearity analysis of SOD genes in *B. napus*, *B. rapa*, *B. oleracea*, and *A. thaliana* chromosomes. Grey lines in the background show the collinear blocks within *B. napus* and other plant genomes, whereas the red and green lines represent the syntenic SOD gene pairs. Genes located on the *B. napus* A subgenome are syntenic with *B. rapa*, *B. oleracea*, and *A. thaliana*, whereas genes located on *B. napus* C subgenome are mainly syntenic with *B. oleracea* and *A. thaliana* except for one gene that showed a syntenic relationship with *B. rapa* and *A. thaliana*.

The Ka/Ks ratio is considered a significant index in evaluating the duplication events and selection pressures [57]. Therefore, to understand the evolutionary story of the *Bn-SODs*, the Ka, Ks, and Ka/Ks ratio was calculated for *B. napus* and the other three plant species (Table S3). Our results revealed that all of the duplicated *BnSOD* gene pairs had a Ka/Ks ratio of <1 (Table S3), signifying that the *BnSOD* genes might have handled intense purifying selective pressure through their evolution (Table S3). Similar findings were observed in *B. rapa*, *B. oleracea*, and *A. thaliana* (Table S3).

3.4. BnSOD Gene Structures and Conserved Motifs Investigation

The exons–introns structures and the introns number play a vital role in gene family evolution [12,58]. The comprehensive examination of the phylogenetic relationships and gene structure illustration supported our knowledge of *BnSODs* gene structures (Figure 4). The findings exposed that exon and intron numbers of *BnSOD* genes had high inconsistency and were steady with the evolutionary hierarchy outcomes, i.e., the numbers of introns and exons were found to be relatively similar within the same group (Figure 4a,b). Briefly, the number of exons and introns varied from three to nine and two to eight in the individual gene sequences, respectively (Table 1; Figure 4b). A group with the Fe-SOD domain had five–nine exons and four–eight introns. A group with the Mn-SOD domain had six exons and five introns (five genes), and only one gene (*BnMSD3*) had nine exons and eight introns. Whereas, the group with Cu/Zn-SOD had five–nine exons and two–eight introns (Table 1; Figure 4b). Notably, group Fe-SOD and Cu/Zn-SOD groups have complex structures compared to the Mn-SOD group.

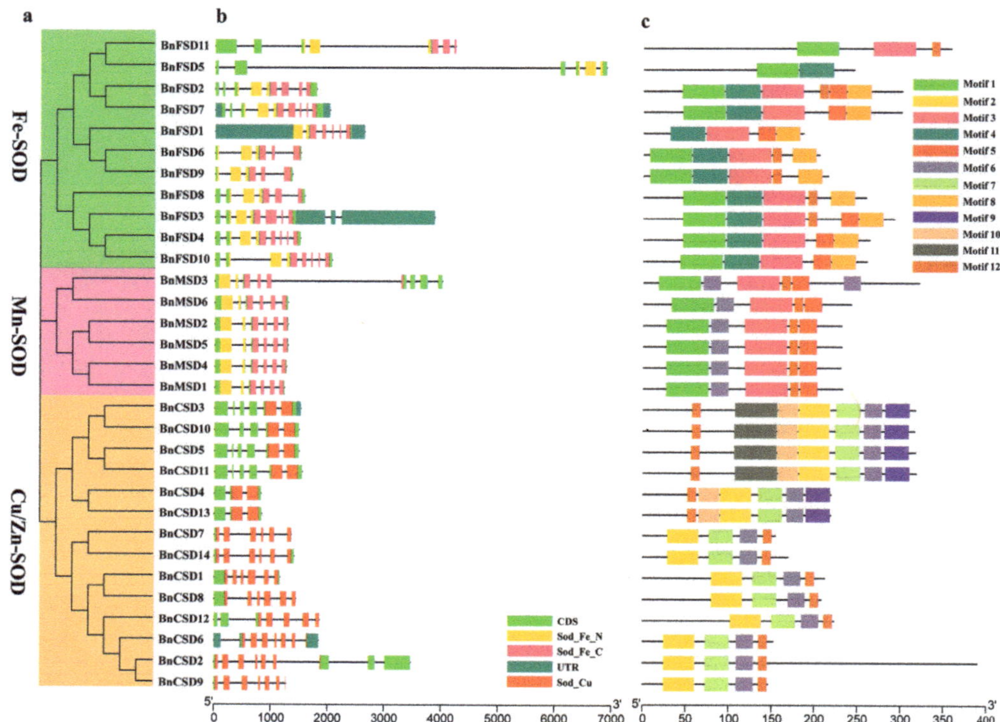

Figure 4. The gene structure and motif analysis of *BnSODs*. (**a**) Based on phylogenetic relationships and domain identification, the *BnSODs* were grouped into three groups. (**b**) Gene structure of *BnSODs*. The blue-gray color represents UTR regions, the light green color represents CDS or exons, the yellow color represents Sod_Fe_N, the pink color represents Sod_Fe_C, the red color represents Sod_Cu, and the black horizontal line represents introns. (**c**) Conserved motif compositions identified in *BnSODs*. Different color boxes denote different motifs.

To systematically appraise BnSODs' protein structure diversity and envisage their functions, we explored the full-length protein sequences of 31 BnSODs by MEME software to identify their conserved motifs. The investigation outcomes displayed that a total of 12 conserved motifs were found (Figure 4c). The detailed information (name, sequence, width, and E-value) of identified motifs is presented in Table S4. In short, the conserved motifs of SOD proteins varied from two to seven, and the motifs distributions were in agreement with the groups. In the Fe-SOD group, only two proteins (BnFSD11 and BnFSD5) had three and two motifs, respectively (Figure 4c). Interestingly, motifs 1, 3, 4, 5, and 8 were predicted to be specific to Fe-SOD and Mn-SOD groups. Motif 6 was specific to Mn-SOD and Cu/Zn groups. Motifs 2, 9, 10, and 11 were specific to the Cu/Zn group. Only motif 12 was largely distributed among all of the domain groups (Figure 4c). In summary, the group arrangements' reliability was mightily sustained by investigating conserved motif arrangements, gene structures, and phylogenetic relationships, suggesting that BnSOD proteins have tremendously well-maintained amino acid remains and members within a group. Thus, it can be assumed that proteins with similar structures and motifs might play similar functional roles.

3.5. Examination of Cis-Elements in Promoters of BnSOD Genes

To distinguish the gene functions and regulatory roles, *cis*-regulatory elements in *BnSODs* promoter regions were examined by searching a 2000 bp upstream region from each gene's transcriptional activation site against the PlantCARE database. The detailed information of *cis*-elements is presented in Table S5. As revealed in Figure 5, five phytohormone-correlated [(abscisic acid (ABA), auxin, methyl jasmonate (MeJA), gibberellin (GA), and salicylic acid (SA)] responsive elements including TCA-element, CGTCA-motif, ABRE, TGACG-motif, TATC-box, GARE-motif, P-box, etc. were recognized (Figure 5; Table S5). Particularly, numerous phytohormone-correlated elements were predicted to be specific to some genes and widely distributed (Figure 5), signifying the crucial role of these genes in phytohormone-mediation.

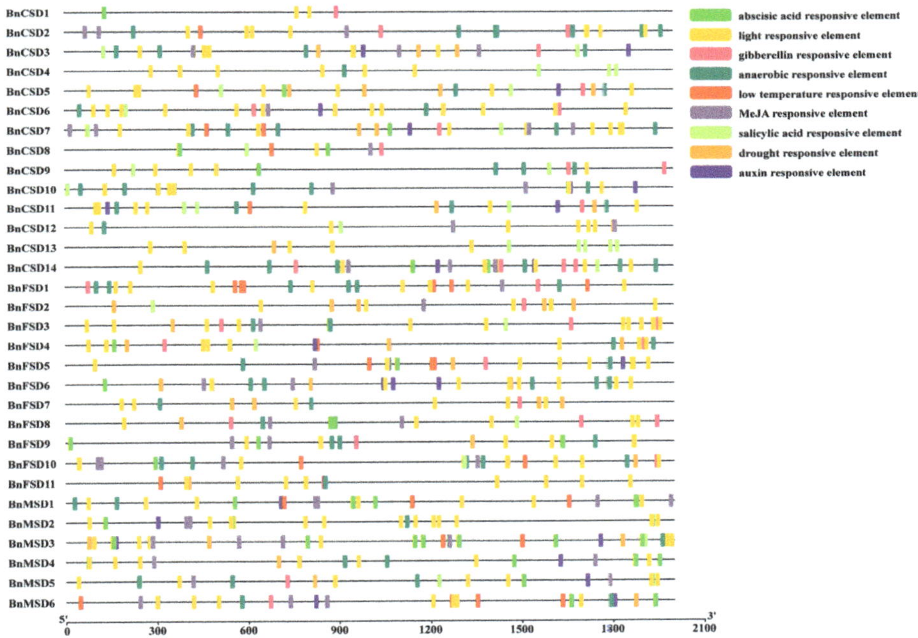

Figure 5. Analysis of *cis*-regulatory elements in the *BnSODs* promoter regions. Different *cis*-elements with functional similarity are denoted by similar colors.

Furthermore, four stress-responsive (drought, low-temperature, anaerobic, and light) elements, including ARE, LAMP-element, LTR, TCT-motif, chs-CMA1a, MBS, G-box, GT1-motif, MBS, etc. were identified (Figure 5; Table S5). Mainly, numerous light-responsive elements were found to be widely distributed among all of the genes (Figure 5; Table S5), signifying the substantial role of *BnSODs* in response to light stress. Overall, results advised that *BnSODs* expression levels may diverge under phytohormone and abiotic stress conditions.

3.6. Genome-Wide Analysis of miRNA Targeting BnSOD Genes

In the recent past, several studies have revealed that miRNA-mediated regulation accompanies the stress responses in plants. Thus, for a deep understanding of miRNA-mediated post-transcriptional regulation of *BnSODs*, we predicted 30 miRNAs targeting 13 genes (Figure 6a; Table S6). Some of the miRNA-targeted sites are presented in Figure 6b, whereas the detailed information of all miRNAs targeted genes/sites is presented in Table S6. Our results showed that four members of the bna-miR159 family targeted four genes (*BnCSD7*, *BnCSD14*, *BnMSD1*, and *BnMSD4*). Six members of the bna-miR166 family targeted one gene (*BnCSD10*). Fourteen members of the bna-miR169 family targeted one gene (*BnFSD11*). Four members of the bna-miR172 family targeted one gene (*BnCSD2*). Two members of the bna-miR394 family targeted four genes (*BnFSD11*, *BnFSD2*, *BnFSD7*, and *BnFSD5*). Two members of the bna-miR397 family targeted one gene (*BnCSD10*). One member of the bna-miR6033 family targeted three genes (*BnFSD4*, *BnFSD8*, and *BnFSD3*) (Figure 6a; Table S6). Mainly, *BnFSD11*, *BnCSD10*, and *BnCSD2* were prophesied to be targeted by a greater number of miRNAs (Figure 6a; Table S6). The expression levels of these miRNAs and their targeted genes requires validation in additional research to govern their biological roles in the rapeseed genome.

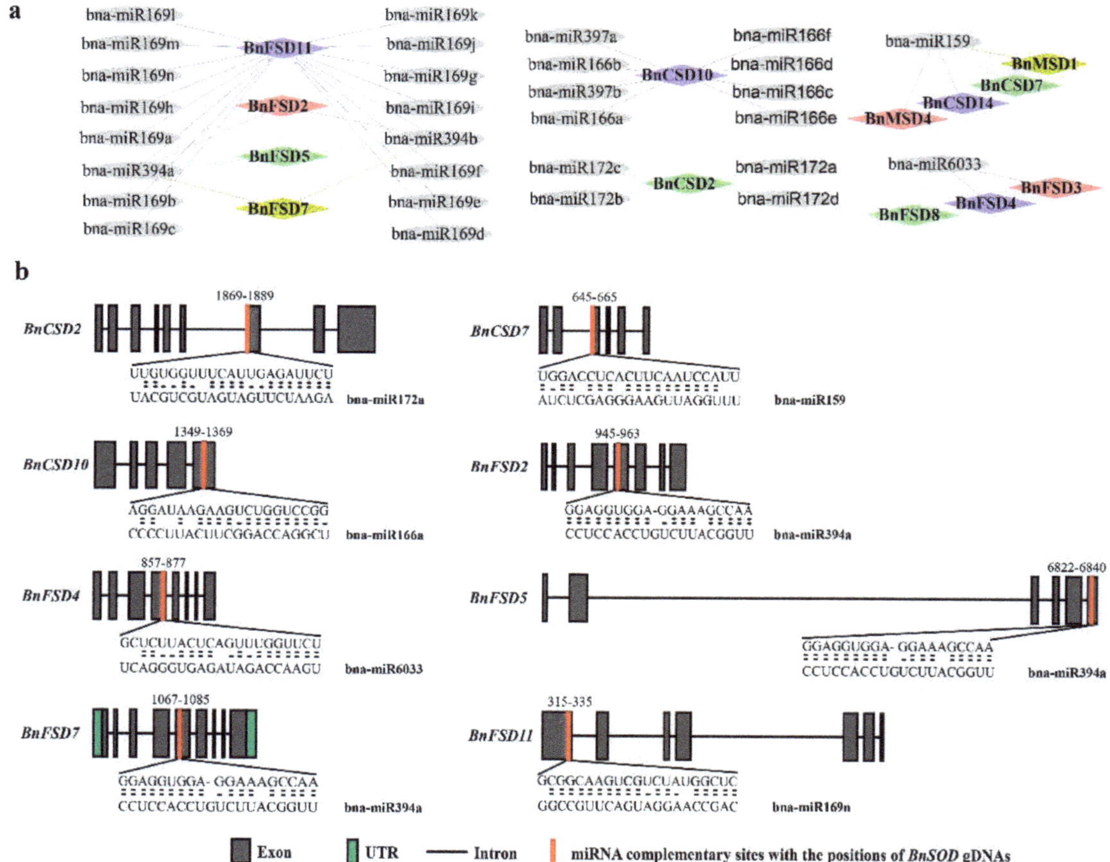

Figure 6. miRNA targeting *BnSOD* genes. (**a**) Network diagram of predicted miRNA targeting *BnSODs* genes. Different diamond colors represent *BnSODs* genes, and gray ellipse shapes represent miRNAs. (**b**) Schematic illustration indicates the *BnSODs* targeted by miRNAs. The RNA sequence of each complementary site from 5′-3′ and the predicted miRNA sequence from 3′-5′ are exposed in the long-drawn-out areas. See Table S6 for the detailed data of all predicted miRNAs.

3.7. Functional Annotation Study of BnSOD Genes

The functions of the *BnSODs* genes were prophesied through GO annotation investigation based on biological process (BP), molecular function (MF), and cellular component (CC) classes. The GO annotation results revealed numerous significantly enriched terms (Table S7). Analysis of BP annotations indicated that these genes were mainly involved in responses to stimulus (GO:0050896), responses to chemicals (GO:0042221), responses to stress (GO:0006950), responses to inorganic substances (GO:0010035), responses to abiotic stimulus (GO:0009628), cellular oxidant detoxification (GO:0098869), cellular response to oxygen-containing compound (GO:1901701), etc. (Table S7). Analysis of MF annotations revealed that these genes were primarily involved in ion binding (GO:0043167), copper ion binding (GO:0005507), zinc ion binding (GO:0008270), oxidoreductase activity (GO:0016491), superoxide dismutase activity (GO:0004784), antioxidant activity (GO:0016209), etc. (Table S7). A study of CC annotations showed that these genes are mainly involved in cellular anatomical entity (GO:0110165), cytoplasm (GO:0005737), obsolete intracellular part (GO:0044424), cellular component (GO:0005575), intracellular anatomical structure (GO:0005622), membrane-bounded organelle (GO:0043227), etc. (Table S7). In conclusion, GO annotation results confirmed the *BnSODs* role in response to different stress stimulus, cellular oxidant detoxification processes, metal ion binding activities, SOD activity, and different cellular components.

3.8. Expression Analysis of BnSOD Genes in Several Developmental Tissues

Tissue-specific expression levels of *BnSODs* genes were appraised in six different tissues and organs (including roots, stems, leaves, flowers, seeds, and silique) using RNA-seq data from *B. napus* (ZS11 variety) (BioProject ID: PRJCA001495). As shown in Figure 7, the group I genes display higher expression in all of the tissues except *BnFSD1, BnCSD3, BnCSD7, BnCSD14, BnMSD2*, that showed relatively lower expression than other genes (Figure 7). Meanwhile, genes clustered in group II exhibited considerably low expression, whereas *BnMSD3* and *BnMSD6* showed high expression at the seed_45d stage (Figure 7). Overall, results showed that genes from group I may play significant roles in *B. napus* growth and development.

3.9. Expression Profiles of BnSOD Genes under Hormones and Abiotic Stress Treatments

To examine the *BnSOD* genes' expression levels under different hormones (ABA, GA, IAA, and KT) and abiotic stress (salinity, cold, waterlogging, and drought) treatments, the qRT-PCR analysis of 10 randomly selected *BnSOD* genes was applied to govern the transcription profile (Figure 8). Under all of these stress conditions, *BnCSD6* and *BnFSD1* were down-regulated and showed relatively low expression levels. Likewise, *BnMSD10* showed a considerable response to ABA, GA, IAA, salinity, cold, and drought stress at different time points. The other eight genes were significantly up-regulated and showed higher expression to all hormones and abiotic treatments (Figure 8), signifying that these genes may play a significant role in mitigating various hormone and abiotic stresses.

3.10. Prediction of the 3D Structures of BnSODs

The 3D structures of BnSODs proteins were predicted using the 3DLigandSite tool with the default Search probability threshold (80.0%). The created models were downloaded to view the 3D structures (Figure 9). The detailed information of predicted 3D structures is presented in Table S8. The predicted Cu/Zn-SODs (BnCSD1-BnCSD14) model possessed a primarily highly conserved β-barrel structure, including a few short α-helices (Figure 9). The distinctive quaternary model of eukaryotic Cu/ZnSODs proteins contained a β-barrel domain comprised of eight antiparallel β-strands and Cu/Zn binding sites positioned at the exterior of the β-barrel in the active site network [13]. An individual subunit of Cu/ZnSODs proteins, the binding site possessed the Cu/Zn ion ligated by four histidines and important enzymes (such as arginine and alanine) with different conserved residues (Table S8). The conserved disulfide bond residues were examined in the active site channel of Cu/ZnSODs, including Cys202, Cys110, Cys201, and Cys110 in BnCSD3, BnCSD4, BnCSD10, BnCSD13, respectively (Table S8). The remaining Cu/ZnSODs proteins do not possess the Cys–Cys conserved disulfide bonds.

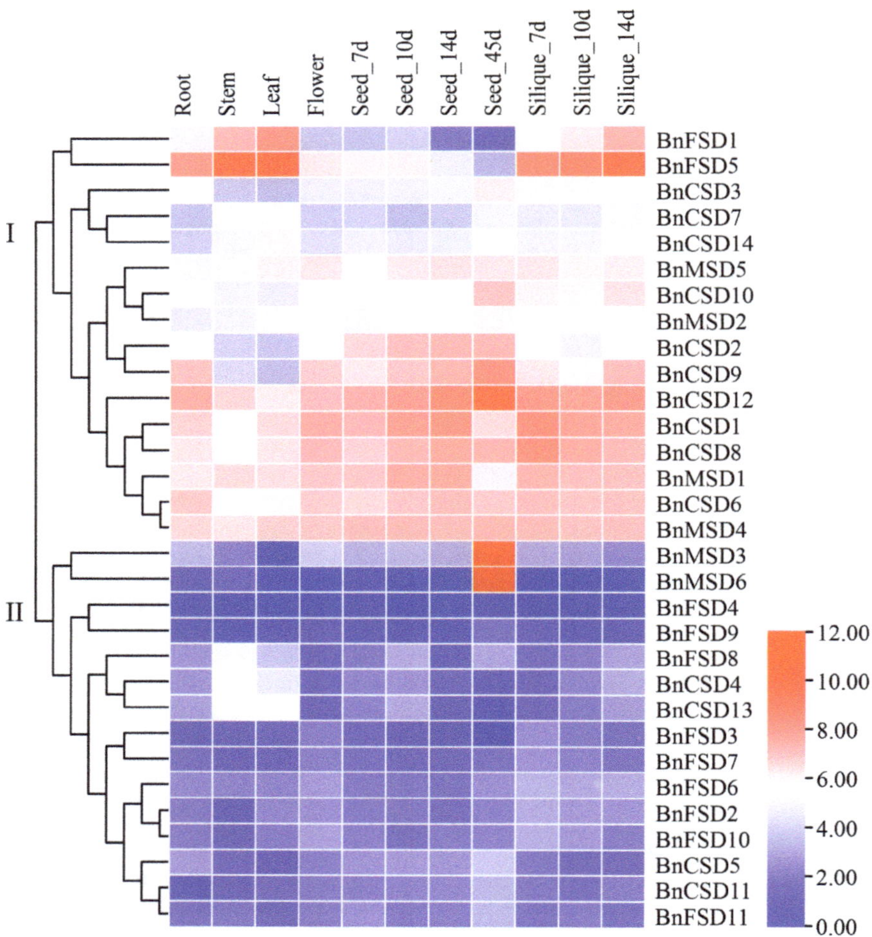

Figure 7. Expression analysis of *BnSOD* genes in several developmental organs. The 7d, 10d, 14d, and 45d tags indicate the time-points when the samples were harvested. In the expression bar, the red color shows high, and the blue color shows low, expression levels. The expression heat map was created using transcripts per million (TPM) values.

Likewise, the examination of *B. napus* Fe/MnSODs discovered that the dominant α-helices and β-sheet structures were detected (Figure 9). However, considerable variance in the sum of α-helices was observed and missing the disulfide bonds (exclusive features of Cu/ZnSODs). For instance, the active spot of Mn/FeSODs was positioned among the N- and C-terminal domains, and it varied from that of Cu/ZnSODs as it was comprised of a single metal ion. The metal ion is harmonized in a stressed trigonal bipyramidal geometry through eight–ten amino acid side-chains (including HIS, HIS, TYR, TRP, GLN, ASP, ALA) (Table S8) with solvent molecules. These Mn/FeSODs active sites also comprise hydrogen bonds that spread from the metal-bound solvent molecule to solvent-visible residues at the edge among subunits (Table S8). Additional structural investigation of these amino acids can boost our understanding of the catalytic process and metal ion binding.

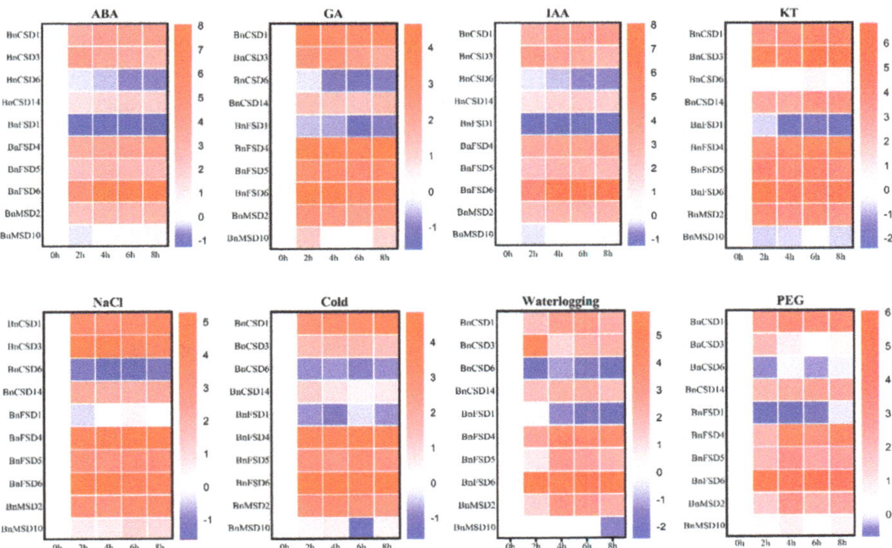

Figure 8. *BnSOD* gene expression levels under different hormone and abiotic stress conditions at different time points (0 (CK), 2, 4, 6, and 8 h). The expression bars display the comparative gene expression trends based on the log2 fold change values. In the expression bar, the red color displays high, and the blue color displays low, expression levels.

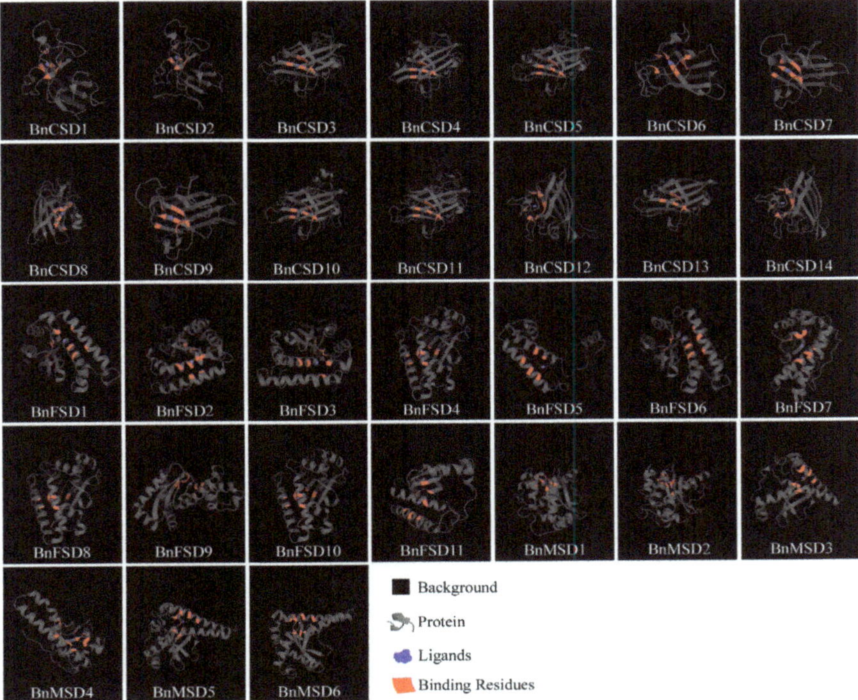

Figure 9. Predicted 3D structures and binding sites of BnSODs proteins using a 3DLigandSite tool. See Table 1 for the complete protein information, including subcellular location. See Table S8 for the complete information of the predicted 3D models, including predicted binding sites, conserved residues, and probability levels.

4. Discussion

Rapeseed is an allotetraploid crop that experiences extensive genome repetition and integration actions [59]. Nonetheless, rapeseed yield is affected by several abiotic stresses, including cold, heat, salinity, drought, and heavy metals [33,34]. ROS production under typical and stress environments is measured and scavenged through SODs, which act as the primary markers of an enzyme complicated in ROS scavenging, and play a crucial role in plants' physiological and biochemical procedures to survive with environmental cues [3]. Over the last few years, *SOD* family genes have been documented in different plant species, such as five *SOD* genes in *Zostera marina* [60], seven genes in *Medicago truncatula* [12] and barley [61], eight genes in sorghum [62], nine genes in tomato [63], ten genes in grapevine [64], 18 genes in cotton [28], 25 genes in banana [65], 26 genes in wheat [66], and 29 genes in *B. juncea* [67]. Thus far, there is no wide-ranging investigation of the *SOD* family genes in rapeseed. The obtainability of the whole rapeseed genome permits the genome-wide characterization of the SOD family genes, which may be used for future rapeseed improvement.

In the current study, we identified 31 *BnSOD* genes, including 14 *Cu/Zn-SODs*, 11 *Fe-SODs*, and six *Mn-SODs* genes (Table 1), which were clustered into three major sets based on the binding domain (Figure 1). So far, this is the largest *SOD* gene family identified in the rapeseed genome. Changes in the SOD gene numbers among plant species may be accredited to gene repetition events, including tandem and segmental repetitions, and act as a key factor in extending *SODs* for divergence. Gene doubling of SOD genes was also detected in various plant species [67,68]. Our results showed that *BnSODs* had experienced segmental duplications (Table S3). Consequently, these outcomes suggested that *BnSODs* duplication actions might play an important role in gene progression.

Gene structure analysis discovered that the number of exons varied from three to nine, and the number of introns varied from two to eight (Table 1; Figure 4). The Fe-SOD group had five to nine exons and four to eight introns. The Mn-SOD group had six exons and five introns (five genes), and only one gene (*BnMSD3*) had nine exons and eight introns. In comparison, the group with Cu/Zn-SOD had five to nine exons and two to eight introns (Table 1; Figure 4b). In wheat, seven *TaSODs* genes were found to have seven introns [66]. In sorghum, intron numbers varied from five to seven [62]. The exon–intron assemblies' difference was proficient by three critical apparatuses (exon/intron gain/loss, exonization/pseudoexonizationand, insertion/deletion), and they are directly subsidized to structural discrepancy [62,69]. Interestingly, the *SOD* genes in each cluster presented similar exon–intron association and conserved motifs (Figure 4), signifying that these genes may participate in the same roles connected to numerous abiotic stressors. Comparable findings have also been reported in sorghum [62], tomato [63], cotton [28], and wheat [66].

To better understand the *BnSOD* genes' role against several environmental strains, *cis*-elements in the promoter regions were prophesied. Our results exhibited that three types of *cis*-elements were recognized: stress-, hormones-, and light-responsive (Figure 5; Table S5). Most of the identified *cis*-elements were connected with ABA, MeJA, GA, SA, drought, low-temperature, light, and anaerobic induction. According to previous reports, *cis*-elements subsidize to plant stress responses [70,71]. These consequences were further validated by the GO annotation analysis (Table S7). Moreover, several researchers reported similar findings in different crop plants where *SOD* genes were found to play a significant part under various stress conditions [62,66,67]. These outcomes can boost our understanding of *BnSOD* genes under diverse environmental conditions.

Previous investigations have described that *SOD* genes may show diverse expression levels in tissues and under stress conditions [62,63,72]. Thus, in the current study, tissue-specific expression levels of *BnSODs* genes were appraised in six different developmental tissues using RNA-seq data (Figure 7), which were in agreement with earlier findings [12,63,66]. Several genes displayed higher expression in all evaluated tissues, indicating that these genes may contribute to rapeseed growth and development. Similarly, the expression levels of 10 *BnSOD* genes were evaluated under different hormones and abiotic stress treatments (Figure 8). Almost all genes were up-regulated under stress treatments, excluding *BnCSD6* and *BnFSD1* genes, whose expression levels were down-regulated. These results agreed with previous findings where several *SOD* genes showed higher expression in response to stresses. For instance, in grapevines, many *SODs* were up-regulated under cold, heat, salinity, and drought treatment [72]. All *SODs* were significantly up-regulated in tomatoes under salt and drought stress [63]. In wheat, almost all *SODs* showed higher expression in response to mannitol, salinity, and drought stress than control conditions [66]. Under cold stress conditions, SOD activity was significantly increased in rapeseed [34]. These findings provided strong evidence that *SOD* genes play a conserved role in vindicating abiotic stresses in different plant species.

Over the past few years, abundant miRNAs have been recognized through genome-wide examination in rapeseed to participate in diverse environmental factors [73–77]. In the present study, we identified 30 miRNAs from seven families targeting 13 *BnSODs* genes (Figure 6a; Table S6). In a recent study, 20 miRNAs were predicted to target 14 *SOD* genes in cotton [28]. Previously, miR398 targeted two Cu/ZnSOD genes in *Arabidopsis thaliana* [27]. miR166 has been stated to be pointedly up-regulated against UV-B radiation in maize [78]; in cassava against cold and drought stresses [79]; in Chinese cabbage under heat stress conditions [79]. According to recent reports, bna-miR159 plays a vital role in fatty acid metabolism in rapeseed [80,81]. Several members of miR172 have been detected while participating in plant development, including regulation of flowering time and floral patterning [82], and developmental control in *A. thaliana* [83]. miR394 has been reported to be involved in drought and salinity tolerance in *A. thaliana* via an ABA-dependent pathway [84]. In another study, miR394 was associated with cold stress responses in *A. thaliana* [85]. Shortly, these reports support our results and recommend that bna-miRNAs might play pivotal roles against several stresses by altering the transcript levels of *SOD* genes in rapeseed.

To obtain further insight, we also predicted the 3D structures of BnSODs proteins (Figure 9). The Cu/Zn-SODs (BnCSD1-BnCSD14) models had primarily highly conserved β-barrel structures, including a few short α-helices (Figure 9). According to a recent report, eukaryotic Cu/ZnSODs proteins contain a β-barrel domain comprised of eight antiparallel β-strands and Cu/Zn binding sites positioned at the exterior of the β-barrel in the active site network [13]. In Cu/ZnSODs proteins, the binding site possessed Cu/Zn ions ligated by four histidines and important enzymes (such as arginine and alanine) with different conserved residues (Table S8). These results were in agreement with previously predicted 3D models of SODs in sorghum [62], in *Gossypium raimondii*, and *G. arboretum* [86], and in rice [87]. The previous study proved that the binding active site of metal ions and the generation of the conserved disulfide bond in individual subunits would participate in the protein constancy, specificity, and dimer gathering [62,88]. Similarly, conserved residues (including arginine residue) also supported our results, and these residues were considered to be vital players for normal enzymatic movement (Table S8) [88]. Similarly, the Fe/MnSODs models of rapeseed suggested that the β-sheets were conquered by α-helices (Figure 9), and these results were in agreement with recently predicted SOD models in rice [87], soybean [89], and sorghum [62].

5. Conclusions

In the current study, we identified 31 *BnSOD* genes in the rapeseed genome via genome-wide comprehensive investigation. To boost our understanding, gene structure, phylogenetic and synteny, conserved motifs, *cis*-elements, GO annotation, miRNA prediction, 3D protein structure, tissue-specific expression, and expression profiling under different hormones and abiotic stress treatments have been performed. The results showed that some genes significantly responded to both hormone and abiotic stress stimuli, enhancing our understanding of the *BnSOD* genes. Thus, additional investigations are required to confirm the purposeful role of *BnSOD* genes in rapeseed growth, development, and response to numerous environmental cues.

Supplementary Materials: The following are available online at https://www.mdpi.com/article/10.3390/antiox10081182/s1, Table S1: the list primer was used for gene expression analysis by qRT-PCR, Table S2: the protein sequences of SOD family genes in *Arabidopsis thaliana*, *Brassica napus*, *Brassica oleracea*, and *Brassica rapa*, Table S3: results of segmentally and tandemly duplicated relationships, and Ka, Ks, and Ka/Ks ratio values between *B. napus*, *B. rapa*, *B. oleracea*, and *A. thaliana* gene pairs, Table S4: the information of identified 12 motifs in BnSOD proteins, Table S5: information of hormone- and stress-related *cis*-elements detected in the promoter regions of BnSODs, Table S6: information of predicted miRNA targeting *BnSOD* genes, Table S7: the GO enrichment analysis of *BnSOD* genes, Table S8: the specifics of predicted BnSODs 3D structures including predicted binding sites, conserved residues, and probability level.

Author Contributions: Conceptualization, A.R. and W.S.; methodology, A.R. and W.S.; software, A.R. and W.S.; formal analysis, A.R., W.S., A.G., Y.Z., Z.J., M.A.H., and S.S.M.; writing—original draft preparation, A.R., W.S., A.G., M.A.H., and S.S.M.; writing—review and editing, A.R., W.S., Y.L., and X.Z.; supervision, Y.C., Y.L. and X.Z.; funding acquisition, Y.L. and X.Z. All authors have read and agreed to the published version of the manuscript.

Funding: This work was supported by the Science and Technology Innovation Project of the Chinese Academy of Agricultural Sciences.

Institutional Review Board Statement: Not applicable.

Informed Consent Statement: Not applicable.

Data Availability Statement: The datasets used and/or analyzed during the current study are shown in the supplementary files. For tissue-specific expression profiling, we downloaded RNA-seq data of rapeseed (BioProject ID: PRJCA001495) from the National Genomics Data Center. The rapeseed genome sequence was downloaded from BnPIR database (http://cbi.hzau.edu.cn/bnapus/index.php, accessed on 1 April 2021). The sequences of *Brassica rapa* and *Brassica oleracea* are available at Phytozome database (https://phytozome.jgi.doe.gov/pz/portal.html, accessed on 1 April 2021).

Acknowledgments: We thank all of the members of The Key Lab of Biology and Genetic Improvement of Oil Crops, Oil Crops Research Institute, Chinese Academy of Agricultural Sciences (CAAS), Wuhan, China for their support throughout the study. Further, W.S., A.R., M.A.H., and S.S.M., want to say special thanks to CAAS for providing a Ph.D. Scholarship and research environment.

Conflicts of Interest: The authors declare no conflict of interest.

References

1. Raza, A.; Razzaq, A.; Mehmood, S.S.; Zou, X.; Zhang, X.; Lv, Y.; Xu, J. Impact of climate change on crops adaptation and strategies to tackle its outcome: A review. *Plants* **2019**, *8*, 34. [CrossRef]
2. Raza, A.; Ashraf, F.; Zou, X.; Zhang, X.; Tosif, H. Plant Adaptation and Tolerance to Environmental Stresses: Mechanisms and Perspectives. In *Plant Ecophysiology and Adaptation under Climate Change: Mechanisms and Perspectives I*; Springer: Singapore, 2020; pp. 117–145.
3. Hasanuzzaman, M.; Bhuyan, M.; Zulfiqar, F.; Raza, A.; Mohsin, S.M.; Mahmud, J.A.; Fujita, M.; Fotopoulos, V. Reactive Oxygen Species and Antioxidant Defense in Plants under Abiotic Stress: Revisiting the Crucial Role of a Universal Defense Regulator. *Antioxidants* **2020**, *9*, 681. [CrossRef]
4. Mittler, R. ROS are good. *Trends Plant Sci.* **2017**, *22*, 11–19. [CrossRef]
5. Hodgson, E.K.; Fridovich, I. Reversal of the superoxide dismutase reaction. *Biochem. Biophys. Res. Commun.* **1973**, *54*, 270–274. [CrossRef]
6. Brawn, K.; Fridovich, I. Superoxide radical and superoxide dismutases: Threat and defense. In *Autoxidation in Food and Biological Systems*; Springer: Boston, MA, USA, 1980; pp. 429–446.
7. Fink, R.C.; Scandalios, J.G. Molecular evolution and structure–function relationships of the superoxide dismutase gene families in angiosperms and their relationship to other eukaryotic and prokaryotic superoxide dismutases. *Arch. Biochem. Biophys.* **2002**, *399*, 19–36. [CrossRef]
8. Abreu, I.A.; Cabelli, D.E. Superoxide dismutases—a review of the metal-associated mechanistic variations. *Biochim. Biophys. Acta (BBA)-Proteins Proteom.* **2010**, *1804*, 263–274. [CrossRef] [PubMed]
9. Dupont, C.; Neupane, K.; Shearer, J.; Palenik, B. Diversity, function and evolution of genes coding for putative Ni-containing superoxide dismutases. *Environ. Microbiol.* **2008**, *10*, 1831–1843. [CrossRef] [PubMed]
10. Xia, M.; Wang, W.; Yuan, R.; Den, F.; Shen, F. Superoxide dismutase and its research in plant stress-tolerance. *Mol. Plant. Breed.* **2015**, *13*, 2633–2646.

11. Zeng, X.; Liu, Z.; Shi, P.; Xu, Y.; Sun, J.; Fang, Y.; Yang, G.; Wu, J.; Kong, D.; Sun, W. Cloning and expression analysis of copper and zinc superoxide dismutase (Cu/Zn-SOD) gene from *Brassica campestris* L. *Acta Agron. Sin.* **2014**, *40*, 636–643. [CrossRef]
12. Song, J.; Zeng, L.; Chen, R.; Wang, Y.; Zhou, Y. In silico identification and expression analysis of superoxide dismutase (SOD) gene family in *Medicago truncatula*. 3 Biotech. **2018**, *8*, 1–12. [CrossRef]
13. Perry, J.; Shin, D.; Getzoff, E.; Tainer, J. The structural biochemistry of the superoxide dismutases. *Biochim. Biophys. Acta (BBA) Proteins Proteom.* **2010**, *1804*, 245–262. [CrossRef] [PubMed]
14. Alamri, S.; Hu, Y.; Mukherjee, S.; Aftab, T.; Fahad, S.; Raza, A.; Ahmad, M.; Siddiqui, M.H. Silicon-induced postponement of leaf senescence is accompanied by modulation of antioxidative defense and ion homeostasis in mustard (*Brassica juncea*) seedlings exposed to salinity and drought stress. *Plant Physiol. Biochem.* **2020**, *157*, 47–59. [CrossRef] [PubMed]
15. Chokshi, K.; Pancha, I.; Trivedi, K.; Maurya, R.; Ghosh, A.; Mishra, S. Physiological responses of the green microalga Acutodesmus dimorphus to temperature induced oxidative stress conditions. *Physiol. Plant* **2020**, *170*, 462–473. [CrossRef]
16. Liu, T.; Ye, X.; Li, M.; Li, J.; Qi, H.; Hu, X. H_2O_2 and NO are involved in trehalose-regulated oxidative stress tolerance in cold-stressed tomato plants. *Environ. Exp. Bot.* **2020**, *171*, 103961. [CrossRef]
17. Mosa, K.A.; El-Naggar, M.; Ramamoorthy, K.; Alawadhi, H.; Elnaggar, A.; Wartanian, S.; Ibrahim, E.; Hani, H. Copper nanoparticles induced genotoxicty, oxidative stress, and changes in Superoxide Dismutase (SOD) gene expression in cucumber (*Cucumis sativus*) plants. *Front. Plant Sci.* **2018**, *9*, 872. [CrossRef]
18. Zhang, L.; Tian, W.; Huang, G.; Liu, B.; Wang, A.; Zhu, J.; Guo, X. The *SikCuZnSOD3* gene improves abiotic stress resistance in transgenic cotton. *Mol. Breed.* **2021**, *41*, 1–17. [CrossRef]
19. Pour-Aboughadareh, A.; Omidi, M.; Naghavi, M.R.; Etminan, A.; Mehrabi, A.A.; Poczai, P. Wild Relatives of Wheat Respond Well to Water Deficit Stress: A Comparative Study of Antioxidant Enzyme Activities and Their Encoding Gene Expression. *Agriculture* **2020**, *10*, 415. [CrossRef]
20. Lv, J.; Zhang, J.; Han, X.; Bai, L.; Xu, D.; Ding, S.; Ge, Y.; Li, C.; Li, J. Genome wide identification of superoxide dismutase (SOD) genes and their expression profiles under 1-methylcyclopropene (1-MCP) treatment during ripening of apple fruit. *Sci. Hortic.* **2020**, *271*, 109471. [CrossRef]
21. Cui, C.; Wang, J.-J.; Zhao, J.-H.; Fang, Y.-Y.; He, X.-F.; Guo, H.-S.; Duan, C.-G. A Brassica miRNA Regulates Plant Growth and Immunity through Distinct Modes of Action. *Mol. Plant* **2020**, *13*, 231–245. [CrossRef]
22. Khandal, H.; Parween, S.; Roy, R.; Meena, M.K.; Chattopadhyay, D. MicroRNA profiling provides insights into post-transcriptional regulation of gene expression in chickpea root apex under salinity and water deficiency. *Sci. Rep.* **2017**, *7*, 1–14. [CrossRef]
23. Ding, Y.; Ding, L.; Xia, Y.; Wang, F.; Zhu, C. Emerging roles of microRNAs in plant heavy metal tolerance and homeostasis. *J. Agric. Food Chem.* **2020**, *68*, 1958–1965. [CrossRef]
24. Ravichandran, S.; Ragupathy, R.; Edwards, T.; Domaratzki, M.; Cloutier, S. MicroRNA-guided regulation of heat stress response in wheat. *BMC Genom.* **2019**, *20*, 488. [CrossRef] [PubMed]
25. Park, S.-Y.; Grabau, E. Bypassing miRNA-mediated gene regulation under drought stress: Alternative splicing affects *CSD1* gene expression. *Plant Mol. Biol.* **2017**, *95*, 243–252. [CrossRef]
26. Shi, G.-Q.; Fu, J.-Y.; Rong, L.-J.; Zhang, P.-Y.; Guo, C.-J.; Kai, X. TaMIR1119, a miRNA family member of wheat (*Triticum aestivum*), is essential in the regulation of plant drought tolerance. *J. Integr. Agric.* **2018**, *17*, 2369–2378. [CrossRef]
27. Beauclair, L.; Yu, A.; Bouché, N. microRNA-directed cleavage and translational repression of the copper chaperone for superoxide dismutase mRNA in Arabidopsis. *Plant J.* **2010**, *62*, 454–462. [CrossRef]
28. Wang, W.; Zhang, X.; Deng, F.; Yuan, R.; Shen, F. Genome-wide characterization and expression analyses of superoxide dismutase (SOD) genes in *Gossypium hirsutum*. *BMC Genom.* **2017**, *18*, 1–25. [CrossRef]
29. Li, Q.; Jin, X.; Zhu, Y.-X. Identification and analyses of miRNA genes in allotetraploid *Gossypium hirsutum* fiber cells based on the sequenced diploid *G. raimondii* genome. *J. Genet. Genom.* **2012**, *39*, 351–360. [CrossRef]
30. Zhang, B.; Wang, Q.; Wang, K.; Pan, X.; Liu, F.; Guo, T.; Cobb, G.P.; Anderson, T.A. Identification of cotton microRNAs and their targets. *Gene* **2007**, *397*, 26–37. [CrossRef] [PubMed]
31. Yang, X.; Wang, L.; Yuan, D.; Lindsey, K.; Zhang, X. Small RNA and degradome sequencing reveal complex miRNA regulation during cotton somatic embryogenesis. *J. Exp. Bot.* **2013**, *64*, 1521–1536. [CrossRef]
32. Sunkar, R.; Kapoor, A.; Zhu, J.-K. Posttranscriptional induction of two Cu/Zn superoxide dismutase genes in Arabidopsis is mediated by downregulation of miR398 and important for oxidative stress tolerance. *Plant Cell* **2006**, *18*, 2051–2065. [CrossRef]
33. Raza, A. Eco-physiological and Biochemical Responses of Rapeseed (*Brassica napus* L.) to Abiotic Stresses: Consequences and Mitigation Strategies. *J. Plant Growth Regul.* **2020**, *40*, 1368–1388. [CrossRef]
34. He, H.; Lei, Y.; Yi, Z.; Raza, A.; Zeng, L.; Yan, L.; Xiaoyu, D.; Yong, C.; Xiling, Z. Study on the mechanism of exogenous serotonin improving cold tolerance of rapeseed (*Brassica napus* L.) seedlings. *Plant Growth Regul.* **2021**, *94*, 161–170. [CrossRef]
35. Mehmood, S.S.; Lu, G.; Luo, D.; Hussain, M.A.; Raza, A.; Zafar, Z.; Zhang, X.; Cheng, Y.; Zou, X.; Lv, Y. Integrated Analysis of Transcriptomics and Proteomics provides insights into the molecular regulation of cold response in *Brassica napus*. *Environ. Exp. Bot.* **2021**, *187*, 104480. [CrossRef]
36. Raza, A.; Razzaq, A.; Mehmood, S.S.; Hussain, M.A.; Wei, S.; He, H.; Zaman, Q.U.; Xuekun, Z.; Hasanuzzaman, M. Omics: The way forward to enhance abiotic stress tolerance in *Brassica napus* L. *GM Crop. Food* **2021**, *12*, 251–281. [CrossRef]

37. Raza, A.; Su, W.; Gao, A.; Mehmood, S.S.; Hussain, M.A.; Nie, W.; Lv, Y.; Zou, X.; Zhang, X. Catalase (CAT) Gene Family in Rapeseed (*Brassica napus* L.): Genome-Wide Analysis, Identification, and Expression Pattern in Response to Multiple Hormones and Abiotic Stress Conditions. *Inter. J. Mol. Sci.* **2021**, *22*, 4281. [CrossRef]

38. Song, J.M.; Liu, D.X.; Xie, W.Z.; Yang, Z.; Guo, L.; Liu, K.; Yang, Q.Y.; Chen, L.L. BnPIR: Brassica napus Pan-genome Information Resource for 1,689 accessions. *Plant Biotechnol. J.* **2021**, *19*, 412. [CrossRef]

39. Rhee, S.Y.; Beavis, W.; Berardini, T.Z.; Chen, G.; Dixon, D.; Doyle, A.; Garcia-Hernandez, M.; Huala, E.; Lander, G.; Montoya, M. The Arabidopsis Information Resource (TAIR): A model organism database providing a centralized, curated gateway to Arabidopsis biology, research materials and community. *Nucleic Acids Res.* **2003**, *31*, 224–228. [CrossRef] [PubMed]

40. Finn, R.D.; Clements, J.; Arndt, W.; Miller, B.L.; Wheeler, T.J.; Schreiber, F.; Bateman, A.; Eddy, S.R. HMMER web server: 2015 update. *Nucleic Acids Res.* **2015**, *43*, W30–W38. [CrossRef] [PubMed]

41. El-Gebali, S.; Mistry, J.; Bateman, A.; Eddy, S.R.; Luciani, A.; Potter, S.C.; Qureshi, M.; Richardson, L.J.; Salazar, G.A.; Smart, A. The Pfam protein families database in 2019. *Nucleic Acids Res.* **2019**, *47*, D427–D432. [CrossRef] [PubMed]

42. Goodstein, D.M.; Shu, S.; Howson, R.; Neupane, R.; Hayes, R.D.; Fazo, J.; Mitros, T.; Dirks, W.; Hellsten, U.; Putnam, N. Phytozome: A comparative platform for green plant genomics. *Nucleic Acids Res.* **2012**, *40*, D1178–D1186. [CrossRef] [PubMed]

43. Gasteiger, E.; Hoogland, C.; Gattiker, A.; Wilkins, M.R.; Appel, R.D.; Bairoch, A. Protein identification and analysis tools on the ExPASy server. In *The Proteomics Protocols Handbook*; Humana Press Inc.: Totowa, NJ, USA, 2005; pp. 571–607.

44. Horton, P.; Park, K.-J.; Obayashi, T.; Fujita, N.; Harada, H.; Adams-Collier, C.; Nakai, K. WoLF PSORT: Protein localization predictor. *Nucleic Acids Res.* **2007**, *35*, W585–W587. [CrossRef]

45. Chen, C.; Chen, H.; Zhang, Y.; Thomas, H.R.; Frank, M.H.; He, Y.; Xia, R. TBtools: An integrative toolkit developed for interactive analyses of big biological data. *Mol. Plant* **2020**, *13*, 1194–1202. [CrossRef]

46. Bailey, T.L.; Boden, M.; Buske, F.A.; Frith, M.; Grant, C.E.; Clementi, L.; Ren, J.; Li, W.W.; Noble, W.S. MEME SUITE: Tools for motif discovery and searching. *Nucleic Acids Res.* **2009**, *37*, W202–W208. [CrossRef] [PubMed]

47. Kumar, S.; Stecher, G.; Li, M.; Knyaz, C.; Tamura, K. MEGA X: Molecular evolutionary genetics analysis across computing platforms. *Mol. Biol. Evol.* **2018**, *35*, 1547–1549. [CrossRef] [PubMed]

48. Subramanian, B.; Gao, S.; Lercher, M.J.; Hu, S.; Chen, W.-H. Evolview v3: A webserver for visualization, annotation, and management of phylogenetic trees. *Nucleic Acids Res.* **2019**, *47*, W270–W275. [CrossRef]

49. Tang, H.; Bowers, J.E.; Wang, X.; Ming, R.; Alam, M.; Paterson, A.H. Synteny and collinearity in plant genomes. *Science* **2008**, *320*, 486–488. [CrossRef] [PubMed]

50. Wang, D.; Zhang, Y.; Zhang, Z.; Zhu, J.; Yu, J. KaKs_Calculator 2.0: A toolkit incorporating gamma-series methods and sliding window strategies. *Genom. Proteom. Bioinform.* **2010**, *8*, 77–80. [CrossRef]

51. Lescot, M.; Déhais, P.; Thijs, G.; Marchal, K.; Moreau, Y.; Van de Peer, Y.; Rouzé, P.; Rombauts, S. PlantCARE, a database of plant cis-acting regulatory elements and a portal to tools for in silico analysis of promoter sequences. *Nucleic Acids Res.* **2002**, *30*, 325–327. [CrossRef] [PubMed]

52. Dai, X.; Zhuang, Z.; Zhao, P.X. psRNATarget: A plant small RNA target analysis server (2017 release). *Nucleic Acids Res.* **2018**, *46*, W49–W54. [CrossRef]

53. Powell, S.; Forslund, K.; Szklarczyk, D.; Trachana, K.; Roth, A.; Huerta-Cepas, J.; Gabaldon, T.; Rattei, T.; Creevey, C.; Kuhn, M. eggNOG v4. 0: Nested orthology inference across 3686 organisms. *Nucleic Acids Res.* **2014**, *42*, D231–D239. [CrossRef]

54. Swift, M.L. GraphPad prism, data analysis, and scientific graphing. *J. Chem. Inform. Comput. Sci.* **1997**, *37*, 411–412. [CrossRef]

55. Wass, M.N.; Kelley, L.A.; Sternberg, M.J. 3DLigandSite: Predicting ligand-binding sites using similar structures. *Nucleic Acids Res.* **2010**, *38*, W469–W473. [CrossRef]

56. Cannon, S.B.; Mitra, A.; Baumgarten, A.; Young, N.D.; May, G. The roles of segmental and tandem gene duplication in the evolution of large gene families in Arabidopsis thaliana. *BMC Plant Biol.* **2004**, *4*, 1–21. [CrossRef]

57. Hurst, L.D. The Ka/Ks ratio: Diagnosing the form of sequence evolution. *Trends Genet.* **2002**, *18*, 486. [CrossRef]

58. Xu, X.; Yang, Y.; Liu, C.; Sun, Y.; Zhang, T.; Hou, M.; Huang, S.; Yuan, H. The evolutionary history of the sucrose synthase gene family in higher plants. *BMC Plant Biol.* **2019**, *19*, 1–14. [CrossRef]

59. Song, J.-M.; Guan, Z.; Hu, J.; Guo, C.; Yang, Z.; Wang, S.; Liu, D.; Wang, B.; Lu, S.; Zhou, R. Eight high-quality genomes reveal pan-genome architecture and ecotype differentiation of *Brassica napus*. *Nat. Plants* **2020**, *6*, 34–45. [CrossRef] [PubMed]

60. Zang, Y.; Chen, J.; Li, R.; Shang, S.; Tang, X. Genome-wide analysis of the superoxide dismutase (SOD) gene family in *Zostera marina* and expression profile analysis under temperature stress. *PeerJ* **2020**, *8*, e9063. [CrossRef] [PubMed]

61. Zhang, X.; Zhang, L.; Chen, Y.; Wang, S.; Fang, Y.; Zhang, X.; Wu, Y.; Xue, D. Genome-wide identification of the SOD gene family and expression analysis under drought and salt stress in barley. *Plant Growth Regul.* **2021**, *94*, 49–60. [CrossRef]

62. Filiz, E.; Tombuloğlu, H. Genome-wide distribution of superoxide dismutase (SOD) gene families in *Sorghum bicolor*. *Turk. J. Biol.* **2015**, *39*, 49–59. [CrossRef]

63. Feng, K.; Yu, J.; Cheng, Y.; Ruan, M.; Wang, R.; Ye, Q.; Zhou, G.; Li, Z.; Yao, Z.; Yang, Y. The SOD gene family in tomato: Identification, phylogenetic relationships, and expression patterns. *Front. Plant Sci.* **2016**, *7*, 1279. [CrossRef]

64. Hu, X.; Hao, C.; Cheng, Z.-M.; Zhong, Y. Genome-wide identification, characterization, and expression analysis of the grapevine superoxide dismutase (SOD) family. *Inter. J. Genom.* **2019**, *2019*, 7350414. [CrossRef]

65. Feng, X.; Lai, Z.; Lin, Y.; Lai, G.; Lian, C. Genome-wide identification and characterization of the superoxide dismutase gene family in *Musa acuminata* cv. Tianbaojiao (AAA group). *BMC Genom.* **2015**, *16*, 1–16. [CrossRef] [PubMed]

66. Jiang, W.; Yang, L.; He, Y.; Zhang, H.; Li, W.; Chen, H.; Ma, D.; Yin, J. Genome-wide identification and transcriptional expression analysis of superoxide dismutase (SOD) family in wheat (*Triticum aestivum*). *PeerJ* **2019**, *7*, e8062. [CrossRef] [PubMed]

67. Verma, D.; Lakhanpal, N.; Singh, K. Genome-wide identification and characterization of abiotic-stress responsive SOD (superoxide dismutase) gene family in *Brassica juncea* and *B. rapa*. *BMC Genom.* **2019**, *20*, 1–18. [CrossRef] [PubMed]

68. Wang, W.; Xia, M.; Chen, J.; Deng, F.; Yuan, R.; Zhang, X.; Shen, F. Data set for phylogenetic tree and RAMPAGE Ramachandran plot analysis of SODs in *Gossypium raimondii* and *G. arboreum*. *Data Brief.* **2016**, *9*, 345–348. [CrossRef]

69. Xu, G.; Guo, C.; Shan, H.; Kong, H. Divergence of duplicate genes in exon–intron structure. *Proc. Nati. Acad. Sci. USA* **2012**, *109*, 1187–1192. [CrossRef]

70. Osakabe, Y.; Yamaguchi-Shinozaki, K.; Shinozaki, K.; Tran, L.S.P. ABA control of plant macroelement membrane transport systems in response to water deficit and high salinity. *New Phytol.* **2014**, *202*, 35–49. [CrossRef]

71. Maruyama-Nakashita, A.; Nakamura, Y.; Watanabe-Takahashi, A.; Inoue, E.; Yamaya, T.; Takahashi, H. Identification of a novel cis-acting element conferring sulfur deficiency response in Arabidopsis roots. *Plant J.* **2005**, *42*, 305–314. [CrossRef]

72. Zhou, Y.; Hu, L.; Wu, H.; Jiang, L.; Liu, S. Genome-wide identification and transcriptional expression analysis of cucumber superoxide dismutase (SOD) family in response to various abiotic stresses. *Inter. J. Genom.* **2017**, *2017*, 7243973. [CrossRef]

73. Chen, L.; Chen, L.; Zhang, X.; Liu, T.; Niu, S.; Wen, J.; Yi, B.; Ma, C.; Tu, J.; Fu, T. Identification of miRNAs that regulate silique development in *Brassica napus*. *Plant Sci.* **2018**, *269*, 106–117. [CrossRef]

74. Buhtz, A.; Springer, F.; Chappell, L.; Baulcombe, D.C.; Kehr, J. Identification and characterization of small RNAs from the phloem of *Brassica napus*. *Plant J.* **2008**, *53*, 739–749. [CrossRef]

75. Fu, Y.; Mason, A.S.; Zhang, Y.; Lin, B.; Xiao, M.; Fu, D.; Yu, H. MicroRNA-mRNA expression profiles and their potential role in cadmium stress response in *Brassica napus*. *BMC Plant Biol.* **2019**, *19*, 1–20. [CrossRef]

76. Körbes, A.P.; Machado, R.D.; Guzman, F.; Almerao, M.P.; de Oliveira, L.F.V.; Loss-Morais, G.; Turchetto-Zolet, A.C.; Cagliari, A.; dos Santos Maraschin, F.; Margis-Pinheiro, M. Identifying conserved and novel microRNAs in developing seeds of *Brassica napus* using deep sequencing. *PLoS ONE* **2012**, *7*, e50663. [CrossRef]

77. Shen, E.; Zou, J.; Hubertus Behrens, F.; Chen, L.; Ye, C.; Dai, S.; Li, R.; Ni, M.; Jiang, X.; Qiu, J. Identification, evolution, and expression partitioning of miRNAs in allopolyploid *Brassica napus*. *J. Exp. Bot.* **2015**, *66*, 7241–7253. [CrossRef]

78. Casati, P. Analysis of UV-B regulated miRNAs and their targets in maize leaves. *Plant Signal. Behav.* **2013**, *8*, e26758. [CrossRef] [PubMed]

79. Li, S.; Yu, X.; Lei, N.; Cheng, Z.; Zhao, P.; He, Y.; Wang, W.; Peng, M. Genome-wide identification and functional prediction of cold and/or drought-responsive lncRNAs in cassava. *Sci. Rep.* **2017**, *7*, 45981. [CrossRef] [PubMed]

80. Wang, Z.; Qiao, Y.; Zhang, J.; Shi, W.; Zhang, J. Genome wide identification of microRNAs involved in fatty acid and lipid metabolism of *Brassica napus* by small RNA and degradome sequencing. *Gene* **2017**, *619*, 61–70. [CrossRef] [PubMed]

81. Wang, J.; Jian, H.; Wang, T.; Wei, L.; Li, J.; Li, C.; Liu, L. Identification of microRNAs actively involved in fatty acid biosynthesis in developing Brassica napus seeds using high-throughput sequencing. *Front. Plant Sci.* **2016**, *7*, 1570. [CrossRef]

82. Zhu, Q.-H.; Helliwell, C.A. Regulation of flowering time and floral patterning by miR172. *J. Exp. Bot.* **2011**, *62*, 487–495. [CrossRef]

83. Wu, G.; Park, M.Y.; Conway, S.R.; Wang, J.-W.; Weigel, D.; Poethig, R.S. The sequential action of miR156 and miR172 regulates developmental timing in Arabidopsis. *Cell* **2009**, *138*, 750–759. [CrossRef] [PubMed]

84. Song, J.B.; Gao, S.; Sun, D.; Li, H.; Shu, X.X.; Yang, Z.M. miR394 and LCR are involved in Arabidopsis salt and drought stress responses in an abscisic acid-dependent manner. *BMC Plant Biol.* **2013**, *13*, 1–16. [CrossRef] [PubMed]

85. Song, J.B.; Gao, S.; Wang, Y.; Li, B.W.; Zhang, Y.L.; Yang, Z.M. miR394 and its target gene LCR are involved in cold stress response in Arabidopsis. *Plant Gene* **2016**, *5*, 56–64. [CrossRef]

86. Wang, W.; Xia, M.; Chen, J.; Deng, F.; Yuan, R.; Zhang, X.; Shen, F. Genome-wide analysis of superoxide dismutase gene family in *Gossypium raimondii* and *G. arboreum*. *Plant Gene* **2016**, *6*, 18–29. [CrossRef]

87. Dehury, B.; Sarma, K.; Sarmah, R.; Sahu, J.; Sahoo, S.; Sahu, M.; Sen, P.; Modi, M.K.; Sharma, G.D.; Choudhury, M.D. In silico analyses of superoxide dismutases (SODs) of rice (*Oryza sativa* L.). *J. Plant Biochem. Biotechnol.* **2013**, *22*, 150–156. [CrossRef]

88. Fisher, C.L.; Cabelli, D.E.; Tainer, J.A.; Hallewell, R.A.; Getzoff, E.D. The role of arginine 143 in the electrostatics and mechanism of Cu, Zn superoxide dismutase: Computational and experimental evaluation by mutational analysis. *Proteins Struct. Funct. Bioinform.* **1994**, *19*, 24–34. [CrossRef] [PubMed]

89. Ramana Gopavajhula, V.; Viswanatha Chaitanya, K.; Akbar Ali Khan, P.; Shaik, J.P.; Narasimha Reddy, P.; Alanazi, M. Modeling and analysis of soybean (*Glycine max*. L) Cu/Zn, Mn and Fe superoxide dismutases. *Genet. Mol. Biol.* **2013**, *36*, 225–236. [CrossRef]

MDPI

St. Alban-Anlage 66

4052 Basel

Switzerland

Tel. +41 61 683 77 34

Fax +41 61 302 89 18

www.mdpi.com

Antioxidants Editorial Office

E-mail: antioxidants@mdpi.com

www.mdpi.com/journal/antioxidants